T0210775

Lecture Notes in Computer Science   9314

Commenced Publication in 1973
Founding and Former Series Editors:
Gerhard Goos, Juris Hartmanis, and Jan van Leeuwen

More information about this series at http://www.springer.com/series/7409

Yo-Sung Ho · Jitao Sang
Yong Man Ro · Junmo Kim
Fei Wu (Eds.)

Advances in Multimedia Information Processing – PCM 2015

16th Pacific-Rim Conference on Multimedia
Gwangju, South Korea, September 16–18, 2015
Proceedings, Part I

 Springer

Editors
Yo-Sung Ho
Gwangju Institute of Science
    and Technology
Gwangju
Korea (Republic of)

Jitao Sang
Chinese Academy of Sciences
Institute of Automation
Beijing
China

Yong Man Ro
KAIST
Daejeon
Korea (Republic of)

Junmo Kim
KAIST
Daejeon
Korea (Republic of)

Fei Wu
College of Computer Science
Zhejiang University
Hangzhou
China

ISSN 0302-9743   ISSN 1611-3349 (electronic)
Lecture Notes in Computer Science
ISBN 978-3-319-24074-9   ISBN 978-3-319-24075-6 (eBook)
DOI 10.1007/978-3-319-24075-6

Library of Congress Control Number: 2015948170

LNCS Sublibrary: SL3 – Information Systems and Applications, incl. Internet/Web, and HCI

Springer Cham Heidelberg New York Dordrecht London

Printed on acid-free paper

Springer International Publishing AG Switzerland is part of Springer Science+Business Media
(www.springer.com)

Preface

We are delighted to welcome readers to the proceedings of the 16th Pacific-Rim Conference on Multimedia (PCM 2015), held in Gwangju, South Korea, September 16–18, 2015. The Pacific-Rim Conference on Multimedia is a leading international conference for researchers and industry practitioners to share and showcase their new ideas, original research results, and engineering development experiences from areas related to multimedia. The 2015 edition of the PCM marked its 16th anniversary. The longevity of the conference would not be possible without the strong support of the research community, and we take this opportunity to thank everyone who has contributed to the growth of the conference in one way or another over the last 16 years.

PCM 2015 was held in Gwangju, South Korea, which is known as one of the most beautiful and democratic cities in the country. The conference venue was Gwangju Institute of Science and Technology (GIST), which is one of the world's top research-oriented universities. Despite its short history of 22 years, GIST has already established its position as an educational institution of huge potential, as it ranked the fourth in the world in citations per faculty in the 2014 QS World University Rankings.

At PCM 2015, we held regular and special sessions of oral and poster presentations. We received 224 paper submissions, covering topics of multimedia content analysis, multimedia signal processing and communications, as well as multimedia applications and services. The submitted papers were reviewed by the Technical Program Committee, consisting of 143 reviewers. Each paper was reviewed by at least two reviewers. The program chairs carefully considered the input and feedback from the reviewers and accepted 138 papers for presentation at the conference. The acceptance rate of 62 % indicates our commitment to ensuring a very high-quality conference. Out of these accepted papers, 68 were presented orally and 70 papers were presented as posters.

PCM 2015 was organized by the Realistic Broadcasting Research Center (RBRC) at Gwangju Institute of Science and Technology (GIST) in South Korea. We gratefully thank the Gwangju Convention and Visitors Bureau for its generous support of PCM 2015.

We are heavily indebted to many individuals for their significant contributions. Firstly, we are very grateful to all the authors who contributed their high-quality research and shared their knowledge with our scientific community. Finally, we wish to thank all Organizing and Program Committee members, reviewers, session chairs, student volunteers, and supporters. Their contributions are much appreciated. We hope you all enjoy the proceedings of the 2015 Conference on Multimedia.

September 2015

Yo-Sung Ho
Jitao Sang
Yong Man Ro
Junmo Kim
Fei Wu

Organization

Organizing Committee

General Chair

Yo-Sung Ho   Gwangju Institute of Science and Technology,
South Korea

Program Chairs

Jitao Sang   Chinese Academy of Sciences, China
Yong Man Ro   Korea Advanced Institute of Science and Technology,
South Korea

Special Session Chairs

Shang-Hong Lai   National Tsinghua University, Taiwan
Chao Liang   Wuhan University, China
Yue Gao   National University of Singapore, Singapore

Tutorial Chairs

Weisi Lin   Nanyang Technological University, Singapore
Chang-Su Kim   Korea University, South Korea

Demo/Poster Chairs

Xirong Li   Renmin University of China, China
Lu Yang   University of Electronic Science and Technology of China,
China

Publication Chairs

Junmo Kim   Korea Advanced Institute of Science and Technology,
South Korea
Fei Wu   Zhejiang University, China

Publicity Chairs

Chin-Kuan Ho   Multimedia University, Malaysia
Gangyi Jiang   Ningbo University, China
Sam Kwong   City University of Hong Kong, Hong Kong
Yoshikazu Miyanaga   Hokkaido University, Japan
Daranee Hormdee   Khon Kaen University, Thailand

Thanh-Sach Le   Ho Chi Minh City University, Vietnam
Ki Ryong Kwon   Pukyong National University, South Korea

Web Chair

Eunsang Ko   Gwangju Institute of Science and Technology, South Korea

Registration Chairs

Young-Ki Jung   Honam University, South Korea
Youngho Lee   Mokpo National University, South Korea

Local Arrangement Chairs

Young Chul Kim   Chonnam National University, South Korea
Pankoo Kim   Chosun University, South Korea

Technical Program Committee

Sungjun Bae	Shoko Imaizumi	Youngho Lee
Hang Bo	Byeungwoo Jeon	Haiwei Lei
Xiaochun Cao	Zhong Ji	Donghong Li
Kosin Chamnongthai	Yu-Gang Jiang	Guanyi Li
Wen-Huang Cheng	Jian Jin	Haojie Li
Nam Ik Cho	Xin Jin	Houqiang Li
Jae Young Choi	Zhi Jin	Leida Li
Wei-Ta Chu	SoonHeung Jung	Liang Li
Peng Cui	YongJu Jung	Songnan Li
Wesley De Neve	Yun-Suk Kang	Xirong Li
Cheng Deng	Hisakazu Kikuchi	Yongbo Li
Weisheng Dong	Byung-Gyu Kim	Chunyu Lin
Yao-Chung Fan	Changik Kim	Weisi Lin
Yuming Fang	ChangKi Kim	Weifeng Liu
Sheng Fang	Chang-Su Kim	Bo Liu
Toshiaki Fujii	Hakil Kim	Qiegen Liu
Masaaki Fujiyoshi	Hyoungseop Kim	Qiong Liu
Yue Gao	Jaegon Kim	Wei Liu
Yanlei Gu	Min H. Kim	Xianglong Liu
Shijie Hao	Seon Joo Kim	Yebin Liu
Lihuo He	Su Young Kwak	Dongyuan Lu
Ran He	Shang-Hong Lai	Yadong Mu
Min Chul Hong	Duy-Dinh Le	Shogo Muramatsu
Richang Hong	Chan-Su Lee	Chong-Wah Ngo
Dekun Hu	Sang-Beom Lee	Byung Tae Oh
Min-Chun Hu	Sanghoon Lee	Lei Pan
Ruimin Hu	Sangkeun Lee	Yanwei Pang
Lei Huang	Seokhan Lee	Jinah Park

Tongwei Ren
Yong Man Ro
Jitao Sang
Klaus Schoffmann
Kwang-Deok Seo
Jialie Shen
Guangming Shi
Hyunjung Shim
Jitae Shin
Donggyu Sim
Kwanghoon Sohn
Doug Young Suh
Yu-Wing Tai
Xinmei Tian
Cong Thang Truong
Anhong Wang
Liang Wang
Ling Wang
Lynn Wilcox
KokSheik Wong

Jinjian Wu
Xiao Wu
Yingchun Wu
Chen Xia
Jimin Xiao
Jing Xiao
Zhiwei Xing
Zixiang Xiong
Min Xu
Xinshun Xu
Long Xu
Toshihiko Yamasaki
Ming Yan
Keiji Yanai
Haichuan Yang
Huan Yang
Lili Yang
Lu Yang
Yang Yang
Yanhua Yang

Yuhong Yang
Jar-Ferr Yang
Chao Yao
Fuliang Yin
Lantao Yu
Li Yu
Lu Yu
Hui Yuan
Junsong Yuan
Zhaoquan Yuan
Zheng-Jun Zha
Guangtao Zhai
Cong Zhang
Qin Zhang
Xue Zhang
Yazhong Zhang
Yongdong Zhang
Lijun Zhao
Nan Zheng

Sponsoring Institutions

Realistic Broadcasting Research Center at GIST
Gwangju Convention and Visitors Bureau

Contents – Part I

Multimedia Applications and Services

Video Coding and Processing

Multimedia Representation Learning

Regular Poster Session

Visual Understanding and Recognition on Big Data

Coding and Reconstruction of Multimedia Data with Spatial-Temporal Information

Contents – Part II

Social Media Computing

Human Action Recognition in Social Robotics and Video Surveillance

Recent Advances in Image/Video Processing

**New Media Representation and Transmission Technologies for Emerging
UHD Services**

Special Poster Sessions

Image and Audio Processing

Internal Generative Mechanism Based Otsu Multilevel Thresholding Segmentation for Medical Brain Images

Yuncong Feng[1,2], Xuanjing Shen[1,2], Haipeng Chen[1,2(✉)], and Xiaoli Zhang[1,2]

[1] Key Laboratory of Symbolic Computation and Knowledge Engineering of Ministry of Education, Jilin University, Changchun 130012, China
chenhp@jlu.edu.cn
[2] College of Computer Science and Technology, Jilin University, Changchun 130012, China

Abstract. Recent brain theories indicate that perceiving an image visually is an active inference procedure of the brain by using the Internal Generative Mechanism (IGM). Inspired by the theory, an IGM based Otsu multilevel thresholding algorithm for medical images is proposed in this paper, in which the Otsu thresholding technique is implemented on both the original image and the predicted version obtained by simulating the IGM on the original image. A regrouping measure is designed to refining the segmentation result. The proposed method takes the predicted visual information generated by the complicated Human Visual System (HVS) into account, as well as the details. Experiments on medical MR-T2 brain images are conducted to demonstrate the effectiveness of the proposed method. The experimental results indicate that the IGM based Otsu multilevel thresholding is superior to the other multilevel thresholdings.

Keywords: Image segmentation · Otsu multilevel thresholding · Internal generative mechanism · Medical images

1 Introduction

Image segmentation is the problem of partitioning an image into some non-overlapping regions with inherent features (such as intensity, texture and color, etc.) in a semantically meaningful way and extracting the objects of interest from the background region [1–3]. Recently, image segmentation has been widely used in the fields of computer vision, pattern recognition and medical image processing 3. Segmentation of medical images including CT (computed tomography) and MRI (magnetic resonance imaging) is still a challenging work. For example, the intensity inhomogeneity and poor image contrast tend to result in missing the boundaries of organ or tissue [6, 7].

In the recent years, various segmentation methods have been proposed for medical image analysis [7–9]. Among them, thresholding is one of the most significant segmentation techniques because of its simplicity and effectiveness [10–12]. Classical thresholding techniques include the Otsu method [13], minimum error method [14] and maximum entropy [15]. They can be easily extended to multilevel thresholding

© Springer International Publishing Switzerland 2015
Y.-S. Ho et al. (Eds.): PCM 2015, Part I, LNCS 9314, pp. 3–12, 2015.
DOI: 10.1007/978-3-319-24075-6_1

problems. Manikandan et al. proposed a multilevel thresholding segmentation algorithm using real coded genetic algorithm with SBX (simulated binary crossover) for medical brain images [7]. In this method, the optimal multilevel threshold is found by maximizing the entropy. Maitra et al. proposed a novel optimal multilevel thresholding based on bacterial foraging for brain MRI segmentation [9], and the performance of the method could comprehensively outperform PSO (particle swarm optimization) [16] based method. However, almost all the traditional thresholding techniques have neglected an essential point, namely Human Visual System (HVS).

Human Visual System (HVS) is very important and complicated. Visual perception is not a direct translation of what we have seen, but a result of complicated psychological inference [17]. Recent developments on Bayesian brain theory and the free-energy principle show that the brain performs as an active inference procedure which is governed by the Internal Generative Mechanism (IGM) [18, 19]. Therefore, it would be meaningless to segment the medical images if we ignore the psychological inference mechanism of HVS. Enlightened by the IGM theory about the visual cognitive procedure, an IGM based Otsu multilevel thresholding for medical brain images was proposed in this paper. The proposed method is superior to the traditional thresholding methods because of its good robustness against the noise, moreover, the segmentation results with IGM is more consistent with the mechanism of HVS perceiving images.

2 Otsu Thresholding

We assume that the gray level of any image is $G = \{0, 1, \ldots, L-1\}$ ($L = 256$). For a grayscale image I with the size of $M * N$. The total number of pixels in it can be depicted as $M \times N = \sum_{i=0}^{L-1} n_i$, where n_i represents the number of pixels with the gray level i, and the probability of such pixels can be defined as $p_i = \frac{n_i}{M \times N}$, $(p_i \geq 0, i \in G)$, which satisfies $\sum_{i=0}^{L-1} p_i = 1$.

Assuming that image I is segmented into $K(K \geq 2)$ classes $(C = \{C_0, C_1, \ldots, C_{K-1}\}$ by $K-1$ thresholds $(\{t_0, t_1, \ldots, t_{K-2}\})$, $C_0 = \{0, 1, \ldots, t_0\}$ denotes the background region of image I, and the object regions are represented as $C_j = \{t_{j-1} + 1, \ldots, t_j\}$, $(j = 1, \ldots, K-1, t_{K-1} = L-1)$. The between-class variance of the image I is defined as follows,

$$\sigma_B^2 = \sum_{j=0}^{K-1} \omega_j (\mu_j - \mu_T)^2 \tag{1}$$

where, ω_j denotes the probability of class j, $\omega_0 = \sum_{i=0}^{t_0} p_i$ and $\omega_j = \sum_{i=t_{j-1}+1}^{t_j} p_i$, $(j = 1, \ldots, K-1)$. μ_j refers to the mean of class j, $\mu_0 = \sum_{i=0}^{t_0} i \frac{p_i}{\omega_0}$ and $\mu_j = \sum_{i=t_{j-1}+1}^{t_j} i \frac{p_i}{\omega_j}$, $(j = 1, \ldots, K-1)$. $\mu_T = \sum_{i=0}^{L-1} i p_i$ represents the total mean of the K classes.

The Otsu thresholding aims at finding the optimal thresholds $\{t_0^*, t_1^*, \ldots, t_{K-2}^*\}$ to maximize the objective function σ_B^2,

$$\{t_0^*, t_1^*, \ldots, t_{K-2}^*\} = \operatorname*{arg\ max}_{0 \leq t_0 < t_1 < \cdots < t_{K-2} \leq L-1} \{\sigma_B^2(t_0, t_1, \ldots, t_{K-2})\} \tag{2}$$

3 The Proposed Segmentation Algorithm

3.1 Segmentation Scheme

Let I denote an original image to be segmented. The framework of the proposed segmentation method is displayed in Fig. 1. Formally, the scheme can be summarized as follows,

- **Step. 1** Use the IGM to obtain the predicted layer P of the image I.
- **Step. 2** Perform the Otsu thresholding on the predicted layer P as well as the original image I, and the obtained segmentation maps are denoted as P_{otsu} and I_{otsu}, respectively.
- **Step. 3** Find out the pixels with different segmentation labels in P_{otsu} and I_{otsu}, which are denoted as controversial pixels. And then regroup them by maximizing the uniformity of each region.
- **Step. 4** Combine the maps of consistent pixels and the regrouped controversial pixels to obtain the final segmentation result.

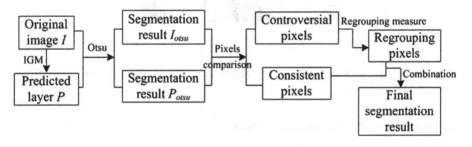

Fig. 1. The framework of the proposed method

3.2 Internal Generative Mechanism

One of attractive features of this method is that it takes the mechanism of HVS perceiving images into account in the process of image segmentation. In this paper, the autoregressive (AR) model is employed to simulate the IGM. Given an original image I with the size of $M * N$, x_{ij} denotes the pixel which is at i_{th} low and j_{th} column. S represents an image block with the size of $V * V$ in image I, and $S(x_{ij})$ denotes the image block in which x_{ij} is the central pixel. The AR model can predict the value of pixel x_{ij} as follows.

$$x'_{ij} = \sum_{x_{pq} \in S(x_{ij})} C_{pq} x_{pq} + \varepsilon, \quad (1 \le p \le V, 1 \le q \le V) \tag{3}$$

where, x_{ij}' is the predicted value of the pixel x_{ij}. x_{pq} denotes the pixel of the image block $S(x_{ij})$. $C_{pq} = \dfrac{I(x_{ij}; x_{pq})}{\sum_{x_{kl} \in S(x_{ij})} I(x_{ij}; x_{kl})}$ refers to the normalized coefficient of the mutual information $I(x_{ij}; x_{pq})$ between x_{ij} and x_{pq}. ε denotes the additive white noise.

According to Eq. (1), it can be noted that the predicted value of each pixel is determined by itself and its surrounding pixels in the image block. The whole predicted layer of image I can be described formally as,

$$P = \{x'_{ij} | 1 \le i \le M, 1 \le j \le N\} \tag{4}$$

Figure 2 gives an example of the predicted image obtained by the IGM. From Fig. 2 (b), it can be seen that the predicted image preserves the primary visual information of the original image and eliminates the weak edge simultaneously. Therefore, the predicted image is conducive to the subsequent image segmentation.

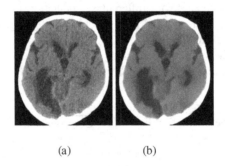

(a)   (b)

Fig. 2. An example of predicted image obtained by the IGM: (a) original image, (b) predicted image.

3.3   Regrouping the Controversial Pixels

After the processing of IGM, the Otsu thresholding is used to segment the predicted image. Figure 3 gives an example of Otsu based multilevel thresholding for a brain image. In Fig. 3(b) and (c) are two segmentation results obtained by performing the Otsu thresholding with $K = 3$ and $K = 4$, respectively. To illustrate the difference between the IGM and the common filtering, Gaussian filter is taken as an example here. (d) and (e) display the segmentation results of the Gaussian filter based Otsu thresholding, respectively. While (f) and (g) are obtained by using the IGM based Otsu thresholding with $K = 3$ and $K = 4$, respectively.

From Fig. 3, it can be observed that the segmentation results in Figs. 3(f) and 2(g) are much better than Fig. 3(b)–(e) on the whole. There are a few of fragmented small regions, namely the misclassified pixels, are existed in Fig. 3(b)–(e), while such a case

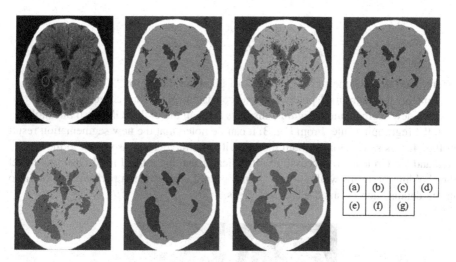

Fig. 3. An example of Otsu based multilevel thresholding for a brain image: (a) original image, (b) Otsu with $K = 3$, (c) Otsu with $K = 4$, (d) Gaussian-Otsu with $K = 3$, (e) Gaussian-Otsu with $K = 4$, (f) IGM-Otsu with $K = 3$, (g) IGM-Otsu with $K = 4$.

does not occur in Fig. 3(f)–(g). It indicates that the IGM can improve the performance of the Otsu multilevel thresholding for the medical image; in addition, it is superior to the traditional filtering. However, the segmentation result obtained by the IGM-Otsu thresholding still has some imperfections due to the complex tissue structure in medical images. For example, in the red arrows area of Fig. 3(g), it can be found that some small edges are not segmented successfully compared with that area in the original image Fig. 3(a). But the edges in the same area of Fig. 3(c) and (e) are segmented clearly. In addition, we can see a very small white area in Fig. 3(a) (see the red circle) if looking closely. Unfortunately, the small area is segmented incorrectly in Fig. 3(g) as well as Fig. 3(e), contrary to the case in Fig. 3(c). It is meaning that the IGM-Otsu thresholding is not very satisfactory in some details.

Therefore, the incorrect segmentation areas need to be regrouped to reduce the misclassification pixels, and a special regrouping rule is designed in the paper to solve this problem. First of all, a new definition is given here. Supposing that an original image I is segmented into K classes, P_{otsu} and I_{otsu} are two segmentation results obtained by IGM-Otsu thresholding and traditional Otsu thresholding, respectively.

Definition. x is a pixel of the original image I, and $x \in R_i^{P_{otsu}} \wedge x \in R_j^{I_{otsu}} (1 \leq i,j \leq K)$, where, $R_i^{P_{otsu}}$ denotes the i_{th} segmented region of P_{otsu}, and $R_j^{I_{otsu}}$ denotes the j_{th} segmented region of I_{otsu}. If $i \neq j$, x is a controversial pixel; otherwise, x is a consistent pixel.

The controversial pixel x will be regrouped based on the discrepancy between the gray level of x and the mean gray level of pixels in the region which x belongs to. The regrouping rule can be denoted formally as,

$$x \in \begin{cases} R_i^{P_{otsu}}, & if\,|x - m(R_i^{P_{otsu}})| < |x - m(R_j^{I_{otsu}})| \\ R_j^{I_{otsu}}, & otherwise \end{cases} \qquad (5)$$

where, $m(\cdot)$ denotes the mean gray level of pixels in the region "$\cdot$".

Finally, the regrouped pixels and the consistent pixels are combined together to obtain a new segmentation. Figure 4 displays an example of the IGM-Otsu thresholding with the regrouping rule. From Fig. 3, it can be noted that the new segmentation result in Fig. 4(b) is satisfactory in both the integrity and the details (as shown in the arrows area and the circle area) successfully. It indicates that the refined IGM-Otsu multilevel thresholding performs well not only on the overall segmentation performance but also in the details such as small edges and areas.

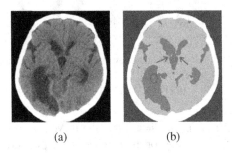

(a)   (b)

Fig. 4. An example of IGM-Otsu thresholding with the regrouping rule: (a) original image, (b) segmentation result with $K = 4$ after regrouping the controversial pixels.

4   Experimental Results and Analysis

4.1   Experimental Settings

To verify the performance of the IGM based Otsu multilevel thresholding, it is compared with some popular multi-threshold segmentation methods including RGA-SBX (Real coded Genetic with Simulated Binary Crossover) [7], BF (Bacterial Foraging), ABF (Adaptive BF) [8], and PSO (Particle Swarm Optimization) [16] with the number of thresholds $p = 2, 3, 4$. Eight MR-T2 brain slices (as shown in Fig. 5) are selected in the experiments. The test images are downloaded freely from a website (http://www.med.harvard.edu/aanlib/home.html) of Harvard Medical School. In addition, the uniformity measure [7, 9] is employed to evaluate the performance of the proposed method precisely.

4.2   Experimental Results

Because of the limitation of the paper space, we only list the segmentation results obtained by the proposed method in the case of $p = 4$ in Fig. 6. From the figure, it can be seen that the segmented images provide clear information and good visual effects. The gray consistency is maintained well in each segmented region.

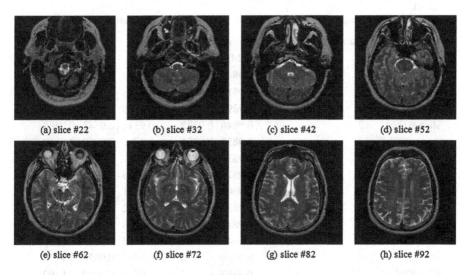

Fig. 5. MR-T2 brain slices.

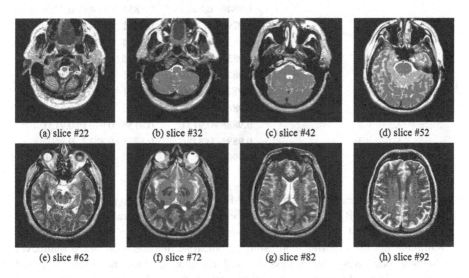

Fig. 6. Segmentation results obtained by the proposed method with $p = 4$.

The comparison of uniformity measure for the proposed method and other segmentation methods is depicted in Table 1. The best results have been marked in bold. From Table 1, it can be observed that, except the slice #042, the proposed algorithm obtains the highest value of the uniformity measure on the remaining seven brain slices, which justifies its superiority to others. Especially for the four testing images, from #062 to #092, all the values obtained by the IGM based thresholding with $p = 2, 3, 4$ are significantly higher than those of others. Comparing to BF, ABF and PSO, the IGM works better for all the testing images with various thresholds. For the slice #042, it can

Table 1. Comparison of the uniformity measure for segmentation methods

Testing images	Number of thresholds	Uniformity measure (u)				
		IGM	RGA-SBX	BF	ABF	PSO
Slice #022	2	**0.9847**	0.9569	0.9569	0.9569	0.9552
	3	**0.9818**	0.9769	0.9708	0.9696	0.9672
	4	**0.9893**	0.9824	0.9765	0.9698	0.9420
Slice #032	2	**0.9846**	0.9342	0.9342	0.9342	0.9368
	3	**0.9857**	0.9801	0.9716	0.9600	0.9619
	4	**0.9854**	0.9848	0.9697	0.9766	0.9144
Slice #042	2	**0.9823**	0.9246	0.9246	0.9246	0.9271
	3	**0.9809**	0.9548	0.9721	0.9689	0.9585
	4	0.9830	**0.9865**	0.9752	0.9821	0.9465
Slice #052	2	**0.9844**	0.9128	0.9128	0.9128	0.9158
	3	**0.9842**	0.9467	0.9713	0.9673	0.9523
	4	**0.9860**	0.9856	0.9764	0.9834	0.9372
Slice #062	2	**0.9809**	0.9015	0.9047	0.9049	0.9192
	3	**0.9822**	0.9030	0.9135	0.9029	0.8777
	4	**0.9840**	0.8989	0.8856	0.8988	0.9236
Slice #072	2	**0.9796**	0.9041	0.9041	0.9041	0.9068
	3	**0.9831**	0.8992	0.9084	0.8985	0.9034
	4	**0.9833**	0.8666	0.8876	0.8804	0.8809
Slice #082	2	**0.9822**	0.9091	0.9091	0.9091	0.9120
	3	**0.9843**	0.8849	0.8621	0.8661	0.8852
	4	**0.9866**	0.8695	0.8479	0.8622	0.8619
Slice #092	2	**0.9888**	0.9131	0.9156	0.9131	0.9131
	3	**0.9895**	0.8786	0.8751	0.8827	0.8607
	4	**0.9905**	0.8641	0.8583	0.8514	0.9490

be noted that the values achieved by the IGM are the highest when $p = 2, 3$. In the case of $p = 4$, the RGA-SBX gives the highest value; however, the value obtained by the IGM is not the lowest, and it ranks in the second place.

Summing up, the experimental results demonstrate that the IGM based multilevel thresholding possesses effectiveness as well as good performance. The proposed method outperforms the other segmentation methods comprehensively with increase in the number of thresholds.

5   Conclusion

In this paper, the problem of automatically segmentation for medical brain images is investigated. Visual perception is not a direct translation of what we have seen, but a result of complicated psychological inference. Recent developments on the brain theory indicate that perceiving an image visually is an active inference procedure of the brain by using the Internal Generative Mechanism (IGM). Inspired by the theory, a novel

Otsu multilevel thresholding based on the IGM is proposed in the paper. In the process of segmentation, the proposed method takes the predicted visual information generated by the complicated Human Visual System (HVS) into account, as well as the details. To demonstrate the performance of the IGM based multilevel thresholding, eight medical brain images are selected in the experiments and four other popular multilevel thresholdings are employed to comparing with the proposed method. The experimental results indicate that the segmented images provide more clear information and the visual effects are much better with the increase of thresholds. According to the values of the uniformity measure, the proposed method performs much better than other multilevel thresholdings.

Acknowledgements. This research is supported by the National Natural Science Foundation of China for Youths (No. 61305046), Jilin Province Science Foundation for Youths (No. 20130522117JH), and the Natural Science Foundation of Jilin Province (No. 20140101193JC).

References

1. Läthén, G. Segmentation Methods for Medical Image Analysis (2010)
2. Farmer, M.E., Jain, A.K.: A wrapper-based approach to image segmentation and classification. IEEE Trans. Image Process. **14**(12), 2060–2072 (2005)
3. Yilmaz, A., Javed, O., Shah, M.: Object tracking: a survey. ACM Comput. Surv. (CSUR) **38**(4), 13 (2006)
4. Sun, C., Lu, H., Zhang, W., Qiu, X., Li, F., Zhang, H.: Lip segmentation based on facial complexion template. In: Ooi, W.T., Snoek, C.G.M., Tan, H.K., Ho, C.-K., Huet, B., Ngo, C.-W. (eds.) PCM 2014. LNCS, vol. 8879, pp. 193–202. Springer, Heidelberg (2014)
5. Peng, B., Zhang, D.: Automatic image segmentation by dynamic region merging. IEEE Trans. Image Process. **20**(12), 3592–3605 (2011)
6. Maulik, U.: Medical image segmentation using genetic algorithms. IEEE Trans. Inf Technol. Biomed. **13**(2), 166–173 (2009)
7. Manikandan, S., Ramar, K., Iruthayarajan, M.W., et al.: Multilevel thresholding for segmentation of medical brain images using real coded genetic algorithm. Measurement **47**, 558–568 (2014)
8. Sathya, P.D., Kayalvizhi, R.: Optimal segmentation of brain MRI based on adaptive bacterial foraging algorithm. Neurocomputing **74**(14), 2299–2313 (2011)
9. Maitra, M., Chatterjee, A.: A novel technique for multilevel optimal magnetic resonance brain image thresholding using bacterial foraging. Measurement **41**(10), 1124–1134 (2008)
10. Sezgin, M.: Survey over image thresholding techniques and quantitative performance evaluation. J. Electron. Imaging **13**(1), 146–168 (2004)
11. Liao, P.S., Chen, T.S., Chung, P.C.: A fast algorithm for multilevel thresholding. J. Inf. Sci. Eng. **17**(5), 713–727 (2001)
12. Huang, D.Y., Wang, C.H.: Optimal multi-level thresholding using a two-stage Otsu optimization approach. Pattern Recogn. Lett. **30**(3), 275–284 (2009)
13. Otsu, N.: A threshold selection method from gray-level histograms. Automatica **11**(285–296), 23–27 (1975)
14. Kittler, J., Illingworth, J.: Minimum error thresholding. Pattern Recogn. **19**(1), 41–47 (1986)

15. Kapur, J.N., Sahoo, P.K., Wong, A.K.C.: A new method for gray-level picture thresholding using the entropy of the histogram. Comput. Vis. Graph. Image Process. **29**(3), 273–285 (1985)
16. Chander, A., Chatterjee, A., Siarry, P.: A new social and momentum component adaptive PSO algorithm for image segmentation. Expert Syst. Appl. **38**(5), 4998–5004 (2011)
17. Sternberg, R.: Cognitive Psychology. Cengage Learning, Belmont (2011)
18. Zhai, G., Wu, X., Yang, X., et al.: A psychovisual quality metric in free-energy principle. IEEE Trans. Image Process. **21**(1), 41–52 (2012)
19. Kersten, D., Mamassian, P., Yuille, A.: Object perception as Bayesian inference. Annu. Rev. Psychol. **55**, 271–304 (2004)
20. Friston, K.: The free-energy principle: a unified brain theory? Nat. Rev. Neurosci. **11**(2), 127–138 (2010)
21. Zhang, X., Li, X., Feng, Y., et al.: Image fusion with internal generative mechanism. Expert Syst. Appl. **42**(5), 2382–2391 (2015)

Efficient Face Image Deblurring via Robust Face Salient Landmark Detection

Yinghao Huang, Hongxun Yao$^{(\boxtimes)}$, Sicheng Zhao, and Yanhao Zhang

School of Computer Science and Technology, Harbin Institute of Technology,
Harbin, China
h.yao@hit.edu.cn

Abstract. Recent years have witnessed great progress in image deblurring. However, as an important application case, the deblurring of face images has not been well studied. Most existing face deblurring methods rely on exemplar set construction and candidate matching, which not only cost much computation time but also are vulnerable to possible complex or exaggerated face variations. To address the aforementioned problems, we propose a novel face deblurring method by integrating classical L_0 deblurring approach with face landmark detection. A carefully tailored landmark detector is used to detect the main face contours. Then the detected contours are used as salient edges to guide the blind image deconvolution. Extensive experimental results demonstrate that the proposed method can better handle various complex face poses while greatly reducing computation time, as compared with state-of-the-art approaches.

Keywords: Efficient · Robust · Face · Image deblurring · Landmark detection

1 Introduction

With the wide popularity of various smart hand-held devices, more and more images and videos are captured and shared by people to record the daily life. Unfortunately, due to various reasons, such as the hand shake of photographer and the motion of target objects, quite a proportion of these images are undesirably blurred, which can degrade ideal sharp images heavily. This predicament catalyzes an important research in the past decade, known as image deblurring, which aims to restore the latent sharp images from the blurred ones.

Mathematically, the underlying process of image degrading due to camera motion can be formulated as:

$$Y = k * X + \varepsilon, \tag{1}$$

where Y refers to the blurred image, k denotes the blur kernel, X and ε correspond to the latent sharp image and the added noise, respectively. Because of the intrinsic illness of this problem, extra information is necessary to constrain

© Springer International Publishing Switzerland 2015
Y.-S. Ho et al. (Eds.): PCM 2015, Part I, LNCS 9314, pp. 13–22, 2015.
DOI: 10.1007/978-3-319-24075-6_2

the solutions. One effective way is to figure out the common and dominating differences between the sharp images and the blurred ones. Then by requiring the restored images to have such properties of natural sharp images, the deblurring process is encouraged to move towards the desired direction. Along this way, various priors have been proposed, such as the sparsity constraints and varient normalized version [2,11], spectral irregularities [7], patch priors [18] and heavy-tailed gradient distributions [6,13]. In terms of solving strategies, most of the state-of-the-art methods are explicitly or implicitly dependent on the intermediate salient edge structures [4,5,10,17,19,20]. The underlying observation is that the original strong edges tend to be salient even after blurring, which can provide useful guiding information for latent sharp image restoration.

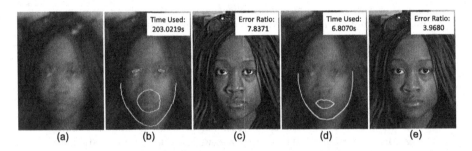

Fig. 1. Illustration of face image deblurring. (a): blurred image; (b) and (d): salient contour achieved by [16] and the proposed method; (c) and (e): corresponding results. Computation cost and error ratios are provided.

Unfortunately, these general-purpose methods fail in some specific cases, such as text images [3,16], low-light images [9] and face images [1,8]. In this paper, we mainly focus on face images. Though the method proposed in [8] can achieve decent results, it requires a sharp reference image sharing similar content with the blurred one, which is unrealistic in reality. Pan et al. [1] extended this approach by constructing a large exemplar set and selecting the best fit candidate as the guidance. Experiments showed that the method can achieve decent generalization ability and satisfying results. However, this method is heavily dependent on exemplar images, which makes it time-consuming and hard to handle complex or exaggerated face poses.

Another research area interesting us is face landmark detection. To accurately locate the key points on a face image, a lot of methods have been proposed [25–28]. Among them, the work of [25] achieved satisfying results in real time, making it feasible to fast detect various kinds of salient edges on face images. Our work is mainly motivated by this observation. As illustrated in Fig. 1, our method can achieve more promising visual quality when processing challenging images, while the computation time is much shorter. Especially when the shape of the testing face is not common or the face pose is exaggerated, the visual quality gain is much obvious.

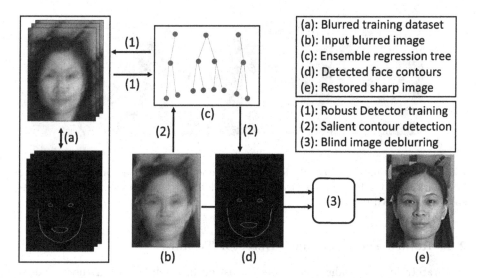

Fig. 2. Framework of the proposed method.

2   The Proposed Method

In this section, we firstly explain the motivation of this work, then give a description of our face deblurring method in detail. As shown in Fig. 2, our method mainly consists of three components: (1) robust face landmark detector training, (2) salient contour detection and (3) blind image deblurring.

2.1   Motivation

As stated previously, most existing face deblurring methods are heavily dependent on exemplar set construction and candidate matching. In consideration of the representation ability, a larger and more complete exemplar set is preferred. Inevitably, much time will be consumed in the candidate matching stage, where the best fit exemplar is chosen by comparing the test image with each candidate image. Taking the method in [1] for example, even in our C++ implementation using a moderate exemplar set with 2245 images, the entire candidate matching process still costs about 198 seconds. Compared with the 5 s required by the remaining procedure, this process is too slow to endure. What's worse, due to the great flexibility of face shapes and poses, it is very hard to construct a complete candidate set. When the best matching exemplar image doesn't fit the test image well, the final restored results will be degraded sharply. This predicament is illustrated in Fig. 1 (b), where the mouse contour matches the real one poorly. To address these two problems, we propose to incorporate face landmark detection with image deblurring, which is elaborated in the following sections.

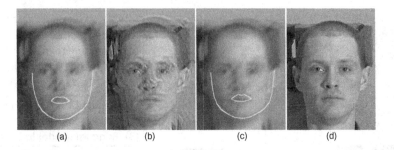

(a)   (b)   (c)   (d)

Fig. 3. Illustration of landmark detection. (a) and (c): detected contours using the landmark detector training on clear images and blurred images; (b) and (d): corresponding restored results.

2.2   Robust Face Landmark Detector Training

Accurately locating salient contours of testing face image within short time is a key step. Due to the difficulty of the problem and its intrinsic pairwise comparison of features, example-based methods cannot fulfill such goal, even with a limited training set. To address the above issues, an efficient and effective face landmark detection method similar to the one proposed in [25] is adopted, which is built upon the ensemble of regression forests.

Ensemble Regression Forests Landmark Detection. Given a training dataset $D = \{(I_1, S_1), \ldots, (I_n, S_n)\}$ with I_i being a face image and S_i its corresponding shape vector, the goal of training ensemble of regression trees is to achieve a series of regression function $r_t(I, S^{(t)})$, which makes the closest landmark prediction to the ground-truth. In the t-th step, the target function can be formulated as

$$\triangle S_i^{(t)} = S_{\pi_i} - \hat{S}_i^{(t-1)}, \tag{2}$$

where $\hat{S}_i^{(t-1)}$ is the prediction result of the previous step. To this end, each regression function r_t is learned using the classical gradient boosting tree algorithm. The intensity difference of two pixels is used as the decision indicator, which is both light-weighted and relatively insensitive to global lighting changes. A piecewise constant function is used to approximate the underlying function for each regression tree, while leaf nodes are represented by constant vectors. To determine node split, we try to minimize the following loss function:

$$E(Q, \theta) = \sum_{s \in \{l,r\}} \sum_{i \in Q_{\theta,s}} \|r_i - \mu_{\theta,s}\|^2, \tag{3}$$

where Q is the set of indices of the training examples at a node, $Q_{\theta,l}$ are those sent to the left node induced by θ, r_i is the current residual vectors and

$$\mu_{\theta,s} = \frac{1}{|Q_{\theta,s}|}, \text{ for } s \in \{l,r\}, \tag{4}$$

The entire procedure is quite complicated and several other tricks introduced in [25] are also adopted, which speeds up the process greately.

Training on Blurred Face Images. As demonstrated in Fig. 3, the landmark detector trained on standard dataset with sharp images often fails blurred ones, which can be a disaster to later processing. To endow the landmark detector with the ability of blur tolearting, we synthesised a blurred landmark detection dataset from Helen [29]. Specifically, 2000 images are randomly chosen from all the 2330 images of the Helen dataset, and then blurred with each of the eight kernels in [13] with their ground-truth labels unchanged. This results in a synthesised blurred face ladmark detection dataset consisting of 160,000 images, which is then used to train the landmark detector. The left 330 images are similarly blurred to generate the test dataset. According to the same evaluation criteria [25], the average test error is 0.4. Though the test error seems large compared with the result reported in the original paper [25], the detection result is accurate enough to achieve good deblurring results in the next step.

2.3   Salient Contour Detection

In this stage, the main face contours are obtained by applying the trained detector on the test image. The experiments in [1] demonstrated that quite decent results can be achieved by using the contours of the lower face, the eyes and the mouth. However we found that in face landmark detection, as a quite difficult part, eyes usually cannot be accurately located to the extend of providing usable guiding information. So only the contours of the lower face and the mouth are singled out and used in the next stage.

2.4   Blind Image Deblurring

Now that the salient contours ∇S have been acquired, the remaining step is to recover the latent sharp image with the guide of the salient edges. A two-stage procedure similar to [1] is used in this paper:

Kernel Estimation. The two target function used to estimate the blur kernel is as follows:

$$\min_X \|X \star k - Y\|_2^2 + \lambda \|\nabla I\|_0, \tag{5}$$

$$\min_k \|\nabla S \star k - \nabla Y\|_2^2 + \gamma \|k\|_2^2, \tag{6}$$

where λ and γ are coefficients of the regularization terms. As shown in [20], some ringing artifacts in X can be effectively removed by using the L_0-norm in Eq. 5. While in Eq. 6, the process of blur kernel estimation can be stabilized by appling the L_2-norm based regularization. The subproblem 5 is solved by employing the half-quadratic splitting L_0 minimization method proposed in [20], while the other subproblem 6 can be solved via the conjugate gradient method.

Latent Image Recovery. After acquiring the blur kernel, quite a number of non-blind deconvolution methods can be used to restore the latent sharp image. Considering the fairness of experimental comparison, we also adopted the method with a hyper-laplacian prior $L_{0.8}$ [15] used in [1].

3   Experimental Results

In this section, extensive experiments are conducted to compare our method with the state-of-the-art in this area. All experiments are conducted on our desktop with an Intel Xeon CPU and 12 GB RAM. To further compare the computation cost with the baseline method [1], we reimplemented our method and the baline method in C++.

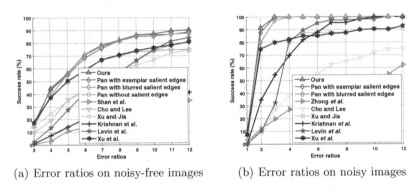

(a) Error ratios on noisy-free images   (b) Error ratios on noisy images

Fig. 4. Quantatitive comparisons with some state-of-the-art single-image deblurring methods: Pan et al. [1], Shan et al. [17], Cho and Lee [4], Xu and Jia [19], Krishnan et al. [11], Levin et al. [14], Zhong et al. [23], and Xu et al. [20].

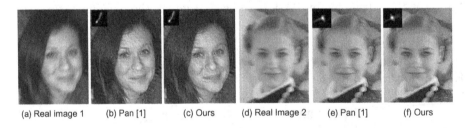

(a) Real image 1   (b) Pan [1]   (c) Ours   (d) Real Image 2   (e) Pan [1]   (f) Ours

Fig. 5. Deblurring results of the real images used in [1]

3.1   Experiments on Synthesised Dataset and Real Images

To validate the effectiveness of our method, experiments are conducted on the synthesised dataset and real images used in [1] with the same settings. Qualitative and qualitative comparison results are provided in Figs. 4 and 5. Due to space constraint, only restored images of our method and the baseline method are shown in Fig. 5. As we can see, our overall performance is comparable or even better than those achieved by the best baseline method.

3.2 Computation Cost Comparison

To further compare the computation cost, we divided the entire procedure of [1] and our method into two stages: salient edge acquisition (SEA) and blind deconvolution (BD). When carrying out the previous experiment on synthesised dataset, for each image the time used in the two stages is recorded. Then we computed the average time of the total 480 trails, as shown in Table 1.

Table 1. Running time comparison. The metric unit is second (s).

	SEA	BD	Total
[1]	198.0210	5.0009	203.0219
Ours	1.5070	5.3000	6.8070

It is easy to observe that, our method is much faster than the baseline method. Note that this result is on a quite small size image with 320 * 240 pixels. When applying on a larger image, the time advantage of our method can be more notable.

3.3 Adaptation to Complex Face Poses

To further compare the ability to process complex face poses, we carefully cropped some face images having exaggerated expression or non-frontal orientation from the Helen dataset. Two experiments are conducted on this new dataset with the same experimental settings, one on the original dataset and another on its noisy version (one percent random noise is added). The result is shown in Fig. 6. Benefiting from the flexibility of the underlying face landmark detection, our method can better handle these difficult situations. Some experimental results are illustrated in Fig. 7. Besides, as shown in Fig. 8, our method is capable of processing images with more than one face, while the baseline method cannot handle this case naturally.

3.4 Rolling Guidance Face Deblurring

If speed is not the main concern, following the loopy style in [30], our method can be further boosted by simply adopting intermediate result image as the guidance and iterating the entire process for several times. Because in this paper algorithm efficiency is also one of the main focuses and the results achieved in a single operation is quite satisfying, this trait of our method is left for future work.

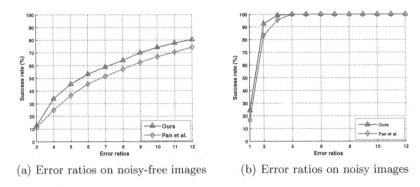

(a) Error ratios on noisy-free images   (b) Error ratios on noisy images

Fig. 6. Quantative comparisionson on the new dataset: Pan et al. [1].

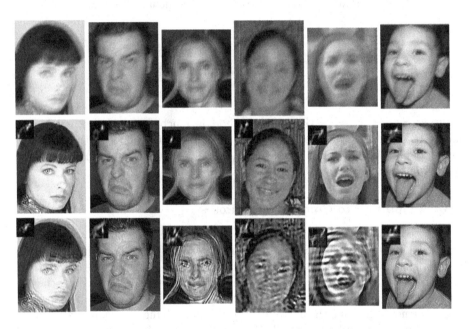

Fig. 7. Deblurring results on complex face poses. From top to bottom: the input blurred images; final restored results by our method; final restored results achieved by [1].

(a) Input   (b) Detected Contours   (c) Final Result

Fig. 8. Our deblurring result on the example image with more than one faces.

4   Conclusions

In this paper, we proposed a novel blind face image deblurring method by incorporating fast landmark detection with classical L_0 method. Benefiting from the accuracy and flexibility of the underlying face landmark detector, the proposed method can better handle challenging face images with various poses and expressions. At the same time, by avoiding time-consuming pair-wise comparison of face features, our method is much more efficient. Extensive experiments wiOur future work will focus on how to improve the deblurring quality further.

Acknowledgement. This work was supported in part by the National Science Foundation of China No. 61472103, and Key Program Grant of National Science Foundation of China No. 61133003.

References

1. Pan, J., Hu, Z., Su, Z., Yang, M.-H.: Deblurring face images with exemplars. In: Fleet, D., Pajdla, T., Schiele, B., Tuytelaars, T. (eds.) ECCV 2014, Part VII. LNCS, vol. 8695, pp. 47–62. Springer, Heidelberg (2014)
2. Cai, J.F., Ji, H., Liu, C., Shen, Z.: Framelet based blind motion deblurring from a single image. IEEE Trans. Image Process. **21**(2), 562–572 (2012)
3. Cho, H., Wang, J., Lee, S.: Text image deblurring using text-specific properties. In: Fitzgibbon, A., Lazebnik, S., Perona, P., Sato, Y., Schmid, C. (eds.) ECCV 2012, Part V. LNCS, vol. 7576, pp. 524–537. Springer, Heidelberg (2012)
4. Cho, S., Lee, S.: Fast motion deblurring. ACM Trans. Graph. **28**(5), 145 (2009)
5. Cho, T.S., Paris, S., Horn, B.K.P., Freeman, W.T.: Blur kernel estimation using the radon transform. In: CVPR, pp. 241–248 (2011)
6. Fergus, R., Singh, B., Hertzmann, A., Roweis, S.T., Freeman, W.T.: Removing camera shake from a single photograph. ACM Trans. Graph. **25**(3), 787–794 (2006)
7. Goldstein, A., Fattal, R.: Blur-Kernel estimation from spectral irregularities. In: Fitzgibbon, A., Lazebnik, S., Perona, P., Sato, Y., Schmid, C. (eds.) ECCV 2012, Part V. LNCS, vol. 7576, pp. 622–635. Springer, Heidelberg (2012)
8. HaCohen, Y., Shechtman, E., Lischinski, D.: Deblurring by example using dense correspondence. In: ICCV, pp. 2384–2391 (2013)
9. Hu, Z., Cho, S., Wang, J., Yang, M.H.: Deblurring low-light images with light streaks. In: CVPR, pp. 3382–3389 (2014)
10. Joshi, N., Szeliski, R., Kriegman, D.J.: PSF estimation using sharp edge prediction. In: CVPR, pp. 1–8 (2008)
11. Krishnan, D., Tay, T., Fergus, R.: Blind deconvolution using a normalized sparsity measure. In: CVPR, pp. 2657–2664 (2011)
12. Levin, A., Weiss, Y., Durand, F., Freeman, W.T.: Understanding and evaluating blind deconvolution algorithms. In: CVPR, pp. 1964–1971 (2009)
13. Levin, A., Weiss, Y., Durand, F., Freeman, W.T.: Efficient marginal likelihood optimization in blind deconvolution. In: CVPR, pp. 2657–2664 (2011)
14. Levin, A., Fergus, R., Durand, F., Freeman, W.T.: Image and depth from a conventional camera with a coded aperture. ACM Trans. Graph. **26**(3), 70 (2007)
15. Nishiyama, M., Hadid, A., Takeshima, H., Shotton, J., Kozakaya, T., Yamaguchi, O.: Facial deblur inference using subspace analysis for recognition of blurred faces. IEEE Trans. Pattern Anal. Mach. Intell. **33**(4), 838–845 (2011)

16. Pan, J., Hu, Z., Su, Z., Yang, M.H.: Deblurring text images via L0-regularized intensity and gradient prior. In: CVPR (2014)
17. Shan, Q., Jia, J., Agarwala, A.: High-quality motion deblurring from a single image. ACM Trans. Graph. **27**(3), 73 (2008)
18. Sun, L., Cho, S., Wang, J., Hays, J.: Edge-based blur kernel estimation using patch priors. In: ICCP, pp. 1–8 (2013)
19. Xu, L., Jia, J.: Two-phase kernel estimation for robust motion deblurring. In: Daniilidis, K., Maragos, P., Paragios, N. (eds.) ECCV 2010, Part I. LNCS, vol. 6311, pp. 157–170. Springer, Heidelberg (2010)
20. Xu, L., Zheng, S., Jia, J.: Unnatural L0 sparse representation for natural image deblurring. In: CVPR, pp. 1107–1114 (2013)
21. Yitzhaky, Y., Mor, I., Lantzman, A., Kopeika, N.S.: Direct method for restoration of motion-blurred images. J. Opt. Soc. Am. A **15**(6), 1512–1519 (1998)
22. Zhang, H., Yang, J., Zhang, Y., Huang, T.S.: Close the loop: joint blind image restoration and recognition with sparse representation prior. In: ICCV, pp. 770–777 (2011)
23. Zhong, L., Cho, S., Metaxas, D., Paris, S., Wang, J.: Handling noise in single image deblurring using directional filters. In: CVPR, pp. 612–619 (2013)
24. He, K., Sun, J., Tang, X.: Guided image filtering. In: Daniilidis, K., Maragos, P., Paragios, N. (eds.) ECCV 2010, Part I. LNCS, vol. 6311, pp. 1–14. Springer, Heidelberg (2010)
25. Kazemi, V., Sullivan, J.: One millisecond face alignment with an ensemble of regression trees. In: CVPR, pp. 1867–1874 (2014)
26. Cao, X., Wei, Y., Wen, F., Sun, J.: Face alignment by explicit shape regression. IJCV **107**(2), 177–190 (2014)
27. Kazemi, V., Sullivan, J.: Face alignment with part-based modeling. In: BMVC, pp. 27–1 (2011)
28. Edwards, G.J., Cootes, T.F., Taylor, C.J.: Advances in active appearance models. In: ICCV, pp. 137–142 (1999)
29. Le, V., Brandt, J., Lin, Z., Bourdev, L., Huang, T.S.: Interactive facial feature localization. In: Fitzgibbon, A., Lazebnik, S., Perona, P., Sato, Y., Schmid, C. (eds.) ECCV 2012, Part III. LNCS, vol. 7574, pp. 679–692. Springer, Heidelberg (2012)
30. Zhang, Q., Shen, X., Xu, L., Jia, J.: Interactive facial feature localization. In: ECCV, pp. 815–830 (2014)

Non-uniform Deblur Using Gyro Sensor and Long/Short Exposure Image Pair

Seung Ji Seo, Ho-hyoung Ryu, Dongyun Choi, and Byung Cheol Song[✉]

Department of Electronic Engineering, Inha University,
Incheon, Republic of Korea
bcsong@inha.ac.kr

Abstract. This paper proposes a deblur algorithm using IMU sensor and long/short exposure-time image pair. First, we derive an initial blur kernel from gyro data of IMU sensor. Second, we refine the blur kernel by applying Lucas-Kanade algorithm to long/short exposure-time image pair. Using residual deconvolution based on the non-uniform blur kernel, we synthesize a final image. Experimental results show that the proposed algorithm is superior to the state-of-the-art methods in terms of subjective/objective visual quality.

Keywords: Deblurring · Non-uniform blur · IMU · Gyro

1 Introduction

Most of the deblur algorithms assume uniform blur kernel for simple computation [1, 2, 6]. However, actual blur kernel is spatially non-uniform because of camera rotation. To resolve the non-uniform blur phenomenon, many approaches have been devised [3–5]. For example, Tai et al. projected three-dimensional (3D) motion into two-dimensional (2D) motion using a finite number of homographies [4]. However, since Tai et al.'s method is a non-blind approach, it does not work when camera motion is unknown. In order to solve this problem, Whyte et al. assumed that non-uniform blur occurs due to camera rotation. They proposed a blur kernel estimation which computes homography by measuring camera rotation [5]. However, their method cannot handle a wide range of camera movement. Cho et al. presented a homography-based algorithm using two blurred images [3]. They computed homographies using Lucas-Kanade algorithm, and estimated a blur kernel from the computed homographies. But, their method is sometimes unstable because it iteratively applies Lucas-Kanade method, thereby its kernel estimation may often diverge.

In order to overcome the drawbacks of the conventional non-uniform deblur algorithms, we present a deblur algorithm using IMU sensor and a pair of long/short exposure-time images. First, camera rotation data is obtained using gyro sensor to improve stability, and calculate an initial homography. Second, the initial blur kernel is refined by applying Lucas-Kanade algorithm to two different exposure-time images. Finally, a deblurred image is produced by residual deconvolution based on the non-uniform blur kernel. Experimental results show that the proposed algorithm provides clearer edges with less ringing

© Springer International Publishing Switzerland 2015
Y.-S. Ho et al. (Eds.): PCM 2015, Part I, LNCS 9314, pp. 23–31, 2015.
DOI: 10.1007/978-3-319-24075-6_3

artifacts than the state-of-the-art algorithms. Also, the proposed algorithm provides higher PSNR of 3.6 dB on average than Xu and Jia's [6] as the state-of-the-art deblur algorithm.

2   The Proposed Algorithm

The proposed algorithm consists of several steps: First, the rotation angle of the moving camera is obtained from the gyro sensor. Second, an initial homography set is calculated using the rotation angle and the camera intrinsic matrix. Third, an initial blur kernel is derived by the homography set. Fourth, the initial kernel is refined by using a pair of long/short exposure-time images. Finally, an input blurred image is deblurred using the refined blur kernel and non-uniform residual deconvolution.

The proposed algorithm has the following contributions: First, accurate blur kernel estimation is possible because the blur kernel can be refined by using near-true camera motion derived from IMU sensor embedded in the camera body. Second, probable artifacts such as ringing are effectively suppressed by employing residual deconvolution based on non-uniform blur kernel.

2.1   Non-uniform Blur Model

In general, uniform blur model is defined assuming that the translation is the only camera motion as in Eq. (1). In other words, a blur image $\mathbf{b}$ is represented as a convolution of a blur kernel $\mathbf{k}$ and a latent image $\mathbf{I}$ as follows:

$$\mathbf{b} = \mathbf{k} \otimes \mathbf{I} \tag{1}$$

where $\otimes$ stands for convolution. However, an instant camera motion includes rotation as well as translation. Actually, blur phenomenon is often dominated by rotation rather than translation. Without loss of generality, the blur kernel by translation can be approximated with that by rotation. Thus, deblur approaches based on non-uniform blur model are required to remove real blur phenomenon completely.

Equation (2) defines the non-uniform blur model because the blur kernel is spatially-variant due to camera rotation [4].

$$\mathbf{b} = \lim_{K \to \infty} \frac{1}{K} \sum_{i=1}^{K} \mathbf{P}_i \mathbf{I} \tag{2}$$

where $\mathbf{P}_i$ indicates the transformation matrix for the i-th homography. Equation (2) can be re-written as follows with a finite number of homographies.

$$\mathbf{b} = \sum_{i=1}^{N} w_i \mathbf{P}_i \mathbf{I} \tag{3}$$

where w_i stands for the weights to compensate for distortion due to the limited number of homographies. Note that w_i is usually set to $1/N$ because IMU sensor data is uniformly sampled on time axis.

2.2 IMU Sensor and Camera Motion

IMU sensor consisting of acceleration and gyro sensors can detect motion and location of an object. Since the cutting-edge digital cameras including smart phones basically embed IMU sensors, we can predict camera motion precisely during shooting period using the IMU sensor.

Rotation angle θ in a general circular movement is represented by $\theta = \int \omega(t)dt$. Here, ω indicates an angular velocity. In discrete form, $\theta = \omega(t)\Delta t$. On the other hand, translational movement distance $\mathbf{T}$ can be obtained from acceleration sensor. $\mathbf{T}$ is represented by $T = \iint a(t)dt^2$. Here, $a(t)$ indicates the acceleration speed. In discrete form, $\mathbf{T} = a(t)\Delta t^2$. However, if $a(t)$ severely changes, the measurement error can increase due to the limited sampling period, too. Also, actual blur phenomenon is usually dominated by rotation rather than translation. So, this paper considers only camera rotation.

2.3 The Initial Kernel Estimation Using Gyro Data

First, rotation matrix $\mathbf{R}$ is computed by θ. Note that θ is normally represented by $\{\theta_x, \theta_y, \theta_z\}$, and it is described by $\theta_i (i = 1 \ldots \ldots N)$ due to sampling. As a result, $\mathbf{R}_i$ is derived from θ_i as follows.

$$\mathbf{R}_i = \mathbf{R}_{ix}\mathbf{R}_{iy}\mathbf{R}_{iz} \tag{4}$$

Where

$$\mathbf{R}_{ix} = \begin{bmatrix} 1 & 0 & 0 \\ 0 & \cos(-\theta_{ix}) & -\sin(-\theta_{ix}) \\ 0 & \sin(-\theta_{ix}) & \cos(-\theta_{ix}) \end{bmatrix}$$

$$\mathbf{R}_{iy} = \begin{bmatrix} \cos(-\theta_{iy}) & 0 & \sin(-\theta_{iy}) \\ 0 & 1 & 0 \\ -\sin(-\theta_{iy}) & 0 & \cos(-\theta_{iy}) \end{bmatrix}$$

$$\mathbf{R}_{iz} = \begin{bmatrix} \cos\theta_{iz} & \sin\theta_{iz} & 0 \\ -\sin\theta_{iz} & \cos\theta_{iz} & 0 \\ 0 & 0 & 1 \end{bmatrix}$$

Here, the minus sign indicates the reversed duality of actual motion and pixel motion.

Second, from $\mathbf{R}_i$ and camera intrinsic matrix $\mathbf{K}$, motion matrix $\mathbf{H}_i$ is computed as follows:

$$\mathbf{H}_i = \mathbf{K}\mathbf{R}_i\mathbf{K}^{-1} \tag{5}$$

Note that $\mathbf{P}_i$ is equivalent to $\mathbf{H}_i$ in transformation matrix form.

2.4   Kernel Refinement

Since $\mathbf{P}_i$ is based on the discretely sampled angular velocity, the measurement error is inevitable. The inaccurate homography may cause ringing artifacts during deconvolution. So, we refine the initial blur kernel by additionally using a pair of long/short exposure-time images.

First, we apply a proper pre-processing to the short exposure time image. Its intensity level is corrected to that of the long exposure-time image, and a specific noise reduction algorithm is applied to the intensity-corrected short exposure-time image. This paper employed a famous BM3D algorithm as the state-of-the-art noise reduction algorithm [8]. Let the pre-processed short exposure-time image be $\mathbf{I}_s$.

Second, we solve Eq. (3) by adopting Cho et al.'s approach [3]. Equation (3) is modified into Eq. (6) as in [3].

$$\mathbf{b} - \frac{1}{N-1} \sum_{j=1, j \neq i}^{N} \mathbf{P}_j \mathbf{I} = \frac{\mathbf{P}_i \mathbf{I}}{N} \tag{6}$$

If the residual image of the left-hand side in Eq. (6) is represented by $\mathbf{e}_i$, the problem to derive $\mathbf{P}_i$ from Eq. (6) can be re-defined as follows:

$$\arg \min_{\mathbf{P}_i} \left\| \mathbf{e}_i - \frac{\mathbf{P}_i \mathbf{I}}{N} \right\| \tag{7}$$

This non-linear problem is solved by using Lucas-Kanade based image registration [7]. In an iterative Lucas-Kanade algorithm, the initial $\mathbf{P}_i$ is very important. Since we already obtained reliable $\mathbf{P}_i$ in Sect. 2.3, we use it as the initial value for Eq. (7). Also, $\mathbf{I}_s$ is employed as an initial $\mathbf{I}$. If the $\mathbf{P}_i$ is updated, we can get the refined blur kernel from Eq. (3).

Figure 1 provides the kernel refinement results. The ground truth kernel in Fig. 1(a) shows that blur kernels are spatially-variant (see two boxes). Figure 1(b) shows that the initial blur kernel is not accurate. Note that the blur kernels in Fig. 1(b) and (c) are discrete because of sampling of IMU data. On the other hand, the refined blur kernel in Fig. 1(c) is very close to the ground truth.

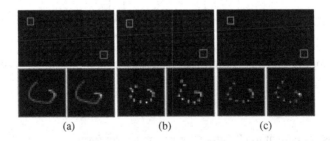

(a)   (b)   (c)

Fig. 1. The kernel refinement result. (a) Ground truth, (b) the initial blur kernel using gyro data only, (c) the refined blur kernel.

2.5 Deconvolution

Now, we obtain a final deblur image **I** by solving Eq. (8).

$$\arg\min_{\mathbf{I}} \left\| \mathbf{b} - \frac{1}{N} \sum_{i=1}^{N} \mathbf{P}_i \mathbf{I} \right\|^2 + \lambda \left| \nabla \mathbf{I} \right|^\alpha \qquad (8)$$

where a blur image **b** indicates a long exposure-time image. Tai et al. solved this problem by applying projective motion concept to conventional Richardson-Lucy deconvolution [4]. However, Richardson-Lucy deconvolution in Tai et al.'s algorithm often suffers from ringing artifacts. Thus, we employed Yuan et al.'s residual deconvolution [2]. Equation (8) is converted to Eq. (9).

$$\arg\min_{\Delta \mathbf{I}} \left\| \Delta \mathbf{b} - \frac{1}{N} \sum_{i=1}^{N} \mathbf{P}_i \Delta \mathbf{I} \right\|^2 + \lambda \left| \nabla \mathbf{I} \right|^\alpha \qquad (9)$$

where a residual image $\Delta \mathbf{b}$ is defined as $\mathbf{b} - (\sum_{i=1}^{N} \mathbf{P}_i \mathbf{I}_s)/N$ and α is a constant. Then, this problem can be solved by Tai et al.'s approach in [4], and $\Delta \mathbf{I}$ is found as a result. Finally, a deblurred image **I** is obtained by adding $\Delta \mathbf{I}$ and $\mathbf{I}_s$.

Figure 2 proves the positive effect of residual deconvolution for a test image. Figure 2(a) results from conventional Richardson-Lucy deconvolution. We can see that it produces ringing artifacts near edges. On the other hand, Fig. 2(b) results from the proposed residual deconvolution. We can observe that edges keep sharpness without ringing artifacts.

(a)   (b)

Fig. 2. The effect of residual deconvolution. (a) Conventional Richardson-Lucy deconvolution, (b) the proposed residual deconvolution.

3   Experimental Results

In order to evaluate the performance of the proposed algorithm, we employed two kinds of test image set: Artificially blurred image set and real blurred image set. The first image set is synthesized as follows. A short exposure-time image is synthesized by lowering the intensity level of a test image and inserting AWGN to it. The corresponding long exposure-time image is obtained by artificially blurring the test image assuming an arbitrary continuous gyro sensor data and an exposure-time of 1/10 s. Here a discrete gyro data which is produced by subsampling the continuous gyro data in 100 Hz. The images of the second test set were taken by Sony α55. For this experiment, RUVA Tech RT $\times$ Q IMU sensor is attached to the top of the digital camera. The sampling rate of the gyro sensor was set to 100 Hz. Long and short exposure-times were set to 1/8 s and 1/40 s, respectively. ISO was set to 100 for both images. The second test images have the resolution of 1280 $\times$ 720, and they were cropped from the center of the original 2448 $\times$ 1376 images for clear comparison.

First, we subjectively compared the proposed algorithm with two deblur algorithms; Xu and Jia's [6] and Whyte et al.'s [5]. Note that Xu et al.'s algorithm is one state-of-the-art algorithm for uniform blur removal, and Whyte et al.'s algorithm is the latest non-uniform deblur algorithm. Figure 3 shows the result for the artificially blurred image set. We can see from Fig. 3(b) that Xu et al's provides good sharpness, but suffers from severe ringing artifacts around image boundary. Since Whyte et al.'s algorithm is inherently weak against large rotation, it cannot provide acceptable visual quality for this test image set having large rotation (see Fig. 3(c)). On the contrary, the proposed algorithm shows outstanding deblur result without ringing artifacts.

Figure 4 shows the deblur results for the real blurred image set. We can observe from Fig. 4(b) and (c) that Xu et al's as well as Whyte et al.'s suffer from significant ringing artifacts near edges.

Table 1. The PSNR comparison

Algorithms	PSNR [dB]
Proposed	33.30
Xu et al.'s	31.14
Whyte et al.'s	20.18

On the other hand, the proposed algorithm successfully preserves edges' sharpness without ringing artifacts.

Table 1 shows that the proposed algorithm provides significantly higher PSNR than Xu et al's.

(a)

(b)

(c)

(d)

Fig. 3. The result for the artificially blurred image set. (a) Input blurred image, (b) Xu et al.'s, (c) Whyte et al.'s, (d) the proposed algorithm.

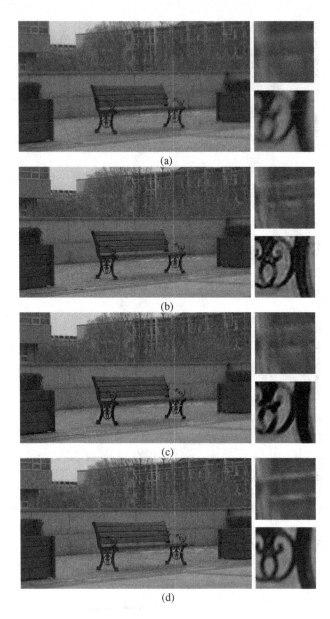

Fig. 4. The result for the real blurred image set. (a) Input blurred image, (b) Xu et al.'s, (c) Whyte et al.'s, (d) the proposed algorithm.

4   Conclusion

This paper proposes a deblur algorithm using gyro sensor and a pair of long/short exposure-time images. First, the proposed algorithm improved the accuracy of the gyro-based kernel estimation by employing a pair of long/short exposure-time images. Second, the proposed non-uniform residual deconvolution preserved edges' sharpness and successfully reduces ringing artifacts. The experimental results show that the proposed algorithm outperforms the existing non-uniform deblur algorithms in terms of objective visual quality as well as subjective visual quality.

Acknowledgement. This work was supported by INHA UNIVERSITY Research Grant.

References

1. Levin, A., Weiss, Y., Durand, F., Freeman, W.: Understanding and evaluating blind deconvolution algorithm. In: IEEE Conference on Computer Vision and Pattern Recognition (2009)
2. Yuan, L., Sun, J., Quan, L., Shum, H.Y.: Image de-blurring with blurred/noisy image Pairs. ACM Trans. Graph. **26**(3), 1–10 (2007)
3. Cho, S., Cho, H., Tai, Y.W., Lee, S.: Registration based non-uniform motion deblurring. Comput. Graph. Forum **31**(7), 2183–2192 (2012)
4. Tai, Y.W., Tan, P., Brown, S.: Richardson-Lucy deblurring for scenes under a projective motion path. IEEE Trans. Pattern Anal. Mach. Intell. **33**(8), 1603–1618 (2011)
5. Whyte, O., Sivic, J., Zisserman, A., Ponce, J.: Non-uniform deblurring for shaken images. Int. J. Comput. Vis. **98**(2), 168–186 (2012)
6. Xu, L., Jia, J.: Two-phase kernel estimation for robust motion deblurring. In: Daniilidis, K., Maragos, P., Paragios, N. (eds.) ECCV 2010, Part I. LNCS, vol. 6311, pp. 157–170. Springer, Heidelberg (2010)
7. Baker, S., Iain, M.: Lucas-Kanade 20 years on: a unifying framework. Int. J. Comput. Vis. **56**(3), 221–255 (2004)
8. Dabov, K., Foi, A., Katkovnik, V., Egiazarian, K.: BM3D Image denoising with shape-adaptive principal component analysis. In: Workshop on Signal Processing with Adaptive Sparse Structured Representations (2009)

Object Searching with Combination
of Template Matching

Wisarut Chantara and Yo-Sung Ho[(⊠)]

School of Information and Communications,
Gwangju Institute of Science and Technology (GIST), 123 Cheomdan-gwagiro,
Buk-gu, Gwangju 500-712, Republic of Korea
{wisarut,hoyo}@gist.ac.kr

Abstract. Object searching is the identification of an object in an image or video. There are several approaches to object detection, including template matching in computer vision. Template matching uses a small image, or template, to find matching regions in a larger image. In this paper, we propose a robust object searching method based on adaptive combination template matching. We apply a partition search to resize the target image properly. During this process, we can make efficiently match each template into the sub-images based on normalized sum of squared differences or zero-mean normalized cross correlation depends on the class of the object location such as corresponding, neighbor, or previous location. Finally, the template image is updated appropriately by an adaptive template algorithm. Experiment results show that the proposed method outperforms in object searching.

Keywords: Object searching · Template matching · Adaptive combination template matching · Normalized sum of squared differences · Zero-mean normalized cross correlation

1 Introduction

Object searching is the process of finding instances of real-world objects such as faces, vehicles, and buildings in images or videos. Object searching algorithms typically use extracted features and learning algorithms to recognize instances of an object category. It is commonly used in applications such as image retrieval, security, surveillance, and automated vehicle parking systems. Moreover, object detection and tracking is considered as an important subject within the area of computer vision. Availability of high definition videos, fast processing computers and exponentially increasing demand for highly reliable automated video analysis have created a new and a great deal for modifying object tracking algorithms. Video analysis has three main steps mainly: detection of interesting moving objects, tracking of such objects from each and every frame to frame and analysis of object tracks to recognize the behavior of the object in the entire video [1]. Several algorithms for object searching have been reported. Mao et al. [2] presented an object tracking approach that integrated two methods consisting of histogram-based template matching method and the mean shift procedure were used to estimate the object location. Choi et al. [3] proposed a vehicle tracking scheme using template matching based on both the scene and vehicle characteristics, including

© Springer International Publishing Switzerland 2015
Y.-S. Ho et al. (Eds.): PCM 2015, Part I, LNCS 9314, pp. 32–41, 2015.
DOI: 10.1007/978-3-319-24075-6_4

background information, local position and size of a moving vehicle. In addition, the template matching [4] is a well-known technique often used in object detection and recognition. This algorithm is a technique in digital image processing for finding small parts of an image which match a template image. It can be used in manufacturing as a part of quality control, a way to navigate a mobile robot, or as a way to detect edges in images. This is also due to the simplicity and efficiency of the method. In the next section, we introduce some brief concepts of conventional template matching.

2 Conventional Methods

The conventional template matching methods have been commonly used as metrics to evaluate the degree of similarity (or dissimilarity) between two compared images. The methods are simple algorithms for measuring the similarity between the template image *(T)* and the sub-images of the target image *(I)*. Then, the process will classify the corresponding object.

(a) *Sum of absolute differences (SAD)*

$$R(x,y) = \sum_{u,v} |(T(u,v) - I(x+u, y+v)|$$ (1)

SAD works by taking the absolute difference between each pixel in T and the corresponding pixel in the small parts of images being used for comparison in I. Absolute differences are summed to create a simple metric of similarity.

(b) *Sum of squared differences (SSD)*

$$R(x,y) = \sum_{u,v} (T(u,v) - I(x+u, y+v))^2$$ (2)

(c) *Normalized sum of squared differences (NSSD)*

$$R(x,y) = \frac{\sum_{u,v}(T(u,v) - I(x+u, y+v))^2}{\sqrt{\sum_{u,v} T^2(u,v) \cdot \sum_{u,v} I^2(x+u, y+v)}}$$ (3)

SSD and NSSD work by taking the squared difference between each pixel in T and the corresponding pixel in the small parts of images being used for comparison in I. Squared differences are summed to create a simple metric of similarity. The normalization process allows for handling linear brightness variation.

(d) *Normalized cross correlation (NCC)*

$$R(x,y) = \frac{\sum_{u,v}(T(u,v) \cdot I(x+u, y+v))}{\sqrt{\sum_{u,v} T^2(u,v) \cdot \sum_{u,v} I^2(x+u, y+v)}}$$ (4)

(e) *Zero-mean normalized cross correlation (ZNCC)*

$$R(x,y) = \frac{\sum_{u,v}(T_*(u,v) \cdot I_*(x+u,y+v))}{\sqrt{\sum_{u,v} T_*^2(u,v) \cdot \sum_{u,v} I_*^2(x+u,y+v)}} \tag{5}$$

where

$$T_*(u,v) = T(u,v) - \bar{T},$$

$$I_*(x+u,y+v) = I(u,v) - \bar{I}$$

NCC works by taking the product of each pixel in T and the corresponding pixel in the small parts of images being used for comparison in I. The normalization process allows for handling linear brightness variation. The main advantage of NCC over the cross correlation is that it is less sensitive to linear changes in the amplitude of illumination in the two compared images. ZNCC is even a more robust solution than NCC since it can also handle uniform brightness variation.

The basic template matching algorithm consists of calculating at each position of the image under examination a distortion function that measures the degree of similarity between the template and image. The minimum distortion or maximum correlation position is then taken to locate the template into the examined image. Many studies on template matching have been reported. Alsaade et al. [5] introduced template matching based on SAD and pyramid structure through compressing both source image and template image. Hager and Belhumeur [6] proposed general illumination models could be incorporated into SSD motion and exhibit a closed-loop formulation for the tracking. Furthermore, Sahani et al. [7] presented object tracking based on two stage search method, whose main application could be tracking aerial target. Maclean and Tsotsos [8] introduced a technique for fast pattern recognition using normalized grey-scale correlation (NCC). Stefano et al. [9] proposed an algorithm for template matching based on the direct computation of the ZNCC function. The algorithm generalized the principle of the BPC technique. Alternative matching algorithms can be found in Refs. [10–12]. However, as far as template matching is concerned, NCC and ZNCC are often adopted for similarity measure as well. The traditional NCC and ZNCC need to compute the numerator and denominator which are very time-consuming. On the contrary, the conventional SAD, SSD, and NSSD are relatively simple.

This paper proposes object searching with combination of template matching. A target image is resized following the object position, then standard robust matching technique NSSD or ZNCC is applied to this image. The applied technique depends on the class of the object location (corresponding, neighbor, or previous location). After that, the object location is identified and the correct positions are updated properly in the whole target image, while the template image is adapted properly. Experiments show that the proposed method outperforms the traditional search algorithms.

3 Proposed Method

The proposed method is initially motivated by Chantara's work [13] based on matching efficiency. Based on Chantara's work, the proposed method adapts an adaptive template matching to enhance the matching accuracy. The contribution of the proposed method is to increase the matching performance.

3.1 Partition Search Area

A partition search area reallocates the target image. A proper size of the target image relates to the previous interested object position as shown in Fig. 1. An algorithm result is illustrated in Fig. 2.

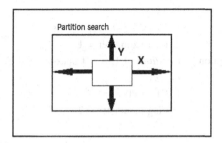

Fig. 1. Partition search area

Fig. 2. (Left) Original target image, (Right) reallocated target image

When the appropriate target image is provided as shown in Fig. 2 (Right), we perform object detection with template matching (NSSD or ZNCC). The result data of matching algorithm provide the location of the corresponding object (CL) in the target image. Other neighbor results and locations (NLs), which are in a limit of a threshold value, can also be found.

3.2 Object Identification

We apply the result data from the previous subsection to locate the suitable location of interested object (OL) in a full target image (Table 1).

Assume: The previous object position (PP) is reserved in a memory buffer.

Table 1. Object identification method

Option	Process
Option 1	*Condition:* CL is the closest of PP.
	Process:
	1. OL = CL
	2. Flag parameter = 0
Option 2	*Condition:* NLs is the closest of PP.
	Process:
	1. OL = the closest of NLs
	2. Flag parameter = 1
Option 3	*Condition:* CL is not option 1 and 2.
	Process:
	1. OL = PP
	2. Flag parameter = 2

Table 2. Adaptive combination template matching method

Flag status	Process
Flag = 0	*Condition:* OL = CL
	Process:
	1. Previous template = Current template
	2. Current template = Original template
Flag = 1	*Condition:* OL = the closest of NLs
	Process:
	1. Previous template = Current template
	2. Current template = Object template
Flag = 2	*Condition:* OL = PP
	Process:
	1. Previous template = Current template
	2. Current template = Original template
	3*. Toggle NSSD
	$\Leftrightarrow$
	ZNCC

3.3   Adaptive Combination Template Matching

The flag parameter is considered to examine a suitable template image. The process also switches the matching algorithms between NSSD and ZNCC when the condition is the previous position of the object uses as the object location (Table 2).

4   Experiment Results

In this section, we present the efficient performance of the proposed method. The experiments are performed to examine the matching accuracy. The system executes on a PC with an Intel (R) Core (TM) i7-3930 K CPU 3.20 GHz, 16.0 GB RAM and operating system of Windows 8.1. A 101 × 98 sized template image is used to match in target image sequences which is a size of 426 × 240 pixels, as shown in Fig. 3. Furthermore, other template image and the related target image contain different sizes and different illuminations is illustrated in Fig. 4.

Fig. 3. (Left) Target image, (Right) template image

Fig. 4. (Left) Target image 768 × 576 pixels, (Right) template image 40 × 75 pixels

Figures 5, 6 and 7 demonstrate the results of each option in Subsects. 3.2 and 3.3. An illumination in each figure shows the outcome of the matching method with the

Fig. 5. Option 1: (Left) object location, (Right) adaptive template image

Fig. 6. Option 2: (Left) object location, (Right) adaptive template image

Fig. 7. Option 3: (Left) object location, (Right) adaptive template image

Table 3. Comparison of the PSNR values

Image Sequence	PSNR Values (dB)				
	SAD	NSSD	ZNCC	Chantara's Method	Proposed Method
Walking	16.48	16.53	16.49	17.25	18.67

Image Sequence	PSNR Gains (dB)			
	$\Delta PSNR_{PS}$	$\Delta PSNR_{PN}$	$\Delta PSNR_{PZ}$	$\Delta PSNR_{PC}$
Walking	+2.19	+2.14	+2.18	+1.42

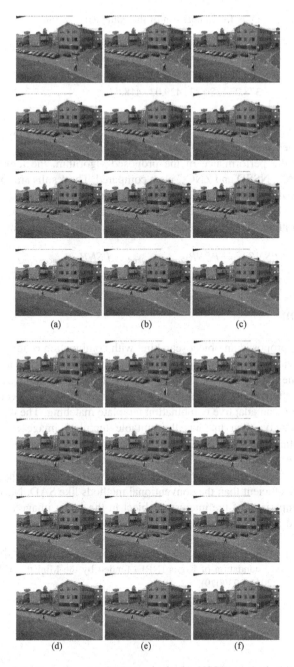

Fig. 8. The experiment results: (a) SAD method, (b) NSSD method, (c) ZNCC method, (d) Chantara's [13] (e) ground truth [14], (f) proposed method, the results are captured in different frames

Table 4. Comparison of the computational time

Image sequence	Computational time (ms)				
	SAD	NSSD	ZNCC	Chantara's method	Proposed method
Face	96.57	97.06	100.72	97.29	90.98
Walking	378.26	380.83	429.03	418.04	57.77

position of involved object and the updated template. In the third option, the matching method is switched between NSSD and ZNCC algorithms.

To illustrate the performance of the proposed algorithm, the template matching algorithm like SAD, NSSD and ZNCC were compared. The results are shown in Fig. 8, Table 3 lists PSNR values for the proposed method and the three above-mentioned methods, the PSNR gain is defined as Eq. 6 and Table 4 presents the computational time of the algorithms. The proposed method outperforms the conventional methods.

$$\Delta PSNR_{PR} = PSNR_{Proposedmethod} - PSNR_{Referencemethod} \tag{6}$$

5   Conclusion

In this paper, we proposed an object searching with combination of template matching. The method provides the proper object location in a target image. We apply a partition search to give the appropriate target image for the template matching algorithm. The process reduces the computing time. The proper target image has been searched the interested object by an adaptive combination template matching. The method identifies the object accurately, then updates the suitable template image and switches the matching algorithms between NSSD and ZNCC when the previous position of the object uses as the object position. This process increases an accurate object location on the target image. Based on the experiment results, we can analyze that the proposed method has more efficient than the conventional methods like SAD, NSSD and ZNCC. Moreover, a comparison of the interested object in the target image provides that the proposed method is outperformance.

Acknowledgement. This research was supported by Basic Science Research Program through the National Research Foundation of Korea (NRF) funded by the Ministry of Science, ICT & Future Planning (No. 2011-0030079).

References

1. Yilmaz, A., Javed, O., Shah, M.: Object tracking: a survey. ACM Comput. Surv. **38**(4), Article 13, 1–45 (2006)
2. Mao, D., Cao, Y.Y., Xu, J.H., Li, K.: Object tracking integrating template matching and mean shift algorithm. In: International Conference on Multimedia Technology, pp. 3583–3586 (2011)

3. Choi, J.H., Lee, K.H., Cha, K.C., Kwon, J.S., Kim, D.W., Song, H.K.: Vehicle tracking using template matching based on feature points. In: IEEE International Conference on Information Reuse and Integration, pp. 573–577 (2006)
4. Brunelli, R.: Template Matching Techniques in Computer Vision: Theory and Practice. Wiley, New York (2009)
5. Alsaade, F., Fouda, Y.M.: Template matching based on SAD and pyramid. Int. J. Comput. Sci. Inf. Secur. **10**(4), 11–16 (2012)
6. Hager, G., Belhumeur, P.: Real-time tracking of image regions with changes in geometry and illumination. In: Proceedings of IEEE Computer Society Conference on Computer Vision Pattern Recognition, pp. 403–410 (1996)
7. Sahani, S., Adhikari, G., Das, B.: A fast template matching algorithm for aerial object tracking. In: International Conference on Image Information Processing, pp. 1–6 (2011)
8. Maclean, J., Tsotsos, J.: Fast pattern recognition using gradient-descent search in an image pyramid. In: International Conference on Pattern Recognition, pp. 873–877 (2000)
9. Stefano, L.D., Mattoccia, S., Tombari, F.: ZNCC-based template matching using bounded partial correlation. Pattern Recogn. Lett. **26**(14), 2129–2134 (2005)
10. Essannouni, F., Oulad Haj Thami, R., Aboutajdine, D., Salam, A.: Adjustable SAD matching algorithm using frequency domain. J. Real-Time Image Process. **1**(4), 257–265 (2007)
11. Hel-Or, Y., Hel-Or, H.: Real-time pattern matching using projection kernels. IEEE Trans. PAMI **27**(9), 1430–1445 (2002)
12. Wei, S., Lai, S.: Fast template matching based on normalized cross correlation with adaptive multilevel winner update. IEEE Trans. Image Process. **17**(11), 227–2235 (2008)
13. Chantara, W., Mun, J.H., Shin, D.W., Ho, Y.S.: Object tracking using adaptive template matching. IEIE Trans. Smart Process. Comput. **4**(1), 1–9 (2015)
14. Wu, Y., Lim, J., Yang, M.H.: Online object tracking: a benchmark. In: IEEE Conference on Computer Vision and Pattern Recognition, pp. 2411–2418 (2013)

Multimedia Content Analysis

Two-Step Greedy Subspace Clustering

Lingxiao Song[(✉)], Man Zhang, Zhenan Sun, Jian Liang, and Ran He

Center for Research on Intelligent Perception and Computing,
Institute of Automation, Chinese Academy of Sciences, Beijing, China
{lingxiao.song,zhangman,znsun,jian.liang,rhe}@nlpr.ia.ac.cn

Abstract. Greedy subspace clustering methods provide an efficient way to cluster large-scale multimedia datasets. However, these methods do not guarantee a global optimum and their clustering performance mainly depends on their initializations. To alleviate this initialization problem, this paper proposes a two-step greedy strategy by exploring proper neighbors that span an initial subspace. Firstly, for each data point, we seek a sparse representation with respect to its nearest neighbors. The data points corresponding to nonzero entries in the learning representation form an initial subspace, which potentially rejects bad or redundant data points. Secondly, the subspace is updated by adding an orthogonal basis involved with the newly added data points. Experimental results on real-world applications demonstrate that our method can significantly improve the clustering accuracy of greedy subspace clustering methods without scarifying much computational time.

Keywords: Greedy subspace clustering · Sparse representation · Subspace neighbor

1 Introduction

Clustering is a classic problem in multimedia and computer vision. As an important branch of clustering, subspace clustering seeks to cluster data into different subspaces and find a low-dimensional subspace fitting each group of points. It is based on the assumption that high-dimensional data often lies on low-dimensional subspaces, which usually holds true for the data acquired in real world. Subspace clustering can be widely applied in image segmentation [1], motion segmentation [2], face clustering [3], image representation, compression [4], and multimedia analysis [5,6]. In these applications, data points of the same class (e.g., pixels belong to the same object, feature points of the same rigid object in a moving video sequence, face images of the same person) lie on same underlying subspace, and the mixture dataset can be modeled by unions of subspaces.

1.1 Related Work on Subspace Clustering

Numerous subspace clustering approaches have been proposed in the past two decades. Existing work on subspace clustering in machine learning and computer

© Springer International Publishing Switzerland 2015
Y.-S. Ho et al. (Eds.): PCM 2015, Part I, LNCS 9314, pp. 45–54, 2015.
DOI: 10.1007/978-3-319-24075-6_5

vision communities can be divided into four main categories: algebraic, iterative, statistical and spectral clustering-based methods [7].

Algebraic methods such as matrix factorization-based algorithms [8,9] segment data points according to a low-rank factorization of the data matrix. But these methods are not effective when the subspaces are dependent. Generalized Principal Component Analysis (GPCA) [10] uses a polynomial function to fit a given point. It can handle both independent and dependent subspaces, but the computational complexity increases exponentially when the dimension of data grows. Such methods are sensitive to noise and outliers, due to their assumption of noise-free data. Iterative methods [3,11] iteratively refine subspaces of each cluster and assign points to the closest subspace. And these methods can be applied to linear as well as affine subspaces, but it is easy to run into a local optimum, thus several restarts are often needed. Statistical approaches such as [12,13], model both data and noise under explicit assumptions of the probabilistic distribution of data in each subspaces and noise. However, these statistical approaches are not suitable for real-world applications due to their sensitivity to outliers.

The standard procedure of spectral clustering-based methods consists of constructing an affinity matrix firstly whose elements measure the similarity between samples, and then applying spectral clustering given the affinity matrix. A number of spectral clustering-based methods spring out in recent years such as Sparse Subspace Clustering (SSC) [14,15], Low-Rank Representation (LRR) [16], Low-Rank Representation via Correntropy (CLRR) [17,18], Low-Rank Sparse Subspace Clustering (LRSSC) [19] and Spectral Curvature Clustering (SCC) [20]. The basic idea of SSC is that a data point can be written as a linear or affine combination of other data points in the same subspace under an l_1-minimization constraint. A similar optimization based method called LRR minimizes nuclear norm instead of the l_1-norm in SSC to guarantee a low-rank affinity matrix. CLRR proposed by Zhang et al. attempt to maximize the correntropy between data points and their reconstruction, and an efficient solution of the optimization problem based on half-quadratic minimization is given in their paper. Wang et al. propose a hybrid algorithm termed LRSSC by combining l_1-norm and nuclear norm, based on the fact that the representation matrix is often not only sparse but also low-rank.

More recently, greedy-like spectral clustering-based approaches gain increasing attention due to their low complexity. Dyer et al. induced a greedy method for sparse signal recovery called orthogonal matching pursuit (OMP) [21], which is used to replace the l_1-minimization in SSC. Heckel and H. Bölcskei proposed a simple algorithm based on thresholding the correlation between data points (TSC) [22]. Park and Caramanis presented Nearest Subspace Neighbor (NSN) in [23], which constructs neighbors via incrementally selecting point that is closest to the subspace. All the greedy approaches share a common advantage of computationally less demanding. However, a major disadvantage of these greedy-like methods is that they can not cope with complex situation where the subspaces intersect or lots of noise exist.

1.2   Paper Contributions

In this paper we propose a two-step greedy subspace clustering (T-GSC) algorithm, which is able to achieve superior clustering performance, comparing to several state-of-the-art methods. A initial subspace construction step is induced aiming to improve the robustness of greedy-like algorithms, especially when the data are not well distributed around their subspaces, e.g. the data is contaminated by outliers, the subspaces are intersected. After the first step, different-class neighbors can be directly ruled out in most cases, resulting in a better spanning of the subspace. Then, in the second step, the nearest subspace neighbor is added to the neighbors set in a greedy way, which requires much less run time than other kinds of algorithms. Numerous experimental results demonstrate that our algorithm can reach state-of-the-art clustering performance, with better robustness and lower computational cost.

The reminder of this paper is organized as follows. Section 2 reviews the Nearest Subspace Neighbor algorithm and then describes the technical details of our method. Then a number of experiments are presented in the Sect. 3. Finally we draw some conclusions in the Sect. 4.

2   Two-Step Greedy Subspace Clustering

As mentioned before, spectral clustering-based methods follow a basic procedure of computing an affinity matrix first and then deriving clusters using spectral clustering. Instead of computing each entry of the affinity matrix, we construct a neighborhood matrix whose entries are either 0 or 1, as in [22,23]. Hence, constructing an affinity matrix is transformed into finding a neighborhood of each data point. The proposed method is mainly inspired by the Nearest Subspace Neighbor (NSN), which iteratively selects neighbor points most likely to be on the same subspace. That is to say, NSN choose the nearest neighbor as the first member of the neighbor set to construct subspace. However, a data point and its nearest neighbor may belong to different true subspaces in many cases. Thus, we propose an initial subspace construction step in our algorithm to solve this problem.

Through this paper, we use uppercase boldface letters to denote matrices, lowercase letters to denote vectors and scalars, and letters that are uppercase but not bold stand for parameters and sets. The given data matrix is denoted as $X \in R^{D \times N}$, with each column representing a data point $x_i \in R^D$. The task of subspace clustering is to recover the union of K subspaces $S = S_1 \cup S_2 \cup ... \cup S_K$ where the data points are belong to, and the dimension of each subspace is supposed to be not higher than P. For each data point x_i, there is a neighbor set Ω_i contains M elements. In addition, U represents subspace spanned by a set of data points: $U = span(\Omega)$; $proj_U(x)$ is defined as the projection of data point x to the subspace U; $\eta_m, m = 1, 2, ..., P$ are orthogonal bases of U; $\mathbb{I}\{.\}$ is the indicator function; $\|.\|$ denote Euclidean norm for vectors; and $\langle . \rangle$ is the inner product operator.

2.1 First Step: Initial Subspace Construction

T-GSC constructs a good initial subspace based on the theory of sparse representation. Considering a data point $x_i \in R^D$ drawn from a union of subspaces, it can be sparsely represented by solving

$$\min \|c\|_0 \qquad s.t. \quad x_i = Xc, \tag{1}$$

where X is the data matrix. However, this optimization problem of l_0-norm is NP-hard and non-convex in general, while it is proved in [24] that the l_1-norm solution is equivalent to l_0-norm solution in certain conditions. So, the l_1-norm is adopted, and Eq. (1) can be reformulated as follows.

$$\min \|c\|_1 \qquad s.t. \quad x_i = Xc. \tag{2}$$

There are a huge number of approaches for extracting the sparse solution of Eq. (2) such as the Basis Pursuit Algorithm [24] and Lasso [25]. The work of SSC [9] proves that the optimal solution $\hat{c}$ has zero entries correspond to data points not lying in the same subspace with x_i. Based on this, neighbors from distinct subspaces can be rejected.

Fortunately, we do not need to find all data points that lie in the same subspace with x_i constructing the initial subspace. We just need to find one neighbor that most likely belongs to the same class with x_i to ensure the reliability of the subspace spanned by them. Therefore, optimizing Eq. (2) among all the data points is undesirable. In our approach, L nearest neighbors around x_i in the ambient space are chosen to be the bases. Then the l_1-optimization problem is simplified to

$$\min \|c\|_1 \qquad s.t. \quad x_i = X_i^{(L)}c. \tag{3}$$

where $X_i^{(L)}$ is comprised of the L nearest neighbors of x_i, and the problem can be efficiently solved when L is not large. We use the glmnet toolbox [25] of Lasso to solve this problem. The data point corresponding to the max entry of c will be the first neighbor added to x_i's neighbor set Ω_i, meanwhile the initial subspace U is spanned by Ω_i.

2.2 Second Step: Greedy Subspace Clustering

Once the initial subspace is built, T-GSC greedily adds the closest point to the subspace into the neighbor set in the following iterations as NSN, until enough neighbors are selected. This step mainly involves two stages: finding new neighbors and updating subspaces.

We use a set of orthogonal bases to represent the subspace. In every iteration, the subspace is updated by adding an orthogonal basis involved with the newly added data points according to the Gram-Schmidt process.

$$\eta_{m+1} = x_{j^*} - \sum_{k=1}^{m} \langle x_{j^*}, \eta_k \rangle \eta_k. \tag{4}$$

Then, $\text{proj}_U(x)$ can be easily obtained through the following equation. Note that we only need to compute the inner product with the newly added orthogonal basis in the k-th iteration.

$$\text{proj}_U(x)^2 = \sum_k \langle x, \eta_k \rangle^2. \tag{5}$$

After all the neighbors are selected, spectral clustering is applied to find the clusters as in other spectral-based algorithms. The following Algorithm 1 summarizes the whole procedure in T-GSC to cluster data points into different subspaces.

Algorithm 1. Two-step Greedy Subspace Clustering(T-GSC)

Input: Data matrix $X \in R^{D \times N}$, maximum subspace dimension P, number of neighbors M, number of subspaces K.

Output: Neighbor matrix $Z \in \{0,1\}^{N \times N}$. Estimated class labels $\hat{c}_1, \hat{c}_2, ..., \hat{c}_N$.

  1. Normalize all data points: $x_i \leftarrow x_i / \|x_i\|$;

for each x_i **do**

    2. (First step) Construct the initial subspace U and initial neighbor set Ω_i by solving Eq.(3);

    3. (Second step) Find M neighbors for x_i:

       (1). Select the closest point to current subspace: $j^* = \arg\max_{j \in [N] \setminus \Omega_i} \text{proj}_U(x_j)$;

       (2). Update the neighbor set: $\Omega_i \leftarrow \Omega_i \cup \{j^*\}$;

       (3). Update the subspace: if $M < P$, then $U \leftarrow span\{x_j : j \in \Omega_i\}$;

    4. Update the neighborhood matrix: $Z_{ij} = 1, \forall j \in \Omega_i$;

end for

  5. Construct the affinity matrix: $W_{ij} = Z_{ij} + Z_{ji}$;

  6. Apply spectral clustering to (W, K).

---

3   Experiments

This section presents experimental results of our study. We compare our method with several state-of-the-art methods on two real-world applications: motion segmentation and face clustering. For the baseline methods, we use the MATLAB codes provided by their authors.

Three evaluation metrics are used in our experiments: clustering error (CE), neighbor selection error (NSE) and run time (CT). CE is defined as

$$CE = \frac{1}{N} \sum_{i=1}^{N} \mathbb{I}(c_i \neq \hat{c}_i), \tag{6}$$

where c_i is the class label, and $\hat{c}_i$ is the estimated class label. Since the definite label index of estimated class may be not consistent with its real index, every permutation of the estimated class label should be calculated in CE, and the minimum among all the permutation is adopted.

$$NSE = \frac{1}{N} \sum_{i=1}^{N} \mathbb{I}(j|W_{ij} \neq 0, c_i \neq \hat{c}_i). \tag{7}$$

NSE is the proportion of points that do not have all correct neighbors, it measures the extent that algorithms misconnect data points from different subspaces in the adjacency matrix. Besides, we compare the average run time (RT) to evaluate our algorithm's efficiency.

3.1   Motion Segmentation

To verify the performance of T-GSC in motion segmentation problem, we evaluate our method on the Hopkins155 motion segmentation database [26],[1] which comprises 155 video sequences of 2 or 3 motions, and the goal of this test is to segment the tracked points in a frame into different motion clusters. All the experiments in this paper are directly done on the raw data downloaded from the database website without any preprocessing.

Choosing the Subspace Dimension P. In theory, the value of subspace dimension P should be set equal to the underlying subspace where the data lies in. However, considering the existence of corruption and noise, choosing the subspace dimension directly equal to its theoretical value is suboptimal. According to the affine projection model, all the trajectories associated with a single rigid motion live in an affine subspace of dimension 3, so we choose the subspace dimension P around 3. The average CE and NSE of T-GSC on Hopkins155 under different P are listed in Table 1: the minimum CE is obtained when $P = 5$. Thus, we set the maximum subspace dimension $P = 5$ in the following motion segmentation experiment.

Table 1. Results on Hopkins 155 under varied P

P	3	4	5	6	7
Mean CE(%)	8.57	5.61	**3.94**	4.73	3.81
Mean NSE(%)	4.71	4.46	3.74	3.54	**3.42**

Motion Segmentation Performance. Table 2 shows the results over all sequences in Hopkins155. Since there is no much corruption or missing data in the Hopkins155 dataset, most baseline algorithms can achieve a really good performance. We can see that T-GSC performs comparable to the state-of-the-art methods, while keeps a relatively low computational time. Moreover, the CE and NSE of T-GSC are much smaller than NSN, which demonstrates the effectiveness of the first step in our method.

---

[1] The Hopkins155 database is available online at http://www.vision.jhu.edu/data/hopkins155/.

Table 2. Results on Hopkins 155 dataset

	Algotithms	SSC	LRR	SCC	OMP	TSC	NSN	T-GSC(Ours)
2 Motions	Mean CE(%)	1.53	**2.13**	2.24	17.25	18.44	3.62	3.07
	Median CE(%)	**0**	**0**	**0**	13.33	16.92	**0**	**0**
	Mean NSE(%)	**1.09**	6.03	–	37.61	2.86	2.91	2.64
	Mean RT(s)	0.50	0.96	0.37	0.17	0.16	**0.06**	0.24
3 Motions	Mean CE(%)	4.40	**4.03**	4.32	27.61	28.58	8.28	6.87
	Median CE(%)	0.56	1.43	**0.21**	23.79	29.67	2.76	1.49
	Mean NSE(%)	**2.44**	10.56	–	78.03	7.42	8.30	7.51
	Mean RT(s)	1.03	1.33	0.68	0.28	0.38	**0.12**	0.38
All	Mean CE(%)	**2.18**	2.56	2.71	19.59	20.73	4.67	3.94
	Median CE(%)	0.13	0.32	**0.05**	15.69	19.80	0.62	0.34
	Mean NSE(%)	**1.39**	7.05	–	46.74	3.89	4.13	3.74
	Mean RT(s)	0.62	1.04	0.44	0.19	0.21	**0.07**	0.27

3.2   Face Clustering

In this section, we evaluate the face clustering performance of T-GSC as well as many state-of-the-art methods on the Extended Yale-B database [27].[2] The Extended Yale-B dataset contains frontal face images of 38 subjects under 64 different illumination conditions.

To reduce the computational time and memory cost, we use the cropped images of the Extended Yale-B dataset and resize them to 4842 pixels, then concatenate the raw pixels value into a 2016-dimensional vector for each data point. We take a series experiments under different number of subjects. All of the experimental results of face clustering are adopted average value under 100 random trials.

Choosing the Subspace Dimension P. Similar to the motion segmentation experiment, we test different P of 9 to 20. The solid line in Fig. 1 shows the influence of P to T-GSC on Extended Yale-B: when the subspace dimension ranges from 9 to 14, the clustering error drops monotonically; while the clustering error keeps almost unchanged when P is increased to 15. This phenomenon implies that T-GSC is relatively robust to the choice of subspace dimension. Results of different P to NSN can be also seen in Fig. 1. Comparing T-GSC with NSN,we can see that T-GSC performs much better in low subspace dimension, and this demonstrates that T-GSC has more powerful ability to recover the subspace. Besides, with almost the same clustering error, T-GSC needs fewer neighbors than NSN, which helps to reduce the computational time.

---

[2] The Extended Yale-B database is available online at http://vision.ucsd.edu/~leekc/ ExtYaleDatabase/ExtYaleB.html.

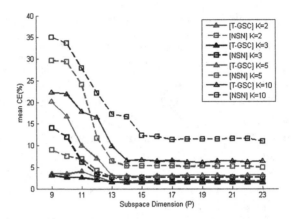

Fig. 1. Influence of P to T-GSC and NSN

Table 3. Results on Yale-B dataset

	Algotithms	SSC	LRR	OMP	TSC	NSN	T-GSC(Ours)
2 Subjects	Mean CE(%)	2.67	4.29	7.41	12.53	1.84	**1.45**
	Median CE(%)	**0**	0.78	1.57	2.36	0.78	0.78
	Mean NSE(%)	8.14	5.59	13.86	10.04	2.63	**1.90**
	Mean RT(s)	17.13	2.59	0.21	**0.15**	0.28	0.50
3 Subjects	Mean CE(%)	4.16	5.60	5.12	20.02	3.32	**2.24**
	Median CE(%)	**1.04**	**1.04**	2.08	13.54	2.6	1.56
	Mean NSE(%)	20.99	9.95	39.07	19.04	4.72	**3.07**
	Mean RT(s)	21.99	4.89	**0.35**	0.36	0.95	1.12
5 Subjects	Mean CE(%)	4.72	5.72	9.26	29.58	6.18	**3.40**
	Median CE(%)	**2.81**	2.90	5.00	31.25	5.31	2.97
	Mean NSE(%)	53.76	17.53	87.00	27.56	7.65	**4.66**
	Mean RT(s)	32.14	10.48	0.76	**0.49**	1.86	2.41
10 Subjects	Mean CE(%)	11.75	11.65	16.15	41.68	13.63	**8.20**
	Median CE(%)	11.02	12.73	17.19	42.66	12.03	**6.41**
	Mean NSE(%)	76.46	33.47	94.54	36.56	12.44	**7.89**
	Mean RT(s)	61.52	41.16	3.24	**1.56**	7.91	9.02

Taking the clustering error and computation time into account, we set $P = 16$ in the following face clustering experiment.

Face Clustering Performance. Table 3 demonstrates the face clustering results of different numbers of subspaces. T-GSC obtains the smallest mean CE as well as mean NSE in most cases, while SSC and LRR perform best on the median CE. One possible reason is that SSC and LRR, which are belong to optimization-based methods, are more suitable to handle general condition but not complex situation, while T-GSC is more robust. Thus the optimization-based methods fail in the small number of difficult cases, resulting in higher mean clustering errors than T-GSC. Besides, the proposed method is about three times faster than SSC, while holds comparable clustering error.

4   Conclusion

This paper studied the initialization problem of greedy subspace clustering methods, and proposed a two-step greedy subspace clustering method to alleviate this problem. First, for a data point, T-GSC constructs an initial subspace by seeking a sparse representation with respect to its nearest neighbors. Second, the subspace is updated by adding an orthogonal basis involved with the newly added data points. A series of experiments of motion segmentation on the Hopkins155 dataset, and face clustering on the Extended Yale-B dataset have been conducted. Experimental results show that T-GSC achieves better performance than other greedy subspace clustering methods, and meanwhile maintains comparable low computational cost.

References

1. Yang, A., Wright, J., Ma, Y., Sastry, S.: Unsupervised segmentation of natural images via lossy data compression. Computer Vis. Image Underst. (CVIU) **110**(2), 212–225 (2008)
2. Vidal, R., Tron, R., Hartley, R.: Multiframe motion segmentation with missing data using powerfactorization and GPCA. Int. J. Comput. Vis. (IJCV) **79**(1), 85–105 (2008)
3. Ho, J., Yang, M.H., Lim, J., Lee, K.C., Kriegman, D.: Clustering appearances of objects under varying illumination conditions. In: Computer Vision and Pattern Recognition (CVPR) (2003)
4. Hong, W., Wright, J., Huang, K., Ma, Y.: Multi-scale hybrid linear models for lossy image representation. IEEE Trans. Image Process. (TIP) **15**(12), 3655–3671 (2006)
5. Chaudhuri, K., Kakade, S.M., Livescu, K., Sridharan, K.: Multi-view clustering via canonical correlation analysis. In: International Conference on Machine Learning (ICML) (2009)
6. Zhao, X., Evans, N., Dugelay, J.L.: A subspace co-training framework for multi-view clustering. Pattern Recogn. Lett. (PRL) **41**, 73–82 (2014)
7. Vidal, R.: Subspace clustering. Sign. Process. Mag. **28**(2), 52–68 (2011)
8. Costeira, J., Kanade, T.: A multibody factorization method for independently moving objects. Int. J. Comput. Vis. (IJCV) **29**(3), 159–179 (1998)
9. Kanatani, K.: Motion segmentation by subspace separation and model selection. In: International Conference on Computer Vision (ICCV) (2001)
10. Vidal, R., Ma, Y., Sastry, S.: Generalized principal component analysis (GPCA). IEEE Trans. Pattern Anal. Mach. Intell. (TPAMI) **27**(12), 1945–1959 (2005)
11. Zhang, T., Szlam, A., Lerman, G.: Median K-flats for hybrid linear modeling with many outliers. In: Proceedings of 2nd IEEE International Workshop on Subspace Methods, pp. 234–241 (2009)

12. Tipping, M., Bishop, C.: Mixtures of probabilistic principal component analyzers. Neural Comput. **11**(2), 443–482 (1999)
13. Yang, A.Y., Rao, S., Ma, Y.: Robust statistical estimation and segmentation of multiple subspaces. In: Computer Vision and Pattern Recognition Workshop (CVPRW) (2006)
14. Elhamifar, E., Vidal, R.: Sparse subspace clustering. In: Computer Vision and Pattern Recognition (CVPR) (2009)
15. Elhamifar, E., Vidal, R.: Sparse subspace clustering: algorithm, theory, and applications. IEEE Trans. Pattern Anal. Mach. Intell. (TPAMI) **35**(11), 2765–2781 (2013)
16. Liu, G., Lin, Z., Yan, S., Sun, J., Yu, Y., Ma, Y.: Robust recovery of subspace structures by low-rank representation. IEEE Trans. Pattern Anal. Mach. Intell. (TPAMI) **35**(1), 171–184 (2013)
17. Zhang, Y., Sun, Z., He, R., Tan, T.: Robust subspace clustering via half-quadratic minimization. In: International Conference on Computer Vision (ICCV) (2013)
18. Zhang, Y., Sun, Z., He, R., Tan, T.: Robust low-rank representation via correntropy. In: Asian Conference on Pattern Recognition (ACPR) (2013)
19. Wang, Y., Xu, H., Leng, C.: Provable subspace clustering: when LRR meets SSC. In: Advances in Neural Information Processing Systems (NIPS) (2013)
20. Chen, G., Lerman, G.: Spectral curvature clustering. Int. J. Comput. Vis. (IJCV) **81**(3), 317–330 (2009)
21. Dyer, E.L., Sankaranarayanan, A.C., Baraniuk, R.G.: Greedy feature selection for subspace clustering. J. Mach. Learn. Res. (JMLR) **14**(1), 2487–2517 (2013)
22. Heckel, R., Bölcskei, H.: Subspace clustering via thresholding and spectral clustering. In: Acoustics, Speech, and Signal Processing (ICASSP) (2013)
23. Park, D., Caramanis, C.: Greedy subspace clustering. In: Advances in Neural Information Processing Systems (NIPS) (2014)
24. Donoho, D.L.: For most large underdetermined systems of linear equations the minimal $l1$-norm solution is also the sparsest solution. Commun. Pure Appl. Aathematics (CPAA) **59**(6), 797–829 (2006)
25. Tibshirani, R.: Regression shrinkage and selection via the lasso. J. R. Stat. Soc. Series B (Methodological) **58**, 267–288 (1996)
26. Tron, R., Vidal, R.: A benchmark for the comparison of 3-D motion segmentation algorithms. In: Computer Vision and Pattern Recognition (CVPR) (2007)
27. Lee, K.C., Ho, J., Kriegman, D.: Acquiring linear subspaces for face recognition under variable lighting. IEEE Trans. Pattern Anal. Mach. Intell. (TPAMI) **27**(5), 684–698 (2005)

Iterative Collection Annotation
for Sketch Recognition

Kai Liu, Zhengxing Sun[✉], Mofei Song, Bo Li, and Ye Tian

State Key Laboratory for Novel Software Technology, Nanjing University,
Nanjing, People's Republic of China
szx@nju.edu.cn

Abstract. Sketch recognition is an important issue in human-computer interaction, especially in sketch-based interface. To provide a scalable and flexible tool for user-driven sketch recognition, this paper proposes an iterative sketch collection annotation method for classifier-training by interleaving online metric learning, semi-supervised clustering and user intervention. It can discover the categories of the collections iteratively by combing online metric learning with semi-supervised clustering, and put the user intervention into the loop of each iteration. The features of our methods lie in three aspects. Firstly, the unlabeled collections are annotated with less effort in a group by group form. Secondly, the users can annotate the collections flexibly and freely to define the sketch recognition personally for different applications. Finally, the scalable collection can be annotated efficiently by combining the dynamically processing and online learning. The experimental results prove the effectiveness of our method.

Keywords: Sketch recognition · Online metric learning · Semi-supervised clustering · Collection labeling

1   Introduction

There have been plenty of works devoted to sketch recognition/classification both in graphics recognition and human-computer interaction [1–3]. Most of them focus on learning a classifier with the pre-labeled samples. As the performance of the classifier is utterly determined by the training set, it is necessary to gather enough number of labeled sketches. However, the manual annotation becomes a bottleneck more and more for the acquisition of labeled data. Thus, with the increasing usage of sketch recognition, the newly annotating tools need to be aroused to flexibly assist the labeling process to fit the needs of different applications while reducing the manual labor.

An appealing annotation tool for sketch collection should meet the requirements at least in three aspects. **Firstly, sketch annotation must be performed online and interactively**. This means the annotation of data collection would be facilitated with the effective machine algorithms. Though active

© Springer International Publishing Switzerland 2015
Y.-S. Ho et al. (Eds.): PCM 2015, Part I, LNCS 9314, pp. 55–65, 2015.
DOI: 10.1007/978-3-319-24075-6_6

learning methods [4] have been used to help select the samples to label, unfortunately they provide the labels for individual data samples one by one according to predefined categories. More efficient way is to put some relatively similar data together and assist user to make samples labeled group by group. Category discovery methods [5] can generate at once the priori number of groups in a collection, but they assume the number and types of categories is known in advance. It is more flexible to discover the potential groups in an online circular manner in the collection, which can assist the user to define the categories continually during the annotation, as recently proposed for image labeling [6–8]. However, as there is no previous works on labeling sketches, it remains unopened to make sketch annotation performed in such a manner. **Secondly, sketch annotation can be done personally with less effort.** The intuitive way of reducing labor effort is to group the collection automatically by clustering, as used in category discovery and image labeling. Unfortunately, most methods only allow adjusting the grouping results by varying the number of clusters, which is not enough for obtaining the personal annotation result. Actually, a personal annotation tool should put the user interaction into the loop of the whole grouping process by interactively identifying the member of each group [8]. But the effort will be heavy when the clustering result is not consistent with the user's requirement. It is more efficient to exploit the extra information to affect the clustering to improve results. Thus, a novel mechanism is required to efficiently integrate user intervention and clustering into a circular annotation. **Thirdly, sketch annotation should be adaptive for scalable collections.** The unlabeled sketches are often added into the collection progressively rather than at once. Though incremental methods have been widely used in image classification [9], the way of sketch annotation in a circular manner not only requires incremental mechanisms to avoid regrouping old and new sketches, but also needs dynamically updating to ensure the online annotation process realtime.

In order to meet the goals described above, we propose an iterative collection annotation method for sketch recognition. It can discover sketch categories iteratively via interleaving the following steps: Online metric learning provides an optimized similarity measuring model. Using the optimized similarity, semi-supervised clustering is introduced to group the unlabeled sketches into clusters, and user intervention is then incorporated to create a group of labeled sketches from the clusters. After each new group is labeled, the supervision information is constructed/updated for the next iteration. Though inspired by previous works [7,8], this paper mainly differs them in two ways. Firstly, this paper exploits semi-supervised clustering to improve results based on single-view sketch feature, while existing works try to include multi-view image features into unsupervised clustering. Secondly, this paper incorporates online metric learning to optimize the similarity measuring incrementally without retraining the whole labeled dataset from scratch each iteration. The features of our method lie in three aspects: (1). the unlabeled collections are annotated with less effort in a group by group form; (2). the users can annotate the sketch collection freely and flexibly to suit for different applications; (3). the scalable collection can be annotated efficiently by combining the dynamically processing and online learning.

2   Overview of Proposed Method

As illustrated in Fig. 1, our learning-based sketch recognition framework is composed of two main components: sketch recognition and iterative annotation. In previous researches, sketch recognition can be done by the incremental SVM classifier as shown in the top of Fig. 1. Here, we focus only in another important component: iterative annotation, which is illustrated in bottom of Fig. 1.

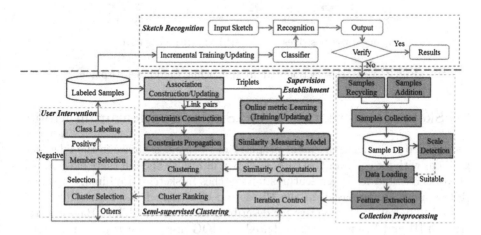

Fig. 1. The framework of sketch recognition based on the iterative annotation

In summary, our iterative annotation method is done followed by four main stages. The first stage, collection preprocessing, adds progressively and incrementally two kinds of sketch samples into the collections: the collected samples and the recycled samples from error recognition results. The loading of sample collections is controlled by scale detection, which measures if the number of samples is enough for iterative annotate. The subsequent three stages perform iteratively the core process of the iterative annotation: based on the feature extraction, **semi-supervised clustering** is introduced to group the loading samples into proper clusters and rank them according to the purity. All the ranked clusters are provided for **user intervention** to create the labeled group. All the remaining unlabeled sketches will be clustered in next iteration. According to the newly labeled group, **supervision establishment** provides the constrained information with the constructed associations (known as triplets and link pairs) for semi-supervised clustering in next iteration. The iteration process stops when there are not enough sketches for grouping in the current collection. The newly labeled sketches are saved into the labeled collection, which can be used to train the classifier of sketch recognition.

Here we briefly sketch how the user intervention is performed sequentially with the three interactions. Firstly, the cluster with highest purity is selected and provided to the user. Secondly, as shown in Fig. 2, the user is then asked to judge the sketches in the cluster through member selection, either (1) mark the positive instances that belong to the major category in the current cluster; or (2) mark the negative instances that do not belong to the major category. In a word, the user always chooses to perform fewest assignments to create the largest group of instances corresponding to a single category. Finally, the user adds a novel or existing category tag for the group of sketches by class labeling. Then, we will describe the main algorithm of the proposed method in details.

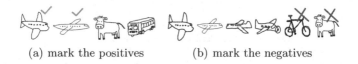

(a) mark the positives   (b) mark the negatives

Fig. 2. Member selection and group labeling (add label:Airplane)

3   Sketch Representation and Similarity Measuring Model

As opposed to images, the main property of freehand sketches is the edge and gradient without color and texture, as shown in Fig. 3. Though there exist plenty of works devoted to design features for sketches in sketch-based image retrieval [10], those features based on the Histogram of Oriented Gradient (HOG) are proved to be relatively effective for describing sketches [11]. Thus, Bag-of-words (BOW) [1] and Fisher Vector [3] based on HOG are chosen here as they have been successfully applied to sketch classification.

In general, it is nontrivial to define an appropriate distance function to measure pairwise affinity for a grouping task, especially when the sketch features based on HOG define a high-dimensional feature space. The bilinear similarity function actually measures the angles rather than distances to compute the similarities between features, so it is more flexible and powerful to discriminate high dimension features [12,17]. Moreover, the bilinear similarity function can be updated online without retraining all the data from scratch [13]. Thus, we exploit a bilinear similarity function to define the similarity measuring model $f(\cdot,\cdot)$ between two sketches p and q: $f(p,q) = S_W(p,q)$, where the positive semi-definite matrix W will be optimized to make the similar sketches more closely during online learning process. Initially, W is identity matrix.

4   Semi-Supervised Clustering

Rather than using unsupervised clustering as previous works [7,8], we use both the learned similarity function and the pairwise constraints as the supervision information to guide the clustering process to improve the results, where the similarity function provide a high quality metric for clustering and the pairwise constraints specify the data relationships (i.e. whether a pair of samples occur together) used readily in spectral clustering.

We firstly use the learned similarity measuring model to compute the similarity matrix $M = \{m_{ij}\}_{N*N}$. The propagated pairwise constraints $F^* =$

$\{f_{ij}^*\}_{N*N}$, which are generated to represent the associated confidence scores between unlabeled sketches, are then used to adjust the similarity matrix M into a constrained matrix $\widetilde{M} = \{\widetilde{m}_{ij}\}_{N*N}$ as: $\widetilde{m}_{ij} = 1 - (1 - f_{ij}^*)(1 - m_{ij})$ if x_j is among the k-nearest neighbors of x_i and $\widetilde{m}_{ij} = (1 + f_{ij}^*)m_{ij}$ otherwise. Then we perform spectral clustering algorithm [16] with the constrained matrix $\widetilde{M}$ to obtain data partition that is consistent with the propagated pairwise constraints.

Meanwhile, the clusters can be ranked according to the average intra-cluster distance. The cluster with the minimum distance (the highest purity) is selected for user judgment, and each instance within the cluster is explicitly confirmed by the user. According to user intervention, a group of newly labeled sketches is generated from the selected cluster. All the remaining unlabeled sketches will be clustered in next iteration. The iteration process can be terminated by detecting whether there are enough sketches for grouping in the current collection (i.e. the size of the remaining sketches is lower than the cluster number).

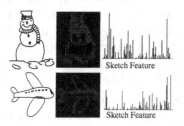

Fig. 3. The illustration of HOG

	x_1	x_2	x_3	x_4	x_5	x_6	x_7
x_1	0	-1	-1	1	0	0	0
x_2	-1	0	1	-1	0	0	0
x_3	-1	1	0	-1	0	0	0
x_4	1	-1	-1	0	0	0	0
x_5	0	0	0	0	0	0	0
x_6	0	0	0	0	0	0	0
x_7	0	0	0	0	0	0	0

Must-link: 1   Cannot-link: -1

Fig. 4. The initial pairwise constraints

5   Supervision Information Establishment

After creating the group of newly labeled sketches, we can use these labeled sketches to construct two kinds of associations: triplets and link pairs. By online metric learn-ing, the triplets are used to train/update the similarity measuring model, which can provide the optimized metric for the clustering process. The link pairs are firstly used to construct the initial pairwise constraints between labeled sketches, and then the propagated pairwise constraints between unlabeled sketches are generated by constraints propagation. The propagated pairwise constraints can be applied to constrain spectral clustering.

Online Metric Learning. The set of triplets $\{(p_i, p_i^+, p_i^-)\}$ are firstly constructed, where p_i^+ denotes the relevant sketches of p_i and p_i^- otherwise. Given the group of newly labeled sketches at each iteration, every two samples in the group are relevant because they belong to the same class. For each relevant pair, we can randomly select an irrelative sample which belongs to different classes in the labeled collection to form a triplet. All the triplets are used to learn the similarity measuring model that assigns higher similarity scores to pairs of more relevant samples. Following the framework of OASIS [13],

W can be updated online by each triplet $(\boldsymbol{p}_i, \boldsymbol{p}_i^+, \boldsymbol{p}_i^-)$: $\boldsymbol{W}_i = \boldsymbol{W}_{i-1} + \tau_i \boldsymbol{V}_i$, where $\tau_i = \min\left\{C, l(\boldsymbol{p}_i, \boldsymbol{p}_i^+, \boldsymbol{p}_i^-)/\|\boldsymbol{V}_i\|^2\right\}$ and $\boldsymbol{V}_i = \boldsymbol{p}_i(\boldsymbol{p}_i^+ - \boldsymbol{p}_i^-)^T$, $l(\cdot, \cdot, \cdot)$ is the hinge loss function according to the Passive-Aggressive algorithm [14]. With the accumulation of labeled sketches, the similar sketches can be drawn closer.

Pairwise Constraints Propagation. Another kind of association contains two types of link pairs: *must-link* pair and *cannot-link* pair. According to the newly and previous labeled sketches, the link pairs can be derived as follows: a pair of sketches with the same label denotes must-link pair and cannot-link pair otherwise. Must-link constraints and cannot-link constraints are two types of pairwise constraints. According to the link pairs, we can separately construct the initial pairwise constraints between labeled sketches [16]. Specially, let matrix $\boldsymbol{Z} = \{z_{ij}\}_{N*N}$ represent the sets of the initial pairwise constraints, where $\boldsymbol{Z}_{ij} = 1$ when (x_i, x_j) is a must-link pair, $\boldsymbol{Z}_{ij} = -1$ when (x_i, x_j) is a cannot-link pair, and $\boldsymbol{Z}_{ij} = 0$ otherwise, as shown in Fig. 4.

In order to get the best set of pairwise constraints between unlabeled sketches ($|z_{ij}| \leq 1$ is viewed as soft constraints), we propagate the initial pairwise constraints between the labeled sketches over the dataset from the initial set $\boldsymbol{Z}$. This can be viewed as a two-class semi-supervised classification problem, which is formulated as minimizing a Laplacian regularized functional [15]:$\min_{\boldsymbol{F}} \|\boldsymbol{F} - \boldsymbol{Z}\|_F^2 + \frac{\psi}{2}(\boldsymbol{F}^T \boldsymbol{L} \boldsymbol{F} + \boldsymbol{F} \boldsymbol{L} \boldsymbol{F}^T)$, where $\boldsymbol{L}$ is a Laplacian matrix. The minimizing problem can be solved by the approximation algorithm [16]. The solution $\boldsymbol{F}^*$ is the final representation of the propagated pairwise constraints. As $\boldsymbol{F}^*$ contains all the pairwise constraints between labeled and unlabeled sketches, we use the propagated constraints between unlabeled sketches as the output.

6   Experiments and Results

Dataset and feature extraction. In our experiments, we evaluated the proposed approach on a public sketch dataset containing 250 categories and each category contains 80 sketches [1]. Our evaluation dataset consists of 20 categories: airplane, bed, bike, bus, car, cat, chair, cow, flying bird, horse, helicopter, motorbike, person, potted plant, sailboat, sheep, table, train, tree, TV. For all sketches in our experiments, we separately exploit the BOW and Fisher Vector representation.All the features are 500-dimension vectors.

The Effectiveness of Single Iteration: We evaluate the effectiveness of single iteration with the quality of clustering in a single iteration. The evaluation criterion is to compute the average purity of the clusters generated from the dataset. The evaluation dataset is equally divided into 4 groups and each group has 20 classes of sketches. Each time, we randomly select 4, 12 and 20 classes of sketches from one group to construct the training set while the other three groups are used as the testing data. We firstly evaluate the effectiveness of similarity measuring model (SMM) and compare it with two methods: the native similarity without metric learning (Native) and the learned similarity by the offline metric learning with rankings (MLR) [8]. Meanwhile, we separately exploit the BOW and Fisher Vector as the representation and the results are illustrated in Figs. 5 and 6. We can see that the average purity achieved by our SMM outperforms the native similarity in all cases. With the number of the training categories increasing, the performance of our SMM is better. The results generated by SMM are little worse than MLR when employing BOW to represent sketches. However, once Fisher Vector is used, the SMM significantly outperforms MLR, which only gets similar results as Native. As SMM measures the angles between the features rather than distances (as MLR does), it has stronger discriminative power to capture the affinities in high-dimension space [17]. According to the results and analysis, SMM could be more adaptive for the existing sketch representations. Based on the learned similarity, the constraints propagation (SMM+CP) can further improve the cluster results, as shown in Figs. 5 and 6. All these can demonstrate our semi-supervised clustering method combining the similarity measuring model with constraint propagation is more efficient for single-modal representation of sketches than the existing methods.

(a) 4 training classes   (b) 12 training classes   (c) 20 training classes

Fig. 5. The curve of the average cluster purity with the BOW representation

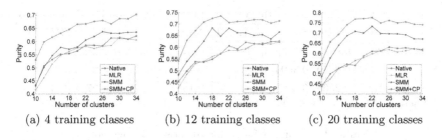

(a) 4 training classes   (b) 12 training classes   (c) 20 training classes

Fig. 6. The curve of the average cluster purity with the fisher vector

The Efficiency of the Iterative Process: In this section, we evaluate the efficiency of the iterative categorization method by computing the average user labor effort required for labeling one sketch:$\mu = U/n$. n is the size of labeled sketches. The user effort U includes two types of interactions: 1) selecting a sketch within the selected cluster, whether positive or negative; and 2) providing a label for the group of sketches. From the evaluation dataset, we randomly select 10 and 20 classes to construct the tow initial unlabeled collections separately, and compute the values of μ with various numbers of clusters on the collections.

Horse	Cow
Bus	Helicopter
Car	Motorbike
Bike	Airplane

	Animal
	Vehicle
	Ridingcar
	Aircraft

(a) Unlabeled sketches   (b) Semantic annotation   (c) Functional annotation

Fig. 7. The annotation results by different taxonomies

Tables 1 and 2 separately show the statistical results with different features. We can see that the proposed semi-supervised clustering method can efficiently reduce the label effort in an iterative framework.

The Diversity of Sketch Annotation: By our interface, users can control the number and categories of the annotated sketches. In order to evaluate the performance of the annotation diversity, we perform the experiments on the whole dataset with two different taxonomies. The first taxonomy is fully consistent with the semantic criterion in benchmark. The second one mainly focuses on the functionality of the sketched objects and includes six categories such as Aircraft (airplane, helicopter), Vehicle (bus, car, sailboat, train), Animal (cat, cow, horse, flying bird, sheep, person), Furniture (bed, chair, table, TV), Riding car (motorbike, bike), and Plant (potted plant, tree). Figure 7 gives an illustration about the semantic and functional annotation. Meanwhile, we test the label effort and the results are shown in Table 3. The results show that our method can produce the labeled collections of different taxonomies with less label effort.

Table 1. The user effort μ for 10 classes of collections

	Sketch representation: BOW				Sketch representation: Fisher Vector			
	Native	MLR	SMM	SMM+CP	Native	MLR	SMM	SMM+CP
NC=10	0.6337	0.6917	0.6055	0.6186	0.3425	0.3022	0.3333	0.2551
NC=20	0.3896	0.3779	0.3694	0.3814	0.2089	0.2122	0.1884	0.1775
NC=30	0.3398	0.2966	0.3006	0.2926	0.1855	0.1826	0.1497	0.1464

Where NC denotes the number of clusters used in the clustering algorithm.

Table 2. The user effort μ for 20 classes of collections

	Sketch representation: BOW				Sketch representation: Fisher Vector			
	Native	MLR	SMM	SMM+CP	Native	MLR	SMM	SMM+CP
NC=10	0.8362	0.7500	0.7895	0.6734	0.7805	0.7327	0.7484	0.6038
NC=20	0.7477	0.6425	0.6850	0.6039	0.5056	0.4990	0.4586	0.4359
NC=30	0.5848	0.5103	0.5321	0.4739	0.4204	0.4552	0.3491	0.3259

Where NC denotes the number of clusters used in the clustering algorithm.

The Labeling Effort Versus Classification Accuracy: This experiment measures how quickly the proposed method collects labeled sketches and how

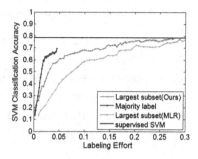

Fig. 8. Comparison of classification accuracy versus labeling effort

effectively the labeled data trains a supervised classifier. The evaluation dataset was first randomly divided into 40/60. The 60 % partition is used to perform the iterative annotation process. At each iteration, a classifier is trained with the new labeled group, and the classification accuracy is measured on the 40 % partition. Note that the 40 % instances don't influence the iterative labeling process. For comparison purposes, we also tested majority labeling used in [7]. Our work and MLR both label the largest subset in the cluster. The results in Fig. 8 show that our work converges to the fully supervised case with faster speed than MLR. Though the majority labeling strategy is more efficient in terms of effort, its accuracy is limited as mislabeled samples exist. This demonstrates that our method can label the collections with less effort while making the generated annotation set more effective.

The Adaptation for Scalable Sketch Collection: In this experiment, we evaluate the efficiency of our method's scalability by two criterions: the training time of metric learning and the average user effort. We partition the initial dataset into four blocks, where Block-1 is the initial collection and the other three groups will arrive one by one as the incremental collections. We separately

Table 3. The user effort for different taxonomies

	Taxonomy 1		Taxonomy 2	
	Native	Ours	Native	Ours
NC=10	0.7366	0.6849	0.1662	0.1496
NC=20	0.4981	0.4163	0.1138	0.0964
NC=30	0.3634	0.3080	0.0977	0.0857

NC denotes the cluster number

Table 4. The training time (s) for the incremental collections

	Block-1	Block-2	Block-3	Block-4
SMM	4.84	5.08	3.87	3.56
MLR	3.54	6.32	11.44	18.98

Table 5. The average label effort μ for the incremental collections

Cluster Number	NC=10		NC=20		NC=30	
	Native	Ours	Native	Ours	Native	Ours
Block-1	0.6954	0.7520	0.3965	0.3021	0.2797	0.2492
Block-2	0.7817	0.5027	0.4298	0.3812	0.3215	0.2596
Block-3	0.7898	0.4382	0.4194	0.3012	0.3173	0.2141
Block-4	0.7143	0.4171	0.4457	0.2698	0.3313	0.2404

treat each block as a new labeled set to evolve the previous training set, and then compute the training time. From the results shown in Table 4, we can see that our SMM is stable with respect to the same number of the incremental training data, while the consuming time of MLR increases more and more. This is intuitive because their offline learning way has to retrain the whole set, but the SMM is relevant to the size of the current labeled set. The SMM can be potentially adapted for the online real-time annotation with the collections increasing. Meanwhile, we also measure whether our methods can improve the labor efficiency with incremental collections addition. The results in Table 5 show that our method needs less effort than the native similarity.

7   Conclusion

This paper proposes an iterative annotation method which can discover the sketch categories iteratively in the dataset. Firstly, we propose an iterative sketch annotation framework for labeling the unlabeled collection with less effort. Secondly, the sketch collection can be annotate freely and flexibly. Finally, the scalable collection can be annotated efficiently and dynamically.

Acknowledgments. This work is supported by the National High Technology Research and Development Program of China (Project No. 2007AA01Z334), National Natural Science Foundation of China (Project No. 61321491 and 61272219), Innovation Fund of State Key Laboratory for Novel Software Technology (Project No. ZZKT2013A12).

References

1. Eitz, M., Hays, J., Alexa, M.: How do human sketch objects? ACM Trans. Graph. **31**(4), 1–10 (2012)
2. Li, Y., Song, Y.Z., Gong, S.G.: Sketch recognition by ensemble matching of structured features. In: BMVC 2013, pp. 35:1–11 (2013)
3. Schneider, R.G., Tuytelaars, T.: Sketch classification and classification-driven analysis using fisher vectors. ACM Trans. Graph. **33**(6), 174:1–174:9 (2014)
4. Li, X., Guo, Y.: Adaptive active learning for image classification. In: CVPR, pp. 859–866 (2013)
5. Tuytelaars, T., Lampert, C.H., Blaschko, M., Buntine, W.: Unsupervised object discovery: a comparison. Springer IJCV **88**(2), 284–302 (2010)

6. Wigness, M., Draper, B.A., Beveride, J.R.: Selectively guiding visual concept discovery. In: WACV, pp. 247–254 (2014)
7. Lee, Y., Grauman, K.: Learning the easy things first: self-paced visual category discovery. In: CVPR, pp. 1721–1728 (2011)
8. Galleguillos, C., McFee, B., Lanckriet, G.R.G.: Iterative category discovery via multiple kernel metric learning. Springer IJCV **108**(1–2), 115–132 (2014)
9. Ristin, M., Guillaumin, M., Gall, J., Van Gool, L.: Incremental learning of ncm forests for large-scale image classification. In: CVPR, pp. 3654–3661 (2014)
10. Eitz, M., Hildebrand, K., Boubekeur, T., et al.: Sketch-based image retrieval: benchmark and bag-of-features descriptors. IEEE TVCG **17**(11), 1624–1636 (2011)
11. Hu, R., Collomosse, J.: A performance evaluation of gradient field hog descriptor for sketch based image retrieval. Elsevier CVIU **117**(7), 790–806 (2013)
12. Kriegel, H.-P., Schubert, M., Zimek, A.: Angle-based outlier detection in high-dimensional data. In: SIGKDD, pp. 444–452 (2008)
13. Chechik, G., Sharma, V., Shalit, U., Bengio, S.: Large scale online learning of image similarity through ranking. JMLR **11**, 1109–1135 (2010)
14. Crammer, K., Dekel, O., Keshet, J., et al.: Online passive-aggressive algorithms. JMLR **7**, 551–585 (2006)
15. Zhou, D., Bousquet, O., Lal, T., et al.: Learning with local and global consistency. NIPS **16**, 321–328 (2004)
16. Lu, Z., Peng, Y.: Exhaustive and efficient constraint propagation: a graph-based learning approach and its applications. Springer IJCV **103**(3), 306–325 (2013)
17. Liu, W., Mu, C., Ji, R.R., et al.: Low-rank similarity metric learning in high dimensions. In: AAAI, pp. 2792–2799 (2015)

Supervised Dictionary Learning Based on Relationship Between Edges and Levels

Qiang Guo[1]($\boxtimes$) and Yahong Han[1,2]

[1] School of Computer Science and Technology, Tianjin University,
Tianjin, China
{qiangguo,yahong}@tju.edu.cn
[2] Tianjin Key Laboratory of Cognitive Computing and Application,
Tianjin, China

Abstract. Categories of images are often arranged in a hierarchical structure based on their semantic meanings. Many existing approaches demonstrate the hierarchical category structure could bolster the learning process for classification, but most of them are designed based on a flat category structure, hence may not be appreciated for dealing with complex category structure and large numbers of categories. In this paper, given the hierarchical category structure, we propose to jointly learn a shared discriminative dictionary and corresponding level classifiers for visual categorization by making use of the relationship between the edges and the relationship between each layer. Specially, we use the graph-guided-fused-lasso penalty to embed the relationship between edges to the dictionary learning process. Besides, our approach not only learns the classifier towards the basic-class level, but also learns the classifier corresponding to the super-class level to embed the relationship between levels to the learning process. Experimental results on Caltech256 dataset and its subset show that the proposed approach yields promising performance improvements over some state-of-the-art methods.

Keywords: Dictionary learning · Hierarchical visual categorization

1 Introduction

Object recognition research has made impressive gains in recent years, with particular success in using discriminative learning algorithms to train classifiers tuned to each category of interest. Some existing models that incorporate richer semantic knowledge about the object categories have achieved more promising results. In recent work, much existing work has shown that using hierarchical category structure to guide the supervised dictionary learning process for classification can bring in improvements in accuracy [4,5,11,12].

A hierarchical category structure is a tree that groups classes together in its nodes according to some human-designed merging or splitting criterion. For example, well-known hierarchical category structure include ImageNet, which

© Springer International Publishing Switzerland 2015
Y.-S. Ho et al. (Eds.): PCM 2015, Part I, LNCS 9314, pp. 66–74, 2015.
DOI: 10.1007/978-3-319-24075-6_7

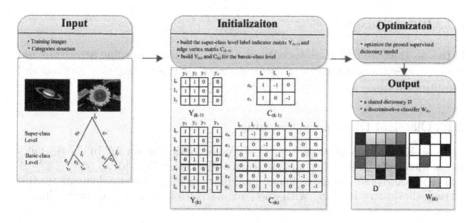

Fig. 1. The flowchart of the proposed method. Firstly, we use the training images and the hierarchical category structure to build the super-class level label indicator matrix $Y_{(k-1)}$ and edge vertex matrix $C_{(k-1)}$, and to build $Y_{(k)}$, $C_{(k)}$ for the basic-class level. $l_i, e_j(i, j = 0, 1, 2, ...)$ represent the labels and the edges respectively. Then we optimize our model. The output of the optimize process is a shared discriminative dictionary D and a classifier $W_{(k)}$. Thus, we can use D and $W_{(k)}$ to predict the categories of the test images.

groups images into sets of cognitive synonyms and their super-subordinate relations, and the phylogenetic tree of life, which groups biological species based on their physical or genetic properties. Critically, such trees implicitly embed cues about human perception of categories, how they relate to one another, and how those relationships vary at different granularities. Thus, for the visual object recognition tasks, such a structure has the potential to guide the learning process. Some models have been made based on this intuition, typically by leveraging the category hierarchy as a prior on iter-class visual similarity [1–3].

In recent years, hierarchical sparse coding algorithms by exploiting the hierarchical semantic relationship among categories have drawn much attention. Zheng et al. [4] proposed to jointly learn multi-layer hierarchical dictionaries and corresponding linear classifiers for region tagging, and bringed in improvements in accuracy. Shen et al. [5] proposed to learn a discriminative dictionary and a set of classification models for each internode of the hierarchical category structure, and the dictionaries in different layers are learnt to exploit the discriminative visual properties of different granularity. Motivated by these algorithms, we present a novel supervised dictionary learning method by taking advantage of the hierarchical category structure. In this paper, the main contributions can be summarized as follows: We present a supervised dictionary learning framework that simultaneously learns a shared discriminative dictionary and a classifier. Our approach not only makes use of the relationship between the edges to help learn the classifiers, but also takes advantage of the relationship between each layer. Our approach is robust to datasets with unbalanced classes.

The rest of the paper is organized as follows. Section 2 presents our approach and model solution. Experimental results and analysis will be provided in Sect. 3. Finally, we conclude this paper in Sect. 4.

2   Our Approach

In this section, we first introduce the classical dictionary learning algorithm and then describe the formulation of our supervised dictionary learning method and its optimization.

2.1   Classical Dictionary Learning

Suppose $\mathbf{X} = [\mathbf{x}^1, ..., \mathbf{x}^N] \in \mathbb{R}^{M \times N}$ of N images of dimension M. In classical dictionary learning tasks, the goal is to learn a dictionary $\mathbf{D} = [\mathbf{d}^1, ..., \mathbf{d}^P] \in \mathbb{R}^{M \times P}$ and represent these images as linear combinations of dictionary elements. In this setting, dictionary learning with an ℓ_1 regularization amounts to computing:

$$\min_{\mathbf{D}, \mathbf{A}} \sum_{i=1}^{N} \left[\frac{1}{2} \left\| \mathbf{x}^i - \mathbf{D}\alpha^i \right\|_2^2 + \lambda \left\| \alpha^i \right\|_1 \right] \tag{1}$$

where, ℓ_1 penalty yields a sparse solution, very few non-zero coefficients in α^i, and λ_1 controls the sparsity of α^i.

2.2   Our Supervised Dictionary Learning

Given the category structure, our goal is to simultaneously take advantages of the relationship between the edges and the relationship between each layer in a unified dictionary learning model.

We consider an image dataset with a k-layer category structure whose levels from the top to the bottom are called: super-class level and basic-class level as shown in Fig. 1. Suppose $\mathbf{X} \in \mathbb{R}^{M \times N}$ denote N training images from J classes. According to the categories taxonomy, images from these J classes in the basic-class level can be merged into S super-classes in the super-class level, e.g., cat and dog belong to the super-class animal. Thus each image region has one class label from the basic-class level and many super-class labels from the super-class level. Matrix $\mathbf{Y}_{(k)} = [\mathbf{y}_{(k)}^1, ..., \mathbf{y}_{(k)}^N] \in \{0,1\}^{J \times N}$, represents the label indicator matrix of image data. $\mathbf{Y}_{(j,i)} = 1$ if the ith image belongs to the jth categories and $\mathbf{Y}_{(j,i)} = 0$ otherwise. Similarly, we use $\mathbf{Y}_{(k-1)} = [\mathbf{y}_{(k-1)}^1, ..., \mathbf{y}_{(k-1)}^N] \in \{0,1\}^{S \times N}$ to denote the super-class label indicator matrix respectively. In order to utilize the relationship among edges and levels, we formulate a supervised dictionary learning process to learn the shared dictionary $\mathbf{D}$ and discriminative matrix $\mathbf{W}$ for images classification.

$$\min_{\mathbf{D}, \mathbf{A}, \mathbf{W}_{(k-1)}, \mathbf{W}_{(k)}} \sum_{i=1}^{N} \left[\left\| \mathbf{y}_{(k-1)}^i - \mathbf{W}_{(k-1)}\alpha^i \right\|_2^2 + \left\| \mathbf{y}_{(k)}^i - \mathbf{W}_{(k)}\alpha^i \right\|_2^2 + \frac{\lambda_0}{2} \left\| \mathbf{x}^i - \mathbf{D}\alpha^i \right\|_2^2 + \lambda_1 \left\| \alpha^i \right\|_1 \right]$$
$$+ \lambda_2 \left[\left\| \mathbf{W}_{(k-1)} \right\|_1 + \left\| \mathbf{W}_{(k)} \right\|_1 \right] + \gamma \left[\Omega_G \left(\mathbf{W}_{(k-1)} \right) + \Omega_G \left(\mathbf{W}_{(k)} \right) \right] \tag{2}$$

We propose to optimize the supervised dictionary learning model in Eq. (2). Where $\Omega_G (\mathbf{W})$ is a graph-guided-fused-lasso penalty. λ_0 controls the importance of the reconstruction term. λ_1, λ_2 control the sparsity of α and $\mathbf{W}$, respectively. γ are regularization parameters that control the fusion effect. G is a graph structure corresponding to the hierarchical categories structure. Assuming that a multi-layer structure over the J categories is given as G with a set of nodes $V = \{1, ..., J\}$ each corresponding to a category and a set of edges E. Thus the graph-guided-fused-lasso penalty is given as:

$$\Omega_G(\mathbf{W}) = \sum_{e=(m,l)\in E, m<l} \omega_{ml} \sum_{p=1}^{M} \left| \mathbf{W}_{(p,m)} - \mathbf{W}_{(p,l)} \right| \tag{3}$$

where $\mathbf{W}_{(p,m)}$ and $\mathbf{W}_{(p,l)}$ are coefficients in $\mathbf{W}$ corresponding to the coefficients of the mth and lth categories for the pth feature, respectively. The weight ω_{ml} measures the fusion penalty for each edge $e = (m, l)$ such that $\mathbf{W}_{(p,m)}$ and $\mathbf{W}_{(p,l)}$ for highly correlated categories with larger ω_{ml} receive a greater fusion effect. Therefore, the graph-guided-fused-penalty in Eq. (2) encourages highly correlated categories corresponding to a densely connected subnetwork in G to be jointly encoded in the dictionary learning process. Because of G encodes correlation information among the edges, and the relationship between the kth level and the $(k-1)$th level is embedded in the formulation, we can expect a better dictionary and a better classifier for image classification.

2.3 Optimization Algorithm

We adopt an alternating optimization method for optimizing the objective function. The steps of our method are as follows:

Optimizing α^i when $\mathbf{D}$, $\mathbf{W}_{(k-1)}$, $\mathbf{W}_{(k)}$ and $\{\mathbf{A}\}_{-i}$ are fixed. The optimization problem to α^i is formulated as:

$$\min_{\alpha^i} \left\| \mathbf{y}^i_{(k-1)} - \mathbf{W}_{(k-1)}\alpha^i \right\|_2^2 + \left\| \mathbf{y}^i_{(k)} - \mathbf{W}_{(k)}\alpha^i \right\|_2^2 + \frac{\lambda_0}{2} \left\| \mathbf{x}^i - \mathbf{D}\alpha^i \right\|_2^2 + \lambda_1 \left\| \alpha^i \right\|_1 \tag{4}$$

We adopt the fast iterative shrinkage-thresholding algorithm (FISTA) [6] to optimize α^i. FISTA can be viewed as an extension of the classical gradient algorithm but with a global rate of convergence. In this algorithm, the Lipschitz constant L is needed. By using $\|x + y\| \leq \|x\| + \|y\|$, we can get the Lipschitz constant of Eq. (4) $L = 2\lambda_{max}\left(\mathbf{W}^T_{(k-1)}\mathbf{W}_{(k-1)}\right) + 2\lambda_{max}\left(\mathbf{W}^T_{(k)}\mathbf{W}_{(k)}\right) + \lambda_{max}\left(\mathbf{D}^T\mathbf{D}\right)$. Then, the procedure of FISTA algorithm for Eq. (4) is shown as:

Optimizing $\mathbf{D}$ when $\mathbf{W}_{(k-1)}$, $\mathbf{W}_{(k)}$ and $\mathbf{A}$ are fixed. The optimization problem to $\mathbf{D}$ is as follows:

$$\min_{\mathbf{D}} \sum_{i=1}^{N} \frac{\lambda_0}{2} \left\| \mathbf{x}^i - \mathbf{D}\alpha^i \right\|_2^2 \tag{5}$$

Obviously, the function is convex, here we adopt gradient descent algorithm to solve it. The gradient of Eq. (5) with $\mathbf{D}$ is:

$$\frac{\partial}{\partial \mathbf{D}} = -\sum_{i=1}^{N} \lambda_0 \left(x^i - \mathbf{D}\alpha^i \right) \alpha^i \tag{6}$$

Algorithm 1. FISTA with backtracking

Input: $y_{(k-1)}^i, y_{(k)}^i, W_{(k-1)}, W_{(k)}, D, x^i, L$.
Initialization: $\alpha_0^i \in \mathbb{R}^P$;
Step 0. Take $s_1 = \alpha_0^i, t_1 = 1$.
Step n. $(n \geq 1)$ Compute until convergence
(1) $\alpha_n^i = p_L(s_n)$,
(2) $t_{n+1} = \frac{1+\sqrt{1+4t_n^2}}{2}$,
(3) $s_{n+1} = \alpha_n^i + \left(\frac{t_n-1}{t_{n+1}}\right)\left(\alpha_n^i - \alpha_{n-1}^i\right)$.

Optimizing W when D and A are fixed. Because of the procedures of optimizing $\mathbf{W}_{(k-1)}$ and $\mathbf{W}_{(k)}$ are the same, we use $\mathbf{W}$ in this part to show how to optimize them. The optimization problem to $\mathbf{W}$ is formulated as:

$$\min_{\mathbf{W}} \sum_{i=1}^N \left\|\mathbf{y}^i - \mathbf{W}\boldsymbol{\alpha}^i\right\|_2^2 + \gamma\Omega_G(\mathbf{W}) + \lambda_2\|\mathbf{W}\|_1 \tag{7}$$

In this paper, we solve Eq. (7) by a general smoothing proximal gradient (SPG) method [7]. The SPG first finds a separable and smooth approximation of $\Omega_G(\mathbf{W})$, and then solves this transformed simple ℓ_1-norm penalized sparse learning problem by the Fast Iterative Shrinkage-Thresholding Algorithm (FISTA) [6]. Accoding to SPG [7], key steps of the approximation process are as follows:

(1) The graph-guided-fused-lasso penalty $\Omega_G(\mathbf{W})$ is reformulated into a linear transformation of $\mathbf{W}$ via the dual norm as follows:

$$\Omega_G(\mathbf{W}) = \max_{\mathbf{B} \in \mathcal{Q}} \left\langle C\mathbf{W}^T, \mathbf{B} \right\rangle \tag{8}$$

where $\langle U, V \rangle \equiv Tr\left(U^T V\right)$ denotes a matrix inter product. $\boldsymbol{\alpha} \in \mathcal{Q} = \{\boldsymbol{\alpha}|\,\|\boldsymbol{\alpha}\|_\infty \leq 1, \boldsymbol{\alpha} \in R^{|E|}\}$ is a column vector in the auxiliary matrix $\mathbf{B}$ and is associated with $\left\|C\mathbf{W}^T\right\|_1$. Matrix $C \in R^{|E| \times J}$ is the edge-vertex incident matrix with the form of

$$C_{e=(m,l),l'} = \begin{cases} \omega_{ml}, & \text{if } l' = m, \\ -\omega_{ml}, & \text{if } l' = l, \\ 0, & \text{otherwise,} \end{cases} \tag{9}$$

where $l' = 1, ..., J$ index the output (i.e., the category) direction of matrix $\mathbf{W}$.
(2) Introduce the smoothing approximation function $f_\mu(\mathbf{W})$ as follows:

$$f_\mu(\mathbf{W}) = \max_{\mathbf{B} \in \mathcal{Q}} \left\langle C\mathbf{W}^T, \mathbf{B} \right\rangle - \mu d(\mathbf{B}) \tag{10}$$

where $\mu \geq 0$ is the smoothness parameter and $d(\mathbf{B}) \equiv \frac{1}{2}\|\mathbf{B}\|_2^2$.
Substitute $\Omega_G(\mathbf{W})$ in Eq. (7) with $f_\mu(\mathbf{W})$, our goal is to solve the following optimization problem:

$$\min_{\mathbf{W}} \sum_{i=1}^N \left\|\mathbf{y}^i - \mathbf{W}D\boldsymbol{\alpha}^i\right\|_2^2 + \gamma f_\mu(\mathbf{W}) + \lambda_2\|\mathbf{W}\|_1 \tag{11}$$

Because the first two terms of Eq. (11) are convex and smooth, Eq. (11) can be efficiently solved by the FISTA algorithm [6].

The steps in Algorithm 2 summarize our alternating optimization algorithm.

Algorithm 2. Our Supervised Dictionary Learning

Part 1: Dictionary Learning
Input: $Y_{(k)}, Y_{(k-1)}, C_{(k)}, C_{(k-1)}, X, P, T(number of iterations), \lambda_0, \lambda_1, \lambda_2, \gamma$, learning rate μ_1.
Initialization: $A, D, W_{(k)}, W_{(k-1)}$;
Output: classifier $W_{(k)}$, dictionary D.
for t=1,...,T **do**
    using Algorithm 1 to get $\{\alpha^i\}$, namely, A.
    $D \leftarrow D - \mu_1 \frac{\partial}{\partial D}$.
    Get $W_{(k-1)}$ by SPG, shown as [7].
    Get $W_{(k)}$ by SPG, shown as [7].
end for
Part 2: Prediction
Input: x.
Output: y (predicted categories)
Evaluate the sparse codes α
The predicted categories for this test data is $W_{(k)}\alpha$.

---

3 Experimental

3.1 Data Set

In this section, we evaluate our dictionary learning approach on two databases: Caltech256 and its subset Caltech90. The Caltech256 database is a large scale database for image classification. We use all the 256 categories and one of its subset with 90 categories to evaluate our method. The caltech256 database has totally 30607 images, every category has least 80 images. We randomly select 30 % of each category to train the model, and the remains for testing. Besides that, we use the hierarchy provided by the Caltech256 dataset as the category structure, which is a tree structure including 318 nodes. Figure 2 shows several subtrees of the hierarchical category structure from Caltech256 dataset. We use 512-dimensional GIST feature as the input descriptors.

3.2 Comparison Methods and Evaluation Criteria

We compare our method with three state-of-the-art algorithms, including MLSVM [8], Least_LASSO [9], SRMTL [9]. In order to demonstrate the effectiveness of using the relationship between the edges and levels, we use DLNE and

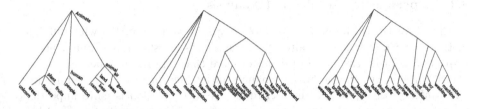

Fig. 2. The hierarchical category structure from Caltech256 dataset contains totally 318 classes, including 62 super-classes, and 256 basic-classes. Here, we show three subtrees of the hierarchical category structure from Caltech256.

Table 1. Comparison of different methods for image classification on Caltech90 dataset

	MLSVM	Least_Lasso	SRML	DLNE	DLNL	Ours
avgprec	0.7050	0.7244	0.7249	0.6987	0.7160	**0.7396**
hloss	0.9707	0.9443	0.9440	0.9481	0.9486	**0.9440**
rloss	0.1276	0.2328	0.2327	0.1317	0.1361	**0.1139**

Table 2. Comparison of different methods for image classification on Caltech256 dataset

	MLSVM	Least_Lasso	SRML	DLNE	DLNL	Ours
avgprec	0.5163	0.5366	0.5358	0.5053	0.5038	**0.5459**
hloss	0.9939	0.9848	0.9848	0.9869	0.9876	**0.9840**
rloss	0.1726	0.2695	0.2695	0.2126	0.2048	**0.1701**

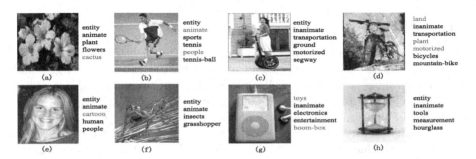

Fig. 3. Some examples of the image classification results on Caltech256 (red fonts denote the false positive predictions).

DLNL as baselines. DLNE is our method without using the relationship between each edge, namely $\gamma = 0$. DLNL is our method without using the relationship between each level. In our experiment, we use three criteria [10] to evaluate the performance of learning with multi-label classification. They are average precision (avgprec), hamming loss (hloss) and ranking loss (rloss).

3.3 Experimental Results and Analysis

We first investigate the performance of comparing our algotithm with different methods on Caltech90 and Caltech256, the results are shown in Tables 1 and 2, respectively. From the results we can observe that: (1) In all experiments, our algorithm achieves the best performances in average precision , hamming loss and ranking loss.(2) In the three criteria, average precision is more important in real application. Our method gains performance improvement of 4.9 %, 2.1 %, 2.0 %, 5.9 %, 3.3 % and 5.7 %, 1.7 %, 1.9 %, 8.0 %, 8.3 % over these algorithms at average precision in these two datasets respectively. Thus the results demonstrate that the effectiveness of the proposed supervised dictionary learning framework.

In our experiments, the column size of D is 32, and γ is equal to 0.0001. In addition, we set $\lambda_0 = 1, \lambda_1 = 3, \lambda_2 = 0.01$ and the learning rate $\mu_1 = 0.01$ on Caltech90. For Caltech256, the learning rate $\mu_1 = 0.001$, other parameters are the same as Caltech90.

In Fig. 3, we show some examples of category prediction on the Caltech256 dataset. We can see the prediction on "segway", "grasshopper", "hourglass" matches the truth perfectly. On the other hand, the category prediction performances on different objects may differ.

4 Conclusions

In this paper, we present a supervised dictionary learning approach. Our approach not only makes use of the relationship between the edges to help learn the classifiers, but also takes advantage of the relationship between each layer. This enables us to simultaneously take advantages of the robust encoding ability of sparse coding as well as the semantic relationship in the category structure. The experimental results on Caltech256 and its subset demonstrate that our model efficiently deal with the classification task with large numbers of categories.

Acknowledgments. This work was partly supported by the NSFC (under Grant 61202166 and 61472276), and Doctoral Fund of Ministry of Education of China (under Grant 20120032120042).

References

1. Zhou, D., Xiao, L., Wu, M.: Hierarchical classification via orthogonal transfer. In: ICML (2011)
2. Marszalek, M., Schmid, C.: Semantic hierarchies for visual object recognition. In: CVPR (2007)
3. Choi, M., Torralba, A., Willsky, A.: A tree-based context model for object recognition. In: TPAMI (2012)
4. Jingjing, Z., Zhuolin, J.: Tag taxonomy aware dictionary learning for region tagging. In: CVPR (2013)
5. Shen, L., Shuhui, W., Gang, S., Shuqiang, J., Qingming, H.: Multi-level discriminative dictionary learning towards hierarchical visual categorization. In: CVPR (2013)
6. Beck, A., Teboulle, M.: A fast iterative shrinkage-thresholding algorithm for linear inverse problems. SIAM J. Imaging Sci. **2**, 183–202 (2009)
7. Chen, X., Lin, Q., Kim, S., Carbonell, J., Xing, E.: Smoothing proximal gradient method for general structured sparse learning. In: Uncertainty in Artificial Intelligence (UAI) (2011)
8. Boutell, M.R., Luo, J., Shen, X., Brown, C.M.: Learning multi-label scene classification. Pattern Recogn. **37**, 1757–1771 (2004)
9. Zhou, J., Chen, J., Ye, J.: MALSAR: Multi-tAsk Learning via StructurAl Regularization. Arizona State University, Tempe (2011)
10. Zhou, Z.-H., Zhang, M.-L., Huang, S.-J., Li, Y.-F.: Multi-instance multi-label learning. Artif. Intell. **176**, 2291–2320 (2012)

11. Wu, F., Han, Y., Tian, Q., Zhuang, Y.: Muti-label boosting for image annotation by structural grouping sparsity. In: ACM MM (2010)
12. Han, Y., Yang, Y., Yan, Y., Ma, Z., Sebe, N., Zhou, X.: Semi-supervised feature selection via spline regression for video semantic recognition. IEEE T-NNLS **26**, 252–264 (2015)

Adaptive Margin Nearest Neighbor for Person Re-Identification

Lei Yao[1], Jun Chen[1,2(✉)], Yi Yu[3], Zheng Wang[1], Wenxin Huang[1], Mang Ye[1], and Ruimin Hu[1,2]

[1] National Engineering Research Center for Multimedia Software, School of Computer, Wuhan University, Wuhan, China
{leiyaoiss,chenj,wangzwhu,wenxin.huang,yemang,hrm}@whu.edu.cn
[2] Collaborative Innovation Center of Geospatial Technology, Wuhan, China
[3] National Institute of Informatics, Chiyoda, Japan
yiyu@nii.ac.jp

Abstract. Person re-identification is a challenging issue due to large visual appearance changes caused by variations in viewpoint, lighting, background clutter and occlusion among different cameras. Recently, Mahalanobis metric learning methods, which aim to find a global, linear transformation of the feature space between cameras [1–4], are widely used in person re-identification. In order to maximize the inter-class variation, general Mahalanobis metric learning methods usually push impostors (i.e., all negative samples that are nearer than the target neighbors) to a fixed threshold distance away, treating all these impostors equally without considering their diversity. However, for person re-identification, the discrepancies among impostors are useful for refining the ranking list. Motivated by this observation, we propose an Adaptive Margin Nearest Neighbor (AMNN) method for person re-identification. AMNN aims to take unequal treatment to each samples impostors by pushing them to adaptive variable margins away. Extensive comparative experiments conducted on two standard datasets have confirmed the superiority of the proposed method.

Keywords: Person re-identification · Metric learning · LMNN · Adaptive margin

1 Introduction

With the rapid development of video surveillance and computer vision techniques, person re-identification (re-id) aiming at recognizing an individual from images observed across non-overlapping cameras, has attracted increasing attentions these years [5–8]. Person re-id researches relying upon biometrics such as face and gait are infeasible or unreliable in the uncontrolled surveillance environment [2,9]. In this condition, most researchers exploit person's appearance to solve the person re-id problem. However, the large variations of appearance, caused by changes on pedestrian pose, camera viewpoint and lighting condition

© Springer International Publishing Switzerland 2015
Y.-S. Ho et al. (Eds.): PCM 2015, Part I, LNCS 9314, pp. 75–84, 2015.
DOI: 10.1007/978-3-319-24075-6_8

Fig. 1. Three rows as three examples from the VIPeR dataset [6], which illustrate that the discrepancies of appearance among different individuals are extremely diverse in person re-id problem. The green box contains the different individuals from camera B compared to each probe image from camera A respectively. For each row, we can see that the impostor on the left is more similar with the probe image than the right one.

between different cameras, often make images of different persons appear even more similar than those of the same person. As a result, person re-id remains a challenging task.

Previous works try to solve this problem mainly from two aspects: (1) Seeking distinctive and stable feature representations for individual appearance, such as color histogram [10,11], graph model [12], spatial co-occurrence representation model [13], principal axis histogram [14], rectangle region histogram [15] and combinations of multiple features [5]. However, designing a set of features that are both distinctive and stable is extremely difficult, specially under realistic conditions where the viewing changes can cause significant intra-object appearance variation. (2) Learning a distance metric or projecting features from different views into a common space to suppress inter-camera variations. Related methods mainly include Large Margin Nearest Neighbor Learning (LMNN) [16,17], Information Theoretic Metric Learning (ITML) [18] , Logistic Discriminant Metric Learning (LDML) [19] and KISS metric Learning [3].

As a kind of metric learning methods, Mahalanobis metric learning methods have recently attracted a lot of attention in person re-id [1–3]. In general, Mahalanobis metric learning methods aim to find a global, linear transformation of the feature space based on the class of Mahalanobis distance functions. In the new feature space, intra-class variation is minimized while inter-class variation is maximized. For maximizing the inter-class variation, general Mahalanobis metric learning methods like LMNN, LMNN-R, ITML push the impostors to a fixed threshold away, which means these methods treat all the impostors equally. For instance, LMNN [16,17] and LMNN-R [1] set one constant margin as the fixed threshold, while ITML makes the distance between dissimilar pairs greater than a fixed lower bound l [18]. However, general Mahalanobis metric learning methods, which treat all the impostors equally, have ignored some important phenomena for person re-id, as follows. First, the discrepancies among different impostors usually exhibit extreme diversity, because the appearance discrepancies between an individual and their impostors are extremely diverse in practice, as shown in Fig. 1. Second, the person re-id problem can be formalized as a content-based image retrieval (CBIR) problem [20], which generates a ranking list where top results are more likely of the same person as the query image. In order to get a more reliable ranking result, it is necessary to consider the discrepancies of different impostors. However, general metric learning methods like LMNN push all impostors of each training sample to a fixed distance away, which can not reflect the diversity among different impostors, as shown in Fig. 2 (a). To solve this problem, we propose an Adaptive Margin Nearest Neighbor (AMNN) method to push different impostors away with adaptive variable margins. With the benefits of treating the impostors differently, our proposed method AMNN achieves a better performance.

The rest of this paper is organized as follows: In Sect. 2, we briefly describe the LMNN method. In Sect. 3, our AMNN method is suggested based on LMNN. Its performance is experimentally evaluated on two benchmark data sets for person re-id and presented in Sect. 4. At last in Sect. 5, we give our conclusion of this paper.

2 Large Margin Nearest Neighbor

LMNN [9] learns a Mahanalobis distance metric, which enforces the k-nearest neighbors to always belong to the same class while samples from different classes are separated by a large margin. Consider a training set of N samples with the corresponding class labels $\{(x_i, y_i)\}_{i=1}^N$. Let $y_{ij} \in \{0, 1\}$ indicate whether y_i and y_j match, and $\eta_{ij} \in \{0, 1\}$ indicate whether x_j is a target neighbor of x_i. The goal of LMNN is to learn a linear transformation $L : \mathbb{R}^d \to \mathbb{R}^d$, to compute the squared distance as:

$$D_L(x_i, x_j) = ||L(x_i - x_j)||^2, \tag{1}$$

which can be rewritten as:

$$D_M(x_i, x_j) = (x_i - x_j)^T M(x_i - x_j), \tag{2}$$

where the Mahalanobis distance metric M is induced by the linear transformation L as $M = L^T L$. Thus, $D^{\frac{1}{2}}$ is a valid distance because M is a symmetric positive-semidefinite matrix. On one hand, we could minimize the distance between each training sample and its K nearest similarly labeled neighbors by minimizing ε_{pull} as follow:

$$\varepsilon_{pull}(M) = \sum_{i,j}^N \eta_{ij} D_M(x_i, x_j). \tag{3}$$

On the other hand, we could maximize the distance between all differently labeled samples which are closer than the aforementioned K nearest neighbors

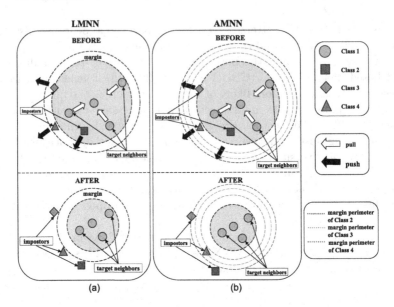

Fig. 2. Schematic illustration of one inputs neighborhood before training (top) versus after training (bottom) in LMNN and AMNN method. (i) LMNN pushes the three impostors to a uniform margin away after training. (ii) AMNN pushes each impostor a different margin away, and the margin is negatively correlated with the distance between a sample and its impostor.

distances plus a constant margin by minimizing ε_{push}.

$$\varepsilon_{push}(M) = \sum_{i,j}^{N} \sum_{l=1}^{N} \eta_{ij}(1 - y_{il})[1 + D_M(x_i, x_j) - D_M(x_i, x_l)]_+. \quad (4)$$

A linear combination of ε_{pull} and ε_{push} could define the overall cost ε_{LMNN}.

$$\varepsilon_{LMNN} = (1 - \mu)\varepsilon_{pull}(M) + \mu\varepsilon_{push}(M), \quad (5)$$

where μ $(0 < \mu < 1)$ is a tuning parameter, $[z]_+ = max(z, 0)$ denotes the standard loss. The cost function consists of two terms, the first term penalizes large distances between each training sample and its target neighbors, while the second term penalizes small distances between each training sample and its impostors.

3   Adaptive Margin Nearest Neighbor

Genaral Mahalanobis metric learning methods treat each impostor equally, which means push the impostors to a fixed threshold away. For the person re-id task, it's not enough to measure the distance between samples using the metric learned by general Mahalanobis metric learning methods. So our proposed method AMNN aims to treat each impostor unequally by setting the fixed distance threshold to be a variable one.

Furthermore, the person re-id task generates a ranking list according to the result similarity to the query image, hence the matching performance is mainly decided by the distance between each sample and its nearest impostor. Namely, the impostor which is nearer to the query sample usually has a more important effect on the final ranking result than the further ones. Therefore, when pushing the impostors away, it's reasonable to push the nearer impostors away from a larger distance than the further ones.

We solve the above problem by modifying the traditional LMNN method. A schematic illustration comparing the LMNN method with our proposed AMNN method is given in Fig. 2. In the figure, LMNN aims to push all impostors, which invade the perimeter surrounding the target neighbors, a constant margin away from the perimeter, as shown in Fig. 2(a). In comparison, our proposed method AMNN pushes all impostors variable margins away from the perimeter surrounding the target neighbors, as shown in Fig. 2(b), a large margin for an impostor (blue rectangle) with a short distance, a small margin for an imposter (purple triangle) with a long distance.

We denote the variable margin mrg_{il} between sample x_i and one of its negative samples x_l as:

$$mrg_{il} = 1 + \alpha \times e^{-\frac{D_M(x_i, x_l)}{Max(D(x_i, x_l))}}, \tag{6}$$

$$D(x_i, x_l) = (x_i - x_l)^T (x_i - x_l), \tag{7}$$

where $D(x_i, x_l)$ denotes the Euclidean distance between x_i and x_l. $Max(D(x_i, x_l))$ is the maximum value of $D(x_i, x_l)$ used to normalize $D_M(x_i, x_l)$. The parameter α reflects the divergence level between different margins, and its value can be tuned by experiments. It is worth noting that our variable margin mrg_{il} is adaptive which has negative correlation with the distance $D_M(x_i, x_l)$.

In the proposed AMNN method, a variable margin mrg_{il} is used, and Eq. 4 can be rewritten as:

$$\varepsilon_{push}(M) = \sum_{i,j}^{N} \sum_{l=1}^{N} \eta_{ij}(1 - y_{il})[mrg_{il} + D_M(x_i, x_j) - D_M(x_i, x_l)]_+. \tag{8}$$

4   Experiment

4.1   Experiment Setting

We evaluated our approach by comparing it with four baseline methods and a number of state-of-the-art methods on two publicly available datasets, the VIPeR dataset [21] and the CUHK dataset [22]. The widely used VIPeR dataset consists of images from 632 pedestrians. Some example images are shown in Fig. 1. The CUHK dataset is a larger dataset recently released by Wang et al.

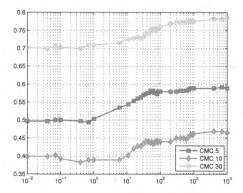

Fig. 3. Matching rate at rank 5, 10, 30 vs. the value of α on the dataset VIPeR.

[22] and contains 971 identities from two disjoint camera views. For each person on these two datasets, a pair of images are taken from cameras with widely inconsistent views. Viewpoint changes of 90 degrees or more as well as huge lighting variations make these datasets as one of the most challenging datasets for re-id. We followed the methodology used in [1,11] for each dataset. The group of pedestrians in each dataset is randomly split into two halves, one for training and the other for testing. The four baseline methods we compared with are ITML [18], LMNN [16], the Mahalanobis distance of the similar pairs and the Euclidean distance. To reduce the bias, we repeated the whole procedure 10 times and the average of the results is given as the final performance. In the evaluation, Cumulative Matching Characteristic (CMC) curve [13], which represents the expectation of finding the true match. will be used as the main metric.

4.2 Parameter Selection

The appropriate selection of parameter α plays an important role in our proposed method. We study this issue experimentally in this section.

Figure 3 shows the matching rate at rank 5, 10, 30 with respect to the value of the α on the dataset VIPeR for our AMNN method. Because the divergence level of the variable margin mrg_{il} is amplified, we can observe that the matching rate increases with the increase of α. However, the improvement of matching rate is getting smaller after α reaches a certain value, such as $\alpha = 10^3$.

4.3 Evaluation on VIPeR and CUHK

To deal with the evaluation, we followed the procedure described in [1,11]. The features were described by HSV color feature and LBP texture feature. We compared our approach with the baseline methods ITML, LMNN, the Mahalanobis distance of the similar pairs and the Euclidean distance. Figure 4 and Fig. 5 show

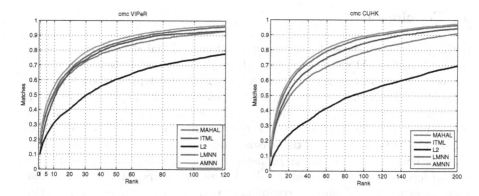

Fig. 4. Average CMC curves of our approaches, ITML, LMNN, Mahal and L2 on the two data sets.

Table 1. Comparisons of matching rates (%) at rank-n on VIPeR. (∗ indicates the best run).

Methods	rank@1	rank@10	rank@25	rank@50
AMNN+SCNCD	**34.8**	**77.2**	**91.5**	93.7
LAFT [24]	29.6	69.3	88.7	**96.8**
KISSME [3]	19.6	62.2	80.7	91.8
SDALF [5]	19.9	49.3	70.5	84.8
SCNCD [27]	20.7	60.6	79.1	90.4
PRDC [2]	15.7	53.9	76	87
LDML [3]	10.4	31.3	44.6	60.4
MCC [2]	15.2	57.6	80	91
MC [25]	12.7	56.1	77	88
LMNN-R∗[1]	23.7	68	84	93

the average CMC curves of these methods on the dataset VIPeR and CUHK, respectively.

As can be seen, our proposed method outperforms the other four baseline methods over the whole range of ranks. It confirms that for metric learning in re-id applications, it's more effective to push the impostors to a variable margin away from the target neighbor perimeter, and the variable margin is negatively correlated with the distance between a training sample and its impostor.

Moreover, we compared our approaches with the state-of-the-art re-id methods on VIPeR. For a fair comparison, the results for these methods are directly taken from the original papers. Table 1 shows that our AMNN method with SCNCD obtains comparable performance with the state-of-the-art.

5 Conclusion

In this paper, we find out the phenomena that general Mahalanobis metric learning methods like LMNN, LMNN-R, ITML push the impostors to a fixed threshold away during metric learning. So we proposed an Adaptive Margin Nearest Neighbor (AMNN) method, aiming to treat each sample's impostors differently by pushing them to adaptive variable margins away during the metric learning. As the impostor which is nearer to the query sample always has a more obvious impact on the performance of the person re-id task, therefore, it is reasonable to push the nearer impostor to a larger distance away than the further ones. Experimental results demonstrate that our AMNN method outperforms the other four baseline methods and obtains comparable performance with the state-of-the-art. In the future, we will design some more advanced adaptive margin model and modify the other traditional Mahalanobis metric learning methods with the adaptive margin idea to further improve the results.

Acknowledgement. The research was supported by National Nature Science Foundation of China (No. 61231015, No. 61170023, No. 61172173, No. 61303114). National High Technology Research and Development Program of China (863 Program, No. 2015AA016306). Technology Research Program of Ministry of Public Security (No. 2014JSYJA016). The EUFP7 QUICK project under Grant Agreement (No. PIRSES-GA-2013-612652). Major Science and Technology Innovation Plan of Hubei Province (No. 2013AAA020). Internet of Things Development Funding Project of Ministry of industry in 2013 (No. 25). China Postdoctoral Science Foundation funded project (2013M530350). Specialized Research Fund for the Doctoral Program of Higher Education (No. 20130141120024). Nature Science Foundation of Hubei Province (2014CFB712). The Fundamental Research Funds for the Central Universities (2042014kf0250, 2014211020203). Jiangxi Youth Science Foundation of China(Grant No. 20151BAB217013).

References

1. Dikmen, M., Akbas, E., Huang, T.S., Ahuja, N.: Pedestrian recognition with a learned metric. In: Kimmel, R., Klette, R., Sugimoto, A. (eds.) ACCV 2010, Part IV. LNCS, vol. 6495, pp. 501–512. Springer, Heidelberg (2011)
2. Zheng, W.-S., Gong, S., Xiang, T.: Person re-identification by probabilistic relative distance comparison. In: IEEE Conference on Computer Vision and Pattern Recognition (CVPR) (2011)
3. Kostinger, M., Hirzer, M., Wohlhart, P., Roth, P.M., Bischof, H.: Large scale metric learning from equivalence constraints. In: IEEE Conference on Computer Vision and Pattern Recognition (CVPR) (2012)
4. Wang, Y., Hu, R., Liang, C., Zhang, C., Leng, Q.: Camera compensation using feature projection matrix for person re-identification. In: IEEE Transactions on Circuits and Systems for Video Technology (TCSVT) (2014)
5. Farenzena, M., Bazzani, L., Perina, A., Murino, V., Cristani, M.: Person re-identification by symmetry-driven accumulation of local features. In: IEEE Conference on Computer Vision and Pattern Recognition (CVPR) (2010)

6. Baltieri, D., Vezzani, R., et al.: Learning articulated body models for people re-identification. In: ACM Multimedia (MM) (2013)
7. Deng, Y., Luo, P., Loy, C.C., Tang, X.: Pedestrian attribute recognition at far distance. In: ACM Multimedia (MM) (2014)
8. Leng, Q., Hu, R., Liang, C., Wang, Y., Chen, J.: Person re-identification with content and context re-ranking. In: Multimedia Tools and Applications (MTA) (2014)
9. Wang, Z., Hu, R., Liang, C., Leng, Q., Sun, K.: Region-based interactive ranking optimization for person re-identification. In: Ooi, W.T., Snoek, C.G.M., Tan, H.K., Ho, C.-K., Huet, B., Ngo, C.-W. (eds.) PCM 2014. LNCS, vol. 8879, pp. 1–10. Springer, Heidelberg (2014)
10. Park, U., Jain, A., Kitahara, I., Kogure, K., Hagita, N.: Vise: visual search engine using multiple networked cameras. In: International Conference on Pattern Recognition (ICPR) (2006)
11. Gray, D., Tao, H.: Viewpoint invariant pedestrian recognition with an ensemble of localized features. In: Forsyth, D., Torr, P., Zisserman, A. (eds.) ECCV 2008, Part I. LNCS, vol. 5302, pp. 262–275. Springer, Heidelberg (2008)
12. Gheissari, N., Sebastian, T., Hartley, R.: Person reidentification using spatiotemporal appearance. In: IEEE Conference on Computer Vision and Pattern Recognition (CVPR) (2006)
13. Wang, X., Doretto, G., Sebastian, T., Rittscher, J., Tu, P.: Shape and appearance context modeling. In: IEEE International Conference on Computer Vision (ICCV) (2007)
14. Hu, W., Hu, M., Zhou, X., Lou, J., Tan, T., Maybank, S.: Principal axis-based correspondence between multiple cameras for people tracking. IEEE Transactions on Pattern Analysis and Machine Intelligence (PAMI) (2006)
15. Dollar, P., Tu, Z., Tao, H., Belongie, S.: Feature mining for image classification. In: IEEE Conference on Computer Vision and Pattern Recognition (CVPR) (2007)
16. Weinberger, K.Q., Blitzer, J., Saul, L.K.: Distance metric learning for large margin nearest neighbor classification. In: Advances in Neural Information Processing Systems (NIPS) (2006)
17. Weinberger, K.Q., Saul, L.K.: Fast solvers and efficient implementations for distance metric learning. In: international Conference on Machine Learning (ICML) (2008)
18. Davis, J.V., Kulis, B., Jain, P., Sra, S., Dhillon, I.S.: Information-theoretic metric learning. In: international conference on Machine learning (ICML) (2007)
19. Guillaumin, M., Verbeek, J., Schmid, C.: Is that you? Metric learning approaches for face identification. In: IEEE International Conference on Computer Vision (ICCV) (2009)
20. Li, X., Tao, D., Jin, L., Wang, Y., Yuan, Y.: Person re-identification by regularized smoothing kiss metric learning. In: IEEE Transactions on Circuits and Systems for Video Technology (TCSVT) (2013)
21. Gray, D., Brennan, S., Tao, H.: Evaluating appearance models for recognition, reacquisition, and tracking. In: IEEE International Workshop on Performance Evaluation for Tracking and Surveillance (PETS) (2007)
22. Li, W., Zhao, R., Wang, X.: Human reidentification with transferred metric learning. In: Lee, K.M., Matsushita, Y., Rehg, J.M., Hu, Z. (eds.) ACCV 2012, Part I. LNCS, vol. 7724, pp. 31–44. Springer, Heidelberg (2013)
23. Hardoon, D.R., Szedmak, S., Shawe-Taylor, J.: An overview with application to learning methods. In: Neural Computation, Canonical Correlation Analysis (2004)

24. Li, W., Wang, X.: Locally aligned feature transforms across views. In: IEEE Conference on Computer Vision and Pattern Recognition (CVPR) (2013)
25. Liu, K., Guo, X., Zhao, Z., Cai, A.: Person re-identification using matrix complex. In: IEEE International Conference on Image Processing (ICIP) (2013)
26. Wang, X., Doretto, G., Sebastian, T., Rittscher, J., Tu, P.: Shape and appearance context modeling. In: IEEE International Conference on Computer Vision (ICCV) (2007)
27. Yang, Y., Yang, J., Yan, J., Liao, S., Yi, D., Li, S.Z.: Salient color names for person re-identification. In: Fleet, D., Pajdla, T., Schiele, B., Tuytelaars, T. (eds.) ECCV 2014, Part I. LNCS, vol. 8689, pp. 536–551. Springer, Heidelberg (2014)

Compressed-Domain Based Camera Motion Estimation for Realtime Action Recognition

Huafeng Chen[1], Jun Chen[1,2(✉)], Hongyang Li[1], Zengmin Xu[1], and Ruimin Hu[1,2]

[1] National Engineering Research Center for Multimedia Software, School of Computer, Wuhan University, Wuhan, China
{chenhuafeng,chenj,lihy,xzm1981,hrm}@whu.edu.cn
[2] Collaborative Innovation Center of Geospatial Technology, Wuhan, China

Abstract. Camera motions seriously affect the accuracy of action recognition. Traditional methods address this issue through estimating and compensating camera motions based on optical flow in pixel-domain. But the high computational complexity of optical flow hinders these methods from applying to realtime scenarios. In this paper, we advance an efficient camera motion estimation and compensation method for realtime action recognition by exploiting motion vectors in video compressed-domain (a.k.a. compressed-domain global motion estimation, CGME). Taking advantage of geometric symmetry and differential theory of motion vectors, we estimate the parameters of camera affine transformation. These parameters are then used to compensate the initial motion vectors to retain crucial object motions. Finally, we extract video features for action recognition based on compensated motion vectors. Experimental results show that our method improves the speed of camera motion estimation by over 100 times with a minor reduction of about 4 % in recognition accuracy compared with iDT.

Keywords: Action recognition · Camera motion estimation · Compressed-domain

1 Introduction

Automatic human action recognition in video is an important and popular research area with potential applications in video analysis, video retrieval, video surveillance and human-computer interaction [6]. Recent research focuses on realistic datasets collected from surveillance videos, web videos, movies, TV shows, etc. These datasets impose significant challenges on action recognition due to camera motions and other fundamental difficulties. Camera motions abound in real-world video and seriously affect the accuracy of action recognition as they fire anywhere in the whole image and easily drown out the object motions.

Local space-time features [1–4,7–14] are shown to be successful on these datasets due to their aggregation of both spatial appearance feature and temporal motion feature. And some approaches [1–4] further consider to separate camera motions from the temporal motions to preserve defining object motions for

© Springer International Publishing Switzerland 2015
Y.-S. Ho et al. (Eds.): PCM 2015, Part I, LNCS 9314, pp. 85–94, 2015.
DOI: 10.1007/978-3-319-24075-6_9

action recognition. Wu et al. [2] apply a low-rank assumption to decompose feature trajectories into camera-induced and object-induced components. Recently, Park et al. [3] perform weak stabilization to remove both camera and object-centric motions using coarse-scale optical flow for pedestrian detection and pose estimation in video. Jain et al. [4] decompose visual motions into dominant and residual motions for extracting trajectories and computing descriptors. Wang et al. [7] introduce motion boundary histograms in Dense Trajectories (DT) to suppress camera motions, and further propose a camera motion estimation (a.k.a. global motion estimation, GME) method in improved Dense Trajectories (iDT) [1] to explicitly rectify the image to remove the camera motions. Benefited from double camera motion inhibition, iDT performs the best in action recognition accuracy among local space-time features.

While these methods [1–4] have improved the recognition accuracy through camera motion estimation and compensation on the basis of optical flow in pixel-domain, they are extremely time-consuming. For example, the speed of iDT ranges in the order of 3-4 frames per second (fps), which absolutely dissatisfies the requirements of realtime application. The main factor of their inefficiency is the pixel-domain based GME algorithm which must performs inefficient operation: OF calculation between adjacent frames. More seriously, some methods [1,3] calculate OF twice: once for GME, once for feature extraction.

To counteract the high computational complexity problem of local space-time feature, Kantorov et al. [5] make an effective attempt to accelerate the method of DT through replacing OF with motion vectors (a.k.a. MPEG flow, MF) which are obtained from video compressed-domain. The replacement of OF with MF for video feature extraction eliminates the calculation process of OF, thus the method of [5] improves the speed of feature extraction by two orders of magnitude at the cost of minor reduction in recognition accuracy compared with DT. Unfortunately, Kantorov et al. have not considered the interference of camera motions in MF.

In order to compensate the influence of camera motion and accelerate the feature extraction process, in this paper we propose a camera motion estimation and compensation method in the compressed-domain (a.k.a. compressed-domain global motion estimation, CGME), avoiding the OF calculation in pixel-domain. Figure 1 presents the comparison of proposed CGME with traditional GME for action recognition. Based on MF, we estimate camera affine transformation parameters by making use of geometric symmetry and differential theory of motion vector [15]. According to the estimated parameters, we compensate initial MF to retain crucial object motions for action recognition. We extract video feature descriptors based on compensated MF by following the method of [5]. Then, we evaluate the speed and accuracy of our approach on UCF50 [16] and HMDB51 [17] benchmarks. Experimental results show that our method improves the speed of GME by over 100 times with a minor reduction of about 4 % in recognition accuracy compared with iDT. It is proved that the proposed approach completely meets the requirements of realtime action recognition.

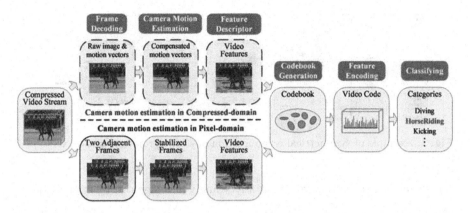

Fig. 1. Comparison of proposed approach to traditional approach for action recognition. **Top**: Pipeline of proposed CGME and feature extraction in compressed-domain. **Bottom**: Pipeline of traditional GME and feature extraction in pixel-domain.

2 Proposed Method

In this section, we first describe the 4-parameter camera affine transformation model for CGME. And then, we estimate the parameters respectively by using initial MF in video compressed-domain. Based on the estimated parameters, we discuss how to rectify the initial MF to eliminate the interference of camera motions. Finally, we extract the video features based on the revised MF by following the method of [5].

2.1 Camera Model

We define the 2D coordinate system of the image for CGME in the first place. The center of the 2D image corresponds to the coordinate origin, the positive direction of **x**-axis to the right, the positive direction of **y**-axis downward, and the image is divided into four quadrants respectively: I quadrant (bottom-right), II quadrant (bottom-left), III quadrant (top-left) and IV quadrant (top-right) (Shown in Fig. 2). Taking any pixel from I quadrant, the spatial coordinates is defined as $z_I = (x, y)^T (x > 0, y > 0)$, it surely determines the symmetry points in other three quadrants: $z_{II} = (-x, y)^T$, $z_{III} = (-x, -y)^T$, and $z_{IV} = (x, -y)^T$.

Based on the 2D coordinate system, we adopt 4-parameter camera affine transformation model for modeling the camera motion [15]. This 4-parameter model can faultlessly model camera translation, scaling, rotation, and their combinations. It is defined by

$$f(z|A, T) = Az + T = \begin{pmatrix} a_1 & -a_2 \\ a_2 & a_1 \end{pmatrix} \begin{pmatrix} x \\ y \end{pmatrix} + \begin{pmatrix} t_x \\ t_y \end{pmatrix}, \tag{1}$$

where a_1 and a_2 are parameters reflecting scaling and rotation changes in motion, t_x and t_y control translation parameters, $(x, y)^T$ is the point in the image.

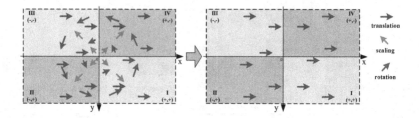

Fig. 2. Example of symmetrical counteraction of scaling/rotation motion vectors.

Thus, a camera motion vector at any point z in image can be expressed as

$$MV(z) = f(z|A, T) - z = (A - I)z + T. \tag{2}$$

2.2   Estimation of T

A camera motion vector can be decomposed into translation, scaling and rotation sub-vectors. The scaling/rotation sub-vectors in image possess the characteristics of symmetrical around the origin. So, any two symmetrical scaling/rotation sub-vectors around the origin can counteract each other (we call it symmetrical counteraction), that is, the sum of them is zero (Shown in Fig. 2).

We make use of the symmetrical counteraction to estimate the parameter T which controls translation sub-vectors. Given two symmetrical around the origin motion vectors $MV(z_I)$ $(z_I = (x, y)^T)$ in I quadrant sector and $MV(z_{III})$ $(z_{III} = (-x, -y)^T)$ in III quadrant sector, we sum them and can get translation parameter T. The sum equation is

$$
\begin{aligned}
MV(z_I) + MV(Z_{III}) &= f(z_I|A, T) - z_I + f(z_{III}|A, T) - z_{III} \\
&= (A - I)(z_I + z_{III}) + 2T \\
&= 2T.
\end{aligned} \tag{3}
$$

Similarly, we can calculate translation parameter T by summing two symmetrical around the origin motion vectors $MV(z_{II})$ $(z_{II} = (-x, y)^T)$ in II quadrant sector and $MV(z_{IV})$ $(z_{IV} = (x, -y)^T)$ in IV quadrant sector,

$$MV(z_{II}) + MV(Z_{IV}) = 2T. \tag{4}$$

By applying Eqs. (3) and (4) on initial MF, we can calculate out a set of initial T parameters $T_{init} = \{T_1, T_2, ..., T_N\}$. Ideally, the values in the set T_{init} are equal ($T_1 = T_2 = ... = T_N$) under the environments without any object motions except camera motions. But generally, camera motions and object motions are mixed together in real videos. So the values in T_{init} are not exactly equal, and we are not sure which truly reflects the real translational motion in T_{init}. To estimate the parameter T from T_{init}, we adopt the mean distance threshold determination algorithm [18]. As can be seen from Algorithm 1, we calculate the mean of all

---

Algorithm 1. Estimation of parameter T from T_{init}.

Require:
$\quad T_{init} = \{T_1, T_2, ..., T_N\};$
Ensure:
$\quad T_{esti};$
1: $T_{mean} \Leftarrow (T_1 + T_2 + \cdots + T_N)/N$
2: **for** $i = 1$ to N **do**
3: $\quad R_i \Leftarrow |T_i - T_{mean}|$
4: **end for**
5: $R_{mean} \Leftarrow (R_1 + R_2 + \cdots + R_N)/N$

6: $count \Leftarrow 0$
7: **for** $i = 1$ to N **do**
8: $\quad$ **if** $R_i > R_{mean}$ **then**
9: $\quad\quad T_i \Leftarrow 0$
10: $\quad\quad count \Leftarrow count + 1$
11: $\quad$ **end if**
12: **end for**
13: $T_{esti} = (T_1 + T_2 + \cdots + T_N)/count$

14: **return** T_{esti}

---

elements in T_{init} firstly, and compute the absolute residuals of all the data based on the mean value. We put the mean of these residuals as a threshold, and weed out the outliers from T_{init} according to whether the element's residual is greater than the threshold. We calculate the mean of rest elements in T_{init}, and put it as the final estimated parameter T_{esti}.

2.3 Estimation of a

We can use the differential principle of motion vectors to calculate the parameter A. Firstly, we deduce general equations of the differences of motion vectors. Given two pixels located on the same line $z_1 = (i_1, c_y)^T$ and $z_2 = (i_2, c_y)^T$ $(i_2 = i_1 + s_x)$. By Eq. (2), we can get

$$
\begin{aligned}
&MV_x(z_2) - MV_x(z_1) \\
&= (f_x(z_2|A, T) - i_2) - (f_x(z_1|A, T) - i_1) \\
&= ((a_1 - 1) \times i_2 - a_2 \times c_y + t_x) - ((a_1 - 1) \times i_1 - a_2 \times c_y + t_x) \quad (5) \\
&= (a_1 - 1)(i_2 - i_1) \\
&= (a_1 - 1) \times s_x.
\end{aligned}
$$

Thus, we can obtain a differential equation on a_1 parameters,

$$
a_1 = \frac{MV_x(z_2) - MV_x(z_1)}{s_x} + 1. \quad (6)
$$

Similarly, we can get a differential equation with respect to a_2 parameters,

$$
a_2 = \frac{MV_y(z_2) - MV_y(z_1)}{s_x}. \quad (7)
$$

Given two pixel coordinates on the same column $z_3 = (c_x, j_1)^T$ and $z_4 = (c_x, j_2)^T$ $(j_2 = j_1 + s_y)$, according to the above mentioned, we can get another set of differential equations on the parameters a_1 and a_2,

$$
a_2 = -\frac{MV_x(z_4) - MV_x(z_3)}{s_y}, \quad (8)
$$

$$a_1 = \frac{MV_y(z_4) - MV_y(z_3)}{s_y} + 1. \tag{9}$$

By applying Eqs. (6), (7), (8) and (9) on initial motion vectors, we can calculate out two sets of initial A parameters $a_{1init} = \{a_{11}, a_{12}, ..., a_{1M}\}$ and $a_{2init} = \{a_{21}, a_{22}, ..., a_{2K}\}$. We estimate parameter A_{esti} based on a_{1init} and a_{2init} by following the algorithm of parameter T estimation in Sect. 2.2.

2.4   Camera Motion Compensation

According to the estimated parameters T_{esti} and A_{esti}, we compensate the initial video motion vectors by

$$\begin{aligned}
MV'(z) &= MV(z) - estiGM(z) \\
&= MV(z) - (Az + T) \\
&= MV(z) - \left(\begin{pmatrix} a_1 & -a_2 \\ a_2 & a_1 \end{pmatrix} \begin{pmatrix} x \\ y \end{pmatrix} + \begin{pmatrix} t_x \\ t_y \end{pmatrix} \right),
\end{aligned} \tag{10}$$

where $MV(z)$ is the initial MF in compressed-domain, $estiGM(z)$ is the estimated camera motion vectors, and $MV'(z)$ is the compensated motion vectors that preserved the defining object motions for action recognition.

2.5   Feature Descriptor Extraction

We follow the design of previously proposed local space-time descriptors [5] and define our descriptor by histograms of the compensated motion vectors in a video patch. We compute HOF descriptors as histograms of compensated motion vectors discretized into eight orientation bins and a non-motion bin. For MBHx and MBHy descriptors the spatial gradients of the v_x and v_y components of the compensated motion vectors are similarly descretized into nine orientation bins. The final descriptor is obtained by concatenating histograms from each cell of the 2×2×3 descriptor grid followed by l_2-normalization of every temporal slice. HOG descriptors are computed at the same sparse set of points.

3   Experimental Results

In this section we evaluate the proposed approach on two publicly available datasets, the UCF50 [16] and HMDB51 [17] (see Fig. 3). The UCF50 dataset has 50 action categories, consisting of real-world videos taken from YouTube. The actions range from general sports to daily life exercises. For all 50 categories, the videos are split into 25 groups. For each group, there are at least 4 action clips. In total, there are 6,618 video clips in UCF50. The HMDB51 dataset is collected from a variety of sources ranging from digitized movies to YouTube videos. There are 51 action categories and 6,766 video sequences in HMDB51.

We compare the speed and action recognition accuracy of the proposed approach to recent methods [1,5]. We follow satandard evaluation setups and report

Fig. 3. Sample frames of standard datasets. Top: UCF50 [16], Bottom: HMDB51 [17].

mean accuracy (Acc) for UCF50 and HMDB51 datasets. The processing speed is reported in frames-per-second (Fps), run at a single-core Intel Xeon X3430 (2.4 GHz) with no multithreading.

To recognize actions, we follow [1,5] to train a GMM model with $K = 256$ Gaussians. Each video is, then, represented by a $2DK$ dimensional Fisher vector for each descriptor type (HOG, HOF, MBHx and MBHy), where D is the descriptor dimension. Finally, we apply l_2-normalization to the Fisher vector. To combine different descriptor types, we concatenate their normalized Fisher vectors. A linear SVM is used for classification.

3.1   GME Evaluation

Table 1 presents action recognition accuracy and speed of the proposed GME approach compared to the GME method adopted by iDT. The performance of iDT (90.9 % in UCF50 and 55.6 % in HMDB51) is approximately four percent higher compared to our proposed approach.

Table 1. Comparison of action classification accuracy and the speed of proposed CGME to GME of iDT. The speed is reported for video of spatial resolution 320×240 pixels on UCF50 and 360×240 pixels on HMDB51.

	Classification (Acc)		Speed (Fps)	
	CGME(Proposed)	GME(iDT)	CGME(Proposed)	GME(iDT)
UCF50	86.3 %	90.9 %	**853.5**	6.5
HMDB51	51.9 %	55.6 %	**912.3**	6.7

When comparing the speed of GME for both methods in UCF50, our CGME method achieves 853.5 fps which is about 24 times faster than real-time and 131 times faster compared to iDT [1]. And when comapring the speed of GME for both methods in HMDB51, our CGME method achieves 912.3 fps which is about 36 times faster than real-time and 136 times faster compared to iDT. From Table 1 we can see, the runtime of proposed CGME method is < 1 % of GME in iDT by avoiding motion vector calculation, and can fully meets the requirements of real-time applications.

3.2   Feature Descriptor Evaluation

We compare our descriptor to iDT [1] and MF [5] on the UCF50 and HMDB51. iDT performs the best in action recognition accuracy, while MF is the fasted algorithm among all existing local space-time descriptor methods.

Proposed method vs iDT [1] - Table 2 presents action recognition accuracy and speed of the proposed approach compared to iDT. The action recognition accuracy of iDT (90.9 % in UCF50 and 55.6 % in HMDB51) is approximately 4 percent higher compared to proposed descriptor (86.3 % in UCF50 and 51.9 % in HMDB51). When comparing the speed of feature extraction for both methods, our method (514.1 fps in UCF50 and 582.2 fps in HMDB51) is far faster than iDT (3.7 fps in UCF50 and 3.9 fps in HMDB51) because proposed approach works in compressed-domain and keeps away from inefficient OF calculation.

Table 2. Comparison of action classification accuracy and speed of proposed feature descriptor to iDT [1].

	Classification (Acc)		Speed (Fps)	
	Proposed	iDT [1]	Proposed	iDT [1]
UCF50	86.3 %	90.9 %	**514.1**	3.7
HMDB51	51.9 %	55.6 %	**582.2**	3.9

Table 3. Comparison of action classification accuracy and speed of proposed feature descriptor to MF [5].

	Classification (Acc)		Speed (Fps)	
	Proposed	MF [5]	Proposed	MF [5]
UCF50	**86.3%**	82.2 %	514.1	698.4
HMDB51	**51.9%**	46.7 %	582.2	752.2

Proposed method vs MF [5] - From Table 3 we can see, the action recognition accuracy of proposed descriptor method (86.3 % in UCF50 and 51.9 % in HMDB51) is approximately 5 percent higher compared to MF feature (82.2 % in UCF50 and 46.7 % in HMDB51). The reason of accuracy increasement between MF and proposed descriptor is that the method of proposed GME significantly inhibit the camera motions (shown in Fig. 4). While the speed of proposed method (514.1 fps in UCF50 and 582.2 fps in HMDB51) is a little slower than MF (698.4 fps in UCF50 and 752.2 fps in HMDB51), it also can meet the needs of realtime processing because it is about 21 times faster than realtime.

Fig. 4. Comparison of initial motion vectors and compensated motion vectors. Left: (a)-UCF50 initial vectors (b)-UCF50 compensated vectors, Right: (a)-HMDB51 initial vectors (b)-HMDB51 compensated vectors. Green Point: motion start point from previous frame, Green Line: motion from the start point to end point in current frame (Color figure online).

4   Conclusion

In this work, we present a method called CGME for realtime action recognition, different from recent mainstream methods. The core idea is: we make full use of motion vectors in compressed domain for GME and feature extraction to avoid inefficient OF calculation in pixel domain. Taking advantage of geometric symmetry and differential theory of motion vectors, we estimate the parameters of camera affine transformation and compensate the initial motion vectors based on the estimated parameters. The proposed method is proved to be more efficient than iDT and completely suitable for realtime action recognition.

Acknowledgement. The research was supported by National Nature Science Foundation of China (No. 61231015), National High Technology Research and Development Program of China (863 Program, No. 2015AA016306), National Nature Science Foundation of China (61170023), Internet of Things Development Funding Project of Ministry of industry in 2013(No. 25), Technology Research Program of Ministry of Public Security (2014JSYJA016), Major Science and Technology Innovation Plan of Hubei Province (2013AAA020), and Nature Science Foundation of Hubei Province (2014CFB712).

References

1. Wang, H., Schmid, C.: Action recognition with improved trajectories. In: IEEE International Conference on Computer Vision (ICCV) (2013)

2. Wu, S., Oreifej, O., Shah, M.: Action recognition in videos acquired by a moving camera using motion decomposition of lagrangian particle trajectories. In: IEEE International Conference on Computer Vision (ICCV) (2011)

3. Park, D., Zitnick, C.L., Ramanan, D., Dollr, P.: Exploring weak stabilization for motion feature extraction. In: IEEE Conference on Computer Vision and Pattern Recognition (CVPR) (2013)

4. Jain, M., Jgou, H., Bouthemy, P.: Better exploiting motion for better action recognition. In: IEEE Conference on Computer Vision and Pattern Recognition (CVPR) (2013)

5. Kantorov, V., Laptev, I.: Efficient feature extraction, encoding, and classification for action recognition. In: IEEE Conference on Computer Vision and Pattern Recognition (CVPR) (2014)

6. Aggarwal, J.K., Ryoo, M.S.: Human activity analysis: A review. ACM Computing Surveys (CSUR) (2011)

7. Wang, H., Klaser, A., Schmid, C., Liu, C.-L.: Action recognition by dense trajectories. In: IEEE Conference on Computer Vision and Pattern Recognition (CVPR) (2011)

8. Wang, H., Klaser, A., Schmid, C., Liu, C.-L.: Dense trajectories and motion boundary descriptors for action recognition. International Journal of Computer Vision (IJCV) (2013)

9. Laptev, I.: On space-time interest points. International Journal of Computer Vision (IJCV) (2005)

10. Scovanner, P., Ali, S., Shah, M.: A 3-dimensional sift descriptor and its application to action recognition. In: ACM International Conference on Multimedia (ACM MM) (2007)

11. Willems, G., Tuytelaars, T., Van Gool, L.: An efficient dense and scale-invariant spatio-temporal interest point detector. In: Forsyth, D., Torr, P., Zisserman, A. (eds.) ECCV 2008, Part II. LNCS, vol. 5303, pp. 650–663. Springer, Heidelberg (2008)

12. Klaser, A., Marszalek, M.: A spatio-temporal descriptor based on 3d-gradients. In: British Machine Vision Conference (BMVC) (2008)

13. Yeffet, L., Wolf, L.: Local trinary patterns for human action recognition. In: IEEE International Conference on Computer Vision (ICCV) (2009)

14. Chen, M., Hauptmann, A.: Mosift: Recognizing human actions in surveillance videos (2009)

15. Zheng, Y., Tian, X., Chen, Y.: Fast global motion estimation based on symmetry elimination and difference of motion vectors. Journal of Electronics & Information Technology (2009)

16. Reddy, K., Shah, M.: Recognizing 50 human action categories of web videos. In: Machine Vision and Applications (MVA) (2012)

17. Kuehne, H., Jhuang, H., Garrote, E., Poggio, T., Serre, T.: Hmdb: A large video database for human motion recognition. In: IEEE International Conference on Computer Vision (ICCV) (2011)

18. Chen, H., Liang, C., Peng, Y., Chang, H.: Integration of digital stabilizer with video codec for digital video cameras. IEEE Transactions on Circuits and Systems for Video Technology (TCSVT) (2007)

Image and Audio Processing

On the Security of Image Manipulation Forensics

Gang Cao[1(✉)], Yongbin Wang[1], Yao Zhao[2], Rongrong Ni[2],
and Chunyu Lin[2]

[1] School of Computer Science, Communication University of China,
Beijing 100024, China
{gangcao,ybwang}@cuc.edu.cn
[2] Institute of Information Science, Beijing Jiaotong University,
Beijing 100044, China
{yzhao,rrni,cylin}@bjtu.edu.cn

Abstract. In this paper, we present a unified understanding on the formal performance evaluation for image manipulation forensics techniques. With hypothesis testing model, security is qualified as the difficulty for defeating an existing forensics system and making it generate two types of forensic errors, i.e., missing and false alarm detection. We point out that the security on false alarm risk, which is rarely addressed in current literatures, is equally significant for evaluating the performance of manipulation forensics techniques. With a case study on resampling-based composition forensics detector, both qualitative analyses and experimental results verify the correctness and rationality of our understanding on manipulation forensics security.

Keywords: Security · Image manipulation forensics · Anti-forensics · Hypothesis testing · Forging attack

1 Introduction

Generally, digital image manipulation refers to any additional alteration on the camera-output images, such as tampering and various typical image operations, i.e., resampling, different types of filtering, contrast adjustment and so on. Such manipulations have been frequently used to create sophisticated image forgeries. In order to verify the authenticity and recover the processing history of digital photograph images, image manipulation forensics techniques have been proposed to detect the suffered manipulations blindly. Existing forensics methods focus on detecting content-changing manipulations, such as splicing [1] and copy-move [2], and non-content-changing manipulations including resampling [3, 4], compression [5], contrast enhancement [6, 7], median filtering [8], sharpening filtering [9], etc. The blind detection of such content-preserving manipulations is also forensically significant [6].

In contrast with cryptography, multimedia forensics remains an inexact science without strict security proofs. As pointed out in [6], although the existing forensics tools are good at uncovering naive manipulations in the scenario without attacks, there is a lack of confidence on their behavior against a sophisticated counterfeiter, who is aware

Y.-S. Ho et al. (Eds.): PCM 2015, Part I, LNCS 9314, pp. 97–105, 2015.
DOI: 10.1007/978-3-319-24075-6_10

of detection techniques in detail. Anti-forensics techniques are developed by forgers to remove or falsify statistical fingerprints of certain image manipulation, aiming to deceive the forensics detectors. Many anti-forensics techniques including undetectable resampling [10], JPEG compression [11] and contrast enhancement [12, 13] etc. have been proposed.

Although the existing anti-forensics techniques have addressed forensics security problem, an objective evaluation framework of security performance is unavailable. In this work, we make efforts to narrow such a gap by addressing the performance evaluation mechanism of image manipulation forensics. With signal detection and hypothesis testing models, reliability and security are characterized with regard to two types of forensic errors: missing and false alarm detection. Manipulation forging attack and manipulation hiding attack are formulated formally. The existing manipulation anti-forensics techniques [10–12], which struggled to hides operation traces, mainly address the security on missing detection risk. There are little concerns on the security on false alarm risk in manipulation forensics. Here we point out that the security on false alarm risk is equally significant for evaluating the performance of manipulation forensics techniques. As a case study, we proposed a resampling forging attack algorithm which could successfully fool the existing resampling-based composition detector [4].

The rest of this paper is organized as follows. In Sect. 2, we present a unified formulation of the image manipulation forensics problem, and address the forensics security evaluation model via two types of attacks. Section 3 gives a case study on resampling forging attack and its application on composition anti-forensics. Experimental results are given in Sect. 4 and the conclusions are drawn in Sect. 5.

2 Understanding and Evaluation of Forensics Security

In this section, we present a unified formulation of the image manipulation forensics problem. Forensics security and attacks would be addressed under such a forensics model.

2.1 Image Manipulation Forensics Model

To address the manipulation forensics problem, first of all, we should make clear about the implication of image manipulation. In this paper, an image manipulation refers to any image operation which can be content-changing or non-content-changing. An image manipulation is denoted as M^θ, in which the type and operation parameter vector of the manipulation are designated by M and θ, respectively. The manipulation space Φ, which is assumed to cover all types of image manipulations with different parameter configurations, is discretized and modeled as

$$\Phi = \left\{ M_i^{\theta_j} \mid i = 1, 2, \ldots, N; j = 1, 2, \ldots, K_i; K_i \in \mathbb{Z}^+ \right\}. \tag{1}$$

Here N types of manipulation, respectively holding K_t kinds of typical parameter configuration, are considered. In real applications, an image may suffer multiple manipulations sequently.

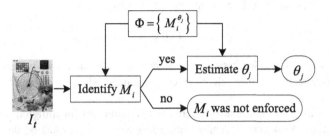

Fig. 1. General image manipulation forensics model

In principle, image manipulation forensics is required to reconstruct the set of processing operations to which the image has been subjected. General manipulation forensics model is proposed and illustrated in Fig. 1. In the manner of passive and blind, the forensics system requires to identify if certain manipulation M_i has been performed on the test image I_t, even to estimate the associated operation parameters when M_i occurs. If we consider the manipulation itself as 'signal', it is appropriate to treat manipulation forensics as series of statistical signal detection and estimation procedure. As a popular statistical analysis tool, hypothesis testing is used to model the image manipulation forensics. In order to detect M_i, two candidate hypothesizes are defined by

$$H_0 : I_t = O(I, \Phi_K), M_i \notin \Phi_K$$

$$H_1 : I_t = O(I, \Phi_K), M_i \in \Phi_K \tag{2}$$

where Φ_K expresses the set of manipulations which the observed image I_t has suffered practically. I denotes the primary unaltered image. The systematic operation function is denoted by $O(\cdot, \cdot)$. As a result, the forensics process can be considered as doing the hypothesis testing. Refined characteristic metric and appropriate decision rule can be used to help make a correct hypothesis testing decision.

2.2 Security Evaluation and Attacks

With respect to the hypothesis testing model, there are two types of detection errors in manipulation forensics [13].

Type I error, often called false positive error, or false alarm, results when the focused manipulation M_i is identified to have been performed while it is actually not. Let $\Re$ be the decision region in the observed signal space, which the receiver uses to assure the occurrence of M_i in the received image signal I_t. Type I error probability is given by

$$P_{fa} = P(I_t \in \Re | H_0). \tag{3}$$

Type II error, often called false negative error, or miss, occurs when the focused manipulation has been performed but the received signal is wrongly verified by the receiver as unaltered. Type II error probability is given by

$$P_m = P(I_t \notin \Re | H_1). \tag{4}$$

Reliability of a forensics system can be measured by such two types of forensics error. It is obvious that higher reliability can be gained when the forensics errors take smaller values.

Security is another significant performance requirement for a digital manipulation forensics system. A successful manipulation forensics system must be designed to be secure against intentional manipulation forging attacks and manipulation hiding attacks, which deceive the forensic detector to yield Type I and Type II error, respectively. If attackers are able to fabricate the images which definitely incur such two types of error, it would be deemed that the forensics system is successfully defeated.

Manipulation forging attack, also called Type I attack, can be defined as

$$\Psi_1 = \left\{ T \mid I_t' = T(I); \; I_t' \in \Re; \; D(I_t', I) < \tau \right\} \tag{5}$$

where T denotes image alteration. D is the visual distortion between the images before and after attack, i.e., I and I_t' correspondingly. τ is the distortion threshold. Manipulation forging attacks intend to counterfeit the inherent statistical fingerprint of targeted manipulation M_i by altering the pixel values of original photograph image, on condition that the image's visual quality is kept invariant and visual effect of the forged manipulation is absent. The attacked image would be falsely detected as manipulated by M_i. We conclude that the evaluation of manipulation forging attacks against digital image forensics should always be benchmarked against the criteria including detectability and image quality.

Manipulation hiding attack, also called Type II attack, can be defined as

$$\Psi_2 = \left\{ T \mid I_t' = T(I_x), I_x \in \{I, I_M\}; I_t' \notin \Re; \; D(I_t', I_M) < \tau \right\} \tag{6}$$

where I_M is the image generated by applying the manipulation M_i to the original image I. And I_t' is the attacked image. Manipulation hiding attacks are often implemented by designing untraceable image manipulation, with the aim to achieve undetectability and preserve as many original image's properties as possible. The evaluation of manipulation hiding attacks against digital image forensics should always be benchmarked against undetectability and image quality. Recently, many research efforts have been devoted to Type II error-related security analysis, in which successful manipulation hiding attacks were proposed to defeat previous forensics systems [10–12].

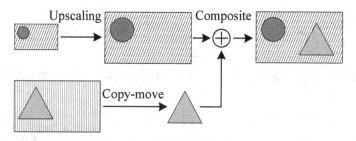

Fig. 2. Resampling involved image composition

In the scenario of targeted manipulation forensics, manipulation forging attacks can also fool the forensic detector by tackling the non-uniqueness of inherent statistical fingerprints [13], which might be counterfeited by attacking alterations or produced by other image manipulations coincidentally.

3 A Case Study with Resampling Forging Attack

As a case study, in this section we propose a resampling forging attack method and show its valuable application in resampling-based composition anti-forensics.

As shown in Fig. 2, in order to create a realistic composite image, resampling may be used to adjust the size of spliced object/background regions and make them matchable. As such, some composition detection methods [3, 4] rely on the detection of locally applied resampling. The inconsistency of detected resampling traces among different image regions is deemed as evidence of composition. In prior works [10], as anti-forensic techniques, Type II attacks have been proposed by hiding the trace of resampling. Here, we show that Type I attack can also be used to defeat the resampling-based composition detectors. Our proposed attack is based on the following two facts: (1) The image would not be taken as composition if the inconsistency of local resampling traces is not detected; (2) Locally forging the resampling trace might be more cost-effective than hiding the trace, especially when the relatively larger spliced region was already resized, as the case shown in Fig. 2.

In order to deceive composition detectors, we can forge resampling trace on unaltered regions to make it be consistent with that on resampled regions. Let the composite image have two source regions, R_1 and R_2, copied from two different source images, respectively. Suppose that R_1 has suffered a certain Geometric Transform (GT) T, while R_2 does not.

We propose to implement the forging attack on the object region R_2 by enforcing inverse geometric transform. Figure 3 shows basic principle of the inverse-GT-based resampling trace forging method. To generate the same operation trace as that incurred by T, a simple but effective way is to enforce the corresponding inverse geometric transform T^{-1}, followed by image quality improvement, and geometric transform T lastly on R_2. Note that the image quality improvement processing can be any post-processing which can improve the quality of the intermediate resampled image,

Fig. 3. Proposed resampling forging attack based on inverse Geometric Transform (GT)

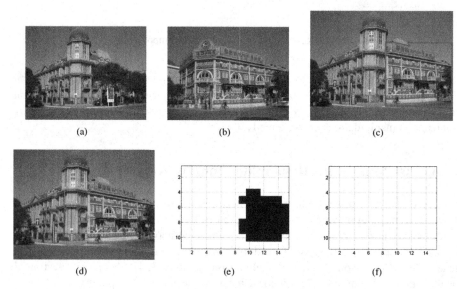

Fig. 4. Resampling forging attack on an example composite image. (a)(b) Source images, both with 3268 × 2448 pixels; (c) Composite image; (d) Attacked composite image; (e)(f) Blockwise resampling detection results of (c)(d) respectively, where 'white' denotes detected while 'black' denotes not detected. Block size in detection is 256 × 256 pixels

such as antialiasing, bilateral filtering and sharpening. Through such enhancement, we endeavor to decrease the quality loss between the attacked and primary versions of R_2. Another benefit is to attenuate and even remove the trace incurred by T^{-1}, which is expected to disappear in resulting attacked images. It should be mentioned that recent advanced resampling techniques such as superresolution [14] can also be introduced to enforce T^{-1}.

4   Experimental Results

Experiments are conducted to verify the efficacy of our proposed resampling forging attack scheme. Antialiasing is performed in the image quality dimprovement step. Bicubic interpolation method is adopted for implementing image scaling. The typical derivative-based resampling detector [4] and its corresponding composition detector are chose as attacking targets.

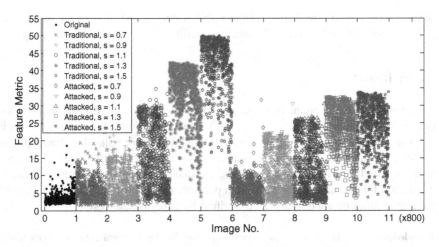

Fig. 5. Feature values of the original, traditionally scaled and attacked versions of 800 image patches (256 × 256 pixels)

Table 1. Average PSNR between attacked and original patches

s	0.7	0.9	1.1	1.3	1.5
PSNR(dB)	46.2	38.9	41.2	41.7	37.4

The test results on an example composite image are shown in Fig. 4. The subfigures (a)(b) indicate two source images. Figure 4(c) shows the composite image, where the region copied from (b) was unaltered but the region from (a) was upscaled by the factor 1.2. The locally scaling is performed to match the size of objects around the boundary, and make the composite image be more realistic. Figure 4(d) shows the tricked image generated by enforcing the inverse-GT-based resampling forging attack on the unaltered source region. The blockwise composition detection results of the forgery images (c)(d) are respectively illustrated in Fig. 4(e)(f), where 'black block' denotes resampling is not detected in corresponding positions and 'white blocks' denotes that is detected. We can see that the image forgery created by our proposed resampling forging attack method can successfully deceive the prior resampling-based composition detector [4]. Image quality has not degraded perceptibly after attack. PSNR between the composite images before and after attack achieves 44.0 dB, and local PSNR on the forged region reaches 38.4 dB.

To evaluate the resampling forging attack, 800 image patches with size of 256 × 256 pixels are collected from natural photographs in raw format. Traditional scaling and the proposed attack are respectively enforced on the original patches. According to the resampling detector [4], the feature value of each patch is computed and shown in Fig. 5. It shows that feature values of attacked samples keep the same high as those of traditionally scaled samples, and both are far above the feature values of original samples. The feature fabrication, namely the resampling trace forging, can be achieved by the proposed attack. High PSNR values shown in Table 1 signify the high quality of attacked images. The attack is verified to have good transparency.

5 Conclusion

We have made two important contributions in this work. First, we propose a formal performance evaluation framework for digital image manipulation forensics. With hypothesis testing model, reliability of a forensic system is formulated by two types of forensics errors, and the security is evaluated via two types of forensics attacks. Second, with a case study on resampling forensics, we demonstrate that the security against manipulation forging attack is significant in performance evaluation. The correctness of our understanding on such an attack is verified by experimental results.

Acknowledgements. This work was supported in part by the National NSF of China under Grants (61401408, 61332012, 61272355), Fundamental Research Funds for the Central Universities (2015JBZ002), Research Founds of CUC (3132015XNG1506), Open Projects Program of NLPR (201306309).

References

1. Zhao, X., Wang, S., Li, S., Li, J., Yuan, Q.: Image splicing detection based on noncausal markov model. In: IEEE International Conference on Image Processing, Melbourne, pp. 4462–4466 (2013)
2. Christlein, V., Riess, C., Jordan, J., Riess, C., Angelopoulou, E.: An evaluation of popular copy-move forgery detection approaches. IEEE Trans. Inf. Forensics Secur. **7**(6), 1841–1854 (2012)
3. Popescu, A.C., Farid, H.: Exposing digital forgeries by detecting traces of resampling. IEEE Trans. Signal Process. **53**(6), 758–767 (2005)
4. Mahdian, B., Saic, S.: Blind authentication using periodic properties of interpolation. IEEE Trans. Inf. Forensics Secur. **3**(3), 529–538 (2008)
5. Luo, W., Huang, J., Qiu, G.: JPEG error analysis and its applications to digital image forensics. IEEE Trans. Inf. Forensics Secur. **5**(3), 480–491 (2010)
6. Stamm, M., Liu, K.J.R.: Forensic detection of image manipulation using statistical intrinsic fingerprints. IEEE Trans. Inf. Forensics Secur. **5**(3), 492–506 (2010)
7. Cao, G., Zhao, Y., Ni, R., Li, X.: Contrast enhancement-based forensics in digital images. IEEE Trans. Inf. Forensics Secur. **9**(3), 515–525 (2014)
8. Chen, C., Ni, J., Huang, J.: Blind detection of median filtering in digital images: a difference domain based approach. IEEE Trans. Image Process. **22**(12), 4699–4710 (2013)
9. Cao, G., Zhao, Y., Ni, R., Kot, A.: Unsharp masking sharpening detection via overshoot artifacts analysis. IEEE Signal Process. Lett. **18**(10), 603–607 (2011)
10. Kirchner, M., Bohme, R.: Hiding traces of resampling in digital images. IEEE Trans. Inf. Forensics Secur. **3**(4), 582–592 (2008)
11. Fan, W., Wang, K., Cayre, F., Xiong, Z.: A variational approach to jpeg anti-forensics. In: IEEE International Conference on Acoustics, Speech and Signal, Vancouver, pp. 3058–3062 (2013)

12. Barni, M., Fontani, M., Tondi, B.: A universal technique to hide traces of histogram-based image manipulations. In: ACM Workshop on Multimedia and Security, Coventry, pp. 97–104 (2012)
13. Cao, G., Zhao, Y., Ni, R.: Attacking contrast enhancement forensics in digital images. Sci. China Inf. Sci. **57**(5), 052110(1)–052110(13) (2014)
14. Glasner, D., Bagon, S., Irani, M.: Super-resolution from a single image. In: IEEE International Conference on Computer Vision, Kyoto, pp. 349–356 (2009)

A Sparse Representation-Based Label Pruning for Image Inpainting Using Global Optimization

Hak Gu Kim and Yong Man Ro[✉]

School of EE, Korea Advanced Institute of Science and Technology (KAIST),
Daejeon, Republic of Korea
ymro@ee.kaist.ac.kr

Abstract. This paper presents a new label pruning based on sparse representation in image inpainting. In this literature, the label indicates a small rectangular patch to fill the missing regions. Global optimization-based image inpainting requires heavy computational cost due to a large number of labels. Therefore, it is necessary to effectively prune redundant labels. Also, inappropriate label pruning could degrade the inpainting quality. In this paper, we adopt the sparse representation of label to obtain a few reliable labels. The sparse representation of label is used to prune the redundant labels. Sparsely represented labels as well as non-zero sparse labels with high similarity to the target region are used as reliable labels in global optimization based image inpainting. Experimental results show that the proposed method can achieve the computational efficiency and structurally consistency.

Keywords: Sparse representation · Label pruning · Image inpainting · Global optimization

1 Introduction

Image inpainting (image completion) is one of interesting research topics in image processing and computer vision field. Image inpainting is a technique that automatically recovers missing region (i.e., hole region) in a visually plausible manner [1]. A variety of image inpainting methods have been presented in order to provide structurally consistent results [2–5]. In particular, a global optimization-based approach is known to be successful [4, 5]. Contrast to the inpainting methods based on local greedy algorithms [2, 3], the global optimization-based image inpainting methods take into account correlation between a target region to-be-filled and neighboring regions in Markov random field (MRF) framework. As a result, in the global optimization process, it is probable to provide structurally consistent and visually plausible results.

In general, the global optimization-based image inpainting methods require heavy computational cost, where combinations of all labels are compared to find the best solution in the global optimization framework [4, 5]. The computational cost is increased with the number of labels in the global optimization framework. Note the labels denote small rectangular patches collected from the known region (i.e., non-hole region). Komodakis and Tzirtas proposed the dynamic label pruning in belief

© Springer International Publishing Switzerland 2015
Y.-S. Ho et al. (Eds.): PCM 2015, part I, LNCS 9314, pp. 106–113, 2015.
DOI: 10.1007/978-3-319-24075-6_11

propagation (BP) [4]. They pruned the unreliable labels with a user specified threshold, which are dis-similar to the target region. They reduced the computational cost by comparing the combinations of remaining reliable labels (active labels). However, many active labels could be duplicated and they were from only known regions so that resultant inpainting quality was limited. Liu et al. proposed the label pruning based on k-mean clustering in coarse-to-fine BP [5]. In the first round BP, they considered a few labels centered in each cluster. In the next round BP, all labels in each cluster were considered. However, the number of label clusters is manually determined by users. Also, the clustering algorithms are not fast in general.

In this paper, we propose a new label pruning method based on sparse representation to maintain the structure consistency as well as to effectively reduce computational cost. The main contribution of this paper is twofold:

(1) We adopt the sparse representation of label to obtain more sparse active labels (see Fig. 1). The sparse representation of label can be useful to prune the redundant labels in the image inpainting [6, 7]. As seen in Fig. 1, a dictionary consists of all labels from known region. The target region is an input of sparse representation of label. Among sparse labels corresponding to nonzero elements, a few labels with high similarity to the target region are selected as active labels.

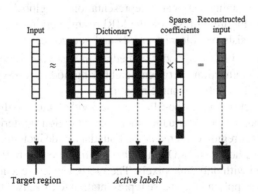

Fig. 1. Illustration of the proposed label pruning based on sparse representation. The black elements in the sparse coefficients indicate the nonzero elements. The black column vectors in the dictionary indicate basis labels (sparse labels) corresponding to the nonzero sparse elements. Active labels in image inpainting are the sparse labels with the high similarity to the target region and the sparsely represented label (i.e., reconstructed input).

(2) We include the sparsely represented label (i.e., a reconstructed input with sparse labels) as active label as well. The sparsely represented label could be one of the best labels in global optimization framework because it had the highest similarity to the target region by sparse representation.

Experimental results demonstrate that the proposed method achieves high computational gain while preserving structural consistency.

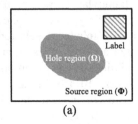
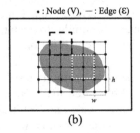

Fig. 2. Illustration of MRF framework for image inpainting. (a) Input image. (b) Nodes and edges of the MRF for the image inpainting. The dashed box represents target region centered at boundary node and the dot box represents target region centered at interior node.

The rest of the paper is organized as follows. In Sect. 2, we present the proposed sparse representation based label pruning. Section 3 presents the validation experiments that evaluate the performance of the proposed label pruning. In Sect. 4, the conclusions are drawn.

2   Proposed Label Pruning

We present a label pruning via sparse representation for global optimization-based inpainting using BP. Figure 2 illustrates the MRF framework for image inpainting [4]. The input image consists of hole region (Ω) and source region (i.e., known region, $\Phi = \Omega^c$). As shown in Fig. 2(b), nodes and edges indicate the sampled points around hole region and the connections between neighboring nodes, respectively. The labels are $w \times h$ patches collected from source region.

In particular, target regions to-be-filled can be divided into two different cases. One is target region at boundary node and the other is target region at interior node, as shown in Fig. 2(b). The target region centered at the boundary node includes the known pixels (non-hole pixels) and hole pixels. On the other hand, the target region centered at the interior node is filled with only hole pixels. Therefore, we devise a novel label pruning method to adaptively applying the sparse representation to these two target region cases in the global optimization-based image inpainting using BP. In the next subsections, we present the proposed label pruning method.

2.1   Dictionary Construction for Two Target Region Cases

To compute the sparse representation of label, it is necessary to construct a dictionary $\mathbf{D}$. For this purpose, labels $L = \{\mathbf{l}_1, \mathbf{l}_2, \ldots, \mathbf{l}_N\}$ obtained from the source region (i.e., known region) are used as basis vectors in the dictionary $\mathbf{D} = [\mathbf{d}_1, \mathbf{d}_2, \ldots, \mathbf{d}_N]$ for sparse representation [6, 7]. Note that N is the number of entire labels L collected from the source region (Φ). Also, the basis vector $\mathbf{d}_n$ represents the n-th reordered column vector of the n-th label $\mathbf{l}_n$.

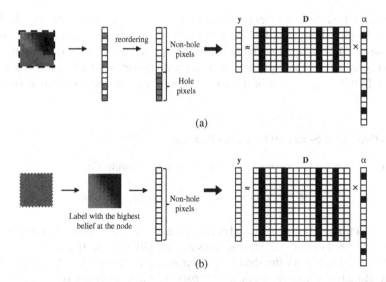

Fig. 3. Illustration of the sparse representation of two target regions of Fig. 2(b): (a) Sparse representation for target region centered at boundary node. (b) Sparse representation for target region centered at interior node. The black elements in the column vectors indicate the non-zero coefficients and basis vectors corresponding to them, respectively.

In this paper, the dictionary **D** is differently constructed according to above mentioned two target region cases. To that end, we differentiate between two types of target region before dictionary construction as,

$$T = \begin{cases} 0, \ if \ (w \times h) > \sum_{i=1}^{w} \sum_{j=1}^{h} \mathbf{M_T}(i,j) \\ 1, \ if \ (w \times h) = \sum_{i=1}^{w} \sum_{j=1}^{h} \mathbf{M_T}(i,j) \end{cases} \tag{1}$$

where T is an indicator, which is zero for the target region at boundary node and one for the target region at interior node. $\mathbf{M_T}$ indicates hole mask of target region, which is one for hole pixel and zero for known pixel. Also, (i, j) indicates the pixel indexes.

In the target region centered at boundary node case ($T = 0$), only known pixels in the target region are used as input **y** (see Fig. 3(a)). Therefore, the basis vectors in the dictionary **D** are made by pixels in labels L corresponding to the known pixels in the target region. In this case, the size of input **y** is '$(w \times h - k) \times 1$' and the size of dictionary **D** is '$(w \times h - k) \times N$'. Note that k indicates the number of hole pixels in the target region.

In the target region centered at interior node case ($T = 1$), the target region does not have any information because there are no known pixels in the target region. In the BP process, the label with the highest belief at the interior node can be used as an input **y**, instead of the hole target region itself (see Fig. 3(b)). The belief is calculated during the

BP process to find the best label combination for image inpainting. Note that the belief indicates the probability that the label is placed to the node. It is reasonable because the label with the highest belief is highly likely to be placed to the node. Because there are no hole pixels in this label, we can construct the dictionary with whole column vectors of labels. In this case, the size of input $\mathbf{y}$ is '$(w \times h) \times 1$' and the size of dictionary $\mathbf{D}$ is '$(w \times h) \times N$'.

2.2 Active Label Selection by Label Pruning

In order to prune the redundant labels we solve the following Eq. (2).

$$\hat{\alpha} = \arg \min_{\alpha} \|\mathbf{D}\alpha - \mathbf{y}\|_2^2 + \lambda \|\alpha\|_1 \tag{2}$$

where the dictionary $\mathbf{D}$ is constructed according to target region cases. In this paper, we used a regularized orthogonal least squares method (ROLS) [8, 9] to obtain the sparse solution of the Eq. (2). By the sparse representation, we can obtain a few sparse labels corresponding to the nonzero elements of sparse coefficients vector α.

The sparse labels $L_S = \{\mathbf{l}_{s1}, \mathbf{l}_{s2}, \ldots, \mathbf{l}_{sM}\}$ are obtained by sparse representation. Note that M indicates the number of sparse labels after the sparse representation ($M \ll N$). We do not use all of sparse labels but select the labels which are highly similar to the target region as active labels. To get active labels in the sparse labels, we calculate the sum of squared difference (SSD) between the sparse labels L_S and the target region. Then, unreliable labels with too high SSD values are pruned. In addition, the sparsely represented label (reconstructed input with sparse labels, i.e., $\mathbf{D}\alpha$) is included in active labels in order to minimize the structure inconsistency in inpainted results. Consequently, we can obtain a few reliable active labels at each target region for an efficient global optimization.

3 Experimental Results

To validate the performance of the proposed method, we performed experiment to evaluate visual effect and computational gain compared with a widely used local greedy inpainting method (Criminisi's method [2]) and global optimization-based image inpainting method (the Komodakis's method [4]).

We implemented a global optimization-based image inpainting method using belief propagation with the proposed label pruning method to evaluate performance. In the sparse representation of label, the parameter λ was 10^{-3}. The threshold of SSD was 10^4 for active label selection in the proposed method. The four public datasets, which are "giraffe", "elephant", "bridge", and "ship", were used [10].

Figures 4 and 5 show the inpainting results for "giraffe" and "ship", respectively. Figures 4(a) and 5(a) represent the original image for each dataset. Figures 4(b) and 5(b) represent the corresponding mask image where black regions indicate the hole regions to-be-filled. Figures 4(c)–(e) and 5(c)–(e) show the results of the Criminisi's method [2], the results of the Komodakis's method [4] and the results of the method using the proposed method for each dataset, respectively. As shown in Figs. 4 and 5, the Criminisi's method and Komodakis's method lead to structure inconsistency in some cases. However, the global optimization-based method using the proposed label pruning provides structure consistency and visually plausible results.

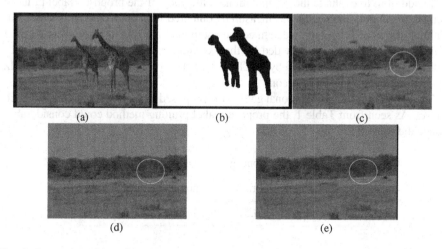

Fig. 4. Image inpainting results for "giraffe". (a) Input image. (b) Mask image (black region indicates the hole region). (c) The result of the Criminisi's method [2]. (d) The result of the Komodakis's result [4]. (e) The result of the inpainting method using a proposed label pruning.

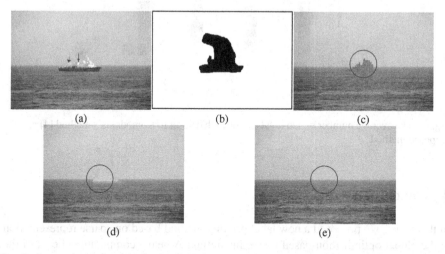

Fig. 5. Image inpainting results for "ship". (a) Input image. (b) Mask image (Black region indicates the hole region). (c) The result of the Criminisi's method [2]. (d) The result of the Komodakis's result [4]. (e) The result of the inpainting method using a proposed label pruning.

To evaluate the performance of the proposed label pruning method, we analyzed the average number of active labels for each dataset. The number of active labels is highly related to the computational cost of the global optimization-based inpainting. Figure 6 shows the average number of active labels for each dataset. As seen in Fig. 6, the average number of active labels of the proposed method is definitely lower than those of the Komodakis's method. It means that the proposed label pruning method effectively pruned the redundant labels.

In addition, to evaluate the computational efficiency of the proposed label pruning method, a computational gain was measured compared with the Komodakis's method. Without of loss of generality, we defined the computational gain as 'the execution time of the Komodakis's method divided by the execution time of the proposed method'. Both of them were implemented by MATLAB 2014a on a 24 GB RAM and a 3.4 GHz quad-core PC for a fair comparison.

Table 1 shows the computational gain of the proposed label pruning method for each dataset. As seen from Table 1, the proposed label pruning method could considerably reduce the execution time.

Table 1. Computational gain of the proposed method.

	Giraffe	Elephant	Bridge	Ship
Computational gain	×6.2	×5.1	×4.2	×5.3

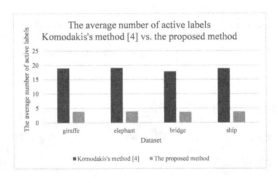

Fig. 6. The average number of active labels for each dataset in Komodakis's method [4] and the proposed method.

4   Conclusion

In this paper, we proposed a new label pruning method based on sparse representation in the global optimization-based image inpainting. A heavy computational cost of the global optimization-based image inpainting methods is challenging. Due to a large number of labels, which include the redundant labels, long computational time was taken for global optimization-based image inpainting. To reduce the number of redundant labels, we developed the sparse representation of labels. We pruned redundant labels using sparse representation and selected active labels as sparse labels with high similarity to the target region. In addition, sparsely represented labels were included in active labels for the image inpainting. The experimental results proved that the proposed label pruning method based on the sparse representation provided both structure consistency and high computational gain in the image inpainting.

Acknowledgement. This work was supported by ICT R&D program of MSIP/IITP [B0101-15-0525, Development of global multi-target tracking and event prediction techniques based on real-time large-scale video analysis].

References

1. Guillemot, C., Meur, O.L.: Image inpainting: overview and recent advances. IEEE Sig. Process. Mag. **31**, 127–144 (2014)
2. Criminisi, A., Pérez, P., Toyama, K.: Region filling and object removal by exemplar-based image inpainting. IEEE Trans. Image Process. **13**(9), 1200–1212 (2004)
3. Wu, Q., Yu, Y.: Feature matching and deformation for texture synthesis. ACM Trans. Graph. **23**(3), 364–367 (2004)
4. Komodakis, N., Tziritas, G.: Image completion using efficient belief propagation via priority scheduling and dynamic pruning. IEEE Trans. Image Process. **16**(11), 2649–2661 (2007)
5. Liu, M., Chen, S., Liu, J., Tang, X.: Video completion via motion guided spatial-temporal global optimization. In: Proceedings of 17th ACM International Conference on Multimedia, pp. 537–540 (2009)
6. Fadili, M.J., Starck, J.L., Murtagh, F.: Inpainting and zooming using sparse representations. Comput. J. **52**(1), 64–79 (2009)
7. Shen, B., Hu, W., Zhang, Y., Zhang, Y.: Image inpainting via sparse representation. In: Proceedings of IEEE International Conference on Acoustics, Speech and Signal Processing, pp. 697–700 (2009)
8. Blumensath, T., Davies, M.E.: On the difference between orthogonal matching pursuit and orthogonal least squares. Technical report, University of Edinburgh, U.K., March 2007
9. Needell, D., Vershynin, R.: Uniform uncertainty principle and signal recovery via regularized orthogonal matching pursuit. Found. Comput. Math. **9**(3), 317–334 (2009)
10. Yan, Q., Xu, L., Shi, J., Jia, J.: Hierarchical saliency detection. In: Proceedings of IEEE CVPR, pp. 1155–1162 (2013)

Interactive RGB-D Image Segmentation Using Hierarchical Graph Cut and Geodesic Distance

Ling Ge, Ran Ju, Tongwei Ren, and Gangshan Wu[✉]

State Key Laboratory for Novel Software Technology,
Collaborative Innovation Center of Novel Software Technology
and Industrialization, Nanjing University, Nanjing 210023, China
gelingnju@gmail.com, juran@smail.nju.edu.cn, {rentw,gswu}@nju.edu.cn

Abstract. In this paper, we propose a novel interactive image segmentation method for RGB-D images using hierarchical Graph Cut. Considering the characteristics of RGB channels and depth channel in RGB-D image, we utilize Euclidean distance on RGB space and geodesic distance on 3D space to measure how likely a pixel belongs to foreground or background in color and depth respectively, and integrate the color cue and depth cue into a unified Graph Cut framework to obtain the optimal segmentation result. Moreover, to overcome the low efficiency problem of Graph Cut in handling high resolution images, we accelerate the proposed method with hierarchical strategy. The experimental results show that our method outperforms the state-of-the-art methods with high efficiency.

Keywords: Interactive image segmentation · RGB-D image · Graph cut · Geodesic distance · Scale space

1 Introduction

Image segmentation aims to partition an image into several parts automatically or with simple interactions. Compared to automatic segmentation [1], interactive image segmentation attracts much attention [2,3] for its advantage in handling complex image content. It is widely used in many applications to simplify further processing, such as object dataset construction [4], image editing [5] and object image retrieval [6]. In interactive image segmentation, the segmentation problem is usually formulated as assigning a binary label to each pixel in an image to denote whether a pixel belongs to foreground or background [7]. Some hints are manual labelled to indicate parts of foreground and background, and the image is segmented by minimizing the defined energy functions.

As a representative of interactive image segmentation technology, Graph Cut [8] converts an image into a graph, in which each pixel is represented as a graph node and the adjacent nodes are connected with weighted edges. Then the segmentation problem is formulated as an energy minimization process, which can be solved using the min-cut algorithm. The main advantage of Graph Cut is that

© Springer International Publishing Switzerland 2015
Y.-S. Ho et al. (Eds.): PCM 2015, Part I, LNCS 9314, pp. 114–124, 2015.
DOI: 10.1007/978-3-319-24075-6_12

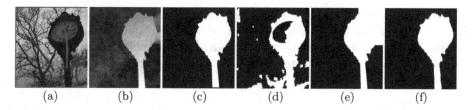

(a)   (b)   (c)   (d)   (e)   (f)

Fig. 1. Comparison of segmentation results using different channels in RGB-D image. (a) Color channels with manual labels. (b) Depth channel. (c) Ground truth. (d) Segmentation result with Graph Cut on color channels. (e) Segmentation result using geodesic distance on depth channel. (f) Our result integrating color and depth cues.

once the energy function is properly defined, it can provide a globally optimal solution with considering of both unary probability and smoothness.

Compared to the prosperity of segmentation research on traditional 2D image, little attention has been paid to interactive segmentation of emerging RGB-D image. Different to RGB image, RGB-D image provides an extra depth channel. Given a pixel in RGB-D image, we can obtain both its color information and its position in 3D spatial space. In this way, the pixels in an RGB-D image form a color point cloud with a certain spatial distribution, in which color channels offer color contrast and texture for distinguishing foreground from background and depth channel describes the geometrical characteristics of objects and scene. It provides quite different cues to RGB image segmentation, and makes interactive RGB-D image segmentation a novel and challenging problem. Figure 1 shows an example of segmentation results comparison using different channels in RGB-D image. We can find that both color channels and depth channel provide partial information to distinguish between foreground and background, which leads to the inaccuracy of segmentation results (Fig. 1(d) and (e)). However, integrating color cue and depth cue can provide more accurate segmentation result, as shown in Fig. 1(f).

In this paper, we propose a novel interactive segmentation method for RGB-D image by integrating color cue and depth cue in a unified hierarchical Graph Cut framework. In the proposed method, RGB-D image is represented with the same graph representation as the original Graph Cut method, but differs in the image properties representation. Specifically, we utilize Euclidean and geodesic distance [9] as the dissimilarity metric for color and depth channels respectively. Compared to directly treating depth as a fourth channel of the input of Graph Cut [10], our method provides a better description of the spatial characteristics of the image content. Moreover, we accelerate our method with hierarchical strategy to efficiently handle high resolution RGB-D images. As the computational complexity of Graph Cut method is in proportion to the square of image pixel number [11], we extend Graph Cut to scale space to generate a primary segmentation result on the coarsest scale and then refine foreground boundary on finer scales, which can obviously reduce the number of graph nodes in segmentation.

We evaluate the proposed method on two public datasets [12,13], and compare it with the state-of-the-art methods as well as manual labelled ground truths. The experiments show that the proposed method obtains better segmentation results with little user interactions, and much more efficient than the other methods.

2   Related Work

We briefly review the relevant researches on image segmentation for RGB images and depth assisted image segmentation.

Interactive Image Segmentation. Interactive image segmentation takes manual labelled certain pixels as input and further segments an image by image content, such as color change and contrast. It can handle the images with complex structure with the assistant of manual interaction, which has been widely used in other techniques, such as mobile search [14] and social media analysis [15]. Graph cut [8] is one of the most representative methods. After converting an image to a graph, in which each pixel in image is represented as a graph node and the adjacent nodes are connected with edges, it formulates segmentation as a min-cut problem on the graph. Based on Graph Cut, GrabCut [16] uses an iterative strategy to improve the quality of segmentation. Random Walks method can also be utilized to perform segmentation [17]. Starting from each unlabelled pixel, the probability of a random walker will first reach one of the pre-labelled pixels is calculated and the pixel is labelled accordingly. This algorithm is also applicable on higher dimension but with lower time performance. Geodesic distance method for interactive segmentation [18] has been previously used in processing color cue. It uses star-convexity prior and replaces Euclidean rays with geodesic path to exploit the structure of shortest paths.

Noted that there are some similar multilevel strategies used in [19,20] to accelerate Graph Cut. However, the application scenes of these methods are different and they are not suitable in processing RGB-D images.

Depth Assisted Image Segmentation. There are several methods proposed to use depth channel to assist segmentation. Kolmogorov et al. [21] proposed two bi-layer segmentation methods for binocular stereo video: Layered Dynamic Programming and Layered Graph Cut. They applied color, contrast and stereo matching information to realize automatic foreground/background separation. It differs with our method in using depth information, namely stereo matching information directly and the usage scenario is limited. Harville et al. [22] proposed a method for modeling the background using time-adaptive, Gaussian mixtures in the combined input space of depth and luminance-invariant color. Ahn et al. [23] used a stereo camera to extract human silhouettes indoors. It also utilized Graph Cut framework. Its contribution focused on object and background seed segmentation and depth assisted Graph Cut. These three algorithms take stereo images as input, which is different from ours, and neglect excavating the geometry of depth information. They all have special limitation for either

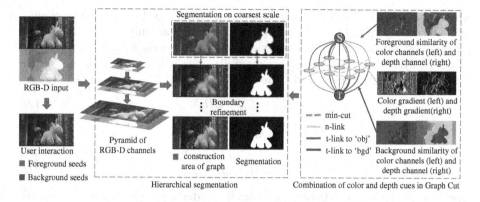

Fig. 2. An overview of the proposed approach.

scene or target scope. As to general segmentation method for RGB-D images, Julia *et al.* proposed a multi-label segmentation method for RGB-D images in [10]. However, they also considered the depth image as an additional data channel to put into Graph Cut framework directly and didn't take further process on depth information.

3   Interactive RGB-D Image Segmentation

Figure 2 shows an overview of the proposed method. For the input RGB-D images and user labels, we first construct the image pyramid on color channels and depth channel. Then Graph Cut is executed on the coarsest scale and the boundary refinements are performed using the processed segmentation result of the previous coarser scale as input. During the segmentation, color and depth are combined to compute the weight of terminal links and neighborhood links. Specifically we employ Euclidean distance and geodesic distance to measure the object likelihood for color and depth channels respectively. Finally, a high quality segmentation result is obtained on the finest scale with full resolution.

3.1   Preliminary of Hierarchical Graph Cut

For further analysis, we briefly review the principle of Graph Cut [8]. In Graph Cut, the original image I is represented as an undirected graph $\mathcal{G} = \langle \mathcal{V}, \mathcal{E} \rangle$. Here, $\mathcal{V}$ is the union of pixel nodes and two additional terminal nodes S and T, and $\mathcal{E}$ is the union of neighborhood links (n-links) and terminal links (t-links), which are the edges between adjacent pixel nodes and between pixel nodes and terminal nodes, respectively. With such representation, the graph $\mathcal{G}$ can be partitioned into disjoint regions by removing edges connecting them, which is formulated as a min-cut problem. The energy function is described as follows:

$$E(L) = \lambda R(L) + B(L), \tag{1}$$

where $R(L)$ is sum of penalties for assigning a certain pixel node p to foreground and background; $B(L)$ is the sum of penalties for discontinuity between adjacent pixel nodes; L is the list of labels assigned to corresponding pixel nodes, whose value is obj(foreground) or bgd(background); λ is a balance coefficient which equals 5 in our experiments; $R(L)$ and $B(L)$ can be further defined as follows:

$$R(L) = \Sigma_{p \in I} R_p(l_p), \tag{2}$$

$$B(L) = \Sigma_{\{p,q\} \in \mathcal{N}} B_{\langle p,q \rangle} \delta(l_p, l_q), \tag{3}$$

$$\delta(l_p, l_q) = \begin{cases} 0, & if \ l_p = l_q \\ 1, & otherwise \end{cases}, \tag{4}$$

where $\mathcal{N}$ is the set of adjacent pixel nodes under a standard 8-neighborhood system; $R_p(L)$ indicates the possibility of pixel p to be labelled as a certain value of L; Boundary penalties $B_{\langle p,q \rangle}$ denotes the cost of cutting off the neighborhood links between adjacent node p and q.

For Graph Cut cannot efficiently handle high resolution images, a hierarchical strategy [19] is utilized to build a pyramid for the input image and construct graph on each scale separately. In this way, the total number of graph nodes and links are obviously reduced and the efficiency of Graph Cut is improved.

3.2 Scale Space Construction

There is a simple but meaningful phenomenon that the segmentation results on different scales of an image have similar appearance, and the difference of segmentation results on different scales occurs on the precision of boundaries. Based on this observation, we construct a scale space $\{I_0, I_1, \ldots, I_n\}$ for each RGB-D image I, which contains the color channel I^c and depth channel I^d. We obtain the primary segmentation result on the coarsest scale I_0 and refine the foreground boundary on the finer scales from I_1 to I_n.

For the original images may have quite different resolutions, it is not suitable to fix the number of scales n for different original images. A better solution is to restrict the scale of coarsest scale and set a proportion to the construction of adjacent scales. In this way, we can control the computation cost of the whole procedure in a acceptable scope. We adopt a self-adaptive strategy to determine the number of scales in scale space construction. A threshold φ is used to constrain the resolution of the coarsest scale. Obviously, too large φ will cause large computational cost to generate the initial segmentation result, but too small φ will result in serious content loss of the original image. In our experiments, we set φ to 50,000. Starting from the original image I, we down sample the current scale at a proportion of ρ to construct the next scale ($\rho = 0.25$ in our experiments). Once the pixel number of current constructed scale is below φ, construction of scale space is finished. We can figure out that $n = \lceil log_\rho \frac{\varphi}{|I|} \rceil$.

3.3 Integration of Color Cue and Depth Cue

On the coarsest scale I_0, we utilize Graph Cut [8] to generate the initial segmentation result s_0. Both RGB and depth cues are applied in Graph Cut.

As mentioned in Sect. 3.1, regional term $R(L)$ and boundary term $B(L)$ in Eq. (1) only takes color cue into consideration in typical Graph Cut framework in RGB segmentation method. In proposed RGB-D segmentation approaches like [10,21–23], they take depth cue as an additional channel to add into Graph Cut directly. To further extract the spatial property of depth cue, we interprete depth information through geodesic distance [9]. Therefore, regional penalty $R_p(L)$ and boundary penalty $B_{\langle p,q \rangle}$ should be redefined.

We first rewrite regional penalty $R_p(L)$ in Eq. (2) as the combination of color penalty $R_p^c(L)$ and depth penalty $R_p^d(L)$ with a balance coefficient α, which equals 1 in our experiments:

$$R_p(L) = R_p^c(L) + \alpha R_p^d(L), L \in \{obj, bgd\}, \quad (5)$$

where color penalty $R_p^c(L)$ of specifying a label l_p to pixel p is the color likelihood between p and the histograms of foreground and background color distributions:

$$R_p^c(L) = P(L = l_p|c_p) = \frac{P(c_p|(L = l_p))}{P(c_p|(L = obj)) + P(c_p|(L = bgd))}. \quad (6)$$

And depth penalty $R_p^d(L)$ is the ratio of geodesic distance, with geodesic distance from p to a specified region as the numerator and the sum of geodesic distance as denominator:

$$R_p^d(L) = \frac{D(p, l_p)}{D(p, obj) + D(p, bgd)}, \quad (7)$$

where $D(x, L)$ indicates the geodesic distance between x and label L, which is formulated as:

$$D(x, L) = \min_{y \in \Omega_L} d(x, y), \quad (8)$$

$$d(x, y) := \min_{C_{x,y}} \int_0^1 |G_d(x, y) \cdot C_{x,y}(p)| dp, \quad (9)$$

where $C_{x,y}(p)$ is a path connecting the pixels x, y; $G_d(x, y)$ is set as the gradient of greyscale.

We also rewrite boundary penalties $B_{\langle p,q \rangle}$ in Eq. (3) with an ad-hoc function as follows:

$$B_{\langle p,q \rangle} \propto exp\left(-\frac{(I_p - I_q)^2}{2\sigma_1^2} - \frac{\beta G_d(p, q)}{2\sigma_2^2} \right), \quad (10)$$

where β is a balance coefficient, which is set to 1; σ_1 and σ_2 are two parameters to adjust the penalty, here $\frac{1}{2\sigma_1^2} = \frac{1}{2\sigma_2^2} = 0.0075$ in our experiments.

3.4   Upscaling Boundary Refinement

The initial segmentation result s_0, after an opening operation with a 3×3 element on it to avoid noise expansion, provides an approximate distribution of foreground and background of the original image. Based on it, we iteratively refine boundary area with Graph Cut from coarse to fine scale.

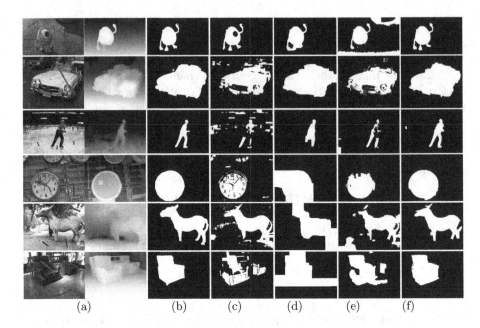

(a)   (b)   (c)   (d)   (e)   (f)

Fig. 3. Examples of segmentation results. (a) Color channels and depth channels of RGB-D images with user labels. (b) Ground truths. (c) Graph Cut. (d) Geodesic distance using depth only. (e) RGB-D segmentation using depth as a fourth channel. (f) our method.

For the procedure of boundary refinement on each scale is similar, without loss of generality, we assume that boundary refinement is performed on the kth scale I_k with the initialization of the segment result $s_{(k-1)}$ of the $(k-1)$th scale. Figure 2(d) shows the procedure of boundary refinement. First, dilation and erosion operations are performed on $s_{(k-1)}$ to determine the inner contour C_{in} and outer contour C_{out} of foreground, respectively. Then the processed $s_{(k-1)}$ is resized to the same size of I_k. The region inside C_{in} is settled as foreground, the region outside C_{out} is settled as background, and the rest part R_c is used to build a new graph for Graph Cut. The size of structuring element used in dilation and erosion operations on each scale is not fixed. A suitable size of structuring element should retain enough pixels in boundary area R_c to generate accurate foreground boundary and avoid too many pixels for efficiency. Here, we set the size of structuring element as $(k+3) \times (k+3)$ when dilate and erode s_k.

Graph Cut has been proved to have a complexity of $O(|C||\mathcal{E}||\mathcal{V}|^2)$ in the worst case [11], $|C|$ denotes the cost of min-cut which equals the total weight of removed edges. According to the result of the experiments, the complexity of refinement is only related with the final min-cut result C and the number of scales n in scale space. Coupled with the fact that the complexity of initial segmentation on I_0 can be seen as a constant, our method is proved practically faster than Graph Cut on one scale. The time cost of our method is presented in Sect. 4.3.

4 Experiments

To validate the performance of our method, we compare it with the methods only using color cue or depth cue and the method directly treating depth as a fourth channel of the input of Graph Cut. We also execute a time evaluation by comparing with other hierarchical methods.

4.1 Datasets and Experimental Settings

To quantitatively evaluate the quality of segmentation results, we use two datasets, NJU400 [12] and RGBD Benchmark [13] in our experiments, which provide 400 stereo image pairs and 1,000 RGB-D images with the corresponding depth maps and pixel-level manual-labelled ground truths, respectively. We treat each left image and its depth map of a stereo image as the color channels and depth channel of a RGB-D image, and compare the segmentation results with the provided ground truths.

The proposed approach is implemented by C++. All the experiments are carried out on a PC with a four-core 3.40 GHz CPU and 8 GB memory.

4.2 Segmentation Accuracy Evaluation

We first compare our method with the methods only using color cue or depth cue. We select three methods using color cue, including Graph Cut (GC) [8], GrabCut (GB) [16] and multi-level Graph Cut (MGC), for their effectiveness, and applying geodesic distance [9] on depth channel (GDD). Figure 3 illustrates some examples of segmentation results generated by different methods. As shown in Table 1, our method (HGG) obtains higher F_β ($\beta = 0.3$) than the compared methods. It demonstrates that both color and depth cues are beneficial to improve segmentation result.

We further compare our method with the method directly treating depth as a fourth channel of the input of Graph Cut method (RGBD) [10]. It shows that our method outperforms than RGBD method in precision, recall and F_β criteria. It shows that geodesic distance can extract geometry attributes of depth channel and provide better distance measurement.

We also compare the proposed method using hierarchical strategy (HGG) or not (GG). It shows that the application of hierarchical strategy slightly influences segmentation performance, but the performance of HGG is still better than other compared methods.

4.3 Running Time Evaluation

We evaluate the efficiency of the above methods on ten randomly selected images with the average resolution of two million pixels, and execute each method ten times to obtain its average running time of segmentation. All the methods are implemented by C++ and executed on the same platform. As shown in Table 2,

Table 1. Comparison of segmentation accuracy of different methods.

	GC	GB	MGC	GDD	RGBD	GG	HGG
precision	0.7163	0.9361	0.7575	0.8542	0.8419	0.9272	0.8946
recall	0.7254	0.5558	0.7360	0.8921	0.7796	0.9032	0.9287
F_β	0.7184	0.8084	0.7524	0.8627	0.8267	0.9215	0.9022

Table 2. Comparison of processing time of different methods.

	GC	GB	MGC	GDD	RGBD	GG	HGG
time(s)	0.4340	5.6015	0.0828	32.0488	0.3423	32.2416	0.1131

out method is only slower than MGC for it contains additional computation for geodesic distance on depth channel in its procedure, but obviously more efficient than other methods. Especially, our method is about 300 times faster than GG method while only having a slight decrease in segmentation accuracy.

5   Conclusions

In this paper, we propose an efficient hierarchical Graph Cut method for interactive RGB-D image segmentation using fusion of color and depth cues, which can generate high quality segmentation results and realtime interaction. Instead of directly using color channels and depth channel in a same mean, we utilize Euclidean distance on color channels and geodesic distance on depth channel, and fuse them in an unified Graph Cut framework. Moreover, to overcome the disadvantage in efficiency of Graph Cut, we accelerate the algorithm by using a hierarchical strategy which improves the efficiency about 300 times. The experiments show that the proposed method can fully utilize the characteristic of color and depth channels, therefore obtain a better performance to the state-of-the-art methods with high efficiency.

Acknowledgments. This work is supported by the National Science Foundation of China (No.61321491, 61202320), Research Project of Excellent State Key Laboratory (No.61223003), and National Special Fund (No.2011ZX05035-004-004HZ).

References

1. Li, S., Ju, R., Ren, T., Wu, G.: Saliency cuts based on adaptive triple threshoding. In: International Conference on Image Processing, pp. 1–4. IEEE (2015)
2. Nguyen, T.N.A., Cai, J., Zhang, J., Zheng, J.: Robust interactive image segmentation using convex active contours. IEEE Trans. Image Process. **21**(8), 3734–3743 (2012)

3. Delgado-Gonzalo, R., Chenouard, N., Unser, M.: Spline-based deforming ellipsoids for interactive 3D bioimage segmentation. IEEE Trans. Image Process. **22**(10), 3926–3940 (2013)

4. Cheng, M.M., Mitra, N.J., Huang, X., Torr, P.H.S., Hu, S.M.: Global contrast based salient region detection. IEEE Trans. Pattern Anal. Mach. Intell. **37**(3), 569–582 (2014)

5. Ren, T., Liu, Y., Wu, G.: Image retargeting based on global energy optimization. In: IEEE International Conference on Multimedia and Expo, pp. 406–409 (2009)

6. Xu, X., Geng, W., Ju, R., Yang, Y., Ren, T., Wu, G.: OBSIR: object-based stereo image retrieval. In: IEEE International Conference on Multimedia and Expo, pp. 1–6 (2014)

7. Greig, D., Porteous, B., Seheult, A.H.: Exact maximum a posteriori estimation for binary images. J. Roy. Stat. Soc. Ser. B (Methodol.) **51**, 271–279 (1989)

8. Boykov, Y.Y., Jolly, M.P.: Interactive graph cuts for optimal boundary & region segmentation of objects in ND images. In: IEEE International Conference on Computer Vision, pp. 105–112 (2001)

9. Yatziv, L., Bartesaghi, A., Sapiro, G.: O(n) implementation of the fast marching algorithm. J. Comput. Phys. **212**(2), 393–399 (2006)

10. Diebold, J., Demmel, N., Hazırbaş, C., Moeller, M., Cremers, D.: Interactive multi-label segmentation of RGB-D images. In: Aujol, J.-F., Nikolova, M., Papadakis, N. (eds.) SSVM 2015. LNCS, vol. 9087, pp. 294–306. Springer, Heidelberg (2015)

11. Boykov, Y., Kolmogorov, V.: An experimental comparison of min-cut/max-flow algorithms for energy minimization in vision. IEEE Trans. Pattern Anal. Mach. Intell. **26**(9), 1124–1137 (2004)

12. Ju, R., Ge, L., Geng, W., Ren, T., Wu, G.: Depth saliency based on anisotropic center-surround difference. In: IEEE International Conference on Image Processing, pp. 1115–1119 (2014)

13. Peng, H., Li, B., Xiong, W., Hu, W., Ji, R.: RGBD salient object detection: a benchmark and algorithms. In: Fleet, D., Pajdla, T., Schiele, B., Tuytelaars, T. (eds.) ECCV 2014, Part III. LNCS, vol. 8691, pp. 92–109. Springer, Heidelberg (2014)

14. Sang, J., Mei, T., Xu, Y.Q., Zhao, C., Xu, C., Li, S.: Interaction design for mobile visual search. IEEE Trans. Multimedia **15**(7), 1665–1676 (2013)

15. Sang, J.: User-centric social multimedia computing. Springer, Heidelberg (2014)

16. Rother, C., Kolmogorov, V., Blake, A.: Grabcut: interactive foreground extraction using iterated graph cuts. ACM Trans. Graph. **23**(3), 309–314 (2004)

17. Grady, L.: Random walks for image segmentation. IEEE Trans. Pattern Anal. Mach. Intell. **28**(11), 1768–1783 (2006)

18. Gulshan, V., Rother, C., Criminisi, A., Blake, A., Zisserman, A.: Geodesic star convexity for interactive image segmentation. In: IEEE Conference on Computer Vision and Pattern Recognition, pp. 3129–3136 (2010)

19. Lombaert, H., Sun, Y., Grady, L., Xu, C.: A multilevel banded graph cuts method for fast image segmentation. In: IEEE International Conference on Computer Vision, pp. 259–265 (2005)

20. Vaudrey, T., Gruber, D., Wedel, A., Klappstein, J.: Space-time multi-resolution banded graph-cut for fast segmentation. In: Rigoll, G. (ed.) DAGM 2008. LNCS, vol. 5096, pp. 203–213. Springer, Heidelberg (2008)

21. Kolmogorov, V., Criminisi, A., Blake, A., Cross, G., Rother, C.: Bi-layer segmentation of binocular stereo video. In: IEEE International Conference on Computer Vision and Pattern Recognition, pp. 407–414 (2005)

22. Harville, M., Gordon, G., Woodfill, J.: Foreground segmentation using adaptive mixture models in color and depth. In: IEEE Workshop on Detection and Recognition of Events in Video, pp. 3–11 (2001)

23. Ahn, J.H., Kim, K., Byun, H.: Robust object segmentation using graph cut with object and background seed estimation. In: International Conference on Pattern Recognition, pp. 361–364. IEEE (2006)

Face Alignment with Two-Layer
Shape Regression

Qilong Zhang and Lei Zhang[✉]

Shenzhen Key Lab of Broadband Network and Multimedia,
Graduate School at Shenzhen, Tsinghua University, Beijing, China
Zhangql13@mails.tsinghua.edu.cn,
zhanglei@sz.tsinghua.edu.cn

Abstract. We present a novel approach to resolve the problem of face alignment with a two-layer shape regression framework. Traditional regression-based
methods [4, 6, 7] regress all landmarks in a single shape without consideration
of the difference between various landmarks in biologic property and texture,
which would lead to a suboptimal prediction. Unlike previous regression-based
approach, we do not regress the entire landmarks in a holistic manner without
any discrimination. We categorize the geometric constraints into two types,
inter-component constraints and intra-component constraints. Corresponding to
these two shape constraints, we design a two-layer shape regression framework
which can be integrated with regression-based methods. We define a term of
"key points" of components to describe inter-component constraints and then
determine the sub-shapes. We verify our two-layer shape regression framework
on two widely used datasets (LFPW [10] and Helen [11]) for face alignment and
experimental results prove its improvements in accuracy.

Keywords: Face alignment · Two-layer shape regression · Inter-component
constraints · Intra-component constraints · Key points of components

1 Introduction

Face alignment is a process of locating facial landmarks or feature points of human face
and plays an important role in many visual tasks like face recognition, face animation
and expression analysis. With massive demand in these domains, the problem of
accurate and effective facial landmarks localization has attracted extensive interests in
the past years.

In general, most facial alignment methods can be classified into two categories: the
first category is the parametric approaches such as the Active Shape Model (ASM) [1],
the Active Appearance Model (AAM) [2] and the Constrained Local Model (CLM) [3],
these methods optimize an energy function consists of facial appearance likelihood and
facial geometric constraints. However, as what we know inside the theory of Point
Distribution Model (PDM), these approaches have limited ability to capture complex
and various facial expression; besides, the parametric methods are sensitive to initial
shape, it would likely fail if the initial shape is far away from the true shape. The
second category is the regression-based methods [4, 5], especially a novel framework

© Springer International Publishing Switzerland 2015
Y.-S. Ho et al. (Eds.): PCM 2015, Part I, LNCS 9314, pp. 125–134, 2015.
DOI: 10.1007/978-3-319-24075-6_13

based on regression approaches [6–9] without using any parametric shape models has been presented in recent years. The Explicit Shape Regression (ESR) [6] and the Robust Cascaded Pose Regression (RCPR) [7] train a global regressor to compute the feature points coordinates. Since the inherent shape constraints and texture information are naturally encoded into the regressor in a cascaded learning framework [12, 13], the novel regression-based methods are more effective and efficient than traditional approaches.

However, regressing all facial landmarks jointly in a single shape is not an optimal method, since the entire shape has too large shape variations to be handled effectively by only one global regressor. We observe that the geometric constraints can be divided mainly into two types: the one is the geometric constraints between facial components such eyes, nose and etc. we define it as inter-component constraints; the other is the interaction of feature points inside a specific facial component, we define it as intra-component constraints. We can see that some feature points have little relationship, such as landmarks of lip and those of eyebrow; it's nearly no use to regress some irrelevant feature points simultaneously; inversely, it makes the regressed shape more various and improves the regression complexity, obviously it hardly leads to an optimal prediction.

In this paper, we present a two-layer regression framework and propose a method to determine suitable sub-shapes. In our framework, we train two-layer shape regressors and enhance the alignment performance. We describe our framework with more details in Sect. 3 and present experiment results in Sect. 4. Our main contributions in this work are three fold: (1) we proposed a two-layer shape regression framework that can be integrated with regression-based methods and boost the face alignment performance; (2) we proposed a method to determine suitable sub-shapes for regression; (3) we test and verify our two-layer shape regression framework on two widely used datasets (LFPW [10] and Helen [11]) for face alignment.

2 Overview

In this section, we describe an overview of the proposed landmark localization framework. Figure 1 displays a brief pipeline of our system.

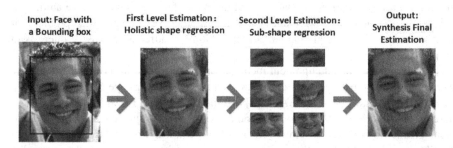

Fig. 1. System overview. Given a face image with a bounding box, we regress all landmarks in a holistic shape in the first layer and get an initial region estimation of each sub-shape; in the second layer, we regress sub-shapes independently, and then we synthesis the final estimation. The face image displayed in the Fig. 1 is from Helen database (68 points).

We divide facial landmarks into six facial components: eyes, eyebrows, nose, mouth, landmarks inside these facial components have strong interaction with each other. Although we can explicitly estimate the local region of each component after the first layer regression in our system, in consideration of strong inter-component constraints, especially when occurs occlusion above a facial component, we choose other suitable sub-shapes for regression instead of regressing the landmarks of each component independently. In our work, we extract two stable landmarks out of each component as the "key points" of a component, the interaction between these key landmarks can be used to describe the inter-component geometric constraints and we determine the sub-shapes with help of "key points". In Sect. 3, we would introduce how to define the stability of landmarks by evaluating an objective function and determine suitable sub-shapes for regression. Although different datasets for face alignment have different numbers of annotated points, they all apply to this method. In our system, we employ the following three-step processing:

(1) In the first layer, we regress the entire facial landmarks in a holistic manner and obtain an initial estimation of coordinates of the feature points.
(2) Based on the initial estimation done by previous step, we can estimate the local region for sub-shapes, then in the second layer we regress the landmarks of each sub-shape independently;
(3) We synthesis the final prediction of landmarks from estimation of each sub-shape.

3 Main Work

In order to clearly discuss our proposed method, we would briefly review the representative regression-based method and introduce some notation. Let $x_i, y_i \in \mathbb{R}^2$ be the x, y-coordinates of the ith facial landmark in an image I. Then the vector $S = [x_1, y_1, \ldots, x_N, y_N]^T$ denotes the coordinates of all the N facial landmarks in I. In this paper we take the vector S as the estimated shape and $\hat{S}$ as the true shape. We define the face alignment error as:

$$\| S - \hat{S} \|_2 \tag{1}$$

In recent relative work [6–9], they adapt a similar shape regression framework. In this framework, it estimates facial shape S in an additive manner. Starting with an initial shape S^0, S is progressively refined by estimating a shape increment ΔS at each stage regressors $(R^1, \ldots, R^T)$:

$$S_i^{(t+1)} = S_i^{(t)} + R^t(I_i, S_i^{(t)}) \tag{2}$$

$$\Delta S_i^{(t+1)} = R^t(I_i, S_i^{(t)}) \tag{3}$$

Where I_i is a face image and the output of the stage regressors $(R^1, \ldots, R^T)$ is a linear combination of all training shapes, so it can be shown that the final regressed shape S is the sum of initial shape S^0 and the linear combination of all training shapes:

$$S = S^0 + \sum_{i=1}^{N} w_i \hat{S}_i \qquad (4)$$

We think that regressing the entire feature points in a holistic manner is not optimal due to two reasons; the first one is that most feature points inside different components have few correlation, for example, the points along the lip have nothing with the points of eyebrows; from Eq. (4) we can know that the higher dimension of shape S is, the more complex to compute coefficients w_i is; regressing these nearly irrespective points in a single shape without discrimination would increase dimensions of shape S and improve the complexity of regression; what's worse, it would likely interfere each other; the second one is that there exists difference in difficulty to locate different landmarks, some salient landmarks like eye corner, nose tip has sufficient local texture information and geometric constraints to locate; inversely, some non-salient landmarks are ambiguous to extract and need help from shape constraints to locate. Although regressing non-salient landmarks with salient landmarks would benefit locating those non-salient landmarks, we conjecture that it also would do harm to locate salient landmarks due to interference from uncertain non-salient points. For verifying our conjecture, we implement the explicit shape regression (ESR) [6] and our implementation is comparable to which was reported by the original authors. We test ESR on LFPW [10] database, as what we conjecture, the experimental results indicates that salient landmarks localization is interfered by non-salient landmarks, which is shown in Tables 2 and 3. For resolving the challenge, we proposed a two-layer shape regression framework. Before specifying how to work out the coordinates of landmarks in our framework, we first clarify a basic term.

3.1   Key Feature Points of a Component

We extract two key feature points out of each component and we consider that inter-component constraints can be represented by the interaction between these key feature points. Intuitively, we may want these key feature points are salient and stable, that is to say, feature points are which shift slightly from their normal location when one make various expression. We select feature points which LFPW [10] defines as the candidate points, because these landmarks are almost biologic landmarks and conclude less interpolation points. In this section, we define how to compute the stability of feature points inside a specific component, and then extract two most stable landmarks from these candidate landmarks as "key feature points" of a component. Let the vector $S_c = (x_1, y_1, \ldots, x_M, y_M)$ denotes the coordinates of all the M feature points inside a facial component. We normalize these true component shapes as what Active Shape Model method [1] do and gain mean shape $\bar{S}_c$ of each component, then aligns true component shape S_c to the mean $\bar{S}_c$ with a similarity transform to minimizes their L2 distance:

$$M_s = \underbrace{argmin}_{M} \parallel \bar{S}_c - M \circ S_c \parallel_2 \qquad (5)$$

Let $\parallel \bar{S}_c - M_s \circ S_c \parallel_2^i$ denotes the displacement of ith feature point from the mean location after aligned and S_c^j denotes the component shape of jth image, then we compute the mean displacement of each feature point inside a specific component as below:

$$P_i = \frac{\sum_{j=1}^{N} \parallel \bar{S}_c - M_s \circ S_c^j \parallel_2^i}{N} \qquad (6)$$

The value of P_i and the stability of ith feature point is negatively correlated and we take the most two stable landmarks as the key feature points of a component. We run this method on LFPW [10] database, Table 1 show the experimental results. Interestingly enough, we observe that the key feature points of a component are almost salient landmarks as well.

Table 1. According to Eq. (6), we compute the stability of every candidate landmarks, and the extract two most stable points out of each component. We use mean width of each component bounding-box to normalize mean displacement of each landmark, the number in brackets is percentage of the mean width of each component. As there are 28 candidate landmarks, we do not display all of them.

Facial components	Key points	
Left eyebrow	Outer corner of left eyebrow (3.45274 %)	Inner corner of left eyebrow (3.0475 %)
Right eyebrow	Outer corner of right eyebrow (3.27516 %)	Inner corner of right eyebrow (2.97617 %)
Left eye	Outer corner of left eye (3.7158 %)	Inner corner of left eye (3.68292 %)
Right eye	Outer corner of right eye (3.82024 %)	Inner corner of right eye (3.63814 %)
Nose	Tip of nose (4.07415 %)	Mid of nostrils (4.75143 %)
Mouth	Outer corner of mouth (6.31268 %)	Inner corner of mouth (6.05473 %)

3.2 Two-Layer Geometric Constraint

We categorize the holistic shape constraints into two parts: inter-component constraints and intra-component constraints. Corresponding to these two shape constraints, we design a two-layer shape regression framework to estimate coordinates of landmarks. We implement this framework for two reasons: Firstly, regressing the sub-shapes independently can avoid the interference from irrespective feature points, and also reduce the complexity of a single regressor as well. Secondly, there exists difference in difficulty to locate different landmarks, we should treat them with distinction. For example, although regressing all landmarks together described in previous work [6, 7] benefits locating those non-salient landmarks, it also do harm to locate salient landmarks due to noisy from uncertain non-salient points. For testifying our conjecture, we implement the explicit shape regression (ESR) [6] and run it in LFPW [10] database. By comparing the alignment error of every landmarks, we observe that those salient landmarks are not better predicted than non-salient landmarks. Table 2 shows fiducial landmarks error and their rankings among all points when regressed in a holistic shape. Inversely, we get a better result which Table 3 shows by regressing the fiducial landmarks in a single sub-shape as display in (H) of Fig. 2.

Table 2. We run ESR in LFPW (29 landmarks) database and compute the error of each landmarks. According to their rankings, we observe that those fiducial points are not better predicted than non-salient landmarks when regressed in a holistic shape.

Fiducial points	Error ($\times 10^{-2}$)	Ranking
Outer corner of left eyebrow	4.71	27
Inner corner of left eyebrow	3.48	13
Outer corner of right eyebrow	5.21	28
Inner corner of right eyebrow	3.44	11
Outer corner of left eye	3.44	10
Inner corner of left eye	2.78	3
Outer corner of right eye	3.77	18
Inner corner of right eye	2.50	1
Tip of nose	4.05	25
Left corner of mouth	3.83	21
Right corner of mouth	3.97	23

3.3   Sub-shape Selection

In first layer regression of our framework, we jointly regress the entire landmarks in a holistic manner and get an initial estimation of coordinates of each landmarks, then we can compute the local region of sub-shapes. However, we don't regress the landmarks of each component independently. In consideration of coarse estimation of local region of component and strong inter-component constraints between two closest facial components, we regress the landmarks of a component with the key feature points of another component which is closest to it. It has been proved useful, especially when occurs occlusion above a component.

We choose the key feature points of a component as what Table 1 shows. Due to the characteristic of contour points, we combine contour points and all the key points as a sub-shape. Figure 2 display the sub-shapes we trained in second layer of our framework.

Table 3. It shows the fiducial landmarks error when regressed in a single sub-shape as display in (H) of Fig. 2. Compared with the result which Table 2 shows, we get a better prediction.

Fiducial points	Error ($\times 10^{-2}$)	Improvement
Outer corner of left eyebrow	4.51	4.2 %
Inner corner of left eyebrow	3.06	12.0 %
Outer corner of right eyebrow	4.29	17.6 %
Inner corner of right eyebrow	2.92	15.1 %
Outer corner of left eye	2.84	17.4 %
Inner corner of left eye	2.20	20.86 %
Outer corner of right eye	2.77	26.5 %
Inner corner of right eye	2.14	14.4 %
Tip of nose	4.02	0.7 %
Left corner of mouth	3.27	14.6 %
Right corner of mouth	3.30	16.9 %

In second layer, we compute the local region for each sub-shape and regress the sub-shapes independently. Finally, we synthesis the final estimation of all facial landmarks. We regress the sub-shapes displayed in Fig. 2 instead of single components for two reasons. Firstly, the estimated local region of each components from the first layer is inaccurate, we need inter-component constraints between a specific components and its nearest components, especially when occurs an occlusion above it. Secondly, the key points are almost salient landmarks, they can offer sufficient appearance clues to help localization and have rare interferences unlike other irrelevant landmarks. We test and verify our method on LFPW [10]. Table 4 shows that it leads to a better prediction when regressed in our sub-shape than in single component independently.

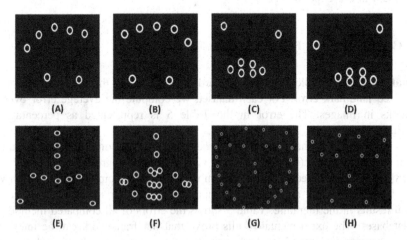

Fig. 2. Sub-Shapes Regressed. (A)–(F) sub-shapes are used to locate landmarks in 6 facial components; G sub-shape is used to locate contour landmarks; H sub-shape is used to locate salient landmarks.

Table 4. We prove that our sub-shape selection is better than single component shape, it also indicates the key points is essential to help localization of other landmarks. The error regressing in our sub-shapes is lower than that regressing in single component.

Sub-shapes	Error ($\times 10^{-2}$)	Facial components	Error ($\times 10^{-2}$)
(A) Sub-shape	3.73	Left eyebrow	4.62
(B) Sub-shape	3.64	Right eyebrow	4.07
(C) Sub-shape	3.03	Left eye	3.36
(D) Sub-shape	3.14	Right eye	3.65
(E) Sub-shape	3.09	Nose	4.04
(F) Sub-shape	3.30	Mouth	4.33

4 Experimental Results

We perform our experiments in two parts. The first part validates our method of selecting sub-shapes and extracting "key points" of components. The Second part compares our approach with previous works. We implemented our approach and ESR in C ++ and tested on four-core Inter Core i5-3407 CPU. For convenience of describing our method, we put the first part experimental results in Sect. 3. In the section, we mainly display the second experimental results.

We briefly introduce two datasets used in the experiments. They present different variations in face shape, appearance, and number of landmarks.

LFPW (29 landmarks) [10] is downloaded from the web. Since some pictures are invalid, we only extract 791 of the 1100 images for training and 221 of the 300 images for testing. We use 29 landmarks in our experiments as what previous work [6] does.

Helen (68 landmarks) [11] is downloaded from the web: 2000 images for training and 330 images for testing. Although they are labeled with 194 landmarks. We use 68 out of them in our experiments.

4.1 Comparison with Previous Works

Evaluation Metric. Following the standard [6, 7], we use the inter-pupil distance normalized landmark error. For each dataset, we compute the average error over all landmarks in images. The error in the Table 5 is represented as percentage of pupil-distance.

We mainly select shape regression-based methods to compete with, including explicit shape regression (ESR) [6], Robust Cascaded Pose Regression (PCPR) [7] and LBF [8]. We implemented ESR and our implementation is comparable to which was reported by the original authors. For comparison with other methods, we used the original results in the literature. Table 5 shows the errors of all compared methods on two databases. The experimental results prove that our framework can be integrated with other regression-based methods and boost the accuracy of original methods (Figs. 3 and 4).

Table 5. We show the error on LFPW and Helen database. The error of ESR is from our implementation, and the errors of RCPR and LBF are from [8].

LFPW (29 landmarks)			Helen (68 landmarks)		
Method	Error	Improvement	Method	Error	Improvement
ESR [6]	3.63	8.8 %	ESR [6]	5.68	7.9 %
RCPR [7]	3.50	5.4 %	RCPR [7]	5.60	6.6 %
LBF [8]	3.35	1.2 %	LBF [8]	4.95	–
Ours	3.31	–	Ours	5.23	–

Fig. 3. Selected results from LFPW

Fig. 4. Selected results from Helen

5   Conclusion and Future Work

We present a two-layer shape regression framework that can be integrated with regression-based methods and boost the face alignment performance. In our system, we propose a term of "key points" and determine suitable sub-shapes for regression. Experiment results with two datasets widely used for face alignment indicate that our framework can improve the performance of original method. In future work, we aim to deeply explore the method of selecting sub-shapes and the application of our framework in other relative scenarios.

Acknowledgments. This work was partially supported by the National High-tech Research and Development Program of China (2015AA015901).

References

1. Cootes, T., Taylor, C., Cooper, D., Graham, J., et al.: Active shape models-their training and application. Comput. Vis. Image Underst. **61**(1), 38–59 (1995)
2. Cootes, T., Edwards, G., Taylor, C.: Active appearance models. IEEE Trans. Pattern Anal. Mach. Intell. **23**(6), 681–685 (2001)

3. Cristinacce, D., Cootes, T.: Feature detection and tracking with constrained local models. In: British Machine Vision Conference (BMVC) (2006)
4. Cristinacce, D., Cootes, T.: Boosted regression active shape models. In: British Machine Vision Conference (BMVC) (2007)
5. Dollar, P., Welinder, P., Perona, P.: Cascaded pose regression. In: IEEE Conference on Computer Vision and Pattern Recognition (CVPR) (2010)
6. Cao, X., Wei, Y., Wen, F., Sun, J.: Face alignment by explicit shape regression. In: IEEE Conference on Computer Vision and Pattern Recognition (CVPR) (2012)
7. Burgos-Artizzu., X.P., Perona, P., Dollar, P.: Robust face landmark estimation under occlusion. In: IEEE International Conference on Computer Vision (ICCV) (2013)
8. Ren, S., Cao, X., Wei, Y., Sun, J.: Face alignment at 3000 FPS via regression local binary features. In: IEEE Conference on Computer Vision and Pattern Recognition (CVPR) (2014)
9. Kazemi, V., Sullivan, J.: One millisecond face alignment with an ensemble of regression trees. In: IEEE Conference on Computer Vision and Pattern Recognition (CVPR) (2014)
10. Belhumeur, P.N., Jacobs, D.W., Kriegman, D.J., Kumar, N.: Localizing parts of faces using a consensus of exemplars. In: IEEE Conference on Computer Vision and Pattern Recognition (CVPR) (2011)
11. Le, V., Brandt, J., Lin, Z., Bourdev, L., Huang, T.S.: Interactive facial feature localization. In: Fitzgibbon, A., Lazebnik, S., Perona, P., Sato, Y., Schmid, C. (eds.) ECCV 2012, Part III. LNCS, vol. 7574, pp. 679–692. Springer, Heidelberg (2012)
12. Duffy, N., Helmbold, D.P.: Boosting methods for regression. Mach. Learn. 47(2–3), 153–200 (2002)
13. Friedman, J.H.: Greedy function approximation: a gradient boosting machine. Ann. Stat. 29 (5), 1189–1232 (2001)
14. Bingham, E., Mannila, H.: Random projection in dimensionality reduction: applications to image and text data. In: ACM SIGKDD Conference on Knowledge Discovery and Data Mining (KDD) (2001)
15. Le, V., Brandt, J., Lin, Z., Bourdev, L., Huang, T.: Interactive facial feature localization. In: European Conference on Computer Vision (2012)
16. Liang, L., Xiao, R., Wen, F., Sun, J.: Face alignment via component-based discriminative search. In: Forsyth, D., Torr, P., Zisserman, A. (eds.) ECCV 2008, Part II. LNCS, vol. 5303, pp. 72–85. Springer, Heidelberg (2008)

3D Panning Based Sound Field Enhancement Method for Ambisonics

Song Wang[1,2], Ruimin Hu[1,2(✉)], Shihong Chen[1], Xiaochen Wang[1,2], Yuhong Yang[1], and Weiping Tu[1]

[1] National Engineering Research Center for Multimedia Software,
School of Computer Science, Wuhan University, Wuhan 430072, China
{wangsongf117,hrm1964,shi_hong_chen,clowang,ahka_yang,echo_tuwp}@163.com
[2] Research Institute of Wuhan University in Shenzhen, Shenzhen 518000, China

Abstract. When conventional first order Ambisonics system uses four loudspeaker with platonic solid layout to reconstruct sound field, the 3D acoustic field effect is limited. A new signal distribution method is proposed to enhance the reproduced field without increasing loudspeakers. First, a platonic solid is extended to get more new vertexes, based on the traditional Ambisonics signal distribution method, original field signal is distributed to loudspeakers at original and new vertexes of platonic solid. Second, signals of loudspeakers at new vertexes are distributed to loudspeakers at original vertexes by a new 3D panning method, then loudspeakers at new vertexes of platonic solid are deleted, only original vertexes of platonic solid are left. The proposed method can improve the quality of the reconstructed sound field and will not increase the complexity of loudspeaker layout in practice. Results are verified through objective and subjective experiments.

Keywords: Ambisonics · 3D panning · Enhancement

1 Introduction

Ambisonics is an important physical acoustic field reconstruction technique in 3D acoustic field technology. It was proposed by Michael Gerzon of Oxford University in the 1970s [1], which uses spherical harmonic function as a way of expressing and reconstructing acoustic field. Ambisonics reconstructs a single point sound field in space efficiently by first-order spherical harmonic function. Then first-order Ambisonics was developed to high order Ambisonics which is used for reconstructing acoustic field in a larger area [2,3]. The principle of Ambisonics is

R. Hu—The research was supported by National Nature Science Foundation of China (61231015, 61201169), National High Technology Research and Development Program of China (863 Program) (2015AA016306), Science and Technology Plan Projects of Shenzhen (ZDSYS2014050916575763), National Nature Science Foundation of China (61201340), the Fundamental Research Funds for the Central Universities (2042015kf0206).

Y.-S. Ho et al. (Eds.): PCM 2015, Part I, LNCS 9314, pp. 135–145, 2015.
DOI: 10.1007/978-3-319-24075-6_14

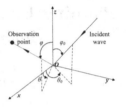

Fig. 1. Tetrahedron, hexahedron, octahedron, dodecahedron, icosahedron.   **Fig. 2.** Spherical coordinate system.

that the desired sound field and sound field of second sound source are expanded into space spheric harmonic function [4–6], which is called the mode matching method. Compared with other 3D acoustic field reconstruction methods such as Wave Field Synthesis [7], Vector Based Amplitude Panning (VBAP) [8] and HRTF [9], Ambisonics represents a practical and asymptotically holographic method to spatialization [10].

In 3D Ambisonics system, the effect of sound field reconstruction is closely related to its order. The mathematical relationship between loudspeaker number L and the order M is: $L \geq (M + 1)^2$. The higher is the order, the better is the sound field effect [10], and more loudspeakers are needed. However, more loudspeakers mean that the loudspeaker layout is more complex, which is not conducive to practical application. The playback system of first order Ambisonics needs four loudspeakers at least. It is the simplest 3D Ambisonics reconstruction system. But the quality of the reconstructed sound field is not high. If too many loudspeakers are used, it will increase complexity in loudspeaker layout. So it needs to improve the quality of the reconstructed sound field with fewer loudspeakers. We research reconstructed sound field enhancement method on first order Ambisonics system in this paper. There are five layouts fulfill the regularity criterion, which are known as platonic solids in Fig. 1. The tetrahedron layout for loudspeaker is researched in this paper, the other four platonic solid layouts can be dealt with similar method.

A new method of signal distribution is proposed in this paper to enhance the reconstructed acoustic field of conventional first order Ambisonics system using four loudspeakers. It can improve the quality of the reconstructed sound field, and it will not increase the complexity of loudspeaker layout in practice use. This paper is outlined as follows: Sect. 2 describes an Ambisonics method, Sect. 3 proposes a 3D panning method, Sect. 4 presents signal distribution new method in detail, Sect. 5 presents the objective and subjective experiments, and compares the results of traditional method and new method.

2   Ambisonics Method

Spherical coordinate system are used in this paper as in Fig. 2. Observation point $\mathbf{x} = (r_x, \varphi, \theta)$, where r_x is the distance to origin, φ is elevation, θ is azimuth. The origin is O. Suppose a plane-wave incident from $\hat{\mathbf{y}} = [cos\theta_0 sin\varphi_0, sin\theta_0 sin\varphi_0, cos\varphi_0]^T$, the sound field at point $\mathbf{x} = x[cos\theta sin\varphi, sin\theta sin\varphi, cos\varphi]^T$ is:

$$S(x, k) = e^{ikx(\hat{\mathbf{y}}^T \hat{\mathbf{x}})} \tag{1}$$

where k is the wave number, $k = 2\pi f/c$, f is the frequency of sound, c is the sound speed in air, T is the transposition of the matrix, $x = |\mathbf{x}|, \hat{\mathbf{x}} = \mathbf{x}/|\mathbf{x}|$ denotes a unit vector in the direction $\mathbf{x}$. Equation (1) also can be approximatively expressed as [11]:

$$S(x, k) = \sum_{n=0}^{M} \sum_{m=-n}^{n} 4\pi i^n j_n(kx) Y_{nm}^*(\hat{\mathbf{y}}) Y_{nm}(\hat{\mathbf{x}}) \tag{2}$$

where $M = \lceil kx_0 \rceil$, x_0 is the radius of considered spherical reproduced region, $Y_{nm}(\hat{\mathbf{x}})$ is spherical harmonics:

$$Y_{nm}(\hat{\mathbf{x}}) = A_{nm} P_{n|m|}(cos\varphi)e^{im\theta}, \quad A_{nm} = \sqrt{\frac{(2n+1)(n-|m|)!}{4\pi(n+|m|)!}} \tag{3}$$

$P_{n|m|}(\cdot)$ is the associated Legendre function, $Y_{nm}^*(\hat{\mathbf{x}})$ and $Y_{nm}(\hat{\mathbf{x}})$ are conjugate, $j_n(\cdot)$ is the nth order spherical Bessel function of the first kind.

Suppose a plain-wave is reproduced in a region by L loudspeakers on a sphere with radius R. See the lth loudspeaker as a point source, which produces sound field at point $\mathbf{x}$ is:

$$T_l(x, k) = |\mathbf{y}_l|e^{ik|\mathbf{y}_l|} \frac{e^{-ik|\mathbf{y}_l - \mathbf{x}|}}{|\mathbf{y}_l - \mathbf{x}|} \tag{4}$$

where $\mathbf{y}_l$ is the loudspeaker position, $|\mathbf{y}_l| = R, \forall\ l$. The sound field produced by L loudspeakers at point $\mathbf{x}$ is:

$$T(x, k) = \sum_{l=1}^{L} a_l T_l(x, k) = \sum_{n=0}^{M} \sum_{m=-n}^{n} 4\pi i^n j_n(kx) R_n(kR) \sum_{l=1}^{L} a_l Y_{nm}^*(\hat{\mathbf{y}}) Y_{nm}(\hat{\mathbf{x}}) \tag{5}$$

where a_l is the weighting coefficient of lth loudspeaker, $R_n(kR) = -ikRe^{ikR}i^{-n} \cdot h_n(kR)$, $h_n(\cdot)$ is the nth order spherical Hankel function of the second kind.

Equate (2) with (5), we could get:

$$P_{n|m|}(cos\varphi_0)e^{-im\theta_0} = R_n(kR) \sum_{l=1}^{L} a_l P_{n|m|}(cos\varphi)e^{-im\theta} \tag{6}$$
$$n = 0, \cdots, M, \quad m = -n, \cdots, n$$

Equation (6) can be expressed as a linear equations, it can be worked out to get the weighting coefficients a_l of L loudspeakers for reproducing the plain-wave in the region $|\mathbf{x}| < R$.

3   3D Panning Method with Sound Pressure Constraint at Two Ears

Our panning method needs to meet the following assumptions: (1) reflected sound is neglected, (2) a loudspeaker can be seen as a point source, (3) only the outgoing sound wave of the loudspeaker is considered, (4) the sound pressure at a unit distance from a loudspeaker is in proportion to the input to the loudspeaker, the proportion coefficient is recorded as G, (5) $k\sigma_{min} \gg 1$, σ_{min} is the minimum distance between the virtual sound source or the loudspeakers and the receiving point. Based on these assumptions, Fourier transform of the sound pressure produced by single loudspeaker at the receiving point is expressed as:

$$p(\mathbf{r}',\omega) = G\frac{e^{-ik|\mathbf{r}'-\boldsymbol{\xi}|}}{|\mathbf{r}'-\boldsymbol{\xi}|}s(\omega) \tag{7}$$

where $\mathbf{r}' = (x',y',z')^T$ is the coordinate of the receiving point, $\boldsymbol{\xi} = (\xi_x,\xi_y,\xi_z)^T$ is the coordinate of single loudspeaker, $s(\omega)$ is the Fourier transform of the input signal to the loudspeaker. Particle velocity produced by single loudspeaker at the receiving point is expressed as [12]:

$$u(\mathbf{r}',\omega) = G(1+\frac{1}{ik|\mathbf{r}'-\boldsymbol{\xi}|})\frac{e^{-ik|\mathbf{r}'-\boldsymbol{\xi}|}}{|\mathbf{r}'-\boldsymbol{\xi}|^2}\begin{pmatrix} x'-\xi_x \\ y'-\xi_y \\ z'-\xi_z \end{pmatrix}s(\omega) \approx G\frac{e^{-ik|\mathbf{r}'-\boldsymbol{\xi}|}}{|\mathbf{r}'-\boldsymbol{\xi}|^2}\begin{pmatrix} x'-\xi_x \\ y'-\xi_y \\ z'-\xi_z \end{pmatrix}s(\omega) \tag{8}$$

The sound pressure or particle velocity of q loudspeakers at the receiving point could be obtained by summing sound pressure or particle velocity of single loudspeakers at the receiving point.

Then we study the case that a virtual sound source (single loudspeaker) is replaced by q loudspeakers. We suppose the virtual sound source and q loudspeakers are on a same sphere, the receiving point is at the center of the sphere which also is the listening point and the origin O. We suppose that the sound pressure produced by a virtual sound source at the receiving point is the same as the sound pressure produced by q loudspeakers at the receiving point, then:

$$G\sum_{j=1}^{q}\frac{e^{-ik\rho}}{\rho}s_j(\omega) = G\frac{e^{-ik\rho}}{\rho}s(\omega) \tag{9}$$

Together with $s_j(\omega) = w_j s(\omega)$, Eq. (9) could be simplified as:

$$w_1 + w_2 + \cdots + w_q = 1 \tag{10}$$

In spherical coordinates, coordinate of the receiving point is $\mathbf{r}' = (0,90°,0°)$, coordinate of the virtual sound source is $\boldsymbol{\xi} = (\rho,\varphi',\theta')$, coordinates of q loudspeakers are $\boldsymbol{\xi}^{(j)} = (\rho,\varphi_j,\theta_j)$, $j = 1,2,\cdots,q$. Then Fourier transform of the particle velocity produced by the virtual sound source at the receiving point is expressed as:

$$u(\mathbf{r}',\omega) = -\frac{G}{c\lambda}\frac{e^{-ik\rho}}{\rho}Hs(\omega), \quad H = \left(\cos\theta'\sin\varphi',\sin\theta'\sin\varphi',\cos\varphi'\right)^T \tag{11}$$

where λ is the density of air. Fourier transform of the particle velocity produced by q loudspeakers at the receiving point is expressed as:

$$\tilde{u}(\mathbf{r}',\omega) = -\frac{G}{c\lambda}\frac{e^{-ik\rho}}{\rho}\tilde{H}Ws(\omega), \quad W = \left(w_1, w_2, \cdots, w_q\right)^T \quad (12)$$

$$\tilde{H} = \begin{pmatrix} \cos\theta_1\sin\varphi_1 & \cos\theta_2\sin\varphi_2 & \cdots & \cos\theta_q\sin\varphi_q \\ \sin\theta_1\sin\varphi_1 & \sin\theta_2\sin\varphi_2 & \cdots & \sin\theta_q\sin\varphi_q \\ \cos\varphi_1 & \cos\varphi_2 & \cdots & \cos\varphi_q \end{pmatrix}$$

Let $u(\mathbf{r}',\omega) = \tilde{u}(\mathbf{r}',\omega)$, then:

$$\tilde{H}\begin{pmatrix} w_1 \\ w_2 \\ \vdots \\ w_q \end{pmatrix} = \begin{pmatrix} \cos\theta'\sin\varphi' \\ \sin\theta'\sin\varphi' \\ \cos\varphi' \end{pmatrix} \quad (13)$$

We divide the first and second rows by the third row in Eq. (13), then:

$$\frac{\sum_{v=1}^{q} w_v\cos\theta_v\sin\varphi_v}{\sum_{v=1}^{q} w_v\cos\varphi_v} = \frac{\cos\theta'\sin\varphi'}{\cos\varphi'}$$
$$\frac{\sum_{v=1}^{q} w_v\sin\theta_v\sin\varphi_v}{\sum_{v=1}^{q} w_v\cos\varphi_v} = \frac{\sin\theta'\sin\varphi'}{\cos\varphi'} \quad (14)$$

Equation (14) could guarantee that direction of particle velocity of virtual sound source and q loudspeakers at the receiving point are the same. From Eq. (14), we could get:

$$\sum_{v=1}^{q} w_v(\cos\theta_v\sin\varphi_v\cos\varphi' - \cos\varphi_v\cos\theta'\sin\varphi') = 0$$
$$\sum_{v=1}^{q} w_v(\sin\theta_v\sin\varphi_v\cos\varphi' - \cos\varphi_v\sin\theta'\sin\varphi') = 0 \quad (15)$$

Together with Eq. (10), we could get:

$$LW = E_1, L = \begin{pmatrix} t_{11} & t_{12} & \cdots & t_{1q} \\ t_{21} & t_{22} & \cdots & t_{2q} \\ 1 & 1 & \cdots & 1 \end{pmatrix}, E_1 = \begin{pmatrix} 0 \\ 0 \\ 1 \end{pmatrix} \quad (16)$$

where for $v = 1, 2, \cdots, q$,

$$t_{1v} = \cos\theta_v\sin\varphi_v\cos\varphi' - \cos\varphi_v\cos\theta'\sin\varphi'$$
$$t_{2v} = \sin\theta_v\sin\varphi_v\cos\varphi' - \cos\varphi_v\sin\theta'\sin\varphi' \quad (17)$$

In realistic listening, listeners keep the center of their head coinciding to the position of spherical center, their ears locates around sphere center, because the radius of human head is about 0.085 m. We suppose the left ear is at $\mathbf{r}'_L = (0.085, 90°, 180°)$, the right ear is at $\mathbf{r}'_R = (0.085, 90°, 0°)$. The sound pressure produced by a virtual sound source at $\mathbf{r}'_L$, $\mathbf{r}'_R$ are:

$$p(\mathbf{r}'_L,\omega) = G\frac{e^{-ik|\mathbf{r}'_L-\xi|}}{|\mathbf{r}'_L - \xi|}s(\omega), \quad p(\mathbf{r}'_R,\omega) = G\frac{e^{-ik|\mathbf{r}'_R-\xi|}}{|\mathbf{r}'_R - \xi|}s(\omega) \quad (18)$$

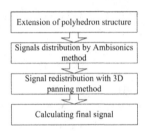

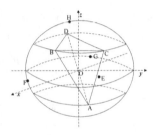

Fig. 3. The overall flow chart.   **Fig. 4.** Structure extension.

the sound pressure produced by q loudspeakers at $\mathbf{r}'_L$, $\mathbf{r}'_R$ are:

$$p_q(\mathbf{r}'_L, \omega) = \sum_{j=1}^{q} G \frac{e^{-ik|\mathbf{r}'_L - \boldsymbol{\xi}^{(j)}|}}{|\mathbf{r}'_L - \boldsymbol{\xi}^{(j)}|} s_j(\omega), \quad p_q(\mathbf{r}'_R, \omega) = \sum_{j=1}^{q} G \frac{e^{-ik|\mathbf{r}'_R - \boldsymbol{\xi}^{(j)}|}}{|\mathbf{r}'_R - \boldsymbol{\xi}^{(j)}|} s_j(\omega)$$

(19)

For a given signal, $\|Gs(\omega)\|$ is a constant. We define:

$$\begin{aligned}
&E(w_1, w_2, w_3, \cdots, w_q) \\
&= \left(\frac{\cos(k|\mathbf{r}'_L - \boldsymbol{\xi}|)}{|\mathbf{r}'_L - \boldsymbol{\xi}|} - \sum_{j=1}^{q} w_j \frac{\cos(k|\mathbf{r}'_L - \boldsymbol{\xi}^{(j)}|)}{|\mathbf{r}'_L - \boldsymbol{\xi}^{(j)}|} \right)^2 + \left(\frac{\sin(k|\mathbf{r}'_L - \boldsymbol{\xi}|)}{|\mathbf{r}'_L - \boldsymbol{\xi}|} \right. \\
&\quad \left. - \sum_{j=1}^{q} w_j \frac{\sin(k|\mathbf{r}'_L - \boldsymbol{\xi}^{(j)}|)}{|\mathbf{r}'_L - \boldsymbol{\xi}^{(j)}|} \right)^2 + \left(\frac{\cos(k|\mathbf{r}'_R - \boldsymbol{\xi}|)}{|\mathbf{r}'_R - \boldsymbol{\xi}|} - \sum_{j=1}^{q} w_j \frac{\cos(k|\mathbf{r}'_R - \boldsymbol{\xi}^{(j)}|)}{|\mathbf{r}'_R - \boldsymbol{\xi}^{(j)}|} \right)^2 \\
&\quad + \left(\frac{\sin(k|\mathbf{r}'_R - \boldsymbol{\xi}|)}{|\mathbf{r}'_R - \boldsymbol{\xi}|} - \sum_{j=1}^{q} w_j \frac{\sin(k|\mathbf{r}'_R - \boldsymbol{\xi}^{(j)}|)}{|\mathbf{r}'_R - \boldsymbol{\xi}^{(j)}|} \right)^2
\end{aligned}$$

(20)

So to make the error of sound pressure at two ears minimum, we should make E minimum. Then the solution of replacing a virtual sound source by q loudspeakers is equivalent to solving the following question:

$$\begin{aligned}
&\min_{W} \tfrac{1}{2} \|\tilde{F}W - F\|_2^2 \\
&s.t. \ LW = E_1 \\
&\quad w_1, w_2, \cdots, w_q \geq 0
\end{aligned}$$

(21)

where

$$\tilde{F} = \begin{pmatrix}
\frac{\cos(k|\mathbf{r}'_L - \boldsymbol{\xi}^{(1)}|)}{|\mathbf{r}'_L - \boldsymbol{\xi}^{(1)}|} & \frac{\cos(k|\mathbf{r}'_L - \boldsymbol{\xi}^{(2)}|)}{|\mathbf{r}'_L - \boldsymbol{\xi}^{(2)}|} & \cdots & \frac{\cos(k|\mathbf{r}'_L - \boldsymbol{\xi}^{(q)}|)}{|\mathbf{r}'_L - \boldsymbol{\xi}^{(q)}|} \\
\frac{\sin(k|\mathbf{r}'_L - \boldsymbol{\xi}^{(1)}|)}{|\mathbf{r}'_L - \boldsymbol{\xi}^{(1)}|} & \frac{\sin(k|\mathbf{r}'_L - \boldsymbol{\xi}^{(2)}|)}{|\mathbf{r}'_L - \boldsymbol{\xi}^{(2)}|} & \cdots & \frac{\sin(k|\mathbf{r}'_L - \boldsymbol{\xi}^{(q)}|)}{|\mathbf{r}'_L - \boldsymbol{\xi}^{(q)}|} \\
\frac{\cos(k|\mathbf{r}'_R - \boldsymbol{\xi}^{(1)}|)}{|\mathbf{r}'_R - \boldsymbol{\xi}^{(1)}|} & \frac{\cos(k|\mathbf{r}'_R - \boldsymbol{\xi}^{(2)}|)}{|\mathbf{r}'_R - \boldsymbol{\xi}^{(2)}|} & \cdots & \frac{\cos(k|\mathbf{r}'_R - \boldsymbol{\xi}^{(q)}|)}{|\mathbf{r}'_R - \boldsymbol{\xi}^{(q)}|} \\
\frac{\sin(k|\mathbf{r}'_R - \boldsymbol{\xi}^{(1)}|)}{|\mathbf{r}'_R - \boldsymbol{\xi}^{(1)}|} & \frac{\sin(k|\mathbf{r}'_R - \boldsymbol{\xi}^{(2)}|)}{|\mathbf{r}'_R - \boldsymbol{\xi}^{(2)}|} & \cdots & \frac{\sin(k|\mathbf{r}'_R - \boldsymbol{\xi}^{(q)}|)}{|\mathbf{r}'_R - \boldsymbol{\xi}^{(q)}|}
\end{pmatrix}, \quad F = \begin{pmatrix}
\frac{\cos(k|\mathbf{r}'_L - \boldsymbol{\xi}|)}{|\mathbf{r}'_L - \boldsymbol{\xi}|} \\
\frac{\sin(k|\mathbf{r}'_L - \boldsymbol{\xi}|)}{|\mathbf{r}'_L - \boldsymbol{\xi}|} \\
\frac{\cos(k|\mathbf{r}'_R - \boldsymbol{\xi}|)}{|\mathbf{r}'_R - \boldsymbol{\xi}|} \\
\frac{\sin(k|\mathbf{r}'_R - \boldsymbol{\xi}|)}{|\mathbf{r}'_R - \boldsymbol{\xi}|}
\end{pmatrix}$$

Equation (21) is a least squares problems with inequality constraints and it could be worked out by existing mature algorithms such as Trust Region Algorithm [13].

(a)   (b)   (c)

Fig. 5. Symmetry center of (a) equilateral triangle; (b) quadrate; (c) regular pentagon.

4 New Signal Distribution Method

The proposed method uses loudspeaker tetrahedron to reproduce a plain-wave, the entire process of our new signal distribution method is shown in Fig. 3.

4.1 Extension of Loudspeakers Structure

First, four loudspeakers will be put into tetrahedron structure, their coordinates are recorded. The origin is $O(0, 90°, 0°)$, as in Fig. 4. Vertexes are denoted by A, B, C, D. Then symmetry center in every face of the tetrahedral should be founded as shown in Fig. 5(a). The origin need to be linked with symmetry centers in every face, and the connections intersect circumsphere surface at four points E, F, G, H respectively. The loudspeakers at A, B, $\cdots$, H are denoted as ld_A, ld_B, $\cdots$, ld_H.

4.2 Calculation of the Input Signal

We suppose the original plain-wave signal is S_o. By traditional Ambisonics method in Sect. 2, the distribution coefficients of ld_A, ld_B, $\cdots$, ld_H can be obtained by Eq. (6) and are denoted as w_A, w_B, $\cdots$, w_H. The initial distribution signal of ld_A, ld_B, $\cdots$, ld_H are:

$$S_I = w_I S_o, \quad I \in \{A, B, \cdots, H\} \tag{22}$$

4.3 Signal Redistribution

Signals of ld_E, ld_F, ld_G, ld_H are assigned to loudspeakers at the nearest three tetrahedron vertices respectively by Eq. (21) in Sect. 3 (set $q = 3$). As in Fig. 4, Signal of ld_E is assigned to ld_A, ld_B, ld_C with distribution coefficients: w_{EA}, w_{EB}, w_{EC}; Signal of ld_F is assigned to ld_A, ld_B, ld_D with distribution coefficients: w_{FA}, w_{FB}, w_{FD}; Signal of ld_G are assigned to ld_A, ld_C, ld_D with distribution coefficients: w_{GA}, w_{GC}, w_{GD}; Signal of ld_H are assigned to ld_B, ld_C, ld_D with distribution coefficients: w_{HB}, w_{HC}, w_{HD}.

4.4 Final Signals

Final signals of ld_A, ld_B, ld_C, ld_D are obtained by adding up the signals they get in Sects. 4.2 and 4.3. Then delete ld_E, ld_F, ld_G, ld_H. At last, there are four loudspeakers at the vertexes of tetrahedron left. Final signals can be expressed as:

$$
\begin{aligned}
S_{fA} &= S_A + w_{EA}S_E + w_{FA}S_F + w_{GA}S_G \\
S_{fB} &= S_B + w_{EB}S_E + w_{FB}S_F + w_{HB}S_H \\
S_{fC} &= S_C + w_{EC}S_E + w_{GC}S_G + w_{HC}S_H \\
S_{fD} &= S_D + w_{FD}S_F + w_{GD}S_G + w_{HD}S_H
\end{aligned}
\tag{23}
$$

5   Experiments

In the following examples, we will illustrate sound field reproduction of a plain-wave by proposed method with loudspeaker tetrahedron layout, loudspeakers' elevation and azimuth are as [14], distances between the origin O and loudspeakers are 2 m. Because there are very less loudspeakers with tetrahedron layout, we just consider the first order case. The original sound field is monochromatic plane wave of frequency of 1000 Hz arriving from $(2, 30°, 120°)$. So we consider a spherical reproduction region of radius $x_0 = 0.05$ m. we assume c is 340 m/s. Then $kx_0 = 0.9240$, the order is $M = \lceil kx_0 \rceil = 1$. According to $L \geq (M + 1)^2$, four loudspeakers are needed at least.

5.1   Objective Tests

Reproduced sound field in x-y plain are compared. The reproduced field is shown in Fig. 6. The considered region is marked by a black solid circle. The sound field reproduced by proposed method is better than traditional Ambisonics method.

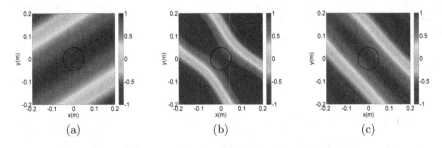

(a)   (b)   (c)

Fig. 6. (a): original field, (b): reproduced field by traditional Ambisonics method, (c): reproduced field by proposed method.

The relative mean square error (RMSE) of the reproduction is used as the error metric:

$$
\varepsilon(kr) = \frac{\int_0^r \int_0^{2\pi} \int_0^\pi |S_r(\mathbf{x}) - S_d(\mathbf{x})|^2 dx}{\int_0^r \int_0^{2\pi} \int_0^\pi |S_d(\mathbf{x})|^2 dx}
\tag{24}
$$

where the integration is over a spherical ball of radius r and $dx = r_{\mathbf{x}}^2 sin\varphi d\varphi d\theta dr_{\mathbf{x}}$, $S_d(\mathbf{x})$ and $S_r(\mathbf{x})$ are the original and reproduced sound fields respectively.

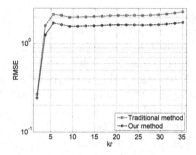

Fig. 7. Reproduced error.

RMSE of the reproduction field is displayed in Fig. 7. RMSE by proposed method is lower than traditional Ambisonics method. When r is 1.9m, the RMSE of our method is 55 % lower than that of traditional method.

5.2   Subjective Tests

Comparison Mean Opinion Score (CMOS) is used to test proposed method and traditional Ambisonics method, the test material consists of Ref/A/B, in which Ref is the original sound source signal, A is signal generated by proposed method, and B is signal generated by traditional Ambisonics method. Ref is played back

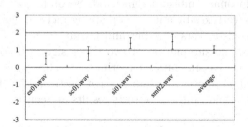

Fig. 8. CMOS score of each item.

Table 1. Levels comparison standard

Comparison of the Stimuli	Score
Sound image of A is much closer to Ref than B	+3
Sound image of A is closer to Ref than B	+2
Sound image of A is slightly closer to Ref than B	+1
Sound image of A to Ref is the same as B	0
Sound image of A is slightly further to Ref than B	−1
Sound image of A is further to Ref than B	−2
Sound image of A is much further to Ref than B	−3

by a loudspeaker at $(2, 30°, 120°)$. A and B are played back with loudspeaker tetrahedron layout,We compare the sound image of A and B which is closer to Ref. The score has 7 levels, which are listed in Table 1. 10 listeners performed the listening test, including 5 males and 5 females. All of them actively work in the audio field. The center of listener's head is at origin O in testing. The test results consist of an average score and a 95 % confidence interval. We use 4 MPEG test sequences: es01, sc01, si01, sm02 including voice and solo, whose sample rate are 48 kHz, bit depth are 16 bits, amplitude are -40 dB, and length are 8s. The test results for each of them are given in Fig. 8. We can see that all these cases are statistically comparable to traditional method in a 95 % confidence interval sense. The average scores of our method are higher than traditional method for all cases, the average value of CMOS of all cases is 1.05. The results mean that location accuracy of our method is better than traditional Ambisonics method and are in accordance with objective test results. The reason is that our method can maintain the spatial information of original sound source signal better, and make sure that the sound pressure error minimum at two ears.

6   Conclusion

In traditional Ambisonics method, for a given loudspeaker number and layout, the signals are directly assigned to loudspeakers. But this paper proposed a new signal distribution method. For a given loudspeaker number and layout, it first extends its structure to get more new vertexes, then it distributes original sound field signal by traditional Ambisonics method. Second, it distribute signals of loudspeakers at new vertexes by a new 3D panning method again, then new vertexes are deleted, loudspeakers' number and layout are still. The signals of the left loudspeakers are the summation of two times signal distributions for each left loudspeaker. Objective simulation experiments with loudspeaker tetrahedron layout show that this proposed method can get better performance than traditional Ambisonics method. Subjective tests results are consistent with objective simulation results.

The faces of five platonic solids are of three kind: equilateral triangle, quadrate, regular pentagon, it's easy to find their symmetry centers as shown in Fig. 5. Then the proposed method could be similarly applied to enhance Ambisonics system with other four kind platonic solids layout.

References

1. Gerson, M.A.: Ambisonics. Part two: Studio Tech. Studio Sound. **17**, 24–30 (1975)
2. Bamford, J.S.: An Analysis of Ambisonics Sound Systems of First and Second Order. Ph.D. thesis, University of Waterloo, Waterloo (1995)
3. Daniel, J., Nicol, R., Moreau, S.: Further investigations of high order ambisonics and wave field synthesis for holophonic sound imaging. In: proceedings of the 114th Audio Engineering Society Convention, Amsterdam, Nertherlands (2003)

4. Daniel, J.: Représentation De Champs Acoustiques, Application à la Transmission et à la Reproduction De Scènes Sonores Complexes Dans Un contexte Multimédia. Ph.D. thesis, University Paris, Paris (2000)
5. Poletti, M.: Three-dimensional surround sound systems based on spherical harmonics. J. Audio Eng. Soc. **53**(11), 1004–1025 (2005)
6. Spors, S., Ahrens, J.: Comparison of Higer-order Ambisonics and Wave Field Synthesis with Respect to Spatial Aliasing Artifacts. In: proceedings of the 19th International Congress on Acoustics. Madrid, Spain (2007)
7. Berkhout, A.J.: A holographic approach to acoustic control. J. Audio Eng. Soc. **36**(12), 977–995 (1998)
8. Pulkki, V.: Virtual sound source positioning using vector base amplitude panning. J. Audio Eng. Soc. **45**(6), 456–466 (1997)
9. Blauert, J.P.: Spatial Hearing. MIT, Revised edition, Cambridge (1997)
10. Kearney, G., et al.: Distance perception in interactive virtual acoustic environments using first and higher order ambisonic sound fields. Acta Acustica United Acustica **98**(1), 68–71 (2012)
11. Ward, D.B., Abhayapala, T.D.: Reproduction of a plane-wave sound field using an array of loudspeakers. IEEE Trans. Speech Audio Process. **9**(6), 697–707 (2001)
12. Ando, A.: Conversion of multichannel sound signal maintaining physical properties of sound in reproduced sound field. IEEE Trans. Audio, Speech, Lang. Process. **19**(6), 1467–1475 (2011)
13. Moré, J.J., Sorensen, D.C.: Computing a trust region step. SIAM J. Sci. Stat. Comput. **3**, 553–572 (1983)
14. Sloane, N.J.A.: Spherical Codes (2006). http://neilsloane.com/packings/

Multimedia Applications and Services

Multi-target Tracking via Max-Entropy Target Selection and Heterogeneous Camera Fusion

Jingjing Wang$^{(\boxtimes)}$ and Nenghai Yu

CAS Key Laboratory of Electromagnetic Space Information,
University of Science and Technology of China, Hefei, China
kkwang@mail.ustc.edu.cn, ynh@ustc.edu.cn

Abstract. Nowadays, dual-camera systems, which consist of a static camera and a pan-tilt-zoom (PTZ) camera, have become popular in video surveillance, since they can offer wide area coverage and highly detailed images of the interesting starget simultaneously. Different from most previous multi-target tracking methods without information fusion, we propose a multi-target tracking framework based on information fusion of the heterogeneous cameras. Specifically, a conservative online multi-target tracking method is introduced to generate reliable tracklets in both cameras in real time. A max-entropy target selection strategy is proposed to determine which target should be observed by the PTZ camera at a higher resolution to reduce the ambiguity of multi-target tracking. Finally, the information from the static camera and the PTZ camera is fused into a tracking-by-detection framework for more robust multi-target tracking. The proposed method is tested in an outdoor scene, and the experimental results show that our method significantly improves the multi-target tracking performance.

1 Introduction

Multi-target tracking is important for activity analysis and anomaly detection in video surveillance systems. Surveillance cameras used in public areas (such as squares, parking lots, railway stations, airports, etc.) usually cover a large area. Therefore, the target size observed from these cameras is small, and the appearance information of the targets is not discriminative enough to distinguish different targets due to the low resolution. If the scene is crowded and long-time occlusion occurs frequently, the number of ID switches would increase significantly. This greatly hampers the performance of multi-target tracking. Increasing the number of cameras can solve this problem, but the costs would also increase. Recently, hybrid camera systems which consist of static cameras and PTZ cameras (which are also referred to as active cameras in this paper) have been widely used in video surveillance [1–10], since they can offer a wide range monitoring and close-up view simultaneously. The static camera can cover a large area providing the motion information of all observed targets, and the active camera is able to zoom in and focus on individual targets to obtain their discriminative appearance information.

© Springer International Publishing Switzerland 2015
Y.-S. Ho et al. (Eds.): PCM 2015, Part I, LNCS 9314, pp. 149–159, 2015.
DOI: 10.1007/978-3-319-24075-6_15

Fig. 1. Examples of tracking results in our dual-camera system. The left images in (a) and (b) are from the static camera, and the right images in (a) and (b) are from the active camera. Same persons moving in static and active cameras are linked with yellow solid lines (Colour figure online).

Previous tracking methods in dual-camera systems mainly focus on the spatial mapping between static cameras and active cameras [1–6], active camera control [7] and scheduling strategies [3]. Most of them use the static camera to detect and track all the targets that appear in the scene, and the active camera to track individual targets or simply capture the images of them at a higher resolution. During tracking, no information fusion from static cameras and active cameras is used to enhance the multi-target tracking performance. Although some of them like [8,9] fuse information from two cameras to improve the tracking accuracy, they focus on improving the single-target tracking accuracy. To the best of our knowledge, there is no research work that focuses on improving the overall multi-target tracking performance in such a dual-camera system.

In multi-target tracking, tracking-by-detection methods, which build long trajectories of targets by associating detection responses or tracklets, have become popular in recent years. Many state-of-the-art trackers [11–14] follow the tracking-by-detection framework. However, they suppose the cameras are homogeneous, and suffer from the low resolution of targets when applied to large-area monitoring systems.

In this paper, we propose a novel framework to improve the multi-target tracking performance in the surveillance systems where static cameras cover large areas and the target size is usually small (some examples are shown in Fig. 1). In our framework, the static camera covers a large area, detects and tracks targets online. The active camera observes one target at each time according to the tracking results in the static camera. The static camera provides the motion information of all observed targets, as the static camera can observe and track all targets simultaneous. The active camera is able to provide discriminative appearance information since it can observe targets at a higher resolution. We first introduce a conservative multi-target tracking method to generate reliable tracklets online in both cameras. Then, a max-entropy target selection method is proposed to choose one target to be observed by the active camera at each time, since there may be more than one target tracked by the static camera. The key idea is that the target with the most ambiguity may lead to the most tracking errors, and should be observed with the highest priority. According to the position of the selected target, the active camera adjusts its parameters to obtain close-up views of the target. The tracking results from both cameras

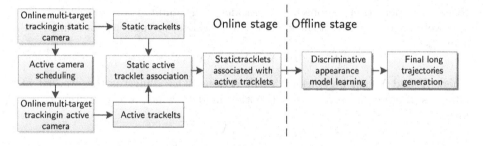

Fig. 2. The overview of our proposed system.

are finally fused into a tracking-by-detection framework to improve the tracking performance. The proposed multi-target tracking system is tested in a challenging outdoor environment, and compared with the state-of-the-art multi-target trackers. The experimental results show that with the cooperation of the active camera, the multi-target tracking performance in the static camera is improved significantly.

2   Our Method

The framework of our proposed method is shown in Fig. 2. It consists of two main stages: the online stage, and the offline stage. During the online stage, an efficient online multi-target tracking algorithm is used to generate tracklets in the static camera which are called *static tracklets*. The tracking results are used to guide the action of the active camera according to the proposed max-entropy target selection strategy. Then, the active camera focuses on the selected target for a short period of time. During this time, the same online multi-target tracking algorithm is used to generate reliable tracklets in the active camera, which are called *active tracklets*. Meanwhile, the active tracklets are associated with static tracklets. During the offline stage, static tracklets which may have associated active tracklets are associated to generate final long trajectories according to the affinity between them. To improve the association accuracy, a discriminative appearance model is learned for each static tracklet using the target image sequences from the static tracklets, which is called *static appearance model*. For each static tracklet which has associated active tracklets, an additional discriminative appearance model, which is called *active appearance model*, is learned using the target image sequences from the associated active tracklets. The appearance affinity between different static tracklets is computed based on the learned static and active appearance models.

2.1   Online Multi-target Tracking

During the online stage, reliable tracklets should be generated efficiently to guide the action of the active camera. We introduce an efficient and reliable online multi-target tracking method. At each frame, pairwise association is performed to associate detection responses with tracklets. The affinity measure to determine how well a detection and a tracklet (or two tracklets) are matched is defined the same as [13]:

$$\Lambda = \Lambda^A(X,Y)\Lambda^S(X,Y)\Lambda^M(X,Y) \tag{1}$$

where X and Y can be tracklets or detections. The affinity is the product of affinities of appearance, shape and motion models, which are computed as follows:

$$\Lambda^A(X,Y) = \sum_{u=1}^{m} \sqrt{h_u(X)h_u(Y)}$$
$$\Lambda^S(X,Y) = \exp\left(-\left\{\left|\frac{h_X-h_Y}{h_X+h_Y}\right| + \left|\frac{w_X-w_Y}{w_X+w_Y}\right|\right\}\right) \tag{2}$$
$$\Lambda^M(X,Y) = \mathcal{N}(P_X + v_X\Delta t; P_Y, \Sigma)$$

The appearance affinity $\Lambda^A(X,Y)$ is the Bhattacharyya coefficient between the color histogram $h_u(X)$ of X and $h_u(Y)$ of Y. The bin number of the histogram is m. For each tracklet, a Kalman Filter is applied to refine the positions and sizes of its detection responses and predict its location. The shape affinity $\Lambda^S(X,Y)$ is computed with the height h and width w of targets. $\Lambda^M(X,Y)$ is the motion affinity between the position P_X of X and the position P_Y of Y with the frame gap Δt and velocity v_X estimated by the Kalman Filter. $\mathcal{N}$ is a Gaussian distribution with covariance matrix Σ. Given the affinity matrix, a conservative strategy is used to link detections with tracklets. A detection response is linked to a tracklet if and only if their affinity is higher than a threshold θ_1, and exceeds their conflicting pairs' affinities by a threshold θ_2. This strategy can avoid ID switches effectively when performing online multi-target tracking. New tracklets are generated from detection responses which are not associated to any tracklet. Tracklets which are not associated with any detection response for a certain period are terminated.

2.2   Active Camera Scheduling

In our dual-camera system, multiple targets are tracked in the static camera. It is necessary for the active camera to observe these targets according to their priority levels at each time. We observe that when performing final tracklet association, a tracklet which can be associated with more than one traklet with similar affinities, may change its ID with a higher probability. The priorities of targets are computed based on the ambiguity of the targets and the earliest deadline first policy. During online multi-target tracking, for each alive static tracklet t_i, we compute its affinity vector A with a tracklet set $T^o = \{t_k^o | k = 1 : m\}$, where t_k^o is the k-th tracklet in T^o. T^o is collected from previous tracklets whose end time is before the start time of t_i with a frame gap less than a certain length. $A = \{a_k | k = 1 : m\}$, where a_k is the affinity between t_i and t_k^o, and a_k is computed using Eq. (1). A is normalized such that a_k ($k = 1 : m$) sum up to one. The ambiguity of target i is defined as $h_i = -\sum_{k=1}^{m} a_k \ln(a_k)$ following the definition of entropy. The priority of the target i is defined as:

$$w_i = \exp(-(t_e)/\sigma) \cdot h_i \cdot \delta(i) \tag{3}$$

where t_e is the predicted time for the target i to exit from the scene or be occluded, and σ controls the importance of t_e. $\delta(i)$ is an indicator function, it equals 0 if the target i has been observed by the active camera, otherwise it equals 1. We predict the future position of the target using the Kalman Filter. If the overlap ratio between the bounding box of the target and the ones of other targets exceeds 0.5, the target is considered in occlusion. The target with the highest priority would be observed by the active camera at each time.

2.3 Static and Active Camera Tracklet Association

Once the target with the highest priority is determined, the active camera is directed to observe the target for a short period of time at a higher resolution. It might be possible to detect multiple targets in the active camera, so we need to associate the targets in the active camera with the tracklets in the static camera. Associating the detection responses in the active camera with the static tracklets would be sensitive to the detection noise. Instead, we suggest to associate the active tracklets with the static tracklets. The score ψ_{ij} between static tracklet i and active tracklet j is defined as:

$$
\psi_{ij} = \begin{cases} \exp(-\frac{\sum_{k=k_1}^{k_2} d(x_s^k, H(x_a^k))}{(k_2 - k_1 + 1) * \gamma}) & \text{if } k_2 - k_1 >= 3 \\ 0 & \text{otherwise} \end{cases}
\tag{4}
$$

where k_1 and k_2 are the start and end frame indexes of the time overlap between the two tracklets. x_s^k and x_a^k are the foot positions of the detection responses in the corresponding tracklets in frame k. d is the Euclidean distance. γ is a parameter controlling the smoothness of the score function. $H(\cdot)$ is the function mapping the position in the active camera to the position in the static camera using a homography, which can be estimated by the calibration method [15].

To obtain the optimal assignment, we use the Hungarian algorithm [17] to assign the active tracklets to the static tracklets. Some association results are shown in Fig. 2.

2.4 Final Trajectory Generation

After the online stage, we obtain the reliable static tracklets from the static camera which may have associated active tracklets. These static tracklets are used to generate final long trajectories based on the affinity between them. To distinguish visually similar targets, discriminative appearance models should be learned for the tracklets. For each static tracklet, we train a static appearance model using the static camera image sequences and an active appearance model using the active camera image sequences if the static tracklet has an associated active tracklet.

Collecting Training Samples. We propose a method to collect positive and negative training samples to train the static appearance model and the active appearance model for each static tracklet t_i.

For the static appearance model, we randomly choose responses in the static tracklet t_i as positive samples. It is intuitive that one target can not appear at different locations at one time and targets in the static camera can not change its positions drastically. Tracklets, that have time overlap with t_i, or are far enough from t_i which makes the targets reach t_i within their time gaps impossibly, represent different targets. Negative samples are collected from these static tracklets.

For the active appearance model, if t_i has associated active tracklet t^a, we randomly choose responses in t^a as positive samples. Any active tracklet which has time overlap with the tracklet t^a can be matched with t^a impossibly and represents a different target. Active tracklets associated with the impossibly matched static tracklets of t_i can also unlikely represent the same target as t_i. Negative samples are collected from these active tracklets.

Appearance Model Learning. The goal of appearance model learning is to learn a model which determines the affinity score $S_{i,j}$ between two tracklets t_i and t_j. In each detection bounding box from tracklets, the color histogram and HOG histogram at different local patches are computed as features. Given a pair of detection responses d_1 and d_2 from two tracklets t_i and t_j, a linear combination of the similarity scores between local patches is learned, which takes the form:

$$S_{i,j}(d_1, d_2) = 1/2(\sum_{k=1}^{n} \alpha_k^i s_k(d_1, d_2) + \sum_{k=1}^{n} \alpha_k^j s_k(d_1, d_2)) \qquad (5)$$

where s_k is the similarity computed at k-th local image region, and we use Bhattachayya distance to measure it. α_k^i and α_k^j are the target specific coefficients which are learned using the Adaboost algorithm [18] similar as [12]. The largest affinity score $S_{i,j}(d_1, d_2)$ between randomly sampled detection responses from t_i and t_j is chosen as the appearance affinity $S_{i,j}$ between t_i and t_j.

For a static tracklet, a static appearance model is learned using training samples from the static camera by Adaboost. If the static tracklet has an associated active tracklet, an active appearance model is learned in the similar way using training samples from the active camera.

Finally, the tracklet affinity is computed using Eq. (1), where $\Lambda^A(t_i, t_j) = S_{i,j}$. For tracklets which have active appearance models, the affinity computed from the active appearance model is chosen as $S_{i,j}$, since it is more discriminative than the static appearance model. Given the affinity matrix of different tracklets, the final trajectories are generated by applying the Hungarian algorithm [17].

It should be noted that, different from [12], we learn active appearance models for static tracklets which have associated active tracklets to enhance the multi-target tacking performance.

3 Experiments

3.1 Experiment Setting

To evaluate the effectiveness of our proposed algorithm, we implement it on a real dual-camera system and test the system in a challenging outdoor environment.

(a)   (b)   (c)   (d)

Fig. 3. Examples of the most discriminative features selected by static appearance models and active appearance models for two targets. (a) and (c) is the most discriminative features selected by active appearance models for target a and target b respectively. (b) and (d) are the corresponding discriminative features selected by static appearance models. The color and HOG descriptors are indicated by red and green bounding boxes, respectively.

We use two off-the-shelf AXIS PTZ Network Cameras Q6032-E in our experiments. One is fixed to serve as the static camera to monitor a wide area, and the other is used as the active camera. The typical height of the target is about 50 pixels in the static camera. Due to the zoom ability of the active camera, the height of the target is about $250 \sim 300$ pixels in the active camera. A detector [19] is trained to detect targets in the scene for its efficiency. The online multi-target tracking algorithm is run in both cameras in real time. The thresholds θ_1 and θ_2 of the online multi-target tracking algorithm are set to 0.4 and 0.1 repetively. We collected 10 videos at different time to evaluate our system.

Standard metrics for multi-target tracking are used to evaluate the proposed method: the multiple object tracking precision (MOTP), the multiple object tracking accuracy (MOTA), the number of fragments (FM) and identity switches (IDS). MOTP and MOTA are the higher the better. FM and IDS are the lower the better. We compare our method with two state-of-the-art multi-target trackers [16] and [12]. Both of them only use the images from the static camera. The difference between them is that [12] trains a discriminative appearance model to compute the appearance similarity, while [16] uses the color histogram to compute the appearance similarity. Different from [16] and [12], our method fuses information from both cameras through learning static discriminative appearance models and active discriminative appearance models.

3.2   Results

The most discriminative features selected by static appearance models and active appearance models for two targets are shown in Fig. 3. The HOG features are selected more by active appearance models, and the color features are selected more by static appearance models. This is reasonable, since the resolution of the target images in the static camera is rather low, and there is little useful shape

Table 1. Multi-target tracking results. Comparison of our method with the state-of-the-art methods.

Methods	MOTP[%]	MOTA[%]	FM	IDS
(a) Huang et al. [16]	76.15	74.33	23	20
(b) Kuo et al. [12]	82.23	81.75	18	15
(c) Our method (hist)	83.89	81.44	16	15
(d) Our method	92.88	93.13	5	4

Table 2. Multi-target tracking results. Comparison of different methods using different features.

Methods	MOTP[%]	MOTA[%]	FM	IDS
Kuo et al. [12] (color)	80.23	78.48	19	17
Kuo et al. [12] (hog)	77.66	75.78	21	19
Kuo et al. [12] (both)	82.23	81.75	18	15
Our method (color)	85.93	84.19	11	12
Our method (hog)	89.37	91.55	8	6
Our method (both)	92.88	93.13	5	4

details on the cloth of people. Compared with color information, the shape details are more discriminative when the color of people's cloth is similar. Therefore, when the resolution is high, more shape features are selected.

Table 1 records the multi-target tracking results. In Table 1, (c) is the result of our method without discriminative model learning, using the similarity of color histogram instead. (d) is the result of our method with discriminative model learning. The results shows that discriminative appearance models can improve the tracking performance, and active discriminative models learned from higher resolution images of the active camera can make the multi-target tracking performance even better, reduce the ID switch errors significantly.

Table 2 records the tracking results using different features when learning discriminative appearance models, which shows that hog features are more discriminative when using active discriminative models, and color features are more discriminative when using static discriminative models. This is consistent with the most features selected by them.

Figure 4 shows some sample results of multi-target tracking. The top row, middle row and bottom row show the results of [12,16] and our method respectively. Person #1 is tracked by [12,16] and our system in frame #441 (left column). Then the person are occluded for a long time, as shown in Frame #469 (middle column). The motion similarity is unreliable when long time occlusion happens. If the appearance information is not discriminative enough, ID switches may happen. When the person #1 reappears, ID switch occurs in results of both [16] and [12]. While our method associates it with the correct person. The discriminative models of person #1 are shown in Fig. 3 (a)-(b). The discriminative

Frame 441   Frame 469   Frame 533

Fig. 4. Some tracking results on our collected video. The top row shows the results of [16], and the middle row shows the results of [12]. The results of our method are shown in the bottom row.

models of the person which the ID of person#1 changed to using method [12], are shown in Fig. 3 (c)-(d). The appearance similarity of the two persons computed using the active appearance model is 0.18, while the score is 0.63 using the static appearance model.

4   Conclusion

In this paper, we have proposed a novel framework for multi-target tracking in dual-camera surveillance systems. Information of the heterogeneous cameras is fused into a tracking-by-detection framework to improve the multi-target tracking performance, based on the learned discriminative appearance models in both cameras. To achieve this goal, an efficient online multi-target tracking algorithm is introduced to generate reliable tracklets. A max-entropy target selection strategy is proposed to reduce the ambiguity of multi-target tracking. Experiments in an outdoor scene show the significant improvement produced by our proposed method compared with the state-of-the-art multi-target trackers.

Acknowledgement. This work is supported by National Natural Science Foundation of China (No. 61371192).

References

1. Horaud, R., Knossow, D., Michaelis, M.: Camera cooperation for achieving visual attention. Mach. Vis. Appl. **16**(6), 1–2 (2006)
2. Cui, Z., Li, A., Feng, G., Jiang, K.: Cooperative object tracking using dual-pan-tilt-zoom cameras based on planar ground assumption. In: IET Computer Vision (2014)
3. Chen, C.H., Yao, Y., Page, D., Abidi, B., Koschan, A., Abidi, M.: Heterogeneous fusion of omnidirectional and PTZ cameras for multiple object tracking. IEEE Trans. Circuits Sys. Video Technol. (TCSVT) **18**(8), 1052–1063 (2008)
4. Senior, A.W., Hampapur, A., Lu, M.: Acquiring multi-scale images by pan-tilt-zoom control and automatic multi-camera calibration. In: IEEE Workshops on Application of Computer Vision (WACV), pp. 433–438 (2005)
5. Zhou, X., Collins, R.T., Kanade, T., Metes, P.: A master-slave system to acquire biometric imagery of humans at distance. In: ACM SIGMM International Workshop on Video Surveillance, pp. 113–120 (2003)
6. Alberto, D.B., Dini, F., Grifoni, A., Pernici, F.: Uncalibrated framework for on-line camera cooperation to acquire human head imagery in wide areas. In: IEEE International Conference on Advanced Video and Signal Based Surveillance (AVSS), pp. 252–258 (2008)
7. Bernardin, K., Van, F., Stiefelhagen, R.: Automatic person detection and tracking using fuzzy controlled active cameras. In: IEEE Conference on Computer Vision and Pattern Recognition (CVPR), pp. 1–8 (2007)
8. Cui, Y., Samarasckera, S., Huang, Q., Greiffenhagen, M.: Indoor monitoring via the collaboration between a peripheral sensor and a foveal sensor. In: IEEE Workshop on Visual Surveillance, pp. 2–9 (1998)
9. Yao, Y., Abidi, B., Abidi, M.: Fusion of omnidirectional and PTZ cameras for accurate cooperative tracking. In: IEEE International Conference on Advanced Video and Signal Based Surveillance (AVSS), pp. 46–46 (2006)
10. Wang, X.: Intelligent multi-camera video surveillance: a review. Pattern Recogn. Lett. **34**(1), 3–19 (2013)
11. Kuo, C.H., Huang, C., Nevatia, R.: Multi-target tracking by on-line learned discriminative appearance models. In: IEEE Conference on Computer Vision and Pattern Recognition (CVPR), pp. 685–692 (2010)
12. Kuo, C.H., Nevatia, R.: How does person identity recognition help multi-person tracking? In: IEEE Conference on Computer Vision and Pattern Recognition (CVPR), PP. 1217–1224 (2011)
13. Bae, S.H., Yoon, K.J.: Robust online multi-object tracking based on tracklet confidence and online discriminative appearance learning. In: IEEE Conference on Computer Vision and Pattern Recognition (CVPR), pp. 1218–1225 (2014)
14. Schindler, K.: Continuous energy minimization for multi-target tracking. IEEE Trans. Pattern Anal. Mach. Intell. (TPAMI) **36**(1), 51–65 (2014)
15. Wu, Z., Radke, J.: Keeping a pan-tilt-zoom camera calibrated. IEEE Trans. Pattern Anal. Mach. Intell. (TPAMI) **35**(8), 1994–2007 (2013)
16. Huang, C., Wu, B., Nevatia, R.: Robust object tracking by hierarchical association of detection responses. In: Forsyth, D., Torr, P., Zisserman, A. (eds.) ECCV 2008, Part II. LNCS, vol. 5303, pp. 788–801. Springer, Heidelberg (2008)
17. Munkres, J.: Algorithms for the assignment and transportation problems. J. Soc. Ind. Appl. Math. **5**(1), 32–38 (1957)

18. Schapire, R.E., Singer, Y.: Improved boosting algorithms using confidence-rated predictions. Mach. Learn. **37**(3), 297–336 (1999)
19. Dollár, P., Appel, R., Belongie, S., Perona, P.: Fast feature pyramids for object detection. IEEE Trans. Pattern Anal. Mach. Intell. (TPAMI) **36**(8), 1532–1545 (2014)

Adaptive Multiple Appearances Model Framework for Long-Term Robust Tracking

Shuo Tang , Longfei Zhang$^{(\boxtimes)}$, Jiapeng Chi, Zhufan Wang, and Gangyi Ding

School of Software, Beijing Institute of Technology, Beijing, China
{tang_shuo,longfeizhang,dgy}@bit.edu.cn

Abstract. Tracking an object in long term is still a great challenge in computer vision. Appearance modeling is one of keys to build a good tracker. Much research attention focuses on building an appearance model by employing special features and learning method, especially online learning. However, one model is not enough to describe all historical appearances of the tracking target during a long term tracking task because of view port exchanging, illuminance varying, camera switching, etc. We propose the Adaptive Multiple Appearance Model (AMAM) framework to maintain not one model but appearance model set to solve this problem. Different appearance representations of the tracking target could be employed and grouped unsupervised and modeled by Dirichlet Process Mixture Model (DPMM) automatically. And tracking result can be selected from candidate targets predicted by trackers based on those appearance models by voting and confidence map. Experimental results on multiple public datasets demonstrate the better performance compared with state-of-the-art methods.

Keywords: Dirichlet process mixture model · Appearance model · Object tracking

1 Introduction

Robust object tracking in a long term is a challenging task in computer vision. There have been many trackers proposed by different researchers [1–3] that employ different types of visual information and learned features to build the appearance models as the base of the tracking, e.g. color histogram in Meanshift, multiple features in particle filter [4], Haar-like in MIL [5], etc.

However, it is still not enough to represent the tracking target with one appearance model, even online updating model or patch dictionary model and so on, while the target in internal and external variations. Internal variation includes pose changing, motion, shape deformation, illumination variation, etc. And External variation includes background changing, covered by foreground objects. Tackle this problem, an appearance model set is needed for describing the historical appearances of tracking target.

Therefore, we propose a novel nonparametric statistical method to model the appearance of the target as combination of multiple appearance models. Each

© Springer International Publishing Switzerland 2015
Y.-S. Ho et al. (Eds.): PCM 2015, Part I, LNCS 9314, pp. 160–170, 2015.
DOI: 10.1007/978-3-319-24075-6_16

model describes an typical appearance character under specific situation, and clustered by Dirichlet Process Mixture Model (DPMM) [7] framework dynamically unsupervised.

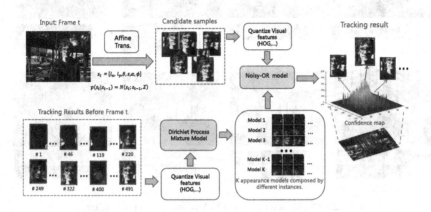

Fig. 1. The framework of the adaptive multiple appearance model tracking.

The Framework of our system is shown in Fig. 1. Experimental results on several public datasets show that AMAM tracking system is applicable to multiple camera system and indoor and outdoor climates tracking system, and outperform several state-of- the-art trackers.

The rest of the paper is organized as follows. Section 2 overviews some of the related works. Section 3 describes the proposed AMAM algorithm. We present experimental validation on several public datasets in Sect. 4 and conclude the paper in Sect. 5.

2   Related Works

In long term tracking task, the biggest challenge is drifting problem. Tackle this problem, appearance models tolerance range need to be enhanced. Ensemble tracking [9] and the Multiple Instance Learning boosting method (MIL) [5] using positive sample and negative samples of tracking targets to train classifiers. Semi-online boosting [10] using both unlabeled and labeled tracking candidate target to train classifiers online. Fragment-based tracking [16] coupled with a voting map can accurately track the partially occluded target. However, historical information is ignored when updating classifiers or models. Dictionary learning [15] was employed to using the linear combination to represent the dynamic appearance of the target and handles the occlusion as a sparse noise component. However, spatial and temporal information are lost when algorithm performing. Appearance representation learned by In our model, we build appearance model set to keep the spatial information of tracking target and tracking system could keep the temporal information as well. at the same time, all efficient

Fig. 2. The procedure of AMAM framework working.

appearance model can be employed in this framework including sparse coding, dictionary learning and learned target descriptions by deep learning or other machine learning methods.

3 The Framework of Adaptive Multiple Appearances Model Tracking

In this section, we describe the common framework of adaptive multiple appearance model. In first, we present the Dirichlet Process Mixture model (DPMM), which are employed to organize the adaptive appearance set. After that, we describe the tracking system based on AMAM framework.

3.1 Dirichlet Process Mixture Model

The Dirichlet process (DP) is parameterized by a base distribution H which has corresponding density $h(\theta)$, and a positive scaling parameter $\alpha > 0$. We denote a DP and suppose we draw a random measure G from a DP, and independently draw N random variables θ_n from G, this can be described as follows:

$$G|\,\{\alpha, H\} \sim DP(\alpha, H) \tag{1}$$

$$\theta_n \sim G, \ n \in \{1, \ldots, N\}$$

As shown by [8], given N independent observations $\theta_i \sim G$, the posterior distribution also follows a DP:

$$p\,(G\,|\,\theta_1, \ldots, \theta_N, \alpha, H) \sim Dir\,(\alpha H\,(A_1) + n_1, \ldots, \alpha H\,(A_r) + n_r)$$

$$\sim DP\left(\alpha + N, \frac{1}{\alpha + N}\left(\alpha H + \sum_{i=1}^{N} \delta_{\theta_i}\right)\right) \tag{2}$$

where $n_1 n_2, ..., n_r$ represent the number of observations falling in each of the partitions $A_1 A_2, ... A_r$ respectively, N is the total number of observations, and δ_{θ_i} represents the delta function at the sample point θ_i.

3.2 Model Inference

Given N observations $X = \{X_i\}_{i=1}^{N} (X_i \in N^d)$, each $X_i = \{x_i\}_{i=1}^{d}$ represents a quantized d-dim HOG feature, and x_i is the histogram quantized bin counts, which is a quantized integer. Let z_i indicate the cluster or appearance model, associated with the ith observation which is represented by quantized HOG feature. As shown in Fig. 1, we would like to infer the number of latent clusters or different appearances underlying those observations, and their parameters θ_k. Since the exact computation of the posterior is infeasible especially when data size is large, we resort to a variant of MCMC algorithms, namely, the collapsed Gibbs sampler [7] for faster approximate inference.

We choose multinomial distribution $F(\theta)$ to describe HOG features of observations, and the cluster prior $H(\lambda)$ is a Dirichlet distribution which is conjugate to $F(\theta)$. Given fixed cluster assignments z_{-i} for other observations, the posterior distribution of z_i factors as follows:

$$p(z_i \mid z_{-i}, X, \alpha, \lambda) \propto p(z_i \mid z_{-i}, \alpha) p(X_i \mid X_{-i}, z, \lambda) \tag{3}$$

The prior $p(z_i \mid z_{-i}, \alpha)$ is given by the Chinese restaurant process (CRP).

$$p(z_i \mid z_{-i}, \alpha) \sim \frac{1}{\alpha + N - 1}\left(\sum_{k=1}^{K} N_k^{-i} \delta(z_i, k) + \alpha \delta(z_i, \bar{k})\right) \tag{4}$$

The $\bar{k}$ denotes one of the infinitely many unoccupied clusters or new appearances. N_k^{-i} is the total number of observations in cluster k except observation i.

For the K clusters to which z_{-i} assigns observations, the likelihood of Eq. (3) is shown as follows:

$$p(X_i \mid z_i = k, z_{-i}, X_{-i}, \lambda) = p(X_i \mid \{X_j \mid z_j = k, j \neq i\}, \lambda) \tag{5}$$

Because dirichlet distribution $H(\lambda)$ is conjugate to multinomial distribution $F(\theta)$, $\theta = (p_1, p_2, ..., p_d)$ and $\{X_i\}_{i=1}^{N} \sim Mult(p1, p2, ..., pd)$, we can get a closed-form of predictive likelihood expression for each cluster or appearance k as follows:

$$p(X_i \mid z_i = k, z_{-i}, X_{-i}, \lambda) = \frac{\Gamma(n+1)}{\prod_{j=1}^{d} \Gamma\left(X_i^{(j)} + 1\right)} \frac{\Gamma\left(\sum_{j=1}^{K} \lambda_j'\right)}{\prod_{j=1}^{K} \Gamma\left(\lambda_j'\right)} \frac{\prod_{j=1}^{K} \Gamma\left(X_i^{(j)} + \lambda_j'\right)}{\Gamma\left(n + \sum_{j=1}^{K} \lambda_j'\right)} \tag{6}$$

Algorithm 1. DPMM algorithm.

Set multiple appearance model assignments $z = z_{t-1}$ of observation at last
iteration $t - 1$. For each $i \in \{1, ..., N\}$ frame targets, resample and rebuild each
appearance model z_i as follows:
repeat

  Step1. Remove a data item X_i from its appearance model k and update its
  model parameter θ_k.

  Step2. For each of the K existing clusters, determine the predictive
  likelihood using Eq. 6, and then determine the likelihood of a potential new
  appearance cluster $\bar{k}$ via Eq. 7.

  Step3.Sample cluster assignment z_i from the $(K + 1)$ - dim multinomial
  Eq. 4.

  Step4. Update appearance model parameter θ_k to reflect the assignment of
  x_i to cluster z_i. If $z_i = \bar{k}$, create a new cluster and increment cluster or
  appearance number K.

  Step5. Repeat the steps shown above until convergence.

until $i = k$;

---

where λ' is the posterior of λ and Γ is the gamma function. Similarly, new
clusters $\bar{k}$ are based upon the predictive likelihood implied by the prior hyper
parameters λ:

$$p\left(X_i \mid z_i = \bar{k}, z_{-i}, X_{-i}, \lambda\right) = p\left(X_i \mid \lambda\right) = \frac{\Gamma(A)}{\Gamma(N + A)} \prod_{k=1}^{K} \frac{\Gamma(n_k + \lambda_k)}{\Gamma(\lambda_k)} \quad (7)$$

where $A = \sum_k \lambda_k$ and $N = \sum_k n_k$, and where n_k = number of x_i's with value k.
Combining these expressions, we employed Gibbs sampler at Algorithm 1.

3.3 AMAM Tracking

Given the observation set of the target$X_{1:t} = [X_1, ..., X_t]$, where each X_t represents a quantized HOG feature, up to time t, the tracking result s_t can be deter-
mined by the Maximum A Posteriori (MAP) estimation, $\hat{s}_t = argmaxp\left(s_t | X_{1:t}\right)$,
where $p\left(s_t | X_{1:t}\right)$ is inferred by the Bayes theorem recursively with

$$p\left(s_t \mid X_{1:t}\right) \propto p\left(X_t \mid s_t\right) \int p\left(s_t \mid s_{t-1}\right) p\left(s_{t-1} \mid X_{1:t-1}\right) ds_{t-1}. \quad (8)$$

Let $s_t = [l_x, l_y, \theta, s, \alpha, \phi]$, where $l_x, l_y, \theta, s, \alpha, \phi$ denote x, y translations,
rotation angle, scale, aspect ratio, and skew respectively.We apply the affine
transformation with those six parameters to model the target motion between
two consecutive frames. The state transition is formulated as $p\left(s_t | s_{t-1}\right) =$
$N\left(s_t; s_{t-1} \sum\right)$, where $\sum$ is the covariance matrix of six affine parameters.

The observation model $p\left(X_t | s_t\right)$ denotes the likelihood of the observation X_t
at state s_t. The Noisy-OR (NOR) model is adopted for doing this:

Algorithm 2. AMAM Tracking Algorithm.

For each $i \in \{1, ..., N\}$, resample z_i as follows:

repeat

    Step1. $L_t(X)$ denote the location of sample X_t at the t-th frame. We have the object location $L_t(X)$ where we assume the corresponding sample is X_t representing the quantized HOG feature.

    Step2. We apply the affine transformation to $L_t(X)$ with six affine parameters to product candidate samples s_{t+1}.

    Step3. For each candidate samples s_{t+1}, we extract quantized HOG feature X_{t+1}, then use NOR model of Eq. 9 and each of the multiple appearance models H^k to compute the likelihood of X_{t+1}.

    Step4. We select the state s_{t+1} which has maximum probability of X_{t+1}

    Step5. Let $X_{1:t+1} = [X_1, ..., X_{t+1}]$ which represents the quantized HOG features set of the target up to time $t + 1$, and then use Algorithm 1 to recreate multiple appearance models.

until $i = N$;

$$p(X_t \mid s_t) = 1 - \prod_k \left(1 - p\left(X_t \mid s_t, H^k\right)\right) \qquad (9)$$

where $H_k, k \in (1, 2, ..., K)$ represents the multiple appearance models learned from Algorithm 1.

The equation above has the desired property that if one of the appearance models has a high probability, the resulting probability will be high as well.

Algorithm 2 illustrates the basic flow of our algorithm.

4 Experiments

In our experiments, we employ 10 challenging public tracking datasets selected from [2] and using the same evaluation methods, the center location error and success rate, to verify the performance of our algorithm. The proposed approach is compared with ten state-of-the-art tracking methods. Table 1 shows all the tracking methods we need to evaluate. In addition, we evaluate the proposed tracker against those methods using the source codes provided by the authors and each tracker is running with adjusted parameters for fair evaluation.

4.1 The AMAM Modeling

Figure 2 shows how the AMAM working. These small face images under the main frame shows the appearance instance belong to each appearance model and the historical instances while tracking. The red rectangle in main frame is the tracking result based on the model in red, and the green one is the ground truth. The instances of each appearance models increasing while long term tracking, and the number of appearance models increasing while the inner and inter distance changing based on the DPMM Algorithm 1.

Table 1. Compare trackers and their representations in our experiment [2]

Trackers	Representation	Trackers	Representation
LOT [17]	L, Color	IVT [18]	H, PCA
ASLA [19]	L, SR, GM	L1ANG [20]	H, SR, GM
MTT [21]	H, SR, SM	VTD [22]	H, SPCA, GM
OAB [23]	H, Haar, DM	MIL [5]	H, Haar, DM
TLD [21]	L, BP, DM	Struck [25]	H, Haar, DM
Ours	H, DPMM		

4.2 Tracking System

Figure 3 shows how the AMAM tracking results based on 2. Bounding boxes in red are our results. We can find that our tracker can track the target very well while most of the other tracker are drifting.

In order to measure the performance of tracking result, we employed two traditional measurement operator. One is the center error (CE), and the other is coverage rate (CR).

The center error is defined as the average Euclidean distance between the center locations of the tracked targets and the manually labeled ground truths, for calculating precision plot. In a general way, the overall performance of one sequence depends on the average center location error over all the frames of one sequence, but when the tracker loses the target, we will only get the random

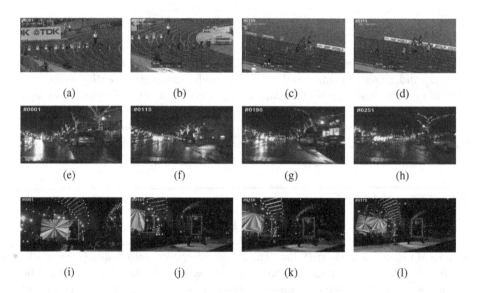

(a)   (b)   (c)   (d)

(e)   (f)   (g)   (h)

(i)   (j)   (k)   (l)

Fig. 3. Three tracking video sequences with all tracking result from the all trackers. The bounding boxes in red are our results.

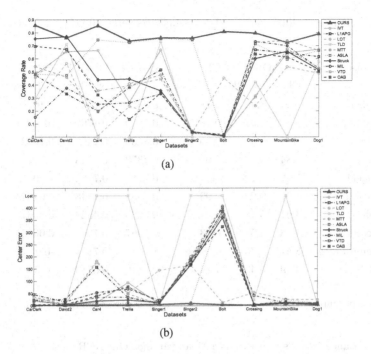

(a)

(b)

Fig. 4. Tracking result compare with the recent appeared 10 trackers listed in Table 1 by coverage rate (a) and center error (b) and measurement operators in public datasets.

output location and the average error value which may not measure the tracking performance correctly [5]. Therefore, we use the precision plot to measure the overall tracking performance. It shows the percentage of frames whose estimated location is within the given threshold distance of the ground truth. Figure 4(b) shows result in our experiment. Since the smaller is better, our AMAM tracker performs good in those public testing videos.

The coverage rate is defined as the bounding box overlapping rate between the tracking target and the ground truth. A higher score means the tracking result is closer to ground truth. The formula of calculating score is $score = \frac{area(ROI_T \cap ROI_G)}{area(ROI_T \cup ROI_G)}$ while the formula of calculating average score is $avrScore = \frac{\sum_1^{frameLength} score}{frameLength}$.

In Table 2, we compare the performance of trackers in each testing dataset with the same testing result shown in Fig. 4. We selected the best performed tracker and the second tracker in each testing data both based on CR and CE operators. We also calculate the differences between them for tracking accuracy measuring. In Table 3, we also import the variation and average CR to measure the robustness and accuracy. Since the inhumations, backgrounds and targets are different at all, if the tracker performing stable with low variation, the tracker can be considered robustness.

Table 2. Compare of all trackers in all datasets by converge rate (CR) and center error (CE).

Dataset	Our CE	Other best CE	CE differences	Our CR	Best CR of others	CR differences
carDark	1.24	3.42	2.18	0.857	0.7549	0.1021
david2	2.89	3.07	0.18	0.7619	0.7708	-0.0089
car4	3.01	6.8	3.79	0.8538	0.7466	0.1072
trellis	3.84	10.86	7.02	0.7372	0.7243	0.0129
singer1	5.22	4.22	-1	0.7617	0.7566	0.0051
singer2	8.97	9.96	0.99	0.7585	0.7448	0.0137
bolt	3.91	13.18	9.27	0.8085	0.4524	0.3561
crossing	1.58	2.51	0.93	0.7967	0.7304	0.0663
mountainBike	8.61	7.75	-0.86	0.7253	0.7391	-0.0138
dog1	3.59	5.48	1.89	0.7916	0.6719	0.1197

Table 3. Compare of trackers by variance and average coverage rate (ACR) in performance.

Our variance	Other's min variance	Mean variance	Our ACR	Best other ACR
0.002031	0.041081246	0.06492	0.78522	0.59107

From the Table 2, we find that our AMAM Tracker is outperform in 8 testing videos. The differences between best and our tracker in CR and CE are less than 1.4 % and 1 pixel in the rest 2 testing videos.

From the Table 3, the variation of our tracker in all videos is 0.002, extremely lower than others both in average and individual.The ACR of our AMAM tracker in all testing videos is 19 % higher than others. That means our tracker can perform more robust and more accurate than others.

5   Conclusion

This paper tackled the drifting problem in tracking and proposed an Adaptive Multiple Appearance Model framework for long term robust tracking. We simply employed HOG to build the basic appearance representation of the tracking target in experiment but all efficient representation of visual objects could be joint in our algorithm framework. Historical appearance descriptions could be employed and grouped unsupervised and modeled by Dirichlet Process Mixture Model automatically. And tracking result can be selected from candidate targets predicted by trackers based on those appearance models by voting and confidence map. Experiment in several public datasets shows that, our tracker has low variation (less than 0.002) and high tracking performance (19 % better than other 10 trackers in average) when compared with the state-of-the-art methods.

Acknowledgments. This material is based upon work supported by the Key Technologies Research and Development Program of China Foundation under Grants No. 2012BAH38F01-5. Any opinions, findings, and conclusions or recommendations expressed in this material are those of the author(s) and do not necessarily reflect the views of the Key Technologies Research and Development Program of China Foundation.

References

1. Li, X., Hu, W., Shen, C., et al.: A survey of appearance models in visual object tracking. ACM Trans. Intell. Syst. Technol. (TIST) **4**(4), 58:1–58:48 (2013)
2. Wu, Y., Lim, J., Yang, M.: Online object tracking: a benchmark. In: Proceedings of the IEEE Conference on Computer Vision and Pattern Recognition (CVPR), pp. 2411–2418 (2013)
3. Smeulders, A., Chu, D., Cucchiara, R.: Visual tracking: an experimental survey. IEEE Trans. Pattern Anal. Mach. Intell. (TPAMI) **36**(7), 1442–1468 (2014)
4. Wang, H., Suter, D., Schindler, K.: Effective appearance model and similarity measure for particle filtering and visual tracking. In: Proceedings of European Conference Computer Vision (ECCV) (2006)
5. Babenko, B., Yang, M., Belongie, S.: Robust object tracking with online multiple instance learning. IEEE Trans. Pattern Anal. Mach. Intell. (TPAMI) **33**(8), 1619–1632 (2011)
6. Wang, N., Yeung, D.: Learning a deep compact image representation for visual tracking. In: Proceedings of the NIPS, (5192), pp. 809–817 (2013)
7. Neal, R.M.: Markov chain sampling methods for Dirichlet process mixture models. J. Comput. Graph. Stat. **9**, 249–265 (2000)
8. Ferguson, T.: A Bayesian analysis of some nonparametric problems. Ann. Stat. **1**(2), 209–230 (1973)
9. Avidan, S.: Ensemble tracking. IEEE Trans. Pattern Anal. Mach. Intell. **29**(2), 261–271 (2007). IEEE society
10. Grabner, H., Bischof, H.: On-line boosting and vision. In: Proceedings of Computer Vision and Pattern Recognition, **1**, pp. 260–267 (2006)
11. Stenger, B., Woodley, T., Cipolla, R.: Learning to track with multiple observers. In: Proceedings of Computer Vision and Pattern Recognition, pp. 2647–2654 (2009)
12. Yu, Q., Dinh, T.B., Medioni, G.: Online tracking and reacquisition using co-trained generative and discriminative trackers. In: ECCV (2008)
13. Gao, Y., Ji, R., Zhang, L., Hauptmann, A.: Symbiotic tracker ensemble toward a unified tracking framework. IEEE Trans. Circuits Syst. Video Technol. (TCSVT) **24**(7), 1122–1131 (2014)
14. Zhang, L., Gao, Y., Hauptmann, A., Ji, R., Ding, G., Super, B.: Symbiotic black-box tracker. In: Proceedings of the Advances on Multimedia modeling (MMM), pp. 126–137 (2012)
15. Wang, N., Wang, J., Yeung, D.: Online robust non-negative dictionary learning for visual tracking. In: Proceedings of the IEEE International Conference on Computer Vision (ICCV 2013), pp. 657–664 (2013)
16. Adam, A., Rivlin, E., Shimshoni, I.: Robust fragments-based tracking using the integral histogram. In: Proceedings of the CVPR (2006)
17. Oron, S., Bar-Hillel, A., Levi, D., Avidan, S.: Locally orderless tracking. In: CVPR (2012)

18. Ross, D., Lim, J., Lin, R., Yang, M.: Incremental learning for robust visual tracking. IJCV **77**(1), 125–141 (2008)
19. Jia, X., Lu, H., Yang, M.: Visual tracking via adaptive structural local sparse appearance model. In: CVPR (2012)
20. Bao, C., Wu, Y., Ling, H., Ji, H.: Real time robust L1 tracker using accelerated proximal gradient approach. In: CVPR (2012)
21. Zhang, T., Ghanem, B., Liu, S., Ahuja, N.: Robust visual tracking via multi-task sparse learning. In: CVPR (2012)
22. Kwon, J., Lee, K.M.: Visual tracking decomposition. In: CVPR (2010)
23. Grabner, H., Grabner, M., Bischof, H.: Real-time tracking via online boosting. In: BMVC (2006)
24. Kalal, Z., Matas, J., Mikolajczyk, K.: P-N learning: bootstrapping binary classifiers by structural constraints. In: CVPR (2010)
25. Hare, S., Saffari, A., Torr, P.H.S.: Struck: structured output tracking with kernels. In: ICCV(2011)

On-line Sample Generation for In-air Written Chinese Character Recognition Based on Leap Motion Controller

Ning Xu[✉], Weiqiang Wang, and Xiwen Qu

School of Computer and Control Engineering, University of Chinese Academy of Sciences,
Beijing, China
{xuning12,wqwang,queiwen13b}@ucas.ac.cn

Abstract. As intelligent devices and human-computer interaction ways become diverse, the in-air writing is becoming popular as a very natural interaction way. Compared with online handwritten Chinese character recognition (OHCCR) based on touch screen or writing board, the research of in-air handwritten Chinese character recognition (IAHCCR) is still in the start-up phase. In this paper, we present an on-line sample generation method to enlarge the number of training instances in an automatic synthesis way. In our system, the in-air writing trajectory of fingertip is first captured by a Leap Motion Controller. Then corner points are detected. Finally, the corner points as well as the sampling points between corner points are distorted to generate artificial patterns. Compared with the previous sample generation methods, the proposed method focuses on distorting the inner structure of character patterns. We evaluate the proposed method on our in-air handwritten Chinese character dataset IAHCC-UCAS2014 which covers 3755 classes of Chinese characters. The experimental results demonstrate that proposed approach achieves higher recognition accuracies and lower computational cost.

Keywords: In-air handwritten character recognition · Sample generation · Sample distortion · Leap motion controller

1 Introduction

Online handwritten Chinese character recognition (OHCCR) has been investigated extensively during the last several decades [1]. So far, the researchers have overcome many challenges in OHCCR and satisfying experimental results have been obtained on the existing datasets [2, 3]. In recent years, with the advancement of human computer interaction (HCI) devices, the writing interaction is no longer restricted to the plane-based way, e.g., touch screen, writing board. Several writing-in-the-air systems [4, 5] have been developed, and it seems that the 3D in-air writing interaction system is coming as a very natural Human-computer interaction way.

For now, the works of IAHCCR are still very few. Feng *et al.* [4] proposed a finger-writing character recognition system using Kinect. By using the depth information and clustering algorithms, the fingertip is located and then its trajectory is captured. They have gained high tracking accuracy in a dataset including digit characters and a limited

© Springer International Publishing Switzerland 2015
Y.-S. Ho et al. (Eds.): PCM 2015, Part I, LNCS 9314, pp. 171–180, 2015.
DOI: 10.1007/978-3-319-24075-6_17

number of Chinese characters. Jin *et al.* [5] proposed a digit string recognition method using Kinect to capture fingertip trajectory method. Digit strings with length of 2–6 are over-segmented and recognized by a path-searching method. Vikram *et al.* [9] applied the Leap Motion Controller to handwriting recognition problem and exploited the dynamic time warping (DTW) algorithm to search the nearest matches for English characters and a limited number of words.

Compared with traditional OHCCR, the in-air handwritten Chinese character recognition is technically more difficult due to two aspects of reasons. First, when a user is writing with his fingertip in the air, the motion of his fingertip is uncontrolled, since it is very difficult for the user to guarantee the validity of relative spatial location among strokes. Thus, the casual writing behavior generally results in the great variation of character structure. Second, there is no pen-up or pen-down information, since each written character always consists of one single stroke, since all the strokes are connected together when writing a Chinese character.

Some examples from handwritten dataset SCUT-COUCH2009 [2] and some from in-air handwritten characters of our dataset IAHCC-UCAS2014 are shown in Fig. 1. The difficulty of the IAHCCR can be seen from it owing to the great variation in the structure of characters.

Fig. 1. Examples of handwritten and in-air hand written data for Chinese "Shi"

As a new robust interaction device, the Leap Motion Controller has been applied into several fancy fields [12–14]. Since the Leap Motion Controller can provide accurate fingertip position in 3D space and its projection on a screen, we apply it to capture the writing trajectory of fingertip.

Since the ICHCCR is challenging when compared with OHCCR, a more robust classifier is needed. The Chinese language contains thousands of characters, so millions of training instances or even more are needed to construct a high-performance recognizer. Apparently, it is a tedious and heavy work to collect enough training instances for thousands of Chinese characters. In the OHCCR, the sample generation methods have been proved to be an efficient and effective way to automatically collect a large volume of training data to construct the recognizer with high accuracy. Leung *et al.* [6] proposed a sample-generation method for off-line handwritten Chinese character recognition. Chen *et al.* [7] apply this method to OHCCR and combine their sample-generation method with Leung's. These two methods generate the samples by distorting off-line patterns, which means they distort the binary images of the samples and neglect the writing sequence of the samples.

This paper presents a sample generation method for handwritten Chinese character samples. Our observation shows that the handwriting differences within a character class result from the inner structure variation. This variation can be reflected in the positions of corner points. Therefore, if we want to mimic a character written by a writer, we can distort the corner points of the same character written by another writer to get a rather approximate match. Under such assumption, distorting the corner points of the original character is parallel to mimicking the writing behavior of the same character by other writers. Accordingly, we propose a sample generation method by changing the positions of corner points of original pattern.

The rest of this paper is organized as follows. First, we introduce how the in-air writing trajectory is captured using our writing-in-the-air system. Then, we present the proposed sample generation approach in Sect. 3. Finally, the evaluation experiments and results are reported in Sect. 4. Section 5 concludes the paper.

2   Writing Trajectory Capturing

In our writing-in-the-air system [21], users conduct their writing behavior in a customized 3D space to write a Chinese character. During the writing process, user can move his writing fingertip above the Leap Motion Controller and the 3D moving trajectory of the fingertip can be captured using APIs of the controller. By projecting the 3D trajectory to the screen plane, the 2D writing trajectory can be obtained and as the input of IAHCCR. A demo of our system can be seen from Fig. 2, where a user is writing the Chinese character of "Shi" in the 3D space.

Fig. 2. A demo of our writing-in-the-air system.

In practice, the stability and accuracy of writing trajectory can be guaranteed by tuning the parameters of the controller. To the distortion of character structure caused by slight shake during the in-air writing process, we smooth 2D tracking trajectory by using the Kalman Filter. Compared with using the Kinect sensor, writing with our writing-in-the-air system can be more natural. Owing to the high accuracy and real-time performance of the Leap Motion Controller, the writing process can be relatively fast and the writing can be done only by moving the writing finger with little body movement and.

3   Proposed Method

In this section, we first introduce Leung's [6] distortion method to get an overview of previous off-line sample generation methods (Subsect. 3.1). Then, an on-line sample generation method is presented in details (Subsect. 3.2).

3.1   Off-line Sample Generation

The off-line sample generation method distorts a pattern on a given binary image, e.g., a 64×64 binary image, by a coordinate mapping function. The mapping function defines a new location for each foreground pixel in the input pattern. Concretely, let $f(x, y)$ denotes a given handwritten pattern, $u = h_x(x, y)$ and $v = h_y(x, y)$ denote the mapping function. In the Leung's method, the mapping function is defined as

$$u = h_x(x, y) = w_n\left(a_1, b_1(x)\right) + k_1 y + c_1,$$
$$v = h_y(x, y) = w_n\left(a_2, b_2(y)\right) + k_2 x + c_2,$$
(1)

where w_n is the warping function, a_1 and a_2 are distortion parameters of the warping function, b_1 and b_2 are used to normalize x and y to the interval $[0, 1]$, k_1 and k_2 are the shearing slopes, c_1 and c_2 are the constants to align the centroid position of the distorted pattern back to the original position. The warping function has two implementations to produce more kinds of variations.

$$w_1(a, t) = \frac{1 - e^{-at}}{1 - e^{-a}}, \quad w_2(a, t) = \begin{cases} 0.5w_1(a, t), 0 \leq t \leq 0.5 \\ 0.5 + 0.5w_1(-a, 2(t - 0.5)), 0.5 < t \leq 1 \end{cases}$$
(2)

In practice, the choice between w_1 and w_2 is selected randomly with a certain pre-defined probability. The distorted patterns are generated by a set of different parameters. Concretely, the warping function is first selected, and then the parameters a_1, a_2, k_1, k_2 are randomly sampled in a constrained parameter space. Some examples of the off-line sample generation are illustrated in Fig. 3. The left-most image is the original binary pattern, and the rest four images are the instances generated by the Leung's off-line sample generation method [6].

Fig. 3.   Instances generated by Leung's algorithm for Chinese character "De".

3.2   On-line Sample Generation

Usually, the on-line handwritten Chinese character pattern consists of many time-ordered sampling points, and these sampling points form the writing trajectory. The proposed method intends to change the positions of these sampling points to generate more synthetic patterns.

The proposed on-line sample generation method utilizes the new mapping function to map each sampling point to its new position in the distorted pattern. Since the proposed generation method is based on corner points of the original pattern, we first introduce the detection of corner points, and then present the details of the distortion computation on these sampling points.

Corner Points Detection. In our method, corner points refer to those sampling points which have local maximal curvature in the writing trajectory. The detection of corner points from a given pattern image is important, since it directly decides the effectiveness of the distortion. We exploit Wu's method [8] to locate the corner points from time-ordered sampling points.

The potential corner points among sampling points are first located by calculating the direction change corresponding to each sampling points. Since the trajectory curve can be represented by a sequence of Freeman's chain codes [15], which uses an integer varying from 0 to 7 to denote eight directions respectively. If the current sampling points lie in the same straight stroke as its preceding sampling points, they have the same chain code, and otherwise their chain codes will be different. By tracking the chain code changes, the points with no direction change can be removed, and the remaining corner points are chosen as potential corner points for further verification.

Then we determine the region of support for each potential corner point. The region of support is defined as the neighborhood region to calculate the curvature. A large region of support will smooth out fine corner points and a small one will generate too many dummy corner points. Here, we determine the region of support adaptively based on the bending value. Let S_i denote the ith sampling point of the original trajectory with coordinate (x_i, y_i) and k_i be the length of region of support for S_i. For a given sampling point S_i, its corresponding k is computed according to the following procedure. Starting with $k = 1$, we constantly compute the bending value $b_{ik}, k = 1, 2, 3, \dots$ by

$$b_{i,k} = max\left(\left|(x_{i-k} - x_i) + (x_{i+k} - x_i)\right|, \left|(y_{i-k} - y_i) + (y_{i+k} - y_i)\right|\right) \qquad (3)$$

until $b_{i,k} < b_{i,k+1}$. Then $k_i = k$.

For a given potential corner point S_i, its region of support is defined as $R_i = \{S_{i-k_i}, \dots, S_{i+k_i}\}$.

Further, the curvature ρ_i of candidate corner point S_i is evaluated by the average bending value of its region of support, i.e., $\rho_i = \frac{1}{k_i} \sum_{j=1}^{k_i} b_{i,j}$.

Finally, a candidate corner point S_i will be regarded as a non-corner point and removed from the candidate list, if it does not meet one of the following four conditions:

1. $\rho_i < \varepsilon$
2. $\rho_i < \rho_{i-1}$ or $\rho_i < \rho_{i+1}$
3. $\rho_i = \rho_{i-1}$ and $k_i < k_{i-1}$
4. $\rho_i = \rho_{i+1}$ and $k_i \leq k_{i+1}$

Through the above computation, all the corner points are identified and located. Figure 4 shows some examples of corner point detection, where the red circles are used to label the locations of corner points detected. It is very significant to robustly detect corner points for the subsequent distortion computation.

Fig. 4. Examples of results of corner point detection.

Sampling Points Distortion. In the proposed distortion algorithm, corner points are first distorted, and then the remaining sampling points are processed. It should be noted that the first and the last sampling points are always taken as corner points for distortion processing. For all corner points, we distort them by the mapping function

$$u_i = x_i + w \cdot t_1,$$
$$v_i = y_i + w \cdot t_2 \tag{4}$$

Where t_1 and t_2 are distortion parameters varying in the interval $[-1, 1]$, and w is defined as the distortion scale which is determined by experiments. By selecting appropriate parameters, the character structure of the original pattern can be well maintained, and the writing differences at the same time can also be reflected.

For each non-corner point, its position is aligned by maintaining its relative position with respect to the closest corner points before and after it. Concretely, for a given non-corner point S_k with coordinate (x_k, y_k), if the coordinates of corner points before and after it, S_i and S_j are (x_i, y_i) and (x_j, y_j), respectively, the corresponding mapping function for S_k is defined as

$$u_k = x_i + \left(x_k - x_i\right) \frac{d(i,k)}{d(i,j)},$$
$$v_k = y_i + (y_k - y_i) \frac{d(i,k)}{d(i,j)} \tag{5}$$

Where $d(i,j)$ denotes the Euclidean distance between point S_i and point S_j.

By randomly sampling different values for parameters t_1 and t_2, the corner points can be mapped to different positions, and correspondingly, we can obtain different instance patterns. Figure 5 shows an example of instances generated by the proposed algorithm, where the binary images are obtained by simply connecting neighboring sampling points into straight lines. The leftmost image is the original binary pattern, and the remaining four images are the generated artificial patterns, where red points labeled in each image are the detected corner points.

Fig. 5. Instances generated by the proposed algorithm for Chinese character "De".

4   Experimental Results

We evaluate the performance of the proposed algorithm on our dataset constructed using our writing-in-the-air system. Our dataset contains 3755 classes, including all classes in the GB2312-80 level-1 set for Chinese characters. In our dataset, each class includes 65 samples and we randomly select 10 samples from them as the testing samples and the rest are used for training.

In the experiments, we adopt feature extraction and classification model in [2], since OHCCR and IAHCCR share a lot of similarities. Concretely, we apply a series of pre-processing steps including the dot density shape normalization method [19] to normalize the input patterns. Then we extract 8-directional feature [16] from each pattern, and adopt a three-level classifier to recognize the input pattern. For the first two levels, the features are obtained by reducing the 512-dimensional feature vector to 20 dimensional and 160 dimensional using Linear Discriminant Analysis (LDA). By the first two-level classifiers, the top 50 candidates are determined. At the third level, the Modified Quadratic Discriminant Function (MQDF) classifier [18] is applied to determine the top 1 and top 5 candidates for each pattern, using the 160-dimensional feature generated by LDA.

Figure 6 shows some examples of the instances generated by Leung's algorithm [6] as well as ours. The original patterns are listed at the first row, and the synthesized instances are shown below. The instances generated by Leung's algorithm are listed at the left column beneath each original pattern, and the results generated by ours are listed at the right column. As shown in Fig. 6 although both methods can relatively change the structure of the original pattern, but the instances generated by ours seems more natural, i.e., they have higher similarity with the patterns written by real writers, since the distortion computation only results in the variation of character skeleton without changing the basic character structure. These examples justify the rationality of the proposed method to some extent.

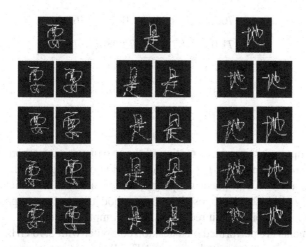

Fig. 6. Comparison of instances generated by Leung's method and ours.

It should be noted that the proposed method can also generate some over-distorted instances, and some of them cannot well conform to character structure of original patterns to some extent. However, in IAHCCR, the in-air written Chinese characters themselves are highly distorted, so those over-distorted instances can also conform to reality and can bring the diversity of instances.

Further, we quantitatively compare the performance of Leung's method and ours based on the classification accuracy of in-air Chinese character recognizers trained on the instances generated by them respectively.

In the experiments, we uniformly sample distorting parameters t_1 and t_2 from interval $[-1, 1]$, and the value of the distortion scale w is empirically set to w = 3 to generate the instances based on the proposed algorithm. We generate N_g new distorted instances for each instance based on our method and compare the recognition results with Leung's method. The N_g is chosen as 0, 19, 39, 59 and 79. We compare the accuracy of top 1 and top 5, which are metrics widely used to compare Chinese character recognition performance. The recognition results are summarized in Table 1. In the experiments, the proposed algorithm obtains much better performance compared with Leung's method on top 1 and top 5 metrics. The experimental results show that the validity of the proposed algorithm. It can also be seen that although the framework of recognition system in [2] has obtained high recognition accuracy on the OHCCR dataset, it achieves rather low accuracy on our IAHCCR dataset. This demonstrates that the IAHCCR is very challenging and needs much research efforts in the future.

Table 1. Comparison of recognition accuracy of Leung's algorithm vs. ours on our in-air written Chinese character dataset

N_g	Leung		Ours	
	Top1	Top5	Top1	Top5
0	70.52	86.00	70.52	86.00
19	71.07	87.22	72.28	87.47
39	71.09	87.27	72.31	87.58
59	71.10	87.29	72.36	87.60
79	71.10	87.30	72.37	87.60

We also compare the time consumption of the two distortion method. As many off-line sample generation methods, the mapping function of Leung's method is also relatively complex, and two non-linear warping functions are used to achieve more kinds of distortions. By contrast, the proposed on-line method just need to compute the mapped position for sampling points, so it reduces a lot of computation cost.

The experiments are performed on a desktop computer with 2.40 GHz CPU, and the recognition system is implemented using MATLAB. In our experiments, the Leung's method is implemented by a fast self-generation framework [20]: eight pattern images are calculated from the original sample and then two nonlinear mapping functions generated by a set of parameters are used to distort the eight pattern images. In our experiments, we compare the average time cost of the two methods when $N_g = 19$ and the experimental results are shown in Table 2. Compared with Leung's method, the proposed method needs lower time consumption during the sample-generation period.

Table 2. Comparison of time cost of Leung's algorithm vs. ours on our dataset

	Leung	Ours
Time consumption (s)	59259.5	27667.7

5 Conclusion

In this paper, we present an on-line sample generation method for in-air handwritten Chinese character recognition. First, we introduce a challenging problem, i.e., in-air handwritten Chinese character recognition (IAHCCR). In the proposed instance generation method, one instance can be distorted to generate dozens of artificial instances, which is very helpful to train robust classifiers. The proposed instance generation method distorts the inner structure of the original instance using detected corner points. The experimental results show that the proposed approach achieves higher recognition accuracy and lower computation cost compared with Leung's algorithm.

Acknowledgments. This work is supported by the National Science Foundation of China (NSFC) under Grant No. 61232013, No. 61271434 and No. 61175115.

References

1. Liu, C., Jaeger, S., Nakagawa, M.: Online recognition of Chinese characters: the state-of-the-art. IEEE Trans. Pattern Anal. Mach. Intell. **26**(2), 198–213 (2004)
2. Jin, L., Gao, Y., Liu, G., Li, Y., Ding, K.: SCUT–COUCH2009–a comprehensive online unconstrained Chinese handwriting database and benchmark evaluation. Int. J. Doc. Anal. Recogn. (IJDAR) **14**(1), 53–64 (2011)
3. Liu, C., Yin, F., Wang, D., Wang, Q.: Online and offline handwritten Chinese character recognition: benchmarking on new databases. Pattern Recogn. **46**(1), 155–162 (2013)
4. Feng, Z., Xu, S., Jin, L., Ye, Z., Yang, W.: Real-time fingertip tracking and detection using Kinect depth sensor for a new writing-in-the air system. In: Proceedings of the 4th International Conference on Internet Multimedia Computing and Service, pp. 70–74 (2012)
5. Jin, X., Wang, Q., Liu, C.: Visual gesture character string recognition by classification-based segmentation with stroke deletion. In: Proceedings of the 2nd IAPR Asian Conference on Pattern Recognition (ACPR), pp. 120–124 (2013)
6. Leung, K.C., Leung, C.H.: Recognition of handwritten Chinese characters by combining regularization, fisher's discriminant and distorted sample generation. In: Proceedings of the 10th International Conference on Document Analysis and Recognition, pp. 1026–1030 (2009)
7. Chen, B., Zhu, B., Nakagawa, M.: Training of an on-line handwritten Japanese character recognizer by artificial patterns. Pattern Recogn. Lett. **35**, 178–185 (2014)

8. Wu, W.Y.: Dominant point detection using adaptive bending value. Image Vis. Comput. **21**(6), 517–525 (2003)

9. Vikram, S., Li, L., Russell, S.J.: Writing and sketching in the air, recognizing and controlling on the fly. In: CHI 2013 Extended Abstracts on Human Factors in Computing Systems, pp. 1179–1184 (2013)

10. Hodson, H.: Leap motion hacks show potential of new gesture tech. New Sci. **218**(2911), 21 (2013)

11. Weichert, F., Bachmann, D., Rudak, B., Fisseler, D.: Analysis of the accuracy and robustness of the leap motion controller. Sensors **13**(5), 6380–6393 (2013)

12. Sutton, J.: Air painting with corel painter freestyle and the leap motion controller: a revolutionary new way to paint. In: ACM SIGGRAPH 2013 Studio Talks, no. 21 (2013)

13. Hantrakul, L., Kaczmarek, K.: Implementations of the leap motion device in sound synthesis and interactive live performance. In: Proceedings of the 2014 International Workshop on Movement and Computing, pp. 142 (2014)

14. Khademi, M., Hondori, H.M., McKenzie, A.: Free-hand interaction with leap motion controller for stroke rehabilitation. In: CHI 2014 Extended Abstracts on Human Factors in Computing Systems, pp. 1663–1668 (2014)

15. Freeman, H.: On the encoding of arbitrary geometric configurations. IRE Trans. Electron. Comput. **EC-10**(2), 260–268 (1961)

16. Bai, Z.-L., Huo, Q.: A study on the use of 8-directional features for online handwritten Chinese character recognition. In: Proceedings of 8th International Conference on Document Analysis and Recognition, pp. 262–266 (2005)

17. Kimura, F., Takashina, K., Tsuruoka, S., Miyake, Y.: Modified quadratic discriminant functions and the application to Chinese character recognition. IEEE Trans. Pattern Anal. Mach. Intell. **PAMI-9**(1), 149–153 (1987)

18. Long, T., Jin, L.: Building compact MQDF classifier for large character set recognition by subspace distribution sharing. Pattern Recogn. **41**(9), 2916–2925 (2008)

19. Bai, Z.L., Huo, Q.: A study of nonlinear shape normalization for online handwritten chinese character recognition: dot density vs. line density equalization. In: The 18th International Conference on Pattern Recognition, vol. 2, pp. 921–924 (2006)

20. Shao, Y., Wang, C., Xiao, B.: Fast self-generation voting for handwritten Chinese character recognition. Int. J. Doc. Anal. Recogn. (IJDAR) **16**(4), 413–424 (2013)

21. Xu, N., Wang, W., Qu, X.: Recognition of in-air handwritten Chinese character based on leap motion controller. In: Proceedings of the 8th International Conference on Image and Graphics, 13–16 August 2015

Progressive Image Segmentation Using Online Learning

Jiagao Hu, Zhengxing Sun[✉], Kewei Yang, and Yiwen Chen

State Key Laboratory for Novel Software Technology, Nanjing University,
Nanjing, People's Republic of China
`szx@nju.edu.cn`

Abstract. This article proposed a progressive image segmentation, which allow users to segment images according to their preferences without any boring pre-labeling or training stages. We use an online learning method to train/update the segmentation model progressively. User can scribble on the image to label initial samples or correct the false-labeled regions of the result. To efficiently integrate the interaction with the learning and updating process, a three-level representation of images is built. The proposed method has three advantages. Firstly, the segmentation model can be learned online along with user's manipulation without any pre-labeling. Secondly, the diversity of segmentation accord with user's preferences can be met flexibly, and the more use the more accurate the segmentation could be. Finally, the segmentation model can be updated online to meet the needs of users. The experimental results demonstrate these advantages.

Keywords: Progressive segmentation · Image processing · Online learning

1 Introduction

Images segmentation is a fundamental problem in computer vision. Most of existing methods focus on training a classifier by a large number of labelled samples based on supervised or semi-supervised learning [1–3]. A newly tendency attempt to co-segment a set of related images based on unsupervised learning [4–6], which can explore the coherence of their appearance and segment them simultaneously. As the training dataset or the coherent images utterly restrict the extent of segmentation in the fixed scale and granularity, there becomes necessary for users to correct the results interactively to fit their needs of the diversity and variety of the different applications. Accordingly, it remains a challenge to make image segmentation flexible and susceptible for user with their own efforts and personalized preferences to accommodate dynamic variation of requirements, as it is usually a context-aware, application-related and task-driven effort.

From our knowledge, there should be at least three aspects of demands for a flexible user-centered image segmentation tool. **Firstly, user could train**

© Springer International Publishing Switzerland 2015
Y.-S. Ho et al. (Eds.): PCM 2015, Part I, LNCS 9314, pp. 181–191, 2015.
DOI: 10.1007/978-3-319-24075-6_18

the segmentation model online along with their manipulation without any pre-labelled samples or pre-trained classifiers. Intuitively, the co-segmentation [4] is one of the priority choices, as it works in a batch process without any pre-labeling or pre-training. Unfortunately, they also limit user's intervention for segmentation criterion. On the other extreme, the supervised methods [1,2] train the classifier heavily dependent on user's labelled samples. But it is a trouble and time-consuming burden for users to pre-label a large amount of images. Accordingly, it would be a best option to allow user doing segmentation while labelling images according to their practical requirements. The problem is *how to introduce online training into image segmentation to integrate segmentation learning with user interaction adaptively.* **Secondly, user could enable and enrich the intelligent image segmentation progressively and accumulatively on their own initiatives.** Almost all of the existing learning-based image segmentation [1,2] are offline trained in a batch process. It seems impossible for any user to determine at once the comprehensive category and context of images what would be faced. A more practical solution is to let them make decisions progressively. Moreover, there is always a hope for users to take the corrected results back to training accumulatively. That is to say, segmentation tool must not only link training process with its working process in a loop, but do segmentation training progressively and accumulatively accompanied by user's correction preferences, as done in 3D shape segmentation [7]. However, *how to make image segmentation in such a loop manner effectively remains unopened.* **Thirdly, user could update the segmentation model dynamically and incrementally to fit their diversity of applications.** The incremental learning and updating strategies have been widely used, such as image classification [8] and object tracking [9]. Most of the existing strategies focus on learning a more adaptive classifier with the offline trained classes from newly added samples. As the training samples are added online and the categories are extending dynamically and randomly in user-centered manner, some additional requirements arises. For example, when segmenting an object never appeared before, users may extend it as a new rather than the most similar one from exist classes. Especially, the online increment and update of segmentation must be integrated with the classifier training progressively and accumulatively. That is, it is necessary for user-centered image segmentation to *design a novel incremental and updating mechanism to avoid re-training on all images.*

This paper introduces an image segmentation method for user-centered application. A progressive segmentation framework is established based on the classifier named Online Multi-Class LPBoost (OMCLP) [10], which uses several Online Random Forests(ORF) [11] as the weak learners and can be online trained and updated. User intervention can be interposed to label initial samples or correct the false-labeled regions. The advantages of our methods are threefold: Firstly, the segmentation model can be learned online along with user's manipulation without any pre-labeling. Secondly, the diversity of segmentation accord with user's preferences can be flexible and the more use the more accurate the segmentation could be. Finally, the segmentation model can be updated online to handle the dynamic class to meet the needs of users.

2   Overview of Progressive Segmentation Method

Our method starts with a multi-level analysis of the image. The next step is an online iterative stage for progressive segmentation. The segmentation process is composed of two steps: segmentation prediction and segmentation optimization. The segmentation prediction model can learn the appearance of object categories progressively by online learning. The segmentation optimization process is used to get an efficient inference and smooth boundaries of segmented regions by taking into account local appearance signals. After user's interactions, we utilize a label augmenting process to make this interaction more efficient.

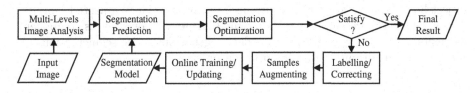

Fig. 1. The pipeline of proposed method

Figure 1 illustrates the pipeline of our method. At the very beginning, the segmentation model is empty, so there is no segmentation prediction and optimization process for the first input image. The user need firstly to scribble some labels to initially train the segmentation model. As soon as the model has been initialized, it can segment the image to get an initial result. If the user was not satisfied with the result, he can scribble to correct some false-labeled regions, and then the model would be updated to re-segment the image. Otherwise, the system would output it as a final result.

After the first image has been segmented, the model can be used to segment subsequent images directly without any scribbles. The segmentation model can be formed and become more and more accurate gradually along with the increase of segmented images. Figure 4 demonstrates a whole segmentation flow of images.

3   Multi-level Image Representation

To integrate the interaction with the training/updating process, a multi-level representation is built based on the pixel-level original image. The progressive segmentation should be based on online learning. But it's too time expensive to train the model directly in pixel-level because an image may have too many pixels. Accordingly, a fine-grained abstraction of the image must be considered to reduce the training time of the segmentation model. On the other hand, a user-centered method should make the user intervention as efficient as possible. In our implementation, a coarse-grained segmentation is used to maximum utilize the information of user's interaction. Thus, we build a three levels representation of the image: a pixel-level representation, a fine-grained level representation and a coarse-grained level representation, as shown in Fig. 2.

Three-Level Image Analysis. The fine-grained representation is a finer scale segmentation of the image, and we call it superpixel here, as shown in Fig. 2(b). Similar with many other works [2,6,12], we use a superpixel, and regard the superpixels as samples instead of pixels in the segmentation prediction. In our implementation, we use the SLIC [13] which has superior performance and efficiency. The coarse-grained one is a coarse over-segmentation of the image, and we call it over-segmentation here, as shown in Fig. 2(c). In order to makes the interaction more efficient, we use the over-segmentation to augment labelled samples. We choose the graph-based segmentation [14] which adheres well to object boundaries to avoid the occurrence of error labels in augmented samples. The pixel-level representation is the original image with pixel as its component elements, as shown in Fig. 2(a). In order to get a pixel-wise segmentation, we use pixel as the primitives in the optimization step.

(a) pixel-level   (b) superpixel   (c) over-segme-   (a) labelled 145 (b) augment to
                                     ntation   superpixels   689 superpixels

Fig. 2. Three levels representation of an image   **Fig. 3.** Sample augmenting

Samples Augmenting. In our system, user can scribble on images to label different objects. User's scribble can be used as initial labelling and correcting. For the first image, user scribbles to label some seed points to initialize the segmentation model. If the user is not satisfied with the segmentation, he can scribble on some false-labeled regions to correct them. Then the model will be updated. To make the interaction more efficient, we design a label augmenting strategy to expand the number of labeled samples. After user's scribble, the system can find out the over-segmentation regions those contain labeled superpixels, and mark un-labeled ones as the same with labeled ones in the same region. Figure 3 shows the augmented superpixels from user's scribbles.

Feature Descriptors. The OMCLP is performed based on the feature descriptors of images. We use a 22-dimensional description vector incorporating texture, color and location. According to the study on feature selections in the supervised approach [1], we use the textons (17-dimensional) as texture description. In addition, we use color information (3-dimensional) and pixel location in the image (2-dimensional). The mean values of all pixels in a superpixel is used to form the feature vector.

4   Online Segmentation

Our segmentation model is composed of two steps: segmentation prediction and segmentation optimization. The segmentation prediction model is used to learn the appearance of object class progressively by online learning/updating. And the segmentation optimization helps to get an efficient inference and smooth boundaries of segmented regions.

Segmentation Prediction. We use online learning method to learn the appearance of object class according to user's interaction to make it capable to segment the following submitted images correctly accord with user's intention.

The OMCLP [10], which uses several ORF [11] as weak learners, is adopted as the classifier to be online trained and incremental updated in the progressive segmentation. The labeled superpixels by user are regarded as samples to sequentially train/update the model. Using OMCLP, we can obtain the confidence of the ith superpixel labeled as the kth object class. The prediction model can be defined as

$$f(x_i, k) = \sum_{m=1}^{M} w_m \cdot g_m(k|x_i) \tag{1}$$

where $f(x_i, k)$ represents the confidence of the ith superpixel labelled as the kth object class, $g_m(k|x_i)$ is the probability of mth ORF, M is the number of ORFs, and w_m is the current weight of mth ORF.

Random forest is the ensemble of random decision trees. To achieve the goal of online learning, Saffari et al. [11] combine the online bagging and online decision trees with random feature selection to build the ORF. After training the ORF, the probability of superpixel i labeled as the kth object class in this ORF can be calculated as $g(k|x_i) = \frac{1}{T} \sum_{t=1}^{T} p_t(k|x_i)$, where $p_t(k|x_i)$ is the density of the kth object class in the leaf of the tree where x_i falls, and T is the number of trees in an ORF.

After training M ORFs, the prediction model can be updating by solving a linear programming problem to update the weights w of ORFs. For more details of the update strategy and the OMCLP algorithms, please refer to [10]. With the weights of all the ORFs are updated, the online updating of the model can be completed. And the model then can be used to predict the label of superpixels.

Optimization of Segmentation. The label likelihood from the learning algorithm is typically noisy. A common approach is to employ some kind of regularization to obtain spatially coherent labels. Similar to many segmentation methods [1,2,12], we use the graph cuts algorithm [15] to solve this problem.

For an image, we construct an undirected graph $G = <V, E>$, with a set of nodes (vertices V) to represent the pixels in the image, and a set of undirected edges (E) link the adjacent pixels. Each edge $e \in E$ in the graph is assigned a non-negative cost w_e. The final segmentation result is settled by minimizing the following energy:

$$E(L) = \lambda \sum_{p \in V} R(l_p) + \sum_{\langle p,q \rangle \in E} B_{\{p,q\}} \cdot \delta(l_p \neq l_q) \tag{2}$$

where l_p and l_q indicate the label of pixel p and q respectively. The first term of Eq. (2) measures the penalty of assigning the label l_p to the pth pixel, while the second term measures the penalty for assigning different labels to adjacent pixels, and λ specifies a relative importance of this two terms. The δ is a 0–1 indicator function which returns 1 if $l_p \neq l_q$ else returns 0.

The $R(l_p)$ is defined as $R(l_p) = -log(f(x_p, l_p)/\sum_{k=1}^{K} f(x_p, k))$, where $f(x_p, k)$ is the confidence that the pth sample labeled as the kth object class, which can get from Eq. (1), and K is the number of all exist classes in the model at present. We set $B_{\{p,q\}} = \exp(-(I_p - I_q)^2/2\sigma^2)/dist(p, q)$ just as [16] did.

Here, we use pixels instead of superpixels to improve the segmentation accuracy, as superpixels may leak across objects as described in [12].

5   Experimental Result

Dataset. We used two datasets in our experiments: the CMU-Cornell iCoseg dataset [12] (38 groups with 643 total images with associated pixel-level groundtruth indicating the foreground and background) and a subset of MSRC

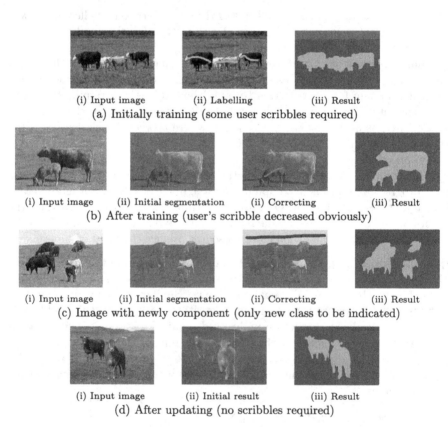

(i) Input image   (ii) Labelling   (iii) Result
(a) Initially training (some user scribbles required)

(i) Input image   (ii) Initial segmentation   (ii) Correcting   (iii) Result
(b) After training (user's scribble decreased obviously)

(i) Input image   (ii) Initial segmentation   (ii) Correcting   (iii) Result
(c) Image with newly component (only new class to be indicated)

(i) Input image   (ii) Initial result   (iii) Result
(d) After updating (no scribbles required)

Fig. 4. An example of the segmentation process flow

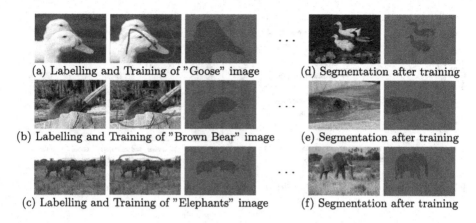

(a) Labelling and Training of "Goose" image   (d) Segmentation after training

(b) Labelling and Training of "Brown Bear" image   (e) Segmentation after training

(c) Labelling and Training of "Elephants" image   (f) Segmentation after training

Fig. 5. Other examples of progressive segmentation

21-class imageset [1] (72 images contain 5 classes: 'cow', 'grass', 'tree', 'water' and 'sky').

Progressive Segmentation. The first experiment is to validate the effectiveness of our progressive segmentation framework. Figure 4 (also the accompanying video) demonstrates the whole segmentation process.

The process of image segmentation in Fig. 4(a) from (i) to (iii) is described as follow: user need firstly to scribble to initialize the segmentation model, as the model is empty; and the model can then segment the image successfully after initialized. The process of image segmentation after initial training is illustrated in Fig. 4(b) from (i) to (iv): the input image can firstly be segmented without any interactions; then the user scribbles to correct some false-labelled pixels as the result is not so satisfactory; the model will further be updated to get more precise result. From (i) to (iv) in Fig. 4(c) illustrates the process when the user submit an image with an object has never seen before (i.e. the 'Sky'). It can't be segmented at first. The segmentation model can be updated after user labeled the new object, then it can correctly segment the current image. Figure 4(d) shows the process of segmentation after updated in Fig. 4(c) from (i) to (iii).

Figure 5 demonstrates some other example of the progressive segmentation, where user scribbles on the first images for initially training are shown from (a) to (c), and the results after stable training are shown from (d) to (f).

From the above, we can see that our method do not need any training data to be prepared in advance or submit all images at once. And the segmentation model can be updated easily if added a new object class.

The Convergence of the Online Learning. To confirm that the online learning is sure to convergence, we conduct this experiment. We regard the groundtruth as user's labels, and submit images one by one to train the model. After each submission, the model is used to segment the remaining images in that group to observe the accuracies changes. The accuracy is measured by the

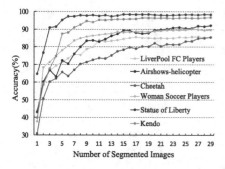
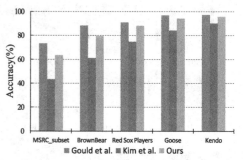

Fig. 6. Accuracies for gradually added   **Fig. 7.** Comparison with related method

intersection-over-union metric ($\frac{GT_i \cap R_i}{GT_i \cup R_i}$), the standard metric of PASCAL challenges. Figure 6 illustrates the result of several groups of images. It shows that the accuracies increase with the accumulation of submitted images until reaching a maximum.

Compare to the State-of-the-Art. We conduct an experiment to compare the segmentation quality of our method with the supervised [2] and co-segmentation [6] methods roughly using accuracy measurement. For our progressive segmentation, we use the segmentation accuracy after submitting half of a group images. For the supervised method, half of a group images is used to train the model to segment the remaining images. And for the unsupervised method, we run it by changing the number of segments from two to ten, and report the best results. Figure 7 shows the result. It can be seen that our method gets a performance close to the supervised one [2], and outperform the unsupervised co-segmentation [6].

Accumulative Learning. With more images segmented, the progressive model will be more reliable to get a more accurate result. Figure 8 shows the segmentation of some given images with the accumulation of segmented images in the same group. The first column are the given images, and the second column are groundtruth. The last four column are results from models trained with different number of images. We can see that the segmentation of the images will become more and more accurate as the accumulation of segmented images.

Diversity Segmentation Accord with User's Intention. Different users may get different segmentation of same images. With the help of learning-based segmentation prediction model, our method can easily achieve this by building a corresponding appearance model of objects, and segment the subsequent images according to the learned model to embody user's intention. Figure 9 demonstrates the diversity of segmentation. The two rows show two different work flows by different users. The users scribble on the same image according to their intention as in Fig. 9(a)(ii) and (v), and get the corresponding results as in Fig. 9(a)(iii) and (vi). The next image can be automatically segmented to corresponding results which is in accord with the users' intention as shown in Fig. 9(b).

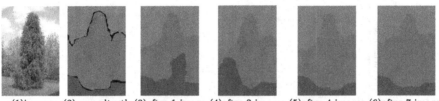

(1)image   (2)groundtruth (3)after 1 image (4)after 2 images (5)after 4 images (6)after 7 images

(a) Image composed of "Tree", "Grass" and "Sky"

(1)image   (2)groundtruth (3)after 1 image (4)after 2 images (5)after 4 images (6)after 7 images

(b) Image composed of "Cows" and "Grass"

(1)image   (2)groundtruth (3)after 1 image (4)after 2 images (5)after 4 images (6)after 7 images

(c) Image composed of "Stonehenge" and "background"

Fig. 8. Segmentation of given images after segmented different number of images

User Study. In order to further evaluate the proposed method, we conduct a user study to verify our hypothesis that the proposed approach can help real users to segment a set of images with increasing efficiency. We invite 15 participants and record the time cost and the number of interactions to segment a group of 31 total images from the CMU-Cornell iCoseg dataset (the 'Goose' group) using our method. The average time cost and interaction numbers of each image is shown in Fig. 10. From it, we can find that the time cost is getting less and less as also the number of interactions. With the accumulation

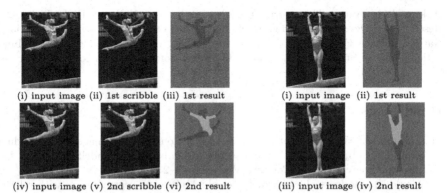

(i) input image (ii) 1st scribble (iii) 1st result   (i) input image (ii) 1st result

(iv) input image (v) 2nd scribble (vi) 2nd result   (iii) input image (iv) 2nd result

(a)Labelling and training with different scribbles   (b)Segmentation after training

Fig. 9. Diversity segmentation accord with different user's intervention

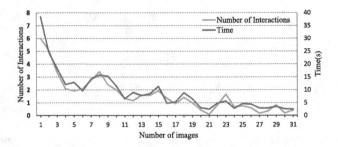

Fig. 10. The time cost and number of interactions for each image in the user study

of segmentation, our model becomes more and more accuracy, so that it will take less and less time and interactions to correct the false-labeled regions. It demonstrated that the proposed method can obvious improve efficiency while segmenting a set of images because of the characteristics of progressive learning. And as the participants say, our tool is getting smarter and smarter through the segmentation.

6   Conclusion

We present a novel framework in this paper which can segment images progressively using online learning algorithms accord with user's preferences. Our method overcomes several challenges in existing methods, and have three advantages. Firstly, the segmentation model can be learned gradually along with user's manipulation without any pre-labeling. Secondly, the diversity of segmentation accord with user's preferences can be met flexibly, and the more use the more accurate the segmentation could be. Thirdly, the segmentation model can be updated online to fit the needs of users and the diversity of applications.

Acknowledgments. This work is supported by the National High Technology Research and Development Program of China (Project No. 2007AA01Z334), National Natural Science Foundation of China (Project No. 61321491 and 61272219), Innovation Fund of State Key Laboratory for Novel Software Technology (Project No. ZZKT2013A12).

References

1. Shotton, J., Winn, J., Rother, C., et al.: Textonboost for image understanding: multi-class object recognition and segmentation by jointly modelling texture, layout, and context. Int. J. Comput. Vis. **81**(1), 2–23 (2009)
2. Gould, S., Rodgers, J., Cohen, D., et al.: Multi-class segmentation with relative location prior. Int. J. Comput. Vis. **80**(3), 300–316 (2008)
3. Socher, R., Fei-Fei, L.: Connecting modalities: semi-supervised segmentation and annotation of images using unaligned text corpora. In: IEEE CVPR, pp. 966–973 (2010)

4. Rother, C., Minka, T., et al.: Cosegmentation of image pairs by histogram matching-incorporating a global constraint into MRFs. In: IEEE CVPR, pp. 993–1000 (2006)
5. Joulin, A., Bach, F., Ponce, J.: Discriminative clustering for image co-segmentation. In: CVPR (2010)
6. Kim, G., Xing, E.P., Fei-Fei, L., et al.: Distributed cosegmentation via submodular optimization on anisotropic diffusion. In: IEEE ICCV, pp. 169–176 (2011)
7. Zhang, F.Q., Sun, Z.X., Song, M.F., Lang, X.F.: Progressive 3D shape segmentation using online learning. Comput. Aided Des. **58**(1), 2–12 (2015)
8. Ristin, M., Guillaumin, M., Gall, J., Van Gool, L.: Incremental learning of NCM forests for large-scale image classification. In: IEEE CVPR, pp. 3654–3661 (2014)
9. Collins, R., Liu, Y., Leordeanu, M.: Online selection of discriminative tracking features. In: IEEE TPAMI, pp. 1631–1643 (2005)
10. Saffari, A., Godec, M., Pock, T., et al.: Online multi-class LPBoost. In: IEEE CVPR, pp. 3570–3577 (2010)
11. Saffari, A., Leistner, C., Santner, J., et al.: Online random forests. In: IEEE ICCV Workshops, pp. 1393–1400 (2009)
12. Batra, D., Kowdle, A., et al.: Interactively co-segmentation topically related images with intelligent scribble guidance. Int. J. Comput. Vis. **93**(3), 273–292 (2011)
13. Achanta, R., Shaji, A., Smith, K., et al.: SLIC superpixels compared to state-of-the-art superpixel methods. IEEE TPAMI **34**(11), 2274–2282 (2012)
14. Felzenszwalb, P.F., Huttenlocher, D.P.: Efficient graph-based image segmentation. Int. J. Comput. Vis. **59**(2), 167–181 (2004)
15. Boykov, Y., Veksler, O., Zabih, R.: Fast approximate energy minimization via graph cuts. IEEE TPAMI **23**(11), 1222–1239 (2001)
16. Boykov, Y., Jolly, M.P.: Interactive graph cuts for optimal boundary & region segmentation of objects in ND images. In: Eighth IEEE ICCV, pp. 105–112 (2001)

A Study of Interactive Digital Multimedia Applications

Chutisant Kerdvibulvech$^{(\boxtimes)}$

Graduate School of Communication Arts and Management Innovation,
National Institute of Development Administration, 118 SeriThai Rd.,
Klong-chan, Bangkapi, Bangkok 10240, Thailand
chutisant.ker@nida.ac.th

Abstract. Many communication models for communication arts and numerous interactive multimedia applications for computer science were discussed over many decades ago. However, there has been little work giving an overview of recent integrated research of digital media and emerging trends, such as interactive multimedia experience in an interdisciplinary aspect. In this paper, we review and study recently interactive digital multimedia applications using and applying the aforementioned emerging trends. We provide a short blueprint for interactive digital multimedia research when applying virtual reality, image processing, computer vision, real-time augmented reality, and interactive media into the senses of hearing and vision for virtual environments. A SMCR (Source-Message-Channel-Receiver) model for communicating via all human senses is also explained and linked to some interactive digital multimedia applications presented recently. After that, the senses of hearing and vision are discussed using related-technologies. It will be of good value to the new researchers in this integrated emerging field of interactive digital multimedia.

Keywords: Interactive media · Interactive digital multimedia · Real-time · Mobile-based · Digital multimedia · Virtual reality · SMCR model

1 Background

Over past several decades, computer with internet has evolved from traditional tele-typewriters to great memorizers to today's powerful tools for connecting people to people, even people to limitless opportunity. This modern technology is changing the way we live, the way we work and, of course, the way we communicate. Today, digital multimedia is not restricted only to old style dimensions [1, 2, 3], but it is also enjoying broader and wider use every moment, especially interactivity. It has even changed the method of old-fashioned communications and conventional art works. More specifically, one of very classical communication models, although it was presented very long time ago, is Berlo's model [4]. This is sometimes called as SMCR (Source-Message-Channel-Receiver) model. This model is defined as four related-elements of communication: Source, Message, Channel, and Receiver. The first element is the start of the communication for encoding the message. The second element is basically the package or packages of meaning that contain generally the intent from the first element.

© Springer International Publishing Switzerland 2015
Y.-S. Ho et al. (Eds.): PCM 2015, part I, LNCS 9314, pp. 192–199, 2015.
DOI: 10.1007/978-3-319-24075-6_19

The third element which is our main element in this paper to be discussed is the medium for transmitting the message. It includes a wide range of things from cell phones to television adverts and from newspaper articles to interactive digital multimedia applications. The fourth element is the person at the end of the communication for seeking to receive the message. Nevertheless, since the third element has to plug into the receiver's sensory system, it suggests that every traditional sense (i.e., hearing, vision, somatosensation, olfactory, and gustatory systems) is the channel which helps human beings to communicate with each other. However in the 20th century, we basically relied only on one traditional sense separately. For instance, when we used the world's first hand-held cell phone, created by Mitchell and Cooper [5] from Motorola in 1973, to communicate people to people via the hearing sense individually, they can transfer only one traditional sense at the same time. In other words, the cell phone technology was mainly used to connect people to people via the sense of hearing. Other senses (such as the sense of vision) cannot be transferred simultaneously at all. More recently in the 21st century with the rise of the smartphone technology, it allows us to transfer more than only one individual sense. Vision sense uses modern technologies such as virtual reality, image processing, real-time augmented reality, interactive multimedia, and smartphone-based digital multimedia to communicate people to people, including but not limited to people to machines, machines to people, and also machines to machines. Figure 1 depicts and illustrates the explanation.

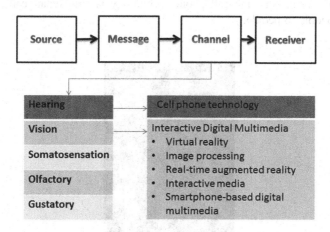

Fig. 1. The third element of SMCR Model [4] does not reply solely on any one individual sense anymore, e.g., the sense of vision uses modern vision-based technologies to connect people.

In this paper, we present a brief blueprint for future interactively digital media research when applying these aforementioned technologies into the sense of vision innovatively. For other three senses (i.e., somatosensation, olfactory, and gustatory systems) such as [6] for affective touch, although it is very important for digital communication, we do not focus in the paper. We structure this paper as follows. Section 2 provides and discusses about unlimited channel for communication.

We focus in this section on the senses of hearing and sight, respectively. We also order according to the technologies used. Section 3 gives finally some conclusion and outlines some possible future works.

2   Unlimited Channel for Communication

Over many decades in the 20th century, communication was only limited through human hearing system, such as cell phone or mobile phone. In that time, there was little work giving an integrated research for additional human senses in an interdisciplinary way. However, voice use is recently still implemented for some intelligent personal assistant and knowledge navigators, such as Apple Inc.'s Siri, and voice user interactions (UIs). We have found the work of embarrassing voice interactions presented recently in [7] from Google, Inc. As depicted in Fig. 2, Siri used in this work for designing voice user interactions to avoid some embarrassing interactions. In his experiments for voice use in search in Google, Siri is one of very few voice user interactions for creating embarrassing situations publicly. Some people using Siri feel quite awkward to interact publicly with a smartphone when other people are around in non-private places. This is because sometimes Siri gives unpredictable answers. In other words, voice user interactions can be particularly embarrassing if implementing without careful evaluation and development, as discussed in CHI workshop in 2015. However, this means that the cell phone technology is still dynamically active for connecting people to people.

Fig. 2. Siri used for designing voice user interactions to avoid some embarrassing interactions as discussed in 2015 CHI Workshop [7].

More interestingly in 21st century, how to connecting people to people expands horizons into new sensory modalities, from real-time augmented reality to interactive multimedia. This means that communication has gone through human vision system. Over time, many interactive digital multimedia applications applying technologies of computer science such as image processing, augmented reality, and computer vision have been found more and more every day. One of very good examples of technologies is image processing, often abbreviated with the acronym IP. For instance, it is used to the related-application of digital Steganography and Steganalysis for digital multimedia as presented in [8]. Similarly to signal processing techniques, image processing applies mathematical operations for which the input is usually an image, including but not limited to only a static image. Video frame or some dynamic images can also be the input of image processing. The output includes more broadly, from an image to a set of parameters related to the static image. This technique for digital Steganography and Steganalysis tried to differentiate between stego-objects and non-stego objects. The differences between stego and non-stego objects are if they have or do not have a hidden message, respectively. Figure 3 gives an example of the process of hiding a message they used in a still image. Despite its effectiveness, this implementation is limited which is able to only detect messages without prior knowledge of the hiding algorithm. In this case, they called this limitation as blind detection.

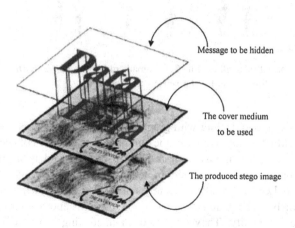

Fig. 3. An example of the process of data hiding in a still image was implemented and described by Rocha and Goldenstein [8] using image processing.

Other similar field of image processing for communicating via the sense of vision is computer vision [9]. It is a computer science related-dynamic branch dealing with trying to understand and interpret images from the physical world. This branch of computer science is utilized in many related-branches. For instance, an algorithm for 3D human motion analysis was recognized for reconstruction [10]. Figure 4 shows some examples for 3D gait signatures computed from 3D data. The data used in this process are obtained and calculated from a triangulation-based projector-camera system using computer vision technology. The 3D human body data used in this implementation are composed of representative poses that occur during the gait cycle of a real walking human. According to the vision-based works, it can be seen that they are able to help human for 3D human motion analysis to communicate more dimensions for the sense of vision.

This means that communication opens up more broadly. It is not limited to only people to people, but also to people to machines, machines to machines, and also machines to people. Furthermore, interactive media has been interestingly defined. In short, it is a method of communication on digital computer-based systems in which the output comes generally from the input of the users. The output of this interactive media includes a wide range of things from text and audio to video and graphics.

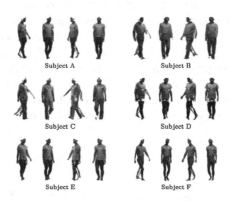

Fig. 4. A vision-based algorithm for 3D human motion analysis [10] helps human to communicate in more dimensions for the sense of vision.

In other words, due to dealing with graphics, interactive media is usually linked to the concepts of virtual reality and real-time augmented reality for connecting people to people through the sense of vision. It is able to be achieved in a great variety of platforms encompassing every branch of technology virtually and physically. For instance, Lee and Lee [11] studied recently about interactive media art for emotion recognition, namely "The Universe", based on the concepts of virtual reality and real-time augmented reality. They presented an interesting framework of media art works. This work changes an emotion recognition prototype for developing in some art works in real-time. In fact, virtual reality has played an essential part of interactive media designs over time. The term of immersion has been seriously discussed for the communication between people to people in the environment of virtual world and also the level of physical submergence of people in the physical world. For example, Emma-Ogbangwo et al. [12] selected representative methods for analyzing to enhance user immersion, co-presence, and also virtual presence. They tested in the application of Social Immersive Virtual Environment (SIVE) through the combination of a virtual reality-based Natural User Interface (NUI). Thus, the system is able to interact smoothly with the environment using the body gestures virtually and naturally. This concept is able to utilize for recognition of body gestures to communicate people to people through the sense of vision. Also, there are indeed many researches towards augmented reality for communicating people in some related-strategies such as the renovation of old buildings using real-time augmented reality. Landrieu et al. evaluated the general benefit of using a mock-up digitally in an augmented reality context [13].

It is done for renovation operations in an old building. Figure 5 illustrates the system scenario when a user is holding the tablet. At the same time, the augmented information is overlaid onto this view for renovation of buildings. For example, the system may automatically suggest renovating a window by designing additionally a double skin facade in windowed partition wall of a building. The goal of this system is to build bidirectional schedule and at the same time manage related-costs after obtaining the results. More interestingly, it is suggestable that the system could be playable into some virtual reality and real-time augmented reality applications usable onsite.

Fig. 5. A user is holding the tablet display with the augmented information for renovating an old building for the work of Landrieu et al. proposed in [13].

Interactive media was also extended to the term of interactive digital multimedia. More specifically, this term was defined as applications and technologies focusing on the creation, transmission, analysis and presentation of multimedia data interactively in an interdisciplinary way through the sense of vision. One of the most important features of interactivity is mutuality between human and machine in an active role automatically. Many applications of interactive digital multimedia in learning, arts, entertainment and communication were presented. For example, there is a work for interactive digital multimedia in an interactive multimedia learning platform in the field of power and heat innovations presented by Fedulov [14]. It developed an interactive evaluation of a repository of multimedia content digitally and interactively for learning in Computerized Educational Platform (CompEdu HPT). In addition, Cochrane [15] evaluated how constructing multimedia learning objects can interactively enhance conventional teaching methodologies during online process virtually. It is developed for both distance e-learning and face-to-face environments. The application of the multimedia architecture 'QuickTime' is used for evaluating. This is since the cross platform capability, multi-platform scalability, and high level of interactivity possible are quite flexible. Similarly, it is then used for designing perspectives, environment, and emerging learning systems in [16] presented by Vartiainen and Enkenberg. Due to this multidisciplinary definition of interactive digital multimedia, it usually overlaps with human-computer interaction, but refers to more about new media with multimedia systems and the study of human interaction in virtual immersive environments. In addition, Starcic et al. [17] built an instructional design method for communicating people and people. An assessment criterion was also designed. In their case, they implemented an integrative system for digital storytelling and scientific problem solving used in a college/university coursework for instructors-students. They concluded that some modern innovations allow multimodal design digitally in communication to be influential practically in shaping the identities of people instructionally. Furthermore, with the rise of smartphones, many digital multimedia applications are often built in mobile-based platforms. For instance, [18] constructed an augmented reality-based system for designing the smartphone application service. Their goal is to convey additional information on advertisements for marketing to users. It is done and implemented on smartphones without any marker. However, despite its interesting features, this system still does not analyze its effectiveness deeply.

3   Conclusion

In this paper, we review some representative of interactive digital multimedia applications. We first explain about a classical communication model (SMCR model) for communicating via all human senses. Then, the hearing sense is discussed using cell phone technology. After that, the sense of vision is reviewed with some related fields. These include virtual reality, image processing, computer vision, real-time augmented reality, interactive media, and interactive digital multimedia. The discussion of vision-based systems for communicating people to people, people to machines, machines to people, and also machines to machines is the main contribution of this paper. While the senses of somatosensation, olfactory and gustatory are directly and importantly related to communicate between people and people, this paper does not focus on these aspects. Future recommendations for the study include studying, researching and implementing for the three senses from a wider range of users. For future work, we plan to build a system to communicate people to people via all five senses accurately, simultaneously and robustly.

Acknowledgments. This research presented herein was partially supported by a research grant from the Research Center, NIDA (National Institute of Development Administration).

References

1. Stevens, S.M.: Multimedia computing: applications, designs, and human factors. User Interface Software, pp. 175–193. Wiley, New York (1993)
2. Wang, Z., Crowcroft, J.: Quality of service routing for supporting multimedia applications. IEEE J. Sel. Areas Commun. **14**, 1228–1234 (1996)
3. Kiene, B.W.: Multimediality, intermediality, and medially complex digital poetry. Revue des littératures de l'Union Européenne (Review of Literatures of the European Union), N. 5 - Juillet, 18, 1–18, Bologna (2006)

4. Berlo, D.: The Process of Communication: An Introduction to Theory and Practice. Holt, Rinehart and Winston, New York (1960)
5. Heeks, R.: Meet Marty Cooper – the inventor of the mobile phone. BBC **41**(6), 26–33 (2008). doi:10.1109/MC.2008.192
6. Stiehl, W.D., Lieberman, J., Breazeal, C., Basel, L., Wolf, M.: The design of the huggable: a therapeutic robotic companion for relational, affective touch. In: AAAI Fall Symposium on Caring Machines, Washington D.C. (2005)
7. Sharma, N.: 'Siri thinks i have two wives' & other embarrassing voice interactions. In: 33rd Annual ACM Conference Extended Abstracts on Human Factors in Computing Systems (CHI) Workshop "Embarrassing Interactions" Collected Papers, Seoul, Korea (2015)
8. Rocha, A., Goldenstein, S.: Steganography and steganalysis in digital multimedia: Hype or Hallelujah?. J. Theor. Appl. Comput. (RITA) **15**, 83–110 (2008)
9. Li, Z., Drew, M.S., Liu, J.: A taste of multimedia. Fundamentals of Multimedia. Texts in Computer Science, p. 25. Springer International Publishing, Switzerland (2014)
10. Kerdvibulvech, C., Yamauchi, K.: 3D human motion analysis for reconstruction and recognition. In: Perales, F.J., Santos-Victor, J. (eds.) AMDO 2014. LNCS, vol. 8563, pp. 118–127. Springer, Heidelberg (2014)
11. Lee, H.Y., Lee, W.H.: A study on interactive media art to apply emotion recognition. Int. J. Multimedia Ubiquit. Eng. (IJMUE) **9**(12), 431–442 (2014)
12. Emma-Ogbangwo, C., Cope, N., Behringer, R., Fabri, M.: Enhancing user immersion and virtual presence in interactive multiuser virtual environments through the development and integration of a gesture-centric natural user interface developed from existing virtual reality technologies. In: Stephanidis, C. (ed.) HCI 2014, Part I. CCIS, vol. 434, pp. 410–414. Springer, Heidelberg (2014)
13. Landrieu, J., Nugraha, Y., Pere, C., Merienne, F., Garbaya, S., Nicolle, C.: Bringing building data on construction site for virtual renovation works. Int. J. Electr. Electr. Comput. Syst. (IJEECS) **14**(1), 737–747 (2013)
14. Fedulov, V.: Educational evaluation of an interactive multimedia learning platform: computerized educational platform in heat and power technology. KTH Skolan för industriell teknik och management (ITM). Energiteknik. Series: Trita-KRV, 1100–7990:06 (2005)
15. Cochrane, T.: Interactive quicktime: developing and evaluating multimedia learning objects to enhance both face to face and distance e-learning environments. Interdisc. J. E-Learn. Learn. Objects **1**(1), 33–54 (2005)
16. Vartiainen, H., Enkenberg, J.: Learning from and with museum objects: design perspectives, environment, and emerging learning systems. Educ. Technol. Res. Dev. (Springer US) **61**(5), 841–862 (2013)
17. Starcic, A.I., Cotic, M., Solomonides, I., Volk, M.: Engaging preservice primary and preprimary school teachers in digital storytelling for the teaching and learning of mathematics. British Journal of Educational Technology, British Educational Research Association (2015). doi:10.1111/bjet.12253
18. Kim, Y., Kim, W.: Implementation of augmented reality system for smartphone advertisements. Int. J. Multimedia Ubiquit. Eng. **9**(2), 385–392 (2014)

Video Coding and Processing

Particle Filter with Ball Size Adaptive Tracking Window and Ball Feature Likelihood Model for Ball's 3D Position Tracking in Volleyball Analysis

Xina Cheng[1]([✉]), Xizhou Zhuang[1], Yuan Wang[1], Masaaki Honda[2], and Takeshi Ikenaga[1]

[1] Graduate School of Information, Production and Systems,
Waseda University, Tokyo, Japan
cheng-c@waseda.jp
[2] Faculty of Sport Sciences, Waseda University, Tokyo, Japan

Abstract. 3D position tracking of the ball plays a crucial role in professional volleyball analysis. In volleyball games, the constraint conditions that limit the performance of the ball tracking include the fast irregular movement of the ball, the small-size of the ball, the complex background as well as the occlusion problem caused by players. This paper proposes a ball size adaptive (BSA) tracking window, a ball feature likelihood model and an anti-occlusion likelihood measurement (AOLM) base on Particle Filter for improving the accuracy. By adaptively changing the tracking windows according to the ball size, it is possible to track the ball with changing size in different video images. On the other hand, the ball feature likelihood enables to track stably even in complex background. Furthermore, AOLM based on a multiple-camera system solves the occlusion problems since it can eliminate the low likelihood caused by occlusion. Experimental results which are based on the HDTV video sequences (2014 Inter High School Games of Men's Volleyball) captured by four cameras located at the corners of the court show that the success rate of the ball's 3D position tracking achieves 93.39 %.

Keywords: Volleyball analysis · Particle filter · 3D position tracking · Ball feature · Likelihood model

1 Introduction

Sports analysis has gained significant development thanks to the computer vision technologies and vast available high-quality videos. In volleyball games, the 3D trajectory and movement of the ball are involved in analysis of the behaviors of players and the performances of teams. The reliability of the analysis is affected by the accuracy of the ball's 3D position. Chen [1] has proposed an automated system to detect ball candidates and find the trajectory fitted from a single camera. However, due to the lack of 3D space information, the approximated result is not reliable. Therefore, ball's 3D position tracking with high accuracy becomes one of the most important issues in volleyball game analysis.

© Springer International Publishing Switzerland 2015
Y.-S. Ho et al. (Eds.): PCM 2015, part I, LNCS 9314, pp. 203–211, 2015.
DOI: 10.1007/978-3-319-24075-6_20

However, in consideration of the unique environment of volleyball games and the limitation of the shooting conditions, the accurate tracking is difficult to be obtained by general tracking methods. There are some problems in ball tracking shown as following: (1) the fast and irregular movement of the ball; (2) the small and changing size of the ball in video; (3) the constant change of the ball's texture caused by the rotation of the ball in each video frame; (4) the complex background including audiences and staff in the scene; (5) dozens of players leading to occlusion problems.

There have been many tracking algorithms such as Mean-Shift [2], Cam-Shift [3], SIFT [4], Kalman Filter [5] and Extended Kalman Filter [6]. However, none of these algorithms can solve all problems. Mean-Shift is not suitable for tracking fast objects, SIFT needs adequate feature points and Kalman Filter is not flexible for tracking irregular moving targets. Nowadays, Particle Filter [7] is widely employed in non-linear and non-Gaussian system and whose likelihood model can be adapted to the tracking target's features. Hess [8] has put forward a discriminatively trained Particle Filter with the HSV likelihood model. Currently, this algorithm has been the basis of Particle Filter and been improved by many researchers.

Huang [9] has proposed a method of small object detection and tracking based on Particle Filter. The used strong and weak detectors based on foreground-background segmentation cannot work well on complex background. Guo [10] has proposed a likelihood model for Particle Filter which combines the HSV color likelihood and the orientation gradient likelihood. However, the orientation gradient likelihood is weak in tracking the objects with changing texture.

In view of the statement above, our research makes use of a multiple-camera system and proposes the BSA tracking window, the ball feature likelihood model and the AOLM method based on Particle Filter.

This paper is organized as follows. Section 2 introduces our 3 proposals in detail. Then, the experiment results and the conclusions are described in the Sects. 3 and 4, respectively.

2   Proposal

The flowchart of our proposals based on Particle Filter is shown in Fig. 1. The state vector of Particle Filter is established by the coordinate system of physical world. In the likelihood estimation part, the BSA tracking window is employed for adjusting the size of tracking window; the ball feature likelihood model and the AOLM method are presented for obtaining the precise likelihood of particles.

2.1   Ball Size Adaptive Tracking Window

The size and shape of optimal tracking window should be the same as tracking target. However, the small and constantly changing size of the ball makes it difficult to estimate the size of tracking window. The undesirable tracking window would bring noise from background or lose edge information. Fortunately, in our research, the BSA tracking window is adjusted to fit the ball's size in video as Fig. 2 shows.

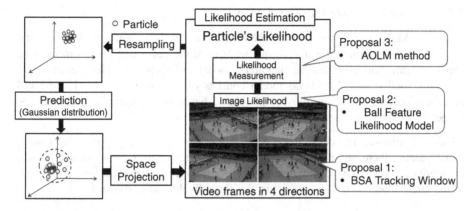

Fig. 1. Overview of proposals in Particle Filter

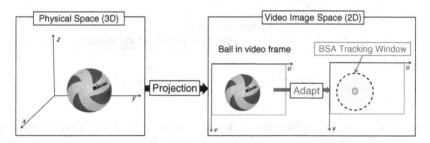

Fig. 2. Concept of BSA tracking window

At the beginning of the likelihood estimation step, both the ball and all the particles are projected from the physical world space to the image space depending on the projection principle as Eq. (1) shows.

$$(\omega x, \omega y, \omega z)^T = A \cdot (X, Y, Z, 1)^T \tag{1}$$

Here, $(\omega x, \omega y, \omega z)^T$ is the 2D homogeneous coordinate of the points in image space, ω is any nonzero constant; $(X, Y, Z, 1)^T$ is the 3D homogeneous coordinate of the points in real world; Matrix A is a 3×4 matrix derived by camera calibration [11].

Since the ball is a generalized rigid sphere, as long as the physical position of the ball and the projection relationship are firmed, the projected radius of the ball in the video image can be obtained. In consideration of the aberration and coma of camera and other effect during the process of the video shooting, the projected radius is just a mathematical result under the ideal condition. So we multiply this value with an adapted coefficient to estimate the radius of the tracking window. Thus, the BSA tracking window is defined as a circle region with the obtained radius.

2.2   Volleyball Feature Likelihood Model

In ball tracking, in order to eliminate the interference of the complex background and solve the rotation problem, some unique features of the volleyball are abstracted then converted into volleyball feature likelihood model to improve tracking performance. The volleyball feature likelihood model consists of the circle likelihood, the SVM detection likelihood and the moving likelihood based on 3 features of the ball: the shape, the texture and the moving feature.

Circle Likelihood. The circle-shape is one of the most obvious features of the ball. In game videos, there are a few circle objects among the mass of noise. Hence, the circularity is effective for distinguishing circle objects from other items and contributes to the circle likelihood. The concept of the circle likelihood is shown below.

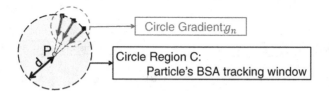

Fig. 3. Concept of the circle likelihood

To calculate the possibility that the tracking window locates at a circle object in video images, the definition of circularity is given. As Fig. 3 shows, a circle region C can be defined by a given position P and a given distance d. We define the component of the gradient which pointing to center P of each pixel i on the edge of C as the circle gradient g_i. The circularity of the region C is the sum value of all pixels' circle gradients on the edge which is described in Eq. (2).

$$circularity = \sum\nolimits_{i=0}^{i=n} g_i \qquad (2)$$

In this equation, n means the amount of pixels on the edge of C. When a particle's BSA tracking window is located at a circle object in an image, the circularity of the image region determined by this particle's tracking window is large. On the contrary, the circularity is small. Thus, this circularity devotes to the circle likelihood of this particle by multiplying it with a coefficient. The value of the coefficient is decided by experiments.

SVM Detection Likelihood. Ball's unique texture is an important feature to make it prominent among all objects in the video image. However, the constantly rotating texture of the volleyball makes it difficult to find an appropriate texture likelihood for volleyball tracking. The Support Vector Machine (SVM) [12] is used to filter out some ball candidates in this paper.

We abstract gray-scale histogram of volleyball samples and apply them as the learning features to train SVM beforehand. During the process of likelihood estimation, every particle is computed in SVM detection. The SVM detection likelihood of one particle is depending on the output of SVM. If the output is positive, the value of SVM detection likelihood is bigger than 1; otherwise it is smaller than 1.

Moving Likelihood. The moving feature of the ball can be used to distinguish moving objects from static ones and background. In this proposal, background subtraction is performed so that a binary mask image is filtered out in which the black pixels stand for static objects. The moving likelihood L_{moving} of the particles is calculated as Eq. (3) shows.

$$L_{moving} = \sum_{j=0}^{j=m} \frac{v_j * d_j}{R} \tag{3}$$

Here, d_i means the distance between the j_{th} pixel and the tracking center of the window, v_j means the j_{th} pixel's value, m means the amount of pixels in tracking window and R means the radius of tracking window. Due to this likelihood, tracking failures is reduced greatly since the non-moving objects in video image are removed.

2.3 Anti-occlusion Likelihood Measurement Method

The AOLM method is proposed by using multiple-camera system. The game video is captured from different directions of the volleyball court, so even the ball occluded in certain direction, the 3D position can still be obtained by information from other directions.

Likelihood measurement is processed to generate the likelihood of every particle after obtaining the image likelihood in different directions' video frame separately. In general, if occlusion occurs in certain camera, this direction's image likelihood will be low and decrease the likelihood of this particle leading to tracking failure. The AOLM method is to sort the likelihood of all directions by numeric value and assign the smallest one the value 1 which can eliminate the low likelihood caused by occlusion. The likelihood of one particle is described in Eq. (4).

$$L_i = L_{i,1} \cdot L_{i,2} \cdot L_{i,3} \cdot \cdots \cdot L_{i,N} \tag{4}$$

Here L_i represents the likelihood of the i_{th} particle and $L_{i,N}$ is the image likelihood in camera N of the i_{th} particle.

3 Experiment

3.1 Tracking Example and Evaluation Method

The experiment is based on videos recording the total 3 sets of the final game of an official volleyball match (2014 Inter High school Games of Men's Volleyball held in

Tokyo Metropolitan Gymnasium in Aug. 2014) by four cameras located at corners of the court. The video's resolution is 1920 × 1080, the frame rate is 60 frames per second and the camera's shutter speed is 1000 per second. The algorithm is implemented by C ++ and OpenCV 2.4.8.

To evaluate the performance of our proposals, we give a definition of hit time (HT) which is a time period between two consecutive ball hitting. If during an *HT*, the tracking is successful, this *HT* becomes a successful hit time (*SHT*). The success rate is described in Eq. (5).

$$\text{success rate} = \frac{\sum SHT}{\sum HT} \times 100\% \tag{5}$$

3.2   Result and Comparison Analysis

Ball Size Adaptive Tracking Window. Firstly, we implement this proposal combining with Hess's [8] HSV color likelihood model and the Gaussian transition model. With the 1st set of the game which contains 218 *HTs* as the test sequence, the results in Table 1 indicate that the proposal of BSA tracking window gains 16.13 % improvement comparing with Hess's work.

Table 1.   Experimental result of BSA tracking window

	Hess's PF[8]	BSA tracking window PF
SHT	93	108
Success rate	42.66 %	49.54 %
Improvement ratio	16.13 %	

Ball Feature Likelihood Model. Then the proposal is evaluated by comparing with Guo's [10] work under the same environment combining with the BSA tracking window. The 1st set of the game is still used as the test sequence. The results shown in Table 2 show that the success rate of the proposed likelihood model rises up to 67.89 % while that of Guo's work is only 47.25 %. In addition, compared with the former result 49.54 % in Table 1, the success rate is increased by 37.04 %. By analyzing the tracking failed sequences, the tracker always loses the target when the ball is occluded by players in some directions. So the AOLM method is the key to deal with the remained over 30 % failures.

Table 2.   Comparison between ball feature likelihood model and Guo's likelihood model

	Guo's likelihood model [10]	Ball Feature Likelihood Model
SHT	103	148
Success rate	47.25 %	67.89 %
Improvement ratio	43.68 %	

Anti-occlusion Likelihood Measurement. At last, the AOLM method is added to test the integral performance. We use all 3 sets of the game as the test sequences. The total success rate is 93.39 % and it is raised by 37.57 % greatly compared to the best result in Table 2 since that we just utilize the position information in non-occlusion directions. The Table 3 gives the comparison data. While the increasing of the tracking accuracy, there are still about 6.4 % tracking failures in the experiment occurring when the ball is occluded in more than two directions or swerves by hitting.

Table 3. The success rate of the integrated proposals implemented on the entire game

	Set 1	Set 2	Set 3	Total
HT	218	223	225	666
SHT	203	204	215	622
Success rate	93.12 %	91.48 %	95.56 %	93.39 %

The experiment results of our proposal are sequences of 3D positions of the ball in physical world. To show the result visualized, the obtained position sequences is plotted in a 3D coordinate system by using Matlab. The Fig. 4 shows the ball's positions tracked from 6 different rounds in the game of the experimental video. Figure 4(a), (b), (c) show three normal rounds which just consist of one serve, one receive, one pass and one spike and finish. Figure 4(d), (e), (f) show the results of long

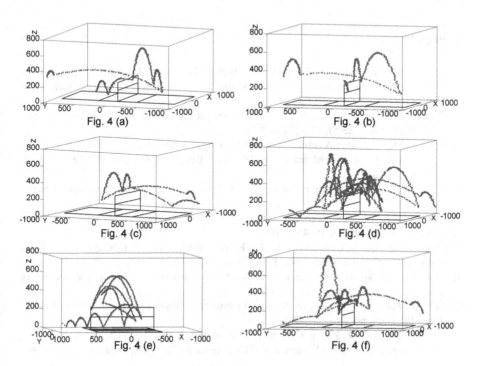

Fig. 4. 3D trajectories of experimental results

rounds. Especially in Fig. 4 (d), this round consists of 26 HTs and can be tracked by our proposal successfully. In every figure, the 3D trajectory of the ball can be observed.

4 Conclusion

This paper presents 3 proposals to achieve high accuracy of ball's 3D position tracking in volleyball game analysis. The BSA tracking window adapts the tracking window for tracking the round ball with changing size in different frames. The ball feature likelihood model which is based on the unique features of the ball consists of the circle likelihood, the SVM detection likelihood and the moving likelihood. By using this likelihood model, the noise caused by complex background and players are removed and the tracking accuracy rises up. At last, the proposal of the AOLM method eliminates the interference of the occlusion problems that occur in certain direction. The proposals have been implemented on prepared test sequences. Comparing with Hess's algorithm, the proposal of the BSA tracking window improved the accuracy by 16.13 %. After adding the ball feature likelihood model, the tracking success rate reaches 67.89 % while the conventional work is only 47.25 %. What's more, the success rate of the combination of all the 3 proposals increases to 93.39 %.

Acknowledgment. This work was supported by KAKENHI (26280016) and Waseda University Grant for Special Research Projects (2015 K-222).

References

1. Chen, H.T., Tsai, W.J., Lee, S.Y., Yu, J.Y.: Ball tracking and 3D trajectory approximation with applications to tactics analysis from single-camera volleyball sequences. Multimedia Tools Appl. **60**(3), 641–667 (2012)
2. Comaniciu, D., Ramesh, V., Meer, P.: Kernel-based object tracking. IEEE Trans. Pattern Anal. Mach. Intell. **25**(5), 564–577 (2003)
3. Dong, X.M., Yuan, K.: A robust CamShift tracking algorithm based on multi-cues fusion. In: 2nd International Conference on Advanced Computer Control (ICACC), vol. 1, pp. 521–524 (2010)
4. Lowe, D.G.: Distinctive image features from scale invariant key points. J. Comput. Vis. **60**, 91–110 (2004)
5. Kalman, R.E.: A new approach to linear filtering and prediction problems. Trans. ASME J. Basic Eng. **85**, 35–45 (1960)
6. Ndiour, I.J., Vela, P.A.: A local extended Kalman filter for visual tracking. In: 49th IEEE Conference on Decision and Control (CDC), pp. 2498–2504 (2010)
7. Kitagawa, G.: Monte Carlo filter and smoother for non-Gaussian nonlinear state space models. J. Comput. Graph. Stat. **5**(1), 1–25 (1996)
8. Hess, R., Fern, A.: Discriminatively trained particle filter for complex multi-object tracking. In: CVPR 2009, pp. 240–247 (2009)
9. Huang, T.S., Llach, J., Zhang, C.: A method of small object detection and tracking based on particle filters. In: 19th IEEE International Conference on Pattern Recognition, ICPR (2008)

10. Guo, C., Lu, Y., Ikenaga, T.: Robust online tracking using orientation and color incorporated adaptive models in particle filter. In: 4th International Conference on New Trends in Information Science and Service Science, pp. 281–286 (2010)
11. Hartley, R., Zisserman, A.: Multiple View Geometry in Computer Vision. Cambridge University Press, Cambridge (2003)
12. Campbell, C., Ying, Y.: Learning with support vector machines. Synth. Lect. Artif. Intell. Mach. Learn. $5(1)$, 1–95 (2011)

Block-Based Global and Multiple-Reference Scheme for Surveillance Video Coding

Liming Yin[1,3], Ruimin Hu[1,2(✉)], Shihong Chen[1], Jing Xiao[1], and Minsheng Ma[1]

[1] NERCMS, School of Computer, Wuhan University, Wuhan 430072, China
{yinlm,hrm}@whu.edu.cn
[2] Collaborative Innovation Center of Geospatial Technology, Wuhan 430072, China
[3] School of Computer Science, Hubei University of Science and Technology, Xianning 437100, China

Abstract. There are often lots of periodic motions in the background of surveillance videos, such as countdown traffic lights, LED billboards and etc. The conventional motion-compensation scheme and the existing frame-based single background reference scheme cannot eliminate this kind of redundancies efficiently, especially when the cycle time exceeds the maximum GOP size. In this paper, we propose a block-based global and multiple-reference scheme to solve this problem. Firstly, the background is modeled on the basis of co-located blocks but not frames, which makes it possible to realize an adaptive block-level background updating. Secondly, multiple background blocks can be kept for one block location, which makes it suitable for modeling periodic background. Thirdly, the scheme enables global reference, which further eliminates the extensively existed redundancies among GOPs in surveillance videos. Experimental results show that the proposed scheme achieves better rate-distortion performance over the existing frame-based single background reference scheme in most cases.

Keywords: Surveillance video coding · Background modeling · Long-term reference

1 Introduction

The widely deployed surveillance cameras generate huge amount of video data every day. To minimize the cost of storage and transmission, many approaches have been proposed to improve the compression efficiency of the existing video coding standards. Background-modeling-based coding scheme is one of the useful facilities for efficient surveillance video coding, because the surveillance videos are often captured from stationary cameras.

The original motivation of background-modeling-based coding scheme is as following. In surveillance videos or other videos from stationary cameras, there usually exist a number of exposed background regions (EBR) [1, 2]. An EBR is a static background region which being occluded by some foreground objects for a while, then

© Springer International Publishing Switzerland 2015
Y.-S. Ho et al. (Eds.): PCM 2015, Part I, LNCS 9314, pp. 212–222, 2015.
DOI: 10.1007/978-3-319-24075-6_21

reappears after the foreground objects moving away. In general, the occlusion interval is too long for a typical H.264/AVC encoder to keep a frame in its short-term reference frame (STR) lists, so if without long-term reference (LTR) frames, the EBRs tend to be encoded again without any appropriate reference when reappearing, which results in a loss of compression efficiency. To achieve better compression efficiency, we should recognize the EBRs as early as possible, and keep the corresponding reconstructed pictures in the LTR frame lists for later referencing.

One important task of background-modeling-based coding scheme is background generation, in which the EBRs are collected. Many algorithms such as GMM [3] and Mean-Shift [4] have been used for this task, but the overhead on memory and computation is not affordable for most practical systems. To avoid this problem, Zhang et al. proposed a segment-and-weight based running average (SWRA) method [5], which can perform background generation with much less cost of memory and modeling time.

Another important task of background-modeling-based coding scheme is background updating. Background may keep static for a certain period, but change occurs often, so the background frame should be updated in time for better prediction. The key problem in background updating is not the updating operation itself, but the determination of appropriate time to update the background. A straight and simple solution for this task is to perform updating every N frames, e.g. N = 900 in Zhang's experiments [2]. A more flexible solution is to perform updating only when the change of the background exceeds some limits, e.g. 70 % in Paul's work [6].

The state-of-the-art works such as BMAP [2] and McFIS [6] are mostly frame-based, and keep only one background frame for each GOP at a time. Frame-based means that the result of background generation is a complete frame, and the background frame should be updated as a whole. For convenience in this paper, we term this kind of scheme as frame-based single background reference scheme.

An illustration of the frame-based single background reference scheme is shown in Fig. 1.

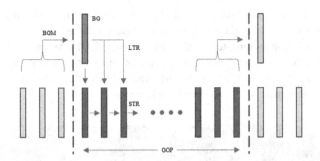

Fig. 1. Illustration of frame-based single background reference scheme, bi-directional references are ignored for clarity. It is notable that there is one and only one background frame modeled as long-term reference for each GOP.

2   Analysis

The existing frame-based single background reference scheme is efficient in many situations, but there are at least three problems.

Firstly, there usually does not exist a common background updating frequency suitable for all regions of a surveillance video. For example, we can see from the top row of Fig. 2 that a low updating frequency is appropriate for those static regions, while for those dynamic regions, it is not the case. Moreover, as the bottom row of Fig. 2 shows, for those rotating regions, a useful modeling result is usually difficult to achieve, but we still have to perform background modeling for these regions and coding the modeling result, which leads to an extra bits occupation. Apparently, background modeling and updating on the basis of frame is not an ideal scheme.

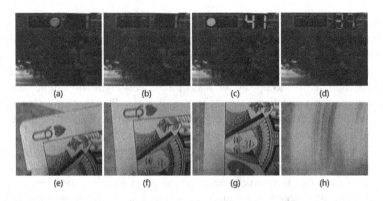

Fig. 2. Illustration of frames with regions of different motions. (a)–(c) are co-located parts of *Intersection*, in which there are several state transition regions. (e)–(g) are co-located parts of *Cactus*, in which there exist a large rotating region. (d) and (h) are the corresponding modeling results using SWRA [5].

Secondly, the reappeared background may be coded and transmitted repeatedly in the existing frame-based single background reference scheme, because it keeps only one background frame as long-term reference for each GOP. As shown in Fig. 3, the vehicles replace the road as the new background at some time, and at some later time, the reappeared road replaces the vehicles vice versa. In this case, the road is coded and transmitted twice as part of the modeled background.

Fig. 3. Illustration of periodic background (from left to right). We can see that if only one background is kept at a time, the reappeared background will be coded and transmitted repeatedly.

Thirdly, when a new GOP starts because of random access (size limitation), the background must be coded and transmitted again, even if the new background is almost the same with the previous one, as Fig. 4 shows.

Fig. 4. Illustration of the similarities between modeled background frames. We can see that only a small part of the background frame changes while the rest is almost the same when a new GOP starts.

Some other common periodic motions in surveillance videos are shown in Fig. 5, which cannot be compressed efficiently using the conventional motion-compensation scheme and the existing frame-based single reference scheme.

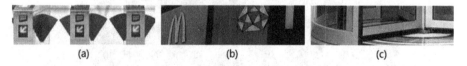

(a)   (b)   (c)

Fig. 5. Illustration of surveillance scenes with periodic motions: (a) ticket gate, (b) LED billboard, (c) revolving door (images are all from internet).

3   The Proposed Scheme

As discussed above, the existing frame-based single background reference scheme (FSR) is inefficient in many situations. In this paper we propose a block-based global and multiple-reference scheme (BGMR). There are **three major improvements** compare to the existing methods. Firstly, the background is modeled on the basis of co-located blocks but not frames, which makes it possible to realize an adaptive block-level background updating. Secondly, multiple background blocks are kept for one block location, which makes it suitable for modeling periodic background. Thirdly, the scheme enables global reference, which eliminates the extensively existed redundancies among GOPs in surveillance videos.

BGMR is a further improvement of our previous work BBM [7]. In the proposed scheme, we focus on the general periodic motions, such as periodic state transition and continuous rotation, but not only EBRs, which keep stationary for several consecutive frames in BBM. Besides, the redundancies among GOPs are eliminated to a large degree by introducing a new global reference scheme.

Since background modeling is the basis of background referencing, for simplicity in the following of this paper, we use the term *reference* to denote the background modeling and referencing.

3.1 Block-Based Reference Scheme

In the frame-based background modeling and updating scheme, the differences of motion properties among regions are ignored completely, thus we have to choose a common background updating frequency suitable for all regions, which is impossible in most situations. To solve this problem, we propose a block-based background modeling and updating scheme. The proposed scheme enables background modeling and updating on the basis of blocks independently, which is shown in Fig. 6.

In block-based reference scheme, background modeling and updating can be performed at a higher frequency for those dynamic regions, while for those static regions, a lower frequency is enough. We can even skip the regions which is difficult to achieve a meaningful background modeling result. In this way, a more efficient background prediction can be realized at the similar cost of frame-based background reference scheme. Besides, this block-based scheme is also the basis for the following two improvements.

Fig. 6. Illustration of block-based background modeling. In this scheme a frame is firstly divided into blocks, background modeling and updating are then performed on the basis of blocks independently. The blocks in green indicate that there is no background block modeled for this block location currently, and the others are the modeled background blocks.

3.2 Multiple-Reference Scheme

To avoid repeated modeling and coding for periodic background, we propose a multiple background reference scheme, which is block-based, and can keep multiple background modeled for one block location at a time. Because the previously modeled background is kept but not replaced, it is unnecessary to perform background modeling and coding again when the periodic background reappears.

In frame-based scheme, multiple reference frames usually consume a large amount of memory, and need much time to find the best matched reference block for all block locations, therefore the maximum number of reference frames is relatively small, e.g. 16 in H.264 standard.

In block-based scheme however, multiple reference blocks are modeled and kept only for those dynamic or periodic regions, which is usually a small portion of the whole scene, therefore the overall memory consumption will be much lower.

Besides, the block-based nature of the proposed scheme makes it possible to save enough consecutive co-located blocks to detect periodic motions. For example, the memory needed for 1800 (corresponding to 1 min at 30 fps) consecutive 16×16 YUV420 blocks is only 675 KB, which is affordable for many practical systems. Additionally, based on the assumption that the camera is stationary, we also have enough time to traverse all of the block locations. In this way, the most frequently reappeared blocks can be selected as reference blocks, thus not only the EBRs but also the general periodic motions can be effectively detected, e.g. the ticket gates and revolving doors. An illustration of multiple background modeling result and multiple-reference scheme is shown in Fig. 7.

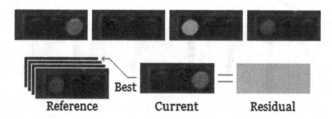

Reference   Current   Residual

Fig. 7. Illustration of multiple background modeling results for one block location and the multiple-reference scheme, which is suitable for periodic background. In the top row, there are four background modeling results for one block location of *Intersection*, where each result represents a background state. In the bottom row, all of the four result images are kept, and the best matched one is chosen as reference when coding.

To find the best reference blocks efficiently, the search process can be well optimized based on the assumption that the camera is stationary. On the one hand, the background search range of each block location can be considered as zero, on the other hand, the features, e.g. mean value of the block, can be extracted when a background block is modeled, and a pre-filter process will further improve the search performance.

3.3   Global Reference Scheme

For random access, the sequences to be encoded are firstly divided into GOPs (Group of Pictures). If the size of the current GOP exceeds the preset maximum size, or the differences between two frames exceed a certain limit, a new GOP should start. In general video coding, the reference lists should be cleared to ensure correct decoding.

In surveillance video coding however, there are lots of similar regions between two successive GOPs as discussed above. If the modeled background blocks in reference lists are cleared when a new GOP starts, a similar modeling result will be generated and a number of bits should be allocated again to represent this result. If the modeling result requires a large number of bits, e.g., the multiple modeling results for periodic background, the redundancies from repeated background modeling will not be ignored. To eliminate the redundancies from periodic motions when the cycle time is longer than the GOP size, the cross-GOP background reference is necessary.

To solve these problems we propose a global reference scheme, which is also block-based. In the proposed scheme, the background blocks are modeled and used as common references among GOPs. That is, the background blocks will not be cleared when a new GOP starts. For random access in global reference scheme, e.g. copy a video segment to a movable disk, the referenced global background blocks should also be extracted along with it. An illustration of global reference scheme is shown in Fig. 8.

The LRU (Least Recently Used) strategy is adopted to eliminate the excessive global background blocks when the reference lists overflow.

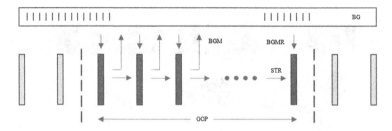

Fig. 8. Illustration of global reference scheme, bi-directional references are ignored for clarity. To avoid the repeated background modeling and coding, the reference lists (modeled background blocks) are not cleared when new GOP starts. It is notable that the first frame of GOP may also reference the background blocks.

3.4   Costs

The proposed scheme will add complexity to both encoder and decoder, which mainly comes from the multiple background modeling and reference. Since the block-based background modeling is relatively simple and effective [7] (only to determine if two blocks are similar or not), and the regions which are difficult to model have been skipped, the costs is affordable for many practical systems.

Memory consumption is another issue because we have to save the global reference blocks in the proposed scheme. For a specific scene however, the periodic motions are often negligible with respect to the long-term encoding, which makes the proposed scheme applicable for surveillance video.

Besides, some extra bits has to be allocated to indicate the best background blocks for reference. In our experiments, these bits can be compressed efficiently using arithmetic coding because of the strong correlation of the reference blocks between two consecutive frames.

4   Experimental Results

To evaluate the overall performance of BGMR, we conducted a series of experiments on the following three encoders: (1) encoder without background reference extension, (2) encoder with FSR extension, and (3) encoder with BGMR extension. We choose H.264/AVC reference software JM18.6 [8] as the standard encoder, and also as the basis to implement FSR and BGMR extensions. To simplify the experiments, the dual-frame motion compensation structure is adopted, and only the IPPP bitstream structure is considered. Some other features such as high quality background image [2, 6] and foreground-background-hybrid block [2] are also ignored for simplicity and clarity, because they are not the fundamental differences between FSR and BGMR, and we presume that each of them has the similar effects on compression efficiency for both models. In the following we also use FSR to denote our implementation of BMAP with these simplifications.

Based on the default settings of JM18.6 and the simplifications mentioned above, some common parameters applied for these three encoders are determined and listed in Table 1. We also turn on the LTR option of the standard H.264/AVC encoder and use the first frame as its content for a fair comparison. The test sequences used in our experiments are listed in Table 2 and shown in Fig. 9, which are all public datasets. Some of them are scaled or cropped to speed up the testing process. Column *Frames* in Table 2 is the frames to be encoded, while the first 120 frames of each sequence are reserved as training set for FSR. In this paper, all experiments are conducted on a computer with an Intel(R) Xeon(TM) Processor E31220 @3.10 GHz and 16 GB memory.

Table 1. Common Parameters of the Encoders

Parameter	Value	Parameter	Value
ProfileIDC	High	NumberReferenceFrames	2
LevelIDC	40	NumberBFrames	0
IntraPeriod	0	SymbolMode	CABAC
IDRPeriod	0	SearchMode	EPZS
RDOptimization	High	SearchRange	32

Table 2. Test Sequences Used in Our Experiments

Sequence	Resolution	Frames	Motion
Silent	352 × 288	180	Low
Hall	352 × 288	180	Medium
Office	720 × 576	300	Medium
Crossroad	720 × 576	300	High
BasketballDrill	832 × 480	300	Medium
Intersection	176 × 144	880	Periodic

The compression performance of BGMR is evaluated using the Bjontegaard-Delta measurement. The BD-Rate of BGMR compared with AVC and FSR for all test sequences are listed in Table 3. The rate-distortion curves for all test sequences are shown in Fig. 10. From Table 3 and Fig. 10 we can see that BGMR achieves better rate-distortion performance over FSR in most cases. For sequence *silent*, all three encoders show almost the same performance, because there are few motions therefore few EBRs and almost no periodic motion. For sequence *Intersection*, BGMR outperforms the other two encoders significantly, because there are lots of periodic motions, and the redundancies from periodic motions are eliminated in a large degree. The other sequences are public surveillance videos with few periodic motions, we can see that BGMR also performs better than FSR especially when there are more background changes, which is consistent with the previous analysis.

It is notable that the overheads of background, such as the background modeling result and the extra background block indicators are all included in the result data for those short sequences (all except *Intersection*). For sequence *Intersection* the global reference blocks (constructed based on their frequencies of recurring) are not included, because the overheads are often negligible since they are shared in a long period.

Fig. 9. Test sequences used in our experiments.

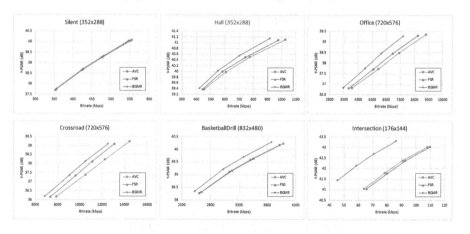

Fig. 10. RD-curves for test sequences.

Table 3. BD-Rate of BGMR Compared with AVC and FSR

Sequence	AVC	FSR	Sequence	AVC	FSR
Silent	−0.54	−0.10	Hall	−11.20	−7.53
Office	−16.98	−12.23	Crossroad	−13.00	−6.13
BasketBallDrill	−6.10	−5.68	Intersection	−31.33	−32.90

5 Conclusion

In this paper, we propose a block-based global and multiple-reference scheme to improve the compression efficiency for surveillance video coding. Since most of the surveillance videos are captured from stationary cameras, and there are often lots of periodic motions in the background, we try to improve the compression efficiency by eliminating the redundancies not dealt with by the existing coding schemes: (1) the redundancies come from frame-based background reference scheme, (2) the redundancies come from single background reference scheme, and (3) the redundancies come from repeated background modeling when new GOP starts. Experimental results show that the proposed scheme achieves better rate-distortion performance over the existing frame-based single background reference scheme in many cases, especially when there are lots of periodic motions. In the proposed scheme however, more memory is required in general and a little more complexity will be added to both encoder and decoder.

Acknowledgements. This research was partly supported by the National Nature Science Foundation of China (No. 61231015), National High Technology Research and Development Program of China (863 Program) No. 2015AA016306, EU FP7 QUICK project (PIRSES-GA-2013-612652), China Postdoctoral Science Foundation (2014M562058), Fundamental Research Funds for the Central Universities (2042014kf0025), National NSFC (No. 61271256).

References

1. Ding, R., Dai, Q., Xu, W., Zhu, D., Yin, H.: Background-frame based motion compensation for video compression. In: Proceedings of IEEE International Conference on Multimedia Expo, pp. 1487–1490 (June 2004)
2. Zhang, X.G., Huang, T.J., Tian, Y.H., Gao, W.: Background-modeling-based adaptive prediction for surveillance video coding. IEEE Trans. Image Process. **23**(2), 769–784 (2014)
3. Haque, M., Murshed, M., Paul, M.: Improved Gaussian mixtures for robust object detection by adaptive multi-background generation. In: Proceedings of IEEE 19th International Conference on Pattern Recgnition, pp. 14 (December 2008)
4. Liu, Y., Yao, H., Gao, W., Chen, X.L., Zhao, D.B.: Nonparametric background generation. J. Vis. Commun. Image Represent. **18**(3), 253263 (2007)
5. Zhang, X.G., Tian, Y.H., Huang, T.J., Gao, W.: Low-complexity and high-efficiency background modeling for surveillance video coding. In: Proceedings of IEEE International Conference on Visual Communication and Image Processing, pp.1–6 (November 2012)

6. Paul, M., Lin, W., Lau, C.T., Lee, B.-S.: Explore and model better I-frame for video coding. IEEE Trans. Circ. Syst. Video Technol. **21**(9), 12421254 (2011)
7. Yin, L., Hu, R., Chen, S., Xiao, J., Hu, J.: A block-based background model for surveillance video coding. In: Proceedings of Data Compression Conference, pp. 476 (April 2015)
8. Shring,K.: Joint model reference software JM18.6 2014. http://iphome.hhi.de/suehring/tml/download/jm18.6.zip

Global Object Representation of Scene Surveillance Video Based on Model and Feature Parameters

Minsheng Ma[1,2], Ruimin Hu[1,2(✉)], Shihong Chen[1], Jing Xiao[1], Zhongyuan Wang[1], and Shenming Qu[1,3]

[1] National Engineering Research Center for Multimedia Software, Computer School of Wuhan University, Wuhan, China
maminsheng@whu.edu.cn, hrm1964@163.com,
chen_lei0605@sina.com, jing@whu.edu.com
[2] Collaborative Innovation Center of Geospatial Technology, Wuhan, China
[3] School of Software, Henan University, Kaifeng, China

Abstract. Scene surveillance video is a kind of video which are captured by stationary camera for a long time in specific surveillance scene. Due to regular movement of vehicles with similarity structures, models and appearances, surveillance video produce amounts of redundancy and needs to be efficiently coded for transmission and storage. In this study, we investigated the video redundancy generation mechanism of scene surveillance, exploit and presents a new redundancy type-Global Object Redundancy (GOR), it is proven that the vehicles occupy the mostly proportion which caused by amounts of vehicles movement. Secondly, aiming at global vehicle objects representation and GOR elimination, a global object representation scheme of scene surveillance video based on model and feature parameters is introduced, by establish a global knowledge dictionary and feature parameter sets, low bitrate with high quality compression can be achieved due to only few vehicle objects individual semantic and feature parametric be transfer and coded. Finally, we carried out preliminary experiments in simulation environment and shows that the object representation scheme can effectively improve the compression of long-term archive surveillance video which with a certain of image quality assurance.

Keywords: Scene surveillance video · Global objects redundancy · Global objects representation · Global texture dictionary · Model-based coding

1 Introduction

Surveillance video data volume increasing dramatically in big data era, needs to efficiently coded for storage and transmission. Scene surveillance video is defined as a kind of surveillance video captured by stationary cameras for a long time in specific scene, amounts of redundancy produces and needs to efficiently coded for transfer and stored due to vehicles moving which with similarity structure, model and appearances. The vehicle in the videos is named global object because of its global and commonality

© Springer International Publishing Switzerland 2015
Y.-S. Ho et al. (Eds.): PCM 2015, Part I, LNCS 9314, pp. 223–232, 2015.
DOI: 10.1007/978-3-319-24075-6_22

characteristics. This paper aims at global objects representation and coding scheme of scene surveillance video, to explore its redundancy generation mechanism, presents the concept of global object redundancy (GOR).

On the basis of the commonality and characteristics analysis of scene surveillance video, a global object representation scheme based on model and feature parameter is proposed. This scheme split the vehicle object from original video firstly, then coding the background video and vehicle object video individually, eliminating the scene redundancy with conventional coding pattern, by utilize the proposed global presentation scheme to encoding global vehicle objects video, the final bitstream is made up of the bitstream of the background video, the bitstream from vehicle parameters and the bitstream from the residuals, the vehicles and original video can be reconstructed and recovered at the decoder side. Due to only few vehicle objects individual semantic and feature parametric information be transfer and coded at the encoder side, dramatically bitrate descend with a certain of image quality assurance. Preliminary experiments carried out in simulation environment and results shown that our coding scheme can effectively improve the compression of long-term archive surveillance video.

Model-based video coding (MBC) focuses on modeling encoding of the structural visual information in the image or video. The representative work is proposed in [1]. Since only semantic high-level semantic and parametric information are transmitted, high compression can be achieved. A series of methods are proposed for model-based localization, recognition and tracking of road vehicles [2–5]. In Musmann's viewpoint [6], MBC is extended to be a more generalized coding system which covers object-based video coding, knowledge-based video coding, semantic-based video coding, etc. Feature based method through SURF (Speeded Up Robust Features), PCA (Principal Components Analysis) and LDA (Linear Discriminate Analysis) to reduce the dimension, can effectively reduce the features amount of image and video presentation [7]. In [8], cloud-based distributed image coding scheme was proposed to explore inter correlation in cloud to further reduce the redundancy in images, spatial correlation in the cloud. But different from image compression, the main compression gain in video coding comes from reducing the inter frame redundancy. Therefore, it is difficult to directly apply cloud image compression techniques to inter frame video compression. The existing video coding scheme can't adapt the characteristics of specific scene surveillance videos, encounter bottleneck in big data era and needs put forwards new representing method.

The rest of this paper is organized as follows: In Sect. 2, we analyze the redundancy generation mechanism and characteristics of scene surveillance video, proposed a global object representation framework for scene surveillance video coding. In Sect. 3, a framework of global objects representation is proposed for scene surveillance video coding. In Sect. 4, preliminary experiments were conducted to prove that the global objects redundancy removal can effectively improve the compression efficiency. And the conclusions and discussion are indicated in Sect. 5.

2   Global Coding Scheme of Scene Surveillance Video

2.1   The Generation Mechanism and Features of Global Redundancy

Scene surveillance video generally captured by a camera which fixed on a stationary place. The video background maintains relatively stable in short-term and slowly changing over time. The unchanged objects like buildings, the regularity of light changes with time, and the seasonal changes of trees, leaves, and flowers, etc. This regular changing of relatively fixed background video information generates scene redundancy. Traffic sequence showing the intersection in Karlsruhe respectively reflect the intersection states under different season and weather conditions (as shown in Fig. 1). The latest AVS-S2 standard removed the scene redundancy by background modeling and added predictive coding technology [9], coding efficiency was nearly reached four times compared with H.264 standard. In this paper, season and weather conditions based background modeling and scene redundancy elimination are not the key problem we considering.

Fig. 1. The intersection states under different season and weather conditions. (a) Normal conditions, (b) heavy snowfall, (c) heavy fog (d) winter, snow on lanes

The vehicles and pedestrian are the main moving objects of scene surveillance videos. Compared with pedestrian and other moving objects, vehicle objects which with strong structure, similar in appearance and texture homogeneity, occupy a large of proportion of the video under the specific situations (such as crossroad with large traffic flow). Meanwhile, different vehicle model with texture similarity, same type vehicles have 3D characteristics similarity. Texture similarity exists between different moving object. Semantic consistency exists between the same type objects. Texture similarity and semantic consistency constitute the main source of global object redundancy (GOR).

Scene redundancy, and the global object redundancy which generated by vehicle moving with similarity appearance and structures composed the scene surveillance video global redundancy (GR). There are large amount of global redundancy in long-term archive surveillance videos, which provide a huge space to further improve the efficiency of video compression.

2.2   Scene Surveillance Video Global Coding Scheme

Based on the composition analysis and generation mechanism of GOR, we propose a global object representation framework for scene surveillance video coding, concentrates the global vehicle model and feature parameters representation. The work will assume the camera parameter is known to us, not considering the influence of pedestrian and other moving objects. Vehicle object 3D CAD models and global texture information preformed and stored in cloud [5], resorts to feature matching to find the corresponding model ID, due to only transfer the object personality semantic and feature parameter information, high compression efficiency can be achieved. In the previous works of Xiao et al. [10], some efforts have been get effective verification, but there are still many problems have been unsolved, we will continue to carry on the beneficial exploration in this research area.

The schematic diagram of scene surveillance video global coding is shown in Fig. 2. By background modeling and movement detection, we detect and split the vehicles from the input original video firstly, then separately encoding the background video and vehicle video with conventional coding pattern and our proposed global object representation scheme individually. The final bitstream is made up of the bitstream of the background video, the bitstream from vehicle parameters and the bitstream from the residuals. The vehicles and original video can be reconstructed and recovered at the decoder side.

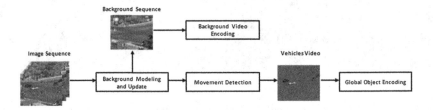

Fig. 2. Schematic diagram of scene surveillance video global coding

3   Global Object Representation Based on Model and Feature Parameters

In this section, we introduce a global object representation scheme based on model and feature parameters, the flowchart of vehicle video global objects representation is shown in Fig. 3. (1) The vehicles video as input get the vehicles 2D location and types information by vehicle localization and discrimination process; (2) The CAD models database for all types vehicle have been established in advance, then create global texture over-complete dictionary for all types vehicle texture information by utilize sparse coding technology, build the global vehicle object database and stored in cloud for retrieval, which including vehicles 3D model database and global texture dictionary; (3) Matching the vehicle 2D object and 3D model by 2D-3D correspondence and pose determination, and get the vehicle pose parameters and corresponding model ID number; (4) Constitute the global feature parameter sets which including vehicles location, pose, texture description parameter and model ID number. (5) Utilize the vehicle global feature parameter sets to conduct motion estimation and motion compensation, then transmit generated parameter residual information to encoder; (6) Parameter residual information and illumination compensation parameters together through the entropy coding are transmitted to the decoder for vehicle and video reconstruction.

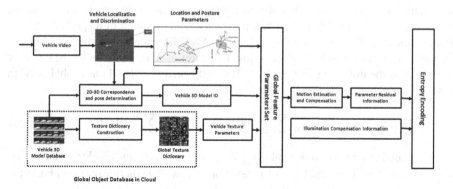

Fig. 3. Flowchart of vehicle video global objects representation

3.1 Model and Shape Representation

We have collected coarse and fine 3D vehicle models and stored in cloud for retrieval, including most types of vehicles in reality [4]. A 3D wire frame model is described by 12 shape parameters. Generic 3D vehicle model can be deformed to fit with different vehicles. Most vehicles should be belongs to a subset of the 12-D shape parameters space.

3.2 Location and Pose Representation

The geometric descriptions (for example, 3D wireframe models) of vehicles will be established in advance. With the ground-plane constraint (GPC) adopted, the pose of a vehicle can be determined only by its position (X, Y) on the ground plane and its orientation θ about the vertical axis of the world coordinate system (WCS) [4, 5]. With these three pose parameters, we can project the model into images. In practice, for model-based vehicle localization and recognition, we need to confirm first an initialized pose to project the 3D model into the image plane for matching and optimization. For moving vehicles, we can exact ROI by motion detection and obtain vehicle regions by motion and shape information. The pose parameters can be described as follows:

$$\rho = [X, Y, \theta]^{T} \tag{1}$$

The vehicles will be respectively detected in each images, we predict the parameters of these in subsequent frames and make associations between detections in consecutive frames. For tracking, each vehicle is represented by a vector (id, cx, cy, vx, vy, w, h), representing the id, center position, speed, and size of the bounding rectangle in image. The predicted vehicle descriptions are matched with the new detections and the position parameters are updated.

3.3   Texture Parameters Representation

By utilize sparse coding technology, we can learnt a global texture over-complete dictionary for all types vehicle texture information from captured vehicles in the videos, which used for texture description and parameter representation of a specific vehicle. The dictionary will occupy extra storage due to store the prior knowledge. A two-level dictionary is created: one level is the common textures of all the vehicles whilst the other level records the difference information. The cost function of level one global moving vehicle texture dictionary can be described as follows:

$$(D_1, \alpha_1) = argmin_{D_1, \alpha_1} \sum_{c=1}^{C} ||y_c - D_1 \alpha_1^c||_2^2 + \tau \sum_{c=1}^{C} ||\alpha_1^c||_1 \tag{2}$$

After obtained the difference value $r_{c,m}$ of the original moving vehicle and the reconstructed global vehicle which with the level one knowledge dictionary, then build the level two texture knowledge dictionary, the cost function is described as:

$$(D_2^c, \alpha_{2,c}) = argmin_{D_2^c, \alpha_{2,c}} \sum_{m=1}^{M} ||r_c^m - D_2^c \alpha_{2,c}^m||_2^2 + \tau \sum_{m=1}^{M} ||\alpha_{2,c}^m||_1 \tag{3}$$

Where D_1 is level one knowledge dictionary, C is types number of global moving vehicles, c is serial number of global moving vehicle types; y_c is the texture information of all types global moving vehicles; α_1 is coding coefficient; τ is balance factor, according to the actual situation and experience to set it, the bigger value indicate the coding coefficients more sparse; D_2^c is level two knowledge dictionary, M is the number of specific type vehicle, m is the serial number of specific one vehicle; $\alpha_{2,c}$ is coding coefficient. Based on the difference value r_c, the level two knowledge dictionary can be obtained, which including all types global moving vehicles 3D structure and personality texture knowledge dictionary.

3.4   Illumination Parameters Representation

Complex lighting condition is a key adverse factor in the process of vehicle reconstruction. Inspired by the method of estimating the natural illumination conditions from a single outdoor image [11], we improved it and used for illumination and shadow parameter estimation of scene surveillance video. For an input singe image, we render the most likely sky appearance using the sun position, which relies on a combination of the sky, the vertical surfaces, the ground, and the convex objects in the image, then fitting the sky parameters using the methods. We can realistically insert a 3D vehicle into the image. The shadows on the ground, and shading and reflections on the vehicle are consistent with the image. When we placed the vehicle object to the corresponding location and adjust its pose, the realistic illumination and shadow will be reflect in the image.

When coding moving vehicles video which is based on the global objects representation framework, only vehicles texture and model, location and pose description parameters are described, and the vehicle model ID, side information, residual information and illumination compensation information are transmitted to reconstruct the video, the global motion vehicles of video data information are transformed into only contained slight information description, that can remove the global redundancy of moving vehicles effectively. After removing the scene redundancy, the framework further removes global objects redundancy in surveillance videos and improves the coding efficiency. Compared with the works of Xiao et al. [10], we further improved the original coding and representation framework, meanwhile added the specific description of the model and feature parameters, make it more operational.

4 Experiments and Results

4.1 Experiment 1

We firstly conducted preliminary experiments to measure the causation of global objects redundancy, to demonstrate the potential of eliminating global redundancy on the bitrate savings in scene surveillance video coding. All video clips are comes from public data sets and vehicles are the mainly moving objects, the experiments carried out on a personal computer with a P4 2.6G central processing unit and 2G RAM. For the each original video, we eliminate all the vehicles by background modeling technologies and built the background video. Four representative scene surveillance clips and its corresponding background images are shown in Fig. 4.

Fig. 4. Scene surveillance video global coding experiment

We encoding the above the scene surveillance video sequence and its background videos (Bg) with HEVC reference code HM16.2, to illuminate how much bitrate are used to encode moving vehicles, the detailed configuration information was included in encoder_lowdelya_P_main file. Set the quantization parameter (QP) individually as 22, 27, 32, 37. Table 1 lists the test sequences information and encoding performance gains of the video clips.

The experiments result shows that compared with encoding original scene surveillance video (Ori), on average around 85.31 % bitrate can be saved when encoding the background video, which is used to record moving vehicles. Although the improvement will be lower if pedestrian and other moving objects involved, it still able to show that

Table 1. Encoding performance gains of videos

Video sequence	Resolution	Coded frames	Bg vs. Ori
A	320 × 240	150	82.64 %
B	720 × 576	150	81.65 %
C	360 × 288	122	85.64 %
D	352 × 288	1000	88.32 %
Average	–	–	85.31 %

the newly discovered global redundancy can be a new chance in the compression of scene surveillance video. Therefore, further eliminating the global redundancy caused by moving objects will lead to a considerable improvement of the compression rate.

4.2   Experiment 2

Two global coding experiments are carried out to verify the global coding performance based on the latest HEVC reference code HM16.2 in simulated environment. The test sequences are taken from our own cameras with two remote controlled toy cars moving repeatedly on a flat ground. Figure 5 shows the reconstructed results of two global coding test sequences, from left to right are respectively is original video frame, background frame and reconstructed frame. The reconstructed frame is generated by the overlay of the background frame and the reconstructed vehicle objects. The specific process can be described as: Firstly find the corresponding vehicle 3D model in the 3D model library, then placed the 3D vehicle object to the corresponding location and adjust its pose according to its location and pose parameters; By utilize the two-level global texture dictionary to generate the personality texture parameters of specific vehicle object, descript and represent the vehicle texture information; Finally the realistic illumination and shadow will be reflect in the image by using the mentioned method in this paper.

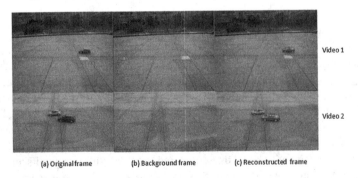

(a) Original frame   (b) Background frame   (c) Reconstructed frame

Fig. 5. Reconstructed result of global coding test sequence

Table 2 shows that compared with encoding original scene surveillance video (Ori), respectively 82.1 % and 85.3 % bitrate can be saved when encoding the background video. The vehicles 3D model library and the global texture dictionary are established in advance and size is fixed, can be repeatedly used for the all types scene surveillance video. And illumination simulation parameters are applicable to all vehicle objects in the video frames. When coding the global vehicle objects, vehicles can be recovered and reconstructed based on 3D model ID, location and pose parameters, texture presentation parameters, illumination compensation parameters, side information and residual information, etc. Since only object semantic parametric information are transmission and coded, low bitrate with high quality compression can be achieved. According to our experimental results, the average bitrate can be descent about 85 % when encoding the background video, consider the various parameters factor in global coding (such as pedestrian and other moving objects), a better coding efficiency can be still achieved which compared with the HEVC standard.

Table 2. Encoding performance gains of videos

Test Seq	Resolution	Coded frame	Average coding bits	Bg coding average bits	Average coding saving
Video1	640 × 480	10	72772	13026.19	82.1 %
Video2	1280 × 720	10	110431.25	16233.39	85.3 %

5   Conclusions

In this paper, we explore the redundancy generation mechanism and characteristics that exists in the specific scene surveillance video, put forward the concept of GOR and global objects representation framework based on model and feature parameters. By establish a global knowledge dictionary and feature parameter sets, low bitrate with high quality compression can be achieved due to only few vehicle objects individual semantic and feature parametric information be transfer and coded at the encoder side. Preliminary global representation verification experiments are carried out and show that the compression efficiency of surveillance video can be greatly improved. In particular, our methods will have obvious efficiency enhancement for long-term archive surveillance video which have large vehicles flow. At present, our method does not fully dig the common information among large scale spatio and temporal objects. Our global coding framework based on plenty of preconditions, it will be a time-consuming process but very effective, although many problems have been unsolved restricted with the current technologies, we still carried on the beneficial exploration and get some inspirations in this research area.

For future work, we plan to implement the framework on top of state-of-the-art high performance computing platforms. The proposed global objects representation method not only provide an efficient coding scheme, but also reserves the important information of moving objects for the further utility of surveillance videos. Global object joint representation based on multiple source scene surveillance video is another focus in the future research.

Acknowledgments. The National Nature Science Foundation of China (No. 61231015), National High Technology Research and Development Program of China (863 Program) No. 2015AA016306, EU FP7 QUICK project under Grant Agreement No. PIRSES-GA-2013-612652, China Postdoctoral Science Foundation funded project (2013M530350, 2014M562058), Fundamental Research Funds for the Central Universities (2042014kf0025), Internet of Things Development Funding Project of Ministry of industry in 2013 (No. 25).

References

1. Yao, Z.: Model based coding: initialization, parameter extraction and evaluation. Tillämpad fysik och elektronik, Umeå, p. 164 (2005)
2. Tan, T.-N., Sullivan, G.D., Baker, K.D.: Model-based localisation and recognition of road vehicles. Int. J. Comput. Vis. **27**(1), 5–25 (1998)
3. Zhang, Z., et al.: Eda approach for model based localization and recognition of vehicles. In: IEEE Conference on Computer Vision and Pattern Recognition, CVPR 2007. IEEE (2007)
4. Zhang, Z., et al.: Three-dimensional deformable-model-based localization and recognition of road vehicles. IEEE Trans. Image Process. **21**(1), 1–13 (2012)
5. Lou, J., et al.: 3-D model-based vehicle tracking. IEEE Trans. Image Process. **14**(10), 1561–1569 (2005)
6. Musmann, H.G.: Object-oriented analysis-synthesis coding based on source models of moving 2D-and 3D-objects. In: 1993 IEEE International Conference on Acoustics, Speech, and Signal Processing (ICASSP 1993), IEEE (1993)
7. Bay, H., Tuytelaars, T., Van Gool, L.: SURF: speeded up robust features. In: Leonardis, A., Bischof, H., Pinz, A. (eds.) ECCV 2006, Part I. LNCS, vol. 3951, pp. 404–417. Springer, Heidelberg (2006)
8. Song, X., et al.: Cloud-based distributed image coding. In: 2014 IEEE International Conference on Image Processing (ICIP)
9. Siwei, M., Shiqi, W., Wen, G.: Overview of IEEE 1857 video coding standard. In: 2013 20th IEEE International Conference on Image Processing (ICIP) (2013)
10. Xiao, J., et al.: Exploiting global redundancy in big surveillance video data for efficient coding. Cluster Comput. **18**(2), 531–540 (2015)
11. Lalonde, J.-F., Efros, A.A., Narasimhan, S.G.: Estimating the natural illumination conditions from a single outdoor image. Int. J. Comput. Vis. **98**(2), 123–145 (2012)

A Sparse Error Compensation Based Incremental Principal Component Analysis Method for Foreground Detection

Ming Qin, Yao Lu$^{(\boxtimes)}$, Huijun Di, and Tianfei Zhou

Beijing Laboratory of Intelligent Information Technology, School of Computer Science, Beijing Institute of Technology, Beijing 100081, China
vis_yl@bit.edu.cn

Abstract. Foreground detection is a fundamental task in video processing. Recently, many background subspace estimation based foreground detection methods have been proposed. In this paper, a sparse error compensation based incremental principal component analysis method, which robustly updates background subspace and estimates foreground, is proposed for foreground detection. There are mainly two notable features in our method. First, a sparse error compensation process via a probability sampling procedure is designed for subspace updating, which reduces the interference of undesirable foreground signal. Second, the proposed foreground detection method could operate without an initial background subspace estimation, which enlarges the application scope of our method. Extensive experiments on multiple real video sequences show the superiority of our method.

Keywords: Foreground detection · Sparse error compensation · Incremental principal component analysis

1 Introduction

Foreground detection often plays an important role in surveillance video analysis. Given a n frame video sequence: o_t ($t \in \{1, 2, \cdots, n\}$), foreground detection aims to separate the foreground from the background: $o_t = b_t + f_t$, where b_t is the relatively stable background signal, and f_t is the foreground signal. Oliver *et al.* [13] first adopt Principal Component Analysis (PCA) to estimate the subspace of background signal for foreground detection. From then on, a great number of achievements have been made by various subspace estimation based foreground detection methods [1].

Robust PCA (RPCA) based methods [4,5,20] show promising results in foreground detection field. Many elegant algorithms [10–12] have been developed to solve the sparse and low rank matrix decomposition problem. While RPCA based methods are both novel and effective, many of them work in batch rather than incrementally, which may limit their range of application.

© Springer International Publishing Switzerland 2015
Y.-S. Ho et al. (Eds.): PCM 2015, Part I, LNCS 9314, pp. 233–242, 2015.
DOI: 10.1007/978-3-319-24075-6_23

Recently, some incremental subspace estimation based methods [7,14,15, 17,19] and robust regression based method [18] have also been proposed for foreground detection in an incremental way. However, there still exist a few limitations among them. Some methods in [17,19] implement subspace estimation directly on the original observations, which may bring undesirable foreground information to background subspace estimation. Some other methods in [7,14,15] rely on an initial subspace estimation or require a training phase to obtain the initial subspace estimation, which may limit their ranges of application. Robust regression based method in [18] utilizes evenly sampled historical frames to recover background signal, which may also suffer from the undesirable foreground information in the historical frames.

In this paper, a Sparse error Compensation based Incremental PCA (SCIPCA) method is proposed for foreground detection. In our method, the subspace updating process is implemented with the compensated frames instead of the original ones, which helps to reduce the chance of introducing the undesirable foreground corruptions and therefore supplies robust detection results. Furthermore, the incremental subspace update process could operate in the absence of an initial subspace estimation, enlarging its application scope.

2   Foreground Detection via Sparse Error Compensation Based Incremental PCA

In this section, we first introduce the proposed foreground detection method, and cast it as an optimization problem. Then we propose an optimization algorithm

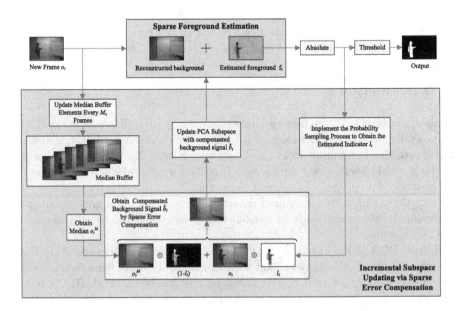

Fig. 1. Block schematic for an overview of the proposed method.

which incrementally detects foreground and updates subspace. A block schematic for an overview of the proposed approach is shown in Fig. 1.

2.1   The Proposed Subspace Based Foreground Detection Model

Given an observation matrix $O = [o_1, o_2, \cdots, o_t, \cdots, o_n] \in R^{d \times n}$ where the column o_t is a d dimension observation vector (o_t is transformed from the t-th frame image), foreground detection aims to separate the foreground signal f_t from the relatively stable background signal b_t in the t-th frame. Therefore, the observation $o_t \in R^d$ can be represented as:

$$o_t = b_t + f_t. \tag{1}$$

The background signal b_t is usually believed to be lying on a low dimension subspace [7,14,15,19] because of its relative stability. In this paper, we adopt PCA subspace to model the background signal, therefore, b_t in Eq. (1) can be further expressed as:

$$b_t = \bar{b}_t + U_t x_t, \tag{2}$$

where $\bar{b}_t = \frac{\sum_{j=1}^{t} b_j}{t}$ is the mean vector of the background for the first t frames, U_t is the $d \times r$ dimension orthogonal basis matrix of the background subspace, and x_t is the reconstruction coefficient.

Because the foreground signal is usually believed to be sparse [4,5,7,14,15, 19,20], the proposed foreground detection model can be formulated as follows:

$$\underset{\bar{b}_t, U_t, x_t, f_t}{argmin} \ \|f_t\|_1 \quad \text{s.t.} \quad o_t = \bar{b}_t + U_t x_t + f_t \tag{3}$$

Given $o_t, \bar{b}_{t-p}, U_{t-p}, x_{t-p}, f_{t-p}$ where $p \in \{1, 2, \cdots, t-1\}$, incremental foreground detection can be done by solving the optimization problem in Eq. (3). However, it's difficult to update $\bar{b}_t, U_t, x_t, f_t$ in Eq. (3) simultaneously.

In this paper, a two-step optimization algorithm is designed to solve the problem by making a reasonable assumption that the subspace change little in q frames (q is a very small number).

In Sect. 2.2, we will elaborate the two-step optimization algorithm in detail. The whole procedure of the proposed algorithm is shown in Algorithm 1.

2.2   Two-Step Optimization Algorithm

The optimization problem in Eq. (3) can be solved by two steps:

(1). Sparse error (foreground) estimation: compute f_t and x_t in Eq. (3) by assuming that the subspace change little in q frames (q is a very small number).

(2). Subspace updating via sparse error compensation: update $\bar{b}_t$ and U_t from the estimated background signal $\hat{b}_t$, which is obtained from a sparse error compensation process.

Algorithm 1. Sparse Error Compensation based Incremental PCA Algorithm

---

Input: The new observation o_t, the background mean estimation $\bar{b}_{t-q}$ and the orthogonal basis matrix estimation U_{t-q} for the past $(t-q)$ frames.

1: Compute f_t by solving Eq. (4) with ADMM method shown in Eq. (6).

2: Obtain the binary foreground detection result $Mask$ with Eq. (7).

3: Compute the Bernoulli distribution p_{ti} defined in Eq. (8) for the i-th pixel in the t-th frame.

4: Sampling from the Bernoulli distribution p_{ti}, obtaining the estimated foreground indicator $\hat{l}_{ti}$.

5: Estimate the background signal $\hat{b}_t$ by the compensation process described in Eq. (11).

6: Update $\bar{b}_t$ and U_t by incrementally computing the mean and SVD of matrix $B = [\hat{b}_1, \hat{b}_2, \cdots, \hat{b}_t]$.

Output: $Mask, \bar{b}_t, U_t, x_t, f_t$

Sparse Error (Foreground) Estimation. As we know, the background signal is relatively stable, therefore we could reasonably assume that the background subspace is nearly unchanged in consecutive q (q is a very small number) frames: $\bar{b}_t \approx \bar{b}_{t-q}$, $U_t \approx U_{t-q}$. Under this assumption, Eq. (3) can be expressed as:

$$\underset{f_t, x_t}{argmin} \|f_t\|_1 \quad \text{s.t.} \quad o_t = \bar{b}_{t-q} + U_{t-q}x_t + f_t \tag{4}$$

If $\bar{b}_{t-q}$ and U_{t-q} are given, foreground f_t and coefficient x_t can be obtained by optimizing the augmented lagrangian form [2] of Eq. (4):

$$\underset{f_t, x_t}{argmin} \|f_t\|_1 + \lambda^T(\bar{b}_{t-q} + U_{t-q}x_t + f_t - o_t) \\ + \frac{\mu}{2}\|\bar{b}_{t-q} + U_{t-q}x_t + f_t - o_t\|_F^2. \tag{5}$$

Equation (5) could be efficiently solved with Alternating Direction Method of Multipliers (ADMM) algorithm [2] in an iteration way:

$$\begin{cases} x_t^{k+1} = U_{t-q}^T(o_t - \bar{b}_{t-q} - f_t^k - \dfrac{\lambda^k}{\mu^k}) \\[2mm] f_t^{k+1} = \text{soft}(o_t - \bar{b}_{t-q} - U_{t-q}x_t^{k+1} - \dfrac{\lambda^k}{\mu^k}, \dfrac{1}{\mu^k}) \\[2mm] \lambda^{k+1} = \lambda^k + \mu^k(U_{t-q}x_t^{k+1} + f_t^{k+1} - o_t + \bar{b}_{t-q}) \\[2mm] \mu^{k+1} = \rho\mu^k \end{cases} \tag{6}$$

where k is iteration number, $\text{soft}(\alpha, \beta) = \max(\alpha - \beta, 0) - \max(-\alpha - \beta, 0)$ is the soft-thresholding operator, and $\rho > 1$ is a parameter in ADMM method [2].

Then the binary foreground detection result $Mask_t$ for the t-th observation o_t can be obtain as follows:

$$Mask_{ti} = \begin{cases} 1, & \text{if } |f_{ti}| > th \quad (i \in \{1, 2, \cdots, d\}) \\ 0, & \text{else} \end{cases} \tag{7}$$

where th is the threshold, f_{ti} denotes the i-th dimension of the estimated foreground f_t.

Subspace Updating via Sparse Error Compensation. With f_t and x_t obtained, a subspace updating algorithm shall be designed for estimating $\bar{b}_t, U_t$.

Some subspace updating methods [16,17,19] directly estimate the subspace on observation o_t, which may introduce the undesirable foreground information and interfere with the accurate estimation on the subspace. To reduce the interference of foreground corruption, we propose a sparse error compensation based incremental subspace updating method, where the subspace estimation is implemented on a compensated background signal $\hat{b}_t$, instead of observation o_t.

To compensate the foreground information of o_t, an binary indicator l_t which estimates the foreground positions in observation o_t shall be first obtained. The binary indicator l_t for the t-th frame can be defined as:

$$l_{ti} = \begin{cases} 0, \text{if the } i\text{-th pixel belongs to foreground} \\ 1, \text{if the } i\text{-th pixel belongs to background} \end{cases} \tag{8}$$

The binary indicator l_t can be estimated by the aid of the estimated sparse error f_t. The absolute value of f_{ti}, which denotes the estimated sparse error on the i-th pixel of the t-th frame, could indicate the possibility of the i-th pixel belonging to the background. The smaller the absolute value of f_{ti} is, the more likely that the i-th pixel belongs to the background.

Based on the above observation, the probability distribution of the i-th pixel belonging to the background could be defined as a Bernoulli distribution:

$$p_{ti} = \begin{cases} \frac{|f_{ti}|}{\max |f_t|} & \text{if } l_{ti} = 0 \\ 1 - \frac{|f_{ti}|}{\max |f_t|} & \text{if } l_{ti} = 1 \end{cases} \tag{9}$$

where $\max |f_t| = \max\{|f_{t1}|, |f_{t2}|, \cdots, |f_{ti}|, \cdots, |f_{td}|\}$ is the maximum absolute value among all the $|f_{ti}|, i \in \{1, 2, \ldots, d\}$.

If we implement a probability sampling process on the Bernoulli distribution described in Eq. (9), an estimated indicator $\hat{l}_{ti}$ for each pixel could be obtained. The estimated indicator $\hat{l}_{ti}$ supplies a position estimation for the foreground and the background signal.

With the position estimation $\hat{l}_{ti}$ of observation o_t, the foreground signal which is indicated by $\hat{l}_{ti} = 0$ shall be compensated with estimated background signal. In this paper, we adopt observation median o_t^M to estimate the background signal which is covered by the foreground signal. The observation median is obtained by taking the median value of a observation vector buffer $O_M = \{o_{M_1}, o_{M_2}, \cdots, o_{M_k}\}$, where $o_{M_h}(h \in \{1, 2, \cdots, k\})$ are observations selected from the columns of the observation matrix $O = [o_1, o_2, \cdots, o_t, \cdots, o_n]$. The observation median o_t^M is obtained as follows:

$$o_{t,i}^M = \text{median}\{o_{M_1,i}, o_{M_2,i}, \cdots, o_{M_k,i}\}, i \in \{1, 2, \cdots, d\}. \tag{10}$$

To reduce the influence of slow moving foreground, the selected observations $o_{M_1}, o_{M_2}, \cdots, o_{M_k}$ in O_M are successively substituted every M_c frames.

With the help of the estimated indicator $\hat{l}_{ti}$ and the observation median o_t^M, the compensated background signal $\hat{b}_t$ could be computed by the following compensation process:

$$\hat{b}_t = o_t^M \odot (1 - \hat{l}_t) + o_t \odot \hat{l}_t, \tag{11}$$

where $\odot$ is the element-wise multiplication operator, $\hat{l}_t = [\hat{l}_{t1}, \hat{l}_{t2}, \cdots, \hat{l}_{td}]$ is the estimated indicator vector.

Now, we discuss the background subspace updating process. The incremental subspace updating task is often solved by singular value decomposition (SVD) technique [3,6,8,16]. In this paper, we adopt the modified Sequential Karhumen-Loeve (SKL) algorithm proposed in [16] to update the subspace, which could operate in the absence of an initial subspace estimation.

Supposed that the subspace is updated every q frames and the compensated signal $\hat{b}_1, \hat{b}_2, \cdots, \hat{b}_{t-q}$ are given (If we set $\hat{b}_t = o_t, t \in \{1, 2, \cdots, q\}$ for the first q observations, the algorithm operates without an explicit training process for initial subspace estimation.), we can construct the old compensated signal matrix $B_{old} = [\hat{b}_1, \hat{b}_2, \cdots, \hat{b}_{t-q}]$ and obtain its mean $\bar{b}_{old} = \frac{1}{t-q} \sum_{i=1}^{t-q} \hat{b}_i$. Then the singular value decomposition (SVD) of the matrix $[B_{old} - \bar{B}_{old}] = [\hat{b}_1 - \bar{b}_{old}, \hat{b}_2 - \bar{b}_{old} \cdots, \hat{b}_{t-q} - \bar{b}_{old}] = [U \Sigma V^T]$ can also be obtained.

When new compensated signal $B_{new} = [\hat{b}_{t-q+1}, \hat{b}_{t-q+2}, \cdots, \hat{b}_t]$ arrives, the compensated signal matrix becomes $B = [B_{old}, B_{new}]$. The incremental subspace updating can be done efficiently by computing the mean and SVD of matrix B with follows steps:

1. Compute the mean vector for matrix B_{new}: $\bar{b}_{new} = \frac{1}{q} \sum_{i=t-q-1}^{t} \hat{b}_i$ and the mean vector for matrix B: $\bar{b} = \frac{t-q}{t} \bar{b}_{old} + \frac{q}{t} \bar{b}_{new}$.

2. Compute $\hat{B}_{new} = [(\hat{b}_{t-q+1} - \bar{b}_{new}), \cdots, (\hat{b}_t - \bar{b}_{new}), \sqrt{\frac{(t-q)q}{t}}(\bar{b}_{new} - \bar{b}_{old})]$.

3. Compute $\tilde{B} = orth(\hat{B}_{new} - UU^T \hat{B}_{new})$, where the $orth()$ is the orthogonalization operation.

4. Construct $R = \begin{bmatrix} \Sigma & U^T \hat{B}_{new} \\ 0 & \tilde{B}_{new}(\hat{B}_{new} - UU^T \hat{B}_{new}) \end{bmatrix}$.

5. Compute the SVD of $R = \hat{U} \hat{\Sigma} \hat{V}^T$. Then $\Sigma' = \hat{\Sigma}$ and $U' = [U \ \tilde{B}_{new}]\tilde{U}$.

6. Update PCA subspace by setting $\bar{b}_t = \bar{b}$ and $U_t = U'$.

3   Experiments

In this section, the proposed method is compared with RPCA [4], GoDec [20], GRASTA [7] and LRFSO [18] methods. The experiments are implemented on nine real video sequences of I2R dataset [9], which contains both crowd scenario videos and dynamic background videos.

To evaluate the performance of different algorithms, F-score is employed:

$$F = \frac{2 \cdot Recall \cdot Precision}{Recall + Precision} = \frac{2TP}{2TP + FN + FP}, \tag{12}$$

where TP, FP and FN denote true foreground (true positive), false foreground (false positive) and false background (false negative) respectively.

For the proposed algorithm, the same parameters are used across all the videos. The column number of the basis matrix r is set to 1. The median buffer size M_k is set to 5. The median buffer is updated every $M_c = 9$ frames. The ρ in Eq. (6) is set to 2.2. The subspace update for $\bar{b}_t$ and U_t is implemented every $q = 5$ frames. The finally binary classification results for all methods are obtained by setting the threshold $th = 25$ for the estimated f_t signal, as done in [18].

Codes for all contrast methods are downloaded from the authors websites without any change. All codes are written in MATLAB and implemented in the same laptop with Intel Core CPU i7-4760HQ 2.10GHz and 16GB memory configuration. We show the F-score results and Frames Per Second (FPS) for all methods in Table 1. For RPCA and GoDec methods which operate in batch, we set the batch size as 150 frames. For GRASTA method, we initialize the subspace by training 100 frames from 100 % entries for 10 times, then we implement the GRASTA method from 10 % entries for foreground detection.

Table 1. The F-score (%) the comparison experiments. The best results are shown in **bold** and the average results are show in blue

Method Video	RPCA	GoDec	GRASTA	LRFSO	IPCA	SCIPCA(ours)
Bootstrap	61.19	60.91	59.45	56.68	58.27	**61.43**
Campus	29.19	24.26	18.38	**36.08**	17.94	18.24
Curtain	54.97	63.03	78.91	79.35	**83.46**	**83.46**
Escalator	**56.77**	47.33	38.24	48.01	35.02	37.59
Fountain	68.81	70.31	63.86	**70.58**	56.32	62.49
Lobby	57.09	46.04	58.42	44.75	65.11	**65.71**
ShoppingMall	**70.03**	66.53	68.96	54.46	68.02	69.25
WaterSurface	48.48	66.56	78.34	76.69	75.49	**85.60**
hall	51.75	51.15	53.49	49.08	**61.96**	61.10
Average	55.36	55.13	57.68	57.30	57.96	**60.54**

As shown, the proposed SCIPCA method exceeds the other comparison methods under F-score evaluation criteria, showing its superiority. We also show the results of the proposed method without the sparse error compensation process (named IPCA in Table 1). The sparse error compensation process boosts the basis IPCA method by 2.58 % points, indicating the effectiveness of the sparse error compensation process.

To further demonstrate the advantages of the proposed method, the ROC curves is depicted in Fig. 2. The ROC curve of the proposed method (the red line

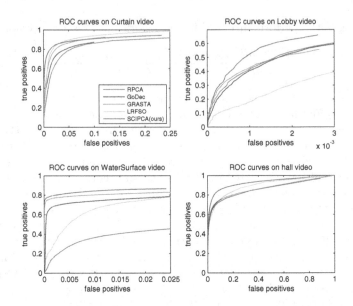

Fig. 2. The ROC curves of RPCA, GoDec, GRASTA, LRFSO and our methods for Curtain, Lobby, WaterSurface and hall videos.

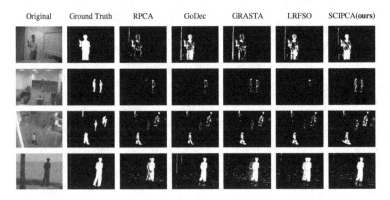

Fig. 3. Foreground detection results for Curtain, Lobby, ShoppingMall and WaterSurface videos respectively.

in Fig. 2) shows the stability and superiority of the proposed method on multiple videos, which is in accordance with the results demonstrated in Table 1. Some foreground detection results are also shown in Fig. 3. All experiments together show the effectiveness of the proposed method.

In terms of algorithm speed, RPCA, GoDec, GRASTA and our method process 10.33, 19.46, 99.77, 0.02 and 54.27 frames per second respectively in foreground detection phase. Our method operates faster than other comparison method except for GRASTA method. However, Considering that GRASTA method need an extra training phase (which averagely cost 17.75 second in our experiment), our method is also time efficient because it could operate without a training process.

4 Conclusion

In this paper, we proposed a novel incremental algorithm for foreground detection. The proposed method is implemented via a sparse error compensation based incremental PCA updating process, which offers a robust estimation on the background subspace and helps to supply a accurate estimation on the foreground signal. Furthermore, our proposed foreground detection method can operate without an explicit training process, which broadens its application scope. Extensive experiments show effectiveness and efficacy of our method. One future direction is to add spatial connexity to the sparse compensation process to take advantage of the spatial information.

Acknowledgements. This work was supported in part by the National Natural Science Foundation of China under Grant No.61273273, 61175096 and 61271374, the Specialized Fund for Joint Bulding Program of Beijing Municipal Education Commission, and the Research Fund for Doctoral Program of Higher Education of China under Grant No. 20121101110043.

References

1. Bouwmans, T.: Traditional and recent approaches in background modeling for foreground detection: an overview. Comput. Sci. Rev. **11**, 31–66 (2014)
2. Boyd, S., Parikh, N., Chu, E., Peleato, B., Eckstein, J.: Distributed optimization and statistical learning via the alternating direction method of multipliers. Found. Trends Mach. Learn. **3**(1), 1–122 (2011)
3. Brand, M.: Incremental singular value decomposition of uncertain data with missing values. In: Heyden, A., Sparr, G., Nielsen, M., Johansen, P. (eds.) ECCV 2002, Part I. LNCS, vol. 2350, pp. 707–720. Springer, Heidelberg (2002)
4. Candès, E., Li, X., Ma, Y., Wright, J.: Robust principal component analysis? J. ACM (JACM) **58**(3), 11 (2011)
5. Guo, X., Cao, X.: Speeding up low rank matrix recovery for foreground separation in surveillance videos. In: IEEE International Conference on Multimedia and Expo (ICME 2014), pp. 1–6 July 2014
6. Hall, P., Marshall, D., Martin, R.: Adding and subtracting eigenspaces with eigenvalue decomposition and singular value decomposition. Image Vis. Comput. **20**(13), 1009–1016 (2002)
7. He, J., Balzano, L., Szlam, A.: Incremental gradient on the grassmannian for online foreground and background separation in subsampled video. In: IEEE Conference on Computer Vision and Pattern Recognition (CVPR 2012) pp. 1568–1575. IEEE (2012)
8. Levey, A., Lindenbaum, M.: Sequential karhunen-loeve basis extraction and its application to images. IEEE Trans. Image Process. **9**(8), 1371–1374 (2000)

9. Li, L., Huang, W., Gu, I.H., Tian, Q.: Statistical modeling of complex backgrounds for foreground object detection. IEEE Trans. Image Process. **13**(11), 1459–1472 (2004)

10. Lin, Z., Chen, M., Ma, Y.: The augmented lagrange multiplier method for exact recovery of corrupted low-rank matrices. arXiv preprint. arXiv:1009.5055 (2010)

11. Lin, Z., Liu, R., Su, Z.: Linearized alternating direction method with adaptive penalty for low-rank representation. Adv. Neural Inf. Process. Sys. **24**, 612–620 (2011). Curran Associates, Inc

12. Lin, Z., Ganesh, A., Wright, J., Wu, L., Chen, M., Ma, Y.: Fast convex optimization algorithms for exact recovery of a corrupted low-rank matrix. Comput. Adv. Multi Sens. Adapt. Process. (CAMSAP) **61**, 199 (2009)

13. Oliver, N., Rosario, B., Pentland, A.: A bayesian computer vision system for modeling human interactions. In: International Conference on Vision Systems (1999)

14. Qiu, C., Vaswani, N.: Real-time robust principal components pursuit. In: 2010 48th Annual Allerton Conference on Communication, Control, and Computing (Allerton), pp. 591–598. IEEE (2010)

15. Qiu, C., Vaswani, N.: Reprocs: A missing link between recursive robust pca and recursive sparse recovery in large but correlated noise. arXiv preprint. arXiv:1106.3286 (2011)

16. Ross, D.A., Lim, J., Lin, R.S., Yang, M.H.: Incremental learning for robust visual tracking. Int. J. Comput. Vision **77**(1–3), 125–141 (2008)

17. Wang, L., Wang, L., Wen, M., Zhuo, Q., Wang, W.: Background subtraction using incremental subspace learning. In: IEEE International Conference on Image Processing (ICIP 2007), vol. 5, pp. V - 45–V - 48, September 2007

18. Xue, G., Song, L., Sun, J.: Foreground estimation based on linear regression model with fused sparsity on outliers. IEEE Trans. Circuits Syst. Video Techn. **23**(8), 1346–1357 (2013)

19. Xue, G., Song, L., Sun, J., Zhou, J.: Foreground detection: Combining background subspace learning with object smoothing model. In: IEEE International Conference on Multimedia and Expo (ICME 2013), pp. 1–6. IEEE (2013)

20. Zhou, T., Tao, D.: Godec: Randomized low-rank and sparse matrix decomposition in noisy case. In: Proceedings of the 28th International Conference on Machine Learning (ICML-2011), pp. 33–40 (2011)

Multimedia Representation Learning

Convolutional Neural Networks Features: Principal Pyramidal Convolution

Yanming Guo[1,2(✉)], Songyang Lao[2], Yu Liu[1], Liang Bai[2], Shi Liu[3], and Michael S. Lew[1]

[1] LIACS Media Lab, Leiden University, Niels Bohrweg, 1, Leiden, The Netherlands
{y.guo,y.liu,m.s.lew}@liacs.leidenuniv.nl
[2] Science and Technology on Information Systems Engineering Laboratory, National University of Defense Technology, Changsha, China
laosongyang@vip.sina.com, xabpz@163.com
[3] School of Arts and Media, Beijing Normal University, Beijing, China
candylstt@163.com

Abstract. The features extracted from convolutional neural networks (CNNs) are able to capture the discriminative part of an image and have shown superior performance in visual recognition. Furthermore, it has been verified that the CNN activations trained from large and diverse datasets can act as generic features and be transferred to other visual recognition tasks. In this paper, we aim to learn more from an image and present an effective method called Principal Pyramidal Convolution (PPC). The scheme first partitions the image into two levels, and extracts CNN activations for each sub-region along with the whole image, and then aggregates them together. The concatenated feature is later reduced to the standard dimension using Principal Component Analysis (PCA) algorithm, generating the refined CNN feature. When applied in image classification and retrieval tasks, the PPC feature consistently outperforms the conventional CNN feature, regardless of the network type where they derive from. Specifically, PPC achieves state-of-the-art result on the MIT Indoor67 dataset, utilizing the activations from Places-CNN.

Keywords: Convolutional neural networks · Concatenate · Principal component analysis · Image classification · Image retrieval

1 Introduction

Convolutional neural networks (CNN) have achieved breakthrough achievements in various visual recognition tasks and have been extensively studied in recent years [1–3, 10].There are several brilliant properties for CNN feature: (1) CNN feature is highly discriminative. Related research has analyzed the behavior of the intermediate layers of CNN and demonstrated that it can capture the most obvious features [1], thus could achieve considerably better results in a number of applications [1, 2]; (2) Unlike the hand-crafted features such as SIFT [5], HOG [6], CNN feature is generated from end-to-end, which eliminates the human intervention; (3) CNN feature can be achieved efficiently. In contrast to the standard feedforward neural networks with similarly-sized layers, CNN has fewer connections and parameters, which reduces the time cost of the

© Springer International Publishing Switzerland 2015
Y.-S. Ho et al. (Eds.): PCM 2015, Part I, LNCS 9314, pp. 245–253, 2015.
DOI: 10.1007/978-3-319-24075-6_24

feature extraction; 4) CNN feature is transferrable. Some works [10, 12] have demonstrated that CNN features trained on large and diverse datasets, such as ImageNet [7] and Places [8], could be transferred to other visual recognition tasks, even there are substantial differences between the datasets.

Owing to those notable characters, our research focuses on the reusing of off-the-shelf CNN feature. But, instead of computing the CNN feature over the full image, we ask whether we could get more information from an image and achieve a refined version of the CNN feature?

An intuitive way to achieve more knowledge is to extract multiple CNN features from one image and organize them in a proper way. In recent years, there are a number of works attempt to extract multiple features from one image, either in region proposals [3] or sliding windows [13]. But most of those methods are used for object detection, not for the refinement of CNN features. Besides, the extraction of numerous CNN features from overlapping regions is quite inefficient.

Related work has been done in the past [13, 14]. In the work by Gong, et al. [13], they extract CNN activations at multiple scale levels, perform orderless VLAD pooling separately, and concatenates them together, forming a high dimensional feature vector which is more robust to the global deformations. Koskela, et al. [14] splits one image into nine regions and calculates the average of the CNN activations, concatenating with the activation of the entire image. The resulting spatial pyramid features are certificated to be more effective in scene recognition.

Different from previous works, we show in this paper that the concatenating of the CNN features from one image could also improve the performance, without further calculation or other time-consuming processes. To avoid increasing the complexity during the test phase and keep the key components meanwhile, we compress the dimension to the normal one (4096-D) using PCA scheme after the concatenation and get the refined feature: Principal Pyramidal Convolution (PPC).

The idea of concatenating features has ever been done in the literatures. The most famous application is in the spatial pyramid matching (SPM) [15] algorithm, which concatenates the BOF vectors of the sub-regions as well as the whole image to import the global spatial information. SPM achieves a substantial improvement over the traditional BOF and has long been a key component in the competition-winning systems for visual recognition before the surge of CNN [16, 17].

In this paper, the BOF vector of the SPM algorithm is replaced by the discriminative CNN feature. Therefore, besides preserving the discrimination of CNN, PPC also introduces some spatial information as well as preserving the most important components. What's more, the strategy is portable, experiments show that whichever network the CNN activations derive from, PPC strategy could continuously improve the performance.

2   Principal Pyramidal Convolution

Inspired by SPM, which extracts features at multiple levels and aggregates them together, we propose the Principal Pyramidal Convolution (PPC) method. It divides the image into two levels and generates the final feature for the image by concatenating and extracting principal components for the features at all resolutions. The basic idea is illustrated in Fig. 1.

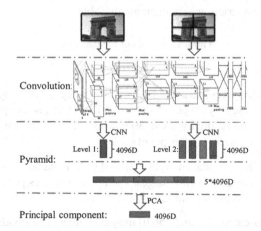

Fig. 1. The procedure of PPC algorithm

We extract CNN features from two scale levels. The first level corresponds to the full image, and the second level is consists of 2 * 2 regions by equally partitioning the full image. Therefore, we need to extract five CNN features for each image: C_0, C_1, C_2, C_3, C_4. Afterwards, we concatenate the five CNN features in an intuitive scheme: $C = [C_0, C_1, C_2, C_3, C_4]$. The resulted C is a 5 * 4096-dimensional vector. The CNN activations are achieved using the Caffe implementation [19]. Here, we select the 4096-dimensional output of the seventh layer (i.e. the last fully-connected layer) and L2-normalize it as the baseline CNN feature.

To address the increasing computational costs during the test phase, we compress the resulted feature vector to 4096-D in the last step. For the dimension reduction, we utilize the well-known PCA method [18], which could reduce the dimensionality of a data set and retain as much as possible of the variation at the same time.

In addition, we also reduce the dimension to other various sizes and compare the performance between conventional CNN and PPC for different visual tasks, including the supervised image classification and unsupervised image retrieval.

3 Experiment

In this part, we make some comparisons between the conventional CNN feature and PPC feature on various image classification and image retrieval [4] databases. The experimental environment is a CPU i7 at 2.67 GHZ with 12 GB RAM and a 2 GB NVIDIA GTX 660.

3.1 Datasets

We present the results on four widely used datasets: Caltech-101 [20], fifteen scene categories (Scene15) [15], MIT Indoor67 database [21] and INRIA Holidays [22].

The details of the datasets are summarized in Table 1.

Table 1. Details of the datasets

Datasets	Details
Caltech-101	102 categories and a total of 9144 images, the image number per category ranges from 31 to 800. We follow the procedure of [15] and randomly select 30 images per class for training and test on up to 50 images per class
Scene15	4485 greyscale images assigned to 15 categories, and each category has 200 to 400 images. We use 100 images per class for training and the rest for testing
Indoors67	67 categories and 15620 images in total. The standard training/test split consists of 80 images for training and 20 images for testing per class
Holidays	1491 images corresponding to 500 image instances. Each instance has 2–3 images describing the same object or location. The images have been rectified to a natural orientation. There are 500 images of them are used as queries

In the databases described above, the first three datasets are used for image classification, on which we train linear SVM classifiers (s = 0, t = 0) to recognize the test images, using the LIBSVM tool [23]. The last dataset is a standard benchmark for image retrieval, and the accuracy is measured by the mean Average Precision (mAP), as defined in [11].

3.2 Comparisons on Different Networks

According to which database the CNN is trained on, the CNN features can be categorized into two types: ImageNet-CNN and Places-CNN. ImageNet-CNN is the most commonly used model which is trained on the well-known database: ImageNet [7]. This database contains 1000 categories with around 1.3 million images and most of the images are object-centric. Places-CNN is another model which is trained on the recently proposed Places database and is scene-centric [8]. This database contains about 2.5 million images assigned to 205 scene categories.

In this paper, we utilize the off-the-shelf CNN features of ImageNet and Places respectively, and compare the performance of CNN and PPC on the four datasets. The results are shown in Table 2.

Table 2. Comparison of SPM and CNN, PPC on different networks

Datasets	ImageNet-CNN	ImageNet-PPC	Places-CNN	Places-PPC	SPM [15]
Caltech-101	86.44 %	**87.45 %**	61.07 %	67.41 %	64.6 %
Scene15	84.49 %	86.4 %	89.11 %	**89.88 %**	81.4 %
Indoors67	59.18 %	64.4 %	72.16 %	**73.36 %**	–
Holidays	73.95 %	**74.9 %**	71.71 %	73.43 %	–

From the table, we can see that the improvements of PPC over CNN vary from about 1 % to 6 %, depending on the network and dataset. We can further conclude that: (1) the features generated from CNN are more distinctive than SIFT in image classification, and this inherent merit brings about the improvement of PPC in contrast to SPM. (2) The features derived from ImageNet-CNN are more discriminative in classifying objects, thus perform better on the Caltech-101 dataset. In contrast, the features achieved from the Places-CNN are better at classifying scenes, and accordingly perform better on Scene15 and MIT Indoor67 datasets. On one hand, choosing a suitable network (i.e. choose ImageNet-CNN for object recognition, or choose Places-CNN for scene recognition) could bring a significant improvement in the performance. As is shown by the experiment on Caltech-101 database, the advantage of ImageNet-CNN feature over the Places-CNN feature is more than 25 % (86.44 % and 61.07 % respectively). Specially, Places-PPC achieves state-of-the-art result on the MIT Indoor67 database. On the other hand, choosing an unsuitable network could highlight the benefit of PPC over CNN. For instance, when we utilize Places-CNN features on the Caltech-101 database, the improvement of PPC over CNN is more than 6 %, rising from 61.07 % to 67.41 %. Similarly, when the ImageNet-CNN features are tested on the Indoor67 dataset, the refinement of PPC over CNN could also be more than 5 % (from 59.18 % to 64.4 %). But no matter which type of networks is applied on the datasets, PPC features consistently outperform the holistic CNN features, demonstrating the effectiveness of the strategy.

For both the CNN and PPC algorithms on the MIT Indoor67 dataset, we visualize the distance between the features of the top performing categories in 3-dimensional space using classical multidimensional scaling technique [9]. As is shown in Fig. 2, the axes correspond to the coordinates in the 3-dimensional space and the categories are buffet, cloister, florist, inside bus.

From the Fig. 2, we can notice that the PPC features are more distinguishable than the holistic CNN features. The advantage is particularly evident on the comparisons between "florist" and "inside bus".

For ImageNet-CNN model, we compare the accuracy of CNN and PPC on each category of Scene15 database, as is demonstrated in Fig. 3. The X-axe details the categories and the Y-axe corresponds to the accuracies of this category.

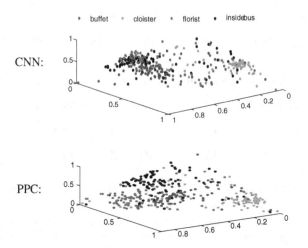

Fig. 2. Top performing feature visualization of CNN and PPC

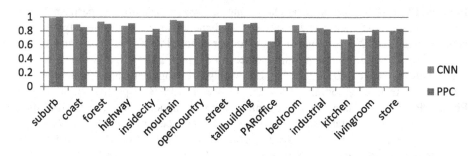

Fig. 3. Comparison of CNN and PPC on each category of Scene15

It can be observed that for most categories of Scene15 (ten of the fifteen categories), PPC performs better than CNN.

Furthermore, we display the confusion matrix of PPC derived from ImageNet in Fig. 4.

The classification accuracies for each class are listed along the diagonal. The element in the ith row and jth column is the percentage of images from class i that were misclassified as class j. Most of the categories rate more than 80 % in precision, and the most easily confused categories are bedroom and living room.

3.3   Comparisons on Different Dimensions

The improvement of PPC over CNN is not limited to 4096-D. To verify this, we further reduce the dimensionality to other sizes and compare the performance of PPC and CNN on different datasets, the results are shown in Fig. 5.

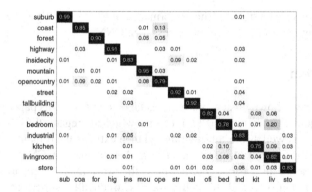

Fig. 4. Confusion matrix of ImageNet-PPC on Scene15

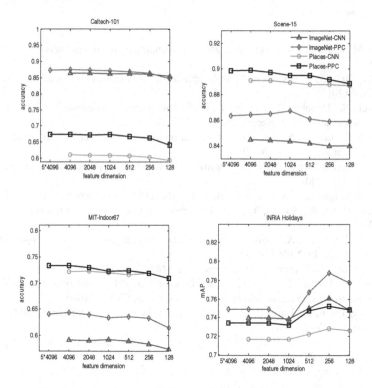

Fig. 5. Comparisons of CNN and PPC on different dimensions

It is noticeable that the performance does not decrease even when the dimensionality of features are reduced to 128-D (most of the accuracies drift within 2 %). On the contrary, the mAP of ImageNet-PPC on the INRIA Holidays dataset even rises to 78.76 %, when the dimensionality is reduced to 256-D. This indicates that the discriminatory power of both CNN and PPC will not be greatly affected with the reduction of the dimensionality. Nevertheless, the performance of PPC is mostly better than that of CNN, indicating that PPC is more robustness than CNN.

4 Conclusion

CNN features have shown great promise in visual recognition. This paper proposed the Principal Pyramidal Convolution (PPC) scheme, which aggregates the CNN features of the whole image as well as the sub-regions and then extracts the principal components. The representation from our strategy outperforms the conventional CNN feature without enlarging the feature dimensions. Furthermore, this work makes comparisons of CNN and PPC on different sizes and shows that the PPC frequently outperforms CNN.

References

1. Zeiler, M.D., Fergus, R.: Visualizing and understanding convolutional networks. In: ECCV (2014)
2. Liu, Y., Guo, Y., Wu, S., Lew, M.S.: DeepIndex for accurate and efficient image retrieval. In: ICMR (2015)
3. Girshick, R., Donahue, J., Darrell, T., et al.: Rich feature hierarchies for accurate object detection and semantic segmentation. In: CVPR (2014)
4. Thomee, B., Lew, M.S.: Interactive search in image retrieval: a survey. Int. J. Multimedia Inf. Retrieval 1(2), 71–86 (2012)
5. Lowe, D.G.: Distinctive image features from scale-invariant keypoints. Int. J. Comput. Vision 60(2), 91–110 (2004)
6. Dalal, N., Triggs, B.: Histograms of oriented gradients for human detection. In: CVPR (2005)
7. Deng, J., Dong, W., Socher, R., et al.: Imagenet: a large-scale hierarchical image database. In: CVPR (2009)
8. Zhou, B., Lapedriza, A., Xiao, J., et al.: Learning deep features for scene recognition using places database. In: NIPS (2014)
9. Seber, G.A.F.: Multivariate observations. Wiley, New York (2009)
10. Oquab, M., Bottou, L., Laptev, I., et al.: Learning and transferring mid-level image representations using convolutional neural networks. In: CVPR (2014)
11. Philbin, J., Chum, O., Isard, M., et al.: Object retrieval with large vocabularies and fast spatial matching. In: CVPR (2007)
12. Yosinski, J., Clune, J., Bengio, Y., et al.: How transferable are features in deep neural networks? In: NIPS (2014)
13. Gong, Y., Wang, L., Guo, R., et al.: Multi-scale orderless pooling of deep convolutional activation features. In: ECCV (2014)
14. Koskela, M., Laaksonen, J.: Convolutional network features for scene recognition. In: ACM Multimedia (2014)
15. Lazebnik, S., Schmid, C., Ponce, J.: Beyond bags of features: spatial pyramid matching for recognizing natural scene categories. In: CVPR (2006)
16. Yang, J., Yu, K., Gong, Y., et al.: Linear spatial pyramid matching using sparse coding for image classification. In: CVPR (2009)
17. Wang, J., Yang, J., Yu, K., et al.: Locality-constrained linear coding for image classification. In: CVPR (2010)
18. Jolliffe, I.: Principal Component Analysis. Wiley Online Library, New York (2005)

19. Jia, Y., Shelhamer, E., Donahue, J., et al.: Caffe: convolutional architecture for fast feature embedding. In: ACM Multimedia (2014)
20. Fei-Fei, L., Fergus, R., Perona, P.: Learning generative visual models from few training examples: an incremental bayesian approach tested on 101 object categories. In: CVIU (2007)
21. Quattoni, A., Torralba, A.: Recognizing indoor scenes. In: CVPR (2009)
22. Jegou, H., Douze, M., Schmid, C.: Hamming embedding and weak geometric consistency for large scale image search. In: ECCV (2008)
23. Chang, C.C., Lin, C.J.: LIBSVM: a library for support vector machines. In: ACM TIST (2011)

Gaze Shifting Kernel: Engineering Perceptually-Aware Features for Scene Categorization

Luming Zhang[(✉)], Richang Hong, and Meng Wang

School of Computer and Information, Hefei University of Technology, Hefei, China
zglumg@gmail.com

Abstract. In this paper, we propose a novel gaze shifting kernel for scene image categorization, focusing on discovering the mechanism of humans perceiving visually/semantically salient regions in a scene. First, a weakly supervised embedding algorithm projects the local image descriptors (*i.e.*, graphlets) into a pre-specified semantic space. Afterward, each graphlet can be represented by multiple visual features at both low-level and high-level. As humans typically attend to a small fraction of regions in a scene, a sparsity-constrained graphlet ranking algorithm is proposed to dynamically integrate both the low-level and the high-level visual cues. The top-ranked graphlets are either visually or semantically salient according to human perception. They are linked into a path to simulate human gaze shifting. Finally, we calculate the gaze shifting kernel (GSK) based on the discovered paths from a set of images. Experiments on the USC scene and the ZJU aerial image data sets demonstrate the competitiveness of our GSK, as well as the high consistency of the predicted path with real human gaze shifting path.

1 Introduction

Scene categorization is a key component in a variety of multimedia and computer vision systems, *e.g.*, scene annotation and video management. It can help to accept likely scene configurations and rule out unlikely ones. For example, a successful scene annotation system should encourage the occurrence of a parking lot in an aerial image belonging to a "downtown area" while suppressing the occurrence of an airport in an aerial image belonging to a "residential area". However, successfully recognizing scene images from various categories is still a challenging task due to the following problems:

- Eye tracking experiments have shown that humans allocate gazes selectively to visually/semantically important regions of an image. As shown on the right of Fig. 1, most people start with viewing the central player, and then shifting their gazes to the rest three players, and finally to the bench and the background trees. Such directed viewing path is informative to distinguish different sceneries, but has not been exploited in the existing categorization models. Current models typically integrate local image descriptors unorderly for recognition, as exemplified on the left of Fig. 1.

© Springer International Publishing Switzerland 2015
Y.-S. Ho et al. (Eds.): PCM 2015, Part I, LNCS 9314, pp. 254–264, 2015.
DOI: 10.1007/978-3-319-24075-6_25

Fig. 1. An illustration of image kernels based on spatial pyramid matching (left) and gaze shifting path (right). The red crosses denote the local image features.

- Psychophysics studies have demonstrated that the bottom-up and top-down visual features collaboratively draw the attention of human eye. An ideal scene categorization system should integrate both the low-level and the high-level visual cues. However, current models typically fuse multiple features in a linear or nonlinear way, where the cross-feature information is not well utilized. Besides, these integration schemes are not designed under the objective of optimally reflecting human visual perception.

To address these problems, we propose gaze shifting kernel (GSK) which encodes human active viewing process. The key technique is a sparsity-constrained ranking algorithm jointly optimizing the weights of graphlets from multiple channels. The pipeline of our method is shown in Fig. 2. By transferring the semantics of image labels into different graphlets using a manifold embedding, we represent each graphlet by multiple low-level and high-level visual features. Then, a sparsity-constrained algorithm is proposed to integrate the multiple features for calculating the saliency of each graphlet. In particular, by constructing the matrices containing the visual/semantic features of graphlets in an image, the proposed algorithm seeks the consistently sparse elements from the joint decompositions of the multiple-feature matrices into pairs of low-rank and sparse matrices. Compared with previous methods that linearly/non-linearly combine multiple global features, our method can seamlessly integrate multiple visual/semantic features for salient graphlets discovery. These discovered graphlets are linked into a path to simulate human gaze shifting process. As the learned paths are 2-D planar features which cannot be fed into a conventional classifier like SVM, we quantize them into a kernel for scene categorization.

The main contributions of this paper are two-fold: (1) GSK: a novel image kernel that encodes human gaze shifting path to discriminate different sceneries; and (2) a sparsity-constrained ranking algorithm which discovers visually/semantically important graphlets that draw the attention of human eye.

2   Related Work

The current gaze estimation includes model-based and appearance-based methods. The former uses 3D eyeball models and estimates gaze direction through the geometric eye features [1–3]. They typically use infrared light sources and a high-resolution camera to locate the 3D eyeball position. Although this approach

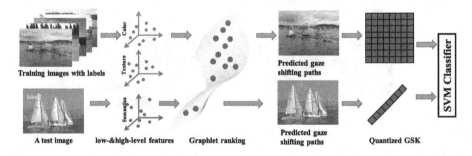

Fig. 2. The pipeline of our approach (The blue and green arrows denote the training and test stages respectively).

can estimate gaze directions accurately, its heavy reliance on the specialized hardware is a limitation. Appearance-based methods compute the non-geometric image features from the input eye image and then estimate gaze directions. The eye position is pre-calculated for estimating the gaze target in the world coordinate system. With the popularity of monocular head pose tracking [4] and RGB-D head-tracking cameras [5], head poses can be captured accurately. Some appearance-based gaze estimation techniques use head poses as an auxiliary input for gaze estimation [6,7]. Appearance variation of the eye images caused by head pose changes is another challenge. Usually it is tackled by a compensation function [6] or warping training images to new head poses [7]. While most of the existing appearance-based models adopt a person-dependent data set, Funes *et al.* proposed a cross subject training method for gaze estimation [8]. An RGB-D camera warps the training and test images to the frontal view. Afterward, an adaptive linear regression is applied to compute gaze directions.

In addition to gaze prediction techniques, our method is also closely related to saliency-guided image recognition. Moosmann *et al.* [9] learned saliency maps for visual search to enhance object categorization. Gao and Vasconscelos [10] formulated discriminative saliency and determined it based on feature selection in the context of object detection [11]. Parikh *et al.* [12] calculated saliency in an unsupervised manner based on how accurate a patch can predict the locations of others. Khan *et al.* [13] modeled color based saliency to weight features. Harada *et al.* [14] learned weights on regions for classification. Noticeably, they learned the class-level weights, where the weights are the same for the entire images. Yao *et al.* [15] learned a classifier based on random forests. Their approach discovers salient patches for the decision trees by randomly sampling patches and selecting the highly discriminative ones.

3 The Proposed Gaze Shifting Kernel

3.1 Low-Level and High-Level Descriptions of Graphlets

There are usually tens of components within a scene. Among them, a few spatially neighboring ones and their interactions capture the local cues. To capture the local cues, we segment a scene image into atomic regions and then construct a number of graphlets. As shown in Fig. 3, each graphlet is a small-sized graph defined as $\mathcal{G} = \{\mathcal{V}, \mathcal{E}\}$.

$\mathcal{V}$ is a set of vertices representing those spatially neighboring atomic regions; and $\mathcal{E}$ is a set of edges, each connecting pairwise adjacent atomic regions. We call a graphlet with t vertices a t-sized graphlet. In this work, we characterize each graphlet in both color and texture channels. Given a t-sized graphlet, each row of matrix $\mathbf{M}_c^r$ and $\mathbf{M}_t^r$ represent the 9-D color moment and 128-D HOG of an atomic region, respectively. To describe the spatial interactions of atomic regions, we employ a $t \times t$ adjacency matrix $\mathbf{M}_s$, as elaborated in [16].

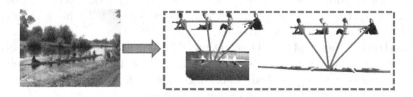

Fig. 3. An example of two graphlets extracted from a photo.

In addition to color and texture channels description, high-level semantic cues should also be exploited for scene modeling. In this paper, the semantic cues are discovered based on a weakly supervised algorithm. We transfer the semantics of image labels into different graphlets in a photo. This is implemented based on a manifold embedding described as follows:

$$\arg \min_{\mathbf{Y}}[\sum_{i,j} ||y_i - y_j||^2 l_s(i,j) - \sum_{i,j} ||y_i - y_j||^2 l_d(i,j)]$$
$$= \arg \min_{\mathbf{Y}} \text{tr}(\mathbf{Y}\mathbf{R}\mathbf{Y}^T), \tag{1}$$

where $\mathbf{Y} = [y_1, y_2, \cdots, y_N]$ denotes a collection of d-D post-embedding graphlets; $l_s(\cdot, \cdot)$ and $l_b(\cdot, \cdot)$ are functions measuring the similarity and discrepancy between graphlets [16]; matrix $\mathbf{R} = [\mathbf{e}_{N-1}^T, -\mathbf{I}_{N-1}]\mathbf{W}_1[\mathbf{e}_{N-1}^T, -\mathbf{I}_{N-1}] + \cdots + [-\mathbf{I}_{N-1}, \mathbf{e}_{N-1}^T]\mathbf{W}_N[-\mathbf{I}_{N-1}, \mathbf{e}_{N-1}^T]$; $\mathbf{W}_i$ is an $N \times N$ diagonal matrix whose h-th diagonal element is $[l_s(h, i) - l_d(h, i)]$.

It is reasonable to assume a linear approximation $\mathbf{Y} = \mathbf{U}^T\mathbf{X}$ of the graphlet embedding process. Thus (1) can be reorganized into:

$$\arg \min_{\mathbf{U}} \text{tr}(\mathbf{U}^T\mathbf{X}\mathbf{R}\mathbf{X}^T\mathbf{U})$$
$$= \arg \min_{\mathbf{U}} \text{tr}(\mathbf{U}^T\mathbf{L}\mathbf{U}) \ \ s.t. \ \ \mathbf{U}^T\mathbf{U} = \mathbf{I}_d, \tag{2}$$

where $\mathbf{X}$ is obtained by row-wise stacking matrix $[\mathbf{M}_r^c, \mathbf{M}_r^t, \mathbf{M}_s]$ into a vector, $\mathbf{U}$ is the linear projection matrix, and matrix $\mathbf{L} = \mathbf{X}\mathbf{R}\mathbf{X}^T$. The above objective function is a basic optimization problem which can be solved using the Lagrangian multiplier. The optimal solution is the d eigenvectors associated with the d smallest eigenvalues of matrix $\mathbf{L}$.

3.2 Sparsity-Constrained Graphlets Ranking

Human vision system typically selects only the distinctive sensory information for further processing. Practically, only a few visually/semantically salient
graphlets within a scene image are perceived by humans. Based on this, we
propose a sparsity-constrained ranking scheme to discover the salient graphlets,
by dynamically tuning the importance of color, texture, and semantic channels.
Mathematically, the ranking algorithm can be formulated as:

Let $\mathbf{X}_1$, $\mathbf{X}_2$, and $\mathbf{X}_3$ be the three feature matrices in color, texture, and
semantic channels respectively, where the columns in different matrices with the
same index correspond to a same graphlet. The size of each $\mathbf{X}i$ is $d_i \times N$, where
d_i is the feature dimension and N is the number of graphlets. Then, the task is
to find a weighting function for each graphlet $S(\mathcal{G}i) \in [0,1]$ by integrating the
three feature matrices $\mathbf{X}_1$, $\mathbf{X}_2$, and $\mathbf{X}_3$.

Based on the theory of visual perception [25], there are usually a strong correlation among the non-salient regions in an image. That is to say, the non-salient
graphlets can be self-represented. This suggests that feature matrix $\mathbf{X}_i$ can be
decomposed into a salient part and a non-salient part:

$$\mathbf{X}_i = \mathbf{X}_i \mathbf{Z}_0 + \mathbf{E}_0, \tag{3}$$

where $\mathbf{X}_i\mathbf{Z}_0$ denotes the non-salient part that can be reconstructed by itself,
$\mathbf{Z}_0$ denotes the reconstruction coefficients, and $\mathbf{E}_0$ denotes the remaining part
corresponding to the salient targets.

According to (3), there is an infinite number of possible decompositions with
respect to $\mathbf{Z}_0$ and $\mathbf{E}_0$. Toward a unique solution, two constraints are adopted:
(1) only a small fraction of graphlets in a scene image are salient, i.e., matrix
$\mathbf{E}_0$ is sparse; and (2) the strong correlation among the background graphlets
suggests that matrix $\mathbf{Z}_0$ has low rankness. Based on these, we can infer the
salient graphlets by adding a sparsity and low-rankness constraint to (3). Then,
the graphlet saliency can be formulated as a low-rankness problem:

$$\min_{\substack{\mathbf{Z}_1,\mathbf{Z}_2,\mathbf{Z}_3 \\ \mathbf{E}_1,\mathbf{E}_2,\mathbf{E}_3}} \sum_{i=1}^{3} ||\mathbf{Z}_i||_* + \lambda ||\mathbf{E}||_{2,1}, \quad s.t. \quad \mathbf{X}_i = \mathbf{X}_i \mathbf{Z}_i + \mathbf{E}_i, \tag{4}$$

where $\mathbf{E} = [\mathbf{E}_1; \mathbf{E}_2; \mathbf{E}_3]$. The integration of multiple features is seamlessly performed by minimizing the $l_{2,1}$ norm of $\mathbf{E}$. That is, we enforce the columns of $\mathbf{E}_1$,
$\mathbf{E}_2$, and $\mathbf{E}_3$ to have jointly consistent magnitude values. It is noticeable that the
above objective function is convex, which can be solved using the inexact ALM
algorithm [17] efficiently.

Denoting $\{\mathbf{E}_1^*, \mathbf{E}_3^*, \mathbf{E}_3^*\}$ as the optimal solution to (4), the saliency of graphlet
$\mathcal{G}_i$ can be quantified as $S(\mathcal{G}_i) = \sum_{j=1}^{3} ||\mathbf{E}_j^*(:,j)||_2$, where $||\mathbf{E}_j^*(:,j)||_2$ denotes the
l_2 norm of the i-th column of $\mathbf{E}_j^*$. A larger score of $S(\mathcal{G}_i)$ reflects that graphlet
$\mathcal{G}_i$ has a higher probability to be salient.

3.3   Gaze Shifting Kernel and SVM Training

As we stated above, an aerial image can be represented by a gaze shifting path, which is notably a planar visual feature in $\mathbb{R}^2$. Unfortunately, conventional classifiers, such as SVM, can only handle 1-D vector form features. Thus, it would be impractical for them to carry out classification directly based on the extracted paths. To tackle this problem, a kernel method is adopted to transform the extracted paths into 1-D vectors.

As shown in (Fig. 4), the proposed kernel method is based on the distances between scene images, which are computed based on the extracted paths. Given a scene image, each of its extracted paths $\mathcal{P}^*$ can be converted into a vector $A = [\alpha_1, \alpha_2, \cdots, \alpha_N]$, where each element of A is calculated as:

$$\alpha_i \propto \exp\left(-\frac{1}{j^2 \cdot K^2} \sum_{j=1}^{K} d(y(\mathcal{P}_j^*), y(\mathcal{P}_j^i))\right), \tag{5}$$

where N is the number of training images, K is the number of graphlets in a path; $\mathcal{P}_j^*$ and $\mathcal{P}_j^i$ denote the j-th graphlets of paths $\mathcal{P}^*$ and $\mathcal{P}^i$ respectively.

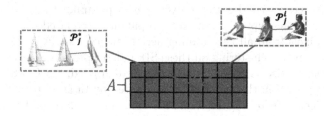

Fig. 4. A graphical illustration of the kernel calculation in (5).

On the basis of the feature vector obtained above, a multi-class SVM is trained. That is, for the training scene images from the p-th and the q-th classes, we construct the following binary SVM classifier:

$$\max_{\alpha \in \mathbb{R}^{N_{pq}}} W(\alpha) = \sum_{i=1}^{N_{pq}} \alpha_i - \frac{1}{2} \sum_{i=1}^{N_{pq}} \alpha_i \alpha_j l_i l_j k(A_i, A_j)$$

$$s.t. \quad 0 \le \alpha_i \le C, \sum_{i=1}^{N_{pq}} \alpha_i l_i = 0, \tag{6}$$

where $A_i \in \mathbb{R}^N$ is the quantized feature vector from the i-th training scene image; l_i is the class label (+1 for the p-th class and -1 for the q-th class) to the i-th training scene image; α determines the hyperplane to separate scene images in the p-th class from those in the q-th class; $C > 0$ trades the complexity of the machine off the number of nonseparable scene images; and N_{pq} is the number of training scene images from either the p-th class or the q-th class.

4   Experimental Results and Analysis

We conduct our experiments on two data sets. The first is the ZJU aerial image [18], which comprises of 20,946 aerial images from ten categories. Most of the aerial images are collected from metropolises, such as New York, Tokyo, and Beijing. The second is the USC scene [19] containing 375 video clips of three USC campus sites, *i.e.*, Ahmanson Center for Biological Science (ACB), Associate and Founders Park (AnF), and Frederick D. Fagg Park (FDF). A number of images are uniformly sampled from the original training and testing video clips for training and testing stages, respectively.

4.1   Comparison with the State-of-the-Art

In our experiments, we compared our GSK with six representative visual descriptors: the walk and tree kernels [20], and four SPM-based generic object recognition models, namely, SPM [21], SC-SPM [23], LLC-SPM [22], and object bank-based SPM (OB-SPM). The parameter settings for the six methods were as follows. For the walk and tree kernels: Their sizes were tuned from one to 10 and the best recognition accuracies were recorded. For SPM, SC-SPM, and LLC-SPM: We constructed a three level spatial pyramid; then extracted over one million SIFT descriptors from 16×16 patches computed over a grid with spacing of 8 pixels from all the training aerial images. Finally, a codebook was generated by k-means clustering on these SIFT descriptors. In our experiments, different codebook sizes: 256, 512, and 1024, were applied. For OB-SPM, we followed the setup used in the experiment conducted by Li *et al.* [24]: The number of generic object detectors is fixed at 200, different regularizers LR1,LRG, and LRG1 were used, and it also has three SPM levels.

To compare the above approaches, on both data sets, we utilized 50 % of the scene images as training images and the rest for testing. As shown in Table 1 and Fig. 5, our approach outperformed the others. This is attributable to four reasons: (1) none of the four SPM-based recognition models incorporate any human gaze shifting information; (2) SPM, SC-SPM, and LLC-SPM are based on SIFT descriptors only. Thus, they ignore other important cues such as semantics and color; (3) the object detectors in OB-SPM are trained for generic objects,

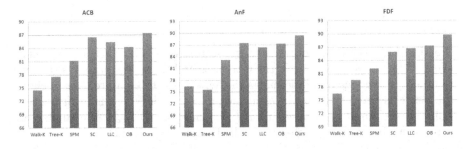

Fig. 5. Category accuracies on the three scenes of the USC data set.

they cannot detect the particular scene components effectively; and (4) both walk and tree kernels are less representative of graph-based descriptors due to the inherent totter phenomenon.

Table 1. Categorization rate on the ZJU aerial image data set (The experiment was repeated 10 times and the average accuracies are reported)

Category	Walk-K	Tree-K	SPM(200)	SC(256)	LLC(256)	OB(LR1)	SPM(400)	SC(512)
Airport	0.882	0.901	0.723	0.721	0.723	0.799	0.811	0.843
Commer.	0.545	0.532	0.441	0.443	0.334	0.517	0.521	0.456
Indust.	0.642	0.611	0.521	0.499	0.413	0.512	0.454	0.576
Inter.	0.645	0.685	0.611	0.643	0.322	0.675	0.674	0.634
Park.	0.523	0.487	0.443	0.512	0.412	0.536	0.512	0.496
Railway	0.556	0.578	0.502	0.511	0.521	0.514	0.521	0.596
Seaport	0.859	0.843	0.774	0.745	0.721	0.766	0.632	0.814
Soccer	0.646	0.655	0.576	0.589	0.578	0.568	0.521	0.624
Temple	0.503	0.454	0.521	0.567	0.511	0.603	0.534	0.565
Univer.	0.241	0.265	0.289	0.301	0.223	0.304	0.498	0.321
Average	0.524	0.601	0.540	0.553	0.4770	0.579	0.568	0.593

Category	LLC(512)	OB(LRG)	SPM(800)	SC(1024)	LLC(1024)	OB(LRG1)	SPM(HC)	SC(HC)
Airport	0.801	0.889	0.799	0.912	0.899	0.872	0.813	0.916
Commer.	0.567	0.565	0.512	0.601	0.521	0.617	0.519	0.584
Indust.	0.521	0.613	0.585	0.557	0.593	0.576	0.598	0.564
Inter.	0.766	0.705	0.644	0.788	0.622	0.676	0.668	0.791
Park.	0.489	0.486	0.503	0.489	0.489	0.512	0.511	0.487
Railway	0.553	0.532	0.6027	0.601	0.599	0.589	0.614	0.609
Seaport	0.751	0.779	0.815	0.745	0.798	0.811	0.822	0.751
Soccer	0.625	0.646	0.634	0.689	0.655	0.668	0.643	0.693
Temple	0.567	0.587	0.577	0.689	0.556	0.612	0.587	0.649
Univer.	0.409	0.389	0.311	0.582	0.281	0.304	0.324	0.537
Average	0.605	0.620	0.606	0.654	0.600	0.636	0.610	0.658

Category	LLC(HC)	Ours
Airport	0.904	0.864
Commer.	0.534	0.646
Indust.	0.598	0.565
Inter.	0.634	0.803
Park.	0.493	0.521
Railway	0.604	0.611
Seaport	0.803	0.732
Soccer	0.659	0.687
Temple	0.574	0.645
Univer.	0.287	0.511
Average	0.609	0.647

4.2   Parameters Analysis

This experiment evaluates the influence of graphlet size T and the number of graphlets K in each path, on the performance of scene categorization.

To analyze the effects of the maximum graphlet size on scene categorization, we set up an experiment by varying T continuously. On the left of Fig. 6, we present the scene classification accuracy (on the ZJU aerial images) when the maximum graphlet size is tuned from one to 10. As can be seen, categorization accuracy increases moderately when $T \in [1, 4]$ but increases slowly when $T \in [5, 10]$. This observation implies that 4-sized graphlets are sufficiently descriptive for capturing the local cues of aerial images in [18]. Considering the time consumption is exponentially increasing with graphlet size, we set $T = 4$ in our experiment.

On the right of Fig. 6, we present the performance when the number of graphlet in a path (K) is tuned from one to 10. As can be seen, the prediction accuracy increases quickly when $K \in [1, 7]$ but remains nearly unchanged when $K \in [7, 10]$. Therefore we set $K = 7$.

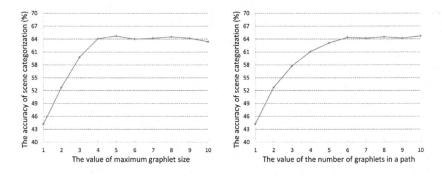

Fig. 6. Scene categorization accuracies under different parameters.

4.3 Visualization Results

In the last experiment, we present the top-ranked graphlets based on our sparsity-constrained ranking algorithm. We experiment on the LHI data set [26]

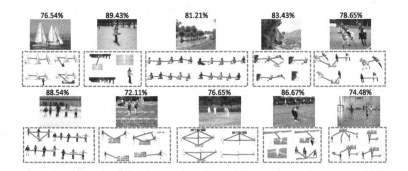

Fig. 7. Top-ranked graphlets selected by our sparsity-constrained ranking algorithm.

because the ground-truth segments are provided. As shown in Fig. 7, the top-ranked graphlets attract humans attention accurately, based on the comparative accuracies with real human gaze shifting paths. This again demonstrates the impressive performance of our categorization model.

5 Conclusion

In this paper, we conduct scene categorization by leveraging human gaze shifting paths. We propose a sparsity-constrained graphlet ranking algorithm. The top-ranked graphlets are visually/semantically important and are linked into a path to mimick human gaze shifting. Finally, the paths are encoded into an image kernel for discriminating different sceneries.

References

1. Guestrin, E.D., Eizenman, M.: General theory of remote gaze estimation using the pupil center and corneal reflections. IEEE T-BE **53**(6), 1124–1133 (2006)
2. Jixu, C., Qiang, J.: Probabilistic gaze estimation without active personal calibration. In: Proceedings of CVPR (2011)
3. Nakazawa, A., Nitschke, C.: Point of gaze estimation through corneal surface reflection in an active illumination environment. In: Fitzgibbon, A., Lazebnik, S., Perona, P., Sato, Y., Schmid, C. (eds.) ECCV 2012, Part II. LNCS, vol. 7573, pp. 159–172. Springer, Heidelberg (2012)
4. Murphy-Chutorian, E., Trivedi, M.M.: Head pose estimation in computer vision: a survey. IEEE T-PAMI **31**(4), 607–626 (2009)
5. Cai, Q., Gallup, D., Zhang, C., Zhang, Z.: Head 3D deformable face tracking with a commodity depth camera. In: Proceeding of ECCV (2010)
6. Lu, F., Okabe, T., Sugano, Y., Sato, Y.: A head pose-free approach for appearance-based gaze estimation. In: Proceedings of BMVC (2011)
7. Mora, K.A.F., Odobez, J.-M.: Gaze estimation from multimodal kinect data. In: CVPR Workshop (2012)
8. Mora, K.A.F., Odobez, J.-M.: Person independent 3D gaze estimation from remote RGB-D camera. In: Proceedings of ICIP (2013)
9. Moosmann, F., Larlus, D., Frederic, J.: Learning saliency maps for object categorization. In: ECCV Workshop (2006)
10. Gao, D., Vasconcelos, N.: Discriminant saliency for visual recognition from cluttered scenes. In: Proceedings of NIPS (2004)
11. Gao, D., Vasconcelos, N.: Integrated learning of saliency, complex features and object detectors from cluttered scenes. In: Proceedings of CVPR (2005)
12. Parikh, D., Zitnick, C.L., Chen, T.: Determining patch saliency using low-level context. In: Forsyth, D., Torr, P., Zisserman, A. (eds.) ECCV 2008, Part II. LNCS, vol. 5303, pp. 446–459. Springer, Heidelberg (2008)
13. Oliva, A., Torralba, A., Castelhano, M.S., Henderson, J.M.: Top-down control of visual attention in object detection. In: Proceedings of ICCV (2009)
14. Harada, T., Ushiku, Y., Yuya Y.: Discriminative spatial pyramid. In: Proceedings of CVPR, Yasuo Kuniyoshi (2011)
15. Yao, B., Khosla, A., Fei-Fei, L.: Combining randomization and discrimination for fine-grained image categorization. In: Proceedings of CVPR (2011)

16. Zhang, L., Song, M., Zhao, Q., Liu, X., Bu, J., Chen, C.: Probabilistic graphlet transfer for photo cropping. IEEE T-IP **21**(5), 803–815 (2013)
17. Lin, Z., Chen, M., Ma, Y.: The augmented lagrange multiplier method for exact recovery of corrupted low-rank matrices, arXiv preprint (2010). arXiv:1009.5055
18. Zhang, L., Han, Y., Yang, Y., Song, M., Yan, S., Tian, Q.: Discovering discriminative graphlets for aerial image categories recognition. IEEE T-IP **22**(12), 5071–5084 (2013)
19. Siagian, C., Itti, L.: Rapid biologically-inspired scene classification using features shared with visual attention. IEEE T-PAMI **29**(2), 300–312 (2007)
20. Harchaoui, Z., Bach, F.: Image classification with segmentation graph kernels. In: Proceedings of ICCV (2007)
21. Lazebnik, S., Schmid, C., Ponce, J.: Beyond bags of features: spatial pyramid matching for recognizing natural scene categories. In: Proceedings of ICCV (2006)
22. Wang, J., Yang, J., Yu, K., Lv, F., Huang, T., Gong, Y.: Locality-constrained linear coding for image classification. In: Proceedings of CVPR (2010)
23. Yang, J., Yu, K., Gong, Y., Huang, T.: Linear spatial pyramid matching using sparse coding for image classification. In: Proceedings of CVPR (2009)
24. Li, L.-J., Su, H., Xing, E.P., Fei-Fei, L.: Object bank: a high-level image representation for scene classification and semantic feature sparsification. In: Proceedings of NIPS (2010)
25. Hou, X., Harel, J., Koch, C., Signature, I.: Highlighting sparse salient regions. IEEE T-PAMI **34**(1), 194–201 (2012)
26. Yao, B., Yang, X., Zhu, S.-C.: Introduction to a large scale general purpose ground truth dataset: methodology, annotation tool, and benchmarks. In: EMMCVPR (2007)

Two-Phase Representation Based Classification

Jianping Gou$^{(\boxtimes)}$, Yongzhao Zhan, Xiangjun Shen, Qirong Mao,
and Liangjun Wang

School of Computer Science and Telecommunication Engineering, JiangSu University,
Zhenjiang 212013, JiangSu, People's Republic of China
goujianping@ujs.edu.cn

Abstract. In this paper, we propose the two-phase representation based classification called the two-phase linear reconstruction measure based classification (TPLRMC). It is inspired from the fact that the linear reconstruction measure (LRM) gauges the similarities among feature samples by decomposing each feature sample as a liner combination of the other feature samples with L_p-norm regularization. Since the linear reconstruction coefficients can fully reveal the feature's neighborhood structure that is hidden in the data, the similarity measures among the training samples and the query sample are well provided in classifier design. In TPLRMC, it first coarsely seeks the K nearest neighbors for the query sample with LRM, and then finely represents the query sample as the linear combination of the determined K nearest neighbors and uses LRM to perform classification. The experimental results on face databases show that TPLRMC can significantly improve the classification performance.

Keywords: Sparse representation · Linear reconstruction measure · Representation based classification

1 Introduction

Representation based classification (RBC) has been one of the research hot spots in the field of computer vision and machine learning. In the RBC methods, how to well evaluate the similarity between a query sample and each class is very crucial to well represent and classify the query sample [1]. Currently, there exist many RBC methods, such as in [2–6].

Nowadays, the RBC methods can be mainly divided into two categories: L_1-norm based representation [4–8] and L_2-norm based representation [1,3,9–12]. Among the L_1-norm based representation methods that are added into by L_1-norm regularization term of the representation coefficients, the earliest and simplest one could almost be the sparse representation based classification (SRC) [4]. In SRC, the sparse linear combination of training samples for the query sample is optimally obtained and then the query sample is classified by its reconstruction residual using the representation coefficients. To enhance the discrimination of sparsity, adaptive sparse representation based classification (ASRC) is proposed in [5] for robust face

© Springer International Publishing Switzerland 2015
Y.-S. Ho et al. (Eds.): PCM 2015, Part I, LNCS 9314, pp. 265–274, 2015.
DOI: 10.1007/978-3-319-24075-6_26

recognition by jointly considering correlation and sparsity. Unlike SRC and ASRC that use all the training samples as the dictionary in sparse representation, the sparse representation based fisher discrimination dictionary learning (FDDL) is proposed in [7] for image classification. In FDDL, the discriminative information in both the representation coefficients and the representation residual is well taken into account in dictionary learning.

L_2-norm based representation with the L_2-norm constraint of the representation coefficients can also well represent and fast classify a query sample in pattern recognition. In these methods, the main characteristic is that the L_2-norm regularization fast offers the closed form solution to the representation coefficients and the solution is stable and robust. The typical L_2-norm based representation is the collaborative representation based classification (CRC) [3]. The CRC focuses on the collaboration between classes to represent the query sample and to improve the classification performance. Since CRC is first introduced, its basic idea has already been successfully borrowed to many new L_2-norm based representation methods [9–11]. Among these algorithms, the two-phase test sample sparse representation (TPTSR) method uses two-phase L_2-norm based representation for good classification, and can be viewed as supervised sparse representation [9]. Besides, the Linear regression classification (LRC) [12] is another classical L_2-norm based representation classification methods that uses a linear combination of class-specific training samples to represent a query.

In many representation based classification problems, the representation coefficients are well employed as a similarity measure among samples [13–20]. On the one hand, the representation coefficients as a similarity measure are directly adopted in the classifier design [13–15]. In [13], the linear reconstruction measure steered nearest neighbor classification (LRMNNC) is designed by using the L_1-norm or L_2-norm based representation coefficients as linear reconstruction measure (LRM) to determine the nearest neighbors of a query. Note that the representation coefficients in LRM are named as the linear reconstruction coefficients. In [14], the sparsity induced similarity measure (SIS) is proposed by using L_1-norm based representation coefficients as a similarity measure for label propagation and action recognition. In [15], sum of L_1-norm based representation coefficients, called sum of coefficients (SoC) is used as the classification decision rule. On the other hand, the representation coefficients as the weights among training samples are used for automatic graph construction of the given data [16–20]. Since the representation coefficients can preserve some intrinsic geometric properties of the original high dimensional data and potential discrimination information, the representation based graph construction can be well applied in dimensionality reduction.

The representation based classification methods perform well in many practical pattern recognition problems. In this paper, based on the idea of TPTSR using LRM, we propose two phase linear reconstruction measure based classification (TPLRMC) for further improving the performance of face recognition. TPLRMC utilizes two-phase L_p-norm based representation with LRM to do classification when $p = 1, 2$. In the first phase, it uses the regularized LRM with

L_p-norm regularization to represent a given query as the combination of the training samples and then chooses the nearest neighbors of the query according to the values of the representation coefficients. The most representative K training samples that correspond to the first K largest representation coefficients are coarsely selected. In the second phase, it further represents the given query as the combination of the chosen K training samples and uses LRM in final classifier design. Through the two phases, the query is classified into the class with the largest sum of representation coefficients among all the classes that the chosen K training samples belong to. In fact, since the first phase of TPLRMC selects partial training samples to represent the query sample, it can be regarded as "supervised" sparse representation, the same as TPTSR. To verify the effectiveness of TPLRMC, the experiments on the face databases are conducted. Experimental results demonstrate the proposed TPLRMC can significantly outperform the state-of-the-art representation based classification methods.

2   The Proposed TPLRMC

In this section, we introduce the novel two-phase representation based classification, called the two-phase linear reconstruction measure based classification (TPLRMC) on the basis of ideas of LRM and TPTSR.

2.1   The Motivation of the TPLRMC

From the perspective of the similarity measure, LRM can well reflect the neighborhood structure of the query sample in terms of the linear reconstruction coefficients [13]. That is to say, LRM uses the representation coefficients to search the nearest neighbors of the query sample by decomposing the query sample as the linear combination of the training samples. The effectiveness of LRM has been intuitively interpreted and mathematically proved in [13]. Furthermore, Cheng *et al.* have also provided sparsity induced similarity measure (SIS), *i.e.*, regularized LRM with L_1-norm regularization, to well gauge similarities among samples [14]. The representation coefficients can be viewed as a good similarity measure, because they can also preserve some intrinsic geometric properties of the data and potential discrimination information [16–20]. In the first phase of TPTSR, the selected nearest neighbors of the test sample are determined by the deviation between neighbors and the test sample. This way to determine neighbors may be unreasonable to some degree, because some training samples may have larger representation coefficients for the test sample (*i.e.* they have larger similarities with the test sample), but they are not chosen as the neighbors due to their corresponding larger deviations.

From the viewpoint of classification, the classification rule of SRC may not be optimal, because it is based on residual that measures the representational ability of the categorical training samples. To enhance the power of the discrimination from sparse coefficients, the SoC classification decision rule has been proposed [15]. In both SoC and LMRNNC, the representation coefficients as similarities of samples are used for good classification, which fully implies that the coefficients potentially preserve more discriminative information. However, in the second phase of TPTSR, reconstruction residual is still utilized in the process of classification just as SRC.

As argued above, motivated by the fact that the representation coefficients can well reflect similarities of samples and contain good pattern discrimination, we develop the two-phase linear reconstruction measure based classification (TPLRMC). In TPLRMC, LRM is adopted to seek nearest neighbors in the first phase and to design the classification rule in the second phase.

2.2   The First Phase of the TPLRMC

For notational convenience in what follows, we first assume a training set $X = [x_1, ..., x_n] \in \mathrm{R}^{d \times n}$ within M classes $\{c_1, ..., c_M\}$ has n training samples in d-dimensional feature space.

In the first phase of TPLRMC, the K nearest neighbors of the test sample are coarsely sought from all training samples using the L_p-norm based representation. It first supposes that the test sample $y \in \mathrm{R}^d$ is approximately represented by a linear combination of X just like TPTSR as follow:

$$y = s_1 x_1 + s_2 x_2 + ... + s_n x_n = Xs , \tag{1}$$

where $s = [s_1, s_2, ..., s_n]^T \in \mathrm{R}^n$ is representation coefficients of all training samples. Obviously, s can be solved by the linear reconstruction process as

$$s^* = \arg\min_s \|y - Xs\|_2^2 , \tag{2}$$

where $s^* = [s_1^*, s_2^*, ..., s_n^*]^T$. Here, in order to enhance discrimination capability, some regularization terms could offen be imposed on s. Equation 2 can be rewritten as

$$s^* = \arg\min_s \|y - Xs\|_2^2 + \gamma \|s\|_p , \tag{3}$$

where γ is a small positive constant, and $p = 1$ or 2 in general. When $p = 2$, the solution of Eq. 3 can be easily achieved by $s^* = (X^T X + \gamma I)^{-1} X^T y$.

After obtaining s^* by Eq. 3, we use s_i^* as the evaluation of the contribution of the ith training sample x_i to represent the test sample y. The larger s_i^* is, the greater the contribution of x_i is, because the coefficient s_i^* well reflects similarity between y and x_i [13,14]. We sort the coefficients in s^* in a decreasing order, and then select first K largest ones $\{s_1^*, s_2^*, ..., s_K^*\}$ that correspond to the K training samples $\bar{X} = [\bar{x}_1, \bar{x}_2, ..., \bar{x}_K]$. The K training samples are viewed as K nearest neighbors of the test sample y, and they are from Q classes denoted as $\bar{C} = \{\bar{c}_1, ..., \bar{c}_Q\}$. $\bar{C}$ must be one of subset of set $\{c_1, ..., c_M\}$, i.e., $\bar{C} \subseteq \{c_1, ..., c_M\}$. Therefore, the test sample will not finally be classified into the rth class, which no chosen neighbors belong to.

2.3   The Second Phase of the TPLRMC

In the second phase of TPLRMC, the test sample is further finely represented by a linear combination of the determined K nearest neighbors, and the representation coefficients of these neighbors are employed for designing the classification decision rule. It still assumes that the test sample y is approximately expressed as follows:

$$y = \bar{s}_1\bar{x}_1 + \bar{s}_2\bar{x}_2 + ... + \bar{s}_n\bar{x}_K = \bar{X}\bar{s} \, , \tag{4}$$

where $\bar{s} = [\bar{s}_1, \bar{s}_2, ..., \bar{s}_K]^T$ are the coefficients. The $\bar{s}$ can be solved by the regularized linear reconstruction process

$$\bar{s}^* = \arg\min_{\bar{s}} \|y - \bar{X}\bar{s}\|_2^2 + \tau\|\bar{s}\|_p \, , \tag{5}$$

where τ is a small positive constant. When $p = 2$, the solution of Eq. 5 can be easily obtained by $\bar{s}^* = (\bar{X}^T\bar{X} + \tau I)^{-1}\bar{X}^T y$.

According to the optimal $\bar{s}^*$, the sum of contribution to represent the test sample y of the neighbors from each class is calculated to do classification by using the linear reconstruction coefficients. The classification decision rule of TPLRMC is defined as

$$\bar{c} = \arg\max_{\bar{c}_i} \sum \delta^{\bar{c}_i}(\bar{s}^*) \, . \tag{6}$$

where $\delta^{\bar{c}_i}(\bar{s}^*) \in R^K$ is a new vector whose only nonzero entries are the entries in $\bar{s}^*$ that are associated with class $\bar{c}_i \in \bar{C}$, and $\sum \delta^{\bar{c}_i}(\bar{s}^*)$ is the sum of coefficients of the neighbors from class $\bar{c}_i$. Consequently, TPLRMC classifies the test sample to the class that has largest sum of coefficients.

2.4   Analysis of the TPLRMC

As stated above, since the main idea of TPLRMC is to use the LRM to seek the nearest neighbors in the first phase and do classification in the second phase, LRM plays an important role in TPLRMC. In order to mathematically verify the efficiency of TPLRMC, we can borrow the theoretical analysis of LRM from [13], which well reveals the representation coefficients can truly reflect the similarities between the nearest neighbors and a given test sample.

Assume that S is a subspace of Euclidean space E and $S = L(x_1, x_2, ..., x_n)$ indicates the subspace S spanned by the basis $x_1, x_2, ..., x_n$. If β is a perpendicular vector from the given test sample y to S, then $\beta \perp S$ and $|\beta|$ is the minimum distance between y and each vector in S, where $\|$ denotes the symbol of the length [13]. It is to be noted that y and each x_i are normalized. According to the LRM function from Eqs. 2 or 3, we can find the vector $y - \sum_{i=1}^{n} s_i x_i$, i.e., the perpendicular vector of the subspace $S = L(x_1, x_2, ..., x_n)$. Through the optimal solution $s^* = [s_1^*, s_2^*, ..., s_n^*]^T$ in the first phase of TPLRMC, we can obtain $y - \sum_{i=1}^{n} s_i^* x_i = 0$ or $y - \sum_{i=1}^{n} s_i^* x_i \perp S$. For any vector x_j in the subspace S, we get

$$< y - \sum_{i=1}^{n} s_i^* x_i, x_j >= 0 \, , \tag{7}$$

where $<,>$ denotes the inner product.

Through Eq. 7, we further discuss the coefficient s_j^* corresponding to x_j from two aspects, in order to reveal the neighborhood structure of y. First, if $s_j^* = 0$, Eq. 7 can be rewritten as

$$< y - \sum_{i=1, i \neq j}^{n} s_i^* x_i, x_j >= 0 . \tag{8}$$

As we know, $s_j^* = 0$ indicates x_j doesn't lie in the best representation linear subspace of y. That is to say, x_j has no contribution to y, and is not nearest neighbor of y.

Generally speaking, we are more concerned about the case that the coefficient s_j^* is not equal to zero. Suppose that $\bar{S} = L(x_1, x_2, ..., x_m)$ is the linear subspace spanned by the subset of all training samples when $s_j^* \neq 0, j = 1, 2, , ..., m, m \leq n$. Equation 7 can be reformulated as

$$< y - \sum_{i=1}^{m} s_i^* x_i, x_j >= 0 . \tag{9}$$

For any sample x_j from $\bar{S}$, we can rewritten the Eq. 9 as

$$s_j^* =< y, x_j > - < \sum_{i=1, i \neq j}^{m} s_i^* x_i, x_j > . \tag{10}$$

Through replacing the inner product with the Euclidean distance, Eq. 10 can be further expressed as follows

$$s_j^* = \frac{1}{2} \left(d^2 (\sum_{i=1, i \neq j}^{m} s_i^* x_i, x_j) - d^2(y, x_j) \right) , \tag{11}$$

where $d(,)$ is the Euclidean distance. It can be observed from Eq. 11 that the value of s_j^* is greater when $d^2(\sum_{i=1, i \neq j}^{m} s_i^* x_i, x_j)$ is larger and $d^2(y, x_j)$ is smaller simultaneously. In other words, the larger s_j^* implies x_j has more representation contribution to y and can be the most likely neighbor of y. Meanwhile, the larger $d^2(\sum_{i=1, i \neq j}^{m} s_i^* x_i, x_j)$ also indicates the other samples except x_j have less contribution to represent y. In addition, if $d^2(y, x_j)$ is greater than $d^2(\sum_{i=1, i \neq j}^{m} s_i^* x_i, x_j)$, i.e., $s_j^* < 0$, x_j has negative representation contribution to y, and cannot become the neighbor of y. Hence, LRM as an effective similarity measure can be convincingly used in TPLRMC for improving the classification performance in theory.

3   Experiments

To well verify the effectiveness of the proposed TPLRMC, the experiments are conducted on the Yale and AR face databases. On each database, we compare

TPLRMC with the state-of-the-art representation based classification methods, such as SRC [4], TPTSR [9], LRC [12], LRMNNC [13] and SoC [15] in terms of classification accuracy. For each database, every image sample is normalized in advance, and l image samples from each class are randomly selected as the training set while the rest of image samples are used as the test set. For a given l, we conduct 10 independent runs in the experiments. Accordingly, the final classification performance evaluation is obtained by averaging recognition accuracy rates from these 10 runs. Note that TPLRMC and LRMNNC with L_1 and L_2 norms is denoted as TPLRMC(L=1) and TPLRMC(L=2), and LRMNNC with L_1 and L_2 norms is denoted as LRMNNC(L=1) and LRMNNC(L=2).

3.1 Databases

The Yale face database (http://cvc.yale.edu/projects/yalefaces/yalefaces.html) contains 165 gray scale images from 15 individuals, each of which has 11 images. All images per subject are taken with different facial expressions and illuminations. The AR face database (http://www2.ece.ohio-state.edu/aleix/ARdatabase.html) comprises over 4,000 color images from 126 people's faces (70 men and 56 women). All images of each subject are taken by varying with different facial expressions, illumination conditions and occlusions (sun glasses and scarf). Here, we choose 1400 face images of 100 individuals, each of which have 14 images under different expressions and illumination conditions. In the experiments, each image is resized to 32×32 pixels. Figure 1 shows image samples of one person on the Yale (the first row) and AR (the second row) face databases.

3.2 Experimental Results

In the experiments, the numbers of training samples from each class are set to be $l = 2, 4, 6, 8$ on Yale and $l = 3, 4, 5, 6$ on AR, and the ratios between K nearest neighbors (*i.e.*, the chosen training samples) and the number of the training samples are varied from 0.1 to 1 by step 0.1 in the first phase of TPLRMC and TPTSR. The experimental comparisons between TPLRMC and TPTSR are conducted by varying the numbers of the chosen training samples K, shown in Figs. 2 and 3. It can be observed from Figs. 2 and 3 that TPLRMC consistently and significantly performs better than TPTSR at different numbers of the chosen training samples. As a consequence, the proposed TPLRMC has more robustness to the chosen training samples in the first phase with satisfactory classification performance.

Fig. 1. The image samples of one subject on the Yale and AR face databases.

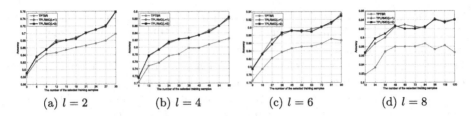

(a) $l = 2$ (b) $l = 4$ (c) $l = 6$ (d) $l = 8$

Fig. 2. The recognition accuracy rates via the chosen training samples K on Yale.

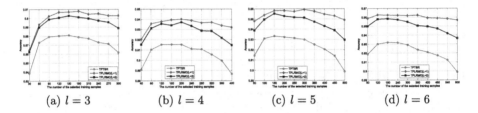

(a) $l = 3$ (b) $l = 4$ (c) $l = 5$ (d) $l = 6$

Fig. 3. The recognition accuracy rates via the chosen training samples K on AR.

The effectiveness of the proposed TPLRMC for classification is further assessed by comparing it with SRC, LRC, TPTSR, SoC and LRMNNC on Yale and AR. The best average classification results of each method are reported in Tables 1 and 2. Note that the numbers in parentheses are the determined training samples K in the first phase in both TPTSR and TPLRMC, and the best performance is indicated in bold-face among the comparative methods. As Tables 1 and 2 displays, the classification performance of each method increases with increasing the number of training samples. Moreover, TPLRMC significantly outperforms the other competing methods, and TPLRMC(L=1) is almost superior to TPLRMC(L=2). Consequently, the promising classification performance of the proposed TPLRMC is well demonstrated on Yale and AR.

Table 1. The average recognition accuracy rates (%) of each method on Yale.

Methods	$l = 2$	$l = 4$	$l = 6$	$l = 8$
SRC	62.22	70.95	82.80	86.67
SoC	61.85	63.71	82.00	86.44
LRC	56.67	67.14	73.07	76.67
LRMNNC(L=2)	62.22	69.52	78.67	82.00
LRMNNC(L=1)	62.22	69.52	77.47	83.11
TPTSR	69.85(30)	82.48(60)	87.07(63)	88.67(84)
TPLRMC(L=2)	**75.85(30)**	**88.48(60)**	92.93(90)	94.00(96)
TPLRMC(L=1)	75.78(30)	88.00(60)	**93.47(90)**	**94.22(96)**

Table 2. The average recognition accuracy rates (%) of each method on AR.

Methods	$l = 3$	$l = 4$	$l = 5$	$l = 6$
SRC	87.57	92.45	94.74	95.90
SoC	87.40	92.21	94.64	95.55
LRC	55.56	67.12	74.58	82.40
LRMNNC(L=2)	88.05	91.13	92.99	93.50
LRMNNC(L=1)	86.92	90.89	92.82	93.91
TPTSR	88.09(150)	91.54(160)	93.33(150)	93.19(180)
TPLRMC(L=2)	90.29(150)	93.72(200)	95.54(150)	95.85(180)
TPLRMC(L=1)	**90.79(180)**	**94.02(200)**	**95.93(300)**	**96.28(120)**

Across the experiments above, three observations can be made as follows: (a) TPLRMC can significantly outperform TPTSR, and has more robustness to the chosen training samples than TPTSR in the first phase. (b) The classification performance of TPLRMC(L=1) is usually better than that of TPLRMC(L=2). (c) TPLRMC almost conformably performs better than the related representation based classification methods significantly.

4 Conclusions

In this paper, we propose the novel two-phase representation based classification, called the two-phase linear reconstruction measure based classification. The proposed TPLRMC utilizes the regularized LRM to choose the nearest neighbors in the first phase and to do classification in the second phase. To well investigate the classification performance of the proposed TPLRMC, the experiments on the face databases are conducted by comparing it with the related representation based classification methods. The experimental results have consistently demonstrated the effectiveness of the proposed TPLRMC method with the satisfactory classification performance.

Acknowledgment. This work was supported by the National Science Foundation of China (Grant Nos. 61170126, 61272211), the Natural Science Foundation of the Jiangsu Higher Education Institutions of China (Grant No. 14KJB520007), China Postdoctoral Science Foundation (Grant No. 2015M570411) and Research Foundation for Talented Scholars of JiangSu University (Grant No. 14JDG037).

References

1. Xu, Y., Li, X., Jian, Y., Lai, Z., Zhang, D.: Integrating conventional and inverse representation for face recognition. IEEE Trans. Cybern. **44**(10), 1738–1746 (2014)
2. Shi, Q., Eriksson, A., van den Hengel, A., Shen, C.: Is face recognition really a compressive sensing problem? In: 2011 IEEE Conference on Date of Conference Computer Vision and Pattern Recognition (CVPR), pp. 553–560, 20–25 June 2011

3. Zhang, L., Yang, M., Feng, X.: Sparse representation or collaborative representation: which helps face recognition. In: Proceedings of IEEE International Conference on Computer Vision, Barcelona, Spain, pp. 471–478, November 2011
4. Wright, J., Yang, A.Y., Ganesh, A., Sastry, S.S., Ma, Y.: Robust face recognition via sparse representation. IEEE Trans. Pattern Anal. Mach. Intell. **31**(2), 210–227 (2009)
5. Wang, J., Lu, C., Wang, M., Li, P., Yan, S., Hu, X.: Robust face recognition via adaptive sparse representation. IEEE Trans. Cybern. **44**(22), 2368–2378 (2014)
6. Deng, W., Hu, J., Guo, J.: Extended SRC: undersampled face recognition via intraclass variant dictionary. IEEE Trans. Pattern Anal. Mach. Intell. **34**(9), 1864–1870 (2012)
7. Yang, M., Zhang, L., Feng, X.: Sparse representation based fisher discrimination dictionary learning for image classification. Int. J. Comput. Vis. **109**(3), 209–232 (2014)
8. Lai, Z.-R., Dai, D.-Q., Ren, C.-X., Huang, K.-K.: Discriminative and compact coding for robust face recognition. IEEE Trans. Cybern. (2015). doi:10.1109/TCYB.2014.2361770
9. Yong, X., Zhang, D., Yang, J., Yang, J.-Y.: A two-phase test sample sparse representation method for use with face recognition. IEEE Trans. Circuits Syst. Video Technol. **21**(9), 1255–1262 (2011)
10. Yang, M., Zhang, L., Zhang, D., Wang, S.: Relaxed collaborative representation for pattern classification. In: IEEE Conference on Computer Vision Pattern Recognition Providence, RI, United states, pp. 2224–2231, June 2012
11. Peng, X., Zhang, L., Yi, Z., Tan, K.K.: Learning locality-constrained collaborative representation for robust face recognition. Pattern Recogn. **47**(9), 2794–2806 (2014)
12. Naseem, I., Togneri, R., Bennamoun, M.: Linear regression for face recognition. IEEE Trans. Pattern Anal. Mach. Intell. **32**(11), 2106–2112 (2010)
13. Zhang, J., Yang, J.: Linear reconstruction measure steered nearest neighbor classification framework. Pattern Recogn. **47**(4), 1709–1720 (2014)
14. Cheng, H., Liu, Z., Hou, L., Yang, J.: Measure, Sparsity Induced Similarity, Applications, Its. IEEE Trans. Circuits Syst. Video Technol. **99**, 1–14 (2012)
15. Li, J., Can-Yi, L.: A new decision rule for sparse representation based classification for face recognition. Neurocomputing **116**, 265–271 (2013)
16. Cheng, B., Yang, J., Yan, S., Fu, Y., Huang, T.S.: Learning with L1-graph for image analysis. IEEE Trans. Image Process. **19**(4), 858–866 (2010)
17. Qiao, L., Chen, S., Tan, X.: Sparsity preserving projections with applications to face recognition. Pattern Recogn. **43**(1), 331–341 (2010)
18. Yang, W., Wang, Z., Sun, C.: A collaborative representation based projections method for feature extraction. Pattern Recogn. **48**, 20–27 (2015)
19. Yan, H., Yang, J.: Sparse discriminative feature selection. Pattern Recogn. **48**, 1827–1835 (2015)
20. Raducanu, B., Dornaika, F.: Embedding new observations via sparse-coding for non-linear manifold learning. Pattern Recogn. **47**, 480–492 (2014)

Deep Feature Representation via Multiple Stack Auto-Encoders

Mingfu Xiong[1], Jun Chen[1,2]([✉]), Zheng Wang[1], Chao Liang[1,2], Qi Zheng[1],
Zhen Han[1,2], and Kaimin Sun[3]

[1] School of Computer, National Engineering Research Center for Multimedia
Software, Wuhan University, Wuhan 430072, China
{xmf2013,chenj,wangzwhu,cliang,zhengq,sunkm}@whu.edu.cn
[2] Collaborative Innovation Center of Geospatial Technology, Wuhan, China
hanzhen_1980@163.com
[3] State Key Laboratory of Information Engineering in Surveying,
Mapping and Remote Sening, Wuhan University, Wuhan 430072, China

Abstract. Recently, deep architectures, such as stack auto-encoders
(SAEs), have been used to learn features from the unlabeled data.
However, it is difficult to get the multi-level visual information from
the traditional deep architectures (such as SAEs). In this paper, a feature representation method which concatenates Multiple Different Stack
Auto-Encoders (MDSAEs) is presented. The proposed method tries to
imitate the human visual cortex to recognize the objects from different
views. The output of the last hidden layer for each SAE can be regarded
as a kind of feature. Several kinds of features are concatenated together
to form a final representation according to their weights (The output
of deep architectures are assigned a high weight, and vice versa). From
this way, the hierarchical structure of the human brain cortex can be
simulated. Experimental results on datasets MNIST and CIRFA10 for
classification have demonstrated the superior performance.

Keywords: Multiple stack auto-encoders · Deep feature learning · Deep
Learning · Classification

1 Introduction

In many tasks of the multimedia systems, a critical problem is how to construct
a good feature representation [1]. Many famous off-the-shelf features (such as
SIFT [2], HOG [3], LBP [4]) have achieved great success in traditional image
recognition. However, they still exist some limitations: (1) they are not adaptive
to data; (2) they can hardly capture the discriminative characteristics of different
classes [5]. In contrast, the deep feature learning techniques, which attract a lot
of attentions in multimedia community, automatically learn intrinsic and subtle
features from the statistics of real data [6–8].

Inspired by the architectural depth of the brain, neural network researchers
have spent for decades in training deep neural networks [9,10]. Other algorithms,

© Springer International Publishing Switzerland 2015
Y.-S. Ho et al. (Eds.): PCM 2015, Part I, LNCS 9314, pp. 275–284, 2015.
DOI: 10.1007/978-3-319-24075-6_27

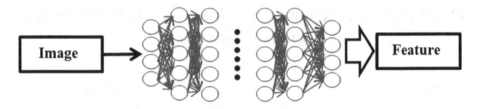

Fig. 1. The traditional auto-encoders for the feature learning. There is only a stack auto-encoder network for images features extraction. When input an image, a feature can be generated at the last hidden layer of the auto-encoder network.

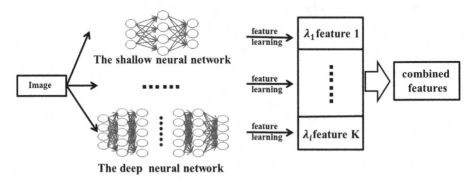

Fig. 2. The combined multiple auto-encoders framework for the feature learning. There are several kinds of features are obtained from different auto-encoders, and denote as feature 1 ,..., feature K (K is the numbers of the auto-encoders. In our experiment, K is 4). Then, the K kinds of features are combined according to their weights to form the final feature. The weight factors are set as $\lambda_1, ..., \lambda_i$ respectively. The more of the hidden layers, the bigger of the λ_i is assigned.

such as hierarchical spatial-temporal features [11], sparse auto-encoder [12], convolutional deep belief network [13], and sum-product network [14], have shown the outstanding performance on many challenging visual recognition tasks. Over the last two years Deep Learning has been choosed for image classification [15,16], attaining even super-human performance levels [17,18], while many other works have shown that the features learned by deep neural networks can be successfully employed in other tasks [19,20]. The key building blocks of deep neural networks for images have been around for many years [21]. The recent success of Deep Neural Networks can be mostly attributed to large datasets, GPU computing and well-engineered choices. Although these algorithms take hierarchical approaches in the feature extraction and provide efficient solution to complex problems, the results in intricate representation of the feature extraction process that is hard to follow intuitively [7]. In addition, the traditional deep architectures (such as SAEs) can only extract a monolayer feature (Fig. 1), which is proved to lose some discriminative information that is helpful to represent the images, accurately [1].

In this paper, the method of combining multi-level features for images representation is presented (Fig. 2), which tries to simulate the human brain visual cortex. In this case, we extend a version of combining several deep neural networks which have same configurations for the feature representation [9]. Note that, in our work, the configurations of the deep neural networks are not the same as each other. The original images are input into each of the SAE for features extraction. The hidden layers of these auto-encoders are set different for each other from top to bottom (Fig. 2). The shallow features are obtained from the auto-encoders that have fewer hidden layers and vice versa. In this way, we can capture different levels features for objects. After that, these features are combined together to form the final feature representation according to their weights.

The contribution of this paper can be summarized in two aspects: Firstly, we propose a method of learning the features representation that capture information of objects at different levels in a layer-wise manner. Secondly, we try to simulate human brain visual cortex from the perspective of bionics. As described above, we can get more robust features for images representation. In addition, the performance of our algorithm has improved greatly on the MNIST datasets. Nearly 2% accuracy rate has improved on the dataset CIRFA 10 comparing with most existing methods.

2 Our Method

2.1 The Basic Auto-Encoder

We recall the basic principles of auto-encoders models, e.g., [22]. An auto-encoder takes an input $x \in R^d$ and first maps the input to the latent representation $h \in R^{d'}$ using a deterministic function of $h = f_\theta = \sigma(Wx + b)$ with parameters $\theta = \{W, b\}$. This "code" is then used to reconstruct the input by a reverse mapping of $f : y = f_{\theta'}(h) = \sigma(W'h + b')$ with $\theta' = \{W', b'\}$. The two parameter sets are usually constrained to be of the form $W' = W^T$, using the same weights for encoding the input and decoding the latent representation y_i. The parameters are optimized, minimizing an appropriate cost function over the training set $D_n = \{(x_0, t_0), ...(x_n, t_n)\}$.

2.2 Building the Multiple Multi-level Auto-Encoders

The traditional auto-encoder is introduced as a dimensionality reduction algorithm. But this paradigm only gets a single feature which can't indicate the traits of the objects accurately. So how to get multiple features is necessary. In this paper, we train several stack auto-encoders which have different hidden layers. And the output of the last hidden layer for each auto-encoder is regarded as a kind of feature. After that, we assign different weights to these features according to their capabilities of representation. If the feature learned from the deep network has a good representation, we will assign a high weight to it, vice versa. This process is unsupervised completely, and simulates the cognitive ability of the human brain, which is from coarse to fine via concatenating different levels features. The framework is described as Fig. 2.

Our framework combine several auto-encoders, and each of them has different configurations. The networks which have fewer hidden layers get the low-level information of the objects, such as edges, and the deep network could get the high-level content, such as object parts or complete objects. The information and content are regarded as different levels features for the objects. Then, these features are combined together to form the final representation. So we can get the features of the objects that is from coarse to fine. When training on natural images, our framework can capture their representation from different views. In this way, to build our MDSAEs, we rely on a novel inference scheme that ensure each layer to reconstruct the input, rather than just the output of the layer directly beneath. So the feature which is obtained in this way is more robust and representative.

2.3   The Layer-Wise Training and Fine Tuning

The basic auto-encoder has been used as a building block to train deep networks in [23], with the representation of the k-th layer used as input for the (k+1)-th, and the k-th layer trained after the (k-1)-th has been trained. However, this major drawback to this paradigm is that the image pixels are discarded after the first layer, thus higher layers of the model have an increasingly diluted connection to the input. This makes learning fragile and impractical for models beyond a few layers. However, in our work, the drawback is handled via combining the multi-level different features, which can compensate for each other.

In addition, the objective for fine-tuning is to obtain the optimized neural network parameters. As we know, the deep model may be over-fitting when there are too many parameters. So the backpropagation is used to train multi-level architectures. Its procedure to compute the gradient of an objective function with respective to the weights of a multi-level stack of modules is nothing more than a practical application of the chain rule for derivatives. Once these derivative have been computed, it is straightforward to compute the gradient with respect to the weights of each model.

2.4   The Weight Assigned for Each Feature

However, how to combine these features is quite important. As described above, the hierarchical features are obtained from different auto-encoders. And the characteristic capability of each feature is different.

$$F = \Sigma_{i=1}^{K} \lambda_i f_i \qquad s.t. \qquad \Sigma_{i=1}^{K} \lambda_i = 1 \qquad (1)$$

We assign different weights to each of the feature to get a good representation. General speaking, for the task of deep neural networks, the more numbers of hidden layers, the better robust feature will be obtained. So we set a high weight to the feature that is learned from the deep network, and vice versa. This process is described as formula (1). F denotes the combined feature, f_i denotes the feature that is learned from the i-th auto-encoder, λ_i is the weight factor that assign to the i-th feature, and K denotes the numbers of the deep auto-encoders. In our work, K values 4, and λ_1, λ_2, λ_3, λ_4 can be set 0.05, 0.25, 0.3, 0.4, respectively.

2.5 Classification

In this work, different features are obtained via multi-level auto-encoders, and these features are preprocessed via weight assigning. At last, the presulfided features are combined together to form the last representation. Then, we can get the final feature followed by a softmax classifier. And the classifier generalizes logistic regression to classification problems where the class label can take on more than two possible values.

3 Experiments

In this section, we evaluate our algorithm on two common objects recognition benchmarks and improve the state-of-the-art on both of them. The output of the last hidden layer for each auto-encoder is used for classification on the MNIST and CIFAR10 datasets.

3.1 The MNIST

The MNIST dataset is a handwritten digits which has a training set of 60,000 examples, and a test set of 10,000 examples. It is a subset of a larger set available from NIST. The digits have been size-normalized and centered in a fixed-size image, which is 784 (28*28). The best performance of reported machine learning methods [24] achieve over 99 % accuracy.

In this work, we trained four kinds of deep framework. The hidden layers of the models were set: 2, 3, 4 and 5, respectively, and the units of each model were set 200, 392, 500, 600, correspondingly. In addition, the number of units for each hidden layer were the same. The networks were indicated as SAE-2 (200), SAE-3 (392), SAE-4 (500) and SAE-5 (600). e.g., SAE-t (M), t denotes hidden layers, M denotes the units of each hidden layer. The input size of each image was set 784 (28*28).

The Single Deep Features: As described in Table 1, The results in the fourth column denote the stack auto-encoders that are not fine-tuned. Generally speaking, the deeper of the neural network, the more necessary is for fine-tuning. After training an auto-encoder, we would like to visualize the weights that were learned by the algorithm, and try to understand what has been learnt (See in Fig. 3).

Table 1. Experimental results of single deep framework. From Table 1, we could see that the performance is related to the dimension of features. Meanwhile, the fine-tuning is also very important for deep neural networks.

Hierarchy	Units	Acc (with fine tuning)	Acc
SAE-2	200	96.72 %	91.71 %
SAE-3	392	96.87 %	89.26 %
SAE-4	500	97.10 %	85.91 %
SAE-5	600	**97.25%**	77.93 %

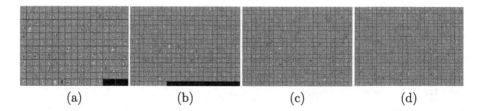

(a)   (b)   (c)   (d)

Fig. 3. Visualization of Features: Visualization of filters learned by single auto-encoder trained on InfiniteMNIST. The weights (called filters) of the first hidden layer for the auto-encoders can be visualized. (a) SAE-2; (b) SAE-3; (c) SAE-4; (d) SAE-5.

The Combined Features: The results are illustrated in Table 2. We implemented four deep auto-encoders as described in above. Then, we combined the four kinds of features together. The combined features improved nearly 1.5 % accuracy rate comparing with a single feature. In addition, the accuracy is related to the dimension of the features. It is worth mentioning that, for the majority of tasks, the models selection procedure chose best performing model with an over-complete first hidden layer representation (typically of size 2000 for the 784-dimension MNIST-derived tasks). It is very different from the traditional"bottleneck" auto-encoders, and make possible by our combined multi-level auto-encoders training procedure. All this suggestions that the proposed procedure is indeed able to produce better feature detectors.

Comparison with Other Algorithms: For the sake of comparison, a variety of other trainable models were trained and tested on the same dataset. The results are illustrated in Table 3. The MDSAEs has improved nearly 2 % accuracy rate comparing with [25].

3.2 The CIFAR 10

The CIFRA 10 is a set of natural images of 32*32 pixels [28]. It contains 10 classes, each with 50000 training samples and 10000 test samples. Images vary greatly within each class. They are not necessarily centered, may contain only parts of the object, and show different backgrounds. Subjects may vary in size by an order of magnitude (i.e., some images show only the head of a bird, other

Table 2. Experimental results of the combined features. In our experiment, there are six kinds of combination for features representation. The results are illustrated in the last two columns of Table 2. The weight fators are assigned as described in the first column of the table. For the last row, the all combined is: 0.05*SAE-2+0.25*SAE-3+0.3*SAE-4+0.4*SAE-5.

Combination	Dimension	Acc(without weighted)	Acc(with weighted)
0.3*SAE-2+0.7*SAE-3	592	98.07 %	98.37 %
0.3*SAE-2+0.7*SAE-4	700	98.39 %	98.42 %
0.3*SAE-2+0.7*SAE-5	800	98.42 %	98.54 %
0.4*SAE-3+0.6*SAE-4	892	98.57 %	98.67 %
0.4*SAE-3+0.6*SAE-5	992	98.70 %	98.89 %
0.5*SAE-4+0.5*SAE-5	1100	99.01 %	99.34 %
All Combined	1692	99.36 %	**99.52**%

Table 3. Comparison with other methods, our algorithm get the best performance.

Methods	Accuracy
RBF Network [26]	96.40 %
Denoising Auto-encoders [25]	98.10 %
Multi-column DNN [27]	99.45 %
Ours	**99.52**%

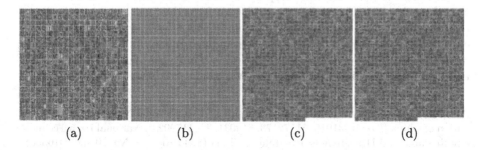

(a) (b) (c) (d)

Fig. 4. Visualization of Features: Visualization of filters learned by single auto-encoder trained on CIFAR. The weights (called filters) of the first hidden layer for the auto-encoders can be visualized. (a) SAE-2. (b) SAE-3. (c) SAE-4. (d) SAE-5.

an entire bird from a distance). Colors and textures of objects/animals also vary greatly.

The input of our MDSAEs have three maps, one for each color channel (RGB), and the size of maps is 3072 (32*32*3). The architecture of the models were set as same as the above did, but the units of each model were 400, 600,800, 1000, correspondingly. The original images and visualization of filters which were learned by single auto-encoder had shown in Fig. 4. We repeated the experiments

Table 4. From the results, our algorithm can show the best performance comparing with other methods.

Methods	Accuracy
LSPR [29]	80.02 %
SERL-Nonnegativity Constraints [30]	82.90 %
Sum-Product Networks [14]	83.96 %
MCDNN [27]	88.79 %
Ours	**89.87%**

with different random initializations and computed mean and standard deviation of the error, which was rather small from original images, showing that our deep model was more robust. Our MDSAEs obtain a quite high accuracy rate of 89.87 %, greatly rising the bar for this benchmark. In addition, our algorithm is much simpler than the other methods. The results have shown in Table 4.

4 Conclusion

In this paper, we have proposed a new method which is named multiple different stack auto-encoders (MDSAEs) for the feature learning. It is trained via various kinds of basic auto-encoders (in this work we have four). Then, the output of the last hidden layers for each deep model is regarded as a kind of feature. At last, these features are combined together to form the final representation, and the experiment results on the MNIST and CIFAR 10 datasets have shown that our model outperformed most of the existing algorithms. As a further work, we would like to apply our proposed method to various types of data for discovering underlying feature hierarchies in complex data.

Acknowledgements. The research was supported by National Nature Science Foundation of China (No. 61231015, 61172173, 61303114, 61170023). National High Technology Research and Development Program of China (863 Program, No. 2015AA016306). Technology Research Program of Ministry of Public Security (No. 2014JSYJA016). The EU FP7 QUICK project under Grant Agreement (No. PIRSES-GA-2013-612652). Major Science and Technology Innovation Plan of Hubei Province (No. 2013AAA020). Internet of Things Development Funding Project of Ministry of industry in 2013 (No. 25). China Postdoctoral Science Foundation funded project (2013M530350, 2014M562058). Specialized Research Fund for the Doctoral Program of Higher Education (No. 20130141120024). Nature Science Foundation of Hubei Province (2014CFB712). The Fundamental Research Funds for the Central Universities (2042014kf0025, 2042014kf0250, 2014211020203). Scientific Research Foundation for the Returned Overseas Chinese Scholars, State Education Ministry ([2014]1685).

References

1. Zeiler, M.D., Taylor, G.W., Fergus, R.: Adaptive deconvolutional networks for mid and high level feature learning. In: IEEE International Conference on Computer Vision (ICCV), pp. 2018–2025. IEEE (2011)
2. Lowe, D.G.: Distinctive image features from scale-invariant keypoints. Int. J. Comput. Vis. **60**, 91–110 (2004)
3. Dalal, N., Triggs, B.: Histograms of oriented gradients for human detection. In: IEEE Computer Society Conference on Computer Vision and Pattern Recognition, vol. 1, pp. 886–893. IEEE (2005)
4. Ojala, T., Pietikainen, M., Maenpaa, T.: Multiresolution gray-scale and rotation invariant texture classification with local binary patterns. IEEE Trans. Pattern Anal. Mach. Intell. **24**, 971–987 (2002)
5. Zuo, Z., Wang, G.: Recognizing trees at a distance with discriminative deep feature learning. In: 2013 9th International Conference on Information, Communications and Signal Processing (ICICS), pp. 1–5. IEEE (2013)
6. Collobert, R.: Deep learning for efficient discriminative parsing. In: International Conference on Artificial Intelligence and Statistics. Number EPFL-CONF-192374 (2011)
7. Song, H.A., Lee, S.-Y.: Hierarchical representation using NMF. In: Lee, M., Hirose, A., Hou, Z.-G., Kil, R.M. (eds.) ICONIP 2013. LNCS, vol. 8226, pp. 466–473. Springer, Heidelberg (2013)
8. Bengio, Y., Courville, A.C., Vincent, P.: Unsupervised feature learning and deep learning: a review and new perspectives. CoRR, abs/1206.5538 1 (2012)
9. Deng, L., Hinton, G., Kingsbury, B.: New types of deep neural network learning for speech recognition and related applications: an overview. In: 2013 IEEE International Conference on Acoustics, Speech and Signal Processing (ICASSP), pp. 8599–8603. IEEE (2013)
10. Jaitly, N., Nguyen, P., Senior, A.W., Vanhoucke, V.: Application of pretrained deep neural networks to large vocabulary speech recognition. In: INTERSPEECH. Citeseer (2012)
11. Le, Q.V., Zou, W.Y., Yeung, S.Y., Ng, A.Y.: Learning hierarchical invariant spatio-temporal features for action recognition with independent subspace analysis. In: 2011 IEEE Conference on Computer Vision and Pattern Recognition (CVPR), pp. 3361–3368. IEEE (2011)
12. Le, Q.V.: Building high-level features using large scale unsupervised learning. In: 2013 IEEE International Conference on Acoustics, Speech and Signal Processing (ICASSP), pp. 8595–8598. IEEE (2013)
13. Huang, G.B., Lee, H., Learned-Miller, E.: Learning hierarchical representations for face verification with convolutional deep belief networks. In: IEEE Conference on Computer Vision and Pattern Recognition (CVPR), pp. 2518–2525. IEEE (2012)
14. Gens, R., Domingos, P.: Discriminative learning of sum-product networks. In: Advances in Neural Information Processing Systems(NIPS), pp. 3248–3256 (2012)
15. Simonyan, K., Zisserman, A.: Very deep convolutional networks for large-scale image recognition. arXiv preprint (2014). arXiv:1409-1556
16. Szegedy, C., Liu, W., Jia, Y., Sermanet, P., Reed, S., Anguelov, D., Erhan, D., Vanhoucke, V., Rabinovich, A.: Going deeper with convolutions. arXiv preprint (2014). arXiv:1409.4842
17. He, K., Zhang, X., Ren, S., Sun, J.: Delving deep into rectifiers: surpassing human-level performance on imagenet classification. arXiv preprint (2015). arXiv:1502.01852

18. Ioffe, S., Szegedy, C.: Batch normalization: accelerating deep network training by reducing internal covariate shift. arXiv preprint (2015). arXiv:1502.03167
19. Razavian, A.S., Azizpour, H., Sullivan, J., Carlsson, S.: CNN features off-the-shelf: an astounding baseline for recognition. In: IEEE Conference on Computer Vision and Pattern Recognition Workshops (CVPRW), pp. 512–519. IEEE (2014)
20. Zeiler, M.D., Fergus, R.: Visualizing and understanding convolutional networks. In: Fleet, D., Pajdla, T., Schiele, B., Tuytelaars, T. (eds.) ECCV 2014, Part I. LNCS, vol. 8689, pp. 818–833. Springer, Heidelberg (2014)
21. LeCun, Y., Bottou, L., Bengio, Y., Haffner, P.: Gradient-based learning applied to document recognition. Proc. IEEE **86**, 2278–2324 (1998)
22. Bengio, Y.: Learning deep architectures for AI. Found. Trends Mach. Learn. **2**, 1–127 (2009)
23. Bengio, Y., LeCun, Y., et al.: Scaling learning algorithms towards AI. Large-scale kernel machines 34 (2007)
24. Labusch, K., Barth, E., Martinetz, T.: Simple method for high-performance digit recognition based on sparse coding. IEEE Trans. Neural Netw. **19**, 1985–1989 (2008)
25. Vincent, P., Larochelle, H., Bengio, Y., Manzagol, P.A.: Extracting and composing robust features with denoising autoencoders. In: Proceedings of the 25th International Conference on Machine Learning, pp. 1096–1103. ACM (2008)
26. Lee, Y.: Handwritten digit recognition using k nearest-neighbor, radial-basis function, and backpropagation neural networks. Neural Comput. **3**, 440–449 (1991)
27. Ciresan, D., Meier, U., Schmidhuber, J.: Multi-column deep neural networks for image classification. In: IEEE Conference on Computer Vision and Pattern Recognition (CVPR), pp. 3642–3649. IEEE (2012)
28. Krizhevsky, A., Hinton, G.: Learning multiple layers of features from tiny images, vol. 1, p. 7. Technical report, Computer Science Department, University of Toronto (2009)
29. Malinowski, M., Fritz, M.: Learning smooth pooling regions for visual recognition. In: 24th British Machine Vision Conference, pp. 1–11. BMVA Press (2013)
30. Lin, T.H., Kung, H.: Stable and efficient representation learning with nonnegativity constraints. In: Proceedings of the 31st International Conference on Machine Learning (ICML), pp. 1323–1331 (2014)

Beyond HOG: Learning Local Parts for Object Detection

Chenjie Huang[✉], Zheng Qin, Kaiping Xu, Guolong Wang,
and Tao Xu

School of Software, TNList, Tsinghua University, Beijing, China
{hcj13,qingzh,xkp13,wanggl13,xut14}@mails.tsinghua.edu.cn

Abstract. Histogram of Oriented Gradients (HOG) features have laid solid foundation for object detection in recent years for its both accuracy and speed. However, the expressivity of HOG is limited because the simple gradient features may ignore some important local information about objects and HOG is actually data-independent. In this paper, we propose to replace HOG by a parts-based representation, Histogram of Local Parts (HLP), for object detection under sliding window framework. HLP can capture richer and larger local patterns of objects and are more expressive than HOG. Specifically, we adopt Sparse Nonnegative Matrix Factorization to learn an over-complete parts-based dictionary from data. Then we can obtain HLP representation for a local patch by aggregating the Local Parts coefficients of pixels in this patch. Like DPM, we can train a supervised model with HLP given the latent positions of roots and parts of objects. Extensive experiments on INRIA and PASCAL datasets verify the superiority of HLP to state-of-the-art HOG-based methods for object detection, which shows that HLP is more effective than HOG.

Keywords: Object detection · Feature learning

1 Introduction

Utilizing sliding windows and local patches for object detection [23] has been widely developed in computer vision community. Instead of the whole image, such methods mainly focus on the local information in small patches, a.k.a. details, of objects [3,9,16]. Their success is build on a celebrated feature, Histogram of Oriented Gradients (HOG), for its accuracy and speed. And the increasing usage of HOG in other computer vision tasks, such as face recognition [28], scene classification [26] and event detection [19], also demonstrated its dramatic power.

In spite of its accuracy and speed, HOG still suffers from some crucial shortcomings. One of the most important is that HOG may fail to directly discover richer and larger local patterns [18] which limits its expressivity. And the other is that HOG is data-independent such that it can't exploit the task specific information in datasets. Though some efforts have been made to design or learn

© Springer International Publishing Switzerland 2015
Y.-S. Ho et al. (Eds.): PCM 2015, Part I, LNCS 9314, pp. 285–295, 2015.
DOI: 10.1007/978-3-319-24075-6_28

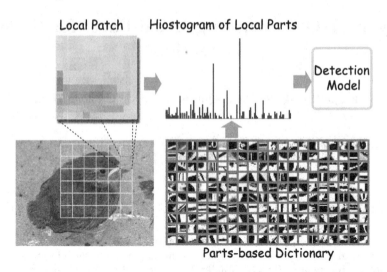

Fig. 1. We focus on learning more expressive features to replace HOG for object detection under sliding window framework. This figure presents the overview of our method.

better local representation to replace HOG [4,13,21], their performance is unsatisfactory and such research topic is still an open and challenging issue.

Recent years have witnessed the rapid development of feature learning [11,14] and its great success in computer vision problems such as recognition [25] and retrieval [5]. So it's straightforward to ask: can we build features for local patches that are more effective by learning from data? Fortunately, the answer is definitely yes [18] and this paper also demonstrates it. In this paper, we focus on learning expressive feature for local patches which can (1) discover richer and larger local patterns of objects, and (2) exploit task specific knowledge in the dataset for certain objects. By fixing such flaws of HOG with two properties, the learned features are expected to be more effective than HOG. To verifies its superiority to HOG, we follow the sliding window framework for object detection.

Specifically, we propose a Histogram of Local Parts (HLP) feature in this paper. At first step, we learn parts-based dictionary from training patches. Research on Nonnegative Matrix Factorization (NMF) [2,14] has suggested that NMF can lead to parts-based representation for images. Motivated by it, we adopt Sparse NMF to learn an over-complete dictionary which can capture the local parts information in patches. Then we can compute the Local Parts coefficients for each pixel, and obtain the HLP representation for a patch by aggregating the coefficients of pixels in it. Next, with the HLP of patches, we can train a supervised model given the latent positions of roots and parts of objects following the Deformable Parts Model (DPM) [9]. In addition, to speed up the detection, we adopt a dimension reduction on the learned model. Different from the data-independent HOG, our parts-based dictionary learned by SNMF from data can capture richer and larger local patterns, such as edges, corners, sharp angles, and etc. And HLP is data-dependent such that it's able to learn some complicated parts for certain objects, which can further promote its expressivity.

We make the following contributions. (1) We propose a parts-based feature, Histogram of Local Parts (HLP) to replace HOG as patch feature for object detection. (2) We adopt SNMF to learn an over-complete dictionary from data. such dictionary can capture richer and larger local patterns of objects and can learn specific parts for certain objects. (3) We adopt dimension reduction for models to guarantee the efficiency for detection. (4) We conducted extensive experiments on INRIA and PASCAL benchmarks and the results demonstrate that our HLP can make large improvements over HOG for object detection.

2 Related Work

Object Detection: The methods developed in recent years for object detection mostly based on the paradigm of classifiers trained on patch features, and the most widely used combination is SVM and HOG [3,9,16] which has achieved promising results on PASCAL benchmark [8]. Upon such paradigm, some efforts have been made to explore other model structures, such as low-dimensional projections [17], compositional grammar structure [10], non-parametric mixtures of exemplars [6], segmentation [22], supervised correspondence [1]. Though different models are adopted, the building block of most of these methods is still HOG feature. In this paper we propose a new feature to replace HOG, hence we can straightforwardly integrate it into any methods above.

Image Features: Extracting effective image features is a long-studied research topic and is still an open issue in computer vision community. It plays a key role for most, if not all computer vision problems, such as detection [23], recognition [25], retrieval [15], and etc. Some hand-designed features are widely used, such as integral channel features [7], local binary patterns [13], RGB covariance features [20], and etc. And combining them by multiple kernels is also adopted in some works [21]. On the other hand, several works focus on learning features from data such as Histogram of Filters [4], mid-level attributes [11], and etc. They demonstrate that the learned features can capture more specific information about data and thus result in better performance. And one of the most closest method to our HLP is the Histogram of Sparse Codes proposed in [18]. However, their sparse coding is much less efficient than our coding method thus their detection is much slower. And their dictionary learning can't discover complicated patterns from large patches. As shown in their experiment, their method performs well when the patch size is 5 to 7, while shows unstable and worse performance when the patch size grows to 9. On the contrary, our dictionary learning method can discover more complicated parts in large patches because we adopt NMF instead of sparse coding [18], and it performs superiorly and stably even with large patches.

Nonnegative Matrix Factorization: NMF [14] is a widely used feature learning method like in image clustering [2]. Unlike sparse coding which mainly focus on edges or corners of objects, NMF can learn parts-based representation because it only allows additive but not subtractive combination of basis.

And the parts-based representation has psychological and physiological interpretation of naturally occurring data [24], especially images. With NMF, we can
discover simple patterns like edges and corners, as well as complicated patterns
such as specific parts of objects. Thus we can utilize NMF to learn more expressive local representation for patches. In addition, the optimization algorithm for
NMF is quite efficient as well, which can guarantee the speed for detection.

3   The Proposed Method

HOG is a gradient-based feature which is long known to be important for object
detection and robust to illumination and appearance changes. However, images
are obviously more than just gradients. In this section, we will introduce our
Histogram of Local Parts (HLP) feature in detail. Our HLP can capture richer
and larger local patterns of objects and exploit some specific information about
certain parts of objects. Here we need to highlight again that the focus of this
paper is on proposing a new feature to replace HOG for object detection, thus
it's very straightforward to integrate it to existing HOG-based frameworks.

An illustration is shown in Fig. 1. At first, we will learn a parts-based dictionary. This is achieved in an unsupervised way by performing Sparse Nonnegative
Matrix Factorization (SNMF) on training patches. With the dictionary, we can
obtain the Local Parts coefficients for each pixel and then obtain the HLP of a
patch by aggregating the coefficients of pixels in it. Now we can replace HOG
with HLP to train a DPM for object detection. Finally, we adopt dimension
reduction on learned models to reduce the complexity for detection efficiency.

Learning Parts-Based Dictionary. First we need to learn an expressive dictionary. We wish that the dictionary can capture complicated local patterns.
Given a set of training patches $\mathbf{X} = [\mathbf{x}_1, ..., \mathbf{x}_n] \in \mathbb{R}^{d \times n}$, where $\mathbf{x}_i$ is the grey-
level representation of patch i, n is the number of patches and d is the size of a
patch. For example, we have $d = 81$ for 9×9 patch. We want to find a nonnegative
over-complete dictionary $\mathbf{D} = [\mathbf{d}_1, ..., \mathbf{d}_m] \in \mathbb{R}^{d \times m}$, where m is the size of dictionary, and corresponding nonnegative coefficients $\mathbf{V} = [\mathbf{v}_1, ..., \mathbf{v}_n] \in \mathbb{R}^{m \times n}$. Since
the dictionary is over-complete, we add sparse regularization to the coefficients.
Now, our dictionary learning is formulated as SNMF [12],

$$\min_{\mathbf{D}, \mathbf{V}} \mathcal{O} = \frac{1}{2}\|\mathbf{X} - \mathbf{D}\mathbf{V}\|_F^2 + \lambda \sum_{i,j} |v_{ij}|, \text{ s.t. } \mathbf{D}, \mathbf{V} \geq 0 \qquad (1)$$

where $\|\cdot\|_F$ is the Frobenius norm of matrix and λ is the regularization parame-
ter. Because of the nonnegative constraints, only additive combination of basis
is allowed, which can lead to parts-based dictionary [14]. As shown in Fig. 1, the
learned parts-based dictionary can indeed capture complicated local patterns
such as parts in large patches (9×9) besides simple patterns such as lines. Furthermore, the learned dictionary are made up of local parts of objects and our
histogram feature is build on it, thus we call it Histogram of *Local Parts*.

Here we'd like to compare to some previous works. In HOG, only gradient information is considered thus they fail to capture rich local patterns. In [4], auxiliary information is required. Sparse coding is adopted in [18] who achieved better performance than HOG. But they can discover simple patterns while shows unstable performance with large patch size, e.g., 9×9. Our dictionary learning algorithm can capture complicated local patterns with large patch size and is unsupervised, hence it is more practical for real-world applications.

The optimization algorithm for Eq. (1) is quite efficient. Actually, we can adopt an iterative strategy to update one matrix while fixing the other. Following [14], the multiplicative updating rules for $\mathbf{D}$ and $\mathbf{V}$ are respectively given as

$$d_{ij} = \frac{(\mathbf{XV'})_{ij}}{(\mathbf{DVV'})_{ij}}, \ v_{ij} = \frac{(\mathbf{D'X})_{ij}}{(\mathbf{D'DV})_{ij} + \lambda} \tag{2}$$

The objective function in Eq. (1) is guaranteed to convergency, whose proof can be found in [14]. The convergency is achieved in hundreds of iterations.

Histogram of Local Parts. With the learned dictionary $\mathbf{D}$, we can compute the Local Parts coefficients $\mathbf{v}$ for each pixel. Firstly we can construct a small patch around this pixel (e.g., 9×9) and represent it by a grey-level vector $\mathbf{x}$. Then we can randomly initialize $\mathbf{v}$ and utilize Eq. (2) iteratively to find the optimal $\mathbf{v}$. Since $\mathbf{D'D}$ can be computed in advance, the overall time complexity for computing $\mathbf{v}$ is $\mathcal{O}(t(m^2 + m) + dm)$ where t is the number of iterations. Generally, 5 iterations are enough for satisfactory result. And because m is small (in our experiment, we found out that a dictionary with hundreds of basis always performs well), computing the Local Parts coefficients can be extremely fast.

Following the sliding window framework, we divide the image into regular cells (e.g., 8×8) and a feature vector (HLP) is computed for each cell. Then the feature vector can be utilized in convolution-based window scanning. Specifically, suppose $\mathbf{v} \in \mathbb{R}^m$ is the Local Parts coefficients computed at a pixel, we can assign its (nonnegative) elements v_i ($i = 1, ..., m$) to the four spatially surrounding cells by bilinear interpolation. Then we can obtain a m-dimension histogram feature H for each cell who averages the coefficients in 16×16 neighborhood. Because each dimension corresponds to a learned local part in the dictionary, we call H the Histogram of Local Parts (HLP). Next we can normalize H to unit Euclidean length. Finally we perform power transform on elements of H,

$$H'_i \leftarrow H^p_i, \ i = 1, ..., m \tag{3}$$

where p is to control the power transform. With proper transform, the distribution of H's elements becomes more uniform and H is more discriminative.

Model Training and Dimension Reduction. Our HLP is for representing local patches in images like HOG, so it's straightforward to apply it to existing HOG-based models like DPM [9]. Denote $p_i = (x_i, y_i, s_i)$ as the location and scale of part i, and $\theta = \{p_0, ..., p_t\}$ as the hypothesis of an object where p_0 represents the root of object. The detection score of θ with a DPM detector is

$$S(\theta) = \sum_{i=0}^{t} w_i' \phi_f(p_i, I) + \sum_{i=1}^{t} c_i' \phi_d(p_i, p_0) + b \qquad (4)$$

where I is the image window, $\phi_f(p_i, I)$ are the local features, $\phi_d(p_i, p_0)$ is deformation cost for part locations, and b is the model bias. When training the detector, we need to learn $\{w_i\}$, $\{c_i\}$ and b. It can be trained in a latent SVM approach. However, because the training needs too much resource which is always unknown, we need to perform model training and latent variables assigning iteratively for final result. Hence it's slow to train such a detector [9]. In addition, we can notice that the complexity to maximize Eq. (4) is linear to the dimension of feature ϕ_f. Thus it could be much slower compared to HOG to train detectors with HLP which is more expressive, redundant and *higher dimensional*.

To speed up training, we adopt injecting supervision from external source like [28] and [1]. Instead of assigning latent variables iteratively, we take advantage of pre-trained detection systems like [9] and treat their outputs as the *pseudo* ground truth. Now with the latent variables fixed, learning the model parameters of the detector can be formulated as the following problem,

$$\min_{\alpha, \xi_i} \frac{1}{2} \|\alpha\|^2 + C \sum_i \xi_i \quad s.t. \quad y_i \alpha' \Phi(\theta) \geq 1 - \xi_i \qquad (5)$$

where ξ_i is a slack variable, α is the model parameter, $\Phi(\theta)$ is the feature for hypothesis θ, and y_i is the label that $y_i = 1$ for positive samples and $y_i = -1$ otherwise. Equation (5) can be efficiently solved by dual-coordinate solver [27]. With such strategy, the learning can be done without any iteration.

To guarantee detection efficiency, we can perform dimension reduction on the HLP features. A common strategy is to utilize PCA. But we found out in our experiment that the performance may degrade severely after PCA. Instead, we adopt supervised dimension reduction. Specifically, we can rewrite w_i as a matrix $\mathbf{W}_i \in \mathbb{R}^{N_i \times m}$ where N_i is the number of spatial cells in the filter. Then by concatenating the matrices of all filters of all classes, we can obtain a large parameter matrix $\mathbf{W} \in \mathbb{R}^{N \times m}$. Then we factorize W to low-rank representation

$$\min_{\mathbf{C}, \mathbf{P}} \|\mathbf{W} - \mathbf{CP}\|_F^2 \quad s.t. \quad \mathbf{PP}' = \mathbf{I}_r \qquad (6)$$

where $\mathbf{P} \in \mathbb{R}^{r \times m}$ is the projection matrix who can project m-dimensional HLP feature H to r-dimensional ($r \ll m$) feature $\hat{H}$ by $\hat{H} \leftarrow \mathbf{P}H$, $\mathbf{C}$ is the filter parameters in reduced space, and r is the target dimensionality. Denote $\mathbf{S} = \mathbf{W}'\mathbf{W}$, Eq. (6) can be optimized by solving a eigenvalue decomposition problem

$$\lambda \psi = \mathbf{S} \psi \qquad (7)$$

where λ is the eigenvalue and ψ is the corresponding eigenvector. By selecting eigenvectors corresponding to the largest r eigenvalues, we can obtain the projection matrix $\mathbf{P}$ which captures the important information in HLP features and results in less redundant features. By incorporating $\mathbf{P}$ in to the feature computation such that it's transparent to the rest of system, we can markedly promote the training and detection efficiency while maintaining satisfactory performance.

Table 1. Average precision (root-only) on INRIA.

HOG	HSC	DPM	HLP$_7$	HLP$_9$	HLP$_{15}$
0.806	0.832	0.849	0.841	0.857	0.860

4 Experiment

4.1 Datasets and Details

We carried out experiments on two benchmark datasets: INRIA person Dataset [3] and PASCAL2007 dataset [8]. The INRIA dataset contains 1208 positive images of standing people cropped and normalized to 64×128, and 1218 negative images for training, and 741 images for testing. This dataset is what HOG was designed and optimized for. PASCAL dataset has 20 different objects and corresponding 9963 images with large variance within and across classes. It's a widely used benchmark for object detection. We utilize the trainval positive and negative images as the training set, and evaluate on the testing images.

When comparing HLP to HOG, we use Eq. (5) to learn detector parameters with respective features for fair comparison. Actually, the difference between detectors can only be the usage of different local features. In addition, we also compare out method to the state-of-the-art DPM [9] who is based on HOG, and HSC [18] which is one of the most related methods to ours. We use the source codes provided by their authors. Average Precision [9,18] is adopted as measure.

We use the following setting for our HLP. The size of local patch is 9×9. The dictionary size is 300 for INRIA and 400 for PASCAL. We set p for power transform in Eq. (3) to 0.4 for INRIA and 0.3 for PASCAL. With parts-based model, the target dimensionality of supervised dimension reduction (SDR) is 100. And we set the parameter $\lambda = 0.1$ in Eq. (1). Other settings follow DPM.

4.2 Result and Discussion

Detection Results. For INRIA, the detection results are summarized in Table 1. Here HLP$_9$ (HLP$_7$, HLP$_{15}$) refers to 9×9 (7×7, 15×15) patch size and 300 (200, 400) basis in dictionary. We can observe that HLP can significantly outperform HOG which demonstrates the superiority of HLP and validates that HLP is indeed expressive than HOG. Our HLP is slightly better DPM. And compared to HSC, our HLP also shows large improvement, especially with larger patch size, since HLP can capture more complicated patterns than HSC.

The quantitative comparison between root-only models on PASCAL is presented in Table 2. We can observe that HLP markedly outperforms HOG, improving mAP from 0.215 to 0.274, and over 0.05 for 12 classes. In Fig. 2, we detailedly exhibit the detection results of 12 examples from 4 classes. This result qualitatively verifies the effectiveness of HLP. We can see HLP do help detect objects in complicated scenes, such as noisy background and overlapping objects.

Table 2. Average precision (root-only) on PASCAL

	aero	bike	bird	boat	btl	bus	car	cat	chair	cow	avg
HOG	.207	.467	.099	.112	.180	.364	.403	.042	.117	.238	.215
HSC	.241	.487	.056	.132	.217	.409	.448	.095	.165	.261	.247
DPM	.251	**.514**	.070	.130	.242	.378	.459	.108	.154	.243	.251
HLP	**.253**	.499	**.112**	**.145**	**.260**	**.419**	**.513**	**.113**	**.187**	**.264**	**.274**

Table 3. Average precision (parts-based) on PASCAL

	table	dog	hors	mbik	prsn	plnt	shep	sofa	train	tv	avg
HOG	.204	.122	.542	.478	.411	.117	.182	.291	.422	.384	.305
HSC	.301	.141	.588	.492	.382	.105	.216	.319	.471	.462	.331
DPM	.249	.121	.597	.494	**.429**	.126	.189	.343	**.492**	.437	.328
HLP	**.307**	**.207**	**.637**	**.577**	.414	**.140**	**.229**	**.358**	.454	**.473**	**.356**

HLP is also much better than HOG-based DPM and HSC. Actually, HLP can outperform the other models in 17 out of 20 classes.

When training with parts by Eq. (5), we use the parts detected by parts-based DPM as the pseudo ground truth. And we reduce the dimension of HLP to 100 for detection efficiency. The results of parts-based models are shown in Table 3. Similar to the results in root-only case, our HLP can significantly outperform HOG and promote mAP by 0.051, and HLP consistently outperforms HOG for all 20 classes. In addition, our HLP also presents observable improvement over DPM and HSC and is better than them in 16 out of 20 classes.

Effect of Patch Size Dictionary Size. The patch size and dictionary size are two important parameters for HLP. In Fig. 3(a) and (d), we plot the effect of them on HLP for root-only detection in both datasets. We have two observations from the figures. First, properly increasing dictionary size results in better performance. Because our SNMF can discover complicated patterns like specific local parts of objects, larger dictionary can contain more patterns such that the feature is more expressive. But the performance turns stable when the dictionary is large enough since it's too redundant with too many basis. Second, HLP performs well with large patch size. This is reasonable because large patch contains more complicated patterns such that our parts-based dictionary may include more parts with larger size. Comparing to HSC whose performance goes down with 9×9 patch, HLP works well even with 15×15 patch. This result validates the scalability of HLP and it can discover richer and larger local patterns.

Effect of Power Transform. We apply power transform to the normalized HLP features in our experiment. We investigate its effect and the results are presented in Fig. 3(b) and (e). By comparing against the case $p = 1$ where no transform is performed, we can see the power transform does make a crucial difference. Generally, selecting $p \in [0.25, 0.5]$ can lead to satisfactory performance.

Fig. 2. Detection (root-only) results of HLP and HOG, showing top three candidates.

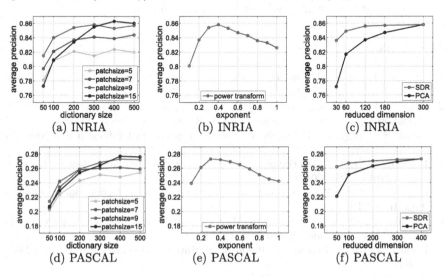

Fig. 3. Effect of patch size, dictionary size, power transform, and dimension reduction.

Effect of Supervised Dimension Reduction. Instead of performing unsupervised dimension reduction like PCA, we learn a projection matrix from the learned model for supervised dimension reduction (SDR) for HLP. To demonstrate the rationality, we compare the effect of SRD and PCA on root-only detection in Fig. 3(c) and (f). Obviously, SDR can outperform PCA and show state-of-the-art performance even with 50 dimensions. However, the effectiveness of HLP degrades significantly after PCA because it may break important discriminative information in features since it's totally unsupervised.

Efficiency. At last, we'd like to investigate the efficiency of extracting HLP. The efficiency is compared on a computer with Intel Xeon E5520 2.27 GHz CPU, 16 GB RAM. The average time for extracting HLP (HOG, HSC) on PASCAL is 0.92 (0.64, 5.87) seconds. HLP is comparable to HOG and faster than HSC.

5   Conclusion

In this paper we focus on learning more expressive features to replace HOG for object detection under sliding window framework. Specifically, we propose a Histogram of Local Parts feature. An over-complete parts-based dictionary is learned from training patches which can capture larger and richer local patterns by SNMF. The Local Parts coefficients at each pixel can be efficiently computed. By aggregating the coefficients of its pixels, we can obtain the HLP representation for a patch. We apply power transform to HLP for more discriminability. Then we can train a supervised model with HLP. For detection efficiency, a supervised dimension reduction is adopted. We carried extensive experiments on INRIA and PASCAL2007 benchmark datasets. The experimental results validate the superiority of HLP to HOG, and demonstrate that our method can outperform state-of-the-art methods for object detection.

References

1. Bourdev, L., Malik, J.: Poselets: body part detectors trained using 3d human pose annotations. In: ICCV (2009)
2. Cai, D., He, X., Han, J., Huang, T.S.: Graph regularized nonnegative matrix factorization for data representation. TPAMI **33**(8), 1548–1560 (2011)
3. Dalal, N., Triggs, B.: Histograms of oriented gradients for human detection. In: CVPR (2005)
4. Dikmen, M., Hoiem, D., Huang, T.S.: A data-driven method for feature transformation. In: CVPR (2012)
5. Ding, G., Guo, Y., Zhou, J.: Collective matrix factorization hashing for multimodal data. In: CVPR (2014)
6. Divvala, S., Efros, A., Hebert, M.: How important are deformable parts in the deformable parts model? In: ECCV (2012)
7. Dollar, P., Tu, Z., Perona, P., Belongie, S.: Integral channel features. In: BMVC (2009)
8. Everingham, M., Gool, L.V., Williams, C., Winn, J., Zisserman, A.: The pascal visual object classes (voc) challenge. IJCV **88**(2), 303–338 (2010)
9. Felzenszwalb, P.F., Girshick, R.B., McAllester, D.A., Ramanan, D.: Object detection with discriminatively trained part-based models. TPAMI **32**(9), 1627–1645 (2010)
10. Girshick, R., Felzenszwalb, P., McAllester, D.: Object detection with grammar models. In: NIPS (2011)
11. Guo, Y., Ding, G., Jin, X., Wang, J.: Learning predictable and discriminative attributes for visual recognition. In: AAAI (2015)
12. Hoyer, P.O.: Non-negative matrix factorization with sparseness constraints. JMLR **5**, 1457–1469 (2004)
13. Hussain, S., Kuntzmann, L., Triggs, B.: Feature sets and dimensionality reduction for visual object detection. In: BMVC (2010)
14. Lee, D.D., Seung, H.S.: Learning the parts of objects by nonnegative matrix factorization. Nature **401**(6755), 788–791 (1999)
15. Lowe, D.G.: Object recognition from local scale-invariant features. In: ICCV (1999)
16. Malisiewicz, T., Gupta, A., Efros, A.: Ensemble of exemplar-svms for object detection and beyond. In: ICCV (2011)
17. Pirsiavash, H., Ramanan, D., Fowlkes, C.: Bilinear classifiers for visual recognition. In: NIPS (2009)
18. Ren, X., Ramanan, D.: Histograms of sparse codes for object detection. In: CVPR (2013)

19. Roshtkhari, M.J., Levine, M.D.: Online dominant and anomalous behavior detection in videos. In: CVPR (2013)
20. Schwartz, W., Kembhavi, A., Harwood, D., Davis, L.: Human detection using partial least squares analysis. In: ICCV (2009)
21. Vedaldi, A., Gulshan, V., Varma, M., Zisserman, A.: Multiple kernels for object detection. In: ICCV (2009)
22. Vijayanarasimhan, S., Grauman, K.: Efficient region search for object detection. In: CVPR (2011)
23. Viola, P.A., Jones, M.J.: Rapid object detection using a boosted cascade of simple features. In: CVPR (2001)
24. Wachsmuth, M.W.O.E., Perrett, D.I.: Recognition of objects and their component parts: responses of single units in the temporal cortex of the macaque. Cereb. Cortex 4(5), 509–522 (1994)
25. Wang, J., Yang, J., Yu, K., Lv, F., Huang, T.S., Gong, Y.: Locality-constrained linear coding for image classification. In: CVPR 2010 (2010)
26. Xiao, J., Hays, J., Ehinger, K., Oliva, A., Torralba, A.: Sun database: large-scale scene recognition from abbey to zoo. In: CVPR (2010)
27. Yang, Y., Ramanan, D.: Articulated pose estimation with flexible mixtures-of-parts. In: CVPR (2011)
28. Zhu, X., Ramanan, D.: Face detection, pose estimation, and landmark localization in the wild. In: CVPR (2012)

Regular Poster Session

Tuning Sparsity for Face Hallucination Representation

Zhongyuan Wang[1]([⊠]), Jing Xiao[1], Tao Lu[2], Zhenfeng Shao[3],
and Ruimin Hu[1]

[1] NERCMS, School of Computer, Wuhan University, Wuhan, China
{wzy_hope,xiaojing,hrml964}@163.com
[2] School of Computer, Wuhan Institute of Technology, Wuhan, China
lutxyl@163.com
[3] LIESMARS, Wuhan University, Wuhan, China
shaozhenfeng@163.com

Abstract. Due to the under-sparsity or over-sparsity, the widely used regularization methods, such as ridge regression and sparse representation, lead to poor hallucination performance in the presence of noise. In addition, the regularized penalty function fails to consider the locality constraint within the observed image and training images, thus reducing the accuracy and stability of optimal solution. This paper proposes a locally weighted sparse regularization method by incorporating distance-inducing weights into the penalty function. This method accounts for heteroskedasticity of representation coefficients and can be theoretically justified from Bayesian inference perspective. Further, in terms of the reduced sparseness of noisy images, a moderately sparse regularization method with a mixture of ℓ_1 and ℓ_2 norms is introduced to deal with noise robust face hallucination. Various experimental results on public face database validate the effectiveness of proposed method.

Keywords: Face hallucination · Sparse regularization · Locally weighted penalty · Mixed norms

1 Introduction

Face super-resolution, or face hallucination, refers to the technique of estimating a high-resolution (HR) face image from low-resolution (LR) face image sequences or a single LR one. A large number of theoretical and applicable works on face hallucination have been carried out. Among them, learning based methods have aroused great concerns provided that they can provide high magnification factors.

Learning based methods can date back to the early work proposed by Freeman et al. [1], who employed a patch-wise Markov network to model the relationship between LR images and the HR counterparts. Afterwards, Baker and Kanade [2] developed a Bayesian approach to infer the missing high-frequency components from a parent structure training samples, and first coined the term "face hallucination". Motivated by their pioneering work, a class of dictionary learning methods has gained popularization recently, in which image patches can be well-approximated as a linear combination of elements from an appropriately designed over-complete dictionary.

© Springer International Publishing Switzerland 2015
Y.-S. Ho et al. (Eds.): PCM 2015, Part I, LNCS 9314, pp. 299–309, 2015.
DOI: 10.1007/978-3-319-24075-6_29

All dictionary learning based methods are equipped with representation methods to resolve optimal coefficients in terms of accuracy, stability and robustness, for example, least squares (LS), ridge regression (RR) and sparse representation (SR). Chang et al. [3] proposed a neighbor embedding (NE) based face hallucination method using LS. It uses a fixed number of neighbors for reconstruction, thus usually resulting in blurring due to under- or over-fitting. Lately, a technical modification of NE [4] is proposed to enhance visual results and reduce complexity, with a nonnegative constraint imposed on solution. By incorporating the position priors of image patch, Ma et al. [5] introduced a position-patch based method to estimate a HR image patch using the same position patches of all training face images. To reduce residuals, Park et al. [6] proposed an example-based method to recursively update a reconstructed HR image by compensating for HR errors. In [7], the Huber norm was used to replace ordinary squared error to improve robustness to outliers. Since face hallucination is an underdetermined problem, LS hardly provides a stable and unique solution.

In contrast, regularization methods enable a global unique and stable solution by improving the conditioning of the ill-posed problem with the help of prior knowledge. One of the most popular regularization methods is ridge regression. It can produce the solution of minimum energy with the help of a squared $\ell 2$ norm penalty. Liu et al. [8] proposed to integrate a global parametric principal component analysis model along with a local nonparametric Markov random field model for face hallucination, where RR is used to represent global face images. Motivated by [8], Li and Lin [9] also presented a two-step approach for hallucinating faces with global images reconstructed with a maximum a posteriori (MAP) criterion and residual images re-estimated with the MAP criterion as well. To enhance the subset selection functionality, Tang et al. [10] proposed to learn a local regression function over the local training set. The local training set is specified in terms of the distances between training samples and a given test sample. Reference [11] instead used a locality-constrained RR model to formulate face hallucination problem and gained impressive results. As pointed out by Fan and Li [12], penalty functions have to be singular at origin to produce sparse solutions. However, the squared $\ell 2$ norm penalty of RR is differentiable at origin, thus leading to over-smooth and under-sharp edge images.

Owing to the well-established theory foundation, sparse representation becomes more appealing than RR for face hallucination. Yang et al. [13] are the first to introduce $\ell 1$ norm SR to face hallucination. Assuming that natural images can be sparsely decomposed, they proposed a local patch method over coupled over-complete dictionaries to enhance the detailed facial information. In [14], a modified version of Yang's algorithm was shown to be more efficient and much faster. To address the biased estimate resulted from LS [5], Jung et al. [15] proposed to hallucinate face images using convex optimization. In [16], the idea of two-step face hallucination [8] was extended to robust face hallucination for video surveillance, where SR was used to synthesize eigenfaces. Zhang et al. [17] presented a dual-dictionary learning method to recover more image details, in which not only main dictionary but also residual dictionary are learned by sparse representation.

The above SR based methods can capture salient properties of natural images, whereas $\ell 1$ norm SR turns out over-sparse for face hallucination. This is primarily due to the fact that face hallucination is a regression problem (pursuing prediction accuracy) rather than a classification problem (seeking discriminability in sparse features). Fan et al. [12] studied a class of regularization methods and showed that $\ell 1$ norm shrinkage produces biased estimates for the large coefficients and could be suboptimal in terms of estimation risk. Meinshausen and Bühlmann [18] also showed the conflict of optimal prediction and consistent variable selection in $\ell 1$ norm. After comparing $\ell 1$ norm with elastic net (EN) proposed in [20], Li and Lin [19] expressly pointed out that the former is much more aggressive in terms of prediction exclusion. Especially, we observe that $\ell 1$ norm results in considerable degradation of hallucination performance in the presence of noise. In the light of the above discussions, we argue that the underlying representation model in face hallucination should maintain a reasonable balance between subset selection and regression estimation. More specifically, in order to better exploit the salient facial features, it is sufficient and necessary to impose moderate sparse constraints, but sparsity should not be overemphasized.

Besides, with the popularization of sparse representation, its weighted variants are also successively developed, such as iterative reweighted $\ell 1$ minimization [21] and weighted $\ell 1$ minimization with prior support [22]. The main idea is to exploit a priori knowledge about the support of coding coefficients so as to favor desirable properties of solution. In particular, Friedlander et al. [22] theoretically proved that if the partial support estimate is at least 50 % accurate, then weighted $\ell 1$ minimization outperforms the standard one in terms of accuracy, stability, and robustness. Intuitively, the geometric locality in terms of Euclidean distance can be treated as intrinsic prior support regarding the observed LR image and training images. If we enforce such a geometric locality constraint on the coding coefficients to induce a weighted sparse representation, the improved hallucinated results can be expected.

In this paper, we propose a locally weighted sparse regularization (LWSR) to boost face hallucination performance. It incorporates distance-inducing weights into regularization penalty function to favor the involvement of near training bases in face hallucination recovery. Particularly, we suggest a mixture of $\ell 1$ and $\ell 2$ norms to deal with noise robust face hallucination in the presence of noise.

2 Locally Weighted ℓ_1 Regularization

From Bayesian perspective, standard ℓ_1 norm regularization corresponds to regularized LS with Laplace prior imposed on coefficients. All entries in the coefficient vector are assumed to share quite identical variance, so called homoscedasticity in statistics. However, Heteroscedasticity is a major concern of regression related problems. A collection of random variables is heteroskedastic when there are sub-populations that have different variances from others. For example, random variables of larger values often have errors of higher variances, leading to weighted least-square method accounting for the presence of heteroscedasticity on error statistics.

To examine whether heteroscedasticity exists in sparse representation, we order its coefficients with respect to Euclidean distances and then calculate their standard deviations at different distances. As shown in Fig. 1, the coefficients in near bases have larger standard deviations than others with the standard variances monotonically decreasing as growing Euclidean distances. According to experimental results in Fig. 1,

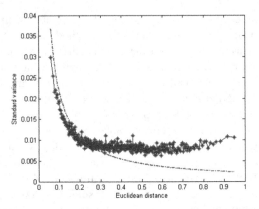

Fig. 1. Standard deviation of sparse representation coefficients declines with increasing Euclidean distance, which can be approximated with the inverse function of the distance as illustrated in dashed red line.

this trend can even be approximated by $f(d) = \frac{0.0022}{d}$, where d denotes distance. Evidently, the solution of linear model in face hallucination problem obeys heteroskedasticity rather than homoskedasticity.

Given the observation vector X, MAP technique is often used to estimate the coefficient vector W:

$$W^* = \arg\max_{W}\{\log P(W|X)\}, \tag{1}$$

where $\log P(W|X)$ is the log-likelihood function. It follows from Bayesian theorem that the following holds:

$$P(W|X) = \frac{P(WX)}{P(X)} = \frac{P(W)P(X|W)}{P(X)}, \tag{2}$$

or,

$$\log P(W|X) = \log P(W) + \log P(X|W) - \log P(X). \tag{3}$$

Since the third term of the log-likelihood function is constant, it can be eliminated from the optimization. Thus, MAP estimator is alternatively given by

$$W^* = \arg\max_{W}\{\log P(W) + \log P(X|W)\}. \tag{4}$$

To solve (4), the conditional probability $P(X|W)$ and the prior probability $P(W)$ must be specified in advance. Because the observation vector X is corrupted by zero-mean i.i.d. Gaussian noise, we have

$$P(X|W) = \frac{1}{(2\pi\sigma^2)^{N/2}} \exp(-\frac{1}{2\sigma^2}\|X - YW\|_2^2), \tag{5}$$

where σ^2 describes noise level. The coefficient vector $W = [w_1, w_2, \ldots, w_M]^T$ is assumed to obey independent zero-mean multivariate Laplace distribution:

$$P(W) = \prod_{i=1}^{M} \{\frac{1}{2\mu_i} \exp(-\frac{|w_i|}{\mu_i})\}, \tag{6}$$

where $\mu_i = \frac{\sigma_i}{\sqrt{2}}$ is a scale parameter indicating the diversity and σ_i describes the standard variance of individual entry. As discussed before, coefficients are highly related to distances and near bases take large magnitudes. Because w_i is viewed as a zero-mean random variable, σ_i or μ_i actually describes the coefficient energy. Therefore, we assume that scale parameter μ_i is inversely proportional to distance. For simplification, let $\mu_i = \frac{\mu}{d_i}$, where d_i denotes Euclidean distance and μ is a common scale parameter, we can rewrite (6) as

$$P(W) = \frac{\prod\limits_{i=1}^{M} d_i}{(2\mu)^M} \exp\{-\frac{\sum\limits_{i=1}^{M} |d_i w_i|}{\mu}\}. \tag{7}$$

According to (5) and (7), we have

$$\log P(W) + \log P(X|W) = N \log \frac{1}{\sqrt{2\pi}\sigma} + M \log \frac{1}{2\mu} + \sum_{i=1}^{M} \log(d_i)$$
$$- \frac{1}{2\sigma^2} \{\|X - YW\|_2^2 + \lambda \sum_{i=1}^{M} |d_i w_i|\}, \tag{8}$$

where $\lambda = \frac{2\sigma^2}{\mu}$. The objective in (4) is thus equivalent to minimizing the cost function $\{\|X - YW\|_2^2 + \lambda \sum\limits_{i=1}^{M} |d_i w_i|\}$. Therefore, the ultimate optimization objective becomes

$$W^* = \arg \min_{W} \left\{ \|X - YW\|_2^2 + \lambda \sum_{i=1}^{M} |d_i w_i| \right\}. \tag{9}$$

If we use D to denote the diagonal matrix with diagonal elements given by $D_{ii} = d_i$, (9) can be rewritten as

$$W^* = \arg \min_{W} \left\{ \|X - YW\|_2^2 + \lambda \|DW\|_1 \right\}, \tag{10}$$

where $\lambda \geq 0$ is an appropriately chosen regularization parameter, controlling the tradeoff between the reconstruction error and the regularization penalty. D is a diagonal weighting matrix with diagonal elements being Euclidean distances (or derivatives from distances). Euclidean distances are measured between the input patch and basis patches in training set.

3   Extension to $\ell_{1,2}$ for Regularizing Noisy Images

Noisy images often contain richer frequency spectrum so that they are difficult to describe with a few representative sparse components. Noisy images behave less sparsely than noiseless ones. Thus, Laplace distribution may insufficiently describe coefficient prior in both noise free and noisy scenarios. To exploit the most approximate prior, we suggest a new prior function:

$$P(W) = \frac{1}{(2\mu)^M} \exp(-\frac{(1-\alpha)\|W\|_1 + \alpha\|W\|_2}{\mu}), \tag{11}$$

called $\ell_{1,2}$ distribution. Parameter α controls the fraction between Laplace and ℓ_2 distributions. The maximum likelihood estimate of scale parameter μ is given by

$$\mu = \frac{(1-\alpha)\sum_{i=1}^{L}\|W_i\|_1 + \alpha\sum_{i=1}^{L}\|W_i\|_2}{LM}. \tag{12}$$

Where $W_i, i = 1, 2, \ldots, L$ be a set of coefficient vectors, with the number of L.

Under MAP framework, the resulting $\ell_{1,2}$ regularization associated with (11) reads

$$W^* = \arg\min_{W}\left\{\|X - YW\|_2^2 + \lambda_1\|W\|_1 + \lambda_2\|W\|_2\right\}, \tag{13}$$

where regularization parameters are given by

$$\lambda_1 = \frac{2(1-\alpha)\sigma^2}{\mu}, \lambda_2 = \frac{2\alpha\sigma^2}{\mu}. \tag{14}$$

Accordingly, the weighted variant reads

$$W^* = \arg\min_{W}\left\{\|X - YW\|_2^2 + \lambda_1\|DW\|_1 + \lambda_2\|DW\|_2\right\}. \tag{15}$$

The above mixture structure enables us to adaptively select ℓ_1 or ℓ_2 penalty and thus overcomes disadvantages using either of them.

4   Experimental Results

This section conducts face hallucination experiments on FEI face database [23] to verify proposed methods. The objective metrics, i.e., PSNR and SSIM index, will be reported along with subjective performance compared with other state-of-the-art methods, like Chang's neighbor embedding (NE) [3] and Yang's sparse representation (SR) [13].

The public FEI face database contains 400 images. Among them, 360 images are randomly chosen as the training set, leaving the remaining 40 images for testing. All the test images are absent completely in the training set. All the HR images are cropped to 120×100 pixels, and then the LR images with 30×25 pixels are generated by smoothing and down-sampling by a factor of 4. Empirically, we set the HR patch size as 12×12 pixels, with the corresponding LR patch size being 3×3 pixels with an overlap of 1 pixel. For the sake of fair competition, we tune the control parameters for all comparative methods to their best results, including the nearest neighborhood number K in NE and regularization parameter λ in all regularized methods.

4.1 Quantitative Evaluation

In Sect. 3, ℓ_2 or $\ell_{1,2}$ is justified to be more favorable than ℓ_1 for face hallucination in the presence of noise. First of all, we conduct an experiment to quantitatively verify this issue. The experiments used both noise free and noisy images, where zero-mean Gaussian noises with a standard variances $\sigma = 5$ and $\sigma = 10$ are added to original noise free images for simulating noisy images.

As results tabulated in Table 1, ℓ_1 gives better results than ℓ_2 under the conditions of noise free, but the outcome is just the opposite in the presence of noise, which exactly coincide with their respective prior approximations to actual distributions. Among all penalty functions, squared ℓ_2 offers the worst results for noise free case. This is partially due to the fact that squared ℓ_2 is a non-sparse model as pointed out by Fan and Li [12]. On the other hand, as the regression coefficients of noisy images contain less sparseness, squared ℓ_2 is slightly better than aggressively sparse ℓ_1, but is still inferior to ℓ_2. Regardless of noisy or noiseless images, $\ell_{1,2}$ always achieves the best results. Especially, it substantially outperforms any counterparts in noisy scenarios. Moreover, different from sole ℓ_2, $\ell_{1,2}$ does not lead to degradation relative to ℓ_1 for noise free case, owning to a compromise of ℓ_1 and ℓ_2.

Table 1. Results by different regularization methods

Images	Metrics	ℓ_1	Squared ℓ_2	ℓ_2	$\ell_{1,2}$
Noise free	PSNR	**32.93**	32.76	32.80	**32.93**
	SSIM	**0.9172**	0.9148	0.9155	**0.9172**
$\sigma = 5$ Gaussian	PSNR	30.15	30.29	30.45	**30.61**
	SSIM	0.8536	0.8601	0.8684	**0.8714**
$\sigma = 10$ Gaussian	PSNR	27.98	28.33	28.56	**28.67**
	SSIM	0.7911	0.8134	0.8224	**0.8253**

4.2 Comparisons of Subjective Results

Our two locally weighted sparse regularization variants such as ℓ_1 norm (WSR hereafter for short) and $\ell_{1,2}$ norm (WL1,2 hereafter for short) get involved in this subjective comparison. Some of randomly selected results are shown in Fig. 2. For the noise free case in Fig. 2, we can discern only slight visual differences among the results of NE, SR, WSR and WL1,2 except that bicubic generates blurring effects. For the cases of

(a)   (b)   (c)   (d)   (e)   (f)

Fig. 2. Comparison of the results of different methods on FEI face database: (a) Bicubic interpolation; (b) Chang's NE; (c) Yang's SR; (d) Proposed WSR; (e) Proposed WL1,2; (f) Original HR faces (ground truth). Top 3 rows for noise free images; middle 3 rows for noisy images with σ = 5; bottom 3 rows for noisy images with σ = 10.

noise, SR based methods (SR, WSR and WL1,2) can remove noise more thoroughly than NE because NE relies on ordinary least squares. Moreover, the hallucinated images by WL1,2 look smooth and clean while SR or WSR still exposes some un-smoothed noisy artifacts, especially on cheeks and noses. Admittedly, WSR does not demonstrate visible improvement on original SR, which again confirms that relatively conservative sparse representation benefits to noise suppression in the process of hallucination.

Robustness against noise is further verified with low-quality images under realistic surveillance imaging conditions. Test faces are captured by a commercial surveillance camera in a low-lighting environment, where the persons are far from the camera and hence unavoidably contain noise and blurring effects. As shown in Fig. 3, W1,2 preserves the same high fidelity for facial features as Bicubic interpolation, while avoids the inherent over-smoothness resulted by up-scaling. The distinct performance of WSR and WL1,2 again confirms that sparseness of the underlying model should not be over-emphasized in practical face hallucination applications. Experimentally, our proposed method can yield acceptable results even though the test images are generated in poor imaging conditions or in the presence of heavy non-Gaussian noise.

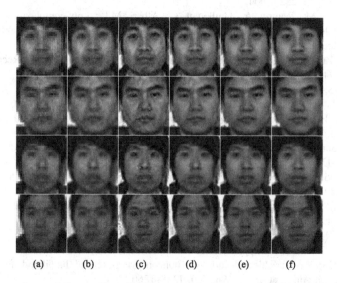

(a)   (b)   (c)   (d)   (e)   (f)

Fig. 3. Comparison of the results of different methods on real-world surveillance images: (a) Input LR faces; (b) Bicubic interpolation; (c) Chang's NE; (d) Yang's SR; (e) Proposed WSR; (f) Proposed WL1,2.

5 Conclusions

Our proposed method takes into account two fundamental statistics of coefficients in sparse domain: heteroskedasticity and under-sparsity of noisy images. Distance-inducing weighting is enforced on penalty term of sparse regularization to favor the locality of dictionary learning. A penalty of ℓ_1 and ℓ_2 mixed norms with conservative sparseness is introduced to characterize the less-sparse nature of noisy images. The resulting weighted $\ell_{1,2}$ norm regularization can significantly promote the accuracy, stability and robustness of solution. Extensive experimental results on FEI face database show its superiority in terms of PSNR, SSIM and subjective quality.

Acknowledgments. This work was supported by the National Natural Science Foundation of China (61231015, 61172173, 61170023, 61303114, 61271256, U1404618), the Fundamental Research Funds for the Central Universities (2042014kf0286, 2042014kf0212, 2042014kf0025, 2042014kf0250), China Postdoctoral Science Foundation (2014M562058), and Natural Science Fund of Hubei Province (2015CFB406).

References

1. Freeman, W., Pasztor, E., Carmichael, O.: Learning low-level vision. Int. J. Comput. Vis. **40** (1), 25–47 (2000)
2. Baker, S., Kanade, T.: Limits on super-resolution and how to break them. IEEE Trans. PAMI **24**(9), 1167–1183 (2002)
3. Chang, H., Yeung, D.Y., Xiong, Y.M.: Super-resolution through neighbor embedding. In: CVPR, pp. 275–282 (2004)
4. Bevilacqua, M., Roumy, A., Guillemot, C., Alberi, M.L.: Low-complexity single-image super-resolution based on nonnegative neighbor embedding. In: British Machine Vision Conference (BMVC), Surrey, pp. 1–10 (2012)
5. Ma, X., Zhang, J., Qi, C.: Hallucinating face by position-patch. Pattern Recogn. **43**(6), 3178–3194 (2010)
6. Park, J.-S, Lee, S.-W: An example-based face hallucination method for single-frame, low-resolution facial images. IEEE Trans. Image Process. **17**(10), 1806–1816 (2008)
7. Suo, F., Hu, F., Zhu, G.: Robust super-resolution reconstruction based on adaptive regularization. In: WCSP, pp. 1–4 (2011)
8. Liu, C., Shum, H., Freeman, W.: Face hallucination: theory and practice. Int. J. Comput. Vis. **7**(1), 15–134 (2007)
9. Li, Y., Lin, X.: An improved two-step approach to hallucinating faces. In: Proceedings of IEEE Conference on Image and Graphics, pp. 298–301 (2004)
10. Tang, Y., Yan, P., Yuan, Y., Li, X.: Single-image super-resolution via local learning. Int. J. Mach. Learn. Cyber. **2**, 15–23 (2011)
11. Jiang, J., Hu, R., Han, Z: Position-patch based face hallucination via locality-constrained representation. In ICME, pp. 212–217 (2012)
12. Fan, J., Li, R.: Variable selection via nonconcave penalized likelihood and its oracle properties. J. Am. Stat. Assoc. **96**(456), 1348–1360 (2001)
13. Yang, J., Tang, H., Ma, Y., Huang, T.: Image super-resolution via sparse representation. IEEE Trans. Image Process. **19**(11), 2861–2873 (2010)
14. Zeyde, R., Elad, M., Protter, M.: On single image scale-up using sparse-representations. In: Curves and Surfaces, pp. 24–30 (2010)
15. Jung, C., Jiao, L., Liu, B., Gong, M.: Position-patch based face hallucination using convex optimization. IEEE Signal Process. Lett. **18**(6), 367–370 (2011)
16. Jia, Z., Wang, H., Xiong, Z.: Fast face hallucination with sparse representation for video surveillance. In: ACPR, pp. 179–183 (2011)
17. Zhang, J., Zhao, C., Xiong, R., Ma, S., Zhao, D.: Image super-resolution via dual-dictionary learning and sparse representation. In: ISCAS, pp. 1688–1691 (2012)
18. Meinshausen, N., Bühlmann, P.: High-dimensional graphs and variable selection with the Lasso. Ann. Statist. **34**(3), 1436–1462 (2006)
19. Li, Q., Lin, N.: The Bayesian elastic net. *Bayesian*. Analysis **5**(1), 151–170 (2010)
20. Zou, H., Hastie, T.: Regularization and variables election via the elastic net. J. R. Stat. Soc. B **67**(2), 301–320 (2005)

21. Candès, E.J., Wakin, M.B., Boyd, S.: Enhancing sparsity by reweighted $\ell 1$ minimization. J. Fourier Anal. Appl. **14**(5), 877–905 (2008)
22. Friedlander, M.P., Mansour, H., Saab, R., Yılmaz, O.: Recovering compressively sampled signals using partial support information. IEEE Trans. Inf. Theory **58**(2), 1122–1134 (2012)
23. FEI Face Database. http://fei.edu.br/~cet/facedatabase.html

Visual Tracking by Assembling Multiple Correlation Filters

Tianyu Yang[✉], Zhongchao Shi, and Gang Wang

Ricoh Software Research Center (Beijing) Co., Ltd, Beijing, China
{tianyu.yang,zhongchao.shi,gang.wang}@srcb.ricoh.com

Abstract. In this paper, we present a robust object tracking method by fusing multiple correlation filters which leads to a weighted sum of these classifier vectors. Different from other learning methods which utilize a sparse sampling mechanism to generate training samples, our method adopts a dense sampling strategy for both training and testing which is more effective yet efficient due to the highly structured kernel matrix. A correlation filter pool is established based on the correlation filters trained by historical frames as tracking goes on. We consider the weighted sum of these correlation filters as the final classifier to locate the position of object. We introduce a coefficients optimization scheme by balancing the test errors for all correlation filters and emphasizing the recent frames. Also, a budget mechanism by removing the one which will result in the smallest change to final correlation filter is illustrated to prevent the unlimited increase of filter number. The experiments compare our method with other three state-of-the-art algorithms, demonstrating a robust and encouraging performance of the proposed algorithm.

Keywords: Visual tracking · Multiple correlation filters · Coefficients optimization

1 Introduction

Visual tracking, as a key component of many applications ranging from surveillance, robotics, vehicle navigation and human computer interaction, remains a challenging problem in spite of numerous algorithms have been proposed recent years. Given the marked object in the first frame, the goal of visual tracking is to determine the position of tracked object in subsequent frames.

Factors like illumination, pose variation, shape deformation and occlusions cause large appearance change for tracked object which easily leads to drift. Thus many efforts have been made to design various appearance models in recent literatures. Several methods [1–3] focus on investigating more robust features to represents object which locate the object position through simple template matching. Sevilla-Lara and Learned-Miller [1] propose an image descriptor called Description of Distribution Fields (DFs) which is a representation that allows smoothing the objective function without

© Springer International Publishing Switzerland 2015
Y.-S. Ho et al. (Eds.): PCM 2015, Part I, LNCS 9314, pp. 310–320, 2015.
DOI: 10.1007/978-3-319-24075-6_30

destroying information about pixel values. Different from the conventional histogram feature, a Locality Sensitive Histogram (LSH) has been presented in [2] which computes the histogram at each pixel location and adds a floating-point value into the corresponding bin for each occurrence of an intensity value. An illumination invariant feature is also constructed based on the LSH. All these algorithms can be categorized as generative approaches since its chief goal is to search for the region most similar to the object appearance model. Other generative algorithms include subspace models [4, 5] and sparse representation [6–8]. Mei and Ling [8] adopt a sparse representation as holistic template by solving minimization problem. However, this process is time-consuming, thereby restricting its usage in real-time situation. Liu et al. [7] utilize a local sparse appearance model with a static dictionary which adopt sparse representation-based voting map and sparse constraint regularized mean-shift for tracking. Admittedly, generative algorithms have demonstrated much success on modeling the appearance of target. However, it does not consider the information of background which is likely to alleviate drifting problem.

Discriminate methods which are widely adopted in tracking recently treat tracking as a classification problem in order to discriminate the object from the background. Lots of machine learning methods like SVM [9], boosting [10], random forests [11] are adopted to tracking problem. However, all these methods treated negative samples equally which means that a negative sample owning a large overlap with object rectangle is weighted the same as the one that overlaps little. This slight inaccuracy during labeling training samples can result in a poor performance in tracking. Many methods are presented such as multiple instance learning [12], loss function [13], semi-supervised learning [14] and continuous labels [15] in order to overcome this problem. To handle occlusions, Adam et al. [16] utilize multiple image fragments to build the template model where every patch can vote on the possible positions of target. In addition, several visual features like HOG [17, 18], Haar-like [19] are popularly used in discriminate methods.

In this paper, we propose an online visual tracking method combing multiple correlation filters [20, 21] which can be computed very fast in Fourier domain. Unlike other learning algorithms which utilize a sparse sampling mechanism to generate training and testing samples, our methods adopt a dense sampling strategy. Contrary to our intuition, it constructs a more efficient learning process because of the highly structured kernel matrix. Different cyclic shift of a base sample is employed to model negative samples which builds a circulant matrix when solving linear regression problem. And this circulant matrix can transfer closed-form solution of regression into Fourier domain due to its property of decomposition. We train a correlation filter based on the tracked target image patch each frame and create a correlation filter pool during tracking. We regard the weighted sum of these correlation filters as the final correlation filter. A coefficient optimization strategy is presented which updates the weights of each correlation filter through an iteration process. Also, a budget mechanism for the correlation filters pool is illustrated by removing the one which will result in the smallest change to final correlation filter.

This paper is organized as follows. The kernelized correlation filter is introduced in Sect. 2. In Sect. 3, we describe the online coefficient optimization mechanism for each correlation filter. Section 4 shows the experimental results on several commonly used datasets and a conclusion is presented in Sect. 5.

2 Kernalized Correlation Filter

In order to make fully use of the negative samples, we adopt dense sampling strategy as is illustrated in [15]. Instead of binary labels, we use the continuous labels which range from 0 to 1 for training samples. Thus, we treat this training process as linear regression problem which has a closed-form solution.

2.1 Linear Regression

Given a set of training samples and their corresponding labels $\{\mathbf{x_0}, y_0\}, \{\mathbf{x_1}, y_1\}, \ldots, \{\mathbf{x_n}, y_n\}$, the goal of training is to find the weight vector $\mathbf{w}$ that can minimize the following objective function,

$$\min_{\mathbf{w}} \sum_i (f(\mathbf{x_i}) - y_i)^2 + \lambda \|\mathbf{w}\|^2 \tag{1}$$

where λ is a regulation factor that control overfitting. This equation has a closed-form solution which is presented in [22],

$$\mathbf{w} = (X^T X + \lambda I)^{-1} X^T \mathbf{y} \tag{2}$$

where X is the data matrix where each sample $\mathbf{x_i}$ in a row. I is an identity matrix and $\mathbf{y}$ stands for the continuous label vector arranged by each sample label y_i. Computing the weight vector directly is time-consuming due to the calculation of inverse matrix. A better way to solve this problem is transfer this equation into Fourier domain using the property of data matrix.

We generate an image patch which is larger than the object box in order to cover some background information. Then this base sample is treated as positive and several other virtual samples obtained by translating it is modeled as negative samples. By using these shifted samples as training samples we can form a circulant data matrix. For simplicity, we utilize an n × 1 vector representing a patch with object of interest.

$$\mathbf{x_0} = \{x_1, x_2, \ldots, x_n\}^T \tag{3}$$

Thus, the data matrix can be written as,

$$X = C(\mathbf{x}) = \begin{bmatrix} x_1 & x_2 & \cdots & x_n \\ x_n & x_1 & \cdots & x_{n-1} \\ \vdots & \vdots & \ddots & \vdots \\ x_2 & x_3 & \cdots & x_1 \end{bmatrix} \tag{4}$$

Circulant matrix has several useful and amazing properties [23] among which is the one that it can be decomposed into the multiplication of Discrete Fourier Transform (DFT) matrix and the diagonal matrix constructed by base vector $\mathbf{x}_0$. This can be presented as,

$$X = F^H diag(\hat{\mathbf{x}}_0)F \tag{5}$$

where F is the DFT matrix which is an unitary matrix satisfying $F^H F = I$. $\hat{\mathbf{x}}_0$ stands for the DFT of vector $\mathbf{x}_0$, that is to say, $\hat{\mathbf{x}}_0 = F\mathbf{x}_0$. Then we can replace X in Eq. (2), thereby get the following equation which is detailed in [21],

$$\hat{\mathbf{w}}^* = \frac{\hat{\mathbf{x}}^* \odot \hat{\mathbf{y}}}{\hat{\mathbf{x}}^* \odot \hat{\mathbf{x}} + \lambda} \tag{6}$$

where $\hat{\mathbf{w}}^*$ and $\hat{\mathbf{x}}^*$ stand for the complex-conjugate of $\hat{\mathbf{w}}$ and $\hat{\mathbf{x}}$ respectively. $\hat{\mathbf{x}}, \hat{\mathbf{y}}, \hat{\mathbf{w}}$ refer to the DFT of these vectors. $\odot$ denotes the element-wise product. This equation can be calculated in Fourier domain through element-wise division which is time-efficient.

2.2 Kernel Regression

To build a more discriminative classifier, kernel trick is a good choice to achieve it. We map the linear problem into a rich high-dimensional feature space $\varphi(\mathbf{x})$. Then the solution $\mathbf{w}$ can be expressed as a linear combination of these samples

$$\mathbf{w} = \sum_i \alpha_i \varphi(\mathbf{x}_i) \tag{7}$$

Then instead of optimizing $\mathbf{w}$, we compute the dual space $\boldsymbol{\alpha} = \{\alpha_1, \alpha_2, \ldots, \alpha_n\}$ to solve the regression problem. Before illustrating the optimization process, we first point out the definition of kernel matrix K with elements

$$K_{ij} = \kappa(\mathbf{x}_i, \mathbf{x}_j) = \langle \varphi(\mathbf{x}_i), \varphi(\mathbf{x}_j) \rangle \tag{8}$$

Then the kernelized regression function thus becomes

$$f(\mathbf{z}) = \mathbf{w}\mathbf{z} = \sum_{i=1}^n \alpha_i \kappa(\mathbf{z}, \mathbf{x}_i) \tag{9}$$

Replacing the regression function in Eq. 1, we obtain the solution of this kernel regression [22],

$$\boldsymbol{\alpha} = (K + \lambda I)^{-1}\mathbf{y} \tag{10}$$

where K is the kernel matrix. Since it is proved [21] that the kernel matrix K is circulant when using radial basic function kernels (e.g. Gaussian) or dot-product kernels (e.g. linear, polynomial), it is easy to transfer the Eq. 10 into a Fourier domain using the property of circulant matrix as illustrated in Eq. 5. Thus, the Fourier version of Eq. 10 is given as,

$$\hat{\alpha}^* = \frac{\hat{y}}{\hat{k}^{xx} + \lambda} \tag{11}$$

where $\mathbf{k}^{xx}$ stands for the first row of kernel matrix K, and the hat $\hat{\cdot}$ denotes the DFT of a vector. Now the calculation of $\boldsymbol{\alpha}$ is determined by the kernel correlation $\mathbf{k}^{xx}$ which can be computed very fast using an inverse DFT and two DFT operations [21].

3 Correlation Filter Fusion

The training process mentioned in the above section is an illustration of learning a correlation filter for only one frame. However, as the tracking goes on, we will train a lot of correlation filters. A linear combination of these correlation filters is regarded as a proper way to represents the variation of object appearances.

3.1 Online Correlation Filter Update

Supposing the correlation filter trained for all frames during tracking are $\{\mathbf{w}_1, \mathbf{w}_2, \dots, \mathbf{w}_n\}$, we integrate these correlation filters into one classifier as $\mathbf{w}_{final}$

$$\mathbf{w}_{final} = \sum_i \beta_i \mathbf{w}_i \quad s.t. \sum_i \beta_i = 1 \tag{12}$$

The core step is to optimize the coefficient β_i monotonically based on the new coming correlation filter trained for new frame. It is hard to optimize all the coefficients one time. However, an update of two coefficients with an iteration process is likely to implement. We define a standard to choose a pair of correlation filter at each iteration of coefficients optimization. We regard the correlation filter whose corresponding image patch has a larger test error on final classifier $\mathbf{w}_{final}$ as the one that should be emphasized and vice versa. Also the recent coming image patches should be assigned more weights than the remote ones due to their best expressions of current tracked object's appearance. Based on these two factors, we proposed an optimization score when choosing the pair of correlation filter,

$$S_i = E(\mathbf{w}_{final}, \mathbf{z}_i) * L(\Delta t_i) \tag{13}$$

where $E(\mathbf{w}_{final}, \mathbf{z}_i)$ refers to the sum of test error for an image patch $\mathbf{z}_i$ which corresponding to the correlation filter $\mathbf{w}_i$. $L(\Delta t_i)$ represents the loss function which assigns a higher weight for image patches near current frame and a lower weight for the ones far away from current frame.

The image patch test error is defined as follows,

$$E(\mathbf{w}_{final}, \mathbf{z}_i) = \sum_{mn} (\mathbf{f}(\mathbf{z}_i) - R_{reg})_{mn} \tag{14}$$

where the output of $\mathbf{f}(\mathbf{z}_i)$ is a $m \times n$ matrix which contains the final classifier responses for all cyclic shift of $\mathbf{z}_i$. R_{reg} is the regression output for the regression function which is also a $m \times n$ matrix. In ideal condition, the responses tested by different shift of $\mathbf{z}_i$ should be fit the regression label perfectly, that is to say, the difference between these two matrix is a zero matrix ideally. However, due to the noise caused by illumination changes, appearances variation or occlusions, the sum of test error is not zero in reality. We believe the correlation filter with a higher image patch test error is caused by the low weight assigned to it. So this correlation filter should be emphasized by increasing its coefficient β_i.

In addition, we also decrease the effect lead by the correlation filter trained by remote frames using a loss function written as,

$$L(\Delta t_i) = G(\Delta t_i) \sim N(0, \sigma^2) \tag{15}$$

where Δt_i is the difference between current frame number and the number of frame which trains i classifier and $G(t)$ is a Gaussian function with 0 as its mean value and σ as its variance. Correlation filter which are trained by recent frames are assigned a higher weight than the ones trained by remote frames.

A correlation filter with maximum optimization score means that it is not emphasized properly. By increasing the coefficient of this correlation filter, we can decrease the test error of final classifier on its corresponding image patch. Similarly, a correlation filter with minimum optimization score means that it is modeled well by current final classifier. We do not need a larger coefficient to emphasize it.

Based on the optimization score of each correlation filter, we propose two schemes of choosing the correlation filter pair to update their coefficient at each iteration. For the new coming image patch, we choose this new trained correlation filter as w_+, and the one with minimum optimization score as w_- For existing correlation filter, we choose the one with maximum optimization score as w_+, and the one with minimum optimization score as w_-.

We calculate an optimal value by an unconstrained optimum

$$\lambda^u = \frac{S_+ - S_-}{2} \tag{16}$$

and then enforcing the constrains

$$\lambda = \min(\beta_-, \min(\theta - \beta_+, \lambda^u) \tag{17}$$

where the θ is the value to limit maximum value of coefficient. β_+ and β_- are the coefficients of two chosen classifiers. Finally, we update the optimization score for each classifier based on the score changes caused by

$$\Delta S_i = \lambda(E(\mathbf{w}_+, \mathbf{z}_i) - E(\mathbf{w}_-, \mathbf{z}_i)) \tag{18}$$

The optimization scores are updated by adding $\Delta S_i * L_i$,

$$S_i = S_i + \Delta S_i * L_i \tag{19}$$

where L_i is the loss function This procedure is summarized in Algorithm 1.

Algorithm 1. Coefficient optimization

Require: w_+, w_-
1. Calculate the optimum value λ

$$\lambda = \min\left(\beta_-, \min\left(\theta - \beta_+, \frac{S_+ - S_-}{2}\right)\right)$$

2. Update coefficient β

$$\beta_+ = \beta_+ + \lambda$$
$$\beta_- = \beta_- - \lambda$$

3. Update optimization score S
   for $w_i \in C$, *do*
   $$\Delta S_i = \lambda(E(w_+, z_i) - E(w_-, z_i))$$
   $$S_i = S_i + \Delta S_i * L_i$$
   end for

3.2 Budgeting on Correlation Filters

Another problem needs to be solved so far is that the number of classifier is not bounded now, that is to say, it will increase without limit as tracking goes on. Obviously, we cannot sustain the increasing memory requirement and runtime. So we introduced a budget mechanism to restrict the number of classifier. We choose to remove the classifier which will result in the smallest change to final classifier weight $\mathbf{w}_{final}$. In order to satisfy the constraint $\sum_i \beta_i = 1$, we add the coefficient of removed classifier to the one with maximum op

$$\Delta\mathbf{w} = -\beta_r\mathbf{w}_r + \beta_r\mathbf{w}_{\max}$$
$$r = \arg\min_r \|\Delta\mathbf{w}\|^2 \tag{20}$$

where β_r is the coefficient of classifier to be removed and $\mathbf{w}_r, \mathbf{w}_{\max}$ are the classifier to be removed and the classifier with maximum coefficient. We choose the one with minimum $\|\Delta\mathbf{w}\|^2$ to remove.

4 Experiments

This section presents the evaluation of our tracking algorithm using 8 challenging video sequences from MIL [24]. It is worth noticing that these sequences covers almost all situations in object tracking like heavy occlusion, illumination changes, pose variation and complex background. To demonstrate the performance of our methods, we compares our results with other three state-of-the-art algorithms which are Struck [13], TLD [25] and CT [26]. An accompanying video is attached to this paper.

Table 1. Mean center location error comparison with other three algorithms. Bold font means best.

Sequence	Struck	TLD	CT	Ours
Coke	11.1	36.3	36.1	**8.4**
David	9.7	**6.6**	12.4	7.4
FaceOcc1	18.6	20.4	26.7	**10.7**
FaceOcc2	7.1	139.8	23.6	**6.4**
Girl	**2.8**	7.8	14.9	6.2
Sylvester	6.1	12.2	19.0	**5.6**
Tiger1	25.6	179.8	28.1	**14.1**
Tiger2	19.6	173.5	28.7	**12.8**

Since our algorithm does not implement the scale changes adaption scheme, it outputs object rectangle with the same size during all tracking frames. So we utilize the mean center error as the standard to evaluate the performance of our method. Table 1 reveals the mean center errors of our methods comparing with other three algorithms.

The results show that our algorithm outperforms other three methods on most of data sequences. In the "David" sequence, TLD performs the best because of its size adaption scheme to handle the size changes during tracking. But for sequences "FaceOcc2", "Tiger1" and "Tiger2", TLD perform poorly due to its failed reacquisition after drifting. Due to the multiple correlation filters fusion scheme, our method is adaptable to various appearance changes including illumination changes and pose variation which are revealed on the last two data sequences. The results on data "FaceOcc1" and "FaceOcc2" show our algorithm's ability to handle occlusion. Struck achieves the best performance in "Girl" sequence because of its appearance changes adaption when the girl turn to his head with her back to us. Our method also obtains a good result on this data set due to our fusion of multiple object appearances during tracking.

To illustrate the details of tracking results for different algorithms, we plot the center location errors per frame in Fig. 1 for all datasets. In the plot of "Coke", we easily find that the center location errors of "TLD" and "CT" are fluctuant because their poor adaption to appearances caused by illumination changes and rotation. For data sets with occlusions such as "FaceOcc1" and "FaceOcc2", it is observed that our methods outperform all other algorithms thanks to the robustness of fusion of multiple correlation filters. Struck also obtains comparable performances due to its online learned structured output support vector machine. From the plot of "Sylvester", we observe that our method shows robust tracking results even when other methods in-crease their center location error around frame 1200. There are lots of appearance changes in sequences "Tiger1" and "Tiger2" which are caused by illumination variants, tiger's mouth movement and pose transformation. The performance of TLD is poor because it easily failed when appearance changes happens. Our method and Struck are more robust to tackle this situation.

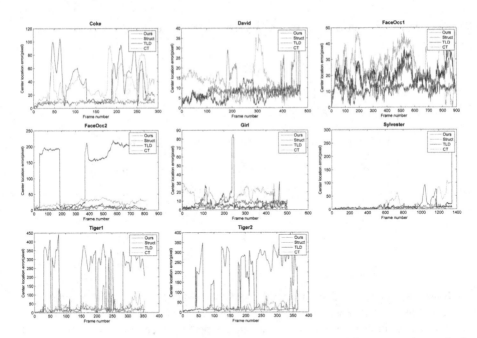

Fig. 1. Center location error per frame for all data sequences. The red line repents the result of our method and the green one is Struck's result. The blue one and cyan one are the results of TLD and CT respectively.

5 Conclusions

In this paper, we present a novel object tracking method using a weighted combination of multiple correlation filters as the classifier. By adopting the property of circulant matrix, we obtain a very efficient learning process in Fourier domain. At each frame we can learn a correlation filter, thus as tracking goes on, a filter pool containing correlation filters trained at different frames is constructed. To adapt the different appearances of object presented in historical frames. We assemble all these correlation filters into a weighted sum of them and propose a coefficient optimization scheme to update the weights assigned for each correlation filter. Since the number of correlation filter in the pool will increase linearly if there is no limit for it, we also illustrated a budget mechanism by removing the one which will result in the smallest change to final correlation filter. Experiments show that our algorithm is robust and effective compared with other three state-of-the-art methods. Our future work is to incorporate the scale changes into current algorithm without increasing the runtime substantially.

References

1. Sevilla-Lara, L., Learned-Miller, E.: Distribution fields for tracking. In: 2012 IEEE Conference on Computer Vision and Pattern Recognition (CVPR), pp. 1910–1917 (2012)

2. He, S., Yang, Q., Lau, R.W., Wang, J., Yang, M.-H.: Visual tracking via locality sensitive histograms. In: CVPR, pp. 2427–2434 (2013)

3. Oron, S., Bar-Hillel, A., Levi, D., Avidan, S.: Locally orderless tracking. In: CVPR, pp. 1940–1947 (2012)

4. Ross, D.A., Lim, J., Lin, R.S., Yang, M.H.: Incremental learning for robust visual tracking. Int. J. Comput. Vision **77**, 125–141 (2008)

5. Black, M.J., Jepson, A.D.: EigenTracking: robust matching and tracking of articulated objects using a view-based representation. IJCV **26**, 63–84 (1998)

6. Jia, X., Lu, H., Yang, M.-H.: Visual tracking via adaptive structural local sparse appearance model. In: Computer Vision and Pattern Recognition, pp. 1822–1829 (2012)

7. Liu, B., Huang, J., Yang, L., Kulikowsk, C.: Robust tracking using local sparse appearance model and k-selection. In: CVPR, pp. 1313–1320 (2011)

8. Mei, X., Ling, H.: Robust visual tracking using $\ell 1$ minimization. In: 2009 IEEE 12th International Conference on Computer Vision, pp. 1436–1443 (2009)

9. Avidan, S.: Support vector tracking. IEEE Trans. Pattern Anal. Mach. Intell. **26**, 1064–1072 (2004)

10. Grabner, H., Bischof, H.: On-line boosting and vision. In: CVPR 2006, June 17, pp. 260–267, New York, NY, USA (2006)

11. Saffari, A., Leistner, C., Santner, J., Godec, M., Bischof, H.: On-line random forests. In: ICCV Workshops 2009, pp. 1393–1400, Kyoto, Japan (2009)

12. Babenko, B., Belongie, S., Yang, M.-H.: Visual tracking with online multiple instance learning. In: CVPR, pp. 983–990, Miami, FL, USA (2009)

13. Hare, S., Saffari, A., Torr, P.H.S.: Struck: structured output tracking with kernels. In: 2011 IEEE International Conference on Computer Vision (ICCV), pp. 263–270 (2011)

14. Grabner, H., Leistner, C., Bischof, H.: Semi-supervised on-line boosting for robust tracking. In: Proceedings of Computer Vision - ECCV Pt I, vol. 5302, pp. 234–247 (2008)

15. Henriques, J.F., Caseiro, R., Martins, P., Batista, J.: Exploiting the circulant structure of tracking-by-detection with kernels. In: Fitzgibbon, A., Lazebnik, S., Perona, P., Sato, Y., Schmid, C. (eds.) ECCV 2012, Part IV. LNCS, vol. 7575, pp. 702–715. Springer, Heidelberg (2012)

16. Adam, A., Rivlin, E., Shimshoni, I.: Robust fragments-based tracking using the integral histogram. In: CVPR, pp. 798–805 (2006)

17. Dalal, N., Triggs, B.: Histograms of oriented gradients for human detection. In: Proceedings of CVPR, vol. 1, pp. 886–893 (2005)

18. Felzenszwalb, P.F., Girshick, R.B., McAllester, D., Ramanan, D.: Object detection with discriminatively trained part-based models. In: PAMI, pp. 1627–1645 (2010)

19. Viola, P., Jones, M.J.: Robust real-time face detection. Int. J. Comput. Vis. **57**, 137–154 (2004)

20. Bolme, D.S., Beveridge, J.R., Draper, B.A., Lui, Y.M.: Visual object tracking using adaptive correlation filters. In: CVPR, pp. 2544–2550 (2010)

21. Henriques, J.F., Caseiro, R., Martins, P., Batista, J.: High-speed tracking with kernelized correlation filters. IEEE Trans. PAMI (2014)

22. Rifkin, R., Yeo, G., Poggio, T.: Regularized least-squares classification. Nato Sci. Ser. Sub Ser. III Comput. Syst. Sci. **190**, 131–154 (2003)

23. Gray, R.M.: Toeplitz and Circulant Matrices: A Review. Now Publishers Inc. (2006)

24. Babenko, B., Yang, M.-H., Belongie, S.: Robust object tracking with online multiple instance learning. IEEE Trans. PAMI **33**, 1619–1632 (2011)

25. Kalal, Z., Mikolajczyk, K., Matas, J.: Tracking-learning-detection. In: IEEE Transactions on Pattern Analysis and Machine Intelligence, pp. 1–1 (2011)
26. Zhang, K., Zhang, L., Yang, M.-H.: Real-time compressive tracking. Presented at the ECCV 2012 (2012)

A Unified Tone Mapping Operation for HDR Images Including Both Floating-Point and Integer Data

Toshiyuki Dobashi[1]([✉]), Masahiro Iwahashi[2], and Hitoshi Kiya[1]

[1] Tokyo Metropolitan University, Tokyo, Japan
dobashi-toshiyuki1@ed.tmu.ac.jp
[2] Nagaoka University of Technology, Niigata, Japan

Abstract. This paper considers a unified tone mapping operation (TMO) for HDR images. This paper includes not only floating-point data but also long-integer (i.e. longer than 8-bit) data as HDR image expression. A TMO generates a low dynamic range (LDR) image from a high dynamic range (HDR) image by compressing its dynamic range. A unified TMO can perform tone mapping for various HDR image formats with a single common TMO. The integer TMO which can perform unified tone mapping by converting an input HDR image into an intermediate format was proposed. This method can be executed efficiently with low memory and low performance processor. However, only floating-point HDR image formats have been considered in the unified TMO. In other words, a long-integer which is one of the HDR image formats has not been considered in the unified TMO. This paper extends the unified TMO to a long-integer format. Thereby, the unified TMO for all possible HDR image formats can be realized. The proposed method ventures to convert a long-integer number into a floating-point number, and treats it as two 8-bit integer numbers which correspond to its exponent part and mantissa part. These two integer numbers are applied the tone mapping separately. The experimental results shows the proposed method is effective for an integer format in terms of the resources such as the computational cost and the memory cost.

Keywords: High dynamic range · Tone mapping · Unified · Integer · Floating-point

1 Introduction

High dynamic range (HDR) images are spreading in many fields: photography, computer graphics, on-vehicle cameras, medical imaging, and more. They have wider dynamic range of pixel values than standard low dynamic range (LDR) images. In contrast, display devices which can express the pixel values of HDR images are not popular yet. Therefore, the importance of a tone mapping operation (TMO) which generates an LDR image from an HDR image by compressing its dynamic range is growing.

© Springer International Publishing Switzerland 2015
Y.-S. Ho et al. (Eds.): PCM 2015, Part I, LNCS 9314, pp. 321–333, 2015.
DOI: 10.1007/978-3-319-24075-6_31

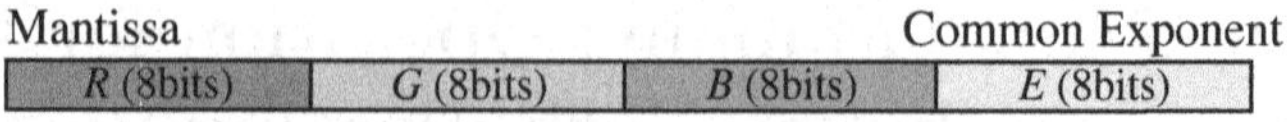

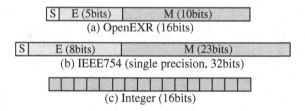

Fig. 1. The bit allocation of the RGBE format.

(a) OpenEXR (16bits)

(b) IEEE754 (single precision, 32bits)

(c) Integer (16bits)

Fig. 2. The bit allocation of the OpenEXR, the IEEE754, and the integer format. Each of RGB channels has this. Let S, E, and M denote a sign, a exponent, and a mantissa, respectively.

Various research works on tone mapping have so far been done [1–11]. Reference [1–9] focus on compression techniques or quality of tone mapped images, and [10,11] focus on speeding-up of a tone mapping function. In [10,11], visibility and contrast are simply controlled with a single parameter. Nevertheless, tone mapping functions for these approach is limited to a specific one. Moreover, the tone mapping function is only one process out of many processes in a TMO.

Unlike these research works, an integer TMO approach which deals with resource reduction was proposed in [12–14]. Considering not only a function itself but also the whole process of a TMO, this method tries to resolve the essential problem on high demand of resources. In these methods, any kind of global tone mapping functions can be used. The method in [12] treats a floating-point number as two 8-bit integer numbers which correspond to a exponent part and a mantissa part, and applies tone mapping to these integer numbers separately. The method reduces the memory cost by using 8-bit integer data instead of 64-bit floating-point data. Moreover, using 8-bit integer data facilitates executing calculations with fixed-point arithmetic because it eases the limitation of the bit length. Fixed-point arithmetic is often utilized in image processing and embedded systems because of the advantages such as low-power consumption, the small circuit size and high-speed computing [15–17]. The method in [13] executes the integer TMO with fixed-point arithmetic, and therefore it reduces the computational cost as well. However, this integer TMO approach is designed for the RGBE format; its performance is not guaranteed for other formats. In [18,19], the intermediate format was introduced, and the integer TMO was extended for it. By using the intermediate format, it can be applied for other formats such as the OpenEXR and the IEEE754. This method can also be implemented with fixed-point arithmetic, and it applies tone mapping with low resources. Nevertheless, in [18,19], only floating-point HDR image formats are used as input images; a long-integer which is one of the HDR image formats is not considered.

This paper extends the method in [19] to an integer format, and confirms its efficacy. The proposed unified TMO can treat all possible HDR image formats including both floating-point data and integer data. There are two patterns of tone mapping for an integer format. One is a method of processing an integer format directly without format conversion. Although this method is simple, it can not be used as a unified TMO because it is exclusive use of the integer format. The other is a method of processing after converting an integer format to an intermediate format. The latter, namely, the proposed method can be used as a unified TMO which can process various HDR image formats in addition to the integer format. In addition, the experimental results shows the proposed method is more effective for an integer format in terms of the resources such as the computational cost and the memory cost.

2   Preliminaries

2.1   Floating-Point HDR Image Formats

This section describes HDR image formats. There are several formats, this paper focuses on the integer format. However, the proposed method is not limited to the integer format. The method is a unified tone mapping operation, and it can be used for other formats as well. The HDR image formats supported by the proposed method are the RGBE [20], the OpenEXR [21], the IEEE754 [22], and the integer. Figure 1 shows the bit allocation of the RGBE format, and Fig. 2 shows that of the others. The RGBE, the OpenEXR, and the IEEE754 are floating-point format.

2.2   Global Tone Mapping Operation

A TMO generates an LDR image from an HDR image by compressing its dynamic range. There are two types of a TMO: global tone mapping and local tone mapping, this paper deals with global tone mapping. A procedure of "Photographic Tone Reproduction" [1], which is one of the well-known global TMOs, is described in this section.

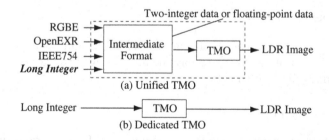

Fig. 3. The scheme of a unified tone mapping operation.

First, the world luminance $L_w(p)$ of the HDR image is calculated from RGB pixel values of the HDR image,

$$L_w(p) = 0.27R(p) + 0.67G(p) + 0.06B(p), \qquad (1)$$

where $R(p), G(p)$, and $B(p)$ are RGB pixel values of the HDR image, respectively.

Next, the geometric mean $\bar{L}_w$ of the world luminance $L_w(p)$ is calculated as follows

$$\bar{L}_w = \exp\left(\tfrac{1}{N}\textstyle\sum_p \log_e\left(L_w(p)\right)\right), \tag{2}$$

where N is the total number of pixels in the input HDR image. Note that Eq. (2) has the singularity due to zero value of $L_w(p)$. It is avoided by introducing a small value as shown in [1]. However, its affection is not negligible for pixel values in a resulting LDR image because a typical HDR image format such as the RGBE can express a small pixel value. Therefore, only non-zero values are used in this calculation.

Then, the scaled luminance $L(p)$ is calculated as

$$L(p) = k \cdot \frac{L_w(p)}{\bar{L}_w}, \tag{3}$$

where $k \in [0, 1]$ is the parameter called "key value".

Next, the display luminance $L_d(p)$ is calculated as follows

$$L_d(p) = \frac{L(p)}{1 + L(p)}, \tag{4}$$

Finally, the 24-bit color RGB values $C_I(p)$ of the LDR image is calculated as follows

$$C_I(p) = \text{round}\left(L_d(p) \cdot \tfrac{C(p)}{L_w(p)} \cdot 255\right), \tag{5}$$

where $C(p) \in \{R(p), G(p), B(p)\}$, $\text{round}(x)$ rounds x to its nearest integer value, and $C_I(p) \in \{R_I(p), G_I(p), B_I(p)\}$.

Despite the resulting LDR image is integer data, the data and arithmetic in the above procedure are both floating-point. Large computational and memory cost is required because of this.

3   Proposed Method

This section describes a scheme of a unified TMO, an intermediate format, an integer TMO for the intermediate format, and the way to execute the integer TMO with fixed-point arithmetic.

3.1   Unified TMO

Figure 3 shows the scheme of a unified tone mapping. In (a), various input HDR image formats are converted to an intermediate format, and then a TMO for the intermediate format is applied. Thus, (a) is a unified method which can process various HDR image formats using a single common TMO. On the other hand, in (b), each input HDR image format is processed by a TMO dedicated to each format. This paper focuses on the TMO (a) including the long-integer as the input HDR image format.

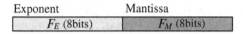

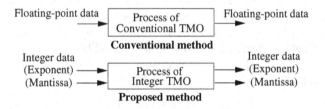

Fig. 4. The bit allocation of the proposed intermediate format.

Fig. 5. The difference between the conventional method [1] and the integer TMO.

3.2 Intermediate Format

An input HDR image is converted to an intermediate format at the first step. This paper considers two types of an intermediate format; the IEEE754 format and the proposed intermediate format. The IEEE754 format is the standard floating-point format as described earlier. In this section, the proposed intermediate format is described.

Figure 4 shows the bit allocation of the proposed intermediate format. Unlike the RGBE format, the exponent part of each RGB channel in this format is independent, and it reduces the error of the format conversion. The format with 8-bit exponent part and 8-bit mantissa part is selected to reduce the memory cost. Because the memory cost is proportional to the format, it is estimated to be 1/2 and 1/4, compared to the IEEE754 single and double precision format. Details of the memory cost are described later. The encode functions which yield the exponent part F_E and the mantissa part F_M of each RGB channel F are defined as

$$F_E = \lceil \log_2 F + 128 \rceil , \quad F_M = \lfloor F \cdot 2^{136-F_E} \rfloor , \tag{6}$$

where $\lceil x \rceil$ rounds x to the nearest integer greater than or equal to x, and $\lfloor x \rfloor$ rounds x to the nearest integer less than or equal to x. On the other hand, the decode function which yields the original RGB value from the intermediate format is defined as

$$F = (F_M + 0.5) \cdot 2^{F_E - 136}. \tag{7}$$

3.3 Integer TMO for the Intermediate Format

The integer TMO converts input and output data of each process to two 8-bit integer data. Using 8-bit integer data facilitates executing calculations with fixed-point arithmetic because it eases the limitation of the bit length. Figure 5 shows the difference between the integer TMO and the conventional method [1] described in Sect. 2.2. Note that this technique of the integer TMO works well by using the proposed intermediate format. The technique does not work well for the IEEE754 format because it has denormalized numbers as well as the OpenEXR [18]. The integer TMO defines new processes and replaces each tone mapping process by them. These new processes are composite functions. Each process of the proposed method is described as follows.

The integer TMO converts RGB values $C(p)$ into the intermediate format described in Sect. 3.2 at the first step. The exponent parts $C_E(p) \in \{R_E(p), G_E(p), B_E(p)\}$ and the mantissa parts $C_M(p) \in \{R_M(p), G_M(p), B_M(p)\}$ are calculated as

$$C_E(p) = \lceil \log_2 C(p) + 128 \rceil, \tag{8}$$

$$C_M(p) = \left\lfloor C(p) \cdot 2^{136-C_E(p)} \right\rfloor. \tag{9}$$

Then, the exponent part $L_{wE}(p)$ and the mantissa part $L_{wM}(p)$ of the world luminance $L_w(p)$ of the HDR image are calculated as

$$L_{wE}(p) = \lceil \log_2 ML(p) - 8 \rceil, \tag{10}$$

$$L_{wM}(p) = \left\lfloor ML(p) \cdot 2^{-L_{wE}(p)} \right\rfloor, \tag{11}$$

$$ML(p) = 0.27(R_M(p) + 0.5) \cdot 2^{R_E(p)} +$$
$$0.67(G_M(p) + 0.5) \cdot 2^{G_E(p)} +$$
$$0.06(B_M(p) + 0.5) \cdot 2^{B_E(p)}, \tag{12}$$

where $0 \leq L_{wE}(p) \leq 255$ and $0 \leq L_{wM}(p) \leq 255$. The method sets $L_{wE}(p) = L_{wM}(p) = 0$ if $C_E(p) = 0$, and the method sets $L_{wM}(p) = 255$ if $L_{wM}(p) = 256$.

Next, the exponent part $\bar{L}_{wE}$ and the mantissa part $\bar{L}_{wM}$ of the geometric mean $\bar{L}_w$ of the HDR image are calculated as

$$\bar{L}_{wE} = \lceil SL_{wM} + SL_{wE} + 128 \rceil, \tag{13}$$

$$\bar{L}_{wM} = \left\lfloor 2^{SL_{wM} + SL_{wE} - \bar{L}_{wE} + 136} \right\rfloor, \tag{14}$$

$$SL_{wE} = \frac{1}{N} \sum_p (L_{wE}(p) - 136), \tag{15}$$

$$SL_{wM} = \frac{1}{N} \sum_p \log_2 (L_{wM}(p) + 0.5), \tag{16}$$

where $0 \leq \bar{L}_{wE} \leq 255$ and $0 \leq \bar{L}_{wM} \leq 255$. Here, $\bar{L}_{wE}$ and $\bar{L}_{wM}$ are computed using only non-zero $L_{wE}(p)$'s.

Then, the exponent part $L_E(p)$ and the mantissa part $L_M(p)$ of the scaled luminance $L(p)$ of the HDR image are calculated as

$$L_E(p) = \lceil \log_2(AL_w(p)) + L_{wE}(p) - \bar{L}_{wE} + 128 \rceil, \tag{17}$$

$$L_M(p) = \left\lfloor AL_w(p) \cdot 2^{136+L_{wE}(p)-L_E(p)-\bar{L}_{wE}} \right\rfloor, \tag{18}$$

$$AL_w(p) = k \cdot \frac{L_{wM}(p) + 0.5}{\bar{L}_{wM} + 0.5}. \tag{19}$$

The method sets $L_E(p) = L_M(p) = 0$ if $L_E(p) < 0$, and $L_E(p) = L_M(p) = 255$ if $L_E(p) > 255$. That is, $0 \le L_E(p) \le 255, 0 \le L_M(p) \le 255$.

Next, the method calculates the exponent part $L_{d_E}(p)$ and the mantissa part $L_{d_M}(p)$ of the display luminance $L_d(p)$. This calculation depends on tone mapping functions. Here, the tone mapping function of Eq. (4) is used as an example,

$$L_{d_E}(p) = \lceil \log_2(FL(p)) + 128 \rceil, \tag{20}$$

$$L_{d_M}(p) = \left\lfloor FL(p) \cdot 2^{136 - L_{d_E}(p)} \right\rfloor, \tag{21}$$

$$FL(p) = \frac{L_M(p) + 0.5}{L_M(p) + 0.5 + 2^{136 - L_E(p)}}. \tag{22}$$

The method sets $L_{d_E}(p) = L_{d_M}(p) = 0$ if $L_{d_E}(p) < 0$, and $L_{d_E}(p) = L_{d_M}(p) = 255$ if $L_{d_E}(p) > 255$. That is, $0 \le L_{d_E}(p) \le 255, 0 \le L_{d_M}(p) \le 255$.

Finally, the 24-bit RGB pixel values $C_I(p)$ of the LDR image is obtained as

$$C_I(p) = \text{round}\left(RL(p) \cdot 2^{C_E(p) + L_{d_E}(p) - L_{w_E}(p) - 136}\right), \tag{23}$$

$$RL(p) = \frac{(L_{d_M}(p) + 0.5)(C_M(p) + 0.5)}{L_{w_M}(p) + 0.5} \cdot 255. \tag{24}$$

In the above processes, the input and output data of each calculation are all 8-bit integer data. The next section describes fixed-point arithmetic in the proposed method.

3.4 Fixed-Point Arithmetic

In the integer TMO, only the data are converted to integer, and the memory cost is reduced. However, the internal arithmetic of the integer TMO is still with floating-point. The proposed method introduces fixed-point arithmetic to reduce the computational cost as well. This section describes the way to execute the internal arithmetic with fixed-point arithmetic. Most of equations can be calculated with fixed-point arithmetic because each variable is expressed in 8-bit integer [13]. Nevertheless, Eq. (22) is difficult to be calculated without floating-point arithmetic because the range of value of the denominator is very wide. Because of this, the method deforms Eq. (22) as follows

$$FL(p) = \frac{1}{1 + \frac{2^{136 - L_E(p)}}{L_M(p) + 0.5}}. \tag{25}$$

Furthermore, the method branches Eq. (25) into three cases and approximates it based on the power of two in the denominator as follows.

Case 1: If $136 - L_E(p) > 15$ in Eq. (25), '1' in the denominator can be ignored because the right part of the denominator is very large, and so it is approximated as

$$FL(p) = \frac{L_M(p) + 0.5}{2^{136 - L_E(p)}}, \tag{26}$$

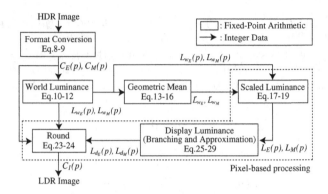

Fig. 6. The outline of the proposed method.

$$L_{d_E}(p) = \lceil \log_2(L_M(p)+0.5) - (136 - L_E(p)) + 128 \rceil, \tag{27}$$

$$L_{d_M}(p) = \left\lfloor (L_M(p) + 0.5) \cdot 2^{L_E(p) - L_{d_E}(p)} \right\rfloor. \tag{28}$$

Case 2: If $136 - L_E(p) < -8$ in Eq. (25), the right part of the denominator can be ignored because it is very small, and so it is approximated as

$$FL(p) = 1, \ L_{d_E}(p) = 128, \ L_{d_M}(p) = 255. \tag{29}$$

Case 3: Otherwise, it can be calculated with fixed-point arithmetic.

In addition, the method uses pre-calculated tables for calculations of 2^x (in Eq. (14)) and $\log_2$ (in Eq. (16)). Each table consists of 16×256 bits. In Eqs. (8)–(29), division operations are simply done by division, not right shift. Moreover, 2^x and $\log_2$ are conducted by using simple bit shift operation except Eqs. (14) and (16). The method can calculate all equations of the TMO with only fixed-point arithmetic by these branching, approximation, and tables.

4   Experimental and Evaluation Results

This paper proposed the unified TMO with the intermediate format. The proposed method can be executed with fixed-point arithmetic to reduce the computational cost. However, errors can occur by these format conversion and fixed-point arithmetic. To confirm the efficacy of the proposed method and the errors involved with it, the experiments and evaluation were carried out. These experiments and evaluation consist of measurements of peak signal-to-noise ratio (PSNR), the structural similarity index (SSIM) [23] of the resulting LDR images and processing time of the TMO, and evaluation of memory usage. Figure 7 shows the block diagram of these. These experiments and evaluation compared the proposed method (with floating-point and fixed-point arithmetic), the IEEE754 methods (double and single precision), and the integer dedicated method. The common conditions in the experiments and the evaluation are listed in Table 1. All of the methods were implemented in C-language.

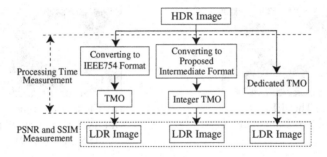

Fig. 7. The block diagram of the experiments.

Table 1. The common conditions in the experiments and the evaluation.

	Arithmetic	Data
IEEE754 (double precision)	64-bit Floating-point	64-bit Floating-point
IEEE754 (single precision)	64-bit Floating-point	32-bit Floating-point
Integer dedicated	64-bit Fixed-point	16-bit Integer
Proposed (floating-point)	64-bit Floating-point	8-bit Integer
Proposed (fixed-point)	32-bit Fixed-point	8-bit Integer

4.1   Comparison of Tone-Mapped LDR Images

This experiment applied tone mapping for 73 HDR images in 16-bit integer format using each method, and measured the PSNR and the SSIM. The tone mapped images $I_{\mathrm{LDR}}(p)$ and $I'_{\mathrm{LDR}}(p)$ are given by

$$I_{\mathrm{LDR}}(p) = \mathrm{T_c}[I_{\mathrm{HDR}}(p)], \quad I'_{\mathrm{LDR}}(p) = \mathrm{T_p}[I_{\mathrm{HDR}}(p)], \tag{30}$$

where $I_{\mathrm{HDR}}(p)$ is an input HDR image, and $\mathrm{T_c}[\cdot]$ and $\mathrm{T_p}[\cdot]$ are TMO of the conventional method and the proposed method, respectively. The PSNR between $A \times B$ sized LDR images is given by

$$PSNR = 10 \log_{10} \frac{255^2}{MSE}, \tag{31}$$

$$MSE = \frac{1}{AB} \sum_{p=1}^{AB} [I_{\mathrm{LDR}}(p) - I'_{\mathrm{LDR}}(p)]^2. \tag{32}$$

The SSIM offers more subjective evaluation than the PSNR. If $I_{\mathrm{LDR}}(p) = I'_{\mathrm{LDR}}(p)$, the PSNR will be ∞ and the SSIM will be 1.0. This experiment used the resulting LDR image of the IEEE754 double precision method as a true value because it was executed with most plenty resources. That is, $\mathrm{T_a}[\cdot]$ was the IEEE754 double precision method and $\mathrm{T_b}[\cdot]$ was the proposed method or the integer dedicated method.

Table 2 shows the maximum, minimum, and average PSNR and the average SSIM. The integer dedicated method gave high PSNR and SSIM values because

(a) The IEEE754
double precision method.

(b) The proposed method
with fixed-point arithmetic.
(SSIM = 0.999)

Fig. 8. LDR images comparison. It is impossible for human eyes to distinguish these images

Table 2. The maximum, minimum, and average PSNR and the average SSIM of the methods.

	PSNR [dB]			SSIM
	Maximum	Minimum	Average	
Integer dedicated	91.76	49.91	60.10	0.9993
Proposed (floating-point)	77.28	50.39	56.24	0.9985
Proposed (fixed-point)	66.56	50.39	55.84	0.9985

it is designed for integer format. On the other hand, the proposed method also gave a high SSIM value. Although the PSNR of the proposed method were slightly lowered, they were still sufficiently high values. Moreover, the proposed method with fixed-point arithmetic also maintains the sufficiently high values. Figure 8 shows example LDR images obtained by this experiment. It indicates that it is impossible for human eyes to distinguish these images. From the above results, this experiment confirmed that the proposed method can execute the TMO with high accuracy.

4.2 Comparison of the Memory Usage

Table 3 shows the memory usage of each calculation when the size of the input HDR image is $A \times B$ pixels. The rest of calculations which is not included

Table 3. The memory usage of the methods.

	Memory usage [bytes]				
	IEEE754 (double precision)	IEEE754 (single precision)	Integer dedicated	Proposed (floating-point)	Proposed (fixed-point)
An HDR image	$A \times B \times 24$	$A \times B \times 12$	$A \times B \times 6$	$A \times B \times 6$	$A \times B \times 6$
World luminance	$A \times B \times 8$	$A \times B \times 4$	$A \times B \times 4$	$A \times B \times 2$	$A \times B \times 2$
Geometric mean	8	4	4	2	2
Table	–	–	393216	-	1024

this table can be conducted per pixel, and it is indicated in Fig. 6. The proposed method and the integer dedicated method used the pre-calculated tables in order to calculate with fixed-point arithmetic. To reduce the table of data, sometimes interpolation is used. However, these implementation did not use interpolation in order to avoid its influence such as errors and load. Therefore, the table corresponding to bit-length of the data used in the method is required. Because the data is 8-bit each in the exponent and the mantissa, the table of the proposed method is smaller than the integer dedicated method. The memory usage of the proposed method which depends on the image size is 75 %, 50 %, and 20 % less than the IEEE754 double precision method, the IEEE754 single precision method, and the integer dedicated method, respectively.

Table 4. The platforms used in the experiment.

	Processor	FPU	RAM
Platform A	Intel Core i7 3930 K 3.2 GHz	Yes	16 GB
Platform B	Marvell PXA270 624 MHz	No	128 MB

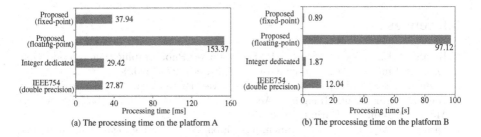

(a) The processing time on the platform A   (b) The processing time on the platform B

Fig. 9. The processing time of the methods.

4.3   Comparison of the Processing Time

This experiment applied tone mapping for HDR images with 512×768 pixels in the integer format using each method, and measured the processing time of the methods. This experiment was carried out on two platforms listed in Table 4.

Figure 9 (a) compares the processing time of the methods on the platform A. This platform has an FPU which can process the IEEE754 floating-point data. Thus, IEEE754 method was processed at the highest speed. The processing time of single precision and double precision were almost same because single precision is treated as double precision in the processor.

On the other hand, Fig. 9 (b) compares the processing time of the methods on the platform B. The methods with fixed-point arithmetic were processed fast on this platform because it does not have an FPU. The proposed method with fixed-point arithmetic was 13.53 and 2.10 times faster than the IEEE754 double precision method and the integer dedicated method, respectively. Therefore, this experiment confirmed that the proposed method with fixed-point arithmetic can perform high speed processing on the processor without an FPU.

5   Conclusion

This paper proposed the unified TMO including both floating-point data and long-integer data. The proposed method ventures to convert a long-integer number into the floating-point number, and treats it as two 8-bit integer numbers which correspond to its exponent part and mantissa part. The memory usage of the method is reduced by using these 8-bit integer numbers. Moreover, the method with fixed-point arithmetic can be executed fast on a processor without an FPU. The method is the unified TMO which can process various HDR image formats including a long-integer, namely a dedicated TMO for each format is not required. The proposed unified TMO can treat all possible HDR image formats including both floating-point data and integer data. The experimental and evaluation results confirmed that the method can be executed with fewer resources than the other methods, while it offers high accuracy of tone mapping.

References

1. Reinhard, E., Stark, M., Shirley, P., Ferwerda, J.: Photographic tone reproduction for digital images. ACM Trans. Graph. **21**(3), 267–276 (2002)
2. Reinhard, E., Ward, G., Pattanaik, S., Debevec, P., Heidrich, W., Myszkowski, K.: High Dynamic Range Imaging - Acquisition, Display and Image based Lighting. Morgan Kaufmann, Burlington (2010)
3. Drago, F., Myszkowski, K., Annen, T., Chiba, N.: Adaptive logarithmic mapping for displaying high contrast scenes. Comput. Graph. Forum **22**(3), 419–426 (2003)
4. Fattal, R., Lischinski, D., Werman, M.: Gradient domain high dynamic range compression. ACM Trans. Graph. **21**(3), 249–256 (2002)
5. Iwahashi, M., Kiya, H.: Efficient lossless bit depth scalable coding for HDR images. In: Asia-Pacific Signal and Information Processing Association Annual Summit and Conference (APSIPA), no.OS.37-IVM.16-4 (2013)
6. Iwahashi, M., Kiya, H.: Two layer lossless coding of HDR images. In: Proceedings of the IEEE International Conference on Acoustics, Speech and Signal Processing (ICASSP), pp. 1340–1344 (2013)
7. Xu, R., Pattanaik, S.N., Hughes, C.E.: High-dynamic-range still image encoding in JPEG2000. IEEE Trans. Comput. Graph. Appl. **25**(6), 57–64 (2005)
8. Zhang, Y., Reinhard, E., Bull, D.: Perception-based high dynamic range video compression with optimal bit-depth transformation. In: Proceedings of the IEEE International Conference on Image Processing (ICIP), pp. 1321–1324 (2011)
9. Iwahashi, M., Yoshida, T., Mokhtar, N.B., Kiya, H.: Bit-depth scalable lossless coding for high dynamic range images. EURASIP J. Adv. Sig. Process. **2015**, 22 (2015)

10. Thakur, S.K., Sivasubramanian, M., Nallaperumal, K., Marappan, K., Vishwanath, N.: Fast tone mapping for high dynamic range images. In: Proceedings of the IEEE International Conference on Computational Intelligence and Computing Research (ICCIC), pp. 1–4 (2013)
11. Duan, J., Qiu, G: Fast tone mapping for high dynamic range images. In: Proceedings of the International Conference on Pattern Recognition (ICPR), pp. 847–850 (2004)
12. Murofushi, T., Iwahashi, M., Kiya, H.: An integer tone mapping operation for HDR images expressed in floating point data. In: Proceedings of the IEEE International Conference on Acoustics, Speech and Signal Processing (ICASSP), pp. 2479–2483 (2013)
13. Dobashi, T., Murofushi, T., Iwahashi, M., Kiya, H.: A fixed-point tone mapping operation for HDR images in the RGBE format. In: Proceedings of the Asia-Pacific Signal and Information Processing Association Annual Summit and Conference (APSIPA), no.OS.37-IVM.16-4 (2013)
14. Dobashi, T., Murofushi, T., Iwahashi, M., Kiya, H.: A fixed-point global tone mapping operation for HDR images in the RGBE format. IEICE Trans. Fundam. **E97–A**(11), 2147–2153 (2014)
15. Lampert, C.H., Wirjadi, O.: Anisotropic gaussian filtering using fixed point arithmetic. In: Proceedings of the IEEE International Conference on Image Processing (ICIP), pp. 1565–1568 (2006)
16. Chang, W.-H., Nguyen, T.Q.: On the fixed-point accuracy analysis of FFT algorithm. IEEE Trans. Sig. Process. **56**(10), 4673–4682 (2008)
17. Rocher, R., Menard, D., Scalart, P., Sentieys, O.: Analytical approach for numerical accuracy estimation of fixed-point systems based on smooth operations. IEEE Trans. Circ. Syst. Part-I **59**(10), 2326–2339 (2012)
18. Murofushi, T., Dobashi, T., Iwahashi, M., Kiya, H.: An integer tone mapping operation for HDR images in OpenEXR with denormalized numbers. In: Proceedings of the IEEE International Conference on Image Processing (ICIP), no.TEC-P10.6 (2014)
19. Dobashi, T., Tashiro, A., Iwahashi, M., Kiya, H.: A fixed-point implementation of tone mapping operation for HDR images expressed in floating-point format. APSIPA Trans. Sig. Inf. Process. **3**(11), 1–11 (2004)
20. Ward, G.: Real pixels. In: Arvo, J. (ed.) Graphic Gems 2, pp. 80–83. Academic Press, San Diego (1992)
21. Kainz, F., Bogart, R., Hess, D.: The OpenEXR image file format. In: ACM SIGGRAPH Technical Sketches & Applications (2003)
22. Information technology - Microprocessor Systems - Floating-Point arithmetic. ISO/IEC/IEEE 60559 (2011)
23. Wang, Z., Bovik, A.C., Seikh, H.R., Simoncelli, E.P.: Image quality assessment: from error visibility to structural similarity. IEEE Trans. Image Process. **13**(4), 600–612 (2004)

Implementation of Human Action Recognition System Using Multiple Kinect Sensors

Beom Kwon, Doyoung Kim, Junghwan Kim, Inwoong Lee, Jongyoo Kim,
Heeseok Oh, Haksub Kim, and Sanghoon Lee[✉]

Department of Electrical and Electronic Engineering, Yonsei University,
Seoul, Korea
{hsm260,tnyffx,junghwan.kim,mayddb100,jongky,angdre5,
khsphillip,slee}@yonsei.ac.kr
http://insight.yonsei.ac.kr/

Abstract. Human action recognition is an important research topic that has many potential applications such as video surveillance, human-computer interaction and virtual reality combat training. However, many researches of human action recognition have been performed in single camera system, and has low performance due to vulnerability to partial occlusion. In this paper, we propose a human action recognition system using multiple Kinect sensors to overcome the limitation of conventional single camera based human action recognition system. To test feasibility of the proposed system, we use the snapshot and temporal features which are extracted from three-dimensional (3D) skeleton data sequences, and apply the support vector machine (SVM) for classification of human action. The experiment results demonstrate the feasibility of the proposed system.

Keywords: Human action recognition · Multiple kinect sensors · Support vector machine

1 Introduction

Human action recognition is one of the actively researched topics in computer vision, because of its potential applications, including the areas of video surveillance, human-computer interaction, and virtual reality combat training. In the past few years, most studies on human action recognition assume that only a single camera is considered for recognizing human action [1–3]. However, it is impractical to apply the methods of [1–3] to human action recognition system because the above methods are vulnerable to partial occlusion due to fixed view point. In addition, the pose ambiguity problem in a two-dimensional (2D) image still remains.

For this reason, the authors in [4,5] proposed a new multi-view human action recognition method by using depth information. The authors in [6] proposed a RGB/depth/skeleton information based multi-view human action recognition

© Springer International Publishing Switzerland 2015
Y.-S. Ho et al. (Eds.): PCM 2015, Part I, LNCS 9314, pp. 334–343, 2015.
DOI: 10.1007/978-3-319-24075-6_32

method. However, it is still not suitable to apply these methods to practical human action recognition system due to its high complexity.

In order to alleviate the complexity problem of the conventional human action recognition method, instead of using depth information, we use three-dimensional (3D) skeleton data which are obtained from Kinect sensor in real-time. Many studies on multi-view human action recognition including [4–6] have been carried out by using Kinect sensor due to its convenience and cheap price. In addition, Kinect sensor provides real-time skeleton data of user without any usage of particular markers, which are attached to human body. However, Kinect sensor captures the user's position and movements under the assumption that the user faces Kinect sensor. Therefore, if the user does not face Kinect sensor, it may provide inaccurate position values. In addition, the inaccurate position values may lead to poor performance of the multi-view skeleton integration.

In order to obtain accurate skeleton data in a multi-view environment, the authors in [7] construct a silhouette of user by using depth images obtained from four Kinect sensors, and then they extract the skeletal representations of user from the silhouette. However, the high complexity of this method impedes its practical implementation. In [8], the integrated skeleton is obtained from four Kinect sensors by using an averaging method. In order to perform the integration, the authors in [8] select points for every joint, which satisfy the criterion of the average distance between them less than a given threshold value. Then, the point of the integrated skeleton is computed as the average of the selected points. However, it is impractical to apply this method to human action recognition system because of the low accuracy of the integrated skeleton. To improve accuracy, we in [9] proposed a weighted integration method. The proposed method enables skeleton to be obtained more accurately by assigning higher weights to skeletons captured by Kinect sensors in which the user faces forward [9].

In this paper, we propose a multi-view human action recognition system that utilizes 3D skeleton data of user to recognize user's actions. In the proposed system, six Kinect sensors are utilized to capture the whole human body and skeleton data obtained from six Kinect sensors are merged by using the multi-view integration method of [9]. In addition, snapshot features and temporal features are extracted from integrated 3D skeleton data, and then utilized as the input of the classifier. To classify the human actions, support vector machine (SVM) is employed as the classifier. The experimental results show that the proposed system achieves high accuracy, so it can be stated that the proposed system is feasible to recognize human actions in practical.

The reminder of this paper is organized as follows. In Sect. 2, we present the proposed human action recognition system. In Sect. 3, the experimental results are demonstrated. Finally, we conclude this paper in Sect. 4.

2 Proposed Human Action Recognition System

In this section, the proposed human action recognition system is explained. Figure 1 shows the block diagram of the proposed human action recognition

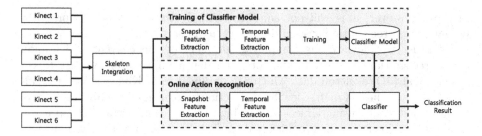

Fig. 1. Block diagram of the proposed human action recognition system.

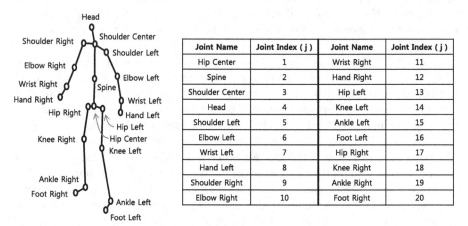

Joint Name	Joint Index (j)	Joint Name	Joint Index (j)
Hip Center	1	Wrist Right	11
Spine	2	Hand Right	12
Shoulder Center	3	Hip Left	13
Head	4	Knee Left	14
Shoulder Left	5	Ankle Left	15
Elbow Left	6	Foot Left	16
Wrist Left	7	Hip Right	17
Hand Left	8	Knee Right	18
Shoulder Right	9	Ankle Right	19
Elbow Right	10	Foot Right	20

Fig. 2. 3D skeleton model derived from Kinect sensor.

system. In the proposed system, 3D skeleton model derived from Kinect sensor is utilized for the recognition of human action. As shown in Fig. 2, this model is composed of a set of 20 joints and includes spatial coordinates information of each joint.

2.1   Multi-view Skeleton Integration

Figure 3 (a) gives the schematic illustration of the omnidirectional six Kinects system. In order to capture the whole human body, six Kinects sensors are arranged in a ring with a radius of 3 m at 60° intervals as shown in Fig. 3 (b). In addition, each Kinect sensor is placed at a height of 1 m above the ground by using a tripod.

Skeleton data captured with a Kinect sensor depends on the coordinate system of the Kinect sensor. Therefore, in order to integrate skeleton data obtained from six Kinect sensors, a camera matrix of each Kinect sensor must first be obtained through calibration. In addition, skeleton data obtained from six Kinect sensors must be transformed into a common coordinate system.

In order to obtain the camera matrix of each Kinect sensor, the calibration method of [10] is employed. Let (X_i, Y_i, Z_i) be a coordinate in the coordinate

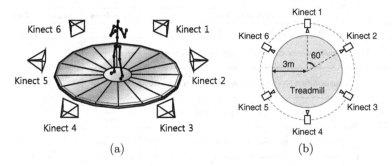

Fig. 3. Schematic illustration (a) and top view (b) of the omnidirectional six Kinects system.

system of ith Kinect sensor and (X_c, Y_c, Z_c) be a coordinate in the common coordinate system. The coordinate of each Kinect sensor is transformed into a common coordinate system as follows:

$$
\begin{bmatrix} X_c \\ Y_c \\ Z_c \end{bmatrix} = \mathbf{R} \begin{bmatrix} X_i \\ Y_i \\ Z_i \end{bmatrix} + \mathbf{T}, \tag{1}
$$

where $\mathbf{R}$ is the rotation matrix and $\mathbf{T}$ is the translation matrix.

Kinect sensor captures the user's position and movements under the assumption that the user faces Kinect sensor. Therefore, Kinect sensor may provide inaccurate position values when the user does not face Kinect sensor. In the omnidirectional six Kinects system, since it is impossible that the user faces all Kinect sensors, the inaccurate position values may lead to poor performance of multi-view skeleton integration. In order to improve performance of multi-view skeleton integration, the weighted integration method of [9] is employed.

2.2 Snapshot Feature Extraction

In the snapshot feature extraction step, various features (such as joint velocities, angles and angular velocities) are extracted from the integrated skeleton data. The features are calculated as follows:

- **Joint Velocities**: the joint velocity is computed from two consecutive records of joint positions and frames. The velocity of the joint j at the frame n can be calculated as follows:

$$
\mathbf{V}_j(n) = \frac{\mathbf{P}_j(n) - \mathbf{P}_j(n-1)}{\Delta n}, \tag{2}
$$

where $\mathbf{P}_j(n) = [x_j(n) \ \ y_j(n) \ \ z_j(n)]^T$ is a 3D coordinate vector expressing the position of the joint j at the frame n, the superscript $[\cdot]^T$ indicates the tranpose of a vector, and Δn is a interval of time between frame n and frame $(n-1)$. Since the frame rate of Kinect sensor is set to 30 fps, Δn is $1/30 = 33.33$ ms [11].

Notation	Kinematic Chain (index)	Euler Angle	Parent Joint Index
A_1	Hip Center (1) - Spine (2) - Shoulder Center (3)	Yaw	1
A_2	Shoulder Center (3) - Shoulder Left (5) - Elbow Left (6)	Roll	3
A_3		Yaw	
A_4	Shoulder Center (3) - Shoulder Right (9) - Elbow Right (10)	Roll	3
A_5		Yaw	
A_6	Shoulder Left (5) - Elbow Left (6) - Wrist Left (7)	Roll	5
A_7		Yaw	
A_8	Shoulder Right (9) - Elbow Right (10) - Wrist Right (11)	Roll	9
A_9		Yaw	
A_{10}	Hip Center (1) - Hip Left (13) - Knee Left (14)	Roll	1
A_{11}		Yaw	
A_{12}	Hip Center (1) - Hip Right (17) - Knee Right (18)	Roll	1
A_{13}		Yaw	
A_{14}	Hip Left (13) - Knee Left (14) - Ankle Left (15)	Roll	13
A_{15}		Yaw	
A_{16}	Hip Right (17) - Knee Right (18) - Ankle Right (19)	Roll	17
A_{17}		Yaw	

Fig. 4. Description of the angles derived from skeleton model.

- **Angles**: the angle is computed from the positions of three joints by using the method in [12] (see Chap. 5). The description of the angles derived from skeleton model is presented in Fig. 4, where A_k is a value of angle k.
- **Angular Velocities**: the angular velocity is computed from two consecutive records of angles and frames. The velocity of the angle k at the frame n can be calculated as follows:

$$W_k(n) = \frac{A_k(n) - A_k(n-1)}{\Delta n}. \qquad (3)$$

Through the snapshot feature extraction step, the feature vector for each frame contains, 94 float values (3×20 joint velocities, 17 angles, and 17 angular velocities).

2.3 Temporal Feature Extraction

In this step, in order to capture the temporal characteristics of human action, a buffer is used to store the snapshot features over L frames. In addition, by using the stored snapshot features, we calculate the following temporal features:

- **Average of Joint Velocities**: the average velocity of joint j at frame n can be calculated as follows:

$$\widehat{\mathbf{V}}_j(n) = \frac{1}{L} \sum_{l=n-(L-1)}^{n} \mathbf{V}_j(l). \qquad (4)$$

- **Average of Angles**: the average of angle k at frame n can be calculated as follows:

$$\widehat{A}_k(n) = \frac{1}{L} \sum_{l=n-(L-1)}^{n} A_k(l). \tag{5}$$

- **Average of Angular Velocities**: the average velocity of angle k at frame n can be calculated as follows:

$$\widehat{W}_k(n) = \frac{1}{L} \sum_{l=n-(L-1)}^{n} W_k(l). \tag{6}$$

Through the temporal feature extraction step, 94 float values (3×20 average of joint velocities, 17 average of angles, and 17 average of angular velocities) for each frame are added to the feature vector. Then, the input of the classifier is a vector of 188 float values including 94 snapshot features.

2.4 Classification

In this step, SVM with radial basis kernel is used to classify human actions. SVM constructs a maximal-margin hyperplane in a high dimensional feature space, by mapping the original features through a kernel function [13]. Then, by using the maximal-margin hyperplane, SVM classifies the features. In the next section, we evaluate the performance of the proposed system using SVM. In the experiment, we employ the multi-class SVM implemented in OpenCV library [14].

3 Experiment and Results

In the experiment, we test the feasibility of the proposed human action recognition system. Figure 5 shows our experiment environment. To test feasibility of the proposed system, we recorded a database containing 16 types human actions in four different scenarios. In scenario 1(2), user walks clockwise(counter-clockwise) around the semicircle. In scenario 3(4), user walks in a crouching posture clockwise(counter-clockwise) around the semicircle. Figure 6 shows the path used in each scenario. Figure 7 shows the type of human actions which are contained in our database. The database contains 8354 frames (1816 frames of scenario 1, 1827 frames of scenario 2, 2310 frames of scenario 3, and 2401 frames of scenario 4) for each Kinect sensor.

Figure 8 shows the results of our experiments about scenarios 1 and 2. The average accuracy in scenario 1 is 87.75 % and its performance varies between 80 % to 91 %. The average accuracy in scenario 2 is 89 % and its performance varies between 80 % to 98 %.

Figure 9 shows the results of our experiments about scenarios 3 and 4. The average accuracy in scenario 3 is 87 % and its performance varies between 84 % to 93 %. The average accuracy in scenario 4 is 90.75 % and its performance varies between 81 % to 96 %.

As shown in Figs. 8 and 9, the proposed human action recognition system achieves 88.625 % the average accuracy rate, so it can be stated that the proposed system is feasible to recognize human actions.

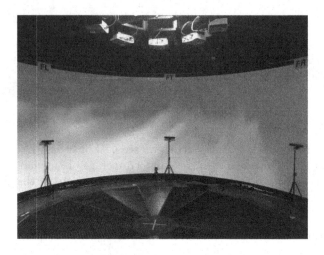

Fig. 5. A partial view of our experiment environment.

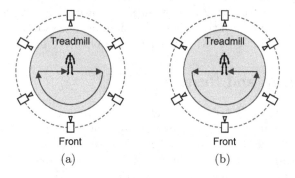

Fig. 6. The path used in the experiment. (a) Clockwise and (b) counter-clockwise.

Action Label	Action Name	Action Label	Action Name
1	Walking Forwarding	9	Crouch Walking Forward
2	Walking Backward	10	Crouch Walking Backward
3	Walking Left	11	Crouch Walking Left
4	Walking Right	12	Crouching Walking Right
5	Walking Forward Left	13	Crouching Walking Forward Left
6	Walking Forward Right	14	Crouching Walking Forward Right
7	Walking Backward Left	15	Crouching Walking Backward Left
8	Walking Backward Right	16	Crouching Walking Backward Right

Fig. 7. Description of the action labels and names.

	Walking Forward	Walking Backward	Walking Left	Walking Right	Walking Forward Left	Walking Forward Right	Walking Backward Left	Walking Backward Right
Walking Forward	91	0	0	1	0	7	1	0
Walking Backward	0	86	1	10	0	0	2	1
Walking Left	0	0	91	4	2	0	3	0
Walking Right	0	0	0	88	0	12	0	0
Walking Forward Left	4	0	1	1	90	3	1	0
Walking Forward Right	2	0	0	10	0	88	0	0
Walking Backward Left	0	5	4	3	0	0	88	0
Walking Backward Right	0	7	0	6	0	7	0	80

(a)

	Walking Forward	Walking Backward	Walking Left	Walking Right	Walking Forward Left	Walking Forward Right	Walking Backward Left	Walking Backward Right
Walking Forward	89	0	1	0	3	7	0	0
Walking Backward	0	85	0	0	0	2	10	3
Walking Left	0	0	96	0	0	1	3	0
Walking Right	0	0	0	90	0	3	0	7
Walking Forward Left	0	0	6	1	80	13	0	0
Walking Forward Right	8	0	0	1	1	90	0	0
Walking Backward Left	0	2	10	1	0	3	84	0
Walking Backward Right	0	2	0	0	0	0	0	98

(b)

Fig. 8. Confusion matrix for walking. (a) Clockwise and (b) counter-clockwise.

	Crouching Forward	Crouching Backward	Crouching Left	Crouching Right	Crouching Forward Left	Crouching Forward Right	Crouching Backward Left	Crouching Backward Right
Crouching Forward	84	1	0	0	7	8	0	0
Crouching Backward	0	85	0	0	0	0	9	6
Crouching Left	1	0	88	0	5	0	5	1
Crouching Right	2	0	0	92	1	2	1	2
Crouching Forward Left	7	0	6	2	85	0	0	0
Crouching Forward Right	0	0	2	3	1	93	0	1
Crouching Backward Left	0	7	7	0	2	0	84	0
Crouching Backward Right	0	7	0	4	0	4	0	85

(a)

	Crouching Forward	Crouching Backward	Crouching Left	Crouching Right	Crouching Forward Left	Crouching Forward Right	Crouching Backward Left	Crouching Backward Right
Crouching Forward	86	0	0	2	7	4	0	1
Crouching Backward	0	91	1	2	0	0	2	4
Crouching Left	1	0	95	1	2	0	1	0
Crouching Right	0	0	0	96	0	2	0	2
Crouching Forward Left	3	0	2	0	93	0	2	0
Crouching Forward Right	0	0	0	2	4	94	0	0
Crouching Backward Left	0	9	4	0	5	0	81	1
Crouching Backward Right	0	6	1	3	0	0	0	90

(b)

Fig. 9. Confusion matrix for walking in a crouching posture. (a) Clockwise and (b) counter-clockwise.

4   Conclusion

In this paper, we proposed a human action recognition system using multiple Kinect sensors. To integrate the multi-view skeleton data, a weighted integration method is used. For recognizing human action, we use joint velocities, angles and angular velocities as the snapshot features. In addition, in order to capture the temporal characteristics of human action, we use average of joint velocities, average of angles and average of angular velocities as the temporal features. We apply SVM to classify human actions. The experiment results demonstrate the feasibility of the proposed system.

Acknowledgments. This work was supported by the ICT R&D program of MSIP/IITP. [R0101-15-0168, Development of ODM-interactive Software Technology supporting Live-Virtual Soldier Exercises]

References

1. Lv, F., Nevatia R.: Single view human action recognition using key pose matching and viterbi path searching. In: Computer Vision and Pattern Recognition, IEEE (2007)
2. Liu, H., Li, L.: Human action recognition using maximum temporal inter-class dissimilarity. In: The Proceedings of the Second International Conference on Communications, Signal Processing, and Systems, pp. 961–969. Springer International Publishing (2014)
3. Papadopoulos, G.T., Axenopoulos, A., Daras, P.: Real-time skeleton-tracking-based human action recognition using kinect data. In: Gurrin, C., Hopfgartner, F., Hurst, W., Johansen, H., Lee, H., O'Connor, N. (eds.) MMM 2014, Part I. LNCS, vol. 8325, pp. 473–483. Springer, Heidelberg (2014)
4. Cheng, Z., Qin, L., Ye, Y., Huang, Q., Tian, Q.: Human daily action analysis with multi-view and color-depth data. In: Fusiello, A., Murino, V., Cucchiara, R. (eds.) ECCV 2012 Ws/Demos, Part II. LNCS, vol. 7584, pp. 52–61. Springer, Heidelberg (2012)
5. Ni, B., Wang, G., Moulin, P.: RGBD-HuDaAct: a color-depth video database for human daily activity recognition. In: Fossati, A., Gall, J., Grabner, H., Ren, X., Konolige, K. (eds.) Consumer Depth Cameras for Computer Vision, pp. 193–208. Springer, London (2013)
6. Liu, A.A., Xu, N., Su, Y.T., Lin, H., Hao, T., Yang, Z.X.: Single/multi-view human action recognition via regularized multi-task learning. Neurocomputing **151**, 544–553 (2015). Elsevier
7. Berger, K., Ruhl, K., Schroeder, Y., Bruemmer, C., Scholz, A., Magnor, M.A.: Markerless motion capture using multiple color-depth sensors. In: Vision Modeling, and Visualization, pp. 317–324 (2011)
8. Haller, E., Scarlat, G., Mocanu, I., Trăscău, M.: Human activity recognition based on multiple kinects. In: Botía, J.A., Álvarez-García, J.A., Fujinami, K., Barsocchi, P., Riedel, T. (eds.) EvAAL 2013. CCIS, vol. 386, pp. 48–59. Springer, Heidelberg (2013)
9. Junghwan, K., Inwoong, L., Jongyoo, K., Sanghoon, L.: Implementation of an omnidirectional human motion capture system using multiple kinect sensors. In: Computer Science and Engineering Conference, Transactions on Fundamentals of Electronics, Communications and Computer Sciences, IEICE (2015) (submitted)
10. Zhang, Z.: A flexible new technique for camera calibration. IEEE Trans. Pattern Anal. Mach. Intell. **22**(11), 1330–1334 (2000). IEEE

11. Parisi, G.I., Weber, C., Wermter, S.: Human action recognition with hierarchical growing neural gas learning. In: Wermter, S., Weber, C., Duch, W., Honkela, T., Koprinkova-Hristova, P., Magg, S., Palm, G., Villa, A.E.P. (eds.) ICANN 2014. LNCS, vol. 8681, pp. 89–96. Springer, Heidelberg (2014)
12. Caillette, F., Howard, T.: Real-time Markerless 3-D Human Body Tracking. University of Manchester (2006)
13. Castellani, U., Perina, A., Murino, V., Bellani, M., Rambaldelli, G., Tansella, M., Brambilla, P.: Brain morphometry by probabilistic latent semantic analysis. In: Jiang, T., Navab, N., Pluim, J.P.W., Viergever, M.A. (eds.) MICCAI 2010, Part II. LNCS, vol. 6362, pp. 177–184. Springer, Heidelberg (2010)
14. Support Vector Machines - OpenCV 2.4.9.0 documentation. http://docs.opencv.org/2.4.9/modules/ml/doc/support_vector_machines.html

Simplification of 3D Multichannel Sound System Based on Multizone Soundfield Reproduction

Bowei Fang[1,2], Xiaochen Wang[1,2(✉)], Song Wang[1,2], Ruimin Hu[1,2], Yuhong Yang[1,2], and Cheng Yang[1,3]

[1] National Engineering Research Center for Multimedia Software, School of Computer, Wuhan University, Wuhan, China
boweifun@sina.com,
{clowang,wangsongf117,hrml1964}@163.com,
ahka_yang@yeah.net,
yangcheng41506@126.com
[2] Research Institute of Wuhan University in Shenzhen, Shenzhen, China
[3] School of Physics and Electronic Science, Guizhou Normal University, Guiyang, China

Abstract. Home sound environments are becoming increasingly important to the entertainment and audio industries. Compared with single zone soundfield reproduction, 3D spatial multizone soundfield reproduction is a more complex and challenging problem with few loudspeakers. In this paper, we introduce a simplification method based on the Least-Squares sound pressure matching method, and two separated zones can be reproduced accurately. For NHK 22.2 system, fourteen kinds of loudspeaker arrangements from 22 to 8 channels are derived. Simulation results demonstrate the favorable performance for two zones soundfield reproduction, and subjective evaluation results show the soundfield of two heads can be reproduced perfectly until 10 channels, and 8-channel systems can keep low distortions at ears. Compared with Ando's multichannel conversion method by subjective evaluation, our proposed method is very close Ando's in terms of sound localization in the center zone, what's more, the performance of sound localization are improved significantly in the other zone of which position off the center.

Keywords: Simplification · NHK 22.2 system · Least-Squares method · 3D multizone soundfield reproduction

The research was supported by National Nature Science Foundation of China(No. 61201169); National High Technology Research and Development Program of China (863 Program) No. 2015AA016306; National Nature Science Foundation of China (No. 61231015,No. 61201340); the Science and Technology Plan Projects of Shenzhen No. ZDSYS2014050916575763; the Fundamental Research Funds for the Central Universities(No. 2042015kf0206).
The research was supported by Science and Technology Foundation of Guizhou Province (No. LKS [2011]1).

Y.-S. Ho et al. (Eds.): PCM 2015, Part I, LNCS 9314, pp. 344–353, 2015.
DOI: 10.1007/978-3-319-24075-6_33

1 Introduction

3D spatial soundfield reproduction enables enhanced immersive acoustic experience for listeners. On one hand, some methods which are used for spatial soundfield reproduction is well founded theoretically, such as the *higher order ambisonics* (HOA) [1], *wave field synthesis* (WFS) approach [2] and the *spherical harmonics-based systems* [3, 4]. But, most of these existing techniques in spatial soundfield reproduction mainly focus on a single zone. On the other hand, practical techniques for spatial reproduction have been developed further in parallel to the theoretical methods. It is worth mentioning that NHK 22.2 multichannel sound system (proposed by NHK laboratory of Japan) considers the perceptive characteristic of human, resulting in good sense of spatial sound impressions. However, the requirements for loudspeakers number make them impractical applied in home environment. Based on single region soundfield reproduction, some simplification methods had been proposed [5, 6]. Thus, in order to meet more people's demand for 3D audio in family at the same time, it's a critical problem to recreate 3D multizone soundfield with few loudspeakers.

1.1 Related Work

Typically, in 2011, Akio Ando proposed a M- to M'-channel conversion method [5] which maintained the physical properties of sound at the center listening point (i.e., the center of head) in the reproduced soundfield. Signals of 22.2 multichannel sound system without two low-frequency channels could be converted into those of 10-, 8- or 6-channel sound systems. Although subjective evaluation showed that this method could reproduce the spatial impression of the original 22-channel sound with 8 loudspeakers, only one person can enjoy 3D audio, what's more, unacceptable distortion will be generated at other listening points.

Some theory and methods which are used for 2D multizone soundfield reproduction had been proposed. In 2008, Poletti proposed a 2D multizone surround sound system using the Least-Squares pressure matching approach. It is considered the first published work in multizone soundfield reproduction [7]. However, the investigations are mainly performed based on simulation results; In 2011, T D. Abhayapala proposed some frameworks to recreate multiple 2D soundfield using the spatial harmonic coefficients translation theorem [8] at different locations within a single circular loudspeaker array. Both of which have made an greatly positive influence on 2D spatial multizone soundfield reproduction. However, realization of 3D multizone soundfield reproduction is a conceptually challenging problem.

The relation between the radius of reproduction soundfield and the number of loudspeakers had been introduced in literature [4], the larger region we recreate, the more loudspeakers we need. Thus, we choose to recreate two arbitrary separated zones instead of a contiguous large zone. The key to this issue is how to calculate the loudspeaker weight coefficients, which can minimize reproduction error between actual and desired reproduction soundfield. In this paper, we introduce a simplification method based on the Least-Squares sound pressure matching technique to find the optimized loudspeaker arrangements automatically. Derivative process, simulation and subjective evaluation results will be demonstrated in following part.

Notation: Throughout this paper, we use the following notations: matrices and vectors are represented by upper and lower bold face respectively, e.g., $\mathbf{H}$ and $\mathbf{p}$. the imaginary unit is denoted by i ($i = \sqrt{-1}$). The superscripts "d" and "a" are used to represent desired soundfield and actual reproduction soundfield respectively.

2 Problem Formulation

We denote soundfield reproduction inside a spherical region Ω, and consider a point source incident from the arbitrary direction, let there be an arbitrary point source a location $\mathbf{y} = (\upsilon, \psi, r)$, the soundfield produced at an arbitrary receiving point $\mathbf{x} = (\theta, \varphi, x) \in \Omega$ is given by

$$S(\mathbf{x}; k) = \frac{e^{ik\|\mathbf{y}-\mathbf{x}\|}}{\|\mathbf{y} - \mathbf{x}\|}, \tag{1}$$

where $k = 2\pi f c^{-1}$ is the wavenumber (with the speed of wave propagation and the frequency). Throughout this paper, we use k instead of f to represent frequency since we assume constant c. We assume no sound sources or scattering objects being present inside the reproduction area. We assume that c is independent of frequency, implying that the wavenumber is a constant multiple of frequency.

2.1 Multizone Soundfield Model

We assume that there are Q nonoverlapping spatial zones and corresponding desired spatial soundfield, as Fig. 1 shown. In the spherical coordinates system, the origin of $zone_q$ is denoted as O_q, which locates at $(\theta_q, \varphi_q, R_q)$ with respect to a global origin O_1. An arbitrary receiving point $\mathbf{x}$ within $zone_q$ is denoted as (θ, φ, x) with respect to the

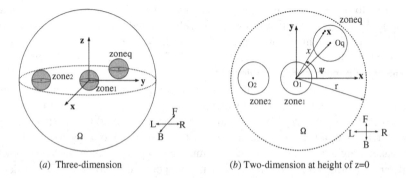

(a) Three-dimension   (b) Two-dimension at height of z=0

Fig. 1. Geometry of the multizone sound reproduction system. (*a*) and (*b*) show the position relation in three-dimension and two-dimension respectively.

global origin O. Loudspeakers are placed on a sphere with radius r outside the Ω, and assume that no sound sources or scattering objects being present inside the reproduction area.

2.2   Formulation of Simplification from *L*- to *(L-1)*-Channel

Let there be L loudspeakers placed outside the region Ω at locations $\mathbf{y}_l = (v_l, \psi_l, r_l), l = 1, \ldots, L$. Similarly, the soundfield produced by the *lth* loudspeaker at receiving point $\mathbf{x}$ is given by

$$T_l(\mathbf{x}; k) = \frac{e^{ik\left\|\mathbf{y}_l - \mathbf{x}\right\|}}{\|\mathbf{y}_l - \mathbf{x}\|}. \tag{2}$$

We denote that the *lth* loudspeaker weight coefficient is $\omega_l(k)$, desired reproduction soundfield produced by L loudspeakers is given by

$$S_L^d(\mathbf{x}; k) = \sum_{l=1}^{L} T_l(k) = \sum_{l=1}^{L} \frac{e^{ik\|\mathbf{y}_l - \mathbf{x}\|}}{\|\mathbf{y}_l - \mathbf{x}\|}. \tag{3}$$

Without the *jth* loudspeaker, actual reproduction soundfield produced by $(L-1)$ loudspeakers, we denote

$$S_{L-1}^a(\mathbf{x}; k) = \sum_{l=1, l \neq j}^{L-1} \omega_l(k) T_l(k) = \sum_{l=1, l \neq j}^{L-1} \omega_l(k) \frac{e^{ik\|\mathbf{y}_l - \mathbf{x}\|}}{\|\mathbf{y}_l - \mathbf{x}\|}, j = 1, \ldots, L. \tag{4}$$

The reproduction error of $(L-1)$ loudspeakers is given by

$$\varepsilon_j = \min \left\| S_L^d(\mathbf{x}; k) - S_{L-1}^a(\mathbf{x}; k) \right\|^2, j = 1, \ldots L. \tag{5}$$

On this basis, we could calculate the minimal squared error ε_j, which means that the *jth* loudspeaker has the lowest distortion to the whole reproduction soundfield among L loudspeakers. Then, we denote

$$\varepsilon_j = \min\{\varepsilon_1, \varepsilon_2, \ldots, \varepsilon_L\}. \tag{6}$$

2.3   Loudspeaker Weight Coefficients

We use Least-Squares method to solve the Eq. (5). The block diagram of Least-Squares sound pressure matching method is shown in Fig. 2, we select M receiving points in each zone. This system can be written in matrix notation as

$$\mathbf{Hw} = \mathbf{S} \tag{7}$$

where $\mathbf{H}$ is a QM by L matrix of sound pressures produced at the M matching points in each zone by the L loudspeakers, $\mathbf{w}$ is the L by 1 vector of loudspeaker weight coefficients and $\mathbf{S}$ is the QM by 1 desired vector of sound pressures produced by L loudspeakers at $\mathbf{y}_l = (\upsilon_l, \psi_l, r_l), l = 1, \cdots, L$.

Assuming that $QM > L$ and including a constraint on the maximum weight power with parameter λ produces the well known least squares error solution [9]

$$\mathbf{w} = \left[\mathbf{H}^H \mathbf{H} + \lambda \mathbf{I}\right]^{-1} \mathbf{H}^H \mathbf{S} \tag{8}$$

where H denotes the conjugate transpose and $\mathbf{I}$ is the L by L identity matrix. In practice, we used 600 randomly distributed points in each zone. In addition, these points had a z coordinate uniformly distributed over [–0.06 m, 0.06 m].

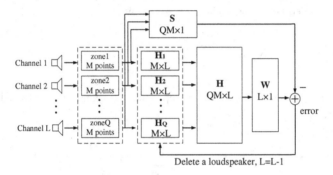

Fig. 2. Block diagram of Least-Squares sound pressure matching method including the calculation of loudspeaker weight coefficients and simplification method.

3   Simulation and Error Analysis

We define the reproduction error as the average squared difference between the entire desired field $\mathbf{S}^d(\mathbf{x}; k)$ and the entire corresponding reproduced field $\mathbf{S}^a(\mathbf{x}; k)$ over the desired reproduction area at different height.

$$\varepsilon = \frac{\int_0^r \int_0^{2\pi} \left| S^d(\mathbf{x}; k) - S^a(\mathbf{x}; k) \right|^2 d\phi x dx}{\int_0^r \int_0^{2\pi} \left| S^d(\mathbf{x}; k) \right|^2 d\phi x dx} \tag{9}$$

3.1   Simplification Results

We implement this proposed simplification method based on NHK 22.2 multichannel system without two low-frequency channels as shown in Fig. 2(*a*), 22 loudspeakers are placed on a sphere of radius $r = 2$ m, and consider two spatial reproduction zones of radius 0.085 m(due to the distance from each ear to the center of head is 0.085 m): $zone_1$ and $zone_2$, the origin O_1 of $zone_1$ locates at $(\pi/2, 0, 0)$, which is the center of Ω, and the origin O_2 of $zone_2$ locates at $(\pi/2, \pi, 0.4)$ is 0.4 m away from O_1, as shown in Fig. 1. The space consisted of these two zones can contain two adults to enjoy the 3D audio, whose heads are in the $zone_1$ and $zone_2$ respectively. We assume that every loudspeaker is a point source and soundfield resulting from a loudspeaker is a spherical wave, desired sound sources are 22-channel *white noise* and the desired soundfield is created by 22-channel system. We delete the selected loudspeaker one by one in the process of simplification.

Ando's method mainly focuses on recreating a center listening point, his simulation results [5] must be symmetrical. Because of the unsymmetrical structure of $zone_1$ and $zone_2$ results in fourteen kinds of unsymmetrical loudspeaker arrangements. The simulation results of 12-, 10-, 8-channel loudspeaker arrangements are shown in Fig. 3. We calculate and denote the reproduction error of $zone_1$, $zone_2$ and entire two zones as $\varepsilon_{zone_1}, \varepsilon_{zone_2}$ and ε. Three reproduction errors are all increased with the decreasing of the number of loudspeakers, as shown in Fig. 4, and we find that $zone_1$ is reproduced more accurately than $zone_2$. For 8-channel system, $\varepsilon_{zone_1} = 0.17\%$, $\varepsilon_{zone_2} = 4.16\%$ and $\varepsilon = 0.37$ %, which are agree very well with the expected error 4 % [4]. Thus, we choose

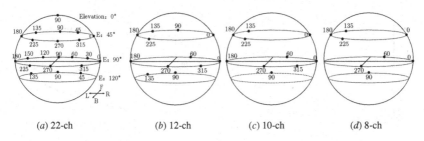

(a) 22-ch	(b) 12-ch	(c) 10-ch	(d) 8-ch

Fig. 3. The results of simplification. (a)–(d) show 22-,12-,10- and 8-channel loudspeaker arrangements respectively.

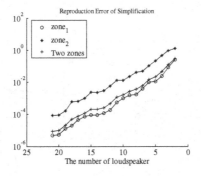

Fig. 4. Reproduction errors of simplification. Reproduction errors corresponding each loudspeaker arrangements from 22- to 2-channel at ears plane.

4 % as the threshold of each reproduction error, an appreciable distortions could be generated when the number of channels under 8.

3.2 Simulation Results and Comparison Analysis

For 10- and 8-channel, we make a comparison between proposed method and Ando's M- to M'-channel conversion method [5] at height of 0 m in terms of soundfield reproduction performance. Let the radius r of $zone_1$ and $zone_2$ range from 0.005 to 0.1 m, we calculate reproduction error corresponding to different radius. Figure 5(a) and (a') show 10-channel soundfield reproduction error in $zone_1$ and $zone_2$ respectively, "o" denotes proposed method and "*" denotes Ando's method. Figure 5(b) and (b') show same case of 8-channel system.

Reproduction error tends to increase with the increasing of radius r. For $zone_1$, Least-Squares method can solve out an global optimal solution to minimize reproduction over given reproduction zone, so reproduction error are increased slowly. Ando's method are mainly to maintain the physical properties of sound at the center listening point instead of a given region, so only the listening point is recreated

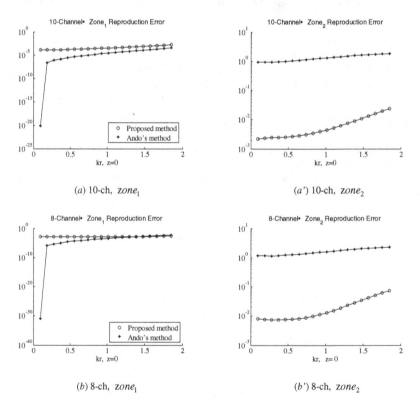

(a) 10-ch, $zone_1$ (a') 10-ch, $zone_2$

(b) 8-ch, $zone_1$ (b') 8-ch, $zone_2$

Fig. 5. Reproduction errors comparison. "o" denotes proposed method and "*" denotes Ando's method.

perfectly, but the reproduction error is closed to proposed method's at last. For $zone_2$, proposed method demonstrates better reproduction performance than Ando's method, the reproduction error of Ando's method even higher 100 %, which means a unacceptable distortion.

Reproduction soundfield are shown in Fig. 6. Figure 6(a) show the real part of desired soundfield produced by 22-channel system. In the following four figures, the left side (b) and (c) show the reproduced soundfield by proposed method corresponding 10- and 8-channel respectively. Correspondingly, Ando's method are show in the right

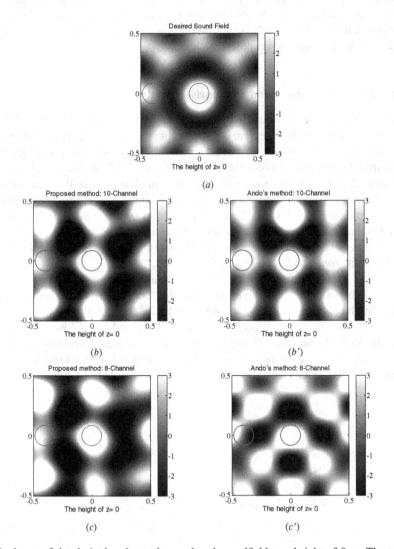

Fig. 6. Real part of the desired and actual reproduced soundfield at a height of 0 m. The two circles denote $zone_1$ and $zone_2$. (a) Desired soundfield produced by 22-channel system. Proposed method: (b) 10-channel reproduction soundfield; (b') 8-channel reproduction soundfield. Ando's method: (c) 10-channel reproduction soundfield; (c') 8-channel reproduction soundfield.

Table 1. Scales used for subjective evaluation

Comparison of the simulation	Much better	Slightly better	Better	The same	Slightly worse	Worse	Much worse
Score	3	2	1	0	−1	−2	−3

side (b') and (c'). The reproduction performance of $zone_1$ agree very well with the desired soundfield by both two methods. The reproduction performance of $zone_2$ is closed to the desired soundfield by proposed method, however, an unacceptable distortions are generated by Ando's method both in 10- and 8-channel. In conclusion, compared with Ando's method, the performance of reproduction are improved significantly in the $zone_2$ by proposed method.

3.3 Subjective Results and Comparison Analysis

Subjective experiments about 10- and 8-channel have been done by the RAB paradigm [10] to evaluate whether they are acceptable in subjective evaluation. Two stimuli (A and B) and a third reference (R) are presented. The reverberation time at 500 Hz in the soundproof room is 0.18 s and background noise is 30 dB. In this experiment, R is the original 22-channel white noise with 10 s, one simulation (A or B) is the sound of 10- (8-) channel of Ando's conversion method in [5] and the other one is the sound of 10- (8-) channel in Fig. 3 (c)−(d). Testers are asked to compare the difference between A and B relative to R in terms of sound localization and give scores according to the continuous seven-grade impairment scale shown in Table 1. Sound localization is evaluated in $zone_1$ and $zone_2$ respectively.

In this experiment, testers are 15 students all major in audio signal processing. Figure 7 shows the Comparative Mean Opinion Score (CMOS) of sound localization given by the subjects with 95 % confidence limits. The CMOS of sound localization reproduced by 10- and 8-channel system are −0.111 and −0.167 in $zone_1$, 1.3125 and 0.75 respectively in $zone_2$. It shows that sound localization performance of 10- and 8-channel systems reproduced in $zone_1$ by our method is very closed to Ando's conversion method, but the performance of our method is improved significantly in $zone_2$.

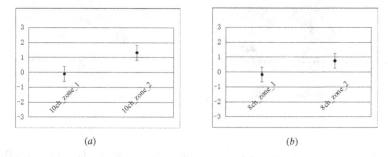

(a)   (b)

Fig. 7. Results of the Comparative Mean opinion score (CMOS) in subjective experiment of 10- and 8-channel between Ando's method and our proposed method. In terms of sound localization in the two zones, (a) 10-channel, (b) 8-channel.

4 Conclusion and Future Work

A practicable method of multichannel system simplification is proposed for reproducing 3D spatial multizone soundfield with few loudspeakers. Based on the Least-Squares sound pressure matching technique, two separated zones can be reproduced accurately. In future, we will explore high-efficiency method to recreate 3D multizone soundfield based on *spherical harmonics-based systems.*

References

1. Gupta, A., Abhayapala, T.D.: Three-dimensional soundfield reproduction using multiple circular loudspeaker arrays. IEEE Trans. Audio Speech Lang. Process. **19**, 1149–1159 (2011)
2. Spors, S., Ahrens, J.: Analysis and improvement of pre-equalization in 2.5-dimensional wave field synthesis. In: Audio Engineering Society Convention 128. Audio Engineering Society (2010)
3. Wu, Y.J., Abhayapala, T.D.: Theory and design of soundfield reproduction using continuous loudspeaker concept. IEEE, Trans. Audio Speech Lang. Process. **17**, 107–116 (2009)
4. Ward, D.B., Abhayapala, T.D.: Reproduction of a plane-wave sound field using an array of loudspeakers. IEEE Trans. Speech Audio Process. **9**, 697–707 (2007)
5. Ando, A.: Conversion of multichannel sound signal maintaining physical properties of sound in reproduced sound field. IEEE Trans. Audio Speech Lang. Process. **19**, 1467–1475 (2011)
6. Yang, S., Wang, X., Li, D., Hu, R., Tu, W.: Reduction of multichannel sound system based on spherical harmonics. In: Ooi, W.T., Snoek, C.G.M., Tan, H.K., Ho, C.-K., Huet, B., Ngo, C.-W. (eds.) PCM 2014. LNCS, vol. 8879, pp. 353–362. Springer, Heidelberg (2014)
7. Poletti, M.: An investigation of 2-D multizone surround sound systems. In: Audio Engineering Society Convention 125. Audio Engineering Society (2008)
8. Wu, Y.J., Abhayapala, T.D.: Spatial multizone soundfield reproduction: theory and design. IEEE Trans. Audio Speech Lang. Process. **19**, 1711–1720 (2011)
9. Sun, H., Yan, S., Svensson, U.P.: Optimal higher order ambisonics encoding with predefined constraints. IEEE Trans. Audio Speech Lang. Process. **20**(3), 742–754 (2012)
10. Recommendation, ITU-R BS. 1284-2: General methods for the subjective assessment of sound quality. International Telecommunications Union (2002)

Multi-channel Object-Based Spatial Parameter Compression Approach for 3D Audio

Cheng Yang[1,3], Ruimin Hu[1,2(✉)], Liuyue Su[1,2], Xiaochen Wang[1,2],
Maosheng Zhang[1,2], and Shenming Qu[1]

[1] National Engineering Research Center for Multimedia Software,
School of Computer, Wuhan University, Wuhan, China
yangcheng41506@126.com, {hrm1964,eterou}@163.com,
{yue.suliu,clowang}@gmail.com, qushenming@foxmail.com
[2] Research Institute of Wuhan University in Shenzhen, Shenzhen, China
[3] School of Physics and Electronic Science,
Guizhou Normal University, Guiyang, China

Abstract. To improve the spatial precision of three-dimensional (3D) audio, the bit rates of spatial parameters are increased sharply. This paper presents a spatial parameters compression approach to decrease the bit rates of spatial parameters for 3D audio. Based on spatial direction filtering and spatial side information clustering, new multi-channel object-based spatial parameters compression approach (MOSPCA) is presented, through which the spatial parameters of intra-frame different frequency bands belonging to the same sound source can be compressed to one spatial parameter. In an experiment it is shown that the compression ratio of spatial parameter can reach 7:1 compared with the 1.4:1 of MPEG Surround and S^3AC (spatial squeeze surround audio coding), while transparent spatial perception is maintained.

Keywords: 3D audio · Dirac · Spatial parameter compression

1 Introduction

In 2013, MPEG has launched a new standard (MPEG-H) and immersive 3D audio is included in the desired sense of audio envelopment, being able to virtualize sound sources at any accurate localization in terms of both direction and distance. With the increasing spatial precision of spatial audio, the bit rates of the spatial parameters are also increasing.

Binaural Cue Coding (BCC) is described in [1] to parameterize the sound image with inter-channel level difference (ICLD), inter-channel time difference (ICTD) and inter-channel coherence (ICC). The sound sources can be expressed in the front within 60° and the bit rates of these spatial parameters are only 3.5 kbps.

MPEG Surround is similar to BCC in expressing the sound image, but the directions of sound sources can be varied from 0° to 360° over the horizontal (2D) plane, and the bit rates of spatial parameters are increased to 14 kbps.

For providing listeners a more immersive feeling and a realistic sound scene within a three-dimensional (3D) space, spatial audio object coding (SAOC), directional audio

© Springer International Publishing Switzerland 2015
Y.-S. Ho et al. (Eds.): PCM 2015, Part I, LNCS 9314, pp. 354–364, 2015.
DOI: 10.1007/978-3-319-24075-6_34

coding (DirAC) [2] and spatial squeeze surround audio coding (S³AC) [3] are proposed. With the increasing spatial resolution in 3D space and the increasing number of channels or objects, the bit rates of spatial parameters are also increased sharply. The bit rates of spatial parameters are 18 kbps/object coding with Spatial Localization Quantization Points (SLQP) method, which presented in S³AC, and for 16 sound sources objects, the bit-rate of 288 kbps for spatial parameters is needed. Therefore, it is very urgent to reduce the bit rates of spatial parameters in 3D audio coding.

BCC, MPEG Surround, S³AC compression methods of spatial parameters consider the features between the adjacent frames, and the bit rates of spatial parameters can be decreased by differential coding. These methods can remove the inter-frame redundancy of spatial parameters between the adjacent frames in the same frequency band, but the intra-frame redundancy of spatial parameters between different frequencies bands of the same sound source still exist.

So, a new multi-channel object-based spatial parameter compression approach for 3D audio recording is proposed. When DirAC analysis is used to record and analyze the sound sources, basing on intra-frame different frequency bands of the same sound source have the same spatial parameters, the intra-frame redundancy of spatial parameters can be removed with higher compression ratio, which is not considered in the earlier spatial parameters compression methods.

2 Background

In this section, the background on the Directional audio coding (DirAC), the 3D audio spatial localization quantization method and the existing compression approaches of spatial parameters are introduced briefly.

2.1 Directional Audio Coding (DirAC)

Directional audio coding (DirAC) is a system for recording and reproducing spatial audio, based on B-format signals. Figure 1 gives a system illustration of the DirAC.

DirAC analysis is performed at a single measurement point based on an energetic analysis of the sound field, which are formulated from the particle velocity and the

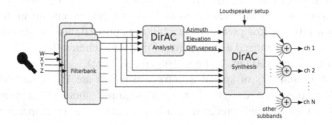

Fig. 1. Analysis and synthesis of DirAC in frequency bands

sound pressure. Both of these values can be derived from a B-format microphone signal (W, X, Y, Z). B-format microphone has four channels: omnidirectional and three figure-of-eight microphones organized orthogonally. The omnidirectional microphone signal is denoted as W. And the three figure-of-eight microphones are denoted as X, Y and Z.

In DirAC, a spatial microphone recording signal is analyzed in frequency domain to derive the sound field information including both localization (azimuth and elevation) and diffuseness information. This information is the side information for a downmix audio channel. DirAC estimates azimuth θ and elevation φ of sound source by analyzing the intensity relation between the 3D Cartesian coordinate axis. This estimation is carried out based on the signal's time-frequency representation.

$$\theta(n,f) = \arg\tan\left[\frac{-I_y(n,f)}{-I_x(n,f)}\right] \tag{1}$$

$$\varphi(n,f) = \arg\tan\left[\frac{-I_z(n,f)}{\sqrt{I_x^2(n,f) + I_y^2(n,f)}}\right] \tag{2}$$

Where n and f are time and frequency indices respectively. The intensity for each axis is denoted with $I_x(n,f)$, $I_y(n,f)$ and $I_z(n,f)$. The diffuseness information is denoted with $\psi(n,f)$.

When synthesizing, the directional sound source in the original sound field is rendered by panning the sound source to the location specified by DirAC directional cues, while the diffuseness cues are used to reproduce surround image with no perceptual localization feature.

2.2   3D Audio Spatial Localization Quantization Method

Different from SLQP [3], the paper's authors propose a new compression method in [5], which extracts the distance and direction information of sound sources as side information, for enhancing the spatial quality of multichannel 3D audio. A 3D audio spatial localization quantization method is designed to describe the azimuth, elevation and distance parameters as Fig. 2.

An azimuth precision of 2° (3° for low precision) is used for front area, and the azimuth precision is gradually decreased to 5° (7° for low precision) in the rear area. A 5° elevation (10° for low precision) resolution is utilized, and the distance r of the sound source can be quantized as the radius of different spheres (10 cm, 20 cm, 20 cm, 40 cm, 50 cm, 75 cm, 100 cm, 130 cm, 160 cm, 320 cm).

The resulting numbers of spatial quantization points are 16740 and 6330 for the high and low precision design. Hence, it requires about 14 kbps/object and 12.6 kbps/object for the high precision and low precision design respectively.

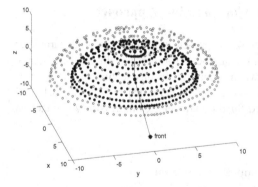

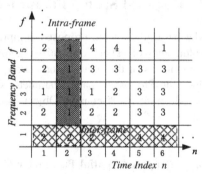

Fig. 2. 3D Audio spatial localization quantization points

Fig. 3. Time-frequency plane of the sound sources

2.3 The Existing Compression Approaches of Spatial Parameters

Binaural Cue Coding (BCC) is a method for multi-channel spatial rendering based on one down-mixed audio channel and side information. The side information includes ICLD, ICTD and ICC. The ICLDs and ICTDs are limited to ranges of ±18 dB and ±800 μs respectively. The ICC is limited to $0 \leq \Gamma_f \leq 1$. The authors of BCC used 7 quantization values for ICLDs and ICTDs respectively and 4 quantization values for ICCs [1]. When BCC is used in stereo coding, the quantization accuracies for ICLDs and ICTDs are 10° respectively, and the quantization accuracy for ICCs is 0.25.

As Fig. 3 shown, the quantizer indices are coded as follows. For each frame, two sets of quantizer index differences are computed. Firstly, for each frequency band f, the difference between the current frame (time index n) and the previous frame ($n-1$) quantizer index is computed. Secondly, for each frame n, the difference between the current frequency band (frequency index f) and the previous frequency band ($f-1$) quantizer index is computed. Both of these sets of index differences are coded with a Huffman code. For transmission, the set is chosen which uses less bits. In this approach, the compression ratio of spatial parameters is about 1.4:1.

The papers [3, 4] investigate the quantization of spatial parameters (azimuth θ and elevation φ) derived from S^3AC. Owing to the band perception property of the human auditory system, each frequency band can be assumed to represent a single sound source. Hence, it is expected that the location of the source varies smoothly over time, resulting in highly correlated cues. So, as Fig. 3 shown, the inter-frame redundancy remaining in the spatial parameters of one frequency band can be removed using frame-wise differential coding. In this approach, the difference between spatial parameter codebook indices derived for the same frequency band f between two adjacent frames is derived as $d_n^f = C_n^f - C_{n-1}^f$ $f = 1, 2, ..., 40; n = 2, 3, ..., N$.

Where C_n^f and C_{n-1}^f are the codebook indices for the f th frequency band of the nth and ($n-1$)th time frame. Here, N represents the differential prediction length. In this approach, the compression ratio of spatial parameters is about 1.4:1.

3   Proposed Spatial Parameter Compression Approach

The basic idea of the proposed compression approach is to reduce spatial parameter redundancy of 3D audio by utilizing spatial parameter similarity of different frequency components in the same sound source when recording. In this section, we will introduce the Proposed Spatial Parameter Compression Scheme [5] in Subsect. 3.1, and introduce the Multi-channel Object-based Spatial Parameter Compression Approach in Subsect. 3.2.

3.1   Proposed Spatial Parameter Compression Scheme

As Fig. 4 shown, a spatial parameter compression scheme of 3D audio is proposed.

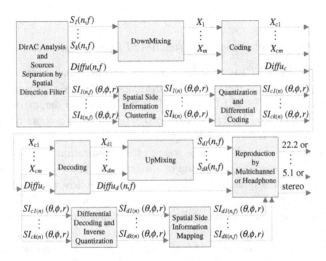

Fig. 4.   Spatial parameter compression scheme of MOSPCA

The inputs of coder are composed of the frequency-domain description of directional sound sources $(S_1(n,f),\dots,S_k(n,f))$, the spatial parameters of directional sound sources $\left(SI_{1(n,f)}(\theta,\varphi,r),\dots,SI_{k(n,f)}(\theta,\varphi,r)\right)$, and the frequency-domain description of diffuse sound source $Diffu(n,f)$. The proposed scheme can be summarized as follows:

- Firstly, we can get the separated signals $(S_1(n,f),\dots,S_k(n,f))$ and the corresponding spatial parameters side information $\left(SI_{1(n,f)}(\theta,\varphi,r),\dots,SI_{k(n,f)}(\theta,\varphi,r)\right)$ by DirAC analysis and spatial direction filtering [6, 7].
- On one hand, downmix k sound sources signals into m signals. And code and decode the m signals. By this way, we can compress the signals of sound sources through the sparsity of audio in frequency domain [8].
- On the other hand, we can remove the intra-frame redundancy of spatial parameters between different frequency bands in the same sound source through spatial side

information clustering. Then differential coding is utilized to remove the inter-frame redundancies of spatial parameters.

- Finally, we can reproduce 3D audio with directional signal $(S_{d1}(n,f),\ldots,S_{dk}(n,f))$, diffuse signal $Diffu_d(n,f)$ and spatial parameter $(SI_{d1(n,f)}(\theta,\varphi,r),\ldots, SI_{dk(n,f)}(\theta,\varphi,r))$.

3.2   Multi-channel Object-Based Spatial Parameter Compression Approach

When we use DirAC analysis to record and analyze the sound sources, the sound sources can be expressed in time-frequency plane as Fig. 5. We can see that there are different frequency bands belonging to the same sound source in the same frame. These different frequency bands have the same spatial parameters, though some few differences would exist between the spatial parameters of them when we record the sound sources. There are many intra-frame redundancies of spatial parameters between these different frequency bands. So, we can express the spatial parameters of different frequency bands, belonging to the same sound source, with one spatial parameter.

For example, when $n = 2$(the second frame) and $f = 1, 2, 3, 4, 5$(in order to describe easily, we only analyses frequency band from the first to the fifth as Fig. 5 shown), there are five spatial parameters, which can be express by $SI_{k(n,f)}(\theta,\varphi,r)$ as $(SI_{2(2,1)}(\theta_2, \varphi_2, r_2), SI_{1(2,2)}(\theta_1, \varphi_1, r_1), SI_{1(2,3)}(\theta_1, \varphi_1, r_1), SI_{1(2,4)}(\theta_1, \varphi_1, r_1), SI_{4(2,5)}(\theta_4, \varphi_4, r_4))$. We can see that $SI_{1(2,2)}(\theta_1, \varphi_1, r_1)$, $SI_{1(2,3)}(\theta_1, \varphi_1, r_1)$ and $SI_{1(2,4)}(\theta_1, \varphi_1, r_1)$ are belonged to the same sound source ($k = 1$, the first sound source), and they are all in the second frame. So their spatial parameter are equal to each other, which mean that their spatial parameter is $(\theta_1, \varphi_1, r_1)$. So, we can use one spatial parameter to express three spatial parameters by intra-frame compression.

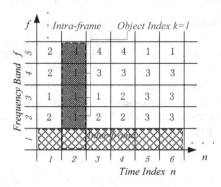

Fig. 5. Time-frequency plane of the sound sources

The intra-frame compression is composed of spatial direction filtering and spatial side information clustering.

Spatial direction filtering includes two steps. The first step is that sound source locations can be obtained by using triangulation based on the DOA estimations in the

3D space when record the sound sources by B-Format array [5]. The second step is that the sound sources are separated by the locations of the sound sources.

Spatial side information clustering is that there is a small measurement error between the spatial parameters of intra-frame different frequency bands belonging to the same sound source, but those spatial parameters of different frequency bands belonging to the same sound source in the same frame can be express with one spatial parameter.

For the spatial parameters of the sound sources changing continuously, there are many inter-frame redundancies of spatial parameters between the adjacent frames in the same frequency band. So we can remove the inter-frame redundancy by differential coding.

4   Performance Evaluation

4.1   Objective Quality Evaluation

The objective quality evaluation was implemented through two methods as follows. One method was that evaluates the decreased bit rate of spatial parameters compared with BCC, MPEG Surround, and S^3AC, and the other method was that calculates the quantization error of spatial parameters after coding.

Our compression method of spatial parameter was similar to BCC, MPEG Surround, and S^3AC in removing the redundancies between the adjacent frames (inter-frame). But our compression method was different from them in removing the redundancies between the frequency bands in the same frame (intra-frame) as shown in Table 1.

Table 1. Compression Feature of Spatial Parameters

	Compression feature of spatial parameters	Compression ratios
BCC, MPEG surround	Inter-frame compression by differential coding or intra-frame compression by differential coding	1.4:1
S^3AC	Inter-frame compression by differential coding	1.4:1
MOSPCA	Inter-frame compression by differential coding and intra-frame compression by spatial side information clustering	7:1

As shown in Fig. 6, the compression ratio of MOSPCA was $p/k \times 1.4$. Where p is the number of frequency bands ($p = 40$), k is the number of sound sources, p/k is intra-frame compression ratio of MOSPCA, and 1.4:1 is the inter-frame compression ratio of MOSPCA as same as the compression ratio of S^3AC. For example, when $p = 40$ and $k = 8$, the compression ratio was 7:1. The compression ratio and the quality of 3D audio decreased as the number of the sound sources increased, so we can control the quality of 3D audio by m (the number of downmixing objects as shown in Fig. 4).

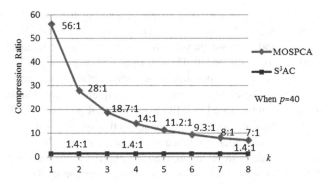

Fig. 6. Compression ratio of spatial parameters

Table 2. Quantization error

	Quantization precision	Quantization error
Azimuth	$2° \sim 5°$	$1° \sim 2.5°$
Elevation	$5°$	$2.5°$
Distance (<=160 cm)	10 cm $\sim$ 30 cm	5 cm $\sim$ 15 cm

The quantization error of spatial parameters for high precision after coding was listed in Table 2. For human auditory system have approximately 1° azimuth resolution and 5° to 10° elevation resolutions respectively [9], the quantization errors of azimuth and elevation cannot be perceived by human. The quantization errors of distance are proved that they also could not be perceived by human through our subjective experiment as below.

4.2   Subjective Quality Evaluation

Eleven persons participated in subjective test. And four test sequences were used to evaluate the performance of the proposed spatial parameters compression method. They were the direction-speech sequence, the distance-speech sequence, the direction-instrument sequence, and the distance-instrument sequence. The durations of the sequences were 7 s respectively. We used two metadata sequences from MPEG to produce the sequences using two different methods. We used the PKU&IOA HRTF database (published in 2013) [10] to produce reproduction sequences with spatial parameters after coding. This database includes the measurements of the proximal region and the distal region, which have approximately 5° azimuth resolution and 10° elevation resolutions respectively. And the distance resolution of the database is about 10 cm $\sim$ 30 cm, which includes the measurement distance of 10 cm, 20 cm, 30 cm, 40 cm, 50 cm, 75 cm, 100 cm, 130 cm and 160 cm.

The first method was to produce four reproduction reference sequences (named as Record-S). We used KEMAR manikin to record two speech/instrument metadata sequences through playing one metadata sequence moving from left 45° to right 45° and at the same time playing the other sequence moving from right 45° to the left 45° (distance = 100 cm), and then we played two speech/instrument metadata sequences moving from 160 cm to 20 cm at left 45° and right 45° respectively.

The second method was to produce four reproduction sequences with HRTF (named as HRTF-S). These four reproduction sequences are similar to Rec-S sequences, but they are reproduced by HRTF with spatial parameters after coding.

MUSHRA test method is used to evaluate subjective quality. Other two sequences were needed, one was the hidden reference sequence (named as Ref-S) which was as same as Record-S but hidden in all test sequences, and the other was anchor sequence (named as Ref-S-3.5 K) which was a low-pass filtered version of Ref-S.

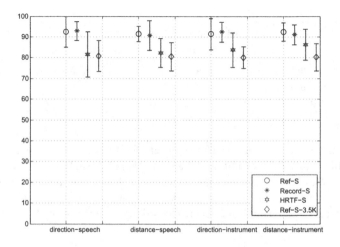

Fig. 7. Subjective evaluation with 95 % confidence interval

In Fig. 7, the direction and distance perception of the HRTF-S sequences (created by spatial parameters after coding) and Record-S sequences was similar to each other, and the Record-S sequences still had a little better quality of direction and distance perception than the HRTF-S sequences. Which mean that the spatial quality 3D audio using our compression method was similar to the original 3D audio recording by KEMAR manikin. The HRTF-S test sequences contained moving sources (including the moving of direction and distance), but we hardly perceive 'jumps' in the location of the source. Which mean that the quantization errors could not be perceived.

Because our method of inter-frame and intra-frame compression was lossless compression and the quantization errors could not be perceived by human, the few different with the original 3D audio was mainly because that HRTF was measured in anechoic chamber and our original 3D audio was recorded in ordinary room. If our HRTF-S sequences add reverb effect of our ordinary room, the few different with the original 3D audio will be narrower.

Therefore, the compression ratio of spatial parameters can reach to 7:1 (when $p = 40$, $k = 8$) comparing to 1.4:1 of BCC and S^3AC, while the spatial perceptual quality of original 3D audio could be still maintained.

5 Conclusions

This paper has presented a Multi-channel Object-based Spatial Parameter Compression Approach (MOSPCA) for 3D audio based on DirAC. Different from other spatial parameter compression approaches, this approach proposed intra-frame compression, which is not taken account in the prior spatial parameter compression approaches. Based on intra-frame different frequency bands of the same sound source have the same spatial parameter, the spatial parameters of different frequency bands can be compressed to one spatial parameter. The results indicate that, when DirAC analysis is used to record and analyze the sound sources, MOSPCA compression ratios of spatial parameters can reach 7:1 (when number of frequency bands $p = 40$ and number of sound sources $k = 8$), while the spatial perceptual quality of original 3D audio could be still maintained.

Acknowledgement. This work is supported by National High Technology Research and Development Program of China (863 Program, No. 2015AA016306), National Nature Science Foundation of China (No. 61231015, 61102127, 61201340, 61201169, 61471271, U1404618), Science and Technology Plan Projects of Shenzhen (No. ZDSYS2014050916575763), Guangdong-Hongkong Key Domain Breakthrough Project of China (No. 2012A090200007), Science and Technology Foundation of Guizhou Province (No. LKS[2011]1).

References

1. Faller, C., Baumgarte, F.: Binaural cue coding-part II: schemes and applications. IEEE Trans. Speech Audio Process. **11**(6), 520–531 (2003)
2. Pulkki, V.: Spatial sound reproduction with directional audio coding. J. Audio Eng. Soc. **55**(6), 503–516 (2007)
3. Cheng, B., Ritz, C., Burnett, I., et al.: A general compression approach to multi-channel three-dimensional audio. IEEE Trans. Audio Speech Lang. Process. **21**(8), 1676–1688 (2013)
4. Cheng, B., Ritz, C.H., Burnett, I.S.: Psychoacoustic-based quantisation of spatial audio cues. Electron. Lett. **44**(18), 1098–1099 (2008)
5. Yang, C., Hu, R., et al.: A 3D audio coding technique based on extracting the distance parameter. In: IEEE International Conference on Multimedia and Expo (ICME) (2014)
6. Kallinger, M., Del, Galdo G., Kuech, F., Mahne, D., Schultz-Amling, R.: Spatial filtering using directional audio coding parameters. In: IEEE International Conference on Acoustics, Speech and Signal Processing (ICASSP), pp. 217–220 (2009)
7. Zheng, X.G., Ritz, C.H., Xi, J.T.: Collaborative blind source separation using location informed spatial microphones. IEEE Signal Proc. Lett. **20**(1), 83–86 (2013)
8. Gorlow, S., Marchand, S.: Informed audio source separation using linearly constrained spatial filters. IEEE Trans. Audio Speech Lang. Process. **21**(1), 3–13 (2013)

9. Blauert, J.: Spatial Hearing: The Psychophysics of Human Sound Localization. MIT Press, Cambridge (1997)

10. Qu, T.S., Xiao, Z., Gong, M., Huang, Y., Li, X.D., Wu, X.H.: Distance-dependent head-related transfer functions measured with high spatial resolution using a spark gap. IEEE Trans. Audio Speech Lang. Process. **17**(6), 1124–1132 (2009)

A FPGA Based High-Speed Binocular Active Vision System for Tracking Circle-Shaped Target

Zhengyang Du[1], Hong Lu[1(✉)], Haowei Yuan[3], Wenqiang Zhang[2], Chen Chen[2], and Kongye Xie[2]

[1] Shanghai Key Laboratory of Intelligent Information Processing,
Fudan University, Shanghai 200433, China
honglu@fudan.edu.cn
[2] Shanghai Engineering Research Center for Video Technology and System
School of Computer Science, Fudan University,
Shanghai, People's Republic of China
[3] Shanghai Electric Group CO., LTD. Central Academe, Shanghai, China

Abstract. With the development of digital image processing technology, computer vision technology has been widely used in various areas. Active vision is one of the main research fields in computer vision and can be used in different scenes, such as airports, ball games, and so on. FPGA (Field Programmable Gate Array) is widely used in computer vision field for its high speed and the ability to process a great amount of data. In this paper, a novel FPGA based high-speed binocular active vision system for tracking circle-shaped target is introduced. Specifically, our active vision system includes three parts: target tracking, coordinate transformation, and pan-tilt control. The system can handle 1000 successive frames in 1 s, track and keep the target at the center of the image for attention.

Keywords: Binocular vision · Active vision system · Target tracking · FPGA

1 Introduction

With the high-speed development of digital image process technology, computer vision technology has been widely used in various areas. Active vision is one of the main research fields in computer vision and can be used in different scenes, such as airports, ball games, and so on.

Active vision [1] was proposed in 1980s. An active observer is the most important part of active vision. An observer is called active when it is engaged in some kind of activities whose purpose is to control the geometric parameters of the sensory apparatus. In the field of computer vision, pan-tilt control is the most common method to make the camera active. Early active vision systems [2, 3] were successively proposed in 1990s. With the high-speed development of hardware and image processing technology, researches on active vision system are still very hot.

Vision system can be divided into monocular vision [4] and multi vision [5]. By using multiple cameras, the 3-D information of the target can be reconstructed from the

© Springer International Publishing Switzerland 2015
Y.-S. Ho et al. (Eds.): PCM 2015, Part I, LNCS 9314, pp. 365–374, 2015.
DOI: 10.1007/978-3-319-24075-6_35

images. However, 3-D reconstruction is also available in monocular vision if some extra knowledge is involved [6].

Most active vision projects relate to robot head [2, 7], driver assistance [8], and target tracking. In the applications of robot head and driver assistance, a simple real-time (30 fps) active vision system may meet the demand. But in some situations of target tracking, a high-speed (1000 fps) active vision system [9, 10] may be necessary, especially when the target is moving at a high speed. When tracking target at such speed, color based features are widely used in most systems. But color based features may involve large noise if the background is complex or the light environment is not good enough.

Nowadays, lots of applications involve complex background. By using color based methods, it is difficult to meet the demand. Thus we take into consideration shape based methods, which may fit the complex background. And circle is one of the most commonly used shapes in our daily life. An active vision system for tracking circle-shaped target is implemented in this paper. It can be used in lots of situations, such as ball games, pupil detection, etc.

Time efficiency is one of the most important parts of a vision system. To improve time efficiency, FPGA (Field Programmable Gate Array) is a more suitable platform for image processing than PC because of its parallel structure. With the efforts of corresponding manufacturers, FPGA becomes faster, cheaper, and widely used in computer vision field [11, 12].

In this paper, a novel FPGA based active vision system is introduced. The system can track circle-shaped target in the complex background at the speed of 1000 fps.

2   Active Vision System

In following sections, a binocular vision system which is used to track circle-shaped target in complex background is presented. A complete binocular active vision system includes 2 cameras, 2 pan-tilts, and a processing software running on a FPGA and a ARM board. The structure of the system is shown as Fig. 1.

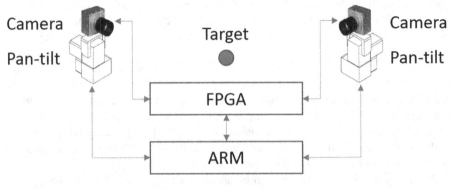

Fig. 1. Structure of our system

Images of the target are transferred from camera to the processing system. The processing system consists of 4 parts: preprocessing, target tracking, 3D localization and pan-tilt control. The overview of the processing system is shown in Fig. 2. In the preprocessing part, the format of the data will be transformed from Bayer to RGB. And the ROI (region of interest) of the current frame will be selected. The selection method of ROI will be introduced in following sections. In the target tracking part, gradients computation and circle fitting methods are used to locate the target in images. Both preprocessing part and tracking part are running on FPGA in parallel. On the ARM board, the image coordinate of the target is transformed to the world coordinate. Then the controlling parameters are computed and sent to the pan-tilt to make the camera fixating at the target.

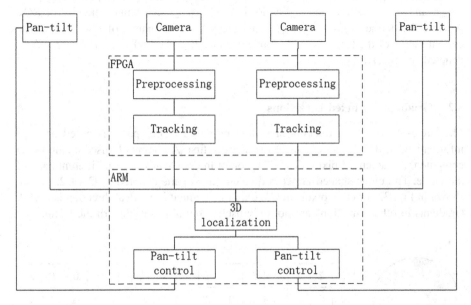

Fig. 2. Overview of the processing system

The whole process is running at the speed of 1000 fps. It means that each part of the process should be finished in 1 ms.

3   Target Tracking

In this section, the circle-shaped target tracking algorithm is presented. The tracking process can be divided into 3 parts. First, by analyzing prior knowledge, the variation of the target's size and position in current frame can be computed. Then for each pixel in ROI (region of interest), its gradients in 4 fixed directions are computed. And the gradients are used to fit the circle. Also some improving methods is presented below.

3.1 Prior Knowledge

According to the whole tracking system, we need to obtain some prior knowledge first. For example, size of the circle-shaped target, the longest and the shortest distance between the target and the camera. Such values can be artificially estimated or measured in specific scene. Through camera calibration, we can also obtain the intrinsic matrix of the camera.

According to the above information, some further knowledge can be easily obtained, such as the range of the target size in image, the maximum variation of the sizes in two adjacent frames, and the maximum variation of the positions in two adjacent frames. So that we can easily obtain candidate target region in each frame.

Assume the target position of the former frame as $P_{pre}(x_{pre}, y_{pre})$, and the radius as r. The current position of the target is in the rectangle with the endpoints of $P_{pre} - \Delta P_{max}$ and $P_{pre} + \Delta P_{max}$, denoted as ROI, where ΔP_{max} determines the size of ROI. Assume the boundary as $[R_{min}, R_{max}]$, and the maximum change of size as Δr . The current radius of the target is in the range of $[\max(R_{min}, r - \Delta r), \quad \min(R_{max}, r + \Delta r)]$, denoted as $[r_{min}, r_{max}]$.

3.2 Gradients in Fixed Directions

The circle-shaped target tracking algorithm proposed in this paper is based on edge information, so the edges need to be extracted at first. A commonly used method is to compute the gradient of pixels. The direction of the gradient varies in different parts of the circle. The circle-shaped target is divided into 8 parts denoted as C1, C2, ..., C8 shown in Fig. 3. For edge pixels in C1, only the gradient in vertical direction is useful. Gradients in other directions are not only useless but also will disturb the result.

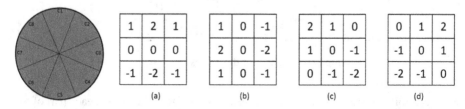

Fig. 3. Division of a circle and gradient operators in different directions

In this paper, gradients in fixed directions are computed for target localization. Considering the time efficiency and complexity of the operator, 4 fixed directions are selected. The operators are shown in Fig. 3. Operator (a) is used in parts C1 and C5, (b) is used in parts C3 and C7, (c) is used in parts C4 and C8, and (d) is used in parts C2 and C6. We have also considered 5 * 5 operators, which perform almost the same as the ones we present above and cost much more time.

According to the prior knowledge, the range of position ROI$[P_{pre} - \Delta P_{max}, P_{pre} + \Delta P_{max}]$ and the range of radius $[r_{min}, r_{max}]$ in current frame are already known.

Obviously, pixels in ROI$'$ $\left[P_{pre} - \Delta P_{max} - (r_{max}, r_{max}), P_{pre} + \Delta P_{max} + (r_{max}, r_{max})\right]$ are involved. Gradients in 4 fixed directions of each pixel in the rectangle ROI$'$ need to be computed.

3.3   Circle Fitting

The circle-shaped target tracking algorithm proposed in this paper finds out the most possible position of target by fitting the circle shape. According to the prior knowledge, the ranges of position and radius are already known. Enumerate different center positions and radii in the range, each group of center and radius determines a circle. Let the average of gradients in fixed direction of pixels on the circumference be the weight of a circle. And find out the circle with the largest weight to be the localization result. The implementation details are as follows.

- Enumerate the center p $\in$ ROI and the radius r $\in$ $[r_{min}, r_{max}]$, which determine a circle.
- Find the pixels on the circumference by using graphics methods, the set of pixels is denoted as $C = p_1, p_2, \ldots, p_N$.
- Compute the weight of each circle $W_{p,r} = \frac{1}{N}\sum_{i=1}^{N} ogd(p_i, dir_i)$.
- Find out the largest $W_{p,r}$, and regard the corresponding circle as the localization result.

Look-up table can be used to improve time efficiency. dir_i represents which part the pixel is belong to. $d(p_i, dir_i)$ represents the gradient in fixed direction of pixel p_i. Log function is used to reduce the influence of noise.

The algorithm proposed in this paper is similar to the inverse process of Hough transformation, and it can achieve higher time efficiency in tracking problems. The performance of tracking is shown in Fig. 4.

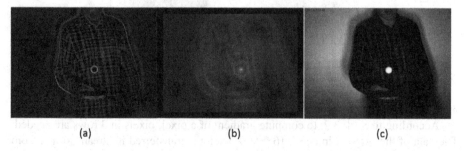

     (a)   (b)   (c)

Fig. 4. Target tracking process: (a) gradient image, (b) circle fitting result, (c) tracking result.

3.4   Improvement

The circle fitting algorithm can also be implemented with progressive accuracy to improve time efficiency. According to the enumerate method mentioned above, the following improvement is available.

First, enumerate the position with the interval of n pixels (n may be 2 or 3, if n is larger than 3, it involves large noise), find out the fuzzy result with the accuracy of n pixels. Then enumerate the position with the interval of 1 pixel near the fuzzy result to find the pixel-level result. By using this method, the enumeration progress is much faster than before.

Also pruning method can be used, 10 % pixels on the circumference are used to quickly compute a fuzzy result. If the fuzzy result is less than a threshold, the process will skip to the next circle. In case the background is not very complex, the pruning method will perform well in improving time efficiency.

4 FPGA Implementation

We implement our system on a programmable board which integrates both FPGA and ARM. The FPGA play the role of computation center while the ARM play the role of center of control. The structure is shown in Fig. 5.

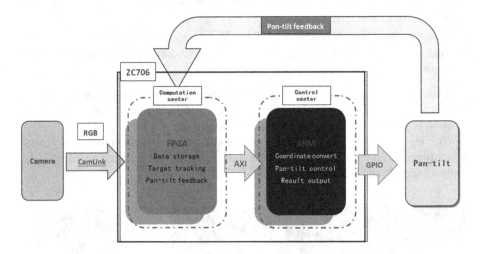

Fig. 5. Structure of our system

The target tracking algorithm, which is the main part of our system, is implemented in FPGA. The FPGA implementation details is presented in this section.

According to Sect. 3.2, to compute gradient of a pixel, pixels in 3 rows are needed. The data of the original image (816 * 600 pixels) is transferred in stream format from camera, so that the original data can be stored in a 4-rows circular array. When finish reading row k of the original image, we can use row $(k - 2)$ %4, row $(k - 1)$ %4 and row k % 4 of the circular array to compute the gradients of row $k - 1$ of the image. At the same time, FPGA is also reading row $k + 1$ of the original image in parallel. So that the gradients computation can be finished when reading original image.

According to Sect. 3.1, only gradients in ROI are needed to be computed and stored. Thus the circle fitting part may start before the whole image has been read. We fit the circle row by row. When we compute the fitting value of row k with radius r, we only need gradient data before row $k + r$. The gradients are stored in a dual port RAM so that the circle fitting part is able to read gradients in parallel.

5   3-D Localization and Pan-Tilt Control

In this section, the world coordinate of target will be computed, and the target statement of pan-tilt, which makes the camera fixating at the target, need to be obtained and sent to the pan-tilt.

In order to compute the world coordinate, some additional values are necessary. First we need the intrinsic matrix of the camera, which can be obtained through camera calibration. Through hand-eye calibration, we can obtain the relationship between pan-tilt and camera. Furthermore, we need the real-time statement of pan-tilt by reading feedback value from the motor.

If the world coordinate is known, it is easy for us to compute the corresponding image coordinate with the values given above:

$$
k \begin{bmatrix} x \\ y \\ 1 \end{bmatrix} = DCBA \begin{bmatrix} X \\ Y \\ Z \\ 1 \end{bmatrix} \tag{1}
$$

Where A is the relationship between world coordinate and the pan-tilt coordinate, B is the rotation matrix of pan-tilt, C is the relationship between pan-tilt and camera, which comes from the hand-eye calibration result, and D is the intrinsic matrix of the camera.

To compute the world coordinate through image coordinate, we should combine the Eq. (1) of both cameras. In the equation set, the world coordinate is the only unknown value. Eliminate the factor k and expand the equation, then we have:

$$
\begin{cases} a_1X + b_1Y + c_1Z + d_1 = x_1 \\ a_2X + b_2Y + c_2Z + d_2 = y_1 \\ a_3X + b_3Y + c_3Z + d_3 = x_2 \\ a_4X + b_4Y + c_4Z + d_4 = y_2 \end{cases} \tag{2}
$$

With the least square method, we can solve the equation set and obtain the world coordinate. Through the hand-eye calibration result and coordinate computed above, it is easy to form the equation to find out the statement that make the camera aiming at the target. And send the statement to both pan-tilt at the same time. So that the system is able to track actively, and the target will always present at the middle of the image.

6   Experiment

6.1   Hardware Environment

2 high-speed cameras and 2 self-made pan-tilt are used in our system. Our processing system is running on Xilinx Zynq-7000 All Programmable SoC ZC706 board, which

integrates both FPGA and ARM. The baud rate of the pan-tilt is 230400 bps. Each instruction consists of 4 Bytes. Thus it can be transferred to pan-tilt within 0.2 ms. The cameras are able to take and transfer at most 1200 high resolution images (816 * 600 pixels) in 1 s.

6.2   Target Tracking Experiment

- Image coordinate error
  To compare with the traditional RANSAC method [13], 100 images with simple background are used to test the target localization algorithm. Each image include a target with a cross mark at the center of the target. Compared with artificial measurement results (may involve 0.5 pixel error), the maximum error of the target localization is 1 pixel, where the maximum error of RANSAC is 3.637 pixels (Fig. 6).

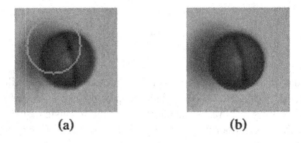

(a)   (b)

Fig. 6. Image coordinate error: (a) RANSAC, (b) our algorithm

In complex background, our algorithm is still able to locate the target while RANSAC may involve large error. The results on complex background is shown in Fig. 7.

Fig. 7. Complex background

- World coordinate error

  It is hard to measure the error in motion state. So 1296 groups of data (the target is 1 m away from the camera) in static state are used to test the error of the processing system. Compared with artificial measurement results (may involve 0.5 mm error), the average error of the result is 3.75337 mm, and the maximum error is 6.919028 mm.

- Speed test

  The process frame rate reaches 1000 fps on FPGA, and the whole control flow is within 2 ms (Fig. 8).

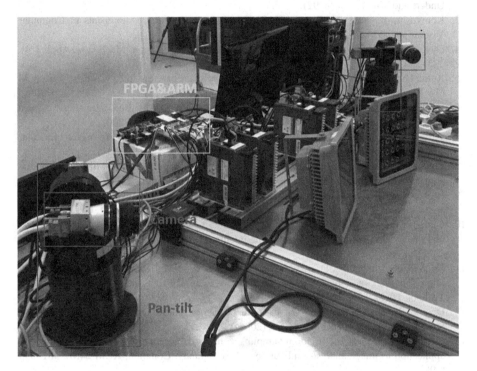

Fig. 8. Experiment setting

7 Conclusions

A binocular active vision system which is able to track circle-shaped target in complex background has been presented in this paper. The core of the system is the processing software which is demonstrated to be robust, fast and accurate.

We finish the active vision system by solving various kinds of difficulties in both hardware and software parts. And we are preparing to apply our system to practical applications, such as ball games and industry scenes.

Acknowledgements. This work was supported in part by the National Natural Science Foundation of China (No. 61170094), 863 Program 2014AA015101, Shanghai Committee of Science and Technology (14JC1402202), and Shanghai Leading Academic Discipline Project (B114).

References

1. Aloimonos, J.Y., Weiss, I., Bandopadhay, A.: Active vision. Int. J. Comput. Vision **1**, 333–356 (1988)
2. Pahlavan, K., Eklundh, J.: A head-eye systemłanalysis and design. CVGIP: Image Understand. **56**, 41–56 (1992)
3. Kuniyoshi, Y., Kita, N., Rougeaux, S., Suehiro, T.: Recent Developments in Computer Vision, pp. 191–200. Springer, Berlin (1996)
4. Dickmanns, E.D., Graefe, V.: Dynamic monocular machine vision. Mach. Vis. Appl. **1**(4), 223–240 (1988)
5. Kutulakos, K.N., Seitz, S.M.: A theory of shape by space carving. Int. J. Comput. Vision **38** (3), 199–218 (2000)
6. Davison, A.J., Reid, I.D., Molton, N.D., Stasse, O.: MonoSLAM: Real-time single camera SLAM. IEEE Trans. Pattern Anal. Mach. Intell. **29**(6), 1052–1067 (2007)
7. Wavering, A.J., Fiala, J.C., Roberts, K.J., Lumia, R.: Triclops: a highperformance trinocular active vision system. IEEE Int. Conf. Rob. Autom. **3**, 410–417 (1993)
8. Clady, X., Collange, F., Jurie, F., Martinet, P.: Object tracking with a pan tilt zoom camera: application to car driving assistance. IEEE Int. Conf. Rob. Autom. **2**, 1653–1658 (2001)
9. Ishikawa, M., Toyoda, H., Mizuno, S.: 1 ms column parallel vision system and its application of high speed target tracking. IEEE Int. Conf. Robot. Autom. **1**, 650–655 (2000)
10. Ishii, I., Taniguchi, T., Sukenobe, R., Yamamoto, K.: Development of high-speed and real-time vision platform, H$_3$ vision. In: Intelligent Robots and Systems, pp. 3671–3678 (2009)
11. Jin, S., Cho, J., Pham, X.D., Lee, K.M., Park, S.-K., Kim, M., Jeon, J.W.: FPGA design and implementation of a real-time stereo vision system. IEEE Trans. Circuits Syst. Video Technol. **20**(1), 15–26 (2010)
12. Diaz, J., Ros, E., Pelayo, F.: FPGA-based real-time opticalflow system. IEEE Trans. Circuits Syst. Video Technol. **16**(2), 274–279 (2006)
13. Fischler, M., Colles, R.: Random sampling consensus: a paradigm for model fitting with application to image analysis and automated cartography. Commun. ACM **24**(6), 381–395 (1981)

The Extraction of Powerful and Attractive Video Contents Based on One Class SVM

Xingchen Liu[1], Xiaonan Song[2], and Jianmin Jiang[3(✉)]

[1] Tianjin University, Tianjin, China
xcliu@tju.edu.cn
[2] Hong Kong Bapitist University, Kowloon, Hong Kong
xnsong@comp.hkbu.edu.hk
[3] Shenzhen University, Shenzhen, Guangdong Province, China
jianmin.jiang@szu.edu.cn

Abstract. With the quick increase of video data, it is difficult for people to find the favorite video to watch quickly. The existing video summarization methods can do a favor for viewers. However, these methods mainly contain the very brief content from the start to the end of the whole video. Viewers may hardly be interested in scanning these kinds of summary videos, and they will want to know the interesting or exciting contents in a shorter time. In this paper, we propose a video summarization approach of powerful and attractive contents based on the extracted deep learning feature and implement our approach on One Class SVM (OCSVM). Extensive experiments demonstrate that our approach is able to extract the powerful and attractive contents effectively and performs well on generating attractive summary videos, and we can provide a benchmark of powerful content extraction at the same time.

Keywords: Video summarization · Powerful content extraction · One-class SVM · Deep learning

1 Introduction

With the development of multimedia technologies, Internet and high speed networking facilities, video data has become a field which draws much attention of human. Facing with so big amount of videos, people find it difficult to choose which video to watch, especially for those long time videos. Thus, how to find the video that is able to attract audiences becomes a vital and challenging problem. To solve this problem, one line of research towards developing feasible solutions is to propose a series of video summarization algorithms to allow users to preview the video content first.

In principle, video summarization consists of two types, static video summarization and dynamic video skimming [1, 2]. Static video summarization extracts several key frames from the original video, which highlight the underlying video content. Taniguchi, Y. et al. proposed an intuitive method of key frame extraction based on sampling [3]. Wafae, S. et al. adopted temporal motion information to extract key frames [4]. The problem of this kind of video summarization methods is that viewers still need

© Springer International Publishing Switzerland 2015
Y.-S. Ho et al. (Eds.): PCM 2015, Part I, LNCS 9314, pp. 375–382, 2015.
DOI: 10.1007/978-3-319-24075-6_36

to take some time to scan all key frame pictures, even for those long time videos with a lot of key frames. Dynamic video skimming, on the other hand, outputs a summarization video sequence with the highlight of events, which can be continuously played back [1]. Viewers can get a brief introduction about the entire content of the video [5]. A representative work along this line is the VAbstract system, which was proposed by Silvia Pfeiffer et al. in 1996 [6]. They defined a series of standards for video abstracts, extracted clips from the original video via several important cues. Omoigui et al. proposed their work to enable users to watch video in a short time with fast play back mode [7]. However, the compressed methods only allow a maximum compression rate of 1.5–2.5 considering the speech speed [8], otherwise the speech will become unrecognizable. Feature plays an important role on video summarization. For example, Smeaton et al. extracted a set of visual and audio features to model the characteristics of shots, and then filter and select them with a Support Vector Machine (SVM) [9]. In addition, more and more researchers focus on visual attention over recent years, and some of whom use their work of visual attention to generate summary videos. A video summarization method based on fMRI-driven visual attention model was proposed by J. Han et al. [10], they combined two fields of brain imaging and bottom-up visual attention model to build a video summarization framework, and reported the natural stimulus of watching videos which are exploited by fMRI brain imaging machine. Similar to [10], N. Ejaz et al. proposed a feature aggregation based on visual attention model to generate summary videos [11].

However, the research we introduced above only show a brief description from the start to the end of a video, viewers often want to see some powerful and special attractive shots. For example, for an action movie, people may want to see the content of fight or explosion. To this end, in this paper, we propose a simple approach which makes use of one class SVM (OCSVM) to obtain the powerful and attractive scenes based on deep learning features. The contributions of our approach are as follows: (a) we abandon the plain plots and extract the powerful and attractive contents from the original video, what make viewers are able to know if they like to watch the whole video or not in a very short time; (b) the sophisticated method, OCSVM is used in our approach, which makes our approach is easy to implement and robust. This research has a promising future on video retrieval, search, data mining and so on based on videos.

The rest of this paper is organized as follows: Sect. 2 describes our proposed algorithm in detail, Sect. 3 presents our experimental results and analysis, and finally Sect. 4 draws conclusions.

2 Video Summarization of Powerful Contents Based on OCSVM

One class SVM (OCSVM) is able to estimate a hyper sphere that can include most of the samples and single out a small number of outliers based on unlabeled data. We utilize this contribution on extracting the powerful contents of videos. Because for a video, only a small part of scenes are special, powerful or magnificent, and the others which account for a large part are relatively plain. Thus, OCSVM fits our intention very well by screening out the outliers which are looked as powerful contents. The flowchart of

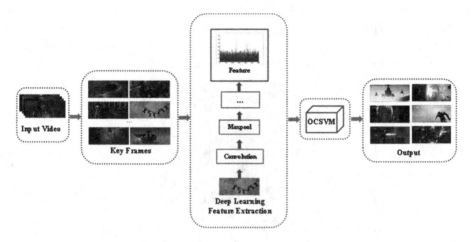

Fig. 1. The flowchart of our approach.

our approach is shown in Fig. 1. We first extract key frames of a video, and describe them with descriptors by deep learning. Then we put these descriptors of key frames in OCSVM, and screen the powerful and attractive frames out.

2.1 Extraction of Key Frames and Features

It is much easier and less time-consuming to process key frames than a video. We extract key frames of the videos by using the technique reported in [12]. Multivariate feature vectors are extracted from the video frames and arranged in a feature matrix. The authors have demonstrated that this approach is suitable for online video summarization and multimedia networking applications. Following the key frame extraction, we then extract two layers of descriptors by deep learning. We can see some simple detail in the third step of Fig. 1. We utilize a very deep convolutional network which was proposed in [13]. By a thorough evaluation of networks of increasing depth using an architecture with very small (3×3) convolution filters, we can get different layers of representations of frames. The frames are passed through a stack of 3×3 convolutional layers, and the padding is 1 pixel for the 3×3 convolution layers. Following some of convolution layers, we put the two representations into one feature matrix by only simple linear combination. Assume that $F_6 \in R^{m \times n}$ is the representation of the output of layer 6, and $F_7 \in R^{m \times n}$ is the representation of layer 7, m is the number of key frames, and n is the dimension of the deep feature. The combination feature F is:

$$F = F_6 + F_7 \tag{1}$$

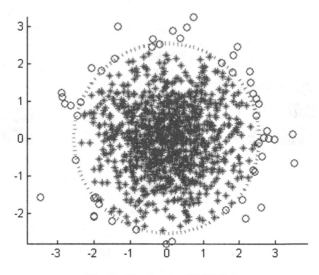

Fig. 2. The sketch of OCSVM.

2.2 Powerful Frames Selection with One Class SVM

We have mentioned above that OCSVM is able to screen the special powerful frames out. There exists another important contribution that it can tune the proportion of outliers by setting the parameter v. As shown in Fig. 2, our intention is to put most of the data into a hyper-sphere (here we use a two-dimensional circle to represent sphere), and make the radius be as small as possible. In this paper, we mainly contribute to tune the value of v to get the best content. This problem can be formulated as an optimization problem:

$$\min_{R \in \Re, \zeta \in \Re^n, m \in F} R^2 + \frac{1}{vn} \sum_{i=1}^{n} \zeta_i$$

$$\left\| \Phi\left(x_i\right) - m \right\|^2 \leq R^2 + \zeta_i, \quad \zeta_i \geq 0, \quad i = 1, \ldots, n \tag{2}$$

where R is the radius of the hyper-sphere which contains all the ordinary frames in mapped feature space, and n is the number of key frames, v sets the proportion of outliers, $\Phi(x_i)$ is the map function, and m is the centroid of the mapped feature space, ζ_i is a relaxation factor.

By using Lagrange multiplier method and kernel function, we can get the decision function:

$$f(x) = R^2 - \sum_{i,j} \alpha_i \alpha_j K\left(x_i, x_j\right) + 2 \sum_{i,j} \alpha_i K\left(x_i, X\right) - K\left(X, X\right) \tag{3}$$

where α_i, α_j are Lagrange parameters. Then the powerful key frames will be screened out by OCSVM.

Fig. 3. The results of a 141 s long online video.

3 Experiments and Discussion

To demonstrate the performance of our approach, we execute our experiments on several online videos which we randomly choose from YouTube and 9 famous movies with the key frames of their official trailers as the ground-truth. These experiments indicate that our approach has an outstanding performance and can be applied on real videos. We use linear kernel in OCSVM, then we train the model to screen out outliers (powerful key frames) with different settings of v who decides the proportion of outliers. Figure 3 shows the outputs frames of an online video. We extract only 4 frames from this video which are more powerful than other frames. We can see the people who are taking kinds of exercises, and the female interviewee. Viewers can scan a few of pictures to know what main and attractive contents a video contains. However, because there is no comparison and ground-truth about online videos, it is difficult to justify the performance of our approach. So we also do experiments on movies, and the official trailers are looked as ground-truth for comparison. We extract key frames of trailers as ground-truth, and evaluate our approach via the number of matched frames between trailers and the output. The number of key frames of movie trailers and our outputs are showed in Table 1. We can see that most number of key frames of outputs are close to trailers'. The evaluation value we adopted is $F1$-value which combines precision and recall, where precision is ratio of the matched frames number and the key frames number of output and recall is ratio of the matched frames number and the key frames number of trailer. The evaluation formulation is:

$$F1 = \frac{2 \times precision \times recall}{precision + recall} \tag{4}$$

Table 1. The number of key frames of movie trailers and outputs and the corresponding occurrence frames.

Movie	Trailer	Output	Occurrence
Les Misérables	76	79	11
The Dark Knight Rises	39	65	5
Transformers 3	41	166	13
Inception	21	55	4
Iron Man 3	55	52	6
Men in Black 3	37	33	8
Pirates of the Caribbean 4	81	123	15
The Twilight Saga	49	49	9
Lucy	55	48	11

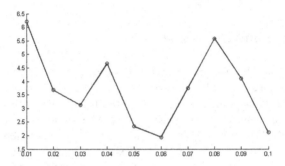

Fig. 4. The F1-value trends with the change of v

We finish extensive experiments, try different settings of parameter v, and choose out the setting is 0.02 finally which can get good performance on all the videos. Table 2 shows the results of this setting on the nine movies. The $F1$-values look not so high, but we should note that the added special effects and the subjectivity of official movie trailers, and our intention is to extract powerful and attractive frames. At the same time, we can find a number of frames which appear in movie trailers, it tells that our approach is able to extract the powerful and attractive contents from videos, and these outputs are very acceptable for human (because trailers contain these frames, too).

Figure 4 shows the plot of F1-value with the change of v on *Les Misérables*, we can see that $F1$-value changes a lot with the different settings of v. Because there are too many frames in a complete film, and a very small variation of v will lead to a big increase or reduce of final extracted frames. Figure 5 shows some output frames of *Les Misérables*, we can see the overall view, the explosion, the scene of fights and the sinking person, these contents usually are exciting for viewers.

Table 2. The F1-value of the setting of $v = 0.02$ (%)

Movie	F1-value
Les Misérables	14.20
The Dark Knight Rises	9.90
Transforrmers 3	12.56
Inception	10.53
Iron Man 3	11.21
Men in Black 3	22.86
Pirates of the Caribbean 4	14.71
The Twilight Saga	18.37
Lucy	10.86
mean	13.91

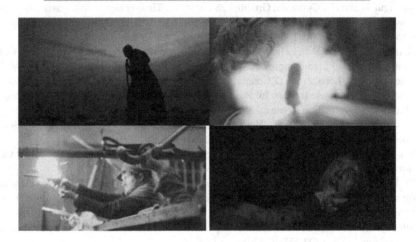

Fig. 5. Some output frames of *Les Misérables*

4   Conclusion

In this paper, we proposed an approach to extract powerful and attractive frames from videos based on One Class SVM (OCSVM), and we also proposed an evaluation method to justify the performance of this kind of research. In addition, we provided a benchmark for the following research about this field. From the theories and experimental results, we can draw the conclusion that our approach is very promising and can be used in many kinds of applications. We will do more further research on this field to get better performance.

Acknowledgment. The authors wish to acknowledge the financial support from the Chinese Natural Science Foundation under the grant No 61373103.

References

1. Li, Y., Zhang, T., Tretter, D.: An overview of video abstraction techniques. HP Laboratories Palo Alto (2001)
2. Lu, S.: Content analysis and summarization for video documents. Doctoral dissertation, The Chinese University of Hong Kong (2004)
3. Taniguchi, Y., Akutsu, A., Tonomura, Y., Hamada, H.: An intuitive and efficient access interface to real-time incoming video based on automatic indexing. In: Proceedings of the Third ACM International Conference on Multimedia, January 2003, pp. 25–33 (1995)
4. Wafae, S., Adil C., Abdelkrim, B.: Video summarization using shot segmentation and local motion estimation. In: Proceedings of the Second International Conference on Innovative Computing Technology, September 2012, pp. 25–33 (2012)
5. Dale, K., Shechtman, E., Avidan, S., Pfister, H.: Multi-video browsing and summarization. Signal Process **89**(12), 2354–2366 (2012). (December 2009)
6. Pfeiffer, S., Lienhart, R., Fischer, S., Effelsberg, W.: Abstracting digital movies automatically. J. Vis. Commun. Image Represent. **7**(4), 345–353 (1996)
7. Omoigui, N., He, L., Gupta, A., Grudin, J., Sanocki, E.: Time-compression: systems concerns, usage, and benefits. In: Proceedings of the SIGCHI Conference on Human Factors in Computing Systems, pp. 136–143 (1999)
8. Heiman, G.W., Leo, R.J., Leighbody, G., Bowler, K.: Word intelligibility decrements and the comprehension of time-compressed speech. Percept. Psychophys. **40**(6), 407–411 (1986)
9. Smeaton, A.F., Lehane, B., O'Connor, N.E., Brady, C., Craig, G.: Automatically selecting shots for action movie trailers. In: Proceedings of the 8th ACM International Workshop on Multimedia Information Retrieval, pp. 231–238 (2006)
10. Han, J., Li, K., Shao, L., Hu, X., He, S., Guo, L., Han, J., Liu, T.: Video abstraction based on fMRI-driven visual attention model. Inf. Sci. **281**, 781–796 (2014)
11. Ejaz, N., Mehmood, I., Baik, S.W.: Feature aggregation based visual attention model for video summarization. Comput Electr. Eng. **40**(3), 993–1005 (2014)
12. Abd-Almageed, W.: Online, simultaneous shot boundary detection and key frame extraction for sports videos using rank tracing. In: Proceedings of the 15th International Conference on Image Processing, pp. 3200–3203 (2008)
13. Simonyan, K., Zisserman, A.: Very deep convolutional networks for large-scale image recognition. arXiv:1409.1556 (2014)

Blur Detection Using Multi-method Fusion

Yinghao Huang, Hongxun Yao$^{(\boxtimes)}$, and Sicheng Zhao

School of Computer Science and Technology, Harbin Institute of Technology,
Harbin, China
h.yao@hit.edu.cn

Abstract. A new methodology for blur detection with multi-method fusion is presented in this paper. The research is motivated by the observation that there is no single method that can give the best performance in all situations. We try to discover the underlying performance complementary patterns of several state-of-the-art methods, then use the pattern specific to each image to get a better overall result. Specifically, a Conditional Random Filed (CRF) framework is adopted for multi-method blur detection that not only models the contribution from individual blur detection result but also the interrelation between neighbouring pixels. Considering the dependence of multi-method fusion on the specific image, we single out a subset of images similar to the input image from a training dataset and train the CRF-based multi-method fusion model only using this subset instead of the whole training dataset. The proposed multi-method fusion approach is shown to stably outperform each individual blur detection method on public blur detection benchmarks.

Keywords: Blur detection · Fusion · Multi-method · CRF

1 Introduction

With various portable hand-held devices like smartphones being widely used, it becomes much easier to capture what interests us with a camera. However, due to some inevitable factors, such as handshake and movement of the target, quite a portion of the captured images are entirely or partially blurred. This can also happen when the photographer pursues some special visual effects like making the foreground persons more salient. Blur detection is one of the research areas around these sort of images. It tries to discriminate the sharp image contents from the blurred, which can benefit many applications such as image segmentation, scene classification and image quality assessment [6–8].

A lot of work has been done on this topic [1–4,9,10,12]. These methods design a variety of discriminative features to compute a blur mask from an input image, like natural image statistics [9,12], average power spectrum [1], gradient histogram span [3] and singular value feature [4]. The scale of blur is also addressed in [1], which proves able to effectively improve the detection performance. Although all being able to achieve promising results, the pros and cons

© Springer International Publishing Switzerland 2015
Y.-S. Ho et al. (Eds.): PCM 2015, Part I, LNCS 9314, pp. 383–392, 2015.
DOI: 10.1007/978-3-319-24075-6_37

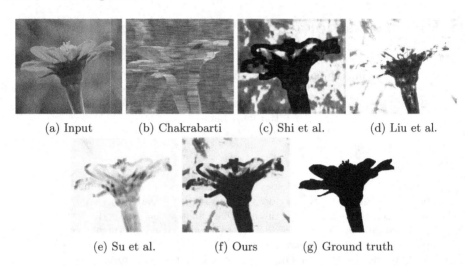

(a) Input   (b) Chakrabarti   (c) Shi et al.   (d) Liu et al.

(e) Su et al.   (f) Ours   (g) Ground truth

Fig. 1. Blur detection via multi-method fusion. Given an input image, results of the individual blur detection methods (b–e) often complement each other. Our method can effectively integrate these results together and generate better blur mask (f) than each of them.

of these methods can be quite different. As shown in Fig. 1, on this specific input image, the methods proposed by Shi et al. [1] makes less false positives, while the methods of Liu et al. [3] and Su et al. [4] get better blur region discrimination.

In this paper we present a multi-method fusion approach to blur detection. Based on the various blur maps generated by different methods, we try to achieve a better overall result. The performance gaps and interrelation of detection results achieved by those methods are modeled by a Conditional Random Field (CRF) framework [5]. Considering the dependence of multi-method fusion on individual image, a subset of similar images to the input image is selected from a training dataset, then the CRF-based fusion model is trained only using this subset instead of the whole training set. Experiments on the public blur detection benchmark validate the effectiveness of our method.

The rest of the paper is organized as follows: firstly we present the motivation of this work in Sect. 2. Then Sect. 3 describes the steps of our proposed method in detail. The experimental results and discussions are covered in Sect. 4. Finally Sect. 5 concludes this paper.

2   Motivation

In the past decade various methods have been proposed to address the problem of blur detection. With the advent of a new method, the global performance of blur detection will usually be improved to a higher level. As shown in Fig. 2 (a), under the criteria of precision-recall, the newly method proposed in [1] performs the best almost in the entire range.

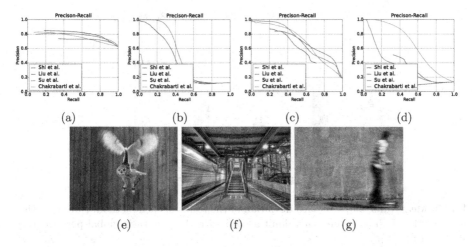

Fig. 2. Not one single method can perform the best on all images. (a): PR (precision-recall) curves of four different methods on the entire dataset used in [1]. (b)–(d): PR curves on three selected images. (e)–(f): The three chosen images: motion0004.jpg, motion0031.jpg and motion0035.jpg.

However, this performance advantage can't stably generalize to every image. Due to the different strengths and weaknesses of the underlying feature engines, how well each method performs on a specific input image is heavily dependent on the image itself. As shown in Fig. 2 (b)–(d), the performance of these four methods fluctuates a lot on the three selected images. To further compare the performance fluctuation of these methods, we rank them according to their F-Score values on all the images in the dataset proposed in [1]. Then the average rank and the standard deviation are computed. The result is shown in Fig. 3. As we can see, the performance range of each method is quite large, and there is an obvious overlapped scope shared by these methods. That means it is more sensible to adjustably combine these methods together than counting on one specific method, which is the main motivation lying the heart of this work.

3   The Methodology

From the perspective of methodology, method-level fusion is a natural extension of feature-level fusion. This idea has been used to address other computer vision tasks. For example, Liu et al. uses SVR (support vector regression) based method aggregation models for image quality assessment [13], and in [11], different saliency detection methods are integrated together to achieve better detection results. Considering the similar setting of saliency detection and blur detection, a method similar to the one proposed in [11] is used in this paper. To the best of our knowledge, this is the first time a multi-method fusion model has been proposed to tackle the problem of blur detection.

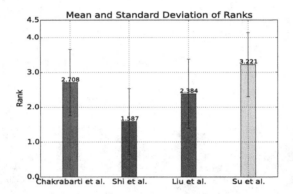

Fig. 3. Mean value and standard deviation of the ranks of each method. Note that the mean ranks of all these methods don't accord with those of their global performance shown in Fig. 2 (a).

3.1   Multi-method Fusion via CRF

Given an input image I, a graph $G = (V, E)$ can be constructed, where V corresponds to all pixels of the image and E is the set of edges connecting pairs of neighbouring vectors. The 8-connection neighbourhood is considered here. Assume we have n different blur detection methods at hand $M = \{M_1, M_2, \ldots, M_n\}$. By applying these methods on the input image, for each pixel p we can get a vector of blur detection values $B(p) = (B_1(p), B_2(p), \ldots, B_n(p))$. Here $B = \{B_1, B_2, \ldots, B_n\}$ is the set of blur masks achieved by the chosen methods, all normalized to the range of $[0, 1]$. In this way, the relationship between the label and features and the interrelation between labels of neighbouring pixels are modeled by the CRF. After setting some potential terms reflecting reasonable priors, we can encourage the fusion model to move towards the way we desire by optimizing the entire objective function. The specific conditional distribution of labels $Y = \{y_p | p \in I\}$ on the features $X = \{\mathbf{x}_p | p \in I\}$ can be formulated as follows:

$$P(Y|X; \theta) = \frac{1}{Z} \exp(\sum_{p \in I} f_a(\mathbf{x}_p, y_p) + \sum_{p \in I} \sum_{q \in N_p} (f_c(\mathbf{x}_\mathbf{p}, \mathbf{x}_\mathbf{q}, y_p, y_q)$$

$$+ f_s(\mathbf{x}_\mathbf{p}, \mathbf{x}_\mathbf{q}, y_p, y_q))) \quad (1)$$

where θ is the parameters of the CRF model, $\mathbf{x}_p$ is the feature vector of the pixel p, y_p is the blur label assigned to the pixel p, and Z is the partition function. $\mathbf{x}_p$ is constructed as follows:

$$\mathbf{x}_p = (s_{p1}, s_{p2}, \ldots, s_{pn}) \quad (2)$$

$$s_{pi} = \begin{cases} B_i(p), & y_p = 1 \\ 1 - B_i(p), & y_p = 0 \end{cases} \quad (3)$$

Intuitively speaking, s_{pi} indicates the probability of the pixel p belonging to the blur area, from the perspective of the ith method. There are three terms in the probability function: $f_a(\mathbf{x}_p, y_p)$, $f_c(\mathbf{x}_p, \mathbf{x}_q, y_p, y_q)$ and $f_s(\mathbf{x}_p, \mathbf{x}_q, y_p, y_q)$. All of them encode the priors we impose on the result labels.

Blur Fusion Term. $f_a(\mathbf{x}_p, y_p)$ is the blur fusion term that defines the relationship between the blur features and the final result label. It is set as follows:

$$f_a(\mathbf{x}_p, y_p) = \sum_{i=1}^{n} \alpha_i s_{pi} + \alpha_{n+1} \tag{4}$$

where $\{\alpha_i\}$ is part of the CRF model parameters θ.

Consistency Constraint Term. We design the consistency constraint term $f_c(\mathbf{x}_p, \mathbf{x}_q, y_p, y_q)$ as:

$$f_c(\mathbf{x}_p, \mathbf{x}_q, y_p, y_q) = \sum_{i=1}^{n} \beta_i((1(y_p = 1, y_q = 0) - 1(y_p = 0, y_q = 1))(s_{pi} - s_{qi})) \tag{5}$$

where $\{\beta_i\}$ is another part of the parameter θ and $1(\cdot)$ is an indicator function. The reason we set this term consistency constraint term is based on the observation that if one of the neighbouring pixels has a high blur feature value, then it usually tends to has a high blur label value. Namely, there is some consistency between the final blur mask and initial blur results achieved by the base methods, at least between neighbouring pixels.

Blur Smooth Term. Then comes the final blur smooth term $f_s(\mathbf{x}_p, \mathbf{x}_q, y_p, y_q)$, which expresses the observation that neighbouring pixels having similar colors should have similar blur labels, the same idea from [14].

$$f_s(\mathbf{x_p}, \mathbf{x_q}, y_p, y_q) = -1(y_p \neq y_q) \exp\left(-\gamma \|I_p - I_q\|\right) \tag{6}$$

where $\|I_p - I_q\|$ is the L_2 norm of the color difference between pixel p and q. γ is set as $\gamma = (2 < \|I_p - I_q\|^2 >)^{-1}$, where $< \cdot >$ denotes the expectation operator. The weight of this term is fixed to 1 to penalize the case where two neighbouring pixels with similar colors are assigned different blur labels.

3.2   Locality-Aware Multi-method Fusion

The above CRF-based fusion method is holistic in the sense that it treats all images as a whole and doesn't discriminate the content of a specific image. Though the performance of different blur detection methods tends to retain some kind of consistency over images, the best fusion parameter of two not so similar images can be quite different. Based on this consideration, we try to further improve this model by endowing it with the locality-aware capability.

To this end, instead of training one holistic model, for each input test image we tailor a specific model to it. This is achieved by selecting the k most similar images of the input image from the entire training dataset, then fit the CRF fusion model on this subset. There are a lot of methods to compare the similarity of images. In this paper we adopted the GIST descriptor proposed in [16], which has proved effective in capturing the spatial configuration of images. After constructing the GIST representation, we use the L_2 distance between them as the metric, as suggested in [16].

4   Experiments

We implement our CRF fusion model on the base of the UGM CRF toolkit[1] from Mark Schmidt, a matlab package for CRF training and inference. Experimental settings and results are described in the following.

Fig. 4. Example images of the dataset.

4.1   Experimental Settings

Dataset. To evaluate the effectiveness of our method, we conduct experiments on the dataset proposed in [1]. This is the largest publicly available benchmark on blur detection, which consists of 1000 images with out-of-focus blur and partial motion blur. All the images have been labeled at a pixel level. This dataset is of quite large diversity, with contents like animals, plants, people and scenes. Several exemplar images with ground-truth masks are shown in Fig. 4.

---

[1] http://www.cs.ubc.ca/~schmidtm/Software/UGM.html.

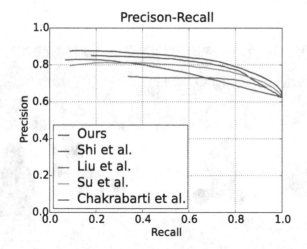

Fig. 5. Quantitative comparison of our fusion method and the base methods.

Base Methods. We adopted the state-of-the-art methods proposed by Chakrabarti et al. [2], Shi et al. [1], Liu et al. [3] and Su et al. [4] as the base methods. We get their blur detection results using existing or our (if the code or executable isn't available) implementation. All blur masks are normalized to the range [0, 1]. The leave-one-out setting is adopted. For each image in the dataset, we treat all the rest images as the training dataset, then train the locality-aware CRF fusion model on the chosen subset. When choosing the training subset for a specific input image, the parameter k is empirically fixed to be 50.

Evaluation Metric. We adopted Precision and Recall (PR) curve as the evaluation metric. Specifically, for each given threshold t, we compute the precision and recall for each image, then adopt the mean value of all the testing images as the final result. The entire PR curve is obtained by varying the threshold t from 1 to 255.

Table 1. Mean rank and standard deviation (Std) comparison.

	[2]	[1]	[3]	[4]	Ours
Mean	3.5550	2.5130	3.2820	4.3230	1.3270
Std	1.0020	1.0164	1.0235	0.8387	0.8042

4.2 Results and Discussion

Detection Accuracy Comparison. The quantitative comparison of our fusion method and the chosen base methods is shown in Fig. 5 and Table 1.

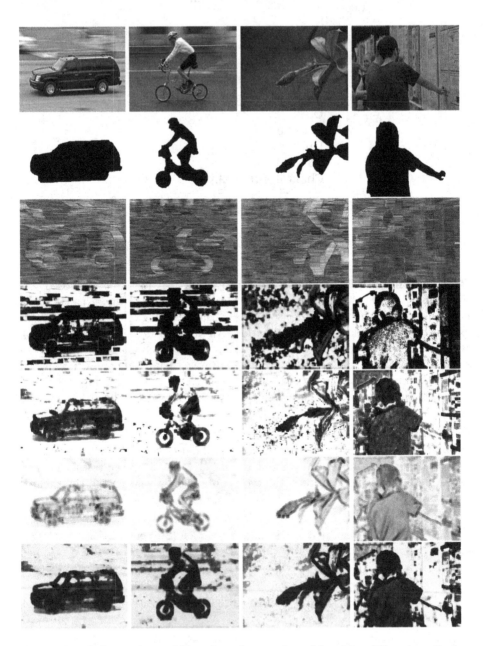

Fig. 6. Visual Comparison of blur detection results achieved by different methods. From top to bottom: the input image; the ground-truth; results of Chakrabarti et al.; results of Shi et al.; results of Liu et al.; results of Su et al.; our results.

It shows that our proposed method consistently outperform each individual base method in the entire recall range $[0, 1]$. This performance improvement is mainly attributed to the powerful modeling capability of the CRF-based fusion method. By taking the performance gap and dependence on specific images of different methods into consideration, the CRF-based fusion model can well figure out the suitable combination weights of the blur masks, then achieve a better final result. Some results of our method are compared with those of the base methods in Fig. 6. As we can see, the proposed fusion method achieves more promising visual results. Compare the final fusion results and the blur masks of each base method and note how correct detection ares are selectively picked out or even enhanced by our method.

Running Time Analysis. Because our method relies on the results of all the base methods, it consumes more running time than each base method. The entire time of blur features acquisition is roughly the sum of those used by the base methods. This procedure can easily be sped up via parallel computing. Besides, on our desktop computer with an i7 3.40 GHZ CPU and 12G RAM, when k is set to 50, for an input image with 640×427 pixels, it takes about 45 s to do model training and inferring.

Failure Case Analysis. The quality of the chosen training subset directly affects our blur detection result. Most failure cases of our method happen when the entire training dataset doesn't contain enough similar images to the input one. Our future work will focus on how to make our fusion method robust to this special situation.

5   Conclusion

In this paper we presented a novel blur detection approach by fusing existing state-of-the-art methods. Different from previous methods focusing on designing discriminative blur features, our proposed method takes blur detection results of the chosen base methods as input and tries to figure out the optimal combination of these blur masks. To this end, a CRF-based framework is used to model the performance gap and interrelation of the base methods. Considering the performance dependence of various methods on specific image content, we make our method sensitive to the input image by endowing it with locality-aware capability. This is achieved by selecting similar images then tailoring the CRF model on this subset. Comparison with the state-of-the-art base methods validate the effectiveness of the proposed method.

Acknowledgement. This work was supported in part by the National Science Foundation of China No. 61472103, and Key Program Grant of National Science Foundation of China No. 61133003.

References

1. Shi, J., Xu, L., Jia, J.: Discriminative blur detection features. In: CVPR, pp. 2965–2972 (2014)
2. Chakrabarti, A., Zickler, T., Freeman, W.T.: Analyzing spatially-varying blur. In: CVPR, pp. 2512–2519 (2010)
3. Liu, R., Li, Z., Jia, J.: Image partial blur detection and classification. In: CVPR, pp. 1–8 (2008)
4. Su, B., Lu, S., Tan, C.L.: Blurred image region detection and classification. In: ACMMM, pp. 1397–1400 (2011)
5. Lafferty, J., McCallum, A., Pereira, F.C.N.: Conditional random fields: Probabilistic models for segmenting and labeling sequence data. In: ICML, pp. 282–289 (2001)
6. Derpanis, K.G., Lecce, M., Daniilidis, K., Wildes, R.P.: Dynamic scene understanding: The role of orientation features in space and time in scene classification. In: CVPR, pp. 1306–1313 (2012)
7. Serre, T., Wolf, L., Bileschi, S., Riesenhuber, M., Poggio, T.: Robust object recognition with cortex-like mechanisms. PAMI **29**(3), 411–426 (2007)
8. Toshev, A., Taskar, B., Daniilidis, K.: Shape-based object detection via boundary structure segmentation. IJCV **99**(2), 123–146 (2012)
9. Levin, A.: Blind motion deblurring using image statistics. In: NIPS, pp. 841–848 (2006)
10. Dai, S., Wu, Y.: Removing partial blur in a single image. In: CVPR, pp. 2544–2551 (2009)
11. Mai, L., Niu, Y., Liu, F.: Saliency aggregation: a data-driven approach. In: CVPR, pp. 1131–1138 (2013)
12. Lin, H.T., Tai, Y.-W., Brown, M.S.: Motion regularization for matting motion blurred objects. PAMI **33**(11), 2329–2336 (2011)
13. Liu, T.-J., Lin, W., Kuo, C.-C.J.: Image quality assessment using multi-method fusion. TIP **22**(5), 1793–1807 (2013)
14. Liu, T., Yuan, Z., Sun, J., Wang, J., Zheng, N., Tang, X., Shum, H.-Y.: Learning to detect a salient object. PAMI **33**(2), 353–367 (2011)
15. Blake, A., Rother, C., Brown, M., Perez, P., Torr, P.: Interactive image segmentation using an adaptive GMMRF model. In: Pajdla, T., Matas, J.G. (eds.) ECCV 2004. LNCS, vol. 3021, pp. 428–441. Springer, Heidelberg (2004)
16. Oliva, A., Torralba, A.: Modeling the shape of the scene: a holistic representation of the spatial envelope. IJCV **42**(3), 145–175 (2001)

Motion Vector and Players' Features Based Particle Filter for Volleyball Players Tracking in 3D Space

Xizhou Zhuang[1(✉)], Xina Cheng[1], Shuyi Huang[1], Masaaki Honda[2], and Takeshi Ikenaga[1]

[1] Graduate School of Information, Production and Systems,
Waseda University, Tokyo, Japan
zhuangxizhou@fuji.waseda.jp
[2] Faculty of Sport Sciences, Waseda University, Tokyo, Japan

Abstract. Multiple players tracking plays a key role in volleyball analysis. Due to the demand of developing effective tactics for professional events, players' 3D information like speed and trajectory is needed. Although, 3D information can solve the occlusion relation problem, complete occlusion and similar feature between players may still reduce the accuracy of tracking. Thus, this paper proposes a motion vector and players' features based particle filter for multiple players tracking in 3D space. For the prediction part, a motion vector prediction model combined with Gaussian window model is proposed to predict player's position after occlusion. For the likelihood estimation part, a 3D distance likelihood model is proposed to avoid error tracking between two players. Also, a number detection likelihood model is used to distinguish players. With the proposed multiple players tracking algorithm, not only occlusion relation problem can be solved, but also physical features of players in the real world can be obtained. Experiment which executed on an official volleyball match video (Final Game of 2014 Japan Inter High School Games of Men's Volleyball in Tokyo Metropolitan Gymnasium) shows that our tracking algorithm can achieve 91.9 % and 92.6 % success rate in the first and third set.

Keywords: Multiple players tracking · Volleyball · Particle filter · 3D space

1 Introduction

Volleyball analysis is arousing more attention due to the increasing demand of professional sports events. To develop more effective tactics, players' movements and trajectories information in 3D space are needed to be extracted by using multiple players tracking algorithm. Traditionally, these data are always estimated manually and then processed using data analysis tools like Data Volley [1], which is very manpower consuming and inefficient. To speed up and make the data more precise, track multiple players in 3D space is necessary in volleyball analysis.

However, tracking multiple players in volleyball match video meets several challenges. Firstly, players always move in irregular directions with variable speeds. In others words, the tracking system is nonlinear and non-Gaussian. Compared with CAM

© Springer International Publishing Switzerland 2015
Y.-S. Ho et al. (Eds.): PCM 2015, Part I, LNCS 9314, pp. 393–401, 2015.
DOI: 10.1007/978-3-319-24075-6_38

shift [2], Kalman Filter [3] and Extend Kalman Filter [4], Particle Filter [5] shows its superiority in handling non-Gaussianity and multimodality by approximating posterior distribution using weighted particles. Secondly, occlusion always occurs during volleyball match. This occlusion problem results in two consequences. One is that it is hard to re-predict players' movements and positions after occlusion. For this consequence, Cheng et al. [6] puts forward a labeling likelihood model, a distance likelihood model and two re-detection models which can track players after occlusion. But that approach can hardly extract players' 3D information. The other consequence is tracking targets might be exchanged or focus on the opposite player when players are occluded by each other. Thus, to distinguish players and track target precisely after occlusion, some unique prediction models and features are needed. Shiina and Ikenaga [7] uses a 2ARDM (second-order auto-regressive dynamical model) model combined with a conventional Gaussian window model to track irregular moving objects. However, as they use only two previous frames to predict the player's current position, once it failed, there is no chance to recover tracking. Lee and Horio [8] adopts a reliable appearance observation model for human tracking. Morais et al. [9] proposes an automatic tracking algorithm by using multiple cameras. Their approach can get 3D positions of soccer players and reduce the error caused by projection. But for volleyball players, they always share the similar features and intersection occurs more often, which makes it difficult to distinguish and track.

To cope with problems mentioned above, this paper proposes a motion vector prediction model and a players' features likelihood model based particle filter in 3D space. Based on this framework, players' 3D coordinates can be obtained. For the first problem of re-predicting player's position after occlusion, because the motion vector prediction model takes advantage of the movement information before occlusion, it can ensure particles still spread closely to the tracking object after occlusion, as long as it distributes the particles according to the previous movement direction when occlusion happens. However, there is still a possibility that those particles may be seized by the opposite player after occlusion. To deal with this problem, a 3D distance likelihood model which assigns weights of particles according to the distance between two players is proposed. Last but certainly not least, as our ultimate aim is to track multiple players belonging to the same team, the jersey number is no doubt a unique feature for distinguishing players, which is employed in number detection likelihood model.

The rest of paper is organized as follows. In Sect. 2, we briefly present the framework of our approach. Then, a proposed motion vector prediction model and a players' features based likelihood model consisted of 3D distance likelihood model and number detection likelihood model are described. Sections 3 and 4 show the experiment result and conclusion, respectively.

2 Proposal

To track players more precisely and obtain players' physical information in the real world, we capture video sequences from different directions to build a tracking system in 3D space. Based on the conventional particle filter, the framework of our proposal is showed in Fig. 1. In the prediction step, a motion vector model is proposed to predict player's position after occlusion. In the likelihood estimation step, the particles in the 3D space are projected to 2D planes to solve the occlusion relation problem, which means when observing two objects from single viewpoint in 3D space, one may occluded another. What's more, a 3D distance likelihood model and a number detection likelihood model are proposed to solve the occlusion and player distinguishing problem respectively.

Assume that the ground in the gym is flat, a 3D coordinate system is built on the court. Thus, the state space can be modified from $\{u, v\}$ (pixel) to $\{x, y, z\}$ (cm). To gain 3D information, camera calibration [10] is applied. For each camera, transformation from image coordinate $p = (u, v, \omega)^T$ to 3D court space coordinate $q = (x, y, z, 1)^T$ is conducted as Fig. 2 shows. Here, A is a 3×4 homogeneous coordinate-transformation matrix.

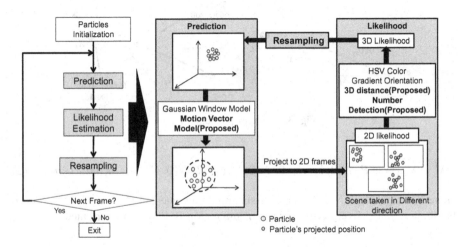

Fig. 1. Framework of particle filter in 3D space

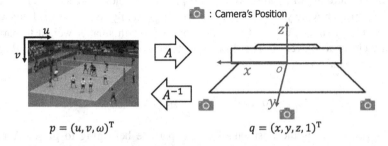

Fig. 2. Image plane and 3D coordinate on the court

2.1   Motion Vector Prediction Model

Although player's movement is irregular, in most cases, we assume the moving direction before and after occlusion remains unchanged. To predict player's position in the t_{th} frame after occlusion, we calculate motion vector using n previous frames like (1):

$$mv_n = X_{t-1}(x, y, z) - X_{t-n-1}(x, y, z) \tag{1}$$

After defining the motion vector model, we distribute particles as Fig. 3 shows. Half of the particles are distributed according to the Gaussian window model to ensure the universality. The other half of particles are distributed in the motion vector direction. So the position of the i_{th} particle can be calculated according to (2).

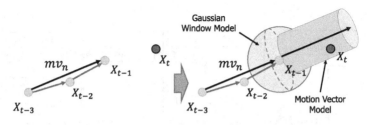

Fig. 3. Particles prediction region

$$X_t^i = X_{t-1}(x + R \cdot N(0, 1), y + R \cdot N(0, 1), z + R \cdot N(0, 1)) + U(0, |mv_n|) \cdot mv_n \tag{2}$$

where R is a constant, $N(0, 1)$ is a Gaussian random value and $U(0, |mv_n|)$ is a uniform distribution value. Every cross section of the region is Gaussian distributed.

2.2   Players' Features Based Likelihood Model

3D Distance Likelihood Model. When two players are close to each other, particles for tracking player A may cover player B, which gradually results in tracking wrong player. As Fig. 4 shows, to avoid this situation, we assign smaller weight to particles which are closer to player B. The distance between one particle and player B in 3D space is defined as (3).

$$D^i = \sqrt{(x_{t,A}^i - x_{t-1,B})^2 + (y_{t,A}^i - y_{t-1,B})^2 + (z_{t,A}^i - z_{t-1,B})^2} \tag{3}$$

where $(x_{t,A}^i, y_{t,A}^i, z_{t,A}^i)$ is the position of the i_{th} particle belonging to player A in the current frame and $(x_{t-1,B}, y_{t-1,B}, z_{t-1,B},)$ is the position of player B in the previous frame.

Based on the 3D distance, we use (4) to calculate the score of this model.

$$L^i_{Distance} = 1 - \exp(-\frac{(D^i)^2}{2000}) \tag{4}$$

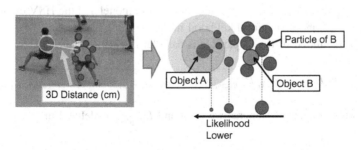

Fig. 4. Concept of 3D distance likelihood model

Number Detection Likelihood Model. For accurately distinguishing and tracking multiple players, jersey number is no doubt a unique feature. The proposed number detection likelihood model consists of two parts, which are described in the follow subsections.

Number Training and Detection. For each number, as shown in Fig. 5, 77 positive samples varying in rotation, deformation and illumination are trained by Ada-boost [10] to get a classifier. The features we extracted are Haar-like feature [12], Histogram of Gradient (HOG) [13] and Local Binary Patterns (LBP) [14].

Normal	Rotation	Illumination Change	Deformation	Total
 1	 1	 1	 1	77

Fig. 5. Example of training sample

Number Detection Likelihood. After detecting number in every 2D frames, the score of number detection likelihood model is calculated as (5):

$$L^i_{Number} = 1 \times Constant_{Haar} \times Constant_{HOG} \times Constant_{LBP} \tag{5}$$

During detection, if the Haar-like classifier considers the particle's sampling region containing a number, $Constant_{Haar}$ is assigned a value larger than 1, otherwise, $Constant_{Haar}$ is assigned as 1, which is same applied to HOG and LBP classifiers.

Likelihood Calculation. The 2D likelihood of the i_{th} particle which is projected on the frame plane taken by one camera is calculated as (6):

$$L_{2D}^i = \sqrt[3]{L_{Gradient}^i \times L_{Color}^i \times L_{Number}^i} \tag{6}$$

where $L_{Gradient}^i$ and L_{Color}^i are the gradient orientation model and the HSV color model. L_{Number}^i is the number detection likelihood model.

The 3D likelihood of the i_{th} particle is calculated as (7):

$$L_{3D}^i = L_{Distance}^i \times \sqrt[N]{L_{2D(camera.1)}^i \times L_{2D(camera.2)}^i \times \ldots \times L_{2D(camera.N)}^i} \tag{7}$$

where N is the amount of cameras we used and $L_{Distance}^i$ is defined in (4).

3   Experiment and Result

Our proposal is implemented using C++ and OpenCV. The input video sequences are 1920 × 1080 in resolution and 60 frames per second, which record an official volleyball match (Final Game of 2014 Japan Inter High School Games of Men's Volleyball in Tokyo Metropolitan Gymnasium). As shown in Fig. 2, three cameras are fixed at right corner, left corner and middle behind the target term to capture the video sequences without zoom in and zoom out. The shutter speed is set at 1000 per second to avoid motion blur.

We use the first and third set of the final game which contains 33 and 44 rounds, respectively. To evaluate our proposal, we define one evaluation criteria by using the concept of Respond Unit (RU). One RU contains the actions like this: one team receives the ball from rivals, passes and then returns the ball to rivals by spiking.

Based on RU, we can calculate the success rate via Eq. (8):

$$Success\_Rate\% = \sum_{1_{st}RU}^{ENDRU} \frac{Tracked\_Players}{6} \times 100\% \tag{8}$$

where 6 is that one team has 6 players on the court and the *Tracked_Players* means the number of players tracked successfully in one *RU*. We sum up the success rate of each *RU* to get the final success rate. Figure 6 shows an example of the results. Both player 2 and player 6 can be tracked by our approach successfully after occlusion, as well as other players:

We compared the experimental results of proposal with conventional works using prepared video sequences. The number of particles we used to track every player is 1000 both for proposal and conventional work. In our proposal, the number of motion vectors we used is 15. For the number detection likelihood, we set the $Constant_{Haar}$ as 1.5, $Constant_{HOG}$ as 2 and $Constant_{LBP}$ as 5. The conventional works we used are particle filter with framework and likelihood model proposed in [9], prediction model proposed in [6].

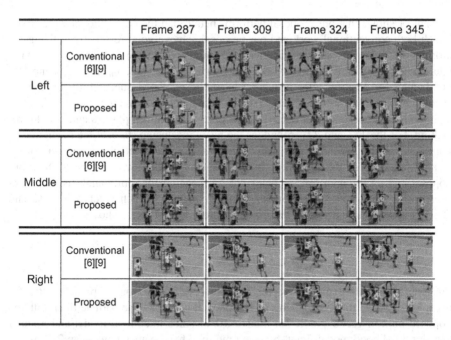

		Frame 287	Frame 309	Frame 324	Frame 345
Left	Conventional [6][9]				
	Proposed				
Middle	Conventional [6][9]				
	Proposed				
Right	Conventional [6][9]				
	Proposed				

Fig. 6. Example of tracking result compared with conventional work

The result of the first set is shown in Table 1.

Table 1. Experimental result of the first set

Round	RU×6	Proposal	Conventional work [6, 9]	Round	RU×6	Proposal	Conventional work [6, 9]	Round	RU×6	Proposal	Conventional work [6, 9]
1	18	16	12	13	18	18	15	25	12	11	7
2	12	12	8	14	12	9	9	26	42	38	39
3	6	5	4	15	18	15	18	27	6	5	5
4	6	5	2	16	6	5	5	28	12	11	11
5	6	6	6	17	12	12	8	29	6	6	4
6	12	10	10	18	6	6	5	30	6	6	6
7	6	6	6	19	6	6	5	31	6	6	6
8	24	23	23	20	6	5	5	32	6	6	5
9	18	15	16	21	6	6	6	33	6	6	5
10	12	12	12	22	12	9	10	Total	348	320	294
11	6	6	5	23	6	6	6	Execution time		20 s/f	8 s/f
12	6	6	6	24	6	6	4	Success_rate		91.9 %	84.5 %

Table 2. Success_rate of the first and third set

	Proposal	Conventional work [6, 9]
First set	91.9 %	84.5 %
Third set	92.6 %	88.6 %

Table 1 shows that our proposal performs better than the conventional work. Even though both conventional work and our proposal achieve good results by utilizing 3D information, some complex conditions with heavy occlusion still lead to tracking failure when using conventional works. For example, in the 1_{st}, 4_{th}, 13_{th}, 17_{th} and 25_{th} round, player's position after heavy occlusion is predicted successfully by motion vector model while the conventional work failed. In most rounds, the 3D distance likelihood model and number detection model can distinguish players and avoid error tracking. For the 15_{th} and 22_{nd} round, as the heavy occlusion starts at the beginning of sequences, motion vector can hardly predict the position precisely, which causes tracking failed. Even the execution time of the conventional work is shorter than our proposal, the results with low accuracy are no meaning for sports analysis. For most cases, our approach is more promising during tracking, which reaches 91.9 % and 92.6 % average success rate in the first and third set as Table 2 shows.

4   Conclusion

This paper presents a motion vector and players' features based particle filter for multiple volleyball players tracking in 3D space. It combines the particle filter with the proposed motion vector prediction model, 3D distance likelihood model and number detection likelihood model. First, by using multiple fixed cameras, we map 2D images to 3D space in the real world which allows us to track players in 3D space. Then, a motion vector prediction model is used to predict players position after occlusion. A 3D distance likelihood model is put forward to avoid error tracking between two players. What's more, the number detection likelihood model is used to distinguish players with unique jersey number. The results show that the success rate of our approach reaches 91.9 % and 92.6 % in the first and third set, respectively.

Acknowledgement. This work was supported by KAKENHI (26280016) and Waseda University Grant for Special Research Projects (2015K-222).

References

1. http://www.dataproject.com/Products/EN/en/Volleyball/DataVolley4
2. Dong, X.M., Yuan, K.: A robust Cam Shift tracking algorithm based on multi-cues fusion. In: 2nd International Conference on Advanced Computer Control (ICACC), vol. 1, pp. 521–524 (2010)
3. Kalman, R.E.: A new approach to linear filtering and prediction problems. J. Fluids Eng. **82**(1), 35–45 (1960)
4. Ndiour, I.J., Vela, P.A.: A local extended Kalman filter for visual tracking. In: IEEE Conference on Decision and Control (CDC), pp. 2498–2504 (2010)
5. Kitagawa, G.: Monte Carlo filter and smoother for non-Gaussian nonlinear state space models. J. Comput. Graph. Stat. **5**(1), 1–25 (1996)

6. Cheng, X., Shiina, Y., Zhuang, X., Ikenaga, T.: Player tracking using prediction after intersection based particle filter for volleyball match video. In: Asia-Pacific Signal and Information Processing Association, Annual Summit and Conference (APSIPA ACS) (2014)
7. Shiina, Y., Ikenaga, T.: Dual model particle filter for irregular moving object tracking in sports scenes. In: Asia-Pacific Signal and Information Processing Association, Annual Summit and Conference (APSIPA ASC) (2011)
8. Lee, S., Horio, K.: Human tracking with particle filter based on locally adaptive appearance model. J. Sig. Process. **18**(4), 229–232 (2014)
9. Morais, E., Goldenstein, S., Ferreira, A., Rocha, A.: Automatic tracking of indoor soccer players using videos from multiple cameras. In: 25th SIBGRAPI Conference on Graphics, Patterns and Images, pp. 174–181 (2012)
10. Hartley, R., Zisserman, A.: Multiple view geometry in computer vision. Cambridge University Press, Cambridge (2003)
11. Freund, Y., Schapire, R.E.: A decision-theoretic generalization of on-line learning and an application to boosting. J. Comput. Syst. Sci. **55**(1), 119–139 (1997)
12. Lienhart, R., Maydt, J.: An extended set of haar-like features for rapid object detection. In: International Conference on Image Processing, Proceedings, vol. 1, pp. I-900–I-903 (2002)
13. Dalal, N., Triggs, B.: Histograms of oriented gradients for human detection. In: IEEE Conference on Computer Vision and Pattern Recognition, vol. 1, pp. 886–893 (2005)
14. Ojala, T., Pietikäinen, M., Harwood, D.: A comparative study of texture measures with classification based on featured distributions. Pattern Recogn. **29**(1), 51–59 (1996)

A Novel Edit Propagation Algorithm via L_0 Gradient Minimization

Zhenyuan Guo[1], Haoqian Wang[1(✉)], Kai Li[1,2], Yongbing Zhang[1],
Xingzheng Wang[1], and Qionghai Dai[1,2]

[1] Graduate School at Shenzhen, Tsinghua University, Beijing, China
wangyizhai@sz.tisnghua.edu.cn
[2] Department of Automation, Tsinghua University, Beijing, China

Abstract. In this paper, we study how to perform edit propagation using L_0 gradient minimization. Existing propagation methods only take simple constraints into consideration and neglects image structure information. We propose a new optimization framework making use of L_0 gradient minimization, which can globally satisfy user-specified edits as well as tackle counts of non-zero gradients. In this process, a modified affinity matrix approximation method which efficiently reduces randomness is raised. We introduce a self-adaptive re-parameterization way to control the counts based on both original image and user inputs. Our approach is demonstrated by image recoloring and tonal values adjustments. Numerous experiments show that our method can significantly improve edit propagation via L_0 gradient minimization.

Keywords: Edit propagation · L_0 gradient minimization · Recoloring · Tonal adjustment

1 Introduction

As media information explodes current days, the need for efficient and intuitive editing methods greatly improved. Photos and videos which contain a huge amount of well-structured information need to be unearthed. In general, traditional editing techniques use commercial or free media editing tools, to construct delicacy mask elaborately, and thus edit using it. However, these methods are both time-consuming and not suitable for practical application. What's more, a new user can hardly get through it. Scribble based editing framework is intuitive and easy to interact. This framework can be applied in colorization [7], matting [8], changing local tonal values [9], editing material [10] and so on. Current edit propagation techniques based on this framework [1,7,12,14] noticed the two major constrain conditions. First, the result will be like to user-specified edit. Second, the final edit is similar for close-by regions of similar appearance. Then edit results such as recoloring and tonal values adjustments can be achieved through post-processing of images. Although edit propagation techniques for digital images and videos developed rapidly, there is still much room for improvement.

© Springer International Publishing Switzerland 2015
Y.-S. Ho et al. (Eds.): PCM 2015, Part I, LNCS 9314, pp. 402–410, 2015.
DOI: 10.1007/978-3-319-24075-6_39

We invoke that L_0 gradient smooth weight term can preserve the original structure better, making the result more continuous and natural. We propose a method using L_0 gradient minimization to constrain the propagated result which refines the final edit. L_0 gradient minimization can globally control how many non-zero gradients are resulted in to apporoximate prominent structure in a sparsity-control manner, make the appearance after edit more continuous and natural. Experiments on a variety of data sets show that our method is effective in remaining the structure of original image when propagating edit.

Propagation methods always observe two principles [1]. First, pixel with user-specified edit should retain that specified amount of edit after propagation. Second, pixels with similar locations (e.g., X, Y coordinates and frame index) and having similar colors are more likely to receive similar amount of edits.

The propagation framework of ours is inspired by previous methods. The previous methods ensure the final edit similar for close-by regions of similar appearance by constraining the gradients to be small for similar samples. They all use the sum of the difference of the squares to constrain the results to be smooth. Different from them, we proposed global L_0 gradient smooth weight term. It can remove low-amplitude structures and globally preserve and enhance salient edges, which is what we expect when propagating. Various experiments demonstrate our result is more continuous and natural.

We further make the following contributions: (1) We integrate commonly used L_0 gradient minimization with edit propagations, which can propagate existing edit to the entire image efficiently and naturally in general scribble-based edit propagation techniques. (2) We propose a self-adaptive re-parameterization way to control the constrained gradient minimization, which is intuitive and convenient. (3) We improve the huge afinity matrix approximation accuracy by select sampled columns carefully with maximizing the interval. In this way, the structure of original afinity matrix is preserved better.

The remainder of this paper is organized as follows. In the next section we discuss various applications of edit propagation techniques, explain different similarity measure and processing methods. Next, in Sects. 3 and 4, we proposed our model and interpret its solution. Finally, we demonstrate the utility of our propagation in the context of several applications.

2   Related Works

We review edit propagation techniques in this section. We categorize them as local propagation and nonlocal ones.

Levin et al. first use scribble based editing framework with optimization methods [7] to colorize photos, which can also be used for recoloring and matting [8]. It is archived by constructing a sparse affinity matrix for neighboring pixel pairs. Using optimization to propagate with enough scribble can get nearly perfect results, but user input is too complicated and need to follow the intricate photo patterns.

An et al. advocate using global method [1] to deal with photos with intricate patterns and spatially discontinuous regions. Using this do not need precise user input. The same pattern in feature space, whether adjacent in distance space, can be propagated properly. Xu et al. use kd-tree to accelerate this method and extend it to video editing [12].

Xu et al. generalize local [7] and global [1] methods and are in favor of a sparse control model [14]. He suggests both local and global methods have their deficiencies and are special cases of sparse control model. Local methods need dense strokes and cannot propagate to far region. Global methods such as AppProp [1] will incur appearance mixing.

Farbman et al. advocate using diffusion distance [2] instead of Euclidean distance when calculating the degree of similarity between pairs of pixels. He proposed that calculate distance from diffusion maps better account for the global distribution of pixels in their feature space.

The last decade has seen the rapid growth of efficient edge-aware filtering, originating in various principles and theories. Recently people have more progress on edge-preserving smoothing [3–6,13,15]. With these methods, we can reserve major edges while eliminate a manageable degree of low-amplitude structures. Especially, L_0 smoothing [13] globally locates important edges instead of depending on local features. Inspired by L_0 smoothing, we integrate the L_0 gradient constrain terms in edit propagation.

3   The L_0 Propagation Method

3.1   Algorithm Framework

We proposed a new framework to propagate the scribbles to the entire image. The framework obey two constraints: first, the final edit is similar for close-by regions of similar appearance by constraining the all-pair difference to be small for similar samples. Second, the final edit should have appropriate gradients. We introduce the gradient counts of edit

$$C(e) = \# \left\{ p \mid |\partial_x e_p| + |\partial_y e_p| \neq 0 \right\} \tag{1}$$

as a constraint. The variable e represents final edit. Using L_0 weight term can remove low-amplitude structures and globally preserve and enhance salient edges of edit. We add the L_0 term to the energy function. With refined e, the edited appearance will be more continuous and natural, then we have

$$\sum_i \sum_j w_j z_{ij} (e_i - g_j)^2 + \lambda C(e). \tag{2}$$

In (2), g means the input edit of user. The form of affinity matrix is

$$z_{ij} = \exp(-\|\mathbf{f}_i - \mathbf{f}_j\|^2 / \sigma_a) \exp(-\|\mathbf{x}_i - \mathbf{x}_j\|^2 / \sigma_s). \tag{3}$$

To solve (2), a special alternating optimization strategy with half quadratic splitting, based on the idea of introducing auxiliary variables to expand the original terms and update them iteratively is adopted. Xu et al. [13] also used this method to iteratively solve the discrete counting mextric. Equation (2) is written as

$$\sum_i \sum_j w_j z_{ij} (e_i - g_j)^2 + \lambda C(h, v) + \beta((\partial_x e_p - h_p)^2 + (\partial_y e_p - v_p)^2). \tag{4}$$

We split (4) into two parts and solve

$$\sum_i \sum_j w_j z_{ij} (e_i - g_j)^2 + \lambda C(h, v) \tag{5}$$

first. To minimize the quadratic energy function (5), we solve the set of linear equations in (6).

$$(Z^* + \beta(\partial_x \partial_x + \partial_y \partial_y))e = ZWg + \beta(h\partial_x + v\partial_y). \tag{6}$$

Obviously, the affinity matrix will be n×n (n indicates the total number of all image pixels) dimensions and is not sparse. To efficiently solve (6), as the affinity matrix has considerable structure and is close to low rank, we write it as

$$Z \approx \tilde{Z} = UA^{-1}U^T. \tag{7}$$

When sampling the matrix, we proposed a method which maximizes the initial distribution in the following section.

Finally, we get

$$e = ((UA^{-1}U^T)^* + \beta(\partial_x \partial_x + \partial_y \partial_y))^{-1}(UA^{-1}U^T Wg + \beta(h\partial_x + v\partial_y)). \tag{8}$$

As for the second half

$$((\partial_x e_p - h_p)^2 + (\partial_y e_p - v_p)^2) + \frac{\lambda}{\beta}C(h, v), \tag{9}$$

we adopt the same threshold method [13], which is

$$(h_p, v_p) = \begin{cases} (0,0) & (\partial_x e_p)^2 + (\partial_y e_p)^2 \leq \lambda/\beta \\ (\partial_x e_p, \partial_y e_p) & \text{otherwise.} \end{cases} \tag{10}$$

We summarize our alternating minimization algorithm in Algorithm 1. Parameter β is automatically adapted in iterations starting from a small value β_0, it is multiplied by κ each time. This scheme is effective to speed up convergence [11].

3.2 Affinity Matrix Approximation

One problem is that computing affinity matrix Z needs both huge amount memory and long time. An et al. suggest the affinity matrix has considerable structure and is close to low rank [1], so they come up with using m linearly independent columns to approximate it. However, the simple approximation way could lead to low accuracy and random results.

Algorithm 1. Edit Propagation via L_0 Gradient Minimization

Input: image I, user-specified scribble g, smoothing weight λ, parameters β_0, β_{max},
   and rate κ
Initialization: $e \leftarrow g, \beta \leftarrow \beta_0, i \leftarrow 0$
1: **repeat**
2:     With $e^{(i)}$, solve for $h_p^{(i)}$ and $v_p^{(i)}$ in (9).
3:     With $h^{(i)}$ and $v^{(i)}$, solve for $e^{(i+1)}$ with (8).
4:     $\beta \leftarrow \kappa\beta, i++$
5: **until** $\beta \geq \beta_{max}$
Output: result edit e

We propose a method which can better approximate the affinity matrix by maximizing interval between column vector. First, k groups of m linearly independent columns are selected. We represent the ith column vector as V_i. Then we choose m columns from $k * m$ columns which maximizes

$$\max \sum_{i=1}^{m} \sum_{j=i}^{m} (V_i - V_j)^2. \tag{11}$$

By maximizing the margins between sampled column, the structure of original affinity matrix is preserved. Compared with appprop [1], it can greatly enhance accuracy and reduce the randomness of approximation result.

3.3   Constrain Parameters

We control the propagation by three parameters, λ, σ_a and σ_s. σ_s indicates the spacial feature weight, and in this paper normalized it with image size. σ_a depends on the types of input image because the appearace's distribution varies. λ is a weight directly controlling the significance of $C(e)$, which is in fact a smoothing parameter. The number of non-zeros gradients is monotone with respect to $1/\lambda$.

We adopt a heuristic way to tune the gradient minimization parameter λ. When the user draws more scribbles, the propagated result will have more non-zeros gradients. In this case the smooth parameter would be small. When the user specifies cribbles, the area which is covered by them will have different scale. On the other hand, λ should be in proportion to the size of the image.

4   Experiments

4.1   Implemention

We do not directly get the final edit results. Instead, edit parameters e are calculated first. After calculate e, we can easily make different appearance changes such as color, intensity, tonality and materials. This will make the framework more general and can be applied more widely Fig. 1.

(a) Input (c) AppProp (e) Edit of AppProp(close-up)

(b) Sparse control model (d) Ours (f) Edit of ours(close-up)

Fig. 1. Comparison with other edit propagation approaches on recoloring. When looking close-up, our method can generate more smooth and natural results. (Parameters of our algorithm: $\sigma_a = 0.8, \sigma_s = 0.2, \lambda = 0.02$)

Lots of experiments demonstrated that our method is able to avoid the noise and make the result much natural and continuous. We tested our algorithm using a variety of edit with images for changing different apearance, partly decribed in Sects. 4.2 and 4.3.

4.2 Recoloring

In recoloring, user generally specifies regions where they want to change color with color strokes. Then they use white strokes to specify region where color should not be changed. We represent the image in YUV color space. The specified color region will propagate to the entire image in UV color space.

We choose AppProp [1] and sparse control model [14] to perform and compare recoloring with our method under the same user input, as indicated in Fig. 2. From results in Fig. 2(b)–(d) we can see our result can achieve more natural and intuitive results. When looking close-up in Fig. 2(e) and (f), the boundary resulted by our method is more clear and have less noise.

A more challenging recoloring example is shown in Fig. 2. In the image, the grapes in background have blurred boundaries and their color is similiar with the stem. All these results validate the superior performance of our propagation model in natural representation and detail noise.

4.3 Tonal Values Adjustments

While photography has touched us more recently, it's still difficult to get satisfactory exposure in the procedure of taking a photograph. We propose that our

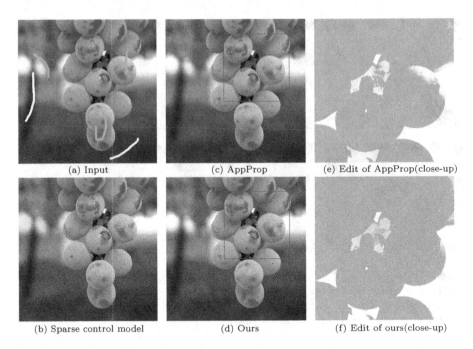

Fig. 2. Comparison with other edit propagation approaches on tonal values adjustments. when looking close-up, our method can generate more smooth and natural results. (Parameters of our algorithm: $\sigma_a = 0.1, \sigma_s = 0.1, \lambda = 0.02$)

algorithm is easy and intuitive for adjusting tonal values of different regions on post-processing. In YUV color space, we use a rough intensity brush to suggest that the gain of the Y component, then the sparse edit will propagate to the entire image, obeying our principles.

We assign scribbles to make the bell tower brighter and the sky darker in Fig. 3, thus simulating scenarios in the evening. We compare our method with mainstream methods in Fig. 3(d)–(f). It is clear that the edit mask of our method can get better results, especially in the high-texture area. After taking the gradient into consideration, the comparison of results and the edit verify that propagation result of our method is more continuous and natural than state-of-art techniques.

5   Discussions and Conclusions

In this paper, we have presented a novel, efficient method to enhance the result in edit propagation. While previous works [1,7,14] only use simple constraints, We integrate with global L_0 gradient terms which can propagate existing edit to the entire image efficiently and naturally in general scribble-based edit propagation techniques. Experiments show that our method effectively make the result more continuous and natural.

(a) Input   (b) Edit of AppProp   (c) Edit of ours

(d) Sparse control model   (e) AppProp   (f) Ours

Fig. 3. Comparison with other edit propagation approaches on recoloring. When looking close-up, our method can generate more smooth and natural results. (Parameters of our algorithm: $\sigma_a = 0.8, \sigma_s = 0.2, \lambda = 0.02$)

In future, we will do more experiments on different propagation methods. While we used L_0 norm weight term in this paper because its global features, we will take more state-of-art edge aware terms into consideration.

References

1. An, X., Pellacini, F.: Appprop: all-pairs appearance-space edit propagation. ACM Trans. Graph. (TOG) **27**, 40 (2008)
2. Farbman, Z., Fattal, R., Lischinski, D.: Diffusion maps for edge-aware image editing. ACM Trans. Graph. (TOG) **29**, 145 (2010)
3. Farbman, Z., Fattal, R., Lischinski, D., Szeliski, R.: Edge-preserving decompositions for multi-scale tone and detail manipulation. ACM Trans. Graph. (TOG) **27**, 67 (2008)
4. Gastal, E.S., Oliveira, M.M.: Adaptive manifolds for real-time high-dimensional filtering. ACM Trans. Graph. (TOG) **31**(4), 33 (2012)

5. He, K., Sun, J., Tang, X.: Guided image filtering. In: Daniilidis, K., Maragos, P., Paragios, N. (eds.) ECCV 2010, Part I. LNCS, vol. 6311, pp. 1–14. Springer, Heidelberg (2010)
6. Kopf, J., Cohen, M.F., Lischinski, D., Uyttendaele, M.: Joint bilateral upsampling. ACM Trans. Graph. (TOG) **26**, 96 (2007)
7. Levin, A., Lischinski, D., Weiss, Y.: Colorization using optimization. ACM Trans. Graph. (TOG) **23**(3), 689–694 (2004)
8. Levin, A., Lischinski, D., Weiss, Y.: A closed-form solution to natural image matting. IEEE Trans. Pattern Anal. Mach. Intell. **30**(2), 228–242 (2008)
9. Lischinski, D., Farbman, Z., Uyttendaele, M., Szeliski, R.: Interactive local adjustment of tonal values. ACM Trans. Graph. (TOG) **25**, 646–653 (2006)
10. Pellacini, F., Lawrence, J.: Appwand: editing measured materials using appearance-driven optimization. ACM Trans. Graph. (TOG) **26**, 54 (2007)
11. Wang, Y., Yang, J., Yin, W., Zhang, Y.: A new alternating minimization algorithm for total variation image reconstruction. SIAM J. Imaging Sci. **1**(3), 248–272 (2008)
12. Xu, K., Li, Y., Ju, T., Hu, S.M., Liu, T.Q.: Efficient affinity-based edit propagation using kd tree. ACM Trans. Graph. (TOG) **28**, 118 (2009)
13. Xu, L., Lu, C., Xu, Y., Jia, J.: Image smoothing via l 0 gradient minimization. ACM Trans. Graph. (TOG) **30**(6), 174 (2011)
14. Xu, L., Yan, Q., Jia, J.: A sparse control model for image and video editing. ACM Trans. Graph. (TOG) **32**(6), 197 (2013)
15. Zhang, Q., Shen, X., Xu, L., Jia, J.: Rolling guidance filter. In: Fleet, D., Pajdla, T., Schiele, B., Tuytelaars, T. (eds.) ECCV 2014, Part III. LNCS, vol. 8691, pp. 815–830. Springer, Heidelberg (2014)

Improved Salient Object Detection
Based on Background Priors

Tao Xi[1][(✉)], Yuming Fang[2], Weisi Lin[1], and Yabin Zhang[1]

[1] Nanyang Technological University,
50 Nanyang Avenue, Singapore 639798, Singapore
xitao1989@gmail.com
[2] Jiangxi University of Finance and Economics, Nanchang, China

Abstract. Recently, many saliency detection models use image boundary as an effective prior of image background for saliency extraction. However, these models may fail when the salient object is overlapped with the boundary. In this paper, we propose a novel saliency detection model by computing the contrast between superpixels with background priors and introducing a refinement method to address the problem in existing studies. Firstly, the SLIC (Simple Linear Iterative Clustering) method is used to segment the input image into superpixels. Then, the feature difference is calculated between superpixels based on the color histogram. The initial saliency value of each superpixel is computed as the sum of feature differences between this superpixel and other ones in image boundary. Finally, a saliency map refinement method is used to reassign the saliency value of each image pixel to obtain the final saliency map for images. Compared with other state-of-the-art saliency detection methods, the proposed saliency detection method can provide better saliency prediction results for images by the measure from precision, recall and F-measure on two widely used datasets.

Keywords: Saliency detection · Background priors · Earth movers distance (EMD)

1 Introduction

Digital images have become an increasingly important part of daily life due to the rapid development of information technologies and applications. Various image processing technologies are much desired for visual content analysis in emerging applications. Salient object detection is crucial in various scenarios, such as image retrieval, object recognition, image classification, etc. To effectively locate the visually attractive or interesting objects in images, saliency detection has been widely explored in the research area of computer vision recently [5,6,8,11].

Salient object detection is regarded as a high-level perception process during observers' viewing of visual scenes [4]. The authors of the study [4] labeled rectangle boxes including objects as the ground truth for performance evaluation of salient object detection. They proposed a saliency detection method

© Springer International Publishing Switzerland 2015
Y.-S. Ho et al. (Eds.): PCM 2015, Part I, LNCS 9314, pp. 411–420, 2015.
DOI: 10.1007/978-3-319-24075-6_40

based on machine learning. A new set of features are extracted for salient object prediction by employing conditional random field (CRF). In [5], Achanta et al. exploited the effectiveness of color and luminance features in salient object detection, and proposed an algorithm to detect the salient objects with well-defined boundaries. In [6], Cheng et al. developed a saliency extraction method by measuring regional contrast to create high-quality segmentation masks. In [7], a novel saliency detection model for JPEG images was proposed by Fang et al., in which the DCT difference is used to calculate the saliency degree of each image block. The authors of [8] exploited two common priors of the background in natural images to provide more clues for saliency detection. In [9], Shen and Wu. proposed a novel low rank matrix recovery based saliency detection method by combining high-level priors and the low-level visual features. In [10], Xie et al. built a Bayesian framework based saliency detection method by employing low and mid level visual cues. In [11], Yang et al. proposed a salient object detection method for saliency prediction in images by using graph-based manifold ranking, which incorporates boundary priors and local cues.

Most existing saliency detection models mentioned above calculate the saliency map by computing the feature differences between center-surround patches (or pixels) in images. Among these models, several studies regard the image boundary as the background priors for saliency detection, because it is commonly accepted that photographers would focus on the objects in natural scenes when taking photos [21]. However, boundary information cannot represent the complete background estimation when some of the salient region is located in the image boundary, and this would result in an inaccurate saliency map.

In this paper, an effective saliency detection method is proposed by employing an image segmentation algorithm to adjust the saliency value of image pixels to address the problem mentioned above. First, the input image is divided into superpixels by using the SLIC (Simple Linear Iterative Clustering) method [14], and the color histogram of each superpixel is extracted as the feature. Then the Earth Mover's Distance (EMD) is adopted to calculate the feature differences between superpixels because it was widely used in computer vision as an effective similarity metric. The saliency degree for each superpixel is computed as the sum of feature differences between this superpixel and other superpixels within the image boundary (which is regarded as the background priors), since saliency is the contrast between current region and background. Finally, a graph-based image segmentation method (GS04) [13] is employed to adjust the saliency value of each image pixel for the final saliency map prediction, since the segmented superpixels by SLIC are almost uniform size and are more suitable to used as the basic processing units for image representation, while in the segment regions from GS04, much more object boundaries are reserved [14]. Experimental results have demonstrated that our saliency detection method can obtain better performance for salient object detection than other existing ones.

2   Improved Saliency Detection Based on Background Priors

We provide the framework of the proposed saliency detection method in Fig. 1. Given an image, we first segment it into superpixels as computational units, and calculate the initial saliency map by measuring the dissimilarity between the superpixel and those in the image boundary. Then, the background prior is updated for the saliency map calculation. Finally, we adjust the saliency value of each image pixel by utilizing a graph-based segmentation method (GS04) [13] to obatain the final saliency map of the input image.

2.1   Pre-processing

In this study, the SLIC method [14] is employed to divide the image into super-pixels. In many vision tasks, superpixels are used to capture image redundancy and reduce the complexity of image processing algorithms. Also, compact and highly uniform superpixels with respect to image color and contour are much desired. The authors of [14] compared five state-of-the-art superpixel extraction methods with respect to speed, image boundaries, memory efficiency, and the performance on segmentation results. Compared with other existing methods (include GS04), the SLIC algorithm is easy to be implemented and it provides compact and highly uniform superpixels [14]. Here, we use the SLIC algorithm to generate superpixels as computational units for saliency calculation. For each superpixel, we extract the color histogram as its feature. To improve the computational efficiency, a sparse histogram representation is adopted for feature extraction of superpixels here. Specifically, we first use a color quantization method [16] to reduce quantization bins of each channel in RGB color space. After that, we transform RGB color space to CIEL*A*B* color space due to the latter's perceptual property.

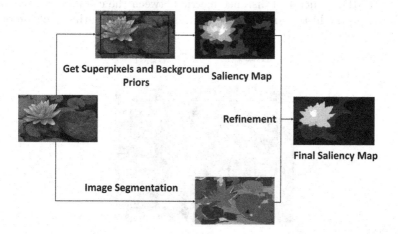

Fig. 1. The framework of the proposed salient object detection method.

2.2   Initial Saliency Map Calculation

Generally, photographers always focus on salient objects when taking photos. Thus, the foreground objects are more likely to be located at the image center. Due to this characteristic, "Center Bias" mechanism can be used as a significant factor in saliency detection. Existing studies have also pointed out that photographers always tends to place interested objects near the center of the shot and to enhance foreground objects relative to the background [12].

In this study, we regard image boundary as the background priors for saliency detection. Although some part of an image boundary might not be background, the image boundary usually captures the representative statistics of background region for an image. For a superpixel S_i where i is the location index of the superpixel, we measure its saliency degree by computing its color difference to all superpixels in the image boundary. Denoting the height and width of the current image as H and W (as shown in Fig. 2), we define the outside of the black rectangle box as $BackgroundPriors$ region (BP region), the inside of the black rectangle box is defined as $Non\text{-}BackgroundPriors$ region ($N\text{-}BP$ region). Specifically, we define the width of four sides of the BP region as $D_{top} = H/n$, $D_{bottom} = H/n$, $D_{left} = W/n$, $D_{right} = W/n$, respectively, n is a parameter which is greater than 1 to control the size of the BP region (here, we set $n=30$). For each superpixel S_i, if there is any one pixel falling into BP region, this superpixel is labeled as $BackgroundPriors$ superpixel (BP superpixel). Otherwise, this superpixel is labeled as $Non\text{-}BackgroundPriors$ superpixel ($N\text{-}BP$ superpixel). For each superpixel S_i, its saliency value $Sal(S_i)$ is defined as:

$$Sal(S_i) = \left(\sum_{j=1}^{M} EMD\,(S_i, S_j) \right)^{\frac{1}{M}}, \quad S_j \in BP \; superpixels \qquad (1)$$

where M is the number of BP superpixels in the image; EMD is earth mover's distance (EMD), which is a distance metric between the color histograms of S_i and S_j. In [6], the histogram distance is measured by computing the sum of the

Fig. 2. The illustration of BP region and N-BP region.

differences between each pair bins. In this study, we use the EMD [16] to calculate the differences between superpixels, because it can obtain better performance for perceptual similarity measure than other similarity metrics [16]. Generally, EMD can be regarded as a solution to the famous transportation problem in linear optimization. For two different color histograms, the EMD is the minimum cost of changing one histogram into the other.

2.3 Saliency Map Refinement

Several recent approaches exploit the background priors to enhance saliency computation and their experimental results demonstrate that the background priors is effective in saliency detection. However, they simply treat all image boundaries as the background region. These models might fail if the salient object is slightly overlapped with the boundary due to the fact that they simply use all image boundary as the background priors [18]. To address this drawback, the initial saliency map is refined by reassigning the saliency value of each image pixel. Specifically, we first segment the input image into different regions by employing a graph-based image segmentation method (GS04) [13], and then reassign saliency value to each pixel by computing the mean value of saliency values in each new region. Compared with SLIC, the segment regions from GS04 are more likely to include the integrated objects than the results from SLIC [14]. Thus, the saliency value in foreground or background from GS04 can be more uniform than that of SLIC. The saliency map with refinement can be computed as follows.

$$S_p = \frac{1}{N_p} \sum Sal_{i,j}, \quad (i,j) \in reg_p \tag{2}$$

where reg_p denotes the p_{th} region from the segmentation result, N_p represents the number of image pixels in reg_p. We can obtain the final saliency map by normalization as follows.

$$Norm(Sal_{i,j}) = \frac{Sal_{i,j} - Sal_{min}}{Sal_{max} - Sal_{min}} \tag{3}$$

where Sal_{min} and Sal_{max} are the minimum and maximum values of $Sal_{i,j}$ over all pixels in the image.

3 Experiments

In this section, we evaluate the performance of our saliency detection method and analyze the effectiveness of the proposed saliency map refinement method on two public datasets: ASD [5] and MSRA [15]. ASD includes 1000 original images with their corresponding binary masks of salient objects, while there are 5000 images and the corresponding ground truth in MSRA. The images in ASD are relatively simpler than the ones in MSRA. The images from MSRA contain complex background and low contrast objects, which make this dataset challenging for saliency detection. For single image saliency detection, we compare our

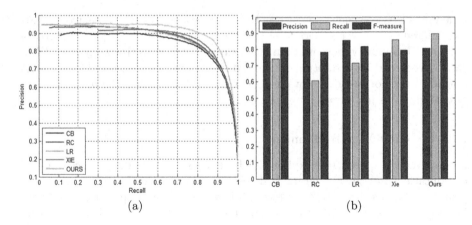

Fig. 3. Comparison results between the proposed saliency detection model and other four state-of-the-art methods on ASD dataset. (a) *P-R* curve. (b) Mean *Precision*, *Recall*, and *F-measure* values of the compared methods.

saliency detection method with four state-of-the-art methods of saliency detection: RC [6], CB [19], XIE [10], and LR [9]. Although there have been dozens of saliency detection methods designed in recent decades since the well-known saliency detection method [20] was proposed by Itti et al., a lot of methods aim to predict humans? eye fixation and cannot get promising results for salient object detection. Therefore, we choose several recent representative methods for salient object detection models to conduct the comparison experiment. Also, for each saliency detection method, we set the default parameters provided by the authors to run the executable source codes.

The *Precision*, *Recall* and *F-measure* are employed to evaluate the matching degree between a saliency map and its corresponding ground truth. *Precision -Recall* curve (*P-R* curve) is commonly applied to measure the performance of the salient object detection model. The *P-R* curve can be drawn as the curve of *Precision* versus *Recall* by means of setting different thresholds. Generally, a higher *P-R* area means a better predication performance for a specified saliency detection method. Furthermore, similar with [5], we use the image dependent adaptive thresholds to evaluate the performance of our model. The image dependent adaptive threshold [5] is defined by as twice of the mean of the saliency map. The *F-measure* is introduced to evaluate the performance of saliency detection models when using image dependent adaptive threshold. The *F-measure* is defined as:

$$F = \frac{(1 + \beta^2) \cdot Precision \cdot Recall}{\beta^2 \cdot Precision + Recall} \tag{4}$$

where the coefficient β^2 is set to 1, indicating the equal importance of *Precision* and *Recall* as [17].

3.1 Performance Evalation on ASD and MSRA Datasets

Images in the ASD dataset contains objects with high contrast and clear contour. Figure 3 (a) shows the P-R curve of RC [6], CB [19], XIE [10], LR [9] and our method, we can see that our method has the highest P-R curve and thus outperforms all the other methods. Also, we compare the performance between our method and all the other methods with the adaptive thresholds for each image. As shown in Fig. 3 (b), our method also obtains the best performance among the compared models on this dataset. We provide some visual comparison samples from the ASD dataset in Fig. 4. From this figure, it is apparent that the saliency map generated by our method on these samples are more consistent with the ground truth compared with other existing models. Specifically, our method outperforms other methods when the visual scene is complex (See the first two rows of Fig. 4).

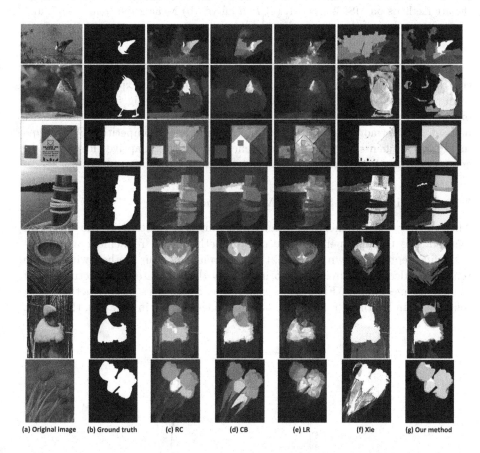

(a) Original image   (b) Ground truth   (c) RC   (d) CB   (e) LR   (f) Xie   (g) Our method

Fig. 4. Visual comparison of our saliency detection method with four other methods.

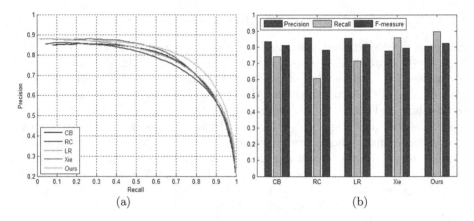

(a)   (b)

Fig. 5. Comparison results between our saliency detection method and four state-of-the-art methods on MSRA dataset. (a) *P-R* curve. (b) Mean *Precision*, *Recall*, and *F-measure* values of the compared methods.

In the MSRA dataset, there are some images including objects with low contrast, which make this dataset more challenging. The experimental results from the compared models on this database are shown in Fig. 5. From Fig. 5(a), our method yields higher *P-R* curve than CB, RC, LR and XIE for the MSRA dataset. Furthermore, we can see that our method has a competitive *F-measure* in Fig. 5(b), when saliency maps are generated by the adaptive thresholds. In general, our method is able to better detect salient objects in images.

3.2   Effectiveness of Saliency Map Refinement

The saliency map refinement is an important contribution in the proposed method. We test the performance of our method with refinement and without refinement. As shown in Fig. 6, the performance of our method without the refinement decreases obviously. To intuitively demonstrate the effectiveness of refinement mechanism, Fig. 7 gives some saliency samples of our method with refinement and without refinement. For the images shown in Fig. 7(a), the salient

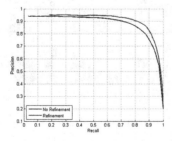

Fig. 6. Evaluation on the effectiveness of refinement.

(a)   (b)

Fig. 7. Visual comparison of our saliency detection method with and without refinement on integrity (a) and uniformity (b) of saliency values on salient object. From left to right (in (a) and (b)): input image, ground truth, saliency map generated by our method without refinement, saliency map generated by our method with refinement.

objects are near to the image boundary, which results in incompleteness of detected object (the third column). By employing refinement, we can obtain an complete salient object in the saliency map (the fourth column), since the segmentation results of GS04 are more likely to include the integrated objects. On the other hand, the refinement can make saliency values of image pixels in the salient objects more uniform, and it tends to highlight the salient object from the image due to the use of GS04, as shown in Fig. 7(b). This strongly indicates that refinement is a significant contributor to the effectiveness of our method.

4   Conclusion

We have presented a new salient object detection method based on background priors, in which the saliency is measured by EMD between current superpixel and boundary superpixels. To address the drawback when regarding the whole image boundary as background priors and salient objects are overlapped with the boundary, we design a saliency map refinement method to reassign saliency value by exploiting the segmentation results of a graph-based image segmentation algorithm. By employing this method, we find that the proposed method can highlight the salient objects more integrally and uniformly. Experimental results on two public test datasets have demonstrated that the proposed saliency detection method outperforms the state-of-the-art saliency detection approaches for salient objects prediction.

Acknowledgements. This research was supported by Singapore MOE Tier 1 funding (RG 36/11: M4010981), and the Rapid-Rich Object Search (ROSE) Lab at the Nanyang Technological University, Singapore. The ROSE Lab is supported by the National Research Foundation, Prime Ministers Office, Singapore, under its IDM Futures Funding Initiative and administered by the Interactive and Digital Media Programme Office.

References

1. Chen, L., Xie, X., Ma, W., Zhang, H., Zhou, H.: Image adaptation based on attention model for small-form-factor devices. In: ICME (2003)
2. Ouerhani, N., Bracamonte, J., Hugli, H., Ansorge, M., Pellandini, F.: Adaptive color image compression based on visual attention. In: ICIAP (2001)
3. Stentiford, F.: A visual attention estimator applied to image subject enhancement and colour and grey Level Compression. In: ICPR (2004)
4. Liu, T., Sun, J., Zheng, N., Tang, X., Shum, H.: Learning to detect a salient object. In: CVPR (2007)
5. Achanta, R., Hemami, S.S., Estrada, F.J., Ssstrunk, S.: Frequency tuned salient region detection. In: CVPR (2009)
6. Cheng, M., Zhang, G., Mitra, N., Huang, X., Hu, S.: Global contrast based salient region detection. In: CVPR (2011)
7. Fang, Y., Chen, Z., Lin, W., Lin, C.: Saliency detection in the compressed domain for adaptive image retargeting. IEEE Trans. Image Process. **21**(9), 3888–3901 (2012)
8. Wei, Y., Wen, F., Zhu, W., Sun, J.: Geodesic saliency using background priors. In: Fitzgibbon, A., Lazebnik, S., Perona, P., Sato, Y., Schmid, C. (eds.) ECCV 2012, Part III. LNCS, vol. 7574, pp. 29–42. Springer, Heidelberg (2012)
9. Shen, X., Wu, Y.: A unified approach to salient object detection via low rank matrix recovery. In: CVPR (2012)
10. Xie, Y., Lu, H., Yang, M.: Bayesian saliency via low and mid level cues. IEEE Trans. Image Process. **22**(5), 1689–1698 (2013)
11. Yang, C., Zhang, L., Lu, H., Ruan, X., Yang, M.: Saliency detection via graph-based manifold ranking. In: CVPR (2013)
12. Tseng, P., Carmi, R., Cameron, I., Munoz, D., Itti, L.: Quantifying center bias of observers in free viewing of dynamic natural scenes. J. Vis. **9**(7:4), 1–16 (2009)
13. Felzenszwalb, P., Huttenlocher, D.: Efficient graph-based image segmentation. Int. J. Comput. Vis. **59**(2), 167–181 (2004)
14. Achanta, R., Smith, K., Lucchi, A., Fua, P., Susstrunk, S.: Slic superpixels. Technical report, EPFL. Technical report: 149300(3) (2010)
15. Jiang, H., Wang, J., Yuan, Z., Wu, Y., Zheng, N., Li, S.: Salient object detection: a discriminative regional feature integration approach. In: CVPR (2013)
16. Grauman, K., Darrell, T.: Fast contour matching using approximate earth movers distance. In: CVPR (2004)
17. Liu, Z., Zou, W., Meur, O.: Saliency tree: a novel saliency detection framework. IEEE Trans. Image Process. **23**(5), 1937–1952 (2014)
18. Zhu, W., Liang, S., Wei, Y., Sun, J.: Saliency optimization from robust background detection. In: CVPR (2014)
19. Jiang, H., Wang, J., Yuan, Z., Liu, T., Zheng, N., Li, S.: Automatic salient object segmentation based on contex and shape prior. In: BMVC (2011)
20. Itti, L., Koch, C., Niebur, E.: A model of saliency-based visual attention for rapid scene analysis. IEEE Trans. Pattern Anal. Mach. Intell. **20**(11), 1254–1259 (1998)
21. Tian, H., Fang, Y., Zhao, Y., Lin, W., Ni, R., Zhu, Z.: Salient region detection by fusing bottom-up and top-down features extracted from a single image. IEEE Trans. Image Process. **23**(10), 4389–4398 (2014)

Position-Patch Based Face Hallucination via High-Resolution Reconstructed-Weights Representation

Danfeng Wan, Yao Lu$^{(\boxtimes)}$, Javaria Ikram, and Jianwu Li

Beijing Laboratory of Intelligent Information Technology,
Beijing Institute of Technology, Beijing, China
{wandanfeng,vis_yl,jikram,ljw}@bit.edu.cn

Abstract. Position-patch based face hallucination methods aim to reconstruct the high-resolution (HR) patch of each low-resolution (LR) input patch independently by the optimal linear combination of the training patches at the same position. Most of current approaches directly use the reconstruction weights learned from LR training set to generate HR face images, without considering the structure difference between LR and the HR feature space. However, it is reasonable to assume that utilizing HR images for weights learning would benefit the reconstruction process, because HR feature space generally contains much more information. Therefore, in this paper, we propose a novel representation scheme, called High-resolution Reconstructed-weights Representation (HRR), that allows us to improve an intermediate HR image into a more accurate one. Here the HR reconstruction weights can be effectively obtained by solving a least square problem. Our evaluations on publicly available face databases demonstrate favorable performance compared to the previous position-patch based methods.

Keywords: Face hallucination · Super resolution · High-resolution reconstruction weights · Position patch · Locality constraints

1 Introduction

Face images captured by surveillance cameras and other equipments are usually of low resolution due to the limits of these devices and some other environmental factors. Further applications are thus obstructed such as face analysis and recognition, 3D facial modeling. Therefore, effective algorithms for face image super resolution, named Face Hallucination [1] are needed.

Since Baker *et al.*'s work [1], many face hallucination methods have been proposed. Wang *et al.* [11] classify these methods into four categories according to the framworks they used: Byesian inference framework [1], subspace learning framework [3,5,12], combination of Bayesian inference and subspace learning framework [15], and sparse representation-based methods [7,8,13]. Here we mainly focus on the subspace learning methods that the proposed algorithm

© Springer International Publishing Switzerland 2015
Y.-S. Ho et al. (Eds.): PCM 2015, Part I, LNCS 9314, pp. 421–430, 2015.
DOI: 10.1007/978-3-319-24075-6_41

belongs to. In subspace learning methods, principal component analysis (PCA) based algorithms [3,12] can well maintain the global property but ignore the neighbor relations which leads to a compensate phrase to add some local facial features. To address this problem and also avoid dimension reduction, Chang *et al.* [2] use Neighbor Embedding to better utilize local features in training samples. Their method, inspired by Local Linear Embedding [10], has the assumption that the low-resolution (LR) and the high-resolution (HR) training images, although in different feature space, hold the same local geometry. After that, several Neighbor Embedding based methods [16] are developed and have achieved promising results.

However, neighborhood preservation for low- and high-resolution patches rarely holds. Ma *et al.* [9] overcome this problem by using position patches rather than neighbor patches to reconstruct HR face images. In [9], each LR patch is reconstructed by the linear combination of the training LR patches at the same position. While the reconstruction weights can be obtained by solving a least square problem, the solution may be not unique when the number of training samples is much larger than the dimension of the patch. Jung *et al.* [6] break the limitation by solving an L1 optimization problem. However, this method emphasizes too much on sparsity rather than locality which is recently demonstrated more improtant [14]. Thus, Jiang *et al.* [4,5] impose locality constraints into the least square problem which achieves better results.

Note that how to get the optimal reconstruction weights is the key for these methods. By solving equations with different constraints, they learn the reconstruction weights in LR feature space and then directly use them to reconstruct the conressponding HR image on HR training data.

However, we observe that the reconstruction weights in HR training data are not totally unavailable. Accordingly, we propose a novel image representation scheme, called High-resolution Reconstructed-weight Representation (HRR), to learn the HR reconstruction weights in HR feature space so that we can obtain more accurately hallucinated images. **The main contributions of our paper are as follows:**

- We are the first to introduce the concept of HR reconstruction weights. It is known that HR space contains much more information compared with the LR one. Hence, instead of directly using the reconstruction weights learned from the LR space [4,6,9], our algorithm optimizes HR reconstruction weights through the intermediate HR patch in HR space by solving the regularized least square problem. Thus, better hallucinated face images can be achieved.
- To better capture local features and reduce the influence of noise, given a position patch, we only search its K nearest neighbor patches in its position patch set for the optimization of HR reconstruction weights, which leads to richer details in final hallucinated images.

The rest of this paper is organized as follows: in Sect. 2, we review the most related work, and introduce the proposed concept of HR reconstruction weights. Section 3 presents the proposed face hallucination algorithm. Section 4 is the experimental results and analysis followed by the conclusion in Sect. 5.

2   High-Resolution Reconstruction Weights

In this section, we firstly review the most representative position-patch based face hallucination method and then introduce the proposed concept of HR reconstruction weights.

In position-patch based algorithms, each image, including the LR input image and the training images, is divided into N overlapping patches. Particularly, let I_L denote the LR input image and $\{X_L^p(i,j)\}_{p=1}^N$ be the patch set divided from I_L. $\{Y^m\}_{m=1}^M$ represents the training samples, where M is the number of training images. For each training image pair, $\{Y_L^{mp}(i,j)\}_{p=1}^N$ and $\{Y_H^{mp}(i,j)\}_{p=1}^N$ are the low- and high- resolution training patches respectively, and $m = 1, 2, \ldots, M$.

Each LR patch $X_L(i,j)$ located at the i-th row, j-th column in I_L is represented as the linear combination of LR training patches at the same position:

$$X_L(i,j) = \sum_{m=1}^M w_m(i,j)Y_L^m(i,j) + e \qquad (1)$$

where e is the reconstruction error.

To get w, least square representation (LSR) [9] optimizes the weights by narrowing down the error e in (1):

$$\min \| X_L(i,j) - \sum_{m=1}^M w_{mL}(i,j)Y_L^m(i,j) \|_2^2 \qquad (2)$$

then the corresponding HR patch is reconstructed with the weights directly:

$$X_H(i,j) = \sum_{m=1}^M w_{mL}(i,j)Y_H^m(i,j) \qquad (3)$$

finally, the whole image is integrated using all the estimated HR patches.

Note that we replace w in [9] with w_L to represent the weights so that the proposed concept of HR reconstruction weights can be distinguished.

2.1   High-Resolution Reconstruction Weights

It is clear that in most patch-based methods [2,4,9], the weights are learned only in LR feature space. Their shared assumption is that the local geometries are the same in low- and high-resolution feature space. However, HR feature space usually contains much more information, and it is reasonable to assume that the local geometries in LR and HR feature space are different to some extent. Thus, the direct usage of the weights learned from LR space can result in imprecise reconstruction. To solve this problem, we propose the concept of HR reconstruction weights.

Let $X_H(i,j)$ be the HR patch, and it can be represented as the linear combination of patches at the same position in HR training data:

$$X_H(i,j) = \sum_{m=1}^{M} w_{mH}(i,j) Y_H^m(i,j) + e \tag{4}$$

where e is the reconstruction error, and w_{mH} is the HR reconstruction weights. The weights can be obtained by solving the following optimization problem:

$$\min \| X_H(i,j) - \sum_{m=1}^{M} w_{mH}(i,j) Y_H^m(i,j) \|_2^2$$

$$\text{s.t.} \sum_{m=1}^{M} w_{mH}(i,j) = 1 \tag{5}$$

here, HR reconstruction weights w_H refer to the weights that learned from HR feature space which are different from the weights w_L used in [4,9].

Face Hallucination aims to get $X_H(i,j)$, but in (5), the HR weights $w_{mH}(i,j)$ is also unknown. Traditional algorithms simply replace $w_{mH}(i,j)$ with $w_{mL}(i,j)$ that is easy to obtain but ignoring the full information in HR feature space for w_H. Hence in this work, rather than directly using w_L, we propose High-resolution Reconstructed-weights Representation (HRR) to approximate w_H. Details are introduced in Sect. 3.

3   High-Resolution Reconstructed-Weights Representation

In this section, we present details of the proposed representation scheme. For simplicity, we first assume HR patch $X_H(i,j)$ is known and then drop this assumption in Sect. 3.1.

To obtain the HR reconstruction weights, we reformulate (5) by imposing it with Local Coordinate Coding [14] as in [4]:

$$w_H^*(i,j) = \underset{w_H(i,j)}{\arg\min} \begin{cases} \|X_H(i,j) - \sum_{m=1}^{K} w_{mH}(i,j) Y_H^m(i,j)\|_2^2 \\ + \lambda \sum_{m=1}^{K} \|d_{mH}(i,j) \circ w_{mH}(i,j)\|_2^2 \end{cases} \tag{6}$$

where $w_H = \{w_{mH}(i,j), m = 1, \ldots, K\}$, and K is the number of nearest neighbor patches in position patch set $\{Y_H^{mp}(i,j)\}_{m=1}^{M}$. λ is a constant to balance the reconstruction error and the local constraints and $\circ$ is the point wise product. $d_{mH}(i,j)$ is the Euclidean distance vector between $X_H(i,j)$ and each neighbor patch:

$$d_{mH}(i,j) = \| X_H(i,j) - Y_H^m(i,j) \|_2^2, 1 \le m \le K \tag{7}$$

It is worth mentioning that, different from [4,9] that take all the training patches, we only choose K nearest patches to optimize reconstruction weights. It will better emphasize the local features and reduces the noise from distant patches. The results are especially pronounced when the training data set is huge, see Sect. 4.3.

3.1   Estimate of the HR Reconstruction Weights

Note that for an LR input image, we actually do not have the conrresponding HR image which makes the calucation of HR reconstruction weights intractable. Thus, an estimated one can be used to optimize w_H.

First, we replace $X_H(i,j)$, $Y_H(i,j)$ with $X_L(i,j)$, $Y_L(i,j)$ respectively and get $w_L(i,j)$ for each patch via (2). Then the estimated HR patch, denoted as $X'_H(i,j)$, can be obtained via (3). After replacing $X_H(i,j)$ with $X'_H(i,j)$, (6) is now a regularized least square problem, and has the following solutions:

$$w'_H(i,j) = (Z^{-1}C)/(C^T Z^{-1}C) + \lambda D \qquad (8)$$

where C is a column vector of ones; D is the distance diagonal matrix:

$$D_{mm} = d_m(i,j), 1 \leq m \leq K \qquad (9)$$

and Z is obtained by:

$$Z = (X'_H(i,j)C^T - Y_H)^T (X'_H(i,j)C^T - Y_H) \qquad (10)$$

where $X'_H(i,j)$ is the estimated HR patch and Y_H is a matrix in which each column is the training HR patch $Y_H^{mp}(i,j)$. Note that this procedure approximates w_H, thus we use w'_H instead of w_H to represent the results.

In our algorithm, however, we do not directly use the estimated HR patch X'_H for calculating $w'_H(i,j)$. On the contrary, we get all the estimated HR patches and integrate them into an intermediate face image , denoted as $\tilde{I}_H$. Then we again divide $\tilde{I}_H$ into overlapping position patches $\{\tilde{X}_H^p(i,j)\}_{p=1}^N$ and use $\tilde{X}_H$ to calculate the estimated HR weights.

The reason for the intermediate synthesis is that by integrating the estimated HR patches into a complete face image and later dividing it again, pixel values in overlapping parts will change. Thus, the overlap parts in each patch can learn from all its adjacent patches so that the HR reconstruction weights can be more precisely estimated.

The process is illustrated in Fig. 1. (a) is two estimated adjacent patches P'_1 and P'_2 with shadows as their overlapping parts. In (b), pixels in overlapping parts of these two patches are averaged after the integration. (c) shows that pathes $\tilde{P}_1$ and $\tilde{P}_2$ for calculating HR reconstruction weights are acctually different from the ones before integration P'_1 and P'_2. Note that for simplicity, we just take two adjacent patches with one overlapping region in each patch. Overlapping areas in other sides are operated similarly with the illustrated ones.

3.2   Face Hallucination via HRR

As we mentioned before, given a LR input image, we first divide it into N overlapping patches and calculate the LR reconstruction weights for each patch to obtain an estimated HR patch. After generating all these estimated HR patches, an intermediate HR face image is synthesized. Then we take it for the division

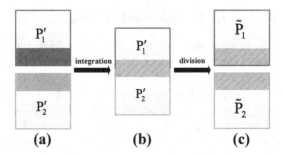

Fig. 1. Illustration on the changes of overlapping regions: (a) two adjacent estimated patches P_1' and P_2'; (b) synthesized patches with averaging pixel values in overlapping part; (c) patches after division: $\tilde{P}_1$ and $\tilde{P}_2$. (Best viewed in color).

of patches and the calculation of w_H'. The final HR patches are reconstructed using the estiamted HR reconstruction weights w_H':

$$X_H(i,j) = \sum_{m=1}^{K} w_{mH}'(i,j)Y_H^m(i,j) \tag{11}$$

with these HR patches, the final face image can be hallucinated.

The detailed description of the proposed algorithm High-resolution Reconstructed-weight Representation is summarized in Algorithm 1.

4   Experimental Results

In this section, we present the details of performed experiments and compare our method with other state-of-the-art algorithms.

4.1   Experiment Settings

To fully and fairly evaluate the proposed method, we run our algorithm on two publicly available databases: CAS-PEAL and FEI.

For CAS-PEAL database, we choose 180 frontal face images for training and other 20 images for testing. These images all have normal lighting and neutral expression. For the FEI database, which includes neutral and smiling expressions, we select more than three hundred images for training and other 27 images for testing. All the training images are aligned to 128×96 and the test LR images are obtained by downsampling the HR images into 32×24.

Each patch divided from the input LR image is 3×3 with 1 overlapping pixels which have been proven to be the best for results [9]. Meanwhile, in the calculation of w_H', we set each HR patch 12×12 with 4 overlapping pixels. In our algorithm, another two important parameters are also involved: the number of nearest neighbor patches K and the regularization parameter λ. In the

---

Algorithm 1. Face hallucination via **HRR**

Input: Training dataset $\{Y_L^m\}_{m=1}^M$ and $\{Y_H^m\}_{m=1}^M$, an LR image I_L, the parameters λ and K.

Output: High-resolution image I_H.

1   Divide I_L and each image in Y_L and Y_H into N small patches $\{X_L^p(i,j)\}_{p=1}^N$, $\{Y_L^{mp}(i,j)\}_{p=1}^N$, $\{Y_H^{mp}(i,j)\}_{p=1}^N$, respectively;

2   **for** *each LR patch* $X_L^p(i,j)$ **do**

3   Compute the LR reconstruction weights $w_L(i,j)$ via (2);

4   Synthesize HR patch $X_H'^p(i,j)$ via (3);

5   **end**

6   Integrate HR patches $\{X_H'^p(i,j)\}_{p=1}^N$ to obtain an intermediate HR image $\tilde{I}_H$;

7   Divide $\tilde{I}_H$ into N patches $\{\tilde{X}_H^p(i,j)\}_{p=1}^N$;

8   **for** *each HR patch* $\tilde{X}_H^p(i,j)$ **do**

9   Calculate Euclidean distance between $\tilde{X}_H^p(i,j)$ and each HR training patch $\{Y_H^{mp}(i,j)\}_{m=1}^M$ by (7);

10   Find its K-nearest patches;

11   Estimate HR reconstruction weights $w_H'(i,j)$ for patch $\tilde{X}_H^p(i,j)$ via (8), (9) and (10);

12   Synthesize the final HR patch X_H^p via (11);

13   **end**

14   Concatenate and integrate all the final HR patches $\{X_H^p\}_{p=1}^N$ according to their position into a facial image I_H by averaging the pixel values in overlapping regions of adjacent patches;

15   **return** I_H.

---

experiments, we set $K = 175$, $\lambda = 0.017$ for CAS-PEAL database and $K = 265$ and $\lambda = 0.027$ for FEI database respectively. Their effects to the final results are discussed in Sect. 4.3.

4.2 Results Comparison

In our experiments, we compare the proposed HRR with two FH algorithms: Least Square Representation (LSR) [9] and Locality-constrained Representation (LcR) [4]. As described in [4,9], LSR and LcR are superior to other patch based face hallucinating algorithms, such as Neighbor Embedding [2], Convex Optimization [6], Eigen-transformation [12]. Especially LcR in [4], the performance improves a lot after imposing the locality constraints. Hence here, we only compare with these two methods: LSR and LcR.

Qualitative Analysis. Some of the hallucinated face images generated by different methods on CAS-PEAL and FEI databases are shown in Fig. 2. In the figures, (a) are the LR input images and (b) are the hallucinated face images by LSR [9]. (c) are the results based on LcR [4] and (d) are the face images hallucinated by our methods. The last column (e) are the original HR images.

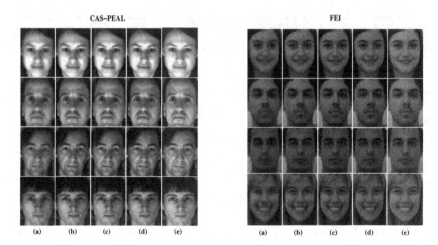

Fig. 2. Some results by different methods on CAS-PEAL and FEI databases respectively: (a) LR input images; (b) LSR's results [9]; (c) LcR's results [4]; (d) Our results; (e) Original HR images.

Quantitative Analysis. For further comparison, we calculate the average PSNR and SSIM values of all the test images. The results are listed in Tables 1 and 2. From Table 1, we can see that in CAS-PEAL database, the proposed algorithm improves average PSNR of all the test images for **0.4533(dB)** and **0.1108(dB)**, and average SSIM for **0.0085** and **0.0036** compared with LSR and LcR respectively. Meanwhile, for FEI database in Table 2, it enhances average PSNR for **1.2475(dB)** and average SSIM for **0.0325** compared with LSR, and **0.0477(dB)** and **0.0022** over LcR. Tables 1 and 2 demonstrate the superiority of our method.

4.3   Influence of Parameters

In this section, we analyze two parameters of our algorithm that are important for good reconstruction results, *e.g.* the nearest neighbor patch number K and the regularization parameter λ. To find out their influence to the results, we run extensive experiments and calculate average PSNR and SSIM values of all the

Table 1. Average PSNR and SSIM on CAS-PEAL Database

Methods	PSNR(dB)	SSIM
LSR [9]	30.9498	0.8579
LcR [4]	31.2923	0.8628
Our EHRR	**31.4031**	**0.8664**

Table 2. Average PSNR and SSIM on FEI Database

Methods	PSNR(dB)	SSIM
LSR [9]	32.5492	0.8062
LcR [4]	33.7490	0.8365
Our EHRR	**33.7946**	**0.8387**

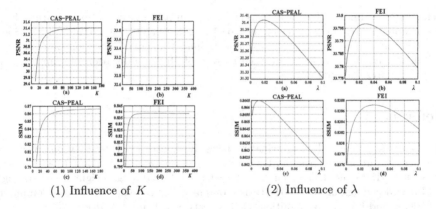

(1) Influence of K (2) Influence of λ

Fig. 3. Influence of K and λ. In both (1) and (2), (a) and (b) are the influence to PSNR on CAS-PEAL and FEI databases; (c) and (d) are the influence to SSIM in CAS-PEAL and FEI two databases.

test images. For each calculation, we have one parameter changed and the other fixed. The results are shown in Fig. 3.

Nearest Neighbor Patch Number K. In Fig. 3 (1), we can see that for K in both data sets, too many position patches from the training data do not enhance the performance. The truth is, although it is not so clear in the figure, with the increase of neighbor patches, PSNR and SSIM of the hallucinated face images decrease slightly after a period of unchangingness. The reason, we assume, is that if K is too large, position patches that are far away from the input patch will be chosen. They contribute noise rather than information which will inevitably lead to worse quanlity.

Regularization Parameter λ. The effect of the regularization parameter λ in Fig. 3 (2) is very clear. For both databases, the range of λ is set from 0.0001 to 0.1 with a step of 0.001 each time. From the average values of PSNR and SSIM, we can see that too small or too large values of λ will decrease the performance which means it penalizes the locality constraints too much or overlearns the locality information.

5 Conclusion

In this paper, we propose a novel algorithm for face hallucination in which the concept of high-resolution reconstruction weights is firstly introduced. By imposing the local constraints into the least square problem in high-resolution feature space, the chosen HR neighbor position patches are fully utilized to reconstruct the HR weights. Experiments show that the proposed method is superior to other state-of-the-art methods.

Acknowledgements. This work was supported in part by the National Natural Science Foundation of China under Grant No.61273273, 61175096 and 61271374, the Specialized Fund for Joint Bulding Program of Beijing Municipal Education Commission, and the Research Fund for Doctoral Program of Higher Education of China under Grant No. 20121101110043.

References

1. Baker, S., Kanade, T.: Hallucinating faces. In: Fourth IEEE International Conference on Automatic Face and Gesture Recognition, pp. 83–88. IEEE (2000)
2. Chang, H., Yeung, D.Y., Xiong, Y.: Super-resolution through neighbor embedding. In: Proceedings of the 2004 IEEE Computer Society Conference on Computer Vision and Pattern Recognition, CVPR 2004, vol. 1, pp. I–I. IEEE (2004)
3. Hsu, C.C., Lin, C.W., Hsu, C.T., Liao, H.Y.: Cooperative face hallucination using multiple references. In: IEEE International Conference on Multimedia and Expo, ICME 2009, pp. 818–821. IEEE (2009)
4. Jiang, J., Hu, R., Han, Z., Lu, T., Huang, K.: Position-patch based face hallucination via locality-constrained representation. In: 2012 IEEE International Conference on Multimedia and Expo (ICME), pp. 212–217. IEEE (2012)
5. Jiang, J., Hu, R., Han, Z., Wang, Z., Lu, T., Chen, J.: Locality-constraint iterative neighbor embedding for face hallucination. In: 2013 IEEE International Conference on Multimedia and Expo (ICME), pp. 1–6. IEEE (2013)
6. Jung, C., Jiao, L., Liu, B., Gong, M.: Position-patch based face hallucination using convex optimization. IEEE Signal Process. Lett. **18**(6), 367–370 (2011)
7. Li, Y., Cai, C., Qiu, G., Lam, K.M.: Face hallucination based on sparse local-pixel structure. Pattern Recogn. **47**(3), 1261–1270 (2014)
8. Liang, Y., Lai, J.H., Yuen, P.C., Zou, W.W., Cai, Z.: Face hallucination with imprecise-alignment using iterative sparse representation. Pattern Recogn. **47**(10), 3327–3342 (2014)
9. Ma, X., Zhang, J., Qi, C.: Hallucinating face by position-patch. Pattern Recogn. **43**(6), 2224–2236 (2010)
10. Roweis, S.T., Saul, L.K.: Nonlinear dimensionality reduction by locally linear embedding. Science **290**(5500), 2323–2326 (2000)
11. Wang, N., Tao, D., Gao, X., Li, X., Li, J.: A comprehensive survey to face hallucination. Int. J. Comput. Vision **106**(1), 9–30 (2014)
12. Wang, X., Tang, X.: Hallucinating face by eigentransformation. IEEE Trans. Syst. Man Cybern. Part C Appl. Rev. **35**(3), 425–434 (2005)
13. Yang, J., Tang, H., Ma, Y., Huang, T.: Face hallucination via sparse coding. In: 15th IEEE International Conference on Image Processing, ICIP 2008, pp. 1264–1267. IEEE (2008)
14. Yu, K., Zhang, T., Gong, Y.: Nonlinear learning using local coordinate coding. In: Advances in Neural Information Processing Systems, pp. 2223–2231 (2009)
15. Zhang, W., Cham, W.K.: Hallucinating face in the dct domain. IEEE Trans. Image Process. **20**(10), 2769–2779 (2011)
16. Zhuang, Y., Zhang, J., Wu, F.: Hallucinating faces: Lph super-resolution and neighbor reconstruction for residue compensation. Pattern Recogn. **40**(11), 3178–3194 (2007)

Real-Time Rendering of Layered Materials with Linearly Filterable Reflectance Model

Jie Guo$^{(\boxtimes)}$, Jinghui Qian, and Jingui Pan

State Key Lab for Novel Software Technology, Nanjing University, Jiangsu, China
guojie_022@163.com

Abstract. This paper proposes a real-time system to render with layered materials by using a linearly filterable reflectance model. This model effectively captures both surface and subsurface reflections, and supports smooth transitions over different resolutions. In a preprocessing stage, we build mip-map structures for both surface and subsurface mesostructures via fitting their bumpiness with mixtures of von Mises Fisher (movMF) distributions. Particularly, a movMF convolution algorithm and a movMF reduction algorithm are provided to well-approximate the visually perceived bumpiness of the subsurface with controllable rendering complexity. Then, both surface and subsurface reflections are implemented on GPUs with real-time performance. Experimental results reveal that our approach enables aliasing-free illumination under environmental lighting at different scales.

Keywords: Layered materials · MovMF reduction · Reflectance filtering · Real-time rendering

1 Introduction

Materials with thin transparent layers are omnipresent in our daily life. Examples include glazed ceramics, waxed floor, metallic car paint, and biological structures like skin or leaves. Although many physically-based rendering techniques have been proposed in recent years, real-time rendering without artifacts is still a delicate task. Furthermore, to enhance the physical realism of surface appearance, normal maps are frequently employed that add a significant amount of details to surfaces via faking the lighting of bumps. Akin to traditional texture mapping, normal mapped surfaces should be correctly filtered for the antialiasing purpose [2]. Currently, the reflectance filtering approaches deal only with the reflection from the outermost surface of an object. However, for layered materials, both the surface scattering and subsurface scattering must be taken into consideration, and this will inevitably complicate the reflectance filtering problem.

This paper presents a solution to this critical problem. We consider a single material layer with two normal mapped interfaces and their roughnesses are characterized by means of the normal distribution functions (NDF) [1,6]. We first build a linearly filterable mip-map structure for each normal map in a

© Springer International Publishing Switzerland 2015
Y.-S. Ho et al. (Eds.): PCM 2015, Part I, LNCS 9314, pp. 431–441, 2015.
DOI: 10.1007/978-3-319-24075-6_42

precomputation stage. Similar to [8], this structure contains the mip-mapped parameters of the NDF which can be fitted by movMF distributions. During the rendering process, we further divide the reflectance model into the surface part and the subsurface part. The difficulty of filtering layered materials mainly reflects in the subsurface part, and we employ a movMF convolution technique to depict the overall roughness of the subsurface. Since the convolution significantly increases the number of vMF lobes, we further propose a movMF reduction algorithm to keep the number of lobe components down to a threshold with minimal loss of fidelity. In brief, our main contributions can be summarized as follows:

- A linearly filterable reflectance model for layered materials with two rough surface boundaries is proposed, including surface and subsurface reflections.
- A movMF convolution algorithm is utilized to predict the appearance of subsurface reflection, and a movMF reduction algorithm is provided to effectively control the number of components in the mixture.
- A practical rendering system for multi-scale layered rough surfaces is presented. The GPU implementation demonstrates that this system supports seamless transitions among different scales at zooming.

2   Related Work

Rendering with Layered Materials: Layered materials are often adopted in computer graphics to describe the complete surface and subsurface scattering. Hanrahan and Krueger [9] proposed an accurate scattering model for layered surfaces in terms of one-dimensional linear transport theory. The efficiency of this model was improved in [14] by deriving the scattering equations in integral form. Gu et al. [7] proposed a customized subsurface scattering model for a specific type of layered material—an optically thin contaminant layer on a transparent surface. Besides, several approximation methods for simulating scattering from layered materials have been proposed. Weidlich and Wilkie [21] proposed a flexible family of layered BRDFs combining several microfacet based surface layers. Dai et al. [4] presented a spatially-varying BTDF model for rough slabs, omitting light transport inside the objects. Recently, a comprehensive and practical framework for computing BSDFs of layered materials is proposed, supporting arbitrary composition of various types of layers [11].

Filtering of Reflectance: To explicitly and efficiently model the small but apparent meso-structures, such as granules or wrinkles on the surface, normal mapping techniques are widely utilized, For normal mapped surfaces, simply averaging texel values of normal maps is not sufficient to perform the reflectance filtering since shading is not linear in the normal [2]. Therefore, specially designed filtering strategies should be developed to obtain physically convincing shading results. Convolution-based normal map filtering methods [6,8,18] build upon the idea that the overall BRDF is the convolution of the base BRDF and the NDF over a pixel's weighted footprint. Since the NDF is linearly interpolable, the

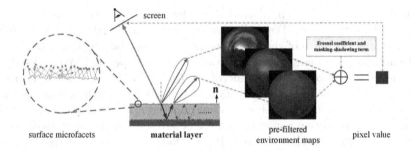

Fig. 1. Illustration of our rendering pipeline with pre-filtered environment maps.

traditional texture filtering methods are applicable to it. LEAN mapping [5,13] also allows linear filtering of the surface reflectance via incorporating normal distribution into the Beckmann shading model. Though efficient, it lacks the ability of capturing multiple modes of the NDF since only one Gaussian lobe is adopted. For better accuracy of the fitted NDF, Tan et al. [16,17] proposed to use a mixture of isotropic Gaussian lobes. However, the planar projection may introduce distortions [8]. Consequently, spherical distributions [12], such as mixture of von Mises-Fisher distributions, are better suited for fitting multi-modal directional data.

Although the filtering of surface reflectance is quite mature, little attention is paid to handle filtering of subsurface reflection which is predominant in many materials. Therefore, our work can be considered as a complement to existing surface reflectance filtering methods.

3 Proposed System

3.1 Overview

When a light impinges on the surface of a layered medium, it is both reflected and refracted depending on its appearance. Light refracted into the layer may be reflected back again and finally exit the medium. For thin transparent layers, multiple rough reflections and rough transmissions happen at the surface boundaries. In this paper, we restrict our consideration to one material layer composed of two smooth or rough boundaries. We further assume the lower boundary of the layer is opaque such that light is only reflected at this boundary. The main stages of our rendering pipeline are illustrated in Fig. 1 and summarized below.

According to the principle of independently propagating of light, the total outgoing radiance $L(\mathbf{o})$ is the sum of the surface and subsurface reflection components (i.e., $L^{sf}(\mathbf{o})$ and $L^{sb}(\mathbf{o})$):

$$L(\mathbf{o}) = L^{sf}(\mathbf{o}) + L^{sb}(\mathbf{o}) = \int_{\Omega} (f_r^{sf}(\mathbf{o}, \mathbf{i}) + f_r^{sb}(\mathbf{o}, \mathbf{i}))L(\mathbf{i})\langle \mathbf{i}, \mathbf{n}\rangle \mathrm{d}\omega_{\mathbf{i}} \quad (1)$$

in which $f_r^{sf}(\mathbf{o}, \mathbf{i})$ and $f_r^{sb}(\mathbf{o}, \mathbf{i})$) are BRDFs for surface and subsurface reflection components, respectively. Depending on the surface material and viewing angle, the relative importance of these two components may varying dramatically.

To account for rough surface reflection from the upper and lower boundaries of the layer, we make use of the microfacet theory [3, 19]. In this theory, rough surfaces can be modeled as a collection of tiny mirrors (with microscopic normal $\mathbf{m}$) whose aggregated behavior determines the material properties. Since the directional reflection pattern is mainly determined by the microfacet normal distribution function (NDF) $D(\mathbf{m})$ [19], we focus on estimating the NDFs for both the surface and subsurface reflection components. In our current implementation, the NDFs are fitted with movMF distributions to facilitate their usage in real-time applications. Mathematically, the NDFs of the upper and lower boundary of the layer can be expressed as $D(\mathbf{m}_1) = \sum_{j=1}^{J} \alpha_j \mathcal{M}(\mathbf{m}_1; \boldsymbol{\mu}_{\mathbf{m}_1,j}, \kappa_{\mathbf{m}_1,j})$ and $D(\mathbf{m}_2) = \sum_{k=1}^{K} \beta_k \mathcal{M}(\mathbf{m}_2; \boldsymbol{\mu}_{\mathbf{m}_2,k}, \kappa_{\mathbf{m}_2,k})$, which are both fitted via the spherical EM algorithm in a pre-processing stage [8]. In these equations,

$$\mathcal{M}(\mathbf{m}; \boldsymbol{\mu}, \kappa) = C_3(\kappa) e^{\kappa \boldsymbol{\mu} \cdot \mathbf{m}} \tag{2}$$

where $C_3(\kappa) = \frac{\kappa}{4\pi \sinh \kappa}$ is the normalization constant, κ and $\boldsymbol{\mu}$ are the concentration parameter and the mean direction, respectively. It is worth noting that both NDFs are linearly filterable, and can be real-time updated according to the viewing resolution. Actually, $D(\mathbf{m}_1)$ directly gives the NDF for surface reflection, while estimating the NDF of subsurface reflection is relatively difficult since both $D(\mathbf{m}_1)$ and $D(\mathbf{m}_2)$ will affect the shape of visually perceived glossiness of subsurface reflection. This will be covered in the following sections.

As sketched in Fig. 1, once we obtained the movMF fitted NDFs for surface reflection (e.g., green lobe in Fig. 1) and subsurface reflection (e.g., blue lobe in Fig. 1), we index into the pre-convolved environment maps with the mean direction served as texture coordinates and the concentration parameter served as mip-map level for each lobe. The returned value is further attenuated by appropriate remaining terms, such as the Fresnel coefficient and the masking-shadowing term. The final pixel value is simply determined by their linear combination.

3.2   Surface Reflection

Let us start by calculating the surface reflection term $L^{sf}(\mathbf{o})$. Given a viewing direction $\mathbf{o}$, we need to obtain the corresponding directional distribution of the incident rays, since the integral of local shading is performed over the incident direction $\mathbf{i}$. Strictly speaking, the directional distribution of $\mathbf{i}$ does not necessarily agree with the movMF distribution, but we can approximate it as a warped movMF distribution according to [20]. Specifically, by preserving the amplitudes of the warped lobes, the directional distribution of $\mathbf{i}$ can be approximated as $D(\mathbf{i}) \approx \sum_{j=1}^{J} \alpha_j \mathcal{M}(\mathbf{i}; \boldsymbol{\mu}_{\mathbf{i},j}, \kappa_{\mathbf{i},j})$, in which $\boldsymbol{\mu}_{\mathbf{i},j} = 2(\boldsymbol{\mu}_{\mathbf{m}_1,j} \cdot \mathbf{o})\boldsymbol{\mu}_{\mathbf{m}_1,j} - \mathbf{o}$ and $\kappa_{\mathbf{i},j} = \frac{\kappa_{\mathbf{m}_1,j}}{4|\mathbf{o} \cdot \boldsymbol{\mu}_{\mathbf{m}_1,j}|}$. Then, surface reflection is found to be

$$L^{sf}(\mathbf{o}) = \int_\Omega L(\mathbf{i}) M_r(\mathbf{o}, \mathbf{i}) D(\mathbf{i}) \langle \mathbf{i}, \mathbf{n} \rangle d\omega_{\mathbf{i}} \approx \sum_{j=1}^{J} \{ \alpha_j M_r(\mathbf{o}, \boldsymbol{\mu}_{\mathbf{i},j}) \langle \boldsymbol{\mu}_{\mathbf{i},j}, \mathbf{n} \rangle E_j[L(\mathbf{i})] \}$$

(3)

where the remaining term $M_r(\mathbf{o}, \mathbf{i}) = \frac{F(\mathbf{i},\mathbf{m})G(\mathbf{o},\mathbf{i})}{4|\mathbf{o}\cdot\mathbf{n}||\mathbf{i}\cdot\mathbf{n}|}$ [19] includes the Fresnel coefficient $F(\mathbf{i}, \mathbf{m})$ and the masking-shadowing term $G(\mathbf{o}, \mathbf{i})$. The approximation in this equation is based on the observation that $M_r(\mathbf{o}, \mathbf{i})$ is very smooth [10,20], therefore we assume this function to be constant across the support of each vMF lobe, i.e., $M_r(\mathbf{o}, \mathbf{i}) \approx M_r(\mathbf{o}, \boldsymbol{\mu}_{\mathbf{i},j})$. In addition, $E_j[L(\mathbf{i})] = \int_\Omega L(\mathbf{i}) \mathcal{M}(\mathbf{i}; \boldsymbol{\mu}_{\mathbf{i},j}, \kappa_{\mathbf{i},j}) d\omega_{\mathbf{i}}$, which can be accomplished by utilizing pre-filtered environment maps. The pre-filtered environment maps are generated as follows. We first convolve the environment map with vMF distributions of decreasing concentration parameter (κ), and then store the results into a mip-map of 2D texture. During the rendering time, we index into this mip-mapped texture according to each vMF lobe ($\mathcal{M}(\mathbf{i}; \boldsymbol{\mu}_{\mathbf{i},j}, \kappa_{\mathbf{i},j})$) to get appropriate shading result.

3.3   Subsurface Reflection

Subsurface reflection is of critical importance in predicting the appearance of layered materials, and its calculation is not straightforward. Since the thickness of the layer is ignored in this paper, the normal distribution for each bounce of subsurface reflection can be closely approximated by a multi-fold spherical convolution of the upper and lower boundaries' NDFs, as inspired by Dai et al. [4]. For instance, the NDF of one-bounce subsurface reflection can be approximated by $D(\mathbf{m}_1) * D(\mathbf{m}_2) * D(\mathbf{m}_1)$, while the NDF of two-bounce subsurface reflection can be approximated by $D(\mathbf{m}_1) * D(\mathbf{m}_2) * D(\mathbf{m}_1) * D(\mathbf{m}_2) * D(\mathbf{m}_1)$. With the increase of iteration, the exitant radiance degrades rapidly attributing to the existence of the Fresnel effects at the upper boundary. Therefore, it can be verified that one-bounce subsurface reflection (whose NDF is given by $D(\mathbf{m}_1) * D(\mathbf{m}_2) * D(\mathbf{m}_1)$) contributes most in the final effect of subsurface reflection.

For the spherical convolution between two vMF lobes, say $\mathcal{M}(\mathbf{x}, \boldsymbol{\mu}_j, \kappa_j)$ and $\mathcal{M}(\mathbf{x}, \boldsymbol{\mu}_k, \kappa_k)$, the result can be approximated by another vMF lobe [12] $\mathcal{M}(\mathbf{x}, \frac{\boldsymbol{\mu}_j + \boldsymbol{\mu}_k}{\|\boldsymbol{\mu}_j + \boldsymbol{\mu}_k\|}, A_3^{-1}(A_3(\kappa_j)A_3(\kappa_k)))$, where $A_3(\kappa) = \coth(\kappa) - \frac{1}{\kappa}$ returns the mean resultant length and $A_3^{-1}(y) = \frac{3y-y^3}{1-y^2}$ [8]. For the spherical convolution between two movMF distributions, similar strategy can be used for each pair o vMF lobes, and the final result is given by a linear combination of several convolved vMF lobes applied with proper weights.

Note that the three-fold convolution for one-bounce subsurface reflection will result in $J^2 K$ vMF lobes, which is excessively complex when J and K are large. Actually, the number of lobes grows exponentially with the number of convolution steps. Therefore, in real-time applications, a movMF reduction process, where the movMF is reduced to a computationally tractable mixture whenever needed, is necessary.

In this paper, we propose a novel movMF reduction algorithm, which includes a pruning step and a merging step. The pruning step simply truncates mixture components that have low weights, e.g., $\alpha_j \beta_k \alpha'_j < PT$, where PT is a pre-defined pruning threshold. For the merging step, we perform the mixture reduction via a top-down merging algorithm. Specifically, the vMF components are merged by analytically minimizing the Kullback-Leibler (KL) divergence between the original and the reduced mixtures. We choose the KL divergence as the cost function for nonlinear optimization because it is considered as an optimal measure of how similar two distributions are [15]. Given two vMF components, say $P_j(\mathbf{x}) = \mathcal{M}(\mathbf{x}; \boldsymbol{\mu}_j, \kappa_j)$ and $P_k(\mathbf{x}) = \mathcal{M}(\mathbf{x}; \boldsymbol{\mu}_k, \kappa_k)$, their KL divergence can be calculated as follows:

$$d_{KL}(P_j, P_k) = \int_\Omega P_j \log\left[\frac{P_j}{P_k}\right] = \log\left[\frac{C_3(\kappa_j)}{C_3(\kappa_k)}\right] + \kappa_j A_3(\kappa_j) - \kappa_k A_3(\kappa_j)(\boldsymbol{\mu}_j \cdot \boldsymbol{\mu}_k)$$

(4)

The pseudocode of our movMF reduction approach is presented in Algorithm 1 . In this algorithm, the merging step starts by picking out the movMF component with maximum weight, say jth, and then merges all other components i for which it satisfies $d_{KL}(P_j, P_i) < MT$. Finally, this algorithm outputs a reduced movMF distribution whose number of components is l.

Algorithm 1. movMF Reduction Algorithm

Require: $\{\alpha_i, P_i(\boldsymbol{\mu}_i, \kappa_i)\}_{i=1,2,...,I}$, a pruning threshold PT, a merging threshold MT.
  1: $l \leftarrow 0$
  2: $\mathbb{I} \leftarrow \{i = 1, 2, ..., I \mid \alpha_i > PT\}$ ▷ The pruning step
  3: **while** $\mathbb{I} \neq \emptyset$ **do**   ▷ The merging step
  4:     $l \leftarrow l + 1$
  5:     $j \leftarrow \underset{i \in \mathbb{I}}{\arg\max}\, \alpha_i$
  6:     $\mathbb{C} \leftarrow \{i \in \mathbb{I} \mid d_{KL}(P_j, P_i) < MT\}$
  7:     $\tilde{\alpha}_l \leftarrow \sum_{i \in \mathbb{C}} \alpha_i$
  8:     $\mathbf{r}_l \leftarrow \frac{1}{\tilde{\alpha}_l} \sum_{i \in \mathbb{C}} \alpha_i A_3(\kappa_i) \boldsymbol{\mu}_i$ ▷ The unnormalized mean direction
  9:     $\tilde{\boldsymbol{\mu}}_l = \frac{\mathbf{r}_l}{\|\mathbf{r}_l\|}$
 10:     $\tilde{\kappa}_l \leftarrow A_3^{-1}\left(\frac{1}{\tilde{\alpha}_l} \sum_{i \in \mathbb{C}} \alpha_i A_3(\kappa_i)(\boldsymbol{\mu}_i \cdot \tilde{\boldsymbol{\mu}}_l)\right)$
 11:     $\mathbb{I} \leftarrow \mathbb{I} \setminus \mathbb{C}$
 12: **end while**

Similar to [15], here we use moment-preserving merging in lines 7-10. Specifically, we replace the vMF lobes to be merged with a single vMF lobe whose zeroth and first-order moments match those of the ordinal ones. In our current implementation, the movMF convolution and the movMF reduction are both performed in a precomputation stage, and the results are stored in 2D textures for further real-time use.

Once we get the simplified NDF of subsurface reflection, we can obtain its corresponding directional distribution of incident rays $\mathbf{i}$ in the same manner as surface reflection: $\tilde{D}(\mathbf{i}) = \sum_{l=1}^{L} \tilde{\alpha}_l \mathcal{M}(\mathbf{i}; \tilde{\boldsymbol{\mu}}_{\mathbf{i},l}, \tilde{\kappa}_{\mathbf{i},l})$. Then, to generate correct subsurface reflection, we use the similar formula as Eq. 3 by replacing $D(\mathbf{i})$ with $\tilde{D}(\mathbf{i})$.

(a) Our method   (b) Ground truth

Fig. 2. Comparison with ground truth across continuous scales. Note that the difference is minimal for each scale.

The remaining term M_r should be modified accordingly, which includes two refraction events and one reflection event. Accurately estimating M_r for subsurface reflection is time-consuming, but since M_r is usually very smooth and has little effect on the shape of BRDF, it can be simply approximated with the mean direction of each vMF lobe. For instance, we use can use $\tilde{\mu}_{i,l}$ and its corresponding refracted ray to get approximated Fresnel effect and masking-shadowing effect.

4   Results

We have implemented the proposed method on a PC with an Intel Core Q8300 CPU and an NVIDIA GeForce GTS 250 graphics card. All the images in this paper are generated using a final output resolution of 512×512. Surface reflection and subsurface reflection are both implemented by GLSL shader programs and run on the GPU. Besides, ground truth images are path traced with 400 samples per pixel. Some of the parameters are set as follows. The refractive index η is always set to 1.5, since it is a good approximation for medium in practical use such as glass, lacquers, or plastics. We choose 4 components for each movMF, i.e., $J = 4$ and $K = 4$. Therefore, 64 vMF lobes are generated for the initial NDF of the subsurface. These lobes are further reduced to 4 lobes ($L = 4$) via the movMF reduction algorithm.

To validate the accuracy of our reflectance model, we show comparisons of our method with ground truth across continuous scales in Fig. 2. Here we display a glazed vase model with both surface reflection and subsurface reflection. Though some subtle differences exist, our results are visually plausible at each scale. This figure also shows that our method can handle both detailed meso-structures at fine resolutions and the aggregated BRDFs at coarse resolutions.

The effectiveness of movMF convolution and movMF reduction is verified in Fig. 3, where we separately render the images with one clustered vMF lobe, first 4 lobes after convolution, total 64 lobes, and 4 lobes after movMF reduction (from Fig. 3(a) to Fig. 3(d)). As seen, our method using reduced 4 lobes (172 fps) is very close to both the rendering method using total 64 lobes (12 fps) and

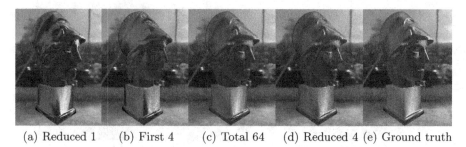

(a) Reduced 1   (b) First 4   (c) Total 64   (d) Reduced 4 (e) Ground truth

Fig. 3. From (a) to (d): comparison of rendering results using different movMF distributions; (e): path traced ground truth. Note that only subsurface reflection is captured here.

ground truth (Fig. 3(e)), while it has much lower rendering time cost. Obviously, using only one lobe after reduction or using the first 4 lobes from the total 64 lobes will produce apparent artifacts, since both of them do not fit well with the real NDF. Note that only subsurface reflection is captured in these images.

LEAN mapping [13] has also been extended to support multiple layers of bumps. However, this method uses only one anisotropic Gaussian lobe, which is not always sufficient for very complex normal distributions, and its original implementation is designed for simple point and directional lighting. We have re-implemented the two-layered LEAN mapping method incorporating environmental light, the masking-shadowing function, and the Fresnel effects. The basic idea of enabling environmental lighting is inspired by [5]. In Fig. 4 we show that our result is much closer to ground truth than the result produced by LEAN mapping, and our linear filter strategy tends to capture more details with the help of multiple vMF lobes.

A simulation of the metallic paint materials is displayed in Fig. 5(a). In this figure, each metallic paint layer is modeled by a randomly distributed normal map (representing metallic flakes) covered with a clear coating. Fig. 5(b) shows the rendering of a tea set. This scene contains several types of layered materials, which indicates the range of materials that can be achieved with our approach.

The performance of rendering different scenes in this section is provided in Table 1. Each row contains timings for: precomputation of the normal maps,

(a) LEAN mapping   (b) Our method   (c) Ground truth

Fig. 4. Visual comparison with two-layered LEAN mapping. Here we only capture subsurface reflection.

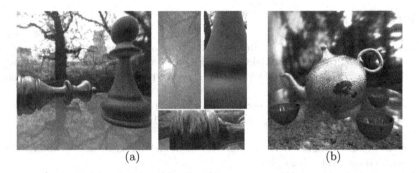

(a)   (b)

Fig. 5. (a): Rendering of metallic paint materials (55K triangles, 42 fps). (b): Real-time rendering of a tea set with various types of layered materials (36K triangles, 35 fps)

surface reflection only (Surface), subsurface reflection only (Subsurface), and the total reflection (Total). The final column also provides the rendering time of ground truth images. Though our approach (Total) is several times slower than rendering of single surface reflection or single subsurface reflection, it still has achieved a speedup of several orders of magnitude compared to ground truth, while maintaining visually comparable image quality.

Table 1. The illustration and comparison of timing performance (in *fps* except for precomputation and ground truth) for different scenes.

Scenes	#Tri	Precomputation	Real-time rendering			Ground truth
			Surface	Subsurface	Total	
Fig. 2	7K	174 s	511	356	**115**	136 s
Fig. 3	15K	201 s	344	172	**50**	155 s
Fig. 4	40K	179 s	243	140	**41**	126 s

5 Conclusions

To summarize, we proposed a linearly filterable reflectance model for layered materials containing two rough or smooth surface boundaries. Both surface and subsurface reflections are accounted for in this model. A movMF convolution algorithm is performed to approximate subsurface roughness and a movMF reduction algorithm is employed to decrease the number of vMF lobes in order to reduce the complexity of rendering. In the future, we would like to enrich our approach by supporting multiple layers and light scattering inside the layer.

References

1. Ashikmin, M., Premože, S., Shirley, P.: A microfacet-based brdf generator. In: Proceedings of the 27th Annual Conference on Computer Graphics and Interactive Techniques, SIGGRAPH 2000, pp. 65–74. ACM Press/Addison-Wesley Publishing Co., New York (2000)
2. Bruneton, E., Neyret, F.: A survey of nonlinear prefiltering methods for efficient and accurate surface shading. IEEE Trans. Visual Comput. Graphics 18(2), 242–260 (2012)
3. Cook, R.L., Torrance, K.E.: A reflectance model for computer graphics. ACM Trans. Graph. 1(1), 7–24 (1982)
4. Dai, Q., Wang, J., Liu, Y., Snyder, J., Wu, E., Guo, B.: The dual-microfacet model for capturing thin transparent slabs. Comput. Graph. Forum 28(7), 1917–1925 (2009)
5. Dupuy, J., Heitz, E., Iehl, J.C., Poulin, P., Neyret, F., Ostromoukhov, V.: Linear efficient antialiased displacement and reflectance mapping. ACM Trans Graph. 32(6), Article No. 211 (2013)
6. Fournier, A.: Normal distribution functions and multiple surfaces. In: Proceedings of Graphics Interface Workshop on Local Illumination, pp. 45–52 (1992)
7. Gu, J., Ramamoorthi, R., Belhumeur, P., Nayar, S.: Dirty glass: Rendering contamination on transparent surfaces. In: EuroGraphics Symposium on Rendering, June 2007
8. Han, C., Sun, B., Ramamoorthi, R., Grinspun, E.: Frequency domain normal map filtering. ACM Trans. Graph. 26(3), July 2007
9. Hanrahan, P., Krueger, W.: Reflection from layered surfaces due to subsurface scattering. In: Proceedings of the 20th Annual Conference on Computer Graphics and Interactive Techniques, SIGGRAPH 1993, pp. 165–174 (1993)
10. Iwasaki, K., Dobashi, Y., Nishita, T.: Interactive bi-scale editing of highly glossy materials. ACM Trans. Graph. 31(6), 144:1–144:7 (2012)
11. Jakob, W., d'Eon, E., Jakob, O., Marschner, S.: A comprehensive framework for rendering layered materials. ACM Trans. Graph. 33(4), 118:1–118:14 (2014)
12. Mardia, K.V., Jupp, P.E.: Directional statistics. Wiley series in probability and statistics. Wiley, Chichester (2000)
13. Olano, M., Baker, D.: Lean mapping. In: Proceedings of I3D 2010, I3D 2010, pp. 181–188 (2010)
14. Pharr, M., Hanrahan, P.: Monte carlo evaluation of non-linear scattering equations for subsurface reflection. In: Proceedings of the 27th Annual Conference on Computer Graphics and Interactive Techniques, SIGGRAPH 2000, pp. 75–84 (2000)
15. Runnalls, A.R.: Kullback-leibler approach to gaussian mixture reduction. IEEE Trans. Aerosp. Electron. Syst., 989–999 (2007)
16. Tan, P., Lin, S., Quan, L., Guo, B., Shum, H.: Filtering and rendering of resolution-dependent reflectance models. IEEE Trans. Visual Comput. Graphics 14(2), 412–425 (2008)
17. Tan, P., Lin, S., Quan, L., Guo, B., Shum, H.Y.: Multiresolution reflectance filtering. In: Proceedings of the Sixteenth Eurographics Conference on Rendering Techniques, EGSR 2005, Aire-la-Ville, Switzerland, Switzerland, pp. 111–116 (2005)
18. Toksvig, M.: Mipmapping normal maps. J. Graph., GPU, Game Tools 10(3), 65–71 (2005)
19. Walter, B., Marschner, S.R., Li, H., Torrance, K.E.: Microfacet models for refraction through rough surfaces. In: Proceedings of the 18th Eurographics Conference on Rendering Techniques, EGSR 2007, pp. 195–206 (2007)

20. Wang, J., Ren, P., Gong, M., Snyder, J., Guo, B.: All-frequency rendering of dynamic, spatially-varying reflectance. ACM Trans. Graph. **28(5)**, 133:1–133:10 (2009)
21. Weidlich, A., Wilkie, A.: Arbitrarily layered micro-facet surfaces. In: GRAPHITE 2007, pp. 171–178 (2007)

Hybrid Lossless-Lossy Compression for Real-Time Depth-Sensor Streams in 3D Telepresence Applications

Yunpeng Liu[1,2(✉)], Stephan Beck[1], Renfang Wang[2], Jin Li[2], Huixia Xu[2],
Shijie Yao[2], Xiaopeng Tong[2], and Bernd Froehlich[1]

[1] Virtual Reality Systems Group, Bauhaus-Universität Weimar, Weimar, Germany
L35633@163.com, {stephan.beck,bernd.froehlich}@uni-weimar.de
[2] College of Computer Science and Information Technology,
Zhejiang Wanli University, Ningbo, China
{wangrenfang,lijin,xuhuixia,yaoshijie,tongxiaopeng}@zwu.edu.cn

Abstract. We developed and evaluated different schemes for the real-time compression of multiple depth image streams. Our analysis suggests that a hybrid lossless-lossy compression approach provides a good trade-off between quality and compression ratio. Lossless compression based on run length encoding is used to preserve the information of the highest bits of the depth image pixels. The lowest 10-bits of a depth pixel value are directly encoded in the Y channel of a YUV image and encoded by a x264 codec. Our experiments show that the proposed method can encode and decode multiple depth image streams in less than 12 ms on average. Depending on the compression level, which can be adjusted during application runtime, we are able to achieve a compression ratio of about 4:1 to 20:1. Initial results indicate that the quality for 3D reconstructions is almost indistinguishable from the original for a compression ratio of up to 10:1.

Keywords: Depth map · Compression · x264 · 3D Tele-immersion

1 Introduction

Color and depth (RGBD) sensors, such as the Microsoft Kinect, are used in many research areas. In 2013, Beck et al. [2] presented an immersive group-to-group telepresence system which is based on a cluster of RGBD-sensors that continuously captures the participants. The color and depth image streams are then sent to a remote site over a network connection where they are used for the real-time reconstruction of the captured participants. Due to the large amount of image data that has to be transmitted, the network bandwidth can become a bottleneck. E.g., a setup of five Kinect V1 sensors running at 30 Hz requires a network connection of 2.46 GBit/s. While this might be feasible with high-speed local area networks, the bandwidth of internet connections is typically in the range of 10 to 100 Mbit/s. To reduce the amount of data that has to be

© Springer International Publishing Switzerland 2015
Y.-S. Ho et al. (Eds.): PCM 2015, Part I, LNCS 9314, pp. 442–452, 2015.
DOI: 10.1007/978-3-319-24075-6_43

transmitted, Beck et al. [2] already suggested the usage of a DXT compression scheme for the color images which compresses the color data by a fixed ratio of 6:1. However, the compression of depth streams was not addressed at all.

Our hybrid lossless-lossy compression method for depth image streams offers different levels of compression quality and compression ratios which can be adjusted during application runtime for adapting to the available bandwidth. The main contributions of our work are:

- Optimal bit assignment from depth values to YUV channels including a lossless encoding mode for the highest bits of the depth values.
- Optimal setting scheme of x264-based encoding parameters for depth maps.
- Efficient encoding and decoding implementation for multiple depth image streams in less than 12ms on average for compression ratios between 4:1 to 20:1.

The integration of our hybrid compression scheme into our immersive telepresence application shows that the quality for 3D reconstructions is almost indistinguishable from the original for a compression ratio of up to 10:1.

2   Related Work

The development of codecs for depth image compression has become an important research topic in recent years. In the context of 3D reconstruction, a class of methods operate on 3D meshes or point clouds [3,12,13,16]. However, the processing costs for 3D compression are typically much higher than for 2D image compression, and, the benefits of segmentation and prediction cannot be used effectively. In addition, geometry-based methods need to maintain special data structures which often rule out high compression ratios.

From another point of view, depth images are "special" gray images, which contain object edges (high frequency areas) and smooth surfaces (low frequency areas). Therefore depth image streams can be compressed by international standard [5,7,10,11,14,15,17–19,21,22] or non-standard [1,4,8,9] video encoding algorithms. Non-standard encoding algorithms focus on retaining the integrity and quality of high frequency areas. On the other side, the compression ratio of above algorithms is not very high and most schemes are not compatible with color video compression.

Pece et al. [15] reported initial results to adapt standard video codec algorithms for depth images. However the error, which is introduced by the compression, is comparably high, even for lower compression ratios. This was mainly caused by the loss of the higher bits of the depth image's pixels.

More recently, the High Efficiency Video Coding (HEVC) standard was introduced and investigated by many researchers [5,10,11,14,17–19]. However, the standardization of 3D Video Coding Extension Development of HEVC is still under development and many parts are in the experimental stage. Besides HEVC, other approaches like JMkta or JM based on H.264/AVC depth map coding [7,21,22] have been proposed. However, neither the JMkta- nor the

JM-reference software can meet real-time application requirements. Our approach is partially inspired by the work of Pece et al. [15]. We therefore further investigated and analyzed the depth bit assignment schemes in order to deduce optimal settings for standard real-time video coding algorithms. In addition, our proposed method uses a hybrid lossless-lossy compression approach, which provides a good tradeoff between quality, processing time and compression ratio.

3   Compression Approach

3.1   System Overview

Our approach is based on the optimal determination of many available encoding parameters. In order to deduce the optimal settings, we recorded a sequence of multiple depth image streams which we used in our analysis. The precision of the depth images is clamped to 12 bit and converted to millimeter. The resulting depth-range of 4.000 mm is sufficient for most practical scenarios. In our analysis, we first investigate several schemes to assign 12 bit depth pixels to YUV-channels. Second, we evaluate different parameters which are used to configure the x246-encoder. Figure 1 gives an overview of our experimental setup. Our analysis is then used to obtain an optimal set of encoding parameters to provide several levels of quality and compression ratios. After the identification of the optimal parameter set we integrated the encoder as well as the decoder into a prototype pipeline for real-time 3D reconstruction (c.f. Fig. 2). Based on this real-time 3D reconstruction pipeline, we were then able to perform a quantitative and qualitative evaluation of our proposed depth compression method.

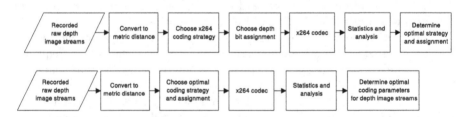

Fig. 1. Pipeline for the analysis and the determination of optimal depth-bit assignment (upper row) and of optimal encoding configurations (lower row).

3.2   Analysis of Depth-Bit Assignment

The x264-codec family provides three different YUV color spaces: 4:2:0, 4:2:2 and 4:4:4. In addition the x264-library can be compiled with an internal precision of 8 bit or 10 bit. In our analysis we tested both precisions. We assign the highest 8 bit/10 bit of a 12 bit depth image's pixel to Y space. The remaining lower bits can be assigned to the UV space using 4 different methods: *Method 1* assigns

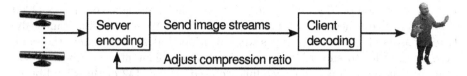

Fig. 2. Overview of the processing and streaming pipeline. Multiple RGB-sensors are connected to a server which encodes the image streams. The client receives the image streams and performs the decoding as well as the 3D-reconstruction.

the remaining lower bits to UV space in byte order. This can be performed for all color spaces (4:2:0, 4:2:2 and 4:4:4). For color spaces 4:2:2 and 4:4:4, *Method 2* assigns the lowest bits to the corresponding lower bits of the U space, while the unassigned space of the byte is set to zero. For color space 4:4:4, *Method 3* interleaves the lowest bits to the corresponding lowest bit of U space and V space, the unassigned space is set to zero. *Method 4* assigns the remaining lower bits of the depth pixel to the lowest bit position in the UV space.

On the one hand, the lower bits of the depth pixels tend to contain noise. As a result, the inter frame prediction that is used in x264-encoding is limited. Therefore, a simple option is to discard the lower 2 bits of the depth pixels. Of course, this will introduce a depth error of 3 mm in the worst case.

On the other hand, the higher bits of the depth pixels are very important for an accurate 3D reconstruction. We therefore investigated a hybrid lossless-lossy compression scheme. The main idea of our hybrid encoding scheme is to encode the highest 2 bits of the depth pixels using a lossless compression algorithm and to assign the lower 10 bits in the Y channel of a YUV image for x264 encoding. For lossless encoding we evaluated the three most commonly used algorithms: Huffman, LZW and run length encoding (RLE). We achieved the best compression ratio and encoding performance with RLE.

3.3   Optimal X264 Encoding Parameters for Depth Image Streams

After the analysis of the depth-bit assignment methods, we had to find the optimal parameter settings for x264-encoding. Because our evaluation revealed that an internal precision of 8 bit results in very poor quality (c.f. Table 1), we focus on the 10 bit version of x264 in the following. First, we decided to set the global encoding parameters to constants: *profile=high10*, *csp=I420*, *framerate=30*, *tune=zerolatency*. Next, we analyzed further encoding parameters, which have a special influence on the encoding quality of x264 for depth image streams. The most important parameters are *preset*, *IDR interval* and *crf*. We will provide details of our analysis in Sect. 4.2.

4   Results and Discussion

We performed our analysis and evaluation for the Kinect V1 and the Kinect V2. The Kinect V1 has a depth resolution of 640 x 480 pixels and the Kinect V2 has

a depth resolution of 512 x 424 pixels. We recorded a sequence of approx. 4000
frames from five Kinect V1 sensors and three Kinect V2 sensors simultaneously.
Three sensors of both Kinect-types were positioned at approximately the same
location in order to capture the same depth image content, actually a person
(standing and gesturing). For encoding we used version 142 of the x246 library
[20], for decoding we used version 2.6.1 of the ffmpeg library [6]. The operating
system was 64-bit Ubuntu 14.04.2 LTS running on a dual six-core Intel® Xeon®
CPU E5-1650 V2@3.50 GHz.

In our evaluation, we use the following notations: *Time* means average encoding (decoding) time in milliseconds per frame, *Ratio* means average compression
ratio, *Error* lists the mean absolute error (MAE) between the decoded depth
image and the original image in millimeters along with the maximum absolute
error in brackets, *Method* ranges from *1* to *4* and indicates the different assignment schemes which were used as described in Sect. 3.2, *Bitrate* is in megabits
per second.

Table 1. Experimental results for different depth-bit assignment methods (1 to 4).

Bit depth	Color space	Method	Time	Ratio	PSNR	Error
8	420	1	13.0	0.43	54.55	3.60 [97]
	422	1	14.5	0.41	53.01	5.41 [112]
		2	14.4	0.39	52.06	7.83 [101]
	444	1	14.6	0.34	53.13	5.11 [89]
		2	14.5	0.34	53.67	4.90 [104]
		3	14.5	0.31	54.23	3.40 [99]
		4	14.7	0.29	56.20	3.49 [103]
10	420	1	14.0	0.21	57.58	3.47 [70]
	422	1	14.5	0.18	58.11	3.51[70]
		2	**13.9**	**0.17**	**58.90**	**2.70 [77]**
	444	1	14.8	0.18	57.30	4.01 [79]
		2	14.7	0.17	57.42	3.88 [88]
		3	14.7	0.17	58.18	3.78 [90]
		4	**14.6**	**0.15**	**58.40**	**3.26 [89]**

4.1   Evaluation of Depth-Bit Assignment-Methods

First, we evaluated our proposed depth-bit assignment schemes. The results are
listed in Tables 1 and 2. It can be seen from Table 1 that method 2 for color space
4:2:2 and method 4 for color space 4:4:4 using 10 bit precision for x264 result in a
very good encoding performance. Both, the achieved compression ratios as well as
the PSNR are relatively high. In Table 2, we compare the encoding performance
for discarding the lowest 2 bits (denoted as *10bit@x264*) with the best results

we achieved for the experiments from Table 1. We can see that the encoding is faster for *10bit@x264*, which is mainly achieved by avoiding the processing of the UV channels. In addition, both, PSNR and compression ratio are very close to *Method 2* and *Method 4*. As a first result, we can deduce that it is feasible to discard the lowest 2 bits instead of assigning them to UV channels. However, the error remains near 3 mm on average, as expected.

Next, we compare our proposed hybrid lossless-lossy compression approach (denoted as *10bit@x264+2bit@RLE*) to the method *10bit@x264*. The results are listed in Table 2. As one can see, the encoding time increases for *10bit@x264+ 2bit@RLE*, which is mainly caused by the RLE encoding time overhead, and, the compression ratio is lower. However, the quality of compression in terms of PSNR, MAE (1.31 mm), as well as maximum error (26 mm) is much better compared to *10bit@x264*. As a second result, we can summarize, that our hybrid approach can achieve a very high quality at the cost of a lower compression ratio and a slightly longer encoding time.

Table 2. Evaluation of discarding the lowest 2 bits per depth pixel and comparison of hybrid lossless-lossy encoding (*10bit@x264+2bit@RLE*) vs. pure x264-encoding (*10bit@x264*).

Color space	Method	Time	Ratio	PSNR	Error
444	4	14.6	0.15	58.40	3.26 [89]
422	2	13.9	0.17	58.90	2.70 [77]
420	10bit@x264	11.8	0.15	58.66	3.07 [80]
420	**10bit@x264 + 2bit@RLE**	**15.1**	**0.23**	**65.17**	**1.31 [26]**

Table 3. Evaluation of different *IDR interval* settings and different *preset* settings.

Preset	IDR interval	Time	Ratio	PSNR	Error
very fast	20	12.0	0.17	57.66	3.40 [98]
very fast	10	12.0	0.17	57.86	3.33 [98]
very fast	1	11.8	0.15	58.66	3.07 [80]
superfast	1	9.7	0.16	59.80	3.00 [73]
ultrafast	1	**8.0**	**0.17**	**60.34**	**3.04 [68]**

4.2 Evaluation of x264 Encoding Parameter Settings

Next, we evaluated the parameter settings for x264-encoding. The results are listed in Tables 3 and 4. In Table 3, *crf* is set to 1, and the depth bit assignment

Table 4. Evaluation of different *crf* settings along with the mapping to 8 compression levels.

Method	crf	Time	Ratio	PSNR	Error	Compr. level
10bit@x264+2bit@RLE	1	11.7	0.25	68.02	1.19 [15]	1
	6	11.0	0.18	62.73	2.17 [28]	2
	12	8.5	0.12	57.92	3.66 [50]	4
	18	7.5	0.06	54.14	5.65 [87]	not used
	24	7.3	0.05	50.30	8.39 [183]	not used
10bit@x264	1	8.0	0.17	60.34	3.04 [68]	3
	6	7.7	0.11	56.50	4.59 [112]	5
	12	7.2	0.08	53.47	7.00 [168]	6
	18	7.2	0.04	49.72	10.16 [281]	7
	24	7.1	0.04	44.90	15.21 [542]	8

scheme is *10bit@x264*. It is obvious that the encoding performance is best for *IDR interval=1*. The reason for this is that depth images typically have areas with silhouette edges and also smooth surfaces, which is more suitable for intra-prediction than inter-prediction. Therefore we set both, *crf* and *IDR interval* to 1 for the evaluation of different *preset* settings (Table 3). In comparison to *veryfast* and *superfast*, *ultrafast* results in smaller depth errors and shorter encoding times with only little loss in the compression ratio.

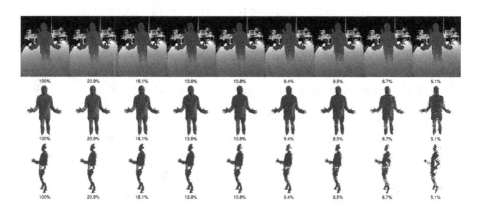

Fig. 3. Surface normal visualizations of 3D reconstructions for different compression ratios, generated from the corresponding depth image (upper row). Note, that the images in the lower row are reconstructed from the corresponding depth image, too.

In our next evaluation, we therefore set *preset* to *ultrafast* and *IDR interval* to 1. We were now interested in a comparison between *10bit@x264+2bit@RLE* and *10bit@x264* for different *crf* settings. The results are listed in Table 4.

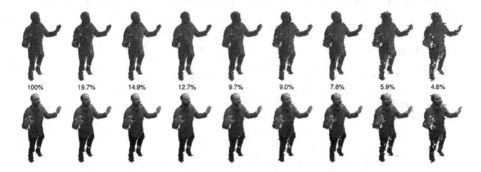

Fig. 4. Comparison of the resulting 3D reconstruction from 2 overlapping Kinect V2 sensors for different compression ratios.

While encoding times are always higher for our hybrid method the quality in terms compression ratio, PSNR and error is much better. However, *10bit@x264* can still be an efficient encoding method if a higher compression ratio (at the cost of quality) is preferred.

4.3   Depth Compression for Real-Time 3D Reconstruction

We evaluated our compression approach in the context of real-time 3D reconstruction. The 3D reconstruction itself is performed using the approach from [2]. We mapped the most suitable configurations to 8 compression levels as listed in Table 4. The compression ratio for depth encoding can then be adjusted during runtime from 1 (lowest) to 8 (highest). Figures 3 and 4 illustrate results for 3D reconstructions for different compression levels. We can see that the subjective 3D reconstruction perception using our compression approach with a compression level from approx. 20 % to 10 % is almost indistinguishable from the original depth image. The perceived quality is still acceptable for a compression ratio of approx. 7 % to 8 %. For the highest compression settings, shown in the two rightmost columns, the quality of the reconstruction is obviously reduced, e.g. edges tend to deteriorate and the reconstructed surface appears blocky.

Table 5 lists the processing times, as well as the compression ratios and the necessary bit rates assuming the devices are running at 30 Hz for different configurations. As one can see, we are able to encode and decode multiple depth image streams in real-time. The encoding as well as the decoding time scales with the compression level. The number of streams has only little influence due to highly parallelized processing. Note, that we do not provide an evaluation for multiple color image streams. In our current system, we use a DXT1-based color compression which is able to encode multiple color image streams in real-time, e.g. less than 30 ms for resolutions up to 1280 x 1080 per color image.

Table 5. Evaluation of different stream configurations and compression levels.

Streams config	Compr. level	Encoding	Decoding	Ratio	Bitrate
1x640x480	1	11.9	6.6	25.7	4.5
	4	7.8	5.2	11.4	2.0
	8	7.2	4.3	4.6	0.8
3x640x480	1	13	6.7	23.6	12.4
	4	9.5	5.5	10.4	5.5
	8	8.6	4.6	4.3	2.2
5x640x480	1	13.8	8.6	24.7	21.8
	4	11	6.6	11.3	10.1
	8	10.2	5.8	4.6	4.1
1x512x424	1	10.2	6.2	21.3	2.6
	4	6.1	4.0	10.7	1.33
	8	5.3	3.4	5.0	0.62
3x512x424	1	11.8	6.3	20.6	7.6
	4	7.8	4.2	10.3	3.8
	8	7.2	3.6	4.9	1.82

5   Conclusion

The real-time compression of depth image streams becomes a basic necessity
if multiple streams have to be sent over wide area networks. We presented an
efficient solution to adapt the x264-video codec, which is designed for 10-bit
YUV color space images, to 12-bit depth maps without modifying the codec
algorithm itself. We deduced and evaluated optimal x264 encoding parameters
for depth images through experimental statistics. Our research suggests that a
hybrid lossless-lossy depth compression scheme provides a good tradeoff between
quality and compression ratio by using x264 for lossy encoding and run length
encoding to preserve the highest bits of 12 bit depth images. Our evaluation
shows that we are able to achieve compression ratios in the range of 4:1 to
20:1 for multiple depth image streams in real-time. The visual quality of 3D
reconstructions is close to the original for a compression ratio of up to 10:1.
Although we did not provide results for real-time color compression our method
can simplify the pipeline of color stream processing. A promising direction for
future research is a silhouette-aware joint color and depth compression which
can further increase both, compression ratio and reconstruction quality.

Acknowledgments. The author gratefully acknowledges the support of K. C. Wong
Education Foundation and DAAD, NSFC (Grant No. 61073074, 61303144), Projects in
Science and Technique of Ningbo Municipal (Grant No. 2012B82003), Zhejiang Higher
Education Reform Project (Grant No. jg2013135), National Students' Innovation and
Entrepreneurship Project (Grant No. 201410876012).

References

1. Bal, C., Nguyen, T.: Multiview video plus depth coding with depth-based prediction mode. IEEE Trans. Circuits Syst. Video Technol. **24**(6), 995–1005 (2014)
2. Beck, S., Kunert, A., Kulik, A., Froehlich, B.: Immersive group-to-group telepresence. IEEE Trans. Visual Comput. Graphics **19**(4), 616–625 (2013)
3. Champawat, Y., Kumar, S.: Online point-cloud transmission for tele-immersion. In: Proceedings of VRCAI 2012, pp. 79–82. ACM, New York (2012)
4. Cho, T., Lee, Y., Shin, J.: A homogenizing filter for depth map compressive sensing using edge-awarded method. In: Proceedings of ICT Convergence 2013, pp. 591–595, October 2013
5. Choi, J.-A., Ho, Y.-S.: Improved near-lossless hevc codec for depth map based on statistical analysis of residual data. In: Proceedings of ISCS 2012, pp. 894–897, May 2012
6. FFMpeg. Ffmpeg (2014, nov. version 2.6.1): audio and video codec library. https://www.ffmpeg.org/
7. Fu, J., Miao, D., Yu, W., Wang, S., Lu, Y., Li, S.: Kinect-like depth data compression. IEEE Trans. Multimedia **15**(6), 1340–1352 (2013)
8. Gautier, J., Le Meur, O., Guillemot, C.: Efficient depth map compression based on lossless edge coding and diffusion. In: Proceedings of PCS 2012, pp. 81–84, May 2012
9. Hoffmann, S., Mainberger, M., Weickert, J., Puhl, M.: Compression of depth maps with segment-based homogeneous diffusion. In: Pack, T. (ed.) SSVM 2013. LNCS, vol. 7893, pp. 319–330. Springer, Heidelberg (2013)
10. Ko, H., Kuo, C.-C.: A new in-loop filter for depth map coding in hevc. In: Proceedings of APSIPA ASC 2012, pp. 1–7, December 2012
11. Lan, C., Xu, J., Wu, F.: Object-based coding for kinect depth and color videos. In: Proceedings of VCIP 2012, pp. 1–6, November 2012
12. Mamou, K., Zaharia, T., Prêteux, F.: Tfan: a low complexity 3d mesh compression algorithm. Comput. Animat. Virtual Worlds **20**(23), 343–354 (2009)
13. Mekuria, R., Sanna, M., Asioli, S., Izquierdo, E., Bulterman, D.C.A., Cesar, P.: A 3d tele-immersion system based on live captured mesh geometry. In: Proceedings of MMSys 2013, pp. 24–35. ACM, New York (2013)
14. Oh, B.T.: An adaptive quantization algorithm without side information for depth map coding. Image Commun. **29**(9), 962–970 (2014)
15. Pece, F., Kautz, J., Weyrich, T.: Adapting standard video codecs for depth streaming. In: Proceedings of EGVE-JVRC 2011, pp. 59–66, Aire-la-Ville, Switzerland. Eurographics Association (2011)
16. Raghuraman, S., Venkatraman, K., Wang, Z., Prabhakaran, B., Guo, X.: A 3d tele-immersion streaming approach using skeleton-based prediction. In: Proceedings of MM 2013, pp. 721–724. ACM, New York (2013)
17. Schwarz, H., Bartnik, C., Bosse, S., Brust, H., Hinz, T., Lakshman, H., Merkle, P., Muller, K., Rhee, H., Tech, G., Winken, M., Marpe, D., Wiegand, T.: Extension of high efficiency video coding (hevc) for multiview video and depth data. In: Proceedings of ICIP 2012, pp. 205–208, September 2012
18. Schwarz, S., Olsson, R., Sjostrom, M., Tourancheau, S.: Adaptive depth filtering for hevc 3d video coding. In: Proceedings of PCS 2012, pp. 49–52, May 2012
19. Van Wallendael, G., Van Leuven, S., De Cock, J., Bruls, F., Van de Walle, R.: Multiview and depth map compression based on hevc. In: Proceedings of ICCE 2012, pp. 168–169, January 2012

20. Videolan. x264 (2014, sep. version 142) : a free h264/avc encoder. http://www. videolan.org/developers/x264.html
21. Yuan, H., Kwong, S., Liu, J., Sun, J.: A novel distortion model and lagrangian multiplier for depth maps coding. IEEE Trans. Circuits Syst. Video Technol. **24**(3), 443–451 (2014)
22. Zamarin, M., Salmistraro, M., Forchhammer, S., Ortega, A.: Edge-preserving intra depth coding based on context-coding and h.264/avc. In: Proceedings of ICME 2013, pp. 1–6, July 2013

Marginal Fisher Regression Classification for Face Recognition

Zhong Ji[1,2(✉)], Yunlong Yu[1], Yanwei Pang[1], Yingming Li[2], and Zhongfei Zhang[2]

[1] School of Electronic Information Engineering, Tianjin University,
Tianjin 300072, China
jizhong@tju.edu.cn
[2] Department of Computer Science, State University of New York,
Binghamton, NY 13902, USA

Abstract. This paper presents a novel *marginal Fisher regression classification* (MFRC) method by incorporating the ideas of *marginal Fisher analysis* (MFA) and *linear regression classification* (LRC). The MFRC aims at minimizing the within-class compactness over the between-class separability to find an optimal embedding matrix for the LRC so that the LRC on that subspace achieves a high discrimination for classification. Specifically, the within-class compactness is measured with the sum of distances between each sample and its neighbors within the same class with the LRC, and the between-class separability is characterized as the sum of distances between margin points and their neighboring points from different classes with the LRC. Therefore, the MFRC embodies the ideas of the LRC, Fisher analysis and manifold learning. Experiments on the FERET, PIE and AR datasets demonstrate the effectiveness of the MFRC.

Keywords: Face recognition · Nearest subspace classification · Linear regression classification · Manifold learning

1 Introduction

Face recognition has been an active research area in computer vision and pattern recognition for decades. In recent years, NSC (nearest subspace classification)-based methods have been drawn considerable attention owing to its efficacy and simplicity [1–3]. These methods aim at seeking the best representation by samples in each class since they assume that a high dimensional face image lies on a low dimensional subspace spanned by the samples from the same subject. Among these methods, linear regression classification (LRC) [2] is one of the representatives. LRC first estimates the regression coefficients on each class by using the least square estimation method for the probe sample, then classifies it to the category with respect to the smallest reconstruction error.

Recently, some efforts have been made to improve LRC. For example, Huang and Yang [4] proposed an improved component regression classification (IPCRC) method to overcome the problem of multicollinearity in the LRC. The IPCRC removes the

© Springer International Publishing Switzerland 2015
Y.-S. Ho et al. (Eds.): PCM 2015, Part I, LNCS 9314, pp. 453–462, 2015.
DOI: 10.1007/978-3-319-24075-6_44

mean of each sample before performing principal component analysis (PCA) and drops the first principal components. The projection coefficients are then obtained by the LRC. To reduce the sensitivity of LSE to outliers, Naseem et al. proposed a robust linear regression classification algorithm (RLRC) [5] to estimate regression parameters by using the robust Huber estimation, which weighs the large residuals more lightly than the least square estimation. In [7], He et al. points out that the LRC fails when the number of samples in the class specific training set is smaller than their dimensionality. They proposed a ridge regression classification (RRC) method to solve this issue. In [6], Huang and Yang proposed a linear discriminant regression classification (LDRC) method by embedding the Fisher criterion into the LRC. The method attempts to maximize the ratio of the between-class reconstruction error to the within-class reconstruction error to find an optimal projection matrix for the LRC. In this way, the LRC on that subspace can achieve a high discrimination for classification.

Although these LRC-based methods are successful in many situations, most of them assume that the face images reside on a linear Euclidean space, which fails to discover the underlying nonlinear structure. Actually, it is well-known that the face images possibly lie in a nonlinear manifold [8–10]. The corresponding technique is manifold learning, which applies a local neighborhood relation to learn the global structure of nonlinear manifolds. Brown et al. [11] made an attempt by introducing the idea of manifold learning to the LRC. They used the k closest samples to the probe image instead of all the samples from the same class. A better performance is obtained against LRC; however, it fails when the training samples are few.

Inspired by the LRC and marginal Fisher analysis (MFA) [9], this paper proposes a novel marginal Fisher regression classification (MFRC) approach by incorporating the idea of LRC into MFA. Particularly, the manifold structure is modeled by applying a within-graph to represent the within-class compactness and a between-graph to represent the between-class separability. The novelty of MFRC is to jointly consider the regression classification information, the manifold structure, and the margin criterion (margin criterion is used to characterize the separability of different classes). Experiments on face recognition achieve a much better performance.

The rest of the paper is organized as follows: Sect. 2 describes the LRC algorithm, and the proposed MFRC method is described in details in Sect. 3. Extensive experiments are presented in Sect. 4, and conclusions are drawn in Sect. 5.

2   Linear Regression Classification

This section gives a brief introduction to linear regression classification (LRC). Assume that we have N training samples containing C different classes with p_c training samples from the cth class in the D dimensional space, $c = 1, 2, \ldots, C$. let $\mathbf{X}_c$ be the training set of the cth class whose data matrix is:

$$\mathbf{X}_c = [\mathbf{x}_c^1, \ldots, \mathbf{x}_c^i, \ldots \mathbf{x}_c^{p_c}] \in \Re^{D \times p_c}, \tag{1}$$

where $\mathbf{x}_c^i$ is the ith sample of class c and $\mathbf{X}_c$ is called the predictor for class c.

According to the subspace assumption, if a probe image $\mathbf{x} \in \Re^{D \times 1}$ belongs to the cth class, it can be represented by a linear combination of the training images from the cth class with an error $\mathbf{e}$ and can be defined as:

$$\mathbf{x} = \mathbf{X}_c \boldsymbol{\beta}_c + \mathbf{e} \quad c = 1, 2, \ldots, C, \tag{2}$$

where $\boldsymbol{\beta}_c \in \Re^{p_c \times 1}$ is the regression coefficient vector of parameters projecting the probe image onto the linear space spanned by the predictor $\mathbf{X}_c$.

The goal of the LRC is to find an optimal $\tilde{\beta}_c$ to minimize the residual error as:

$$\begin{aligned} &\min_{\boldsymbol{\beta}_c} \mathbf{e} \\ &\Rightarrow \tilde{\beta}_c = \arg\min_{\boldsymbol{\beta}_c} \|\mathbf{X}_c \boldsymbol{\beta}_c - \mathbf{x}\|_2^2 \quad c = 1, 2, \ldots, C. \end{aligned} \tag{3}$$

Equation (3) can be solved by using the least square estimation method as follows:

$$\tilde{\beta}_c = (\mathbf{X}_c^T \mathbf{X}_c)^{-1} \mathbf{X}_c^T \mathbf{x}. \tag{4}$$

However, it can be seen that it fails when the number of samples in the class specific training set is smaller than their dimensionality. A common solution to this problem is to add regular terms. The ridge regression method is one of the effective methods, which uses a regularized least square method for regression [6]. Thus, this paper uses it to find the optimal $\tilde{\beta}_c$ by minimizing the residual error as:

$$\tilde{\beta}_c = \arg\min_{\boldsymbol{\beta}_c} (\|\mathbf{X}_c \boldsymbol{\beta}_c - \mathbf{x}\|_2^2 + \alpha \|\boldsymbol{\beta}_c\|_2^2) \quad c = 1, 2, \ldots, C, \tag{5}$$

where α is a regularization parameter. By using the least square estimation method, the regression coefficients can be estimated as:

$$\tilde{\beta}_c = (\mathbf{X}_c^T \mathbf{X}_c + \alpha \mathbf{I})^{-1} \mathbf{X}_c^T \mathbf{x}, \tag{6}$$

where $\mathbf{I} \in \Re^{D \times D}$ is an identity matrix.

The probe image can be represented in terms of the training images as:

$$\tilde{\mathbf{x}}_c = \mathbf{X}_c \tilde{\beta}_c = \mathbf{X}_c (\mathbf{X}_c^T \mathbf{X}_c + \alpha \mathbf{I})^{-1} \mathbf{X}_c^T \mathbf{x} = \mathbf{H}_c \mathbf{x}, \tag{7}$$

where $\tilde{\mathbf{x}}_c$ is the projection of the sample $\mathbf{x}$ on class c, and $\mathbf{H}_c = \mathbf{X}_c (\mathbf{X}_c^T \mathbf{X}_c + \alpha \mathbf{I})^{-1} \mathbf{X}_c^T$ is called the class projection matrix on class c.

The identity of the test image can be determined by calculating the Euclidean distance between the projected vectors and the original vector as:

$$identity(\mathbf{x}) = \arg\min_c \|\tilde{\mathbf{x}}_c - \mathbf{x}\|, \quad c = 1, \ldots, C. \tag{8}$$

3   Marginal Fisher Regression Classification

Given a training data matrix $\mathbf{X} = [\mathbf{x}_1 \ldots \mathbf{x}_N] \in \Re^{D \times N}$, where D is the dimensionality and N is the number of training samples. The proposed marginal Fisher regression classification (MFRC) method aims at finding an optimal transformation matrix $\mathbf{W} = [\mathbf{w}_1 \ldots \mathbf{w}_d] \in \Re^{D \times d}$ such that $\mathbf{X}$ can be transformed to $\mathbf{Y} = [\mathbf{y}_1 \ldots \mathbf{y}_N] \in \Re^{d \times N}$ by $\mathbf{Y} = \mathbf{W}^T \mathbf{X}$, where d denotes the reduced dimensionality.

Since the motivation of the MFRC is to incorporate the idea of LRC into MFA, it is formulated as the following optimization problem by minimizing the objective function given as:

$$\arg\min_{\mathbf{W}} J(\mathbf{W}) = \min_{\mathbf{W}}(\frac{\mathbf{S}_W}{\mathbf{S}_B}), \tag{9}$$

where $\mathbf{S}_W$ and $\mathbf{S}_B$ denote the within-class compactness and between-class separability, respectively. It embodies the idea of Fisher discriminant analysis (FDA). Specifically, $\mathbf{S}_W$ is measured as the sum of distances between each sample and its neighbors within the same class by using the LRC method, which is defined as:

$$\mathbf{S}_W = \sum_{\substack{i,j = 1 \\ k = l(\mathbf{x}_i)}}^{N} \left\| \mathbf{y}_i - \hat{\mathbf{y}}_{kj} \right\|^2 s_{ij}^{(w)}, \tag{10}$$

where $\mathbf{y}_i = \mathbf{W}^T \mathbf{x}_i$ is the transformed vector of $\mathbf{x}_i$, $l(\mathbf{x}_i) \in \{1, 2, \ldots, C\}$ denotes the corresponding class label of $\mathbf{x}_i$, $\hat{\mathbf{y}}_{kj} = \mathbf{W}^T \hat{\mathbf{x}}_{kj}$ is the transformed vector of $\hat{\mathbf{x}}_{kj}$, and $\hat{\mathbf{x}}_{kj}$ denotes the projection of $\mathbf{x}_j$ on class k with Eq. (7). In this way, the idea of LRC is incorporated. Moreover, $s_{ij}^{(w)}$ represents the within-class local graph adjacency relationship, where an edge is connected between nodes i and j if $\mathbf{x}_j$ is among within-class k_w nearest neighbors of $\mathbf{x}_i$, and vice versa, i.e.,

$$s_{ij}^{(w)} = \begin{cases} 1 & \text{if } \mathbf{x}_i \in LN_{k_w}(\mathbf{x}_j) \text{ or } \mathbf{x}_j \in LN_{k_w}(\mathbf{x}_i) \\ 0 & \text{otherwise} \end{cases}, \tag{11}$$

where LN_{k_w} denotes k_w within-class local neighborhood.

Meanwhile, $\mathbf{S}_B$ is characterized as the sum of distances between margin points and their neighboring points from different classes by using the LRC method, which is defined as:

$$\mathbf{S}_B = \sum_{\substack{i,j = 1, \\ k = l(\mathbf{x}_i)}}^{N} \left\| \mathbf{y}_j - \hat{\mathbf{y}}_{kj} \right\|^2 s_{ij}^{(b)}, \tag{12}$$

where $s_{ij}^{(b)}$ is the penalty graph which represents the between-class graph adjacency relationship and connects the margin samples from different classes; each sample is connected to its k_b between-class nearest neighbors, i.e.,

$$s_{ij}^{(b)} = \begin{cases} 1 & \text{if } \mathbf{x}_i \in LN_{k_b}(\mathbf{x}_j) \text{ or } \mathbf{x}_j \in LN_{k_b}(\mathbf{x}_i) \\ 0 & otherwise \end{cases}, \tag{13}$$

where LN_{k_b} denotes k_b between-class local neighborhood.

Therefore, the objective function of the MFRC can be expressed as:

$$\arg \min_{\mathbf{W}} J(\mathbf{W}) = \min_{\mathbf{W}} \left(\frac{\mathbf{S}_W}{\mathbf{S}_B} \right) = \min_{\mathbf{W}} \frac{\sum\limits_{\substack{i,j=1 \\ k=l(\mathbf{y}_i)}}^{N} \left\| \mathbf{y}_i - \hat{\mathbf{y}}_{kj} \right\|^2 s_{ij}^{(w)}}{\sum\limits_{\substack{i,j=1, \\ k=l(\mathbf{y}_i)}}^{N} \left\| \mathbf{y}_j - \hat{\mathbf{y}}_{kj} \right\|^2 s_{ij}^{(b)}}. \tag{14}$$

With some algebraic deduction, we have:

$$J(\mathbf{W}) = \frac{\sum\limits_{\substack{i,j=1 \\ k=l(\mathbf{y}_i)}}^{N} tr[\mathbf{W}^T (\mathbf{x}_i - \hat{\mathbf{x}}_{kj}) s_{ij}^{(w)} (\mathbf{x}_i - \hat{\mathbf{x}}_{kj})^T \mathbf{W}]}{\sum\limits_{\substack{i,j=1 \\ k=l(\mathbf{y}_i)}}^{N} tr[\mathbf{W}^T (\mathbf{x}_i - \hat{\mathbf{x}}_{kj}) s_{ij}^{(b)} (\mathbf{x}_i - \hat{\mathbf{x}}_{kj})^T \mathbf{W}]} \tag{15}$$

$$= \frac{tr(\mathbf{W}^T \mathbf{E}_b \mathbf{W})}{tr(\mathbf{W}^T \mathbf{E}_w \mathbf{W})},$$

where $\mathbf{E}_b = \sum\limits_{\substack{i,j=1 \\ k=l(\mathbf{x}_i)}}^{N} (\mathbf{x}_i - \hat{\mathbf{x}}_{kj}) s_{ij}^{(b)} (\mathbf{x}_i - \hat{\mathbf{x}}_{kj})^T$ and $\mathbf{E}_w = \sum\limits_{\substack{i,j=1 \\ k=l(\mathbf{x}_i)}}^{N} (\mathbf{x}_i - \hat{\mathbf{x}}_{kj}) s_{ij}^{(w)} (\mathbf{x}_i - \hat{\mathbf{x}}_{kj})^T$

denote the between-class and within-class reconstruction error matrices with LRC, respectively.

In addition, the maximum margin criterion (MMC) method [12] is employed by employing the constraint of $\boldsymbol{\omega}_i^T \boldsymbol{\omega}_i = 1$ to avoid the potential small sample size (SSS) problem. The objective function of the MFRC can be reformulated as the following constrained optimization problem:

$$\arg \min_{\mathbf{w}} tr(\mathbf{W}^T \mathbf{E}_w \mathbf{W}) - tr(\mathbf{W}^T \mathbf{E}_b \mathbf{W})$$

$$= \arg \max_{\mathbf{w}} tr(\mathbf{W}^T \mathbf{E}_w - \mathbf{E}_b) \mathbf{W}) \tag{16}$$

$$s.t. \quad \boldsymbol{\omega}_i^T \boldsymbol{\omega}_i \quad i = 1, \dots, d.$$

Equation (16) can be accomplished by using Lagrange multipliers. Thus, the transformation matrix $\mathbf{W}$ can be obtained by solving the generalized eigenvector problem as:

$$(\mathbf{E}_w - \mathbf{E}_b)\omega_i = \lambda_i \omega_i \quad i = 1, 2, \ldots, d, \qquad (17)$$

where λ_i is the eigenvalue vector of $\mathbf{E}_w - \mathbf{E}_b$. Let the column vector $\omega_1, \ldots, \omega_d$ be the solution of Eq. (17), ordered with their eigenvalue $\lambda_1 < \cdots < \lambda_d$. Thus, the embedding is as follows:

$$\mathbf{x}_i \rightarrow \mathbf{y}_i = \mathbf{W}^T \mathbf{x}_i, \quad \mathbf{W} = [\omega_1, \ldots, \omega_i, \ldots, \omega_d], \qquad (18)$$

where $\mathbf{y}_i$ is a d-dimensional vector, and $\mathbf{W} = [\omega_1, \ldots, \omega_i, \ldots, \omega_d]$ is the optimal projection matrix.

Figure 1 illustrates the flowchart of the MFRC, whose main procedures are stated below.

(1) PCA projection: To avoid the singular problem and reduce noise disturbance, principle component analysis (PCA) is first adopted to project $\mathbf{X}$ into a subspace by throwing away the smallest principal components to maintain 99 % of the energy. For convenience, we still use $\mathbf{x}_i$ to denote the data samples in the PCA subspace in the following steps.

(2) Graph construction: Define two affinity graphs G_W and G_B both with N nodes; the ith node corresponds to the sample $\mathbf{x}_i$. In the within-class compactness graph G_W, set the adjacency matrix $s_{ij}^{(w)} = s_{ji}^{(w)} = 1$ if $\mathbf{x}_j$ is among the k_b nearest neighbors of $\mathbf{x}_i$ of the within-class. In the between-class separability graph G_B, set the similarity matrix $s_{ij}^{(b)} = s_{ji}^{(b)} = 1$ if $\mathbf{x}_j$ is among the k_w nearest neighbors of $\mathbf{x}_i$ of the between-class.

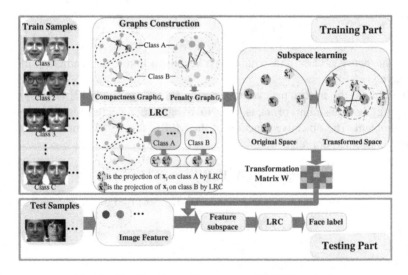

Fig. 1. The flowchart of the proposed MFRC method.

(3) Probe image representation with LRC: Project the probe image $\mathbf{x}_i(i = 1, \ldots, N)$ onto the linear space spanned by the projection matrix $\mathbf{H}_c(c = 1, \ldots, C)$ with the linear regression classification method.

(4) Embedding feature computation: The transformation matrix $\mathbf{W}$ can be obtained with Eq. (17). Then the embedding feature is obtained by:

$$\mathbf{y} = \mathbf{W}^T\mathbf{x}. \tag{19}$$

(5) Classification: First, project the test image into the embedding subspace with Eq. (18). Then, compute the projection vector of the test image on each class by using Eq. (7). At last, the identity is determined with the minimum distance in Eq. (8).

4 Experimental Results

This section demonstrates the effectiveness of the MFRC method on three popular face recognition datasets: FERET [13], PIE [19], and AR [14]. Nine popular methods are used for comparison, which include: (1) subspace learning methods: PCA(Eigenface) [15], LDA (Fisherface) [16], LPP (Laplacianface) [8], KLPP [17], and MFA [9]; (2) LRC-based methods: LRC [2], IPCRC [4], RLRC [5], and LDRC [6]. Specifically, the regularization parameter α in Eq. (5) is set to be 0.005, and the reduced dimensionality d in Eq. (17) is set to be 110. It should be noted that all the experiments are repeated for ten times and the average recognition accuracies are reported.

4.1 Experiment on FERET Dataset

The FERET dataset is one of the largest publicly available datasets. Following [6], we selected 250 people from the dataset, each with four frontal view images from each subject as the four training cases. All the face images contain lighting changes and facial expression variations. All face images were converted to grayscale, cropped manually and resized to 32×32 pixels. Similar to [6], 2 and 3 training samples are employed to investigate the performance in few training samples.

The recognition results are shown in Table 1. It can be observed that the MFRC significantly outperforms the other methods. For example, the MFRC achieves 9.5 % and 5.7 % improvements over the MFA method, and 17.5 % and 10.5 % improvements over the LRC with training samples 2 and 3, respectively. Even for the state-of-the-art LDRC method, the MFRC outperforms it with 1.1 % and 1.8 %, respectively.

4.2 Experiment on PIE Dataset

The CMU PIE face database [19] contains 68 subjects with 41,368 face images. The face images were captured by 13 synchronized cameras and 21 flashes, under varying pose, illumination, and expression. All face images were cropped and resized to

32 × 32 pixels. We take 170 face images for each individual in the experiment, of which 85 images for training and the other 85 images for testing.

The recognition results are also shown in Table 1. The MFRC still performs the best. For example, it outperforms the MFA, LPP, KLPP, LRC, PCRC, RLRC and LDRC approaches with 1.5 %, 1.7 %, 0.5 %, 4.1 %, 9.9 %, 5.6 % and 1.8 %, respectively.

4.3   Experiment on AR Dataset

The AR dataset contains over 4,000 color images corresponding to 126 people's faces (70 men and 56 women) with different facial expressions, lighting changes and partially occlusions. The facial areas were converted to grayscale, cropped manually and resized to 30 × 20 pixels. To evaluate the effectiveness of the proposed method in coping with unseen expressions, the same experimental setup is employed as that in [6], that is four expression variations, including neutral (N), happy (H), angry (A), and scream (S) expressions, are selected from 100 subjects [18] and the single-one-expression training strategy is adopted to evaluate the performance.

The performance comparison is illustrated in Fig. 2. It can be seen that MFRC consistently outperforms all the methods in all the situations except the performance of the LDRC in the case of happy expression as the training set. For example, in the four training cases, the MFRC gets the improvements of 3 %, 2.62 %, 2.34 %, and 2.5 % over the MFA, and gets the improvements of 4.3 %, 1.04 %, 2.17 %, and 4.55 % over the LRC. Compared with LDRC, the MFRC gains 1.94 %, 1.75 % and 1.91 % improvement in the situations when the neutral, angry and screaming expression images are used for training sets, respectively. It is only 0.5 % inferior to the LDRC in the situation of happy expression as the training set.

Table 1. Performance (%) comparison on the FERET and PIE datasets (The number in the bracket represents the dimensionality, note that LDA and LDRC have different dimensionality for FERET and PIE).

Datasets	FERET	FERET	PIE
Training number	2	3	85
PCA (150)	62.4	75.8	79.4
LDA (249/67)	76.2	87.4	84.3
LPP (110)	78.8	90.5	95.4
KLPP (110)	80.6	92.7	96.7
MFA (90)	78.8	88.4	95.7
LRC (1024)	70.8	83.6	93.1
PCRC (150)	63.8	78.8	87.3
RLRC (1024)	72	80	91.6
LDRC (249/67)	87.2	92.3	95.4
MFRC (110)	**88.3**	**94.1**	**97.2**

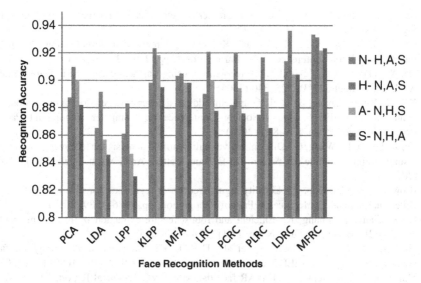

Fig. 2. The performance comparison on the AR dataset.

5 Conclusion and Future Work

A marginal Fisher regression classification approach has been developed for face recognition. The MFRC embeds the linear regression classification into marginal Fisher analysis to seek an optimal projection matrix. Extensive experimental results have validated the effectiveness of MFRC for face recognition. One of our future works is to leverage a proper kernel to make the MFRC nonlinear. In addition, we will employ MFRC in other machine-based recognition applications, such as scene classification.

Acknowledgements. This work was supported by the National Basic Research Program of China (973 Program) under Grant 2014CB340400, the National Natural Science Foundation of China under Grant 61271325, Grant 61472273, and the Elite scholar Program of Tianjin University under Grant 2015XRG-0014.

References

1. Basri, R., Jacobs, D.: Lambertian reflectance and linear subspaces. IEEE Trans. Pattern Anal. Mach. Intell. **25**(2), 218–233 (2003)
2. Naseem, I., Togneri, R., Bennamoun, M.: Linear regression for face recognition. IEEE Trans. Pattern Anal. Mach. Intell. **32**(11), 2106–2112 (2010)
3. Wright, J., Yang, A., Ganesh, A., et al.: Robust face recognition via sparse representation. IEEE Trans. Pattern Anal. Mach. Intell. **31**(2), 210–227 (2009)
4. Huang, S., Yang, J.: Improved principal component regression for face recognition under illumination variations. IEEE Sig. Process. Lett. **19**(4), 179–182 (2012)
5. Naseem, I., Togneri, R., Bennamoun, M.: Robust regression for face recognition. Pattern Recogn. **45**(1), 104–118 (2012)

6. Huang, S., Yang, J.: Linear discriminant regression classification for face recognition. IEEE Sig. Process. Lett. **20**(1), 91–94 (2013)
7. He, J., Ding, L., Jiang, L., et al.: Kernel ridge regression classification. In: IEEE International Joint Conference on Neural Networks, pp. 2263–2267 (2014)
8. He, X., Yan, S., Hu, Y., et al.: Face recognition using Laplacianfaces. IEEE Trans. Pattern Anal. Mach. Intell. **27**(3), 328–340 (2005)
9. Yan, S., Xu, D., Zhang, B., et al.: Graph embedding: a general framework for dimensionality reduction. In: IEEE Computer Society Conference on Computer Vision and Pattern Recognition, vol. 2, pp. 830–837 (2005)
10. Lu, J., Tan, Y.P., Wang, G.: Discriminative multimanifold analysis for face recognition from a single training sample per person. IEEE Trans. Pattern Anal. Mach. Intell. **32**(1), 39–51 (2013)
11. Brown, D., Li, H., Gao, Y.: Locality-regularized linear regression for face recognition. In: IEEE International Conference on Pattern Recognition, pp. 1586–1589 (2012)
12. Li, X., Jiang, T., Zhang, K.: Efficient and robust feature extraction by maximum margin criterion. IEEE Trans. Neural Netw. **17**(1), 157–165 (2006)
13. Phillips, P., Moon, H., Rauss, P., et al.: The FERET evaluation methodology for face recognition algorithms. IEEE Trans. Pattern Anal. Mach. Intell. **22**(10), 1090–1104 (2000)
14. Martinez, A., Benavente, R.: The AR face database. CVC Technical Report, 24 (1998)
15. Turk, M., Pentland, A.: Face recognition using eigenfaces. In: IEEE Conference on Computer Vision and Pattern Recognition, pp. 586–591 (1991)
16. Belhumeur, P., Hespanha, J., Kriegman, D.: Eigenfaces vs. Fisherfaces: recognition using class specific linear projection. IEEE Trans. Pattern Anal. Mach. Intell. **19**(7), 711–720 (1997)
17. Cheng, J., Liu, Q., Lu, H., et al.: Supervised kernel locality preserving projections for face recognition. Neurocomputing **67**, 443–449 (2005)
18. Martinez, A., Kak, A.: PCA versus LDA. IEEE Trans. Pattern Anal. Mach. Intell. **23**(2), 228–233 (2001)
19. Sim, T., Baker, S., Bsat, M.: The CMU pose, illumination, and expression (PIE) database. In: IEEE International Conference on Automatic Face and Gesture Recognition, pp. 46–51 (2002)

Temporally Adaptive Quantization Algorithm in Hybrid Video Encoder

Haibing Yin[✉], Zhongxiao Wang, Zhelei Xia, and Ye Shen

Information Engineering Department, China Jiliang University,
No. 258, Xueyuan Street, Xia Sha, Hangzhou, People's Republic of China
habingyin@163.com

Abstract. In video coder, inter-frame prediction results in distortion propagation among adjacent frames, and this distortion dependency is a crucial factor for rate control and video coding algorithm optimization. The macroblock tree (MBTree) is a typical temporal quantization control algorithm, in which a quantization offset δ is employed for adjustment according to the amount of distortion propagation measured by the relative propagation cost ρ. Appropriate δ-ρ model is the key to the MBTree-like adaptive quantization algorithm. The default δ-ρ model in MBTree algorithm is designed in an empirical way with rough model accuracy and insufficient universality to different input source. This paper focuses on this problem and apply the competitive decision mechanism in exploring optimal δ-ρ model, and then proposes an improved δ-ρ model with rate distortion optimization. The simulation results verify that the improved MBTree algorithm with the proposed model achieves up to 0.14 dB BD-PSNR improvement, and 0.29 dB BD-SSIM improvement. The proposed algorithm achieves better temporal bit allocation and reduces the distortion fluctuation in temporal domain, achieving in adaptive quantization control.

Keywords: Video coding · Rate control · Rate distortion optimization · Distortion propagation · Competitive decision

1 Introduction

Algorithm optimization for video coding is a complicated optimization problem. Video sequence is usually partitioned into multiple coding units at different levels from GOP (group of picture), frame, slice to macroblock etc. Multiple-level joint optimization is ideally desired for rate distortion optimization. However, algorithm optimizations are usually made for the units at different levels individually to simplify the multiple-level optimization, without considering or weakening the dependency between different coding units [1].

Actually, there are complex multilevel spatial and temporal domain dependence in video coding, such as inter prediction resulted temporal distortion propagation effect, which makes the assumption of traditional inter-unit independency invalid anymore. It is meaningful to consider these dependencies to achieve global optimization. Dynamic programming optimization is theoretically suitable for dependent coding optimization. However, dynamic programming method, such as Viterbi algorithm, is ill-suited for

© Springer International Publishing Switzerland 2015
Y.-S. Ho et al. (Eds.): PCM 2015, Part I, LNCS 9314, pp. 463–472, 2015.
DOI: 10.1007/978-3-319-24075-6_45

multilevel simultaneous optimization for multiple parameters mainly due to the huge search space, which is caused by the huge amount of available solutions. As a result, the complexity increases sharply with the increase of graphic trellis states in Viterbi algorithm.

In recent years, global algorithm optimization considering the above dependencies attracted research interests. Liu etc. proposed a model to measure the distortion propagation in temporal domain to optimize bit allocation for scalable video coding [2]. Yu etc. analyzed the relationship between the reconstructed distortion with prediction error, quantization step size, and the impact of reference frame compression [4]. It was generally accepted that there is linear relationship between the frame average distortion with the source prediction error and the reference image distortion. Based on this assumption, some distortion propagation models are proposed, such as Oscar's frame level distortion propagation model [5], and Ma's frame dependent rate distortion model [6]. In addition, Zhu etc. proposed the source distortion propagation model to estimate the impact of the current unit impose on the subsequent units, and to adjust the coding parameters of current unit for algorithm optimization [7].

x264 explorers proposed perceptual rate control algorithm employing perceptual bit allocation using blurred complexity model, and quantization control including temporal MBTree and spatial VAQ (variance adaptive quantization) [3]. These tools improve the rate distortion (RD) performance to some extent. However, they are all based on empirical models and supposed to be improved further. The MBTree algorithm evaluates the degree of distortion propagation marked by the *relative propagation cost* (ρ), which is employed to calculate an offset δ to adjust the frame quantization parameter. The key of the MBTree algorithm is how to measure ρ and how to calculate δ, i.e. the δ-ρ mapping model. The bigger the ρ one coding unit is, the bigger distortion propagate amount is. The unit with larger ρ is supposed to have less distortion to reduce the distortion and decrease the distortion propagation, from the viewpoint of temporal bit allocation optimization. The default δ-ρ model in x264 uses an empirical model with insufficient model accuracy and adaptivity to versatile input sequences, suffering from unsatisfactory RD performance.

This paper focuses on the above problem and tends to apply the competitive decision algorithm (CDA) in exploring optimal δ-ρ model. A large number of samples of optimal (ρ, δ) pairs will be collected for offline statistic analysis for δ-ρ modeling. Then, an improved δ-ρ model is proposed to achieve rate distortion optimization. Intensive simulations will be given to verify the performance of the proposed work.

The rest is organized as follows. Section 2 analyses the deficiency of the traditional temporal quantization control algorithm. Section 3 gives the new δ-ρ model based on CDA, and the proposed quantization control algorithm. Simulation results and conclusions are given in Sects. 4 and 5 in turn.

2   The Temporally Adaptive Quantization Algorithm

The relative propagation cost ρ is measured according to the results of video preanalysis. The preanalysis process is performed within a sliding frame window, which is composed of the current frame and its adjacent frames. The preanalysis is implemented using half-pixel resolution original image in sliding window. The SATD (sum of absolute transformed difference) is used as the criterion for the prediction error to measure the intra-frame and inter-frame prediction costs (ζ_{intra} and ζ_{inter}), as well as the inter-frame reference propagation cost $\Upsilon_{propagate}$. The relative distortion propagation cost ρ of one coding unit is measured as follows.

$$\rho = \Upsilon_{propagate}/\zeta_{intra} \tag{1}$$

The key of the algorithm is how to measure the reference propagation cost $\Upsilon_{propagate}$. Assume that the current coding unit is marked with (s, i, j) which is located at position (i, j) in the current frame s, the flowcharts of estimation and propagation analysis for $\Upsilon_{propagate}$ are shown in Fig. 1. The processing and analysis can be described as follows.

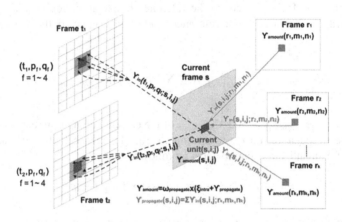

Fig. 1. Diagram of estimation, distribution and transfer process for reference propagation cost ($k = 1, 2, \ldots, K$)

① Estimate the intra-frame and inter-frame prediction costs, ζ_{intra} and ζ_{inter}, they are measured with SATD. If ζ_{inter} is less than ζ_{intra}, then $\zeta_{inter} = \zeta_{intra}$.

② Calculate the weight coefficient ω of the propagation cost $\Upsilon_{propagate}$ for the current unit as follows.

$$\omega_{propagate} = 1 - \zeta_{inter}/\zeta_{intra} \tag{2}$$

③ Estimate the current unit's propagation cost amount Υ_{amout}, which will be distributed to all matched units in the reference frame as follows.

$$\Upsilon_{amout} = (\zeta_{intra} + \Upsilon_{propagate}) \times \omega_{propagate} \tag{3}$$

$\Upsilon_{propagate}$ is the sum of the income reference cost Υ_{in}, which is the propagation cost inputted to the current unit from the inter-predicted blocks (dst, src) in reference frames, and these will be analyzed in step ⑤.

④ The inter-frame displaced block of the current unit may cover four blocks (t_1, p_f, q_f), $f = 1 \sim 4$, due to misalignment in the adjacent frame. Then, estimate the reference weight $\Lambda(dst; src)$ for propagation cost distribution according to area of the referenced pixels, i.e. pixel counts; and Υ_{amout} is distributed to four possible blocks proportionally according to their weight, calculate the income reference cost $\Upsilon_{in}(t_1, p_f, q_f; s, i, j)$ as shown in Eq. (4). These four income cost Υ_{in} are the inputted cost to four blocks (t_1, p_f, q_f) of t_1 frame propagated from the current coding unit (s, i, j) accounting for inter-frame prediction.

$$\Upsilon_{in}(t_1, p_f, q_f; s, i, j) = \Upsilon_{amout}(s, i, j) \times \Lambda(t_1, p_f, q_f; s, i, j) \qquad (4)$$

⑤ As shown in Fig. 1, assume that there are K blocks (r_k, m_k, n_k) in adjacent frame r_k $(k = 1, 2, \ldots, K)$ that will refer to the current unit(s, i, j) in whole or in part. Calculate the income reference cost $\Upsilon_{in}(s, i, j; r_k, m_k, n_k)$ respectively according to (4), and calculate the reference propagation cost $\Upsilon_{propagate}$ that the matched blocks in adjacent reference frames propagate to the current unit as follows,

$$\Upsilon_{propagate}(s, i, j) = \sum_{k=1}^{K} \Upsilon_{in}(s, i, j; r_k, m_k, n_k) \qquad (5)$$

⑥ Calculate the ζ_{intra}, ζ_{inter} and $\Upsilon_{propagate}$ of the current coding unit, and calculate the relative propagation cost ρ according to Eq. (1), then calculate the adjustment offset for quantization parameter δ as follows.

$$\delta = 5(1 - qcompress) \times \log_2(1 + \rho) \qquad (6)$$

The mapping from the propagation cost ρ to quantization offsets δ, called "δ-ρ" model in this paper, is described with logarithmic function in MBTree algorithm as shown in Eq. (6). The default of *qcompress* is 0.6, and it is estimated in an empirical way. The simulation results indicate that the performance improvements of MBTree vary from sequence to sequence, and negative performance may appear in some special sequences as shown in Table 1. The complex dependence among adjacent frames in rate control algorithm appeals to adaptive δ-ρ model to adapt to versatile sequences with different characteristics. However, the traditional model based on logarithmic function is roughly derived in an empirical way. This limits the effectiveness of the MBTree algorithm considerably.

Focusing on this problem, this paper attempts to derive more accurate δ-ρ model to characterize the characteristics of different sequences. This will be helpful for further research for optimization of the MBTree algorithm.

Table 1. PSNR and SSIM performance comparison (compared with MBTree off)

Video sequence	BD-PSNR improvement(dB)				BD-SSIM improvement(dB)			
	Original model	Proposed model			Original model	Proposed model		
		No iteration	One iteration	Two iteration		No iteration	One iteration	Two iteration
Bridge-close	0.6714	0.6624	0.7234	0.7525	0.4322	0.4728	0.5182	0.5454
mobile	0.3872	0.2403	0.0202	−0.1069	0.8552	0.9219	0.9253	0.8824
news	1.2541	1.3027	1.3084	1.2455	1.2022	1.2913	1.3677	1.3248
paris	1.8158	1.8503	1.8741	1.8299	1.4770	1.5783	1.6631	1.6902
highway	−0.0891	−0.0626	−0.0388	−0.0953	0.0772	0.0987	0.1239	0.1256
Bridge-far	−0.2056	−0.0406	0.0183	0.0888	0.0523	0.1329	0.1387	0.1589
forman	0.4230	0.4595	0.4247	0.4691	0.3477	0.4446	0.5052	0.5917
coastguard	0.2129	0.1758	0.1443	0.1067	0.3356	0.3736	0.3489	0.3265
container	0.5893	0.6604	0.7277	0.7229	0.3357	0.4718	0.5517	0.5767
hall	0.2896	0.4220	0.4380	0.4543	0.3005	0.4484	0.5088	0.5920
Average	**0.5348**	**0.567**	**0.564**	**0.5467**	**0.5416**	**0.6234**	**0.6652**	**0.6814**

3 The Proposed CDA Based δ-ρ Model

3.1 The Proposed δ-ρ Model

Intensive simulations are made to collect samples for offline statistical analysis for parameter ρ in the case of different sequences. We find that the samples of ρ are distributed within a local rang with magnitude less than 30. The impact of the samples with magnitude greater than 30 is negligible. Therefore the interval [0, 30] of ρ is selected for model building in this paper.

Competition decision algorithm (CDA) is one general method which is inherited from the natural biological world [8], especially from human competition mechanism and decision theory. In fact, it is an optimization algorithm involving multiple step competitions and decision-making according to the features of candidate competitors. In this work, the CDA is used to derive the optimal δ-ρ model in the rate distortion optimization sense based on the assumption that one specific input ρ only corresponds to a uniquely optimal δ.

In addition, the proposed δ-ρ model is constructed in a piece-wise way to track the optimal relationship as much as possible. Suppose that the whole interval of ρ is [0, T], it is partitioned into N equivalent segments indexed by i, and ρ_i is calculated as follows.

$$\rho_i = T/2N + T \times (i-1)/N \quad (1 \leq i \leq N) \tag{7}$$

The δ in the case of ρ_i is $\kappa_i = f(\rho_i)$, and its initial value is determined by Eq. (6). This is just the traditional suboptimal model with decreased RD performance. The work assigns an offset ω_i to each κ_i to derive an adjusted offset $\kappa_i' = \kappa_i + \omega_i$. The default offsets are used for other segments whose indices are not equal to i in the case of offset adjustment, and the above adjustment is described as follows.

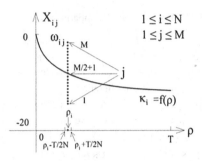

Fig. 2. Competitors in CDA based model building

$$k_i' = \begin{cases} k_i & j = 0 \\ k_i + \omega_{ij} & \text{others} \end{cases} \tag{8}$$

Then, the following task is to determine the adjustment offset for optimal ω_i for κ_i. Suppose that κ_i of Eq. (6) is taken as the initial and the maximum range of ω_i is $\Delta\delta_{max}$, the whole dynamic range is divided into M segments (M is odd), and each segment is indexed by j, then the adjusted offset for the segment j (ω_{ij}) is shown in Fig. 2 and described as follows.

$$\omega_{ij} = \Delta\delta_{max}/M \times j; \quad (-M/2 \leq j \leq M/2) \tag{9}$$

Therefore, there are M competitors for each ρ_i, and this work derives the optimal ω_{ij} shown in Eq. (10) by applying CDA using the RD optimization criterion, $J = D + \lambda \times R$. The optimal ω_{ij} ($\bar{\omega}_i$) is determined by minimizing the rate-distortion cost $J(k_i + \omega_{ij})$.

$$\bar{\omega}_i = \underset{\omega_{ij}}{\text{avg max}} J(\kappa_i + \omega_{ij}) \tag{10}$$

The pair (ρ_i, $\bar{\delta}_i$) for each segment is recorded as follows.

$$\bar{\delta}_i = \kappa_i + \bar{\omega}_i \tag{11}$$

Then, the CDA based model search is performed for the following adjacent segments until all segments have finished the optimization processing with sample points (ρ_i, $\bar{\delta}_i$) recorded. As a result, the new δ-ρ model is actually derived. In above steps, each segment of the new model is determined with optimal offset $\bar{\omega}_i$ respectively. In fact, the adjacent segments with adjustment offsets may have influence on each other. Accounting for this dependency, the obtained offset $\bar{\omega}_i$ for segment i may no longer be the optimal offset. In order to derive more accurate model, the values of κ_i is updated as the new initial to refresh the new model in an iterative way. This iteration may repeat many times until a certain converge condition is satisfied, the model parameters are recorded and refreshed in each iteration. The one with the optimal RD performance is taken as the finally optimal model.

The algorithm flow is shown in Fig. 3, and the details are described as follows:

① Competitor determination: partition ρ into N equivalent segments and determine M candidate competitors ω_{ij} for each segment according to (7) and (6) as shown in Fig. 2. T, N, M, and $\Delta\delta_{max}$ are 30, 12, 21, and 5.25 respectively.

② Determine the states of competitive decision: calculate δ for each competitors, i.e. κ_i, and determine the initial values according to Eq. (6).

③ Calculate the RD performance for each competitors.

④ Competitive decision: calculate the competitor ω_{ij}, record $\bar{\omega}_i = \bar{\delta}_i$ with the optimal RD performance according to the decision function in (10). And record the pair $(\rho_i, \bar{\delta}_i)$.

⑤ Segment iteration: Calculate $(\rho_i, \bar{\delta}_i)$ for all N segments by repeating steps from ② to ④, and derive the new model.

⑥ Refinement iteration: repeat competitive decision using the resource exchange rules $\kappa_i = \bar{\delta}_i$, update the competitors and their initial states to derive the updated new model.

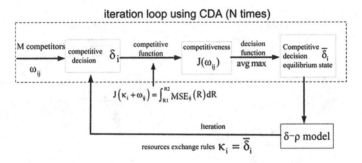

Fig. 3. Flow diagram of CDA based model building

3.2 The Improved Quantization Control Algorithm

The proposed new model is a piecewise function comprised of N segments, and the same δ holds for the whole range in one segment. An array $\delta[n]$ with size of N is used to record the model parameters. Piecewise segmentation is first performed for the current coding unit according to its ρ, and its segment index n is estimated as follows.

$$n = (\rho \times N)/T + T/2N - 1 \tag{12}$$

Then, the array based model is applied in software encoder by table lookup for simulation. The quantization offset δ_{MBTree} is initialized with the array $\delta[n]$, and this offset is taken as the quantization offset to adapt to the temporal propagation, and it works coordinately with the quantization offset δ_{VAQ} for variance adaptive quantization. These two offsets are used to adjust the frame level quantization parameter Qp_{frames}, and the final quantization parameter Qp_{final} is calculated as follows.

$$Qp_{final} = Qp_{frm} + \delta_{VAQ} + \delta_{MBTree} \tag{13}$$

4 Simulation Results and Analysis

The proposed model and the quantization algorithm are verified in x264 video coder. 10 standard test sequences are used for comprehensive evaluation of RD performance.

Figure 4 give the δ-ρ model difference between the logarithmic model and new model derived with CDA based iteration. The absolute distortion comparisons of the MBTree off, MBTree with the traditional model and with the proposed model are shown in Fig. 5. The results of the proposed algorithm derived by two iterations indicate that the distortion fluctuation is smaller than those of the traditional model in the case of equivalent average distortion fluctuation.

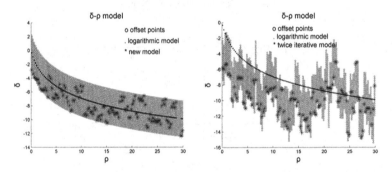

Fig. 4. new δ-ρ model without iteration and δ-ρ model of two iterations

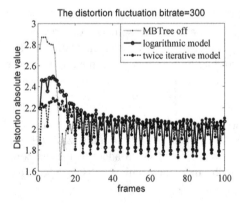

Fig. 5. The frame-level distortion fluctuation

SSIM is a well-accepted perceptual metric for image quality assessment. Similar with PSNR definition, the ssim score is mapped to the dB-scored SSIM using $-10 \times \log_{10}(1\text{-ssim})$, and this SSIM is used for video perceptual quality assessment. The comparisons of the PSNR and SSIM performance are shown in Fig. 6 and in Table 1. Compared with the traditional model, 0.14 dB PSNR and 0.29 dB SSIM PSNR and SSIM performance improvement are observed in the algorithm using the proposed model. In general, the RD performance increase with respect to increment of iterations. For some special sequences in which the traditional MBTree algorithm suffers from performance loss, the proposed algorithm can achieve up to 0.28 dB BD-PSNR improvement. In addition, more iterations result in higher SSIM improvement than PSNR improvement. The simulation results verify that the proposed model based on CDA iteration can achieve more efficient quantization control with better video quality both objectively and subjectively.

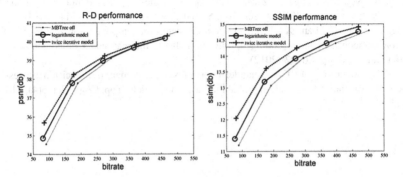

Fig. 6. PSNR and SSIM performances of the hall sequence

5 Conclusions

In video coder, temporal inter-frame prediction results in distortion propagation among adjacent frames, and this temporal distortion dependency is crucial for rate control algorithm optimization. In x264, the MBTree algorithm measure the relative distortion propagation cost ρ to determine the quantization parameter offset δ using an empirical model for perceptual quantization. However, the δ-ρ model is designed in an empirical way with rough accuracy. To solve this deficiency, this paper applies the competition decision algorithm (CDA) to explore the optimal δ-ρ model, and then proposes a new δ-ρ model in the sense of rate distortion optimization. The simulation results verify that the proposed model achieves superior coding performance objectively and subjectively compared with the traditional model.

Acknowledgment. This work is partly supported by the projects ZJNSF Y15F020075, LY12F01011, Y1110114, and LY13H180011.

References

1. Ramchandran, K., Ortega, A., Vetterli, M.: Bit allocation for dependent quantization with applications to multi resolution and MPEG video coders. IEEE Trans. Image Process. **3**(5), 533–545 (1994)
2. Liu, J., Cho, Y., Guo, Z., Kuo, C.C.J.: Bit allocation for spatial scalability coding of H.264/SVC with dependent rate-distortion analysis. IEEE Trans. Circuits. Syst. Video Technol. **20**(7), 967–981 (2010)
3. Garrett-Glaser, J.: A novel macroblock-tree algorithm for high-performance optimization of dependent video coding in H.264/AVC. x264 developer group
4. Jie, C., Yu, L.: The Distortion Model with Reference Frame Decompression for Video Coding. Master thesis of Zhejiang University (2010)
5. Pang, C., Au, O.C., Zou, F., Dai, J., Zhang, X., Dai, W.: An analytic framework for frame-level dependent bit allocation in hybrid video coding. IEEE Trans. Circuits Syst. Video Technol. **23**(6), 990–1002 (2013)
6. Wang, S., Ma, S., Wang, S., Zhao, D., Gao, W.: Rate-GOP based rate control for high efficiency video coding. IEEE J. Sel. Top. Signal Process. **7**(6), 1101–1111 (2013)
7. Yang, T., Zhu, C., Fan, X., Peng, Q.: Source distortion temporal propagation model for motion compensated video coding optimization. IEEE International Conference on Multimedia and Expo, pp. 86–92 (2012)
8. Fuss, I.G., Navarro, D.J.: Open parallel cooperative and competitive decision processes: a potential provenance for quantum probability decision models. Top. Cogn. Sci. **5**(4), 818–843 (2013)

Semi-automatic Labeling with Active Learning for Multi-label Image Classification

Jian Wu[1(✉)], Chen Ye[1], Victor S. Sheng[2], Yufeng Yao[1],
Pengpeng Zhao[1], and Zhiming Cui[1]

[1] The Institute of Intelligent Information Processing and Application,
Soochow University, Suzhou 215006, China
jianwu@suda.edu.cn
[2] Department of Computer Science, University of Central Arkansas,
Conway 72035, USA

Abstract. For multi-label image classification, we use active learning to select example-label pairs to acquire labels from experts. The core of active learning is to select the most informative examples to request their labels. Most previous studies in active learning for multi-label classification have two shortcomings. One is that they didn't pay enough attention on label correlations. The other shortcoming is that existing example-label selection methods predict all the rest labels of the selected example-label pair. This leads to a bad performance for classification when the number of the labels is large. In this paper, we propose a semi-automatic labeling multi-label active learning (SLMAL) algorithm. Firstly, SLMAL integrates uncertainty and label informativeness to select example-label pairs to request labels. Then we choose the most uncertain example-label pair and predict its partial labels using its nearest neighbor. Our empirical results demonstrate that our proposed method SLMAL outperforms the state-of-the-art active learning methods for multi-label classification. It significantly reduces the labeling workloads and improves the performance of a classifier built.

Keywords: Multi-label · Image classification · Semi-automatic labeling · Active learning

1 Introduction

Traditional studies for image classification focus on multi-class settings, where each image only has one label. However, in real-world applications, each image usually has multiple labels. This leads to multi-label classification. Multi-label classification has received broad attentions in recent years [1–4].

Active learning [5] is one of the most effective approaches used in image classification to build an effective classifier with limited resources. Specifically, a learner iteratively selects an example that has the most contribution for the classifier and the classifier is retrained on the training dataset with the new labeled example added. The learning process continues until all the annotation resources are depleted or the obtained classifier is accurate enough. The core of active learning is sampling. In recent decades, most studies of active learning on image classification focus on single-label

© Springer International Publishing Switzerland 2015
Y.-S. Ho et al. (Eds.): PCM 2015, Part I, LNCS 9314, pp. 473–482, 2015.
DOI: 10.1007/978-3-319-24075-6_46

[6–8]. The research on multi-label active learning (MLAL) for image classification is rare. However, there exist many problems when we study active learning to multi-label domains [9].

The studies on MLAL can be divided into two categories: example-based and example-label based MLAL. The example-based MLAL selects examples to query all of their labels simultaneously [10–12]. When the number of labels for each image is large, labeling all labels of each image is expensive or time consuming. Example-based MLAL methods have to require all labels of a selected unlabeled example from human experts. There are a few studies on example-label based MLAL [13–17], which select examples to query its partial labels.

Li et al. [10] is the first research using active learning for multi-label image classification. It calculates the loss only on the most certainly predicted class of this image, which called Max Loss (ML). Li and Guo [11] combined a max-margin prediction uncertainty selection strategy with a label cardinality inconsistency strategy together. An adaptive framework of integrating these two strategies was proposed, based on an approximate generalization error measure. However, its computation cost is so high.

We can see that the above example-based MLAL methods measure the informativeness of an example only based on the example level. When the number of labels is large, those methods cause expensive labeling cost.

Example-label based MLAL methods only require partial labels of a selected unlabeled example from human experts. It selects example-label pairs to request labels. Qi et al. [13, 14] proposed a two-dimensional active learning method (2DAL), which not only considers the example dimension but also the label dimension. It uses entropy to measure the uncertainty of a selected example-label pair, and mutual information to measure the statistical redundancy among a selected label and the rest ones. An Expectation-Maximum (EM) algorithm is used to predict all the rest labels of the example-label pair. Zhang et al. [15] proposed an active learning approach to actively select a batch of informative example-label pairs. In order to discover the informativeness of label correlation, association rule mining is adopted. And then it finds the nearest neighbor example of a selected example-label and considers all the rest labels will be the same. Wu et al. [17] proposed a method to select the most uncertain example-label pairs to decrease the labeling cost. It does not have to predict the rest labels of the selected one, but it neglects the inherent label correlations across all labels of an example.

In order to develop an effective strategy to select example-label pairs, we need to discover the inherent label correlations. And due to the large number of labels in a multi-label dataset, predicting all the rest labels will not be reliable and necessary.

Accordingly, in this paper, we will use cosine similarity to evaluate the inherent label correlations. Then, we select the most informativeness example-label pair to query their labels. After that, we select the most uncertainty example-label pairs to predict its partial labels using its nearest neighbor. This automatical labeling process will decrease the labeling cost. The details of our proposed example-label pair selection strategy will be discussed in Sects. 2 and 3. Our experimental results (in Sect. 4) show that our proposed active learning strategy consistently outperforms the state-of-the-art approaches.

2 Label Correlation Based Sampling Strategy

Previous research shows that label correlation can improve the performance of multi-label classification. For a multi-label domain, its labels are usually correlated with each other. It is common to see that images are labeled by a set of labels simultaneously. Therefore, the correlation information between labels in multi-label domains is very important, which can be utilized to improve the selection strategy of MLAL. So we use cosine similarity to measure the label correlation between all labels. The cosine similarity of two labels y_j and y_k, denoted as C_{jk}, is defined as follows:

$$C_{jk} = cos(y_j, y_k) = \frac{<y_j, y_k>}{||y_j||\,||y_k||} \tag{1}$$

With the cosine similarity of each label pair, we construct a symmetric $l \times l$ matrix C, where l is the number of labels in a multi-label domain. With the matrix C, we will calculate the informativeness of each unknown label of an example x_i.

As we know, the unlabeled labels of an example affect whether an example-label pair would be selected or not. To evaluate the degree of the relationship (informativeness) of an example-label pair on entire labels, we average the cosine similarities of the unlabeled labels in an example. Therefore, the informativeness of the j-th label of the example x_i, denoted as MU_{ij}, is defined as follows.

$$MU_{ij} = \begin{cases} \frac{1}{l_u} \sum_{k=1}^{l} |C_{jk}| \times sign(y_k \in UL(x_i)) & l_u > 1 \\ 0 & l_u = 1 \end{cases} \tag{2}$$

where l_u is the number of unlabeled labels in x_i, l is the number of labels in a multi-label domain, and $UL(x_i)$ is the unlabeled labels of x_i. $sign(y_k \in UL(x_i))$ is a sign function. Its value is 1 if $y_k \in UL(x_i)$, 0 otherwise.

As we know, uncertainty sampling is a simple but effective active learning strategy used for single-label classification. Like Wu et al. [17], we define the uncertainty of each example-label pair as follows.

$$E_{ij} = |\frac{1}{2} - p(y_j = 1|x_i)| \tag{3}$$

where $p(y_j = 1|x_i)$ denotes the probability of the label y_j of the example x_i belonging to the positive class.

After we have the uncertainty evaluation of an unlabeled label of each example and its correlation, we integrate them together to evaluate the general informativeness of an unlabeled label of an example for a learner, using the following equation.

$$U_{ij} = E_{ij} * \alpha - MU_{ij} * (1 - \alpha) \tag{4}$$

where α is a coefficient used to leverage the importance between the uncertainty of an unlabeled label and its correlation. This active learning selection strategy is named as a cosine similarity based multi-label active learning algorithm (CosMAL), defined as follows.

$$x^* = \underset{x_i \in U, j \in \{1,\dots,l\}}{\arg\min} \ U_{ij} \qquad (5)$$

3   Semi-automatic Labeling with Active Learning

3.1   Automatic Labeling Strategy

Previous studies on multi-label active leaning have to automatically label all the rest labels of the selected example-label pairs or do not label them at all. When the number of the labels of a multi-label dataset is great, such as the dataset corel5k with 364 labels, predicting all the rest labels are unreliable. Besides, some labels are correlated to each other. Labeling all of them won't benefit the classifier. Therefore, we don't predict all the rest labels of the selected example-label pairs, but part of them. And we will choose the example-label pair that can benefit the classifier the most to automatically label its rest labels (Fig. 1).

The steps of our semi-automatic labeling process are described as follows:

(1)   We use Eq. 3 to select the most uncertainty example-label pair $x_i - y_j$ as the candidate sets.
(2)   We use the nearest neighbor method to find the nearest example x_k of the selected example-label pair $x_i - y_j$ from the labeled sets $L\{ x|y_j\}$, where $L\{ x|y_j\}$ denotes the labeled example sets with label y_j.
(3)   We compare the value $p(y_j = 1|x_i)$ with $v(x_k - y_j)$, where $v(x_k - y_j)$ is the value of y_j of x_k. If $p(y_j = 1|x_i) \geq 0.5$ and $v(x_k - y_j) = 1$, or $p(y_j = 1|x_i) < 0.5$ and $v(x_k - y_j) = 0$, then the label of the example-label pair $x_i - y_j$ is predicted as $v(x_k - y_j)$ and added into the labeled set. Otherwise, the example-label pair $x_i - y_j$ will not be added.

3.2   Complete Algorithm

The active learning method we propose here is called semi-automatic labeling with active learning (SLMAL). Its steps are described as follows.

We first select the most informative example-label pair from the unlabeled set and label it by a human expert.

Then, we select the most uncertainty example-label pair from the unlabeled set and use its nearest neighbor to determine whether this example-label pair can be predicted and added into the labeled set with this predicted label.

```
Algorithm: SLMAL
Input:
    D: {x₁,···,xₙ} denotes the features of the entire dataset;
    Y_L: {y₁,···,yₗ} denotes the labels of the training set;
    L denotes the labeled set of the example-label pairs,
and U denotes the unlabeled set of the example-label
pairs.
Initialize:
    Train a multi-label classifier F on the labeled data
L;
    Construct the matrix M from the labeled data L;
    Repeat:
    Obtain the predictions for unlabeled examples in U
with F;
        For each unlabeled example-label pair (x*, y*) in U
            Calculate its uncertainty using Equation 3;
            Calculate its label correlation using Equation 2;
            Calculate its informativeness using Equation 4;
        End for
    Select the example-label pair (x*, y*) with the most
informativeness using Equation 5;
    Remove (x*, y*) from U, and add it to L after obtain-
ing its label y*from the oracle;
    Select the example-label pair (x#, y#) with the most
uncertainty using Equation 3;
    Find its nearest neighbor (xª, y#) from the labeled set
L{x|y#};
    Define the value of the example-label pair (x#, y#) us-
ing the method we discussed in Section 3.1, and add the
example-label pair (x#, y#) with the value into L.
    Update the matrix M with (x*, y*)and (x#, y#);
    Update the classifier F;
    Until the number of queries or the required accuracy is
reached
```

4 Experiments

In this section, we investigate the effectiveness of our proposed multi-label active learning method SLMAL by comparing with three existing approaches and one contrast approach proposed as follows:

MML: an example selection strategy proposed in [10]. It calculates the loss value only on the most certain label of an example.

Adaptive: This strategy is to select the example that has the minimize approximate generalization error [11].

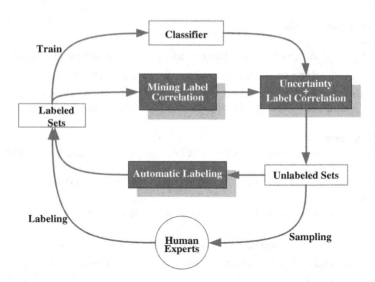

Fig. 1. The framework of the semi-automatic labeling with active learning for multi-label classification

MLAL: an example-label pair selection strategy proposed in [17]. It selects the example-label pairs with the most uncertainty. This uncertainty strategy is the same as the uncertainty we used in our method.

CosMAL: a contrast approach for investigating the performance of SLMAL. The procedure of CosMAL is the same as SLMAL, except that CosMAL does not predict the rest labels of any selected example-label pairs, while SLMAL predicts the partial rest labels of some selected example-label pairs.

To speed up the active learning procedure, we choose a batch of example-label pairs at each iteration of active learning. For each dataset, we use 70 % examples for training (including an initial training set and the rest for active learning) and 30 % examples as a test set. For each active learning approach, we query $10 \times l$ example-label pairs at each iteration. To make the results more confident, we repeat each experiment 10 times and report the average results. In addition, the parameter α in Eq. 4 balancing the uncertainty and label correlation is set as 0.9.

We will conduct experiments on four image datasets and two non-image datasets from MULAN [1]. Three different evaluation metrics are used in our experiments, i.e. accuracy, macro F1 and micro F1.

4.1   On Image Datasets

We first evaluate the performance of our proposed approach on four image datasets from MULAN [1]. The first image dataset we used is Scene, which has been widely used as a benchmark dataset to evaluate multi-label image classification algorithms. It is an image collection for outdoor scene recognition. It has 2407 images tagged with six scenes. The second image dataset we used is Flags. It has 194 images tagged with 7 colors. NUS-WIDE1 is a subset of the image dataset NUS-WIDE. It has almost 5000 examples tagged with 81 labels. Corel16k001 is a subset of the image dataset Corel16k. Corel16k001 has 5188 examples tagged with 153 labels. For the two datasets Flags and Scene, we use 10 examples with all labels as their initial training datasets. For the other two datasets we use 30 examples with all labels as their initial training datasets, since each of them has a large number of labels.

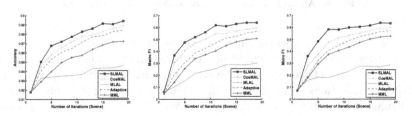

Fig. 2. Average results over 10 runs on the dataset Scene

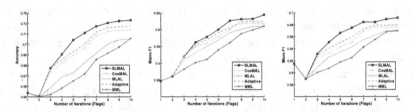

Fig. 3. Average results over 10 runs on the dataset Flags

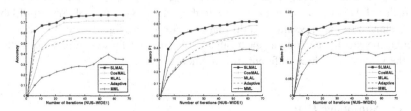

Fig. 4. Average results over 10 runs on the dataset NUS-WIDE1

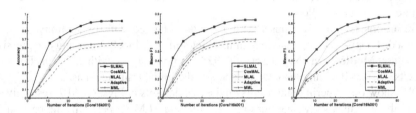

Fig. 5. Average results over 10 runs on the dataset Corel16k001

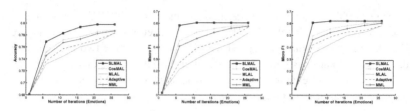

Fig. 6. Average results over 10 runs on the dataset Emotions

Figure 2 shows the experimental results on the dataset Scene, in terms of three common measurements used in multi-label classification, i.e., accuracy, macro F1, and micro F1. The horizontal axis presents the number of iterations. Similarly, Fig. 3 shows the experimental results on the dataset Flags. Figure 4 shows the experimental results on the subset NUS-WIDE1 of the dataset NUS-WIDE. And Fig. 5 shows the experimental results on the subset Corel16k001 of the dataset Corel16k.

From these figures, we can make the following conclusions. Our proposed algorithm SLMAL always performs the best on these datasets, based on the three measurements: accuracy, macro F1, and micro F1. This indicates the effectiveness of automatically labeling the example-label pairs in SLMAL. Among the other comparison methods, it is difficult to make a general conclusion, which one performs better. The ranks among these comparison methods vary. CosMAL is the second best method except on the dataset Flags. This indicates that the label correlations are useful to estimate the informativeness of each sample-label pair in active learning. For the dataset Flags, CosMAL is the third. We attribute this to the label correlations too. It might be that the semantic relationships between those labels are not strong enough. Adaptive performs the second best on the dataset Flags, but the worse on the dataset Corel16k001. This is because this method is based on the label cardinality. The label cardinality of the dataset Corel16k is close to three. However, its total number of labels is 164. Thus, it has no advantage on this dataset. From these figures, we also notice that the three measurements support each other. The ranks of these comparison methods keep the same under three different measurements.

4.2 On Non-image Datasets

We further examine the performance of our proposed approach SLMAL on two non-image datasets (i.e., Emotions and Medical) from MULAN. We briefly describe the three datasets as follows.

Emotions is a music dataset with 593 songs tagged with 6 emotions: amazed-surprised, happy-pleased, relaxing-calm, quiet-still, sad-lonely and angry-aggressive. The dataset **Medical** is a text dataset, with 978 examples tagged with 45 labels. For these datasets, we use 10 examples with all labels as the initial training dataset.

We show our experimental results on the two datasets in Figs. 6 and 7, respectively. Again, these figures show that our proposed method SLMAL performs the best on the two datasets, in terms of all the three different measurements. The comparisons among the other methods are also consistent to the conclusion we made in the above subsection.

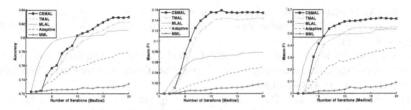

Fig. 7. Average results over 10 runs on the dataset Medical

From the above experiments, we can conclude that SLMAL provides the best performance on diverse datasets. This indicates that it is an effective and robust approach for multi-label classification than the state-of-the-art active learning methods do.

5 Conclusions

In this paper, we proposed a novel semi-automatic labeling with multi-label active learning method (SLMAL). Firstly, we select unlabeled example-label pairs that are uncertain and informative to request their labels from the oracle. Besides, to further reduce the investment of requesting more labels, we proposed semi-automatically labeling to predict the labels for other most uncertainty example-label pairs. Our experimental results on four image datasets and two non-image datasets from MULAN show that our proposed method SLMAL outperforms the state-of-the-art active learning techniques on multi-label classification and can significantly reduce the labeling cost.

Acknowledgement. This research was partially supported by the Natural Science Foundation of China under grant No. 61170020, 61402311 and 61440053, Jiangsu Province Colleges and Universities Natural Science Research Project under grant No. 13KJB520021, and the U.S. National Science Foundation (IIS-1115417).

References

1. Tsoumakas, G., Spyromitros-Xioufis, E., Vilcek, J., et al.: Mulan: a java library for multi-label learning. J. Mach. Learn. Res. **12**, 2411–2414 (2011)
2. Tsoumakas, G., Katakis, I.: Multi-label classification: an overview. Int. J. Data Warehouse. Min. **3**(3), 1–13 (2007)
3. Agrawal, R., Gupta, A., Prabhu, Y., et al.: Multi-label learning with millions of labels: recommending advertiser bid phrases for web pages. In: Proceedings of the 22nd International Conference on World Wide Web. International World Wide Web Conferences Steering Committee, pp. 13–24 (2013)
4. Zhang, M., Zhou, Z.: A review on multi-label learning algorithms. IEEE Trans. Knowl. Data Eng. **26**(8), 1819–1837 (2013)
5. Settles, B.: Active learning literature survey. Computer science technical report 1648, University of Wisconsin-Madison, USA (2010)

6. Holub, A., Perona, P., Burl, M.C.: Entropy-based active learning for object recognition. In: Proceedings of the IEEE Conference on Computer Vision and Pattern Recognition (CVPR), pp. 1–8 (2008)
7. Luo, T., Kramer, K., Goldgof, D.B., Hall, L.O., et al.: Active learning to recognize multiple types of plankton. In: Proceedings of the International Conference on Pattern Recognition (ICPR), vol. 3, pp. 478–481 (2004)
8. Vijayanarasimhan, S., Jain, P., Grauman, K.: Far-sighted active learning on a budget for image and video recognition. In: Proceedings of the IEEE Conference on Computer Vision and Pattern Recognition (CVPR), pp. 3035–3042 (2010)
9. Tsoumakas, G., Zhang, M.L., Zhou, Z.H.: Introduction to the special issue on learning from multi-label data. Mach. Learn. **88**(1–2), 1–4 (2012)
10. Li, X., Wang, L., Sung, E.: Multilabel SVM active learning for image classification. In: Proceedings of the International Conference on Image Processing (ICIP), vol. 4, pp. 2207–2210 (2004)
11. Li, X., Guo, Y.: Active learning with multi-label SVM classification. In: Proceedings of the International Joint Conference on Artificial Intelligence (IJCAI), pp. 1479–1485 (2013)
12. Vasisht, D., Damianou, A., Varma, M., et al.: Active learning for sparse bayesian multilabel classification. In: Proceedings of the 20th ACM SIGKDD International Conference on Knowledge Discovery and Data Mining, pp. 472–481. ACM (2014)
13. Qi, G.-J., Hua, X.-S., Rui, Y., Tang, J., Zhang, H.-J.: Two-dimensional active learning for image classification. In: Proceedings of the IEEE International Conference on Computer Vision (CVPR), pp. 1–8 (2008)
14. Qi, G.J., Hua, X.S., Rui, Y., Tang, J.H., Zhang, H.J.: Two-dimensional multilabel active learning with an efficient online adaptation model for image classification. IEEE Trans. Pattern Anal. Mach. Intell. **31**(10), 1880–1897 (2009)
15. Zhang, B., Wang, Y., Wang, W.: Batch mode active learning for multi-label image classification with informative label correlation mining. In: Proceedings of the IEEE Workshop on Applications of Computer Vision (WACV) (2012)
16. Vasisht, D., Damianou, A., Varma, M., et al.: Active learning for sparse bayesian multilabel classification. In: Proceedings of the 20th ACM SIGKDD International Conference on Knowledge Discovery and Data Mining (KDD), pp. 472–481 (2014)
17. Wu, J., Sheng, V.S., Zhang, J., Zhao, P., Cui, Z.: Multi-label active learning for image classification. In: Proceedings of the 21st International Conference on Image Processing (ICIP), pp. 5227–5231 (2014)

A New Multi-modal Technique for Bib Number/Text Detection in Natural Images

Sangheeta Roy[1], Palaiahnakote Shivakumara[1(✉)], Prabir Mondal[2],
R. Raghavendra[3], Umapada Pal[2], and Tong Lu[4]

[1] Faculty of Computer Science and Information Technology,
University of Malaya, Kuala Lumpur, Malaysia
2sangheetaroy@gmail.com,
shiva@um.edu.my
[2] Computer Vision and Pattern Recognition Unit,
Indian Statistical Institute, Kolkata, India
prrabirmondal@gmail.com, umapada@isical.ac.in
[3] Norwegian Biometric Laboratory, Gjovik University College, Gjovik, Norway
raghu07.mys@gmail.com
[4] National Key Lab for Novel Software Technology,
Nanjing University, Nanjing, China
lutong@nju.edu.cn

Abstract. The detection and recognition of racing bib number/text, which is printed on paper, cardboard tag, or t-shirt in natural images in marathon, race and sports, is challenging due to person movement, non-rigid surface, distortion by non-illumination, severe occlusions, orientation variations etc. In this paper, we present a multi-modal technique that combines both biometric and textual features to achieve good results for bib number/text detection. We explore face and skin features in a new way for identifying text candidate regions from input natural images. For each text candidate region, we propose to use text detection and recognition methods for detecting and recognizing bib numbers/texts, respectively. To validate the usefulness of the proposed multi-modal technique, we conduct text detection and recognition experiments before text candidate region detection and after text candidate region detection in terms of recall, precision and f-measure. Experimental results show that the proposed multi-modal technique outperforms the existing bib number detection method.

Keywords: Face detection · Skin detection · Text detection · Multi-modal text detection · Bib number detection · Bib number recognition

1 Introduction

It is true that lots of marathons have been organized in the whole world nowadays to create awareness and send specific messages to public. There is a need for analyzing and understanding the videos and images, which capture marathon, running races, athletics, etc. in Olympics to identify runners or persons accurately [1]. In addition, it is also useful for security and public safety purpose [2], one example of which we can remember is the recent Boston Marathon bombing [2]. In literature, we can see

© Springer International Publishing Switzerland 2015
Y.-S. Ho et al. (Eds.): PCM 2015, Part I, LNCS 9314, pp. 483–494, 2015.
DOI: 10.1007/978-3-319-24075-6_47

numerous techniques for text detection and recognition in scanned images, degraded document images, video, and natural scene images [3]. However, these techniques may not give satisfactory results for the images or video of Marathon and running race because these images usually contain the combination of numbers and texts, which would be printed on paper, cardboard tag, t-shirts of different colors at different parts of person bodies. Besides, since people always run together with different speeds, background complexity changes frequently with trees, sky, building, moving objects, etc. These result in severe occlusions, the loss of information or quality, and the distortions due to motion blur, non-rigid surfaces and arbitrary orientations [1]. Furthermore, the conventional text detection and recognition techniques [4–13] focus on full text lines but not bib numbers [3]. Therefore, text detection and recognition of bib numbers in such video or images requires a special attention to achieve better results. For instance, the examples for text detection before and after text candidate region detection are respectively shown in Fig. 1, where one can see the text detection method which works based on wavelet and color features [8] fails to detect all the text lines in the image and gives more false positives as shown in Fig. 1(a) when text candidate region detection is not applied, while the same text detection method detects all the text lines with less false positives when applied after text candidate region detection as shown in Fig. 1(b). This shows that a text detection method alone may not give a good solution for detecting bib numbers in marathon and race images. The same conclusion can be drawn from the recognition results before and after text candidate region detection because the recognition results depend on how well a text detection method detects the texts in the image. This motivates us to propose biometric features from face and skin to reduce the background influence and get rough text detection regions, which we call text candidate regions.

 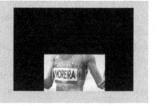

(a). Before candidate region detection   (b) After candidate region detection

Fig. 1. Text detection results obtained before and after candidate region detection by the text detection method [8]

Recently, Shivakumara et al. [14] proposed a technique for arbitrarily oriented text detection in video based on gradient vector flow and grouping. Despite addressing the hard problem of arbitrary orientation, the technique is not tested on Marathon videos, where we can see lots of strings with numerals. Zhang and Kasturi [15] proposed a technique for detecting texts in video and natural scene images using character link and energies. The technique explores the characteristics of text components for detection. The technique is good if text components well preserve their shapes. Minetto et al. [16]

proposed a technique for detecting texts in urban scene images using descriptors and a classifier. It is stated that the technique is good for full text lines and at least words but not single characters and numerals. Kang et al. [17] proposed a robust technique for detecting oriented texts in natural scene images using maximally stable extremal regions. However, it is noted that maximally stable external regions work well for high contrast images but not for low contrast images. In the same way, Yin et al. [18] explored maximally stable extremal regions along with a single link clustering algorithm for detecting texts in natural scene images and born digital images. The technique works well for distortion free images. Yao et al. [19] and Yi et al. [20] proposed techniques for the recognition of video and natural scene texts. Yao et al. explored a forest classifier and dictionary for the recognition of characters, while Yi et al. explored structural features with a simple classifier for recognition. However, both the techniques require high contrast images without illumination effect and blur.

Bib number/text detection and recognition is similar to license plate number detection and recognition, for which we review the recent techniques [21–23] here. Cui and Huang [21] proposed a technique for character extraction from license plates of video based on Markov random field and genetic algorithm. Suresh et al. [22] proposed a technique for license plate recognition using super resolution by fusing sub-pixels shifted and noisy low resolution observations. This technique also explores Markov random field. Yu et al. [23] proposed a technique for license plate location using wavelet transform and empirical mode decomposition analysis. It is noted that the background of license plate numbers usually has homogeneous colors and the number always be in horizontal direction with little tilt. This is the main advantage for license plate detection and recognition techniques. In addition, these techniques are not tested on bib numbers in Marathon video or images.

We have found a paper on racing bib number recognition using stroke width transform and tesseract OCR engine [1]. This technique combines face detection and stroke width transform for bib number detection. We know that since Marathon and race videos have a big crowd and people run with different speeds, it is hard for locating and detecting faces. Moreover, frontal face sometimes may not appear (face be on back side due to body turn). Therefore, this technique has some inherent limitations. Inspired by this work [1] where it is shown that the combination of face detection and text detection is necessary to achieve good results for these images, we propose to explore skin detection along with face detection as skin easily gives hints of an individual person and is robust compared to face for the above situations [24, 25]. The combination of face and skin helps us to identify the regions which can cover bib number/text located at any part of the body. This is the advantage of the proposed technique.

2 Proposed Technique

As discussed in the introduction section, text detection alone may not help us to detect bib numbers accurately, and face or body detection alone may not help us to detect bib numbers/texts properly due to the complexity of the problem. Therefore, we propose a multi-modal technique that combines face, skin and text detections to achieve better results for bib number/text detection and recognition. This is valid because a person who participates in Marathon and Olympic sports generally displays skin at leg, face, neck, etc. With this cue, we can identify person body which can cover bib numbers without missing and irrespective of body turn. This step results in text candidate regions.

We believe the above step eliminates most irrelevant backgrounds and gives the regions that contain bib numbers/text. It is also true that the above step may not detect exact bib numbers/texts since it detects bib numbers/texts with their backgrounds. As a result, we cannot consider the output of the above step as the result of text detection because text detection results are supposed to be produced by closed bounding boxes for text lines. In this way, this step helps text detection and recognition techniques to detect or recognize bib numbers/texts accurately by removing complex background successfully. Therefore, we propose to use a text detection method for detecting texts from the text candidate regions. Then detected text lines are passed to OCR to get recognition results through binarization technique.

2.1   Text Candidate Region Detection

It is true from skin detection that color plays a vital role in identifying skin color in images [24, 25]. Since our inputs are marathon and running race images, definitely the skin of persons appears in the images. Therefore, we propose to explore skin color detection to identify person bodies in the input images. The main advantage of skin color detection is that it does not depend on faces as the existing technique does [1] for text candidate region detection. For a given image, the proposed technique detects faces if faces are present in the image using Open CV implementation [26]. This results in rectangular boxes for the faces in the image. For each pixel in the rectangular box, the proposed technique computes the mean for R, G and B values corresponding to the pixels in each rectangular box. We consider the means of respective color values as a seed, which is represented as the following Eq. (1),

$$Seed = (R_m, G_m, B_m) \qquad (1)$$

where R_m, G_m and B_m are respective seed values for the R, G, B pixels in the rectangular box.

Since these seed color values are extracted from a facial region, we believe the seed values represent the skin color of the person in the image. The proposed technique compares the values of the pixels in the input image with the seed color values, and classifies them into respective seed value clusters when a pixel value is close to a respective seed color value. This results in three clusters as defined in Eqs. (2)–(4).

$$C_R = \{P | P.R \in [Seed.R_m - \Delta, Seed.R_m + \Delta]\} \qquad (2)$$

$$C_G = \{P | P.G \in [Seed.G_m - \Delta, Seed.G_m + \Delta]\} \qquad (3)$$

$$C_B = \{P | P.B \in [Seed.B_m - \Delta, Seed.B_m + \Delta]\} \qquad (4)$$

Where C_R, C_G and C_B denote three color clusters, P is a pixel in the input image, and Δ is a threshold which is set to the minimum distance with the particular cluster. The proposed technique combines all the three cluster values as one cluster to obtain the regions that represent skin of the person in the image. Thus the combined cluster can be defined in Eq. (5).

$$C_C = C_R \cup C_G \cup C_B \tag{5}$$

where C_C represents the combined cluster. When face does not present in the image, we use the existing skin detection method [24] for identifying skin candidates and then employ the above procedure to detect skin regions of persons in the image. The skin detection method [24] adapted in this work is based on a non-parametric histogram model trained by manually annotated skin and non-skin pixels, in which a total of 14,985,845 skin pixels and 304,844,751 non-skin pixels were used. Since the detected region covers the whole body of the person including legs, definitely, the region includes texts or bib number tags which may appear at any part of the body. However, when the image contains several persons, the skin regions detected by the above procedure may be merged. In order to obtain skin regions which cover the bib numbers or texts of each individual person, we propose to use the same hypothesis as used in [1] for torso detection using skin candidate regions, which results in text candidate regions. This procedure is illustrated in Fig. 2, where (a) is the result of face detection and its torso detection, (b) is the result of face and skin detection, (c) is the result of the final text candidate region detection, (d) is the result of skin detection for the image where there is no face, (e) is result of torso detection using skin candidates, and (f) is the final text candidate region for the image that has no faces. In this way, the proposed technique can detect text candidate regions from the image with either face or without face.

(a). Face detection   (b) Face + Skin detection   (c). Text candidate region   (b) Skin detection without face (red patches)

 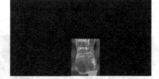

(e). Torso detection using skin candidate   (f) Text candidate region detection

Fig. 2. Text candidate region detection using both face and skin

2.2   Multi-modal Method for Text Detection/Recognition

Figure 2 shows that the technique presented in the previous section detects text candidate regions which contain text or bib number. We noticed from the results that though a text candidate region contains text or bib number, the region covers lots of background information, which cannot be used for recognition. Therefore, we propose to use a text detection method to locate the texts in the candidate regions. Since this section considers text candidate region detected by the biometric features for text detection, we name it multi-modal method. Since there are plenty of text detection techniques available in the literature, we use the technique [8] which works well for marathon and running race images. The technique in [8] works is based on the combination of wavelet and color, and k-means clustering is used for detecting text lines. Since the technique uses color feature as one of the main features for detecting text candidates, we believe the same color plays a vital role in detecting bib numbers/texts in marathon and running race images. In addition, the technique works well for the situations such as low contrast, complex background, low resolution and arbitrary orientation. These characteristics of the text detection technique motivate us to propose the technique for detecting bib numbers or texts in the text candidate regions in this work.

In the same way, we can see lots of techniques for binarizing scanned, camera based and video images in the literature. Since the main objective of this work is to show how the combination of biometric and textual features help in improving text detection and recognition in marathon and running race images compared to text features alone, we prefer to use the existing technique which suits for this work. With this notion, we propose to use the binarization technique proposed in [13], which tunes the parameters automatically for binarizing images. It is known that the main issue of a binarization technique is to control the threshold and parameters used when different data or applications are given as the input. Since it is hard to define parameters or threshold values for the current bib number or text detection in marathon and running race images, we use this binarization method for binarizing bib numbers or texts. Sample results by the above technique for the text candidate regions in Fig. 2 are shown in Fig. 3(a), where we can notice that the text detection technique detects texts well. Similarly, sample binarization results for the text lines detected by the text detection technique are shown in Fig. 3(b). Finally, the binarized results go to the OCR engine [27] to recognize texts or bib numbers as shown in Fig. 3(b), where the recognition results are shown within in quotes.

(a).   (b) "JI\IN\\g"   "JPN"

Fig. 3. Text detection and recognition results of text candidate region: (a) text detection results by [8] and (b) binarized results by [13] and its recognition results by [27]

3 Experimental Results

To evaluate the proposed technique, we create our own dataset of size 200 images of sports, Olympics, Marathon and running race. We also use the standard dataset of size 217 images, which is called the RBNR dataset [1] to evaluate the performance of the proposed technique. In total, the proposed technique has been tested on 417 images. We use well-known measures such as recall (R), precision (P) and F-measure (F) for evaluating the performances of both the proposed and the existing techniques. For evaluating recognition results we use recognition rate (RR). We follow the definition and instructions as proposed in [1] for both text detection and recognition experimentation in this work. Since the RBNR dataset provides the ground truth for calculating measures, we use the same ground truth for the experiments on RBNR data. For our data, we count manually because there is no ground truth.

In order to show the usefulness of the proposed multimodal technique, we conduct text detection and recognition experiments before text candidate region (TCR) detection and after TCR detection. For text detection experiments, we prefer to use the technique due to Shivakumara et al. [8] which is suitable as we discussed in Sect. 2.2 and Epshtein et al. [4] which is well known and the state of the art technique for text detection in scene images. In the same way, we use the method presented by Roy et al. [9] which proposes wavelet and gradient combination for binarizing text lines of video, Moghaddam et al. [12] which proposes a multi-scale adaptive binarization technique for degraded images, Wolf et al. [11] which propose a method for binarizing multimedia documents, Chattopadhyay et al. [10] which propose automatic selection of binarization techniques for improving recognition results, and Howe [13] which proposes an automatic way to tune parameters and threshold values to improve binarization results, for calculating recognition rate using Tesseract OCR engine [27]. The proposed technique in this work consists of face + skin + text detection by [8] + binarization by [13] + Tesseract OCR [27].

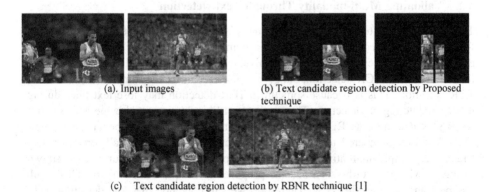

(a). Input images   (b) Text candidate region detection by Proposed technique

(c)   Text candidate region detection by RBNR technique [1]

Fig. 4. Results of text candidate region detection of the proposed and existing technique on our data

3.1   Experiments on Text Candidate Region Detection

To show the effectiveness of the TCR detection step of the proposed technique, we compare TCR detection results with the results of the RBNR technique in terms of recall, precision and f-measure. Sample qualitative results of the proposed and the existing technique are shown in Fig. 4, where it is noted that the proposed technique detects TCR well for the input images in Fig. 4(a) as shown in Fig. 4(b) compared to the results in Fig. 4(c) given by the existing technique. Figure 4(c) shows that the existing technique misses some of the TCR due to low contrast and blur, while the proposed technique detects well. The quantitative results reported in Table 1 show that the proposed technique is better than the existing technique in terms of recall and f-measure for both the two datasets (RBNR and our datasets). However, the precision for the RBNR technique is higher on RBNR than that of the proposed technique due to low false positives. On the other hand, the proposed technique sometimes produces false positives when face + skin combination fails to detect proper TCRs. However, overall, the proposed technique outperforms the existing technique. The main advantage of the proposed technique is that it uses both face and skin while the existing technique uses only face information. As a result, the existing method misses several text candidates. Therefore, the recall for the existing method is lower compared to the proposed method on both the datasets.

Table 1. Performance of the proposed and existing techniques for text candidate region detection

Methods	RBNR data			Our data		
	R	P	F	R	P	F
Proposed method	78.26	58.06	66.66	70.85	84.73	77.17
RBNR method [1]	49.15	97.38	65.32	57.44	76.33	65.55

3.2   Validating Multi-modality Through Text Detection

Table 2 shows that text detection techniques give better results at recall, precision and f-measure when used after TCR detection compared to before TCR detection. However, we can observe from Table 2 that Shivakumara et al.'s method gives a good recall for RBNR before TCR detection, but gives a poor recall for the same RBNR data after TCR detection. This is because sometimes TCR detection may lose text lines during text candidate region detection. The precision of Shivakumara et al.'s method is better for RBNR data after TCR. For our data, both the techniques give better results after TCR detection compared to before TCR detection in terms of recall, precision and f-measure. Sample qualitative results of the proposed technique on our data are shown in Fig. 5, where it can be seen that the proposed technique detects almost all the bib numbers and texts in the TCR. Therefore, we can conclude that text detection techniques alone are not good enough to solve the problems of bib number or text detection in marathon and running race images. Hence, TCR detection is useful to improve the performances of existing text detection techniques.

Fig. 5. Sample text detection results of the proposed method on our data

Table 2. Performance of the text detection techniques before and after text candidate region detection

Text detection methods	Before text candidate region detection						After text candidate region detection					
	RBNR data			Our data			RBNR data			Our data		
	R	P	F	R	P	F	R	P	F	R	P	F
Shivakumara et al. [8]	83.3	17.3	24.9	38.6	37.5	34.8	24.3	40.2	24.5	44.2	46.5	41.0
Epshtein et al. [4]	20.1	1.8	3.0	5.27	1.9	2.61	45.0	31.2	33.3	13.9	7.5	8.9

3.3 Validating Multi-modality Through Recognition

We also conduct experiments on recognition through binarization techniques to show the usefulness of TCR detection using the multimodal way. Table 3 shows the recognition rates of different binarization techniques when used before TCR and after TCR on our data and RBNR data. To show that the conventional OCR engine is not good for text recognition in marathon and running race images before TCR detection, we send the whole input image to OCR engine to calculate recognition rate. Then we pass text lines detected by the text detection technique [8] from the whole input image to OCR engine to calculate recognition rate to show the improvements compared to the whole image. Since Shivakumara et al.'s [8] method gives better results for text detection after TCR detection as shown in Table 2, we choose the same technique for bib number or text detection experiments for recognition. Similarly, we send the whole TCR to Tesseract OCR [27] directly to calculate recognition rate after TCR detection. Again, we send text lines detected by the text detection technique after TCR to Tesseract OCR. Table 3 shows that for both the datasets, different binarization techniques give poor results for the whole image before TCR detection compared to those after TCR detection. It is noted from Table 3 that Howe [13] technique gives a better recognition rate after TCR compared to other binarization techniques for our data. Similarly, Wolf et al.'s method [11] gives a better recognition rate for the RBNR data after TCR compared to the other techniques. However, overall, when we compare recognition results after TCR on both RBNR and our data, we find the recognition rate for our data is lower than that of RBNR data. This means that our data contains much more complex data compared to the existing RBNR data.

Further, according to the results reported in Table 4, we find that the proposed technique gives better results than the existing technique [1] for both the datasets in terms of recall and f-measure. However, the precision of the existing technique on our data is higher than the proposed technique. This is because the way the existing technique converts RGB to binary and uses Tesseract for recognition is different from our technique. The measures recall, precision and f-measure are calculated for recognition results according to instructions given in [1]. The main reason to get better results is the same as discussed in text detection experiments. This is valid because the dataset contains several images where faces are not visible (missing of frontal face), which is common in case of marathon and running race images. In summary, when we look at recognition rates in Tables 3 and 4, the results are far from standard results of document analysis area. Therefore, we are planning to explore machine learning algorithms [28, 29] to use at multi-level in future.

Table 3. Character recognition rate of the binarization techniques on our dataset and RBNR dataset (in %)

Binarization methods	Our data				RBNR data			
	Before-TCR		After-TCR		Before-TCR		After-TCR	
	Whole image	Text	TCR	Text	Whole image	Text	TCR	Text
Wolf et al. [11]	2.1	34.2	4.3	42.2	2.3	20.0	3.9	49.0
Moghaddam et al. [12]	2.4	25.3	4.1	27.2	3.1	17.2	2.1	25.3
Roy et al. [9]	2.7	12.2	3.3	16.3	2.6	10.2	2.7	19.0
Chattopadhyay et al. [10]	1.8	16.2	3.1	21.3	1.3	12.2	2.4	20.2
Howe [13]	3.1	31.6	5.3	43.3	3.3	18.2	4.1	33.8

Table 4. Character recognition rate of the proposed and existing techniques (in %)

Methods	RBNR data			Our data		
	R	P	F	R	P	F
Proposed method	55.3	76.5	64.19	49.69	56.16	52.73
RBNR method [1]	52.0	68.0	58.6	28.48	77.04	41.58

4 Conclusion and Future Work

In this paper, we have proposed a new multimodal technique for bib number and text detection in marathon or running race images. The proposed technique explores the combination of face and skin features for identifying text candidate regions from an input image. We have conducted different ways of experiments before text candidate region detection and after text candidate region detection using text detection and binarization techniques to show the usefulness of text candidate region detection from marathon and running race images. Experimental results show that the performance of text detection and binarization techniques improves much after text candidate region detection compared to before text candidate region detection. Furthermore, experimental results show the proposed multimodal technique outperforms the existing technique. In future, we aim at the identification of the persons who participate in marathon and running race images with the help of text detection and recognition.

Acknowledgment. The work described in this paper was supported by the Natural Science Foundation of China under Grant No. 61272218 and No. 61321491, and the Program for New Century Excellent Talents under NCET-11-0232.

References

1. Ami, B., Basha, T., Avidan, S.: Racing bib number recognition. In: Proceedings of BMCV (2012)
2. Klontz, J.C., Jain, A.K.: A case study of automated face recognition: the boston marathon bombing suspects. Computer **46**, 91–94 (2013)
3. Ye, Q., Doermann, D.: Text detection and recognition in imagery: a survey. IEEE. Trans. PAMI **37**, 1480–1500 (2015)
4. Epshtein, B., Ofek, E., Wexler, Y.: Detecting text in natural scenes with stroke width transform. In: Proceedings of CVPR, pp. 2963–2970 (2010)
5. Rong, L., Suyu, W., Shi, Z.X.: A two level algorithm for text detection in natural scene images. In: Proceedings of DAS, pp. 329–333 (2014)
6. Shivakumara, P., Sreedhar, R.P., Phan, T.Q., Shijian, L., Tan, C.L.: Multi-oriented video scene text detection through bayesian classification and boundary growing. IEEE Trans. CSVT **22**, 1227–1235 (2012)
7. Shivakumara, P., Phan, T.Q., Tan, C.L.: New fourier-statistical features in RGB space for video text detection. IEEE Trans. CSVT **20**, 1520–1532 (2010)
8. Shivakumara, P., Phan, T.Q., Tan, C.L.: New wavelet and color features for text detection in video. In: Proceedings of ICPR, pp. 3996–3999 (2010)
9. Roy, S., Shivakumara, P., Roy, P., Tan, C.L.: Wavelet-gradient-fusion for video text binarization. In: Proceedings of ICPR, pp. 3300–3303 (2012)
10. Chattopadhyay, T., Reddy, V.R., Garain, U.: Automatic selection of binarization method for robust OCR. In: Proceedings of ICDAR, pp. 1170–1174 (2013)
11. Wolf, C., Jolion, J.M., Chassaing, F.: Text localization, enhancement and binarization in multimedia documents. In: Proceedings of ICPR, pp. 1037–1040 (2002)
12. Moghaddam, R.F., Cheriet, M.: A multi-scale framework for adaptive binarization of degraded document images. Pattern Recogn. **43**, 2186–2198 (2010)
13. Howe, N.R.: Document binarization with automatic parameter tuning. IJDAR **16**, 247–258 (2013)
14. Shivakumara, P., Phan, T.Q., Lu, S., Tan, C.L.: Gradient vector flow and grouping based method for arbitrarily-oriented scene text detection in video images. IEEE Trans. CSVT **23**, 1729–1739 (2013)
15. Zhang, J., Kasturi, R.: A novel text detection system based on character and link energies. IEEE Trans. PAMI **23**, 4187–4198 (2013)
16. Minetto, R., Thome, N., Cord, M., Leite, N.J., Stolfi, J.: SnooperText: a text detection system for automatic indexing of urban scenes. CVIU **122**, 92–104 (2014)
17. Kang, L., Li, Y., Doermann, D.: Orientation robust text line detection in natural scene images. In: Proceedings CVPR, pp. 4034–4041 (2014)
18. Yin, X.C., Yin, X., Huang, K., Hao, H.W.: Robust text detection in natural scene images. IEEE Trans. PAMI **36**, 970–983 (2014)

19. Yao, C., Bai, X., Liu, W.: A unified framework for multioriented text detection and recognition. IEEE Trans. IP **23**, 4737–4749 (2014)
20. Yi, C., Tian, Y.: Scene text recognition in mobile application by character descriptor and structure configuration. IEEE Trans. IP **23**, 2972–2982 (2014)
21. Cui, Y., Huang, Q.: Character extraction of license plate from video. In: Proceedings of CVPR, pp. 502–507 (1997)
22. Suresh, K.V., Kumar, G.M., Rajagopalan, A.N.: Super resolution of license plates in real traffic videos. IEEE Trans. ITS **8**, 321–331 (2007)
23. Yu, S., Li, B., Zhang, Q., Liu, C., Meng, M.A.H.: A novel license plate location method based on wavelet transform and EMD analysis. PR **48**, 114–125 (2015)
24. Conaire, C.O., Connor, N.E.O., Smeaton, A.F.: Detector adaptation by maximizing agreement between independent data sources. In: Proceedings of CVPR, pp. 1–6 (2007)
25. Kakumanu, P., Makrogiannis, S., Bourbakis, N.: A survey of skin-color modeling and detection methods. PR **40**, 1106–1122 (2007)
26. Lienhart, R., Maydt, J.: An extended set of haar-like features for rapid object detection. In: Proceedings of ICIP, pp. 900–903 (2002)
27. Tesseract: http://code.google.com/p/tesseract-ocr/
28. Lu, W., Tao, D.: Multiview hessian regularization for image annotation. IEEE Trans. IP **22**, 2676–2687 (2013)
29. Xu, C., Tao, D., Xu, C.: Large-margin multi-view information bottleneck. IEEE Trans. PAMI **36**, 1559–1572 (2014)

A New Multi-spectral Fusion Method
for Degraded Video Text Frame Enhancement

Yangbing Weng[1], Palaiahnakote Shivakumara[2], Tong Lu[1(✉)],
Liang Kim Meng[3], and Hon Hock Woon[3]

[1] National Key Lab for Novel Software Technology, Nanjing University,
Nanjing, China
wengyangbing@gmail.com, lutong@nju.edu.cn
[2] Faculty of Computer Science and Information Technology,
University of Malaya, Kuala Lumpur, Malaysia
shiva@um.edu.my
[3] Advanced Informatics Lab, MIMOS Berhad, Kuala Lumpur, Malaysia
{liang.kimmeng, hockwoon.hon}@mimos.my

Abstract. Text detection and recognition in degraded video is complex and challenging due to lighting effect, sensor and motion blurring. This paper presents a new method that derives multi-spectral images from each input video frame by studying non-linear intensity values in Gray, R, G and B color spaces to increase the contrast of text pixels, which results in four respective multi-spectral images. Then we propose a multiple fusion criteria for the four multi-spectral images to enhance text information in degraded video frames. We propose median operation to obtain a single image from the results of the multiple fusion criteria, which we name fusion-1. We further apply k-means clustering on the fused images obtained by the multiple fusion criteria to classify text clusters, which results in binary images. Then we propose the same median operation to obtain a single image by fusing binary images, which we name fusion-2. We evaluate the enhanced images at fusion-1 and fusion-2 using quality measures, such as Mean Square Error, Peak Signal to Noise Ratio and Structural Symmetry. Furthermore, the enhanced images are validated through text detection and recognition accuracies in video frames to show the effectiveness of enhancement.

Keywords: Multi-spectral images · Text enhancement · Multi-spectral-fusion · Quality measures · Video text detection and video text recognition

1 Introduction

Video text detection and recognition is an emerging area for current researchers in the field of image processing, pattern recognition and video document analysis because it is useful for several real time applications, for example, event retrieval based on semantics, exciting events extraction, assisting blind persons, safe driving, navigation and surveillance [1]. According to the past literature on text detection and recognition in video [2], the applications are divided into two groups based on two types of texts present in video, such as graphics text which is edited and scene text which naturally

© Springer International Publishing Switzerland 2015
Y.-S. Ho et al. (Eds.): PCM 2015, Part I, LNCS 9314, pp. 495–506, 2015.
DOI: 10.1007/978-3-319-24075-6_48

exists. The applications that use graphics text generally aim at indexing and retrieval, while the applications that use scene text aim at navigation, tracking and surveillance. It is noted that most of the methods in literature [1, 2] focus on graphics text detection and recognition in video because graphics text is easy to process compared to scene text. Since graphics text is edited, it has good visibility, clarity and contrast, while scene text suffers from degradations, low contrast, complex background, orientations, etc. One such real time application is license plate detection and recognition in video, for which achieving a good text detection and recognition accuracy is complex and challenging because it severely suffers from the degradations caused by motion blur, lighting, non-uniform illumination, text movements and complex background [3]. Such example can be seen in Fig. 1(a), where the license plate number is even not visible and readable. For the text in Fig. 1(a), the existing text detection method, which uses stroke width (SWT) information for detection [4], fails to detect the number properly because of degradation effects, while the same existing text detection method detects the number successfully for the enhanced image as shown in Fig. 1(c). This shows that we need a robust enhancement method for sharpening text information in degraded video frames. In the same way, the sample illustrations for recognition are show in Fig. 2, where (a) is the text segmented manually from Fig. 1(a), (b) is the result of Otsu thresholding, which is a well-known baseline method for binarizing document images [5], (c) shows the recognition results given by Tesseract available publicly [6], (d) is the text segmented from Fig. 1(c), (e) gives the results of Otsu thresholding, and (f) shows the recognition results by Tesseract. It is observed from Fig. 2(c), (f) that the Tesseract OCR engine recognizes all the characters except "D" for the text segmented from the enhanced image, while it fails to recognize the text segmented from the original input frame. Therefore, in this paper, we propose a novel enhancement method, which fuses multi-spectral images to remove degradation effects such that the current text detection and recognition methods can give good results.

The problem of degradation is familiar to the document analysis field [7–10] and has been explored in several ways, such as through robust binarization methods [11], super-resolution methods [12, 13], and extracting the features that are invariant to degradations [14]. Since the methods aim at the degradations of scanned documents, they expect plain or homogeneous background with high contrast texts. Therefore, these methods may not be used for solving the degradation problem of video text directly because video text suffers from low resolution and complex background.

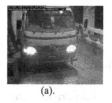 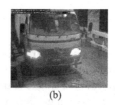 

(a).   (b)   (c)

Fig. 1. Sample text detection experiment for the input frame and the enhanced frame: (a) Degraded License plate, (b) Text detection by existing method [4] in (a) and (c) Text detection on enhanced image

(a). Text in Fig. 1(a)   (b) Otsu thresholding   "REY T0?"
                                                  (c) Recognition

(d). Text in Fig. 1(c)   (e) Otsu thresholding   "OBY S02"
                                                  (f) Recognition

Fig. 2. Sample recognition experiments for the original text and the enhanced text given by the proposed method

Similarly, there are lots of methods proposed in literature for text detection and recognition in video, which work well for both graphics and scene texts of arbitrary orientations [15–17]. In addition, we can also see a few methods for separating graphics and scene texts from video [18]. However, these methods do not consider degraded or blur videos. Recently, Cui and Huang [19] proposed a method for character extraction from license plates using Markov Random Field and multiple frames in video; however, this method does not consider blur and degraded frames. The blur and degradation problem is addressed by Li and Doermann [20] using the projection onto convex sets and sub-pixel concepts. This method requires temporal frames and the performance depends on text tracking. The degradation with a single frame is addressed by Yu et al. [3] by proposing wavelet transform and Empirical Mode decomposition (EMD) analysis, but it consumes more time. Similarly, Suresh et al. [21] proposed super-resolution for enhancing license plate numbers in a blur image using a deblurring method. However, the method is parameter orientated and thus its performance greatly depends on parameters.

Though there are several methods that addressed the issue of degradations and blur in video especially for license plate detection and recognition, none of them gives a satisfactory solution for different situations. In addition, most of the methods assume text lines are detected. It is true that a text detection method may not perform well when the video frame is affected by degradations and blur. Therefore, there is a great demand for developing a novel method for enhancing degraded or blur video text frames. Since the method enhances the whole frame, theoretically it can be used for some other applications and other kind of image enhancement.

2 Proposed Methodology

It is known that degradations are caused due to illumination, lighting, motion blur, etc. [19, 20]. These degradations reflect in the form of noises in background and the losing of sharpness at edges or near edges. In order to restore the sharpness at edges or near edges and to remove background noises, we are inspired by the work proposed in [11, 22] for removing background noises and increasing contrast by studying local minimum and maximum intensity values. We explore the same concept with modifications in gray, R, G and B domains for obtaining an enhanced image in this work. This process results in four multi-spectral images for respectively gray, R, G and B spaces, which reduces degradation effects. To collect useful information from the four multi-spectral images, we propose to explore the fusion concept with five operations, namely, maximum, minimum, sum, average and median as suggested in [23] for predicting OCR accuracy on mobile captured images. This results in five fused images, for which we propose a median operation to fuse all the images to obtain single resultant image. Sometimes, due to background complexity in video, there are chances of enhancing background noises by the fusion operation. Therefore, we propose median operation to fuse the five images as the median operation considers neither high values nor low values, which is called the fusion-1 method. It is valid because both high and low values may not contribute much. We know that the five fused image contain high values for significant information and low values for noisy pixels. Therefore, we propose to apply k-means clustering with k = 2 on the five fused images to classify significant pixels that have high values, rather than performing median operation on the five fused images directly. This process results in five cluster images containing significant information including text pixels. Then we perform the median operation to fuse the five cluster images to obtain the final fused image, which is called the fusion-2 method. This final fusion image is considered as the enhanced image in this work, and its effectiveness would be validated through text detection and recognition experiments in the experimental section.

2.1   Multi-spectral Images for Reducing Degradation Effect

As discussed in Sect. 2, for each input color frame, the method derives gray, R, G and B images. The reason to derive the four channels is to study the non-linear intensity values caused by degradations to suppress background noises and sharpen edges. In order to achieve this, we propose to extract local minimum and maximum values from the respective four images as explained mathematically below. Note that local minimum and maximum have been used for suppressing background noises in the past [11, 22].

Let the four images respectively be $I_{gray}(x,y)$, $I_R(x,y)$, $I_G(x,y)$, and $I_B(x,y)$. Here (x, y) represents any pixel in the channel. For each channel, we obtain its local minimum image, say $I_{min}(x,y)$, and local maximum image, say $I_{max}(x,y)$ as defined in

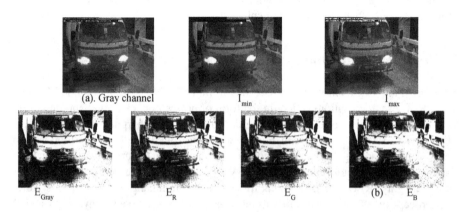

Fig. 3. Multi-spectral images for the input degraded frame

Eqs. (1) and (2) by moving a circular sliding window over the gray image shown in Fig. 3(a), where we can see the contrast looks distributed uniformly compared to the gray image. Similarly, the proposed method derives I_{min} and I_{max} for R, G and B channels, respectively:

$$I_{min} = min(\{I(x_k, y_k) : (x_k, y_k) \in W\}) \tag{1}$$

$$I_{max} = max(\{I(x_k, y_k) : (x_k, y_k) \in W\}) \tag{2}$$

where W denotes a circular window with the radius of two pixels centered at pixel (x, y). The value of the radius is determined empirically. Then we compute the enhanced image, say $E_{gray}(x, y)$ as defined in Eq. (3), which is shown in Fig. 3(b), where one can see significant information is enhanced:

$$E_{gray}(x, y) = \frac{\exp\left(-\frac{\left[I_{gray}(x,y) - \mu_{max}\right]^2}{2\sigma_{max}^2}\right)}{\exp\left(-\frac{\left[I_{gray}(x,y) - \mu_{max}\right]^2}{2\sigma_{max}^2}\right) + \exp\left(-\frac{\left[I_{gray}(x,y) - \mu_{min}\right]^2}{2\sigma_{min}^2}\right)} \tag{3}$$

where μ_{min} and μ_{max} are the mean pixel values of $I_{min}(x, y)$ and $I_{max}(x, y)$ images, respectively. Similarly, σ_{min} and σ_{max} respectively denote the standard deviation of pixel values of $I_{min}(x, y)$ and $I_{max}(x, y)$ images. This process helps us to suppress background noises and increase the sharpness at edges or near edges. In this way, we respectively derive $E_R(x, y)$, $E_G(x, y)$, $E_B(x, y)$ images for $I_R(x, y)$, $I_G(x, y)$, $I_B(x, y)$, which we label as multispectral enhanced images because they are obtained by studying different intensity values. The multi-spectral images (E_{Gray}, E_R, E_G and E_B) are shown respectively for gray, R, G and B channels in Fig. 3(b), where one can notice that the significant information such as text information is sharpened compared to the input frame. However, it also noted that since the proposed enhancement process applies on the whole image, some of the background noise pixels are also sharpened.

fs$_{Max}$ fs$_{Min}$ fs$_{Sum}$ fs$_{Avg}$ fs$_{Median}$ Final Fusion-1

Fig. 4. Five fused images and the final fusion-1 image

2.2 Multi-spectral Fusion-1 for Text Frame Enhancement

It is mentioned in the previous subsection that due to the complexity of video, it is hard to enhance required information as shown in the multi-spectral images, where both text pixels and other background information are also sharpened. Since we are inspired by the work presented in [23] for predicting OCR accuracy using several fusion criteria on document images, we also explore the same fusion criteria to fuse the four multi-spectral images to collect significant information without missing. The fusion criteria are selecting maximum, minimum, sum, mean and median values from the pixel values in the four multi-spectral images. In other words, the proposed method performs the above operations to select one value from the four pixel values corresponding to the four multi-spectral images. Therefore, this process results in five fused images corresponding to the five above mentioned operations as shown in Fig. 4, where one can see the pixels that representing text are more sharpened compared to the multi-spectral images. Further, we fuse again the five fused images through the median operation to produce one enhanced image as shown in Fig. 4, where fusion-1 results are still better than the five fused images when we look at the pixels representing text. The reason to propose the median operation is that according to multi-spectral images, sometimes, background noises may get high values and actual noises get low values. To avoid the noises contributed by background, median is a better operation as it neither considers high nor low values. The mathematics steps of this process are as follows.

After obtaining $E_{gray}(x,y)$, $E_R(x,y)$, $E_G(x,y)$ and $E_B(x,y)$, we combine the four images together by computing the maximum, minimum, sum, average and median values of each pixel in the same position of their own image:

$$fsMax = max\left(\left\{E_j(x_j, y_j) : j = 1, 2, 3, 4\right\}\right) \tag{4}$$

$$fsMin = min\left(\left\{E_j(x_j, y_j) : j = 1, 2, 3, 4\right\}\right) \tag{5}$$

$$fsSum = \sum_{j=1}^{4} \left(\left\{E_j(x_j, y_j)\right\}\right) \tag{6}$$

$$fsAvg = mean(\left\{E_j(x_j, y_j) : j = 1, 2, 3, 4\right) \tag{7}$$

$$fsMed = median\left(\left\{E_j(x_j, y_j) : j = 1, 2, 3, 4\right\}\right) \tag{8}$$

where E is the set of the four enhanced images including $E_{gray}(x,y)$, $E_R(x,y)$, $E_G(x,y)$, $E_B(x,y)$.

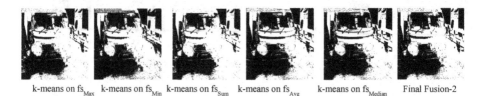

k-means on fs$_{Max}$ k-means on fs$_{Min}$ k-means on fs$_{Sum}$ k-means on fs$_{Avg}$ k-means on fs$_{Median}$ Final Fusion-2

Fig. 5. k-means clustering on five fused images and final fusion-2 image

2.3 Multi-spectral Fusion-2 for Text Frame Enhancement

In the previous subsection, we fuse the five fused images through the median operation in gray domain. There are chances of enhancing background noises caused by degradation effect due to complex background and low resolution of video. It is also true that the five fused images contain both high values representing significant information like text pixels and low values representing noise pixels. Therefore, we propose to use k-means clustering for the five fused images to classify high values into one cluster and low values into another one as shown in Fig. 5, respectively, where these results look brighter than the images in Fig. 4. Since a high value cluster contains text pixels, we consider the clusters that contain high values for fusing through the median operation further. This results in the final enhanced image for the input degraded frame, which is called fusion-2 as shown in Fig. 6, where the fusion-2 image contains enhanced text information. Fusing the five binary fused images can be shown mathematically as follows:

$$F = median\left(\left\{E_j(x_j, y_j) : j = 1, 2, 3, 4, 5\right\}\right) \tag{9}$$

where F denotes the final fusion-2 image, which is nothing but the enhanced image. E is the set of the five binary fused images.

3 Experimental Results

Since this work is new and the problem is not yet addressed properly in the past, there is no standard database for video text frame enhancement. Therefore, we create our own dataset, which is the collection of degraded frames from different sources such as the ICDAR database [24], captured images from our own camera and internet. In total, we collect 200 video images for experimentation. To measure the quality of the images given by the fusion-1 method described in Sect. 2.2 and the fusion-2 method described in Sect. 2.3, we use Mean Square Error (MSE), Peak to Signal to Noise Ratio (PSNR) and Structural Symmetry Index (SSIM) as these measures are well known for evaluating the quality of an image. Definitions and formula can be found in [11]. These measures require a reference image for matching the obtained results. For this purpose, we use the enhanced image given by the local minimum and maximum method proposed in [11] as the reference image because this method is the state of the art for enhancing degraded documents in the literature.

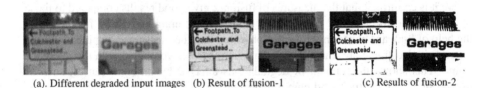

(a). Different degraded input images   (b) Result of fusion-1   (c) Results of fusion-2

Fig. 6. Sample degraded images with corresponding enhanced images given by fusion-1 and fusion-2

To show the effectiveness of enhancement, we conduct experiments on text detection and recognition using several text detection and binarization methods for before enhancement and after enhancement in the subsequent subsections. We implement Stroke Width Transform (SWT) based text detection method [4], which is popular and the state of the art method for detecting texts in natural scene images, and the method proposed by Minetto et al. [15], which is developed for detecting texts in urban scene images where background varies drastically, for experimentations in this work. Yi et al. [16] proposed a method for text detection in natural scene images using structure-based partition and grouping. Similarly, we use classical binarization methods, namely, Otsu thresholding [3], Niblack [7], and Sauvola [8], which are baseline methods for binarizing document images, and one more recent method by Zhou et al. [9] which works well for blurred images and complex background images, for recognition experiments in this work. Su et al. [10] proposed a binarization method for degraded document images based on local contrast and stroke information. The measures used for the text detection methods are recall (R), precision (P) and f-measure (F). Since there is no ground truth for the dataset, we manually count the number of text blocks for calculating recall, precision and f-measure. For recognition, we use recognition rate (RR) before enhancement and after enhancement given by Tesseract OCR, which is available publicly [6].

   (a). Degraded frames   (b). Outputs of fusion-1   (c) outputs of fusion-2

Fig. 7. Text detection results by [4] for the degraded (a), the outputs of fusion-1 (b) and the outputs of fusion-2 (c)

3.1   Experiments on Measuring Quality of the Enhanced Frame

Sample degraded images due to motion blur or low contrast and corresponding enhanced images given by fusion-1 and fusion-2 are shown in Fig. 6(a–c), respectively, where it is noticed that the texts in degraded frames are sharpened well, which can be easily readable from the OCR. Figure 6(b), (c) show that the results of fusion-2 are better than those of fusion-1 because of fusion-2 considers the output of fusion-1 for further fusing. The quality measures for both the methods are reported in Table 1, where it is confirmed that the measures of fusion-2 give good results compared to those of fusion-1 because fusion-2 further fuses the five fused images given by fusion-1 with the help of k-means clustering. It is also observed that the measures computed with the input degraded frame are much higher than the measures computed with the reference frame. This shows the amount of degradations affected by different causes, such as motion blur, low contrast, and complex background.

Table 1. Quality measures for fusion-1 and fusion-2

Methods	Input frame with enhanced frames			Reference frame with enhanced frame		
	MSE	PSNR	SSIM	MSE	PSNR	SSIM
Fusion-1	0.25	61.28	0.83	0.34	55.55	0.73
Fusion-2	0.18	65.41	0.89	0.26	60.23	0.82

3.2 Validating Enhancement Through Text Detection

Sample qualitative results of the text detection method given by Epshtein et al. for degraded, enhanced images by fusion-1 and enhanced image by fusion-2 are shown in Fig. 7(a–c), respectively. Figure 7 shows that the text detection method does not detect texts in the degraded frame, while the same method detects texts well in the enhanced images. Besides, when we compare the text detection results respectively from fusion-1 and fusion-2, we find the results of fusion-2 are better than those of fusion-1. Therefore, we can assert that the enhancement helps in improving text detection performances for degraded frames. The quantitative results of the three text detection methods mentioned in Sect. 3 are reported in Table 2 for before-enhancement and after-enhancement. It is seen from Table 2 that the methods are good at precision for before-enhancement, while the same methods are good at recall and f-measure for after-enhancement. The reason is that due to enhancement, text detection methods produce more false positives after-enhancement. As a result, precision is lower for after-experiment compared to before-experiment. However, when we consider the overall performance of text detection methods, they perform well for enhanced images (after-enhancement) compared to degraded frames (before-enhancement). Therefore, we need to explore some rules or a classifier for the classification of false positives. This would be considered as our future work.

Table 2. Performance of text detection methods before-enhancement and after-enhancement images

| Text detection methods | Before enhancement | | | After enhancement | | | | | |
	Input frame			Fusion-1			Fusion-2		
	R	P	F	R	P	F	R	P	F
Epshtein et al. [4]	27.2	74.0	39.8	47.1	66.5	55.1	53.0	59.0	55.8
Minetto et al. [15]	38.1	51.7	43.8	39.8	47.0	43.1	42.0	47.8	44.7
Yi et al. [16]	31.4	72.1	43.7	38.7	69.3	49.7	45.5	67.8	54.5

3.3 Validating Enhancement Through Recognition

For recognition experiments, we use stroke width transform based method [4] for obtaining text lines from the input degraded frame. According to Table 2, this method is better than other text detection methods in terms of precision for before-enhancement and in terms of f-measure after-enhancement. Therefore, we use this method for text line detection from the degraded input frames and enhanced frames. The text lines detected by text detection method are input for the recognition experiments. We calculate recognition rate (RR) for the text lines with different binarization methods through Tesseract OCR. The results are reported in Table 3, where it is clear that the binarization methods give good recognition rates for after-enhancement (AE) compared to before-enhancement (BE). This shows that in order to achieve a good recognition rate for degraded frames, enhancement is essential. The same conclusions can be drawn from text detection experiments as well. In summary, when we look at the results in Tables 2 and 3, we find the results are far from document analysis. Therefore, we are planning to explore machine learning algorithms [25, 26] to use at multi-level in future.

Table 3. Performance of binarization methods before-enhancement and after-enhancement for text lines in terms character recognition rate

Binarization methods	BE	AE	
	RR	Fusion-1	Fusion-2
		RR	RR
Otsu [5]	28.6	54.9	60.9
Niblack [7]	18.1	31.1	29.7
Sauvola [8]	11.7	33.0	59.7
Zhou et al. [9]	34.8	55.0	59.7
Su et al. [10]	13.6	19.6	23.3

4   Conclusion

In this paper, we have proposed a new method for enhancing text information in degraded video frames caused by distortions, non-uniform illumination, etc. The proposed method derives multi-spectral images from different color channels based on local minimum and maximum values from respective color channels. We perform a new operation to fuse these two images, which results in four multi-spectral images. We propose multiple fusion criteria for each multi-spectral image to obtain fused images. The proposed method further uses the median operation for fusing multiple fused images, which gives a single enhanced image (fusion-1). The same fusion is done at binary level with the help of k-means clustering, which results in another single enhanced image (fusion-2). The experimental results on both text detection and recognition show that the proposed enhancement method is effective and useful to achieve good text detection and recognition accuracies for degraded video frames.

Acknowledgment. The work described in this paper was supported by the Natural Science Foundation of China under Grant No. 61272218 and No. 61321491, and the Program for New Century Excellent Talents under NCET-11-0232.

References

1. Ye, Q., Doermann, D.: Text detection and recognition in imagery: a survey. IEEE. Trans. Pattern Anal. Mach Intell. **1**, 1 (2014)
2. Sharma, N., Pal, U., Blumenstein, M.: Recent advances in video based document processing: a review. In: Proceedings of DAS, pp. 63–68 (2012)
3. Yu, S., Li, B., Zhang, Q., Liu, C., Meng, M.A.H.: A novel license plate location method based on wavelet transform and EMD analysis. Pattern Recogn. **48**, 114–125 (2015)
4. Epshtein, B., Ofek, E., Wexler, Y.: Detecting text in natural scenes with stroke width transform. In: Proceedings of CVPR, pp. 2963–2970 (2010)
5. Otsu, N.: A threshold selection method from gray level histogram. IEEE Trans. Syst. Man Cybern. **11**, 62–66 (1978)
6. Tesseract. http://code.google.com/p/tesseract-ocr/
7. Niblack, W.: An Introduction to Digital Image Processing. Strandberg Publishing Company, Birkeroed (1985)
8. Sauvola, J., Seeppanen, T., Haapakoski, S., Pietikainen, M.: Adaptive document binarization. In: Proceedings of ICDAR, pp. 147–152 (1997)
9. Zhou, Y., Feid, J., Miller, E.L., Wang, R.: Scene text segmentation via inverse rendering. In: Proceedings of ICDAR, pp. 457–461 (2013)
10. Su, B., Lu, S., Tan, C.L.: A robust document image binarization for degraded document images. IEEE Trans. Image Process. **22**, 1408–1417 (2013)
11. Su, B., Lu, S., Tan, C.L.: Binarization of historical document images using the local maximum and minimum. In: Proceedings of DAS, pp. 159–166 (2010)
12. Nayef, N., Chazalon, J., Kramer, P.G., Ogier, J.M.: Efficient example-based super-resolution of single text images based on selective patch processing. In: Proceedings of DAS, pp. 227–231 (2014)
13. Zheng, Y., Li, X.K.S., Sun, Y.H.J.: Real-time document image super-resolution by fast matting. In: Proceedings of DAS, pp. 232–236 (2014)
14. Saleem, S., Hollaus, F., Sablatnig, R.: Recognition of degraded ancient characters based on dense SIFT. In: Proceedings of DATeCH, pp. 15–20 (2014)
15. Minetto, R., Thome, N., Cord, M., Leite, N.J., Stolfi, J.: SnooperText: a text detection system for automatic indexing of urban scenes. In: CVIU, pp. 92–104 (2014)
16. Yi, C., Tian, Y.: Text string detection from natural scenes by structure-based partition and grouping. IEEE Trans. Image Process. **20**, 2594–2605 (2011)
17. Shivakumara, P., Phan, T.Q., Lu, S., Tan, C.L.: Gradient vector flow and grouping based method for arbitrarily-oriented scene text detection in video images. IEEE Trans. Circ. Syst. Video Technol. **23**, 1729–1739 (2013)
18. Xu, J., Shivakumara, P., Lu, T., Phan, T.Q., Tan, C.L.: Graphics and scene text classification in video. In: Proceedings of ICPR, pp. 4714–4719 (2014)
19. Cui, Y., Huang, Q.: Character extraction of license plate from video. In: Proceedings of CVPR, pp. 502–507 (1997)
20. Li, H., Doermann, D.: Super-resolution-based enhancement for text in digital video. In: Proceedings of ICPR, pp 847–850 (2000)
21. Suresh, K.V., Kumar, G.M., Rajagopalan, A.N.: Superresolution of license plates in real traffic videos. IEEE Trans. Intell. Transp. Syst. **8**, 321–331 (2007)
22. Saleeem, S., Sablatnig, R.: A robust SIFT descriptor for multi-spectral images. IEEE Signal Process. Lett. **21**, 400–403 (2014)

23. Rusinol, M., Chazalon, J., Ogier, J. M.: Combining focus measure operators to predict OCR accuracy in mobile-captured document images. In: Proceedings of IWDAS, pp 181–185 (2014)
24. Karatzas, D., Shafait, F., Uchida, S., Iwamura, M., Boorda, L.G.I., Mestre, S.R., Mas, J., Mota, D.F., Almazan, J.A., De las Heras, L.P.: ICDAR 2013 robust reading competition. In: Proceedings of ICDAR, pp. 1115–1124 (2013)
25. Lu, W., Tao, D.: Multiview Hessian regularization for image annotation. IEEE Trans. Image Process. **22**, 2676–2687 (2013)
26. Xu, C., Tao, D., Xu, C.: Large-margin multi-view information bottleneck. IEEE Trans. Pattern Anal. Mach. Intell. **36**, 1559–1572 (2014)

A Robust Video Text Extraction and Recognition Approach Using OCR Feedback Information

Guangyu Gao[1(✉)], He Zhang[2], and Hongting Chen[3]

[1] School of Software, Beijing Institute of Technology, Beijing, China
`guangyu.ryan@gmail.com`
[2] Operations Office (Beijing), People's Bank of China, Beijing, China
`zhanghe86@126.com`
[3] School of Computer, Beijing University of Posts and Telecom, Beijing, China
`hongting.chen@yahoo.com`

Abstract. Video text is very important semantic information, which brings precise and meaningful clues for video indexing and retrieval. However, most previous approaches did video text extraction and recognition separately, while the main difficulty of extraction and recognition with complex background wasn't handled very well. In this paper, these difficulty is investigated by combining text extraction and recognition together as well as using OCR feedback information. The following features are highlighted in our approach: (i) an efficient character image segmentation method is proposed in consideration of most prior knowledge. (ii) text extraction are implemented both on text-row and segmented single character images, since text-row based extraction maintains the color consistency of characters and backgrounds while single character has simpler background. After that, the best binary image is chosen for recognition with OCR feedback. (iii) The K-means algorithm is used for extraction which ensures that the best extraction result is involved, which is the binary image with clear classification of text strokes and background. Finally, extensive experiments and empirical evaluations on several video text images are conducted to demonstrate the satisfying performance of the proposed approach.

Keywords: Text extraction · Text recognition · Character segmentation · *K*-means · OCR feedback

1 Introduction

According to the official statistic-report of *YouTube* over 6 billion hours of video are watched each month and about 300 h of video are uploaded every minute [1]. Thus, it has become a crucial and challenging task to retrieve videos in those large dataset.

This work was supported by National Natural Science Foundation of China (Grant No. 61401023) and Fundamental University Research Fund of BIT (Grand No. 20140842001).

© Springer International Publishing Switzerland 2015
Y.-S. Ho et al. (Eds.): PCM 2015, Part I, LNCS 9314, pp. 507–517, 2015.
DOI: 10.1007/978-3-319-24075-6_49

Video text is one of the most important high-level semantic features for this. Generally, there are two types of video text: the superimposed text (added during the editing process) and the scene text (existing in real scene). Moreover, there are three steps involved before text recognition, i.e., detection, localization and extraction. Here in video text extraction, text pixels remain after the background pixels in the text rows are removed. For video text recognition, the Optical Character Recognition (OCR) is always used to deal with the image pre-processed by text extraction.

Comparing to scene text, the superimposed text offers concise and direct description of the video content. For example, subtitles in sport video highlight the information of scores and players, and captions in movie can be used to summarize the core description of the story [2, 3]. Therefore, in this paper, we mainly discussed the video text extraction combined with recognition for superimposed texts. Actually, both text extraction and recognition problems are handled giving the detected video text image. Compared with previous studies, our main contributions include:

- In extraction, an efficient and accurate character segmentation method is proposed, in which the peak-to-valley eigenvalues is defined to evaluate character width.
- Extraction on row-text images and segmented character images are combined in order to remain both of their advantages. Meanwhile, the best extracted binary image is adaptively chosen for recognition.
- The K-means algorithm with different "K"s is used for text extraction to ensures the best extraction is obtained and chosen by OCR feedback information.

2   Related Work

Recently, video annotation has become a very attractive functionality in video analysis and understanding. i.e., Kim [4] applied manual video tagging to generate text-based metadata for context-based video retrieval. Bhute and Meshram [5] have given a review of text based approach for indexing and retrieval of image and video. However, manual metadata annotation is both time consuming and incapable of preventing new errors. Meanwhile, although automated speech recognition is applied to provided text scripts of spoken language, poor recording quality and noises affect the achieved performance beyond further usefulness [6]. Therefore, video text recognition based video annotation is still efficient and precise for large scale videos.

As a crucial process before recognition, text extraction needs to separate the pixels of a localized text region into categories of text and background. Video text extraction is always difficult because of complex background, unknown text character color, and various stroke widths [7]. Most existing methods can be classified into:

(1) *Threshold-based.* Otsu method [8] was the widely used threshold-based text extraction method due to its simplicity and efficiency. Leedham et al. [9] proposed the automatic selection or combination of appropriate algorithms. Besides, Ngo and Chan [10] used the adaptive thresholding with four sets of different operations for text extraction. In [11], a novel extraction framework with an improved color-based thresholding is proposed. However, thresholding-based methods did

not work well facing complex background, since a strict threshold of text and the background does not always exist.

(2) *Statistic model-based.* The statistic model-based methods deemed that features of text pixels obey some probability distribution. Gao and Yang [12] expressed all the same color regions with a Gaussian kernel function and then to determine the class for each region. Chen et al. [13] used Markov Random Field to determine which Gaussian term each pixel belongs to, and consequently, to segment the text from background. Fu et al. [14] proposed to extract multilingual texts in images, through discriminating characters from non-characters based on the Gaussian mixture modeling of neighbor characters. Meanwhile, Roy et al. [15] presented a statistical model based scheme for automatic extraction of text components from digital images. Statistic model-based methods can handle complex background well since it took consideration of multi-peak distribution of text color, but they always need establish different models for different images, and also owing to simply using color information, it is hard to determine model functions.

(3) *Connected components-based.* These methods considered that character strokes are always preserved connectivity but it is not true for background. Lienhart and Wernicke [16] adapted image segmentation, to cluster the foreground pixels with unsupervised color clustering for text extraction. Lyu et al. [7] proposed a synthetic method with local adaptive threshold and connected components analysis. Li et al. [18] proposed a two-threshold method using stroke edge filter, which can effectively identify stroke edges in subjective evaluation. In addition, the authors of [19] proposed a video text extraction scheme using Key Text Points (KTPs). Liu et al. [20] proposed a novel multi-oriented CC based video text extraction algorithm specifically for Chinese text.

However, none of these works have perfectly solved the problem of complex background especially in videos. Meanwhile, while the goal of text extraction is to get a good result for recognition, it is not reasonable to discuss the performance of text extraction separately. The feedback of OCR will be helpful for selecting and evaluating the extraction results. Besides, while most previous works performed text extraction in a whole text row, Sharma et al. [21] proposed to do character segmentation from detected text before binarization recognition, and to achieve better accuracy even for scene text. Meanwhile, our previous work [22] also got a satisfactory performance by separating the text row into individual characters for text extraction.

Thus, in this paper, text extraction are applied on both whole text and segmented characters considering their own advantages. Meanwhile, we propose a novel way to optimally configure K since the best K can't be easily obtained in the K-means algorithm. Finally, the extraction result with the best K is used to recognize text accurately. The whole framework is shown in Fig. 1.

The rest of the paper is organized as follows. In Sect. 3, Video text segmentation will be discussed. While in Sect. 4, the text extraction in both segmented character and whole text row will be introduced. Meanwhile, the OCR recognition feedback based best scheme choosing is illustrated in Sect. 5. The experimental results are shown in Sect. 6. Finally, conclusion will be included in Sect. 7.

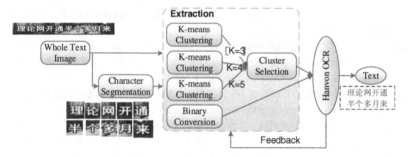

Fig. 1. Framework diagram of the proposed approach

3 Video Text Segmentation

An assumption is made that the video text rows is obtained with state-of-the-art video text detection and localization method [23, 24]. Characters always have strong spatial frequency variations compared with un-character in a text image. Edge features correspond to spatial frequency which synthesizes vertical and horizontal gradient together, where the vertical gradient is the most useful one. Because most of the video texts are horizontal texts (it can be rotated to horizontal one for the vertical text.), and the vertical gradient can strongly discriminate characters from un-character regions or gaps. Besides, horizontal gradient may also introduce noises to segmentation process. Thus, here, only vertical gradient is considered for segmentation.

The gradient projection is performed on the gradient map to get the projection map, namely gradient projection array. Meanwhile, this array is smoothed by the mean filter. Character Divider Lines (DLs) has to be corresponded to the troughs in the gradient projection array. Thus, in the smoothed gradient array, all troughs form the initial DL set, $\mathcal{L} = \{\ell_1, \ell_2, \ldots, \ell_N\}$. And then, the Peak-to-Valley Eigen values (PVE) is defined for each DL ℓ_i as follows.

$$V(i) = \frac{C(i-1) + C(i+1) - 2 \times T(i)}{T(i)} \tag{1}$$

where $V(i)$ is the PVE of ℓ_i, $C(i)$ and $T(i)$ refers to the i_{th} crest and trough respectively. In fact DLs who's PVEs are less than the average of all PVEs are removed. The remains are represented by,

$$L = \{l | V(i) \geq \sum_j \frac{V(j)}{N}\} \tag{2}$$

In the end, character DLs are reserved. They divide characters from each other with the following steps:

- DLs are sorted with their PVEs and the biggest M ones are remained. Generally, given that the characters are near-square word with the same height and width, we set $M = 1.5 \times W/H$, where W and H refer to the width and height of the text image.

The reason of using weight of 1.5 is that characters are only near-square, not really square, especially for English characters. Therefore, it make sure that enough DLs are existed, and also it ensure not to abandon any character DLs.

- Then, we aim to estimate the single Character Width (CW). The gap between adjacent DLs is named as DL interval. Generally, characters always have the same width, so we sorted DL intervals with descending order as $W = \{\omega_1, \omega_2, \ldots, \omega_{M+1}\}$, and then the CW is calculated as,

$$CW = \frac{\sum_{\omega \in W'} \omega}{\|W'\|}, W' = \{\omega_k \| \omega_k - \omega_{k-1} < T, \omega_k - \Omega_{K+1} < T \qquad (3)$$

In fact, DL interval must be big enough compared with CW. Here, while a DL interval is less than 0.2CW, the right DL of this interval is removed as noise.

- For Chinese characters with left-right structure, there could be several false DLs located between character components (including two components with 0.5CW or components with 0.3CW and 0.7CW). Specifically, for the first case, if the width of two adjacent DL intervals are both greater than 0.4CW and less than 0.6CW, and also the PVE of the middle DL is smaller than that of both sides' DLs, we merge this two DL intervals. For the second case, if the width of any DL intervals is smaller than 0.3CW, we check that if its left/right DL interval is smaller than 0.9CW (normal width with 0.1CW offset), we merge these two DL intervals. Otherwise, these three DL intervals are merged together.
- We estimate CW again with the above steps corresponding to the new numbers of DLs. The width of DL intervals between these new DLs may be greater than 1.5CW. Therefore, a second segmentation which is similar to the above steps, is conducted in DL intervals whose width are larger than 1.5CW.

4 Text Extraction

Algorithm 1. Best Extraction Scheme Selection Mechanism.

Input: Text image I and clusters number K.
Output: Character binary images B_1 and B_2.
1: Classify pixels into initial classes based on mod operation of pixel index and K.
2: Select the average gray values of RGB channels in each class as initial centers.
3: **while** Not convergence or iteration smaller than maximum **do**
4:      Calculate the distance between each pixel and the K cluster centers.
5:      Classify pixel into cluster center with the smallest distance.
6:      Refresh the cluster centers with average value calculation.
7: **for** Each cluster do
8:      Calculate mean square deviation cd and the mean cm as in [22].
9:      Choose the two binary images B_1 and B_2 with the two smallest cds.
10: Output character binary images B_1 and B_2

Now, we get the final character DLs which accurately segment a whole video text row into several single characters, as shown in Fig. 2. We perform the K-means clustering [17] to retrieve several candidate binary images as shown in Algorithm 1. K is set to 4 in our previous work since a text image is composed of *text characters, contrast contours around characters, background* and *some noise points*. However, it will introduce false extraction results when the noisy points or the contrast contours around characters are not so obvious, for instance, some character points will forcibly be classified into noises or contours. Actually, we find that the binary conversion or clustering with $K = 3$ and 5 have more satisfactory performance sometime. Consequently, both $K = 5$, $K = 4$ and $K = 3$ based K-means clustering (as shown in Algorithm 1) as well as the binary conversion (namely, $K = 2$) are both applied to each text image.

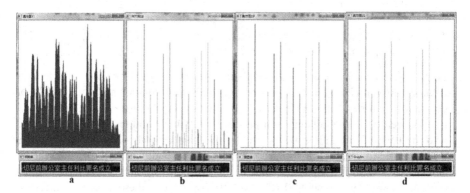

Fig. 2. Segmentation histogram of text segmentation. (a) Gradient Projection map, (b) Initial DLs, (c) First Segmentation Results, and (d) Final Character Divider Lines.

For all the clustering results and binary conversion, dam point labeling and inward filling [22] are used firstly. Then, variable cd [22] is used to select the binary images refer to characters for recognition. Considering the clustering bias, binary images corresponding to the two smallest cds are chose as candidate images.

So far, for each whole text or segmented single character image, two binary images have been got for a specific K. In fact, there are 13 binary images for recognition, including each 2 binary images for $K = 3$, $K = 4$, $K = 5$ on both whole text and segmented single characters, and also one for directly binary conversion.

5   Best Extraction Schemes Choosing

While 13 binary images are generated by different Ks, the confidence value is also given to each binary image. Then the binary image with the biggest confidence will be the best one for recognition. Specifically, as characters in the same row have nearly the same width, an assumption is made that the character width in each text row follows Gaussian distributions. And the confidence values are evaluated as follows. Let a text as character string $C \to c_i C$, where c_i denotes a character, and C means a string. As shown in Sect. 3, we denote the initial DL intervals as $W = \{\omega_1, \omega_2, \ldots, \omega_{M+1}\}$. Then, Gaussian distributions of the character width named C_ω is defined as,

$$C_\omega \sim P_{C_\omega}(x) \sim N(\mu, \sigma^2) \tag{4}$$

$$\mu = [\overline{\omega}] = \frac{1}{M+1} \sum_{i=1}^{M+1} \omega_i, \ \sigma^2 = \frac{1}{M+1} \sum_{i=1}^{M+1} (\omega_i - \overline{\omega})^2 \tag{5}$$

While a binary text image is input into *Hanvon OCR*, the feedback information: the total recognized characters number N and the dubious characters count D, are used. That is to say, the confidence value S of each binary image is calculated as,

$$S = P_{C_\omega}(N)e^{-D/N} = \frac{1}{\sqrt{2\pi}\sigma} e^{\frac{(N-\mu)^2}{2\sigma^2} - D/N} \tag{6}$$

The binary image with the biggest confidence S is chose as the best extraction result. Consequently, the corresponded best extraction binary image is input into OCR to successfully recognize the text.

Table 1. The video set used in our experiments.

Name	Duration (Seconds)	# Text Row
CCTV TV news	1503	208
Liaoning TV news	538	67
Petroleum TV news	311	46
Hunan TV news	476	70
Phoenix TV news	593	75

6 Experimental Results

To assess the performance of our approach, a large set of videos with different resolutions and characteristics is selected whose details are shown in Table 1.

6.1 Performance of Text Segmentation

The text segmentation performance is evaluated with: *Recall* = DT/(DT + LS) and *Precision* = DT/(DT + FA), where DT, LS, FA are the number of correct detects, loss detects and false alarms, respectively. The combination of recall and precision is also evaluated: *RP = Recall × Precision*, as shown in Table 2. Some segmentation examples can be seen in Fig. 3. Actually, the segmentation performance in video of *Phoenix News* are relatively weaker than the others. Because there are more English words and numbers combined with Chinese characters, causing the inconsistent character width.

6.2 Performance of Text Extraction

In order to evaluate the performance of the proposed approach, our approach is compared with: the Otsu method [8], the Lyu's method [7] and the Li's method [19].

Table 2. Text segmentation performance in test videos.

Test Videos	Recall (%)	Precision (%)	PR (%)
CCTV TV news	98.3	100	98.3
Liaoning TV news	97.9	98.1	96
Petroleum TV news	99	100	99
Hunan TV news	98.6	99.5	98.1
Phoenix TV news	95.3	96.4	91.8

The Otsu method is a simple but very classic solution, while Lyu method is robust to various background complexities and text appearances, and also broadly used for comparison in recent works. Meanwhile, the Li method is one of the most recent approach for text extraction. The Character Error Rates (CER) [7] is adopted to evaluate the extraction performance by *Hanvon OCR*. For the total of 4849 characters in 466 text rows, the comparisons on CERs of the four methods are shown in Table 3.

From Table 3 and Fig. 4, we can see that our approach get more satisfactory results than the other three methods. The main advantage of our approach is that not any single result generated by different way (using whole text or segmented characters and different value of K) is adopted as outputs alone, but we applied all these ways on the text image and chose the best one in term of the OCR feedback.

Fig. 3. Text segmentation results. The colorful image is the original one, the first binary image refer to the whole row, while the following binary images means the segmented characters.

Table 3. CERs evaluation of four text extraction methods.

	Chinese CER	English CER
Our method	0.122	0.073
Li's method [19]	0.187	0.151
Lyu's method [7]	0.201	0.160
Otsu's method [8]	0.690	0.411

6.3   Recognition Performance with Best Scheme Choosing

Without considering of the recognition performance, the text extraction evaluation is incomplete. Furthermore, the purpose of extraction is to recognize the text. Therefore, to assess the recognition performance of different methods, we also adopt the definition

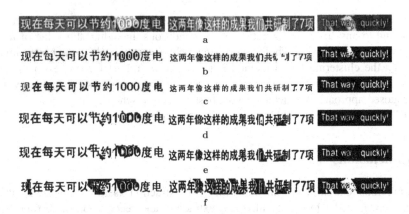

Fig. 4. Comparison of text extraction results: (a) the original text; (b) our whole row results; (c) our single character merged results; (d) Li's results; (e) Lyu's results; (f) Otsu's results.

of *Recall* and *Precision*, where *DT* is the number of correctly recognized characters, *LS* is the number of characters loss to be recognized, and *FA* is the number of false alarm in recognized characters by *Hanvon OCR* toolkit.

First, the recognition performance that applying the Hanvon OCR on the extracted texts of different methods, including the proposed method without using OCR feedback (*Without Feedback*), Otsu method [8], Lyu method [7], Li method [19] as well as our approach with OCR feedback (*With Feedback*), is shown in Table 4. Specifically, in *Without Feedback*, K = 4 is used directly for clustering, and the single character based extraction is adopted. From Table 4 we can see that without OCR feedback, the recognition performance is similar to those robust works of *Li's method* and *Lyu's method*. Besides, with unpredictable complex background, none of the method can extract perfect text for general text recognition.

Table 4. Comparison of text recognition performance.

Methods	Recall (%)	Precision (%)	RP (%)
With feedback	98.3	85.4	83.9
Without feedback	88.7	75.0	66.5
Li's method [19]	89.8	74.1	66.5
Lyu's method [7]	87.1	75.2	65.5
Otsu's method [8]	75.3	72.2	54.4

The recognition performance of our approach with OCR feedback is better than most of the previous methods. Our recognition recall even achieved 0.983, which means almost all of the characters are successfully recognized. However, the recognition precision is not so satisfactory, and the reasons include: the frame resolution is low, several noises are introduced in the detection and extraction steps, and also the OCR's recognizing capability is limited.

In addition, the best scheme with the most satisfactory recognition results does not correspond to any fixed configuration of the conditions, including the value of K and the extraction results using whole text or single characters as well as the binary image. Although the clustering based method is proved to be robust for text extraction, the binary image can sometimes generate more accurate recognition results. In other words, the proposed approach that determine the best extraction scheme with consideration of the OCR feedback is reasonable for satisfactory performance.

7 Conclusions

Most previous text extraction methods focus on extraction itself, but not the results of recognition. In order to make video text extraction and recognition more robust and general for different video conditions, we propose a robust text extraction and recognition approach using OCR feedback information. Specifically, we applied the K-means based color clustering on both whole text row, and segmented single characters. Firstly, we proposed an efficient character segmentation method. Then, we calculated the confidence value with the OCR feedback information. After that, the best extraction scheme is adaptively determined. Finally, we compared our approach with several typical methods, and the result shows that our approach is able to extract and recognize almost all the characters in the test videos efficiently and accurately.

References

1. https://www.youtube.com/yt/press/statistics.html
2. Zhang, D., Chang, S.: Event detection in basketball video using superimposed caption recognition. In: Proceedings of the ACM MM, pp. 315–318 (2002)
3. Zhang, D., Rajendran, R., Chang, S.: General and domain-specific techniques for detecting and recognizing superimposed text in video. In: Proceedings of ICIP, pp. I-593–I-596
4. Kim, H.H.: Toward video semantic search based on a structured folksonomy. J. Am. Soc. Inf. Sci. Technol. 62(3), 478–492 (2011)
5. Bhute, A.N., Meshram, B.B.: Text based approach for indexing and retrieval of image and video: a review. Adv. Vis. Comput. 1(1), 27–38 (2014)
6. Mitra, V., Franco, H., Graciarena, M., Vergyri, D.: Medium-duration modulation cepstral feature for robust speech recognition. In: Proceedings of ICASSP, pp. 1749–1753 (2014)
7. Lyu, M.R., Song, J., Cai, M.: A comprehensive method for multilingual video text detection, localization, and extraction. IEEE Trans. Circ. Syst. Video Technol. 15(2), 243–255 (2005)
8. Otsu, N.: A threshold selection method from gray-level histograms. IEEE Trans. Circ. Syst. Video Technol. 9(1), 62–66 (1979)
9. Leedham, G., Yan, C., Takru, K., Tan, J.H.N., Mian, L.: Comparison of some thresholding algorithms for text/background segmentation in difficult document images. In: Proceedings of ICDAR, pp. 859–864 (2003)
10. Ngo, C.W., Chan, C.K.: Video text detection and segmentation for optical character recognition. Multimedia Syst. 10(3), 261–272 (2005)
11. Kim, W., Kim, C.: A new approach for overlay text detection and extraction from complex video scene. IEEE Trans. Image Process. 18(2), 401–411 (2009)

12. Gao, J., Yang, J.: An adaptive algorithm for text detection from natural scenes. In: Proceedings of CVPR, pp. II-84–II-89 (2001)
13. Chen, D., Olobez, J.M., Bourlard, H.: Text segmentation and recognition in complex background based on Markov random field. In: Proceedings of ICPR, pp. 227–230 (2002)
14. Fu, H., Liu, X., Jia, Y., Deng, H.: Gaussian mixture modeling of neighbor characters for multilingual text extraction in images. In: Proceedings of ICIP, pp. 3321–3324 (2006)
15. Roy, A., Parui, S.K., Roy, U.: A pair-copula based scheme for text extraction from digital images. In: Proceedings of ICDA, pp. 892–896 (2013)
16. Lienhart, R., Wernicke, A.: Localizing and segmenting text in images and videos. IEEE Trans. Circ. Syst. Video Technol. **12**(4), 256–268 (2002)
17. Song, Y., Liu, A., Pang, L., Lin, S., Zhang, Y., Tang, S.: A novel image text extraction method based on k-means clustering. In: Proceedings of ICIS, pp. 185–190 (2008)
18. Li, X., Wang, W., Huang, Q., Gao, W., Qing, L.: A hybrid text segmentation approach. In: Proceedings of ICME, pp. 510–513 (2009)
19. Li, Z., Liu, G., Qian, X., Guo, D., Jiang, H.: Effective and efficient video text extraction using key text points. IET Image Process. **5**(8), 671–683 (2011)
20. Liu, Y., Song, Y., Zhang, Y., Meng, Q.: A novel multi-oriented Chinese text extraction approach from videos. In: Proceedings of ICDAR, pp. 1355–1359 (2013)
21. Sharma, N., Shivakumara, P., Pal, U., Blumenstein, M., Tan, C.L.: A new gradient based character segmentation method for video text recognition. In: ICDAR, pp. 126–130 (2011)
22. Huang, X., Ma, H., Zhang, H.: A new video text extraction approach. In: Proceedings of ICME 2009, pp. 650–653 (2009)
23. Shivakumara, P., Phan, T.Q., Tan, C.L.: A Laplacian approach to multi-oriented text detection in video. IEEE Trans. Pattern Anal. Mach. Intell. **33**(2), 412–419 (2011)
24. Huang, X., Ma, H., Yuan, H.: A novel video text detection and localization approach. In: Huang, Y.-M.R., Xu, C., Cheng, K.-S., Yang, J.-F.K., Swamy, M.N.S., Li, S., Ding, J.-W. (eds.) PCM 2008. LNCS, vol. 5353, pp. 525–534. Springer, Heidelberg (2008)

Color and Active Infrared Vision: Estimate Infrared Vision of Printed Color Using Bayesian Classifier and K-Nearest Neighbor Regression

Thitirat Siriborvornratanakul[✉]

Graduate School of Applied Statistics, National Institute of Development Administration (NIDA), 118 SeriThai Rd., Bangkapi, Bangkok 10240, Thailand
thitirat@as.nida.ac.th

Abstract. Speaking of active infrared vision, its inability to see physical colors has long been considered as one major drawback or something everybody has paid no attention to until very recently. Looking at this color blindness from other perspective, we propose an idea of a novel medium whose visibilities in both visible and active infrared light spectrums can be controlled, enabling vision-based techniques to transform everyday printed media into smart, eco-friendly and sustainable monitor-like interactive displays.

To begin with, this paper observes the most important key success procedure regarding the idea—estimating how physical colors should look like when being seen by an active infrared camera. Two alternative methods are proposed and evaluated here. The first one uses Bayesian classifier to find some color-attribute combinations that can precisely classify our sample data. The second alternative relies on simple weighted average and k-nearest neighbor regression in two color models—RGB and CIE L*a*b*. Suggesting by experimental results, the second method is more practical and consistent at different distances. Besides, it shows likelihoods of the model created in this work being able to estimate infrared vision of colors printed on different material.

Keywords: Color · Active infrared · Machine learning · Bayesian classification · K-nearest neighbor regression

1 Introduction

It is undeniable that digital media, like email, e-brochure, e-newsletter and website, have increasingly been used instead of traditional physical media like printed newspaper and letter. Nevertheless, the more we have got used to the digital media, the more we have realized that the physical printed media do own some unique strengths which cannot be replaced by the digital ones. While recent advances in pure digital media form factors have been proposed continuously, there are also many attempts to introduce an alternative reality combining both digital and physical media in a same space. Creating a real-time hybrid reality

© Springer International Publishing Switzerland 2015
Y.-S. Ho et al. (Eds.): PCM 2015, Part I, LNCS 9314, pp. 518–527, 2015.
DOI: 10.1007/978-3-319-24075-6_50

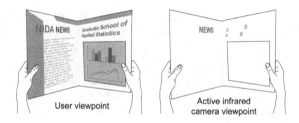

Fig. 1. Example usage scenario of a color printed pamphlet whose active infrared vision is controlled by careful color selection.

is nontrivial, particularly when referring to a self-contained mobile device used in an uncontrolled dynamic environment. One of the most challenging problems is probably to analyze the dynamic environment in real time and in a stealth manner so that the background analysis does not distract or interrupt users from their foreground tasks. In order to do this, camera is a popular choice of sensors as it is a self-contained sensor that can observe rich visual information of an environment regardless of any prior knowledge.

Unfortunately, vision-based analysis of a complete unknown environment is complicated and time consuming. Hence, using some kinds of invisible light to encode an invisible known pattern into the unknown environment is one common trick. Active infrared is an example of this trick that utilizes near-infrared light to invisibly light up or encode an unknown environment. Under active infrared vision, captured images are usually shown in greyscale which represents amount of infrared light reflected back through camera lens. To the best of our knowledge, relationship between colors (as seen by human) and their corresponding active infrared vision (as seen by an infrared camera) has barely been studied as it depends on too many unpredictable factors.

In this paper, we propose a study regarding printed colors and their active infrared visual appearances. Our goal is to overcome the long-time limitation of "color in active infrared vision" and find some mathematical models that can be used to estimate how an active infrared camera sees different printed colors. Results of this study will become useful for designing smart and sustainable hybrid media where traditional colorful printed media, like brochure and letter, can be used to display multimedia digital contents in real time (Fig. 1). In the long run, this study can be extended to the next level of visual data encryption where different printed colors are treated as different layers of visual data whose visibilities regarding human and infrared camera can be designed to be unique.

2   Related Works

Active infrared vision has long been used in commercial products as well as researches, particularly those relating to interactive image processing and real-time computer vision. In commercial products, the most well-known usages of active infrared are probably in night-vision surveillance cameras and in smart devices like Kinect Sensor for Xbox 360, Leap Motion and Smart TV. In researches, [10] fuses active infrared images with RGB images for the purpose of automatic skin enhancement. In [15], active infrared light is used to illuminate either background or foreground, solving the classic problem of background-foreground image segmentation in a highly dynamic background scene. In [16], the proposed goggle attached with near infrared LEDs helps prevent its wearer from being facial recognized by any passerby camera, using infrared light as invisible noises. Moreover, it is an active infrared technique that has been popular among interactive projection researches. Similar to commercial surveillance cameras, interactive projection systems in [7,12,13] use active infrared to illuminate their working environment or target object. In [1,2,4,6,7,14], active infrared is utilized the same way as Kinect sensor for Xbox 360 does—to encode an unknown environment or object with a known infrared pattern.

Using a low-cost infrared retro-reflective material is another trick for active infrared systems. As in [5] and [7] where an infrared retro-reflective material is used to visually emphasize the target object captured by infrared cameras, allowing easy visual detection and segmentation. Another use of active infrared includes light-activated invisible ink that cannot be seen under normal lighting but specific infrared lighting; this technique has been used for marking confidential documents, detecting counterfeit banknotes, marking museum visitors, etc. Researches using this invisible ink trick include interactive knit [9] and interactive storybook based on mobile projector [12].

Despite of a large number of vision-based researches using active infrared, most of them mention nothing about how physical colors behave under active infrared vision. Whenever a physical printed marker is used under active infrared vision, most of the times, the marker is pure binary (i.e., only black and white) as in [13]. When human skin is about to be detected under active infrared vision as in [2,7,13], previous researches show no concrete explanation about how human skin appears in the images; only the work of [10] describes that human skin (consisting of melanin and hemoglobin) barely absorbs infrared light and usually appears bright in infrared images. Recently there is an outstanding progress of color and active infrared exhibited in October 2014 at CEATEC2014. The exhibition named IR Color Vision (Sharp-LZ0P420A) was said to be the world first industrial camera that can see colored objects even in a pitch dark environment [3]. According to the brief detail revealed by developers, this camera uses a high-speed computation engine to analyze colorless infrared reflectance of each pixel and render the computed color back to a display monitor in real time.

In this paper, inspired by the work of Disney Research [12] and the world first color infrared camera [3], we focus on studying visual characteristics of physical colors when being seen by an active infrared camera.

3 Proposed Methods

To estimate how an active infrared camera sees physical colors, we start by narrowing down the scope and investigating only physical colors printed on general A4 papers. The current work consists of two separated methods—one using

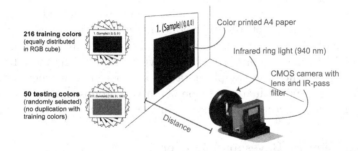

Fig. 2. Experimental setup for measuring active infrared reflectances of 216 training colors and 50 testing colors.

Bayesian classifier (Sect. 3.1) and the other using k-nearest neighbor (KNN) regression (Sect. 3.2). The Bayesian-based method predicts greyscale infrared values of unknown printed colors by pure machine learning of multiclass classification, whereas the KNN-based method tries to approximate those unknown values by referring to the known values of K closest neighbors.

The two proposed methods share the same sets of training and testing data collected at the same time in the same indoor environment. There are six sets of sample data collected and used in this work. Five are from five different distances based on a primary set of 80 gsm A4 papers, including the distances of 40, 57, 75, 92 and 110 cm. The sixth set is collected from the distance of 75 cm based on a secondary set of 80 gsm A4 papers sold by a different manufacturer. Most of our experiments were conducted based on the data collected from the primary set of papers at 75 cm unless specified otherwise. Note that all A4 papers used in our work are printed by the same printer (i.e., Ricoh Aficio SP C430DN color laser printer) using the same software, printing mode and computer.

To collect data used during training and testing, we set up the experiment as shown in Fig. 2. Each test color is printed on a common 80 gsm A4 white paper. The camera, coupling with lens and 940 nm infrared long pass filter, is rigidly fixed with an external 940 nm infrared ring light in a coaxial manner. For each experiment, the fixed module of camera and ring light is placed at a static distance from the test paper. Then for each test paper, our developed software performs the following sampling algorithm:

1. Capture five greyscale infrared images from the camera (this is to avoid transient noises in any image, if any).
2. For each captured image:
   2.1. Crop the region-of-interest (ROI) representing only the target color area.
   2.2. Compute average greyscale values of all pixels in the cropped ROI.
3. The final greyscale value of this printed color regarding 940 nm active infrared vision, is obtained by averaging the five values from step 2.2.

At this moment, there are 266 different RGB colors used in total: 216 colors as training data and 50 colors as testing data. By dividing the unsigned 8-bit RGB cube into 125 equal subcubes, 216 training colors are obtained in an equally distributed manner. The 50 testing colors are then randomly generated. Our reason of using RGB instead of CMYK color model, is for future convenience in developing vision-based interactive systems where RGB is the most frequently used color space. Once the sample data are all collected, another software is used to train, test and evaluate precision of our two proposed methods. Detailed explanation of each method can be found in the following two subsections.

3.1   Prediction by Bayesian Classifier

From Wikipedia [11], Naive Bays classifier or Bayesian classifier is "a simple probabilistic classifier based on applying Bayes' theorem with strong (naive) independence assumptions between the features." Despite of simplicity, Bayesian classifier can outperform other complicated machine learning method in many previous studies and most importantly it is capable of multiclass classification as required by our work. Our assumption here is that there might exist some combinations among popular color attributes where greyscale active infrared values of printed colors strongly depend on, allowing infrared appearance of each physical color to be predicted by Bayesian classifier.

To prove our assumption, three color models (i.e., RGB, HSV and YCrCb) are used, resulting in 9 color attributes: R (red), G (green), B (blue), H (hue), S (saturation), V (value), Y (luminance or greyscale value), Cr (red-difference) and Cb (blue-difference). Apart from R, G and B attributes specified by us, the other six color attributes are computed by OpenCV library [8]. Using these color attributes, 129 combinations are tested; this includes 9 one-attribute combinations, 36 two-attribute combinations, and 84 three-attribute combinations. As for the target classes of classification, ideally, there should be 256 classes representing all possible greyscale values of unsigned 8-bit images (i.e., 0–255). Nevertheless, to allow some flexibility in classification, instead of assigning one greyscale value per one target class, we vary the number of greyscale values per class ($nval$) from 1 to 10, resulting in the number of target classes varied from 26 (when $nval$ equals 10) to 256 (when $nval$ equals 1). Each target class is represented by an integer number, a.k.a. class number, started from 1. The bigger the class number, the bigger the greyscale value(s) in that class.

3.2   Regression by K-Nearest Neighbors

In this method, we apply the concept of KNN regression to locally approximate an unknown value from 8-nearest known training data. As mentioned earlier, our training data consist of 216 RGB colors that are equally distributed in the unsigned 8-bit RGB cube. Therefore, finding the 8-nearest neighbors means finding the RGB subcube in where the unknown value resides. Nevertheless, relying solely on the closest distances computed in RGB color space, which is not perceptually uniform, may not be a good idea. Hence, in our work, the KNN regression is applied in both RGB and perceptually uniform CIE L*a*b* color spaces. In summary, there are six modes of computation used in this method:

1. The unknown value is computed by equally averaging values of the 8-closest neighbors found in RGB color space.
2. The unknown value is computed by equally averaging values of the 8-closest neighbors found in CIE L*a*b* color space.
3. The unknown value is computed by weighted averaging values of the 8-closest neighbors found in RGB color space. The smaller the Euclidean distance in RGB space, the larger the assigned weight.
4. The unknown value is computed by weighted averaging values of the 8-closest neighbors found in CIE L*a*b* color space. The smaller the Euclidean distance in CIE L*a*b* space, the larger the assigned weight.
5. The unknown value is computed by weighted averaging values of the 8-closest neighbors found in RGB color space. The smaller the Euclidean distance in CIE L*a*b* space, the larger the assigned weight.
6. The unknown value is computed by weighted averaging values of the 8-closest neighbors found in CIE L*a*b* color space. The smaller the Euclidean distance in RGB space, the larger the assigned weight.

4   Experimental Results

All experiments were conducted on a HP Pavilion laptop with Intel(R) Core(TM) i7-4510U CPU running at $2\,GHz$ on $8\,GB$ RAM. The main developer tools include Qt 5.4.0 MSVC2013 OpenGL 32bit and OpenCV 2.4.10 libraries, running in Windows 8.1 Enterprise N 64-bit operating system. For each experiment, four indicators are used for evaluation:

- **Correct percentage (%):** this indicator shows the percentage of correct classifications or estimations. Subtracting this indicator from 100 equals to the percentage of incorrect test results whose detailed imprecisions are refined by other two indicators named $missAvg$ and $missSD$.

$$correctPercent = \frac{N_{correct} * 100}{N_{total}}, \tag{1}$$

where the number of all testing data or N_{total} equals $(N_{correct} + N_{miss})$.
- **Average miss (class):** this indicator refines how much the overall errors are in average. The computation is

$$missAvg = \frac{\sum_{i=1}^{N_{total}} |A_i.trueClass - A_i.computeClass|}{N_{miss}}, \tag{2}$$

where A_i represents an ith test datum.
- **SD miss (class):** Similar to the above indicator, this indicator represents a standard deviation of all errors.

$$missSD = \sqrt{\frac{\sum_{i=1}^{N_{total}} (|A_i.trueClass - A_i.computeClass| - (missAvg))^2}{N_{miss}}}, \tag{3}$$

where A_i represents an ith test datum.

– **Time (ms):** this indicator represents an average time used per one prediction or estimation, excluding the time used for model creation.

4.1   Prediction by Bayesian Classifier

To conclude the experimental results regarding each indicator, we start with the *correctPercent* indicator based on data collected from the primary set of papers at 75 cm distance. Among 129*10=1,290 tested combinations, we obtain the highest *correctPercent* of 60 % when G, H and V color attributes are used together at $nval = 9$. Looking more closely at the experimental results (not shown here due to page limitation), it is quite obvious that the V attribute from HSV color model occupies majority of the best results selected from one-, two- and three-attribute tests. Besides, it is noticeable that, most of the times, the bigger the value of $nval$ is, the more likely the *correctPercent* grows; this is neither a coincidence nor a discovery because a big $nval$ allows an internal error within a class (i.e., one target class represents many greyscale values) that cannot be measured by this *correctPercent* indicator.

Having a high value of *correctPercent* definitely represents an accurate classifier. However, results with small *correctPercent* but small errors (i.e., small *missAvg*) can be portrayed as good classification as well. Because the value of 1 in our *missAvg* and *missSD* indicators represents an error of one class whose maximum number of class members is $nval$, we look forward to finding the results with small $(missAvg * nval)$ and $(missSD * nval)$; this is to make the compared data more similar to actual infrared greyscale values. By collecting the minimum $(missAvg * nval)$ values from all 1,290 tests, surprisingly the smallest $(missAvg * nval)$ of 9.18 is when G and V attributes are used together at $nval = 1$. Again, it is the V attribute that contributes to this best result.

After all these experimental results, the V color attribute seems to play an important role in predicting an unknown infrared greyscale value of a printed color using Bayesian classifier. To decide which one of the two selected combinations (i.e., V and GV) to use at the distance of 75 cm, it probably depends on what is a system's first priority. Unfortunately, when we use the two selected combinations to classify data collected from other distances, the results in Table 1 are not impressive as the successfulness at 75 cm does not seem to repeat itself at other distances; there is barely connection or behavior that can be used as norms for future classification. Our conclusion here is that, Bayesian classifier alone is not enough for predicting greyscale values of printed colors observed by an active infrared camera at arbitrary distances.

4.2   Regression by K-Nearest Neighbors

In this method, a greyscale value of active infrared vision regarding an unknown printed color is estimated by KNN regression, which means that $nval$ always equals to 1 and there is no concern about class's internal error. Table 2 shows the experimental results. Comparing to Bayesian method, the *correctPercent* values are small here. But in substitution for that, this regression method produces

Table 1. Bayesian classifier's experimental results at five distances on the primary set of papers. (GHV) and (GV) are the color-attribute combinations that provide the highest *correctPercent* and the lowest (*missAvg* * *nval*) respectively, during previous experiments at distance of 75 cm.

	nval	distance (cm)	correctPer	missAvg	missSD	missAvg * nval	missSD * nval	time (ms)
(GHV)	9	40	78 %	4.61	4.61	37.62	41.49	6.97
		57	10 %	2.64	1.91	23.76	17.19	8.13
		75	* 60 % *	**1.65**	**1.24**	**14.85**	**11.16**	**5.45**
		92	56 %	1.05	0.21	9.45	1.89	3.43
		110	30 %	1.29	0.66	11.61	5.94	2.98
(GV)	1	40	54 %	17.22	34.20	17.22	34.20	14.54
		57	4 %	19.69	23.48	19.69	23.48	27.96
		75	**0 %**	**9.18**	**8.95**	* 9.18 *	**8.95**	**19.41**
		92	10 %	8.13	10.57	8.13	10.57	17.54
		110	6 %	5.38	4.05	5.38	4.05	14.86

small errors as all *missAvg* values are approximately 5 regardless of the color models used for KNN and the Euclidean distance computation. In addition, it can be seen that accuracy results from the six modes are hardly different; only two modes that relatively use more time than the others because they involve color conversion between two color models.

Based on the primary set of papers, we repeat the same experiment (as in Table 2) on data collected from other four different distances. This time, they all show similar behaviors, which is, small values of *missAvg* (maximum values of 12 occur at the distances of 40 and 57 cm). Hence, for the distances with available training data, it should be acceptable to apply this method for estimating active infrared greyscale values regarding unknown printed colors. However, estimating values observed from unknown distances is beyond the scope of this paper and has not been studied yet.

Next, to prove whether the KNN model created by data from the primary set of papers can or cannot be used to estimate active infrared greyscale values of colors printed on another type of similar papers (a.k.a. the secondary set of papers), another experiment was conducted and its results are shown in Table 3. Comparing to Table 2, the *missAvg* values become about three times bigger. When looking more closely at the greyscale values observed from both sets of papers, we found that 99.62 % of the values from secondary papers are larger than their correspondences in primary papers, and absolute differences between the two sets have a mean of 13.97 with standard deviation of 5.28. By simply subtracting all original values observed from the secondary set of papers with 13.97, the absolute differences between the two sets of data are significantly reduced to the mean of 4.34 and the standard deviation of 3.01. Suggesting by these results, our conclusion is that using an unknown type of papers with the prebuilt KNN regression model is possible but an additional compensation is required.

Table 2. KNN regression's experimental results at the distance of 75 cm. The data used for training and testing here are sampled from the primary set of papers.

Weighted by:	None		RGB's distance		CIE L*a*b*'s distance	
Data in RGB space	$correctPer$ (%)	6.00	$correctPer$ (%)	8.00	$correctPer$ (%)	10.00
	$missAvg$	5.30	$missAvg$	5.24	$missAvg$	5.53
	$missSD$	3.50	$missSD$	4.35	$missSD$	4.04
	$time$ (ms)	4.52	$time$ (ms)	4.69	$time$ (ms)	8.56
Data in CIE L*a*b* space	$correctPer$ (%)	4.00	$correctPer$ (%)	8.00	$correctPer$ (%)	6.00
	$missAvg$	5.31	$missAvg$	5.63	$missAvg$	5.53
	$missSD$	3.95	$missSD$	4.28	$missSD$	4.35
	$time$ (ms)	4.37	$time$ (ms)	8.49	$time$ (ms)	4.47

Table 3. KNN regression's experimental results at the distance of 75 cm. The KNN model created by data collected from the primary set of papers are used to perform regression on data collected from the secondary set of papers.

Weighted by:	None		RGB's distance		CIE L*a*b*'s distance	
Data in RGB space	$correctPer$(%)	0.75	$correctPer$(%)	0.38	$correctPer$(%)	0.00
	$missAvg$	14.56	$missAvg$	15.04	$missAvg$	14.65
	$missSD$	6.52	$missSD$	5.74	$missSD$	5.94
	$time(ms)$	4.53	$time(ms)$	4.37	$time(ms)$	8.16
Data in CIE L*a*b* space	$correctPer$(%)	1.13	$correctPer$(%)	0.00	$correctPer$(%)	0.00
	$missAvg$	14.87	$missAvg$	14.83	$missAvg$	14.64
	$missSD$	6.48	$missSD$	5.88	$missSD$	5.87
	$time(ms)$	4.36	$time(ms)$	8.29	$time(ms)$	4.34

5   Conclusion

In this paper, we present two methods whose shared goal is to connect physical colors printed on plain A4 papers with their greyscale values when being seen by an active infrared camera. All experimental data are sampled from two sets of white A4 papers printed with 266 RGB colors and placed at the predefined five distances. The first method of machine learning with Bayesian classifier presents some interesting characteristics around the V attribute of HSV color model; however, these characteristics do not repeat when the distance changed. The second method relies on k-nearest neighbor regression that, although requires more memory storage because of its lazy learner nature, is more consistent and utilizable with additional compensation.

Future works include observing alternative machine learning methods that might possibly provide some usable and consistent results, extending the work on the KNN regression to cover compensation and estimation of those at an unknown distance, and finally, creating an interactive color-printed medium based on the fully developed concept of designable active infrared vision.

References

1. Akasaka, K., Sagawa, R., Yagi, Y.: A sensor for simultaneously capturing texture and shape by projecting structured infrared light. In: Proceedings of the 6th International Conference on 3-D Digital Imaging and Modeling, pp. 375–381 (2007)
2. Harrison, C., Benko, H., Wilson, A.: Omnitouch: wearable multitouch interaction everywhere. In: Proceedings of the 24th Annual ACM Symposium on User Interface Software and Technology (UIST'11), pp. 441–450 (2011)
3. Hornyak, T.: Sharp's security camera captures color video when it's pitch black. http://www.pcworld.com/article/2843032/sharps-color-security-camera-shoots-in-the-dark.html (4 November 2014) Accessed on 10 February 2015
4. Jones, B., Benko, H., Ofek, E., Wilson, A.: Illumiroom: Peripheral projected illusions for interactive experiences. In: Proceedings of the SIGCHI Conference on Human Factors in Computing Systems (CHI 2013), pp. 869–878 (2013)
5. Kim, D., Izadi, S., Dostal, J., Rhemann, C., Keskin, C., Zach, C., Shotton, J., Large, T., Bathiche, S., NieBner, M., Butler, D., Fanello, S., Pradeep, V.: Retrodepth: 3d silhouette sensing for high-precision input on and above physical surface. In: Proceedings of the SIGCHI Conference on Human Factors in Computing Systems (CHI 2014), pp. 1377–1386 (2014)
6. Lee, J., Hudson, S., Dietz, P.: Hybrid infrared and visible light projection for location tracking. In: Proceedings of the 20th Annual ACM Symposium on User Interface Software and Technology (UIST 2007), pp. 57–60 (2007)
7. Molyneaux, D., Izadi, S., Kim, D., Hilliges, O., Hodges, S., Cao, X., Butler, A., Gellersen, H.: Interactive environment-aware handheld projectors for pervasive computing spaces. In: Kay, J., Lukowicz, P., Tokuda, H., Olivier, P., Krüger, A. (eds.) Pervasive 2012. LNCS, vol. 7319, pp. 197–215. Springer, Heidelberg (2012)
8. OpenCV: Open source computer vision. http://opencv.org/
9. Rosner, D., Ryokai, K.: Weaving memories into handcrafted artifacts with spyn. In: Proceedings of the SIGCHI Conference on Human Factors in Computing Systems (CHI 2008), pp. 2331–2336 (2008)
10. Susstrunk, S., Fredembach, C., Tamburrino, D.: Automatic skin enhancement with visible and near-infrared image fusion. In: Proceedings of the 18th ACM International Conference on Multimedia (MM 2010), pp. 1693–1696 (2010)
11. Wikipedia: Naive bayes classifier. http://en.wikipedia.org/wiki/Naive_Bayes_classifier Accessed on 22 April 2015
12. Willis, K., Shiratori, T., Mahler, M.: Hideout: Mobile projector interaction with tangible objects and surfaces. In: Proceedings of the 7th International Conference on Tangible, Embedded and Embodied Interaction (TEI 2013), pp. 331–338 (2013)
13. Wilson, A.: Playanywhere: a compact interactive tabletop projection-vision system. In: Proceedings of the 18th Annual ACM Symposium on User Interface Software and Technology (UIST 2005), pp. 83–92 (2005)
14. Wilson, A., Benko, H., Izadi, S., Hilliges, O.: Steerable augmented reality with the beamatron. In: Proceedings of the 25th Annual ACM Symposium on User Interface Software and Technology (UIST 2012), pp. 413–422 (2012)
15. Wu, Q., Boulanger, P., Bischof, W.: Bi-layer video segmentation with foreground and background infrared illumination. In: Proceedings of the 16th ACM International Conference on Multimedia (MM 2008), pp. 1025–1026 (2008)
16. Yamada, T., Gohshi, S., Echizen, I.: Use of invisible noise signals to prevent privacy invasion through face recognition from camera images. In: Proceedings of the 20th ACM International Conference on Multimedia (MM 2012), pp. 1315–1316 (2012)

Low Bitrates Audio Bandwidth Extension Using a Deep Auto-Encoder

Lin Jiang[1,3], Ruimin Hu[1,2(✉)], Xiaochen Wang[1,2], and Maosheng Zhang[1]

[1] National Engineering Research Center for Multimedia Software,
Computer School of Wuhan University, Wuhan, China
`{jlcdf,hrml964,clowang,eterou}@163.com`
[2] Research Institute of Wuhan University, Shenzhen, China
[3] Software School of East China Institute of Technology, Nanchang, China

Abstract. Modern audio coding technologies apply methods of bandwidth extension (BWE) to efficiently represent audio data at low bitrates. An established method is the well-known spectral band replication (SBR) that can provide the very high sound quality with imperceptible artifact. However, its bitrates and complexity are very high. Another great method is LPC-based BWE, which is part of 3GPP AMR-WB+ codec. Although its bitrates and complexity are reduced distinctly, the sound quality it provided is unsatisfactory for music. In this paper, a novel bandwidth extension method is proposed which provided the high sound quality close to eSBR, with only 0.8 kbps bitrates. The proposed method predicts the fine structure of high frequency band from low frequency band by a deep auto-encoder, and only extracts the envelope of high frequency as side information. The performance evaluation demonstrates the advantage of the proposed method compared to the state of the art. Compared with eSBR, the bitrates drop about 63 %, and the subjective listening quality is close to it. Compared with LPC-based BWE, the subjective listening quality is better than it with the same bitrates.

Keywords: Bandwidth extension · Audio compression · Deep auto-encoder

1 Introduction

Bandwidth extension (BWE) is a standard technique within contemporary audio codecs to efficiently code audio signals at low bitrates. The main idea of bandwidth extension is to exploit correlations between the low frequency band (LF) and the high frequency band (HF) of the signal [1]. There are two main categories of BWE methods: blind bandwidth extension and non-blind bandwidth extension. In blind bandwidth extension, the HF signals are reproduced utilize LF signals on the decoder side without any side information. In non-blind bandwidth extension, the high frequency band is represented by additional side information parameters which are set for the HF band on the encoder side, on the decoder side, the HF band is reproduced by shifting the LF band into the HF band by additional post processing procedure. In this paper, we only discuss the non-blind BWE.

© Springer International Publishing Switzerland 2015
Y.-S. Ho et al. (Eds.): PCM 2015, Part I, LNCS 9314, pp. 528–537, 2015.
DOI: 10.1007/978-3-319-24075-6_51

In non-bandwidth extension, there are two methods: time domain BWE [2, 3] and frequency domain BWE [4–7]. The time domain methods generate the HF signal by up sampling the LF signal or by LPC-based estimation. Both methods double the bandwidth of the LF signal. However, the frequency domain methods, compared with bandwidth of the LF signal, the bandwidth of the generated HF signal is wider or narrower. In addition, the frequency-dependent of the generated signal can be adjusted. Due to these merits, the frequency domain methods are suitable for SWB codecs where the extended bandwidth should be wide. Therefore, we focus on the frequency domain methods in this paper. Spectral Band Replication (SBR) [4, 5] is the most widely used frequency domain BWE method, and is used in MPEG-4 High-Efficiency AAC (HE-AAC) [7] and AAC Enhanced Low Delay (AAC-ELD) [8], and its enhanced version (eSBR) is applied by the MPEG USAC codec [9]. Because the MPEG-SBR module uses a 64 channel QMF decomposition to obtain the higher band parameters, for 32 kHz signal, its algorithmic delay reaches about 30 ms and side information data rate reaches about 3.20 kbps, beyond that, its algorithm complexity is very high. This is too much for many time-critical applications like as real-time speech communications. In fact, SBR technology uses more side information to obtain the high sound quality of construction of high frequency part.

A recent advance in training methods for multilayer neural networks has motivated renewed interest in exploring deep, multilayer networks for a number of machine learning problems including encoding, retrieval [10], as well as the problems associated with classification and regression that involves image, language and speech [11–13]. The deep networks are first pre-trained, one layer at a time, to form a good generative model of the input data. After pre-training has discovered multiple layers of non-linear features that are good at capturing the structure in the input domain, the whole network is discriminatively fine-tuned. The discriminative fine-tuning can make the network perform good classification or regression, but also can make the network be good at reconstructing its input from a compact code. In [14], a deep auto-encoder is used to code the binary speech spectrograms, experimental results demonstrate that this binary codes produce a log-spectral distortion lower than a conventional subband vector quantization technique.

Our goal for this paper is to achieve a BWE system in which the bitrates audio codecs are reduced distinctly while sound quality is held. To achieve this purpose, we use a deep auto-encoder to predict the fine structure of high frequency component, and then extract the spectral envelope of high frequency component, the high frequency signals are reconstructed on the decoder side by the spectral envelope and fine structure.

2   Proposed Methodology

2.1   Motivation

In existed frequency domain BWE, the HF is synthesized with the spectral envelope and fine structure. Generally, the spectral envelope of HF is represented by subband energy benefit from its high quantification performance, the fine structure is represented by LF component because of the correlation between HF and LF, and the generation of HF is shown in Fig. 1. In the decoder the LF are transposed to the HF (Fig. 1(b)), and using the envelope parameters of HF to shape their level so that the level of the new HF is similar to the original HF (Fig. 1(c)). In Fig. 1(b), the reason is, for doing this, that there is correlation between LF and HF. However, in most case, their similarity of tonality (or harmonic) between LF and HF is not high (see the right of Fig. 1(a, b)). Hence, tonality and noise compensation is applied in order to adjust the harmonic structure of HF (Fig. 1(d)).

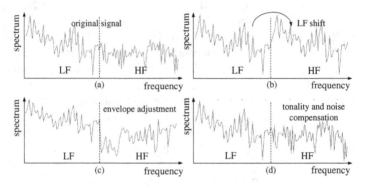

Fig. 1. Principle of HF generation

In Fig. 1(d), the ability of harmonic adjustment is limited, and there are some artifacts, such as tone trembling, tone shift, tone spike and noise overflow [14]. What's more, the tonality and noise parameters resulted in increasing bitrates about 50 %. In fact, in frequency domain, when the envelope of HF is represented by frequency subband energy, the fine structure is important cue for sound perception [15]. Unfortunately, this property is neglected in existed BWE. In this paper, we try to obtain the fine structure of HF replace the LF component in Fig. 1(b), in order to reduce bitrates, we also remove the module of tonality and noise compensation (Fig. 1(d)). For the new fine structure, the level of harmonic or tonality is more accordant with the original. Take account of the correlation between LF and HF, and the power model ability of deep neural network. We predict the fine structure of HF from the LF by a deep auto-encoder.

2.2   Proposed BWE Method

The proposed BWE system is shown in Fig. 2. The input signal sampled at 25.6 kHz is processed in 20 ms per frame. A 1st order IIR high-pass filter is applied to the 25.6 kHz sampled input signal to remove 0-50 Hz components. The pre-processed signal is divided into two 6.4 kHz sampled low frequency band (LF 0 ~ 6.4 kHz) and high frequency band (HF 6.4 ~ 12.8 kHz) using an analysis filterbank.

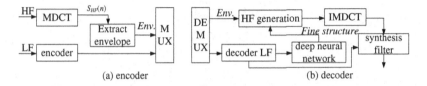

Fig. 2. Block diagram of the proposed method

In encoder side (Fig. 2(a)), the high frequency signals are transformed in the frequency domain by the well-known MDCT using a 50 % overlapped sinusoidal window, $S_{HF}(n)$. The $S_{HF}(n)$ is further divided into 8 subbands uniformly. We extract the spectrum envelope of HF by Eq. 1:

$$Env(i) = \sum_{j=1}^{M} \left(S_{HF}((i-1)M+j)\right)^2, \qquad (1)$$

where $Env(i)$ denotes the energy of ith subband in current frame and M is the subband length. The envelope parameters are quantized by vector quantization method with 16 bits.

In decoder side (Fig. 2(b)), the HF signals are synthesized by decoded envelope parameters and the predicted HF fine structure from the decoded LF by deep neural network. The final HF signals are generated by Eq. 2:

$$\hat{S}_{HF}(i,j) = \hat{E}nv(i) * Fine\_stru(i,j)$$
$$1 \le i \le 8, \quad 1 \le j \le M, \qquad (2)$$

where $\hat{S}_{HF}(i,j)$ denotes generated HF signal the jth point of ith subband in current frame, $Fine\_stru(i,j)$ is the normalized HF fine structure the jth point of ith subband, and $\hat{E}nv(i)$ is the decoded energy of ith subband. The fine structure is important for perception sound quality, and would be described in Sect. 3.

3   Prediction of Fine Structure Using Deep Auto-Encoder

3.1   Auto-Encoders

A basic auto-encoder (AE), which consists of only one hidden layer, is trained to reconstruct an input. So the learner must capture the structure of the input distribution. However, for many complicated signals, the stacked auto-encoder (SAE, also be called deep auto-encoder) is applied to represent the structure of signals. The architecture of a SAE is given in Fig. 3.

Formally, an input example $x \in R^n$ is input to the first hidden layer h_1, the hidden representation $h_1(x) \in R^{m1}$ is

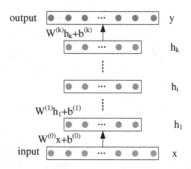

Fig. 3. Stacked auto-encoder (SAE) architecture

$$h_1(x) = f(W^{(0)}x + b^{(0)}), \tag{3}$$

where $f(\cdot)$ is a non-linear activation function, typically a logistic sigmoid function applied component-wise, $w^{(0)} \in R^{m1 \times n}$ is a weight matrix, and $b^{(0)} \in R^{m1}$ is a bias vector, it is easily found that the topology structure of the auto-encoder completely relies on the size of the input layer k and the number of each hidden units m_i. For others layer, the input is the output of previous layer, and the output of current layer is used for the input of next layer. The each hidden layer representation $h_i(x) \in R^{m_i}$ is

$$h_2(h_1) = f(W^{(2)}h_1 + b^{(2)})$$
$$\vdots$$
$$h_i(h_{i-1}) = f(W^{(i)}h_{i-1} + b^{(i)}) \tag{4}$$
$$\vdots$$
$$h_k(h_{k-1}) = f(W^{(k)}h_{k-1} + b^{(k)}),$$

where k denotes the number of hidden layer, and m_i is the number of each layer hidden units.

The network output maps the hidden representation h_k back to a reconstruction $y \in R^n$:

$$y = f(w^{(k)}h_k + b^{(k)}). \tag{5}$$

Given a set of input sample χ, the SAE training consists of finding parameters $\theta = \{W^{(0)}, W^{(1)}, \ldots, W^{(k)}, b^{(0)}, b^{(1)}, \ldots, b^{(k)}\}$ which minimize the reconstruction error. This corresponds to minimizing the following objective function:

$$\zeta(\theta) = \frac{\lambda}{2}\left(\sum_{l=0}^{n}\sum_{j}\left\|w_j^{(l)}\right\|^2\right) + \sum_{x \in \chi}\|x - y\|^2, \tag{6}$$

where λ denotes a regularization parameter. Here, we also include a weight-delay regularization term with its hyper-parameter to the objective function to avoid over-fitting. $w_j^{(l)}$ is the jth column vector of the lth layer weight matrix $W^{(l)}$. The minimization is usually realized either by stochastic gradient descent or more advanced optimization techniques such as L-BFGS or conjugate gradient method.

3.2 Prediction of Fine Structure

In this paper, our main goal is to predict the fine structure of HF by the deep auto-encoder from decoded LF. The motivation is the power model ability of deep neural network and the correlation between LF and HF.

In Fig. 3, input x is replaced by the decoded low band normalized MDCT coefficient LF ($\hat{s}_{LF}$). The output y is just the predicted value of the fine structure of HF (*Fine_stru*). For doing so, the parameters $\theta = \{W^{(0)}, W^{(1)}, \ldots, W^{(k)}, b^{(1)}, b^{(2)}, \ldots, b^{(k)}\}$ must be obtained by training the deep auto-encoder. We used greedy layer-wise training to obtain the parameters. To do this, first train the first layer on raw input ($\hat{s}_{LF}$) to obtain parameters $W^{(1,1)}, W^{(1,2)}, \ldots, W^{(1,n)}, b^{(1,1)}, b^{(1,2)}, \ldots, b^{(1,n)}$, ($W^{(i,j)}$ denotes the weight of the jth input element of the ith layer, $b^{(i,j)}$ denotes the bias item of the jth input element of the ith layer). Use the first layer to transform the raw input into a vector consisting of activation of the hidden units. Train the second layer on this vector to obtain parameters $W^{(2,1)}, W^{(2,2)}, \ldots, W^{(2,m1)}, b^{(2,1)}, b^{(2,2)}, \ldots, b^{(2,m1)}$, ($m_1$ is the number of hidden unit). Repeat for subsequent layers, using the output of each layer as input for the subsequent layer. This method trains the parameters of each layer individually while freezing parameters for the remainder of the model. To produce better results, after this phase of training is complete, fine-tuning using back propagation can be used to improve the results by tuning the parameters of all layers are changed at the same time.

After the parameters are obtained by training, the fine structure of HF can be predicted from the decoded normalized LF signals by the deep auto-encoder. Specifically, the decoded normalized LF MDCT coefficient is input the deep auto-encoder, then the forward propagation algorithm is implemented, the output of the deep auto-encoder is the prediction of fine structure of HF.

4 Experiments and Evaluation

In this section, our goal is to obtain the deep auto-encoder network architecture by training that will help us to predict the fine structure of HF, and we also will evaluate the performance of the proposed BWE system.

4.1 Training Auto-Encoders

In our experiments, the audio database is used in order to obtain the architecture of deep neural network. The audio database including TIMIT speech, natural sounds and music, which in total about 1.2 million frames (0.2 ms per frames, 1 million frames for training, 0.2 million frames for testing). Due to only evaluated our BWE system, so the

MDCT coefficients of original low frequency (not specific low frequency codec) and the original fine structure of HF are extracted. The training and testing dataset was normalized to have a standard deviation of 10. Normalization is essential to prevent saturation of hidden units, which produces small learning signal. A standard deviation of 10 was found to give lower percentage reconstruction error than other values, with the parameter initializations we were using.

The structure of the stacked auto-encoders was (256, 128, 64, 64, 128, 256) with a mini-batch size of 16 training samples. The number of epoch for each layer of single auto-encoder pre-training was 20. Learning rate of pre-training was 0.0005. For the fine-tuning, learning rate was set as 0.1 for the first 10 epochs, and then decreased by 10 % after every epoch. Total number of epoch was 50. During the weight updates, we apply a 0.9 momentum and 0.0001 weight decay.

4.2   The Results of Implement

The results of predicted HF fine structure are shown in Fig. 4(a). In original HF fine structure (red), there are rich harmonic structure (10 ~ 12 kHz). In LF fine structure, the correlation is weak compare with HF fine structure, the predicted HF fine structure more match with original HF fine structure on harmonic structure, which provides the guarantee for get high quality HF generation signals. The HF rebuild signals are shown in Fig. 4(b). For the HF rebuild signal that didn't used DNN (Fig. 4(b) middle), which have distinct distortion on harmonic (10 ~ 12 kHz). For our results (Fig. 4(b) bottom), the property of tonality is well maintained. From our results, this confirmed the validity of the proposed BWE method.

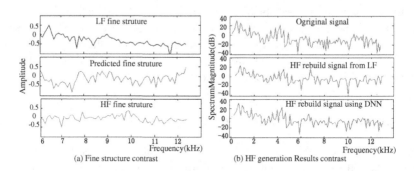

Fig. 4. Results contrast of proposed method

4.3   Performance Evaluation

In order to validate our method, we compare proposed method with the state-of-the-art methods, eSBR [6] and LPC-based BWE [3], where the eSBR is an enhanced version of SBR which is applied by the newest MPEG USAC codec [9], the LPC-based is a famous BWE method at low bitrates. The performance evaluation is implemented, including: bitrate comparison, subjective and objective testing with eSBR and

LPC-based BWE. The proposed method, eSBR and LPC-based BWE were applied to a lossless LF signal, in order to assess the performance of just the BWE part, independent of the performance of the core codec.

In our method, there is only energy parameters (16 bit per 0.2 ms frame, which is same with LPC-based BWE) are transmitted to the bit stream, the bitrate is 0.8 kbps. For eSBR, the envelope and control parameters are extracted, and then the parameters are quantized by Huffman coding, its bitrate is flexible, about 2.2 kbps in our experiments. Compare with eSBR, the bitrate dropt about 63 % ((2.2 − 0.8)/2.2 * 100 %).

We carry out subjective tests to assess the quality of the proposed BWE system using MUSHRA method with a five grade impairment scale ranging from 1.0 (very annoying) to 5.0 (imperceptible) [16]. Fifteen expert listeners participate in the test, which is performed for two categories, that is, speech and music/mixed contents, and test materials derived from MEPG standard. Figure 5 shows a summary of the subjective test results. As shown in the figure, for all cases, the quality provided by the proposed BWE system is slightly lower than eSBR, and is better than LPC-based BWE. Noticeably, the proposed BWE system outperforms the LPC-based BWE, and we think our method is close to eSBR because the difference SDG score is only 0.06 (4.77 − 4.71) at average.

The objective performance of the proposed BWE system is also evaluated by the ITU-R BS.1387 PEAQ test. The average Objective Difference Grade (0 ∼ −5, form imperceptible to very annoying) is −1.71 (LPC), −1.25 (proposed), −1.11 (eSBR) respectively. The test results have been shown to highly correlate with the SDG score (Fig. 5) from a subjective listening test.

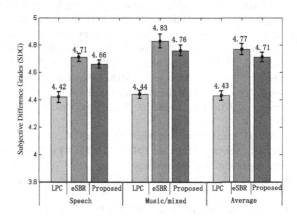

Fig. 5. Subjective test results

5 Conclusions

The work presented here show that performing bandwidth extension under very low bitrate (0.8 kbps) becomes available, and the same time obtains the very high sound quality (close to eSBR). The result is attributed to deep auto-encoder which has good function for predicting the fine structure. From the results of performance evaluation, the proposed method reaches nearly the sound quality of eSBR provided, but the bitrate dropt about 63 % compare with eSBR. Compare with LPC-based BWE, the sound quality of the proposed method is better.

Acknowledgments. The research was supported by National Nature Science Foundation of China (No. 61231015); National High Technology Research and Development Program of China (863 Program) No. 2015AA016306; National Nature Science Foundation of China (No. 61102127, 61201340, 61201169, 61471271), Guangdong-Hongkong Key Domain Breakthrough Project of China (No. 2012A090200007), and The major Science and Technology Innovation Plan of Hubei Province (No. 2013AAA020).

References

1. Larsen, E., Aarts, R.M.: Audio bandwidth extension: application of psychoacoustics. In: Signal Processing and Loudspeaker Design, pp. 113–117. Wiley, Hoboken (2005)
2. Geiser, B., Jax, P., Vary, P., et al.: Bandwidth extension for hierarchical speech and audio coding in ITU-T Rec. G. 729.1. IEEE Trans. Audio Speech Lang. Process. **15**(8), 2496–2509 (2007)
3. Makinen, J., Bessette, B., Bruhn, S., et al.: AMR-WB+: a new audio coding standard for 3rd generation mobile audio services. In: Proceedings of the IEEE International Conference on Acoustics, Speech, and Signal Processing (ICASSP), pp. 1109–1112. IEEE Press, Philadelphia, USA, 19–23 March 2005
4. Dietz, M., Liljeryd, L., Kjorling, K., et al.: Spectral band replication, a novel approach in audio coding. In: Proceedings of the 112th Audio Engineering Society Convention, pp. 1–8. Audio Engineering Society press, Munich, Germany, 10–13 May 2002
5. Ekstrand, P.: Bandwidth extension of audio signals by spectral band replication. In: Proceedings of the 1st IEEE Benelux Workshop on Model Based Processing and Coding of Audio (MPCA), pp. 53–58. IEEE Press, Leuven, Belg, 15 November 2002
6. Neukam, C., Nagel, F., Schuller, G., et al.: A MDCT based harmonic spectral bandwidth extension method. In: Proceedings of IEEE International Conference on Acoustics, Speech and Signal Processing (ICASSP), pp. 566–570. IEEE Press, Vancouver, Canada, 26–31 May 2013
7. ISO/IEC 14496–3:2001/Amd1:2004. Bandwidth extension
8. Schnell, M., Geiger, R., Schmidt, M., et al.: Enhanced MPEG-4 low delay AAC-Low bitrate high quality communication. In: Proceedings of the 122nd Audio Engineering Society Convention, pp. 1211–1223. Audio Engineering Society Press, Vienna, Austria, 5–8 May 2007
9. Max, N., Markus, M., Nikolaus, R., et al.: MPEG unified speech and audio coding – the ISO/MPEG standard for high-efficiency audio coding of all content types. In: 132nd Audio Engineering Society Convention, pp. 248–269. Budapest, Hungary, 26–29 April 2012
10. Hinton, G., Osindero, S., Teh, Y.W.: A fast learning algorithm for deep belief nets. Neural Comput. **18**(7), 1527–1554 (2006)
11. Deng, L., Seltzer, M., et al.: Binary coding of speech spectrograms using a deep auto-encoder. In: Proceedings of the 11th Annual Conference of the International Speech Communication Association (INTERSPEECH), pp. 1692–1695. IEEE Press, Makuhari, Chiba, Japan, 26–30 September 2010

12. Mohamed, A., George, E.D., Hinton, G.: Acoustic modeling using deep belief networks. IEEE Trans. Audio Speech Lang. Process. **20**(1), 14–22 (2012)
13. Ling, Z.-H., Deng, L., Yu, D.: Modeling spectral envelopes using restricted Boltzmann machines and deep belief networks for statistical parametric speech synthesis. IEEE Trans. Audio Speech Lang. Process. **21**(10), 2129–2139 (2013)
14. Liu, C.M., Hsu, H.W., Lee, W.C.: Compression artifacts in perceptual audio coding. IEEE Trans. Audio Speech Lang. Process. **16**(4), 681–695 (2008)
15. Kim, K.T., Choi, J.Y., Kang, H.G.: Perceptual relevance of the temporal envelope to the speech signal in the 4–7 kHz band. J. Acoust. Soc. Am. **122**(3), EL88–EL88 (2007)
16. ITU-R BS.1534-1, MUSHRA. International Telecommunications Union, Geneva, Switzerland (2001–2003)

Part-Aware Segmentation for Fine-Grained Categorization

Cheng Pang, Hongxun Yao[✉], Zhiyuan Yang, Xiaoshuai Sun,
Sicheng Zhao, and Yanhao Zhang

School of Computer Science and Technology, Harbin Institute of Technology,
Harbin, China
{pangcheng3,h.yao,zyyang,xiaoshuaisun,zsc,yhzhang}@hit.edu.cn

Abstract. It is difficult to segment images of fine-grained objects due to the high variation of appearances. Common segmentation methods can hardly separate the part regions of the instance from background with sufficient accuracy. However, these parts are crucial in fine-grained recognition. Observing that fine-grained objects share the same configuration of parts, we present a novel part-aware segmentation method, which can get the foreground segmentation from a bounding box with preservation of semantic parts. We firstly design a hybrid part localization method, which combines parametric and non-parametric models. Then we iteratively update the segmentation outputs and the part proposal, which can get better foreground segmentation results. Experiments demonstrate the superiority of the proposed method, as compared to the state-of-the-art approaches.

Keywords: Image segmentation · Fine-grained visual categorization · GrabCut

1 Introduction

In recent years, fine-grained visual categorization (FGVC) has received increasing attention in computer vision domain and been applied in commercial applications. Due to the high inter-class similarities of the fine-grained objects and the intra-class disparities arising from a large variation of views and object poses, it is challenging to distinguish between sub-categories. Considering the above challenges, recent works [1–6] are dedicated to learn discriminative features coupled with object alignment, which has demonstrated the effectiveness with promising results. Angelova *et al.* [7] firstly introduced segmentation in FGVC and demonstrated that the accuracy of recognition can be boosted with the aid of segmentation.

From the perspective of cognitive psychology, basic-level categories are principally defined by their parts while subordinate-level categories are distinguished by the different properties of these parts [8]. Following these hypotheses, a group of part-based methods have been proposed. Zhang *et al.* [9] presented a

© Springer International Publishing Switzerland 2015
Y.-S. Ho et al. (Eds.): PCM 2015, Part I, LNCS 9314, pp. 538–548, 2015.
DOI: 10.1007/978-3-319-24075-6_52

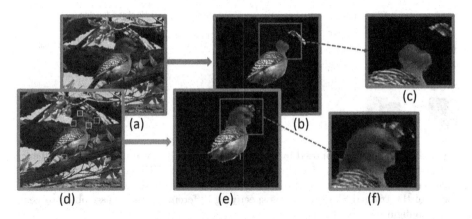

Fig. 1. Glimpse of segmentation result comparison. (a) The original image. (b) Segmentation result of [12]. (c) Details of (b). (d) Part localizations. (e) Outputs of the proposed part-aware segmentation using localized parts indicated by (d). (f) Details of (e) where more parts are preserved.

deformable part descriptors for FGVC and attribute prediction, in which DPM is used to produce coarse part configurations. Xie *et al.* [10] proposed the hierarchical structure learning algorithm to find high level concepts beyond basic parts. Berg *et al.* [11] introduced a part-based, one-vs-one feature where pairwise parts are used for pose normalization and finding discriminative regions.

In addition to the recognition tasks, some methods also employ the response of part detection to direct the process of segmentation [13]. Within the realm of FGVC, Chai *et al.* [12] firstly combined part localization with segmentation, in which DPM is used for part localization and GrabCut is used for segmentation. They found that these two processes, localization and segmentation, could assist each other to get better results. In contrast with their method, the proposed method (Fig. 1) differs in two aspects: (1) DPM only gives a coarse predictions of a few parts while the proposed method gives pixel-wise predictions of more semantic parts. Experiments show that our semantic parts detection leads to better foreground segmentation and thus benefits the final recognition. (2) Instead of re-localize parts, we iteratively update the probabilities of the pixels being the foreground to preserve more discriminative parts.

In order to localize semantic parts, we design a hybrid localization method. Firstly, we search for training images with an object shape similar to the current query object and then transfer the part annotations from them directly. Secondly, in the neighborhood of the transferred parts, we look for the precise locations of parts using several part-specific detectors trained for each individual part. Different from the non-parametric part transfer method [14], the first step adopts the transferred locations as a prior and doesn't rely on ground-truth foreground masks. Compared with global searching, our method has two leading merits: (1) We combine the advantages of parametric and non-parametric methods, yielding relatively more accurate results with less computation time; (2) Our local searching strategy can avoid the confusions caused by different parts with similar appearances.

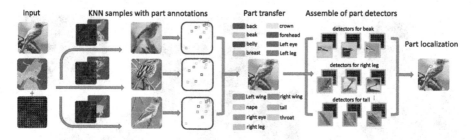

Fig. 2. Pipeline of the proposed hybrid part localization. Given the HOG and saliency map of the query image as descriptors, we first find the KNN samples in training set. Then we average the part annotations and transfer it to the query image. Finally, results of the trained SVM detectors specifying different parts are assembled to get a better configuration of the parts.

We detail our hybrid part localization and part-aware segmentation method in Sects. 2 and 3 respectively. Section 4 presents the main experimental results. Finally, we conclude this paper in Sect. 5.

2 Hybrid Part Localization

Previous works demonstrate that accurate part localization could significantly boost the accuracy of fine-grained recognition [15,16]. DPM and some parametric models designed for FGVC share the assumption that the distribution of parts corresponds to a Gaussian model. However, the assumption is proved incorrect in [14]. Parametric models for detection are neither sufficient for tackling the large variations presented in fine-grained recognition tasks, nor efficient for on-line applications. Therefore, we present a hybrid method combining parametric and non-parametric models which can tackle large pose variations and lead to accurate localizations of semantic parts (Fig. 2).

First, we transfer the part annotations of training data to the query image using saliency map together with HOG instead of ground-truth segmentation masks which are used in [14]. All images are cropped according to the bounding box and resized to 256×256. Then the product of the HOG and saliency map are computed as a shape descriptor for each image which minimizes the background noise interferences. We find the k samples which are most similar to the query instance using Euclidean distance metrics, and average their part locations as an initial prediction for the query instance.

With the transferred locations of parts, we carry out sliding window detection which provides more precise predictions. During the training stage, patches specifying each part are used to train groups of $SVMs$. Specifically, for the i-th part, we get $SVM_1^i, SVM_2^i, \ldots, SVM_m^i$. Due to one certain part belonging to different sub-categories presents large variations, there is a sacrifice of discrimination if we merely have one classifier trained for one part. So the patches are clustered first and then trained for different classifiers individually. The detector of a certain part is an ensemble of these classifiers. The classifiers are trained to be over-fitting, yielding less classification errors. Within the neighborhood of the i-th part, we accept the result of the classifier which gets the highest score among SVM^i. Notice that the method is somewhat like the ensemble of exemplar svm [17], but there are apparently more than one positive examples for training in our method. It can be seen as a follower of the subcategory consistency [16].

3   Part-Aware Segmentation

In this section, we modify GrabCut [18] to obtain a better segmentation by using the detected parts. GrabCut is built on Graph Cut [19], which uses GMM as the color model and updates the model and segmentation outputs iteratively. No additional user interaction is needed except for the bounding box.

3.1   Definitions

Our motivation is that if a part is estimated as the background mistakenly, then increase the probability of it being foreground. We also use a minimum convex polygon computed from the part for segmentation initialization. The proposed cost function can be written as:

$$E(\mathbf{s}, \mathbf{w}; \mathbf{z}) = E^{GC}(\mathbf{s}; \mathbf{z}) + \alpha E^P(\mathbf{w}) + \beta E^C(\mathbf{s}, \mathbf{w}; \mathbf{z}), \tag{1}$$

where $\mathbf{z}$ is an image depicted by an array of RGB values, $\mathbf{s}$ is the foreground mask given by GrabCut, $\mathbf{w} = [w_1, w_2, \ldots, w_n]$ is the part proposal, which denotes the probability of each pixel that it belongs to the foreground, E^{GC} denotes a GrabCut energy, E^P denotes the energy of part proposal, E^C is a consistency term penalizing the cases that the foreground and the part neighborhood do not agree. A hyper-parameter $\mathbf{P} = [p_1, p_2, \ldots, p_n]$ is the part mask denoting the regions of parts and their neighborhood. $\mathbf{w}$ is initialized according to $\mathbf{P}$, that is, if z_i is in the neighborhood of parts, then w_i is set to be 1, otherwise 0. α and β are positive constants controlling the balance between the energy terms. The GrabCut energy is defined by:

$$E^{GC}(\mathbf{s}; \mathbf{z}) = \sum_{n \in I} D_n(s_n; z_n) + \gamma \sum_{m,n \in C} V_{m,n}, \tag{2}$$

where D is the negative logarithm of Gaussian probability distribution and V is the smoothness term, C is the indices of adjacent pixels, γ is a constant, I is the indices of all pixels. The part energy term is defined by:

$$E^P(\mathbf{w}) = \sum_{n \in U} w_n, U = dif(\mathbf{P}, \mathbf{s}). \tag{3}$$

Here, if a pixel is assigned as parts or their neighborhood by $\mathbf{P}$, but estimated as background by $\mathbf{s}$, we term it as inconsistent pixel. $dif(\cdot)$ return the indices of all inconsistent pixels. The consistency term is defined by:

$$E^C(\mathbf{s}, \mathbf{w}; \mathbf{z}) = \sum_{i \in U} (1 - w_i) D_i(s_i; z_i) + \sum_{j \in \overline{U}} D_j(s_j; z_j), \tag{4}$$

where $\overline{U}$ is the complement of U, and their union is I

3.2   Optimization

The cost function Eq. (1) can be optimized in this way, that is, alternating between updating part proposal **w** while fixing the foreground segmentation **s**, and vice versa.

Updating the Part Proposal w. Fix the foreground segmentation **s**, then E^{GC} can be ignored. We optimize Eq. (1):

$$
\min_{\mathbf{w}} \alpha E^P(\mathbf{w}) + \beta E^C(\mathbf{s}, \mathbf{w}; \mathbf{z}) = \min_{\mathbf{w}} \alpha \sum_{n \in U} w_n + \beta \sum_{i \in U} (1 - w_i) D_i(s_i; z_i)
$$
$$
+ \beta \sum_{j \in \overline{U}} D_j(s_j; z_j)
$$
$$
= \min_{\mathbf{w}} w_i \sum_{i \in U} (\alpha - \beta D_i(s_i; z_i)) + \beta \sum_{j \in I} D_j(s_j; z_j), \quad (5)
$$

where the last term does not depend on **w**. Equation (5) can be written as:

$$
\min_{\mathbf{w}} w_i \sum_{i \in U} (\alpha - \beta D_i(s_i; z_i)). \tag{6}
$$

We choose α and β, making $\alpha - \beta D_i(s_i; z_i)$ to be positive for all i. So Eq. (6) is a monotonically increasing function of **w**. We also fix the change of each w_i to be a constant λ or 0 in one iteration. When w_i is reduced to near 0, we set it as 0 and update P_i, so that P_i no longer denotes the part neighborhood. The intuition of (6) is that if the part proposal and the foreground segmentation do not agree, we decrease the probability of the pixels being neighborhood of part or foreground. This ensures that our segmentation is tolerant of imperfect part detections.

Updating the Foreground Segmentation s. Fix **w**, then E^P can be ignored. Optimization of Eq. (1) is turned to:

$$
\min_{\mathbf{s}} E(\mathbf{s}, \mathbf{w}; \mathbf{z}) = E^{GC}(\mathbf{s}) + \beta E^C(\mathbf{s}, \mathbf{w}; \mathbf{z})
$$
$$
= \min_{\mathbf{s}} \sum_{n \in I} D_n(s_n; z_n) + \gamma \sum_{(m,n) \in C} V_{m,n}
$$
$$
+ \beta \sum_{i \in U} (1 - w_i) D_i(s_i; z_i) + \beta \sum_{j \in \overline{U}} D_j(s_j; z_j)
$$
$$
= \min_{\mathbf{s}} \sum_{i \in U} (1 + \beta - \beta w_i) D_i(s_i; z_i)
$$
$$
+ (1 + \beta) \sum_{j \in \overline{U}} D_j(s_j; z_j) + \gamma \sum_{(m,n) \in C} V_{m,n}. \tag{7}
$$

In contrast to the correctly segmented pixels, the modified function Eq. (7) always comparatively minimizes the coefficients of inconsistent part-neighborhood pixels. This encourages the inconsistent parts and their neighborhood to be the foreground. Notice that the Eq. (7) can still be minimized via Graph Cut.

In conclusion, we optimize Eq. (1) between four steps: **(a)** optimizing part proposal **w** according to Eq. (6), **(b)** updating the hyper-parameter **P** according to **w**, **(c)** estimating the GMM color model, **(d)** optimizing the foreground segmentation **s** according to Eq. (7).

4   Experiments

We test the proposed method on Bird 200-2011 dataset [20], which is the benchmark for fine-grained categorization. We consider two aspects of performance: segmentation and recognition.

4.1   Dataset

Bird 200-2011 is a challenging dataset [20] consisting of 11788 images for 200 species of birds. Labels of species, bounding boxes, part annotations and segmentation masks are provided. Although the dataset has been extensively studied by researchers in the field of fine-grained categorization, the highly variations of views, light and object poses make the dataset extremely difficult for recognition and segmentation tasks. In our experiments, we use the bounding boxes and part annotations of the training images to construct the localization models, then the foreground segmentation of all the images can be obtained using our method.

4.2   Part Localization Results

We select three important parts (back, nape and left leg) of a bird to show the effectiveness of our hybrid part localization method. The error of a part is given

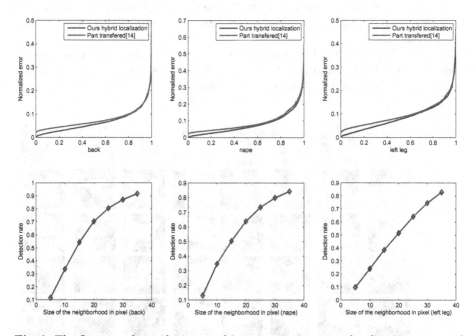

Fig. 3. The first row shows the errors of 3 parts using our part localization compared with that adapted from [14]. The errors for each part are computed from all samples and sorted, our method leads to less errors. The second row shows the detection rate affected by the size of part neighborhood.

by the Euclidean distance between the predicted location and the ground-truth location, and normalized by the size of the bounding box. Then the part errors of all the samples are sorted for evaluation, as shown in the first row of Fig. 3. Our hybrid method leads to less errors compared to [14] using transfer(no masks are used according to our settings). Imperfect part localization will hurt the segmentation. If the ground-truth part is in the neighborhood of the prediction, we can also get a correct segmentation. In this situation, the part can be seen as properly detected. We show the influence of the neighborhood size on the detection rates in the second row of Fig. 3. Notice that larger size leads to higher detection rate. We use 20 pixels in our segmentation settings, which can lead to a competitive detection rate of 0.6.

4.3   Segmentation Results

We first get the part locations using our hybrid localization method. The part annotations of training set are used to train part-specific detectors. We use the recommended training and test split, then segment all the images and evaluate our results across the whole dataset. All images are resized to a maximum width or height of 500 pixels.

Figure 4 illustrates some of the segmentation outputs of different methods. From the top to bottom: GrabCut [18], Chai *et al.* [12] and ours. The ground truth masks and original images with detected parts are also given. It can be seen that some semantic parts have been preserved by our method while others tend to miss them.

Figure 5 illustrates the bad cases, where the failure of part detections couple with confusing background may lead to bad results. In fact, we can see all the segmentation methods have failed in these situations.

Fig. 4. Comparison of segmentation results. More parts (*e.g.* the beaks and crowns of the birds) are correctly segmented by our method with aid of the localized parts.

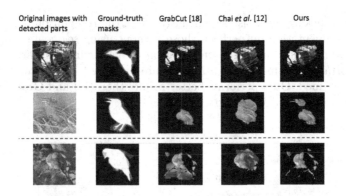

Fig. 5. Some failure cases caused by the confusing background and bad detections. Each row shows the results of one bird.

Table 1. Segmentation results comparison of different methods on Caltech-UCSD Birds-200-2011 dataset.

Methods	SegI (%)	SegII (%)
No segmentation	47.0	47.0
GrabCut [18]	64.5	74.2
Chai *et al.* [12]	73.7	82.9
Ours	**75.6**	**83.6**

Two benchmarked segmentation criteria are used for evaluation. One is the PASCAL VOCC [21] overlap measure $\frac{F_E \bigcap F_G T}{F_E \bigcup F_G T}$ (**Seg.I**), where F_E is the estimated foreground pixels and $F_G T$ is the ground-truth foreground pixels. The other is the percentage of pixels being correctly estimated as foreground and background (**Seg.II**).

Table 1 shows the performance of segmentation on Bird-200-2011 dataset. The proposed method achieves an improvement of 11.1 % and 9.4 % in terms of **Seg.I** and **Seg.II** respectively as compared to [18], and an improvement of 1.9 % and 0.7 % as compared to [12]. Here, we only consider the outputs inside the bounding-boxes.

4.4   Recognition Results

From the intuition that part-based segmentation could preserve discriminative ingredients for recognition. We test our segmentation with the part-based classification method in [6]. The method uses ground-truth part annotations to generate large amount of patches specifying the objects. Then the patches are used to form a codebook and get a kind of mid-level discriminative features via sparse coding.

Figure 6 shows the mean accuracy of recognition within the top r guesses using outputs of different segmentation methods. It is easy to see that, for the rank-1 accuracy, the recognition accuracy benefits from the better segmentation, because the background noise significantly affect the recognition. So the methods using segmentation outperform those without using segmentation(No seg). However, while r increases, the preservation of discriminative parts is getting more important. The more semantic parts are preserved, the better recognition accuracy obtained. Ours performs better results than [12] and GrabCut due to our preservation of semantic parts.

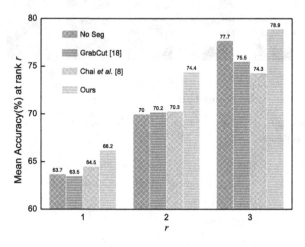

Fig. 6. Comparison of classification results on Birds-200-2011 dataset within the top r guesses. Our method performs significantly better than [12, 18] and the classification without using segmentation (No seg) due to the preservation of more discriminative parts

5   Conclusion

In this paper, we presented a novel segmentation method for fine-grained recognition. We introduced a hybrid part localization method which takes both advantages of parametric and non-parametric models, tackling large variations of poses and getting the locations of parts with less computations. Then the GrabCut is modified to incorporate the detected parts and get better foreground segmentation. By alternating between segmentation and changing the part proposal, our algorithm preserves more discriminative parts, and is tolerant of some imperfect part detections. Experiments demonstrated that our method outperforms the competitors on both segmentation and recognition benchmarks.

Acknowledgements. This work was supported in part by the National Science Foundation of China No. 61472103, and Key Program Grant of National Science Foundation of China No. 61133003.

References

1. Berg, T., Liu, J., Lee, S.W., Alexander, M.L., Jacobs, D.W., Belhumeur, P.N.: Birdsnap: large-scale fine-grained visual categorization of birds. In: 2014 IEEE Conference on Computer Vision and Pattern Recognition (CVPR), pp. 2019–2026. IEEE (2014)
2. Wah, C., Horn, G.V., Branson, S., Maji, S., Perona, P., Belongie, S.: Similarity comparisons for interactive fine-grained categorization. In: 2014 IEEE Conference on Computer Vision and Pattern Recognition (CVPR), pp. 859–866. IEEE (2014)
3. Yang, S., Bo, L., Wang, J., Shapiro, L.G.: Unsupervised template learning for fine-grained object recognition. In: Advances in Neural Information Processing Systems, pp. 3122–3130 (2012)
4. Liu, J., Kanazawa, A., Jacobs, D., Belhumeur, P.: Dog breed classification using part localization. In: Fitzgibbon, A., Lazebnik, S., Perona, P., Sato, Y., Schmid, C. (eds.) ECCV 2012, Part I. LNCS, vol. 7572, pp. 172–185. Springer, Heidelberg (2012)
5. Yao, B., Khosla, A., Fei-Fei, L.: Combining randomization and discrimination for fine-grained image categorization. In: 2011 IEEE Conference on Computer Vision and Pattern Recognition (CVPR), pp. 1577–1584. IEEE (2011)
6. Pang, C., Yao, H., Sun, X.: Discriminative features for bird species classification. In: Proceedings of International Conference on Internet Multimedia Computing and Service, pp. 256. ACM (2014)
7. Angelova, A., Zhu, S.: Efficient object detection and segmentation for fine-grained recognition. In: 2013 IEEE Conference on Computer Vision and Pattern Recognition (CVPR), pp. 811–818. IEEE (2013)
8. Rosch, E., Mervis, C.B., Gray, W.D., Johnson, D.M., Boyes-Braem, P.: Basic objects in natural categories. Cogn. Psychol. 8(3), 382–439 (1976)
9. Zhang, N., Farrell, R., Iandola, F., Darrell, T.: Deformable part descriptors for fine-grained recognition and attribute prediction. In: 2013 IEEE International Conference on Computer Vision (ICCV), pp. 729–736. IEEE (2013)
10. Xie, L., Tian, Q., Hong, R., Yan, S., Zhang, B.: Hierarchical part matching for fine-grained visual categorization. In: 2013 IEEE International Conference on Computer Vision (ICCV), pp. 1641–1648. IEEE (2013)
11. Berg, T., Belhumeur, P.N.: Poof: Part-based one-vs.-one features for fine-grained categorization, face verification, and attribute estimation. In: 2013 IEEE Conference on Computer Vision and Pattern Recognition (CVPR), pp. 955–962. IEEE (2013)
12. Chai, Y., Lempitsky, V., Zisserman, A.: Symbiotic segmentation and part localization for fine-grained categorization. In: 2013 IEEE International Conference on Computer Vision (ICCV), pp. 321–328. IEEE (2013)
13. Wu, B., Nevatia, R., Li, Y.: Segmentation of multiple, partially occluded objects by grouping, merging, assigning part detection responses. In: 2008 IEEE Conference on Computer Vision and Pattern Recognition (CVPR), pp. 1–8. IEEE (2008)
14. Goering, C., Rodner, E., Freytag, A., Denzler, J.: Nonparametric part transfer for fine-grained recognition. In: 2014 IEEE Conference on Computer Vision and Pattern Recognition (CVPR), pp. 2489–2496. IEEE (2014)
15. Liu, J., Li, Y., Belhumeur, P.N.: Part-pair representation for part localization. In: Fleet, D., Pajdla, T., Schiele, B., Tuytelaars, T. (eds.) ECCV 2014, Part II. LNCS, vol. 8690, pp. 456–471. Springer, Heidelberg (2014)

16. Liu, J., Belhumeur, P.N.: Bird part localization using exemplar-based models with enforced pose and subcategory consistency. In: 2013 IEEE International Conference on Computer Vision (ICCV), pp. 2520–2527. IEEE (2013)
17. Malisiewicz, T., Gupta, A., Efros, A.A.: Ensemble of exemplar-SVMs for object detection and beyond. In: 2011 IEEE International Conference on Computer Vision (ICCV), pp. 89–96. IEEE (2011)
18. Rother, C., Kolmogorov, V., Blake, A.: Grabcut: interactive foreground extraction using iterated graph cuts. In: ACM Transactions on Graphics (TOG), vol. 23, pp. 309–314. ACM (2004)
19. Boykov, Y.Y., Jolly, M.P.: Interactive graph cuts for optimal boundary & region segmentation of objects in ND images. In: 2001 IEEE International Conference on Computer Vision (ICCV), pp. 105–112. IEEE (2001)
20. Wah, C., Branson, S., Welinder, P., Perona, P., Belongie, S.: The caltech-UCSD birds-200-2011 dataset (2011)
21. Everingham, M., Van Gool, L., Williams, C.K., Winn, J., Zisserman, A.: The pascal visual object classes (VOC) challenge. Int. J. Comput. Vis. **88**(2), 303–338 (2010)

Improved Compressed Sensing Based 3D Soft Tissue Surface Reconstruction

Sijiao Yu, Zhiyong Yuan[✉], Qianqian Tong, Xiangyun Liao,
and Yaoyi Bai

School of Computer, Wuhan University, Wuhan, China
zhiyongyuan@whu.edu.cn

Abstract. This paper presents a 3D soft tissue surface reconstruction method based on improved compressed sensing and radial basis function interpolation for a small amount of uniform sampling data points on 3D surface. We adopt radial basis function interpolation to obtain the same amount of data points as to be reconstructed and propose an improved compressed sensing method to reconstruct 3D surface: we design a deterministic measurement matrix to signal observation, and then adopt the discrete cosine transform to the 3D coordinate sparse representation and use weak choose regularized orthogonal matching pursuit algorithm to reconstruct. Experimental results show that the proposed algorithm improves the resolution of the surface as well as the accuracy. The average maximum error is less than 0.9012 mm, which is smooth enough to provide accurate surface data model for virtual reality based surgery system.

Keywords: Compressed sensing · Deterministic measurement matrix · Radial basis function interpolation · Sparse 3D discrete point

1 Introduction

3D reconstruction is an important research direction in the field of computer vision and it has been widely used in Virtual Reality (VR), object recognition and visualization. In recent years, 3D reconstruction technology has become a new hotspot in medical research, and VR based surgery is one of its applications. It provides a powerful tool for both surgical training and its evaluation, in which the accurate acquisition of 3D soft tissue structural information is primary premise.

A lot of research has been conducted on 3D surface reconstruction. Different kinds of methods are for different kinds of data and the results produced by the algorithm are highly dependent on the types of data [1]. In typical 3D surface reconstruction methods, data sets are usually obtained from different sources such as medical imagery, laser range scanner and mathematical models [2].

For approaches based on images, Elfarargy et al. [3] presented an image-based modeling technique based on polynomial texture mapping. Qian [4] presented an efficient poisson-based surface reconstruction of 3D model from a non-homogenous sparse point cloud. However, their limitation is that the generated models are only planar nature. For approaches based on point cloud through laser range scanner, lots of techniques usually used dense data set to calculate the connectivity to avoid the

© Springer International Publishing Switzerland 2015
Y.-S. Ho et al. (Eds.): PCM 2015, Part I, LNCS 9314, pp. 549–558, 2015.
DOI: 10.1007/978-3-319-24075-6_53

appearance of holes [5], but there are some data redundancy problems. Amenta et al. [2] is the first that offers a provable guarantee to the surface reconstruction problem from unorganized sample points and they proposed a Voronoi-based surface reconstruction algorithm. Later, Patrick et al. [6] proposed a modular framework for robust 3D reconstruction from unorganized, unoriented, noisy, and outlier ridden geometric data. This approach is scalable while robust to noise, outliers, and holes. But, it is for closed surface and relies on large scale of points. For approaches based on mathematical models, such as implicit surfaces, this approach uses different standards to fit the implicit surfaces toward input points by minimizing the energy that represents different distance functions [7]. To summarize, using existing methods to reconstruct accurate surface data must take advantage of large amounts of data, which will bring data redundancy and affect the speed of reconstruction.

To reduce the amount of data collected and reconstruct the 3D surface with high accuracy, we proposed a method using the recently proposed sampling method, compressed sensing or compressive sampling (CS) theory [7–9] to reconstruct for a small count of uniform point. CS theory has the ability to compress a signal during the process of sampling, which means that CS can collect compressed data at a sampling rate much lower than that needed in Shannon's sampling theorem. Experimental results show that the proposed algorithm improved the resolution of the surface as well as the accuracy.

2   Surface Reconstruction

The goal of this paper is to reconstruct 3D surface with high accuracy and strong robustness. We now describe the model and implementation of our method. Firstly, we collect a small amount of random 3D tissue surface data sets S_1, as described in literature [10]. Secondly, we obtain the same amount of data points S_2 as to be reconstructed by using Radial Basis Function (RBF) interpolation. Then, we adopt the Discrete Cosine Transform (DCT) to the 3D coordinate sparse representation respectively, and design a deterministic measurement matrix to signal observation. Finally, we choose Weak choose Regularized Orthogonal Matching Pursuit (WROMP) algorithm as reconstruction algorithm. The reconstruction process is described in Fig. 1.

2.1   RBF Interpolation

Radial basis functions are commonly used for all kinds of scattered data interpolation problems. RBF interpolation is to find a spline $S(x)$ that passes as close as possible to

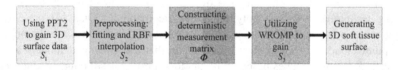

Fig. 1. Reconstruction process

the data points and is as smooth as possible. In this paper, we use Thin Plate Spline (TPS) [11] as basis function. For a set of discrete data sets $\{x_i, f_i\} \in R^d \times R, i = 1, 2, \ldots, n$ and basis function

$$\varphi : R_+ \rightarrow R \begin{cases} \varphi(x) = x^{2k-d} \ln x, & d \in even \\ \varphi(x) = x^{2k-d}, & d \in odd \end{cases} \tag{1}$$

Let $E = (x_j^{\alpha})_{|\alpha| \leq d, j \leq m}, X^T = (1, \ldots, x^{\alpha}, \ldots)$, the solution can be expressed as

$$S(x) = (\varphi^T, X^T) \begin{pmatrix} A & E \\ E^T & 0 \end{pmatrix} \begin{pmatrix} f \\ 0 \end{pmatrix} \tag{2}$$

We first collect only a small amount of data points S_1. In order to ensure the accuracy of the final reconstruction, we combine fitting with RBF interpolation. We get the connectivity information by fitting. When interpolation, we first interpolate in a rough layer to get S_2 so that the data points distribute uniformly, and then interpolate again. The steps are summarized in Table 1.

Table 1. RBF interpolation algorithm

Input: Data set $S_1(x, y, z)$
Output: Data set $S_2(x_2, y_2, z_2)$

1:   Fitting
2:   For each column (x, z)
3:       Construct fitting curves of (x, z) to get R
4:       Compute the inserted x coordinates x_1
5:       Compute the inserted z coordinates z_1 according to R
6:   end
7:   RBF Interpolate
8:   Train RBF neural network N based on S_1
9:   Let (x_1, z_1) as input of N, obtaining the inserted y coordinates y_1
10: Combine $S_1(x, y, z)$ and (x_1, y_1, z_1) to get interpolated data set $S_2(x_2, y_2, z_2)$

2.2   CS Reconstruction Algorithm

Mathematical Model. After preprocessing, we focus on CS reconstruction. CS is a nonlinear theory of sparse signal reconstruction. Suppose $x \in R^n$ is a digital signal, if it is a K sparse or compressible signal, then we can estimate it with few coefficients by linear transformation. We can represent this process mathematically as $y = \Phi x, y \in R^m$. $\Phi(m \times n(m \ll n))$ is a measurement matrix and represents a dimensionality reduction. We can rebuild x from y by solving the optimal problem below:

$$\hat{x} = \arg\min \|x\|_0, \quad s.t \quad y = \Phi x = \Phi \Psi s = \Theta s \tag{3}$$

When $m > K \log(n)$ and Φ has restricted isometry property (RIP) [12]. Our main work is to design a measurement matrix Φ that fulfills the above properties and can recover the original signal x from measured value y.

Deterministic Measurement Matrix Construction. In this paper, we propose an improved CS algorithm. Our main contribution is to propose a practical deterministic measurement matrix construction algorithm. We note that random matrices, such as Gaussian and Bernoulli matrix, generally satisfy RIP. However, due to their uncertainty, random matrices cannot guarantee the reconstruction results are accurate, which is impractical in reconstruction of high accuracy. Thus, we construct deterministic measurement matrix using training method.

The criterion of constructing deterministic measurement matrix [13, 14] is as follows: Let Gram matrix $\mathbf{G} = \Theta^T \Theta$. If $\mathbf{G}$ is a symmetric matrix and nonnegative definite matrix, and the eigenvalues of $\mathbf{G}$ are in $[1 - \delta k, 1 + \delta k]$, then the Θ will meet the conditions of RIP, where δk is constrained equidistant constant.

In work [15], t - averaged mutual-coherence is defined as the average of all absolute and normalized inner products between different columns in Θ (denoted as g_{ij}) that are above t. Put formally,

$$\mu_t(\Theta) = \frac{\sum_{1 \le i,j \le N, i \ne j} (g_{ij} \ge t) \cdot |g_{ij}|}{\sum_{1 \le i,j \le N, i \ne j} (g_{ij} \ge t)} \tag{4}$$

Our goal is to minimize $\mu_t(\Theta)$: the main idea is to construct an adaptive measurement matrix set Φ_B firstly, and then select the matrix with smallest column coherence as the measurement matrix Φ. Based on Gaussian random matrix, we construct adaptive measurement matrix Φ_i according to literature [16]. Then according to the sparsity of signal, we adopt Elad algorithm [15] to modify certain columns of Φ_i. While modifying, we have to make sure that the modified matrix satisfies RIP and can make a small coefficient closer to zero, so that the signal is sparse enough to be better reconstructed. We update Gram matrix $\mathbf{G}$ by formula (5)

$$g_{ij} = \begin{cases} \gamma g_{ij} & |g_{ij}| \ge t \\ \gamma t \cdot sign(g_{ij}) & t \ge |g_{ij}| \ge \gamma t \\ g_{ij} & \gamma t \ge |g_{ij}| \end{cases} \tag{5}$$

Where the value of t, γ is very crucial, which will affect the satisfiability of RIP. The detailed algorithm is explained in Table 2.

Table 2. Deterministic measurement matrix construction algorithm

---

Input: sparse signal s; sparse matrix Ψ; iterations $iter$; output: measurement matrix Φ

---

1. **For** $i = 1 : iter$
2.   **Generate Gaussian matrix** $\Phi_0(m \times n)$;
3.   **Let** $s = (s_1, s_2, \ldots s_m)^T$; **Find the** k_0 **big components** $s_{ir_1}, s_{ir_2}, \ldots, s_{ir_{k_0}}$ **and their positions** $A = \{r_1, r_2, \ldots, r_{k_0}\}$;
4.   **Divide** Ψ **and normalize** $\Psi = (\hat{\Psi}_{m-k_0, n}, \hat{\Psi}_{k_0, n})^T$;
5.   **Calculate threshold matrix** $\Delta = \Psi_1^{-1} \Delta \Psi_1$, **where**

$$\Delta = \begin{cases} 0 & i = j \\ 1 & i = j \,\&\&\, j \in A(i, j \in \{1, 2, \ldots, n\}) \\ 10^q & i = j \,\&\&\, j \notin A \end{cases} ;$$

6.   **Calculate** $\Phi_i = \Phi_0 \tilde{\Delta}$ **and normalize each column;**
7.   **Calculate** $\mu_t(\Theta_i)$ **by formula (5);**
8.   **if** $\mu_t(\Theta_i) < t \,\&\&\, \mu_t(\Theta_i) < \mu_t(\Theta)$
9.   **Let** $\Phi = \Phi_i$;
10.   **else**
11.   **Update** Φ_i **by Elad algorithm;**
12.   **end**
13. **end**

WROMP Algorithm. We found that WROMP algorithm [17] is accurate and stable. The weak selection criterion of each element is as formula (6):

$$\{i \| gg_i | \geq \alpha \cdot \max_j |gg_j| \}, i \in J, \alpha \in (0, 1] \tag{6}$$

Where g_i represents the correlation of iteration residua r and measurement matrix Φ, J represents candidate set and α represents weak choosing factor. In the case of different r and different atomic correlations of Φ, the weak atomic selection criteria can better select the representation of atomic groups of the original signal from candidate set. So WROMP algorithm can reconstruct signal better, enhancing stability.

3 Experiment Results

For comparison purposes, we perform simulation experiments on four algorithms: linear interpolation (LI), RBF interpolation (RBFI), CS and RBF_CS (proposed algorithm), using memory pillow, and then analyze the results. Our experimental platform includes: (1) hardware: Intel Xeon CPU, 2.40 GHz, Operating System windows7-32bit; (2) software: Matlab8.1 (R2013a), Precision Position Tracker with 2 Cameras motion capture system (PPT2); (3) experimental material: Memory pillow ($450 \times 250 \times 100$ mm^3), and the deformation range in Y is 100 mm.

3.1 Data Sets

We sample the memory pillow surface uniformly using the method in literature [10] to obtain its 3D surface data S_1 with $299(23 \times 13)$ points. We collect points by the step of 30 mm in X direction and by the step of 20 mm in Z direction. Three typical surfaces data points are shown in Fig. 2(a)–(c). We suppose the data sets we collect through this method is the actual coordinates of 3D surface.

For comparison, we select points set $S_{11}(5 \times 13)$ from S_1 as reference. We select 5 rows from the third line with the step of 4 as reference to ensure the reliability and the remained data $S_{12}(18 \times 13)$ is selected as the basis of interpolation. On the basis of $S_{12}(18 \times 13)$, we carry out RBFI according to the algorithm in Table 1. After the first interpolation, we get 299 points and then we conduct the second interpolation. During the second interpolation, we first interpolate by row and get $1443(111 \times 13)$ points; then we interpolate by column and get data sets $S_2(111 \times 52)$.

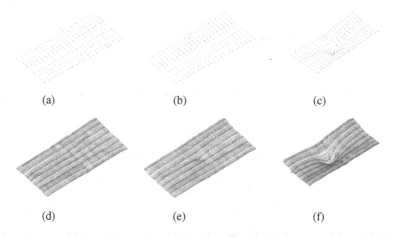

(a)   (b)   (c)

(d)   (e)   (f)

Fig. 2. (a), (b) and (c) are original data point coordinates we collect on conditions of no deformation, rather small deformation and pretty large deformation; and (d), (e) and (f) are the corresponding data point coordinates after RBF interpolation

3.2   Comparison of Different Measurement Matrix

To verify the deterministic measurement matrix we constructed possesse better performance, we performed experiments on three matrices (Gaussian Random matrix, Adaptive matrix and proposed matrix). The experiments are under three kinds of deformation of coordinate Y: no deformation (No), rather small deformation (Small) and pretty large deformation (Large). We mainly performed tests on mean square error (e_{MSE}) and maximum error (e_{MAX}). The reconstruction result along with deformation is shown in Table 3.

Table 3. Reconstruction results of different measurement matrices mm

Deformation	No		Small		Large	
	e_{MSE}	e_{MAX}	e_{MSE}	e_{MAX}	e_{MSE}	e_{MAX}
Gaussian	0.1593	1.0391	0.2033	1.2372	0.403	2.6019
Adaptive	0.1498	1.0219	0.2444	1.6239	0.3939	1.3948
Proposed	0.1071	0.6908	0.1135	0.7854	0.1220	0.8967

In the test, the sampling rate is 0.5. As can be seen, the reconstruction error is small when using CS. However, the error of Gaussian and Adaptive matrix increases apparently when there is larger deformation. While for the proposed matrix, the maximum reconstruction error is within 1 mm and has perfect stability, which fully shows the proposed matrix is better.

3.3   Comparison of LI and RBFI

Table 4 shows the reconstruction results of two kinds of interpolation algorithm, LI and RBFI. We can learn that the reconstruction results are of no difference in small deformation. But as the deformation increases, the error of LI enlarges significantly, while the value of e_{MSE} and e_{MAX} of RBFI is relatively stable. As can be seen from Fig. 2(d)–(f), after RBF interpolation, the resolution of surface is increased by a magnitude. The number of points is increased from $234(18 \times 13)$ points to $5772(111 \times 52)$ points.

Table 4. Reconstruction results of LI and RBFI mm

Deformation	No		Small		Large	
	e_{MSE}	e_{MAX}	e_{MSE}	e_{MAX}	e_{MSE}	e_{MAX}
LI	0.1593	1.0391	0.2033	1.2372	0.403	2.6019
RBF	0.1244	1.0981	0.1635	1.3786	0.1475	1.0054

3.4   Comparison with Other Methods

During the CS reconstruction, we reconstructed the 3D surface coordinates x, y, z respectively. We apply DCT to the sparse representation, choose Fast Fourier Transform (FFT) as sparse base and use the proposed constructing algorithm construct measurement matrix. Finally we adopt WROMP algorithm to reconstruct. In this paper, we combine RBF and CS. CS algorithm is performed on point set S_2 interpolated by RBF interpolation. In the experiment, the sampling rate of y is set to 0.5, and we set $t = 0.2, \gamma = 0.9$ when constructing deterministic matrix. We can see the reconstruction error of memory pillow under 8 kinds of deformation in Figs. 3 and 4.

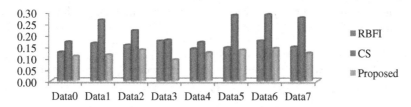

Fig. 3. The comparison of e_{MSE} of three algorithms.

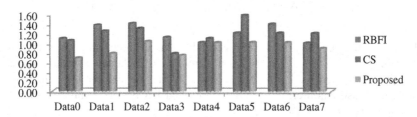

Fig. 4. The comparison of e_{MAX} of three algorithms.

In Figs. 3 and 4, the blue column represents RBFI algorithm, the red column represents CS algorithm and the green column represents the proposed algorithm. As can be seen, the maximum error of CS reconstruction is relatively small and stable, but the value of e_{MSE} is large; for RBF reconstruction, the value of e_{MSE} is small while maximum error is large; and for the proposed algorithm, both the value of e_{MSE} and e_{MAX} are the smallest. And we can see detailed data from Tables 5 and 6, the value of e_{MAX} can be within 1.1 mm under any deformation, less than 1 mm in most cases, and the average maximum error is less than 0.9012 mm. Therefore, it can be seen that the reconstruction results by our algorithm have a smaller degree of error, which fully shows our algorithm is of high accuracy, strong robustness and has obvious advantages. The reconstructed surfaces under three typical deformations are shown in Fig. 5.

Table 5. The comparison of e_{MSE} of three algorithms mm

Deformation	Data0	Data1	Data2	Data3	Data4	Data5	Data6	Data7
RBF	0.1244	0.1635	0.1557	0.1744	0.1404	0.1447	0.1747	0.1475
CS	0.1688	0.2649	0.2186	0.1785	0.1681	0.2869	0.2888	0.2754
Proposed	**0.1071**	**0.1135**	**0.1355**	**0.0925**	**0.123**	**0.1347**	**0.1419**	**0.1220**

Table 6. The comparison of e_{MAX} of three algorithms mm

Deformation	Data0	Data1	Data2	Data3	Data4	Data5	Data6	Data7
RBF	1.0981	1.3786	1.4141	1.1246	1.0163	1.215	1.402	1.0054
CS	1.0575	1.2538	1.3106	0.7849	1.1030	1.5824	1.2152	1.2100
Proposed	**0.6908**	**0.7854**	**1.0383**	**0.7488**	**1.0141**	**1.0192**	**1.0169**	**0.8967**

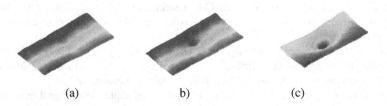

(a)   b)   (c)

Fig. 5. (a), (b) and (c) are final reconstruction surface on conditions of no deformation, rather small deformation and pretty large deformation respectively

4 Conclusions

In this paper, we present a creative 3D surface reconstruction algorithm based on improved CS with RBF interpolation. We put forward a deterministic measurement matrix construction algorithm which can be applied in practical. Experimental results illustrate that the proposed algorithm based on sparse data points satisfies the data precision requirement of tissue parameter measurement in terms of objective evaluation (e_{MSE}, e_{MAX}), and is more realistic from visual effects. Therefore, it can provide rather accurate data model for VR based surgery. In the future, we will investigate how to accelerate reconstruction.

Acknowledgment. The research was supported by the National Nature Science Foundation of China (Grant No. 61372107), the National Basic Research Program of China (Grant No. 2011CB707904), and the Open Funding Project of State Key Laboratory of Virtual Technology and Systems, Beihang University (Grant No. BUAA-VR-13KF-15).

References

1. Lim, S.P., Haron, H.: Surface reconstruction techniques: a review. Artif. Intell. Rev. (AIR) **42**(1), 59–78 (2014)
2. Amenta, N., Bern, M., Kamvysselis, M.: A new voronoi-based surface reconstruction algorithm. In: Proceedings of SIGGRAPH (1998)
3. Elfarargy, M., Rizq, A., Rashwan, M.: 3D surface reconstruction using polynomial texture mapping. In: Bebis, G., Boyle, R., Parvin, B., Koracin, D., Li, B., Porikli, F., Zordan, V., Klosowski, J., Coquillart, S., Luo, X., Chen, M., Gotz, D. (eds.) ISVC 2013, Part I. LNCS, vol. 8033, pp. 353–362. Springer, Heidelberg (2013)

4. Qian, N.: Efficient poisson-based surface reconstruction of 3D model from a non-homogenous sparse point cloud. In: Elmoataz, A., Lezoray, O., Nouboud, F., Mammass, D. (eds.) ICISP 2014. LNCS, vol. 8509, pp. 578–585. Springer, Heidelberg (2014)

5. Gálvez, A., Iglesias, A., Cobo, A., Puig-Pey, J., Espinola, J.: Bézier curve and surface fitting of 3D point clouds through genetic algorithms, functional networks and least-squares approximation. In: Gervasi, O., Gavrilova, M.L. (eds.) ICCSA 2007, Part II. LNCS, vol. 4706, pp. 680–693. Springer, Heidelberg (2007)

6. Mullen, P., De Goes, F., Desbrun, M., Cohen Steiner, D., Alliez, P.: Signing the unsigned: robust surface reconstruction from raw point sets. In: Computer Graphics Forum (CGF) (2010)

7. Donoho, D.L.: Compressed sensing. IEEE Trans. Inf. Theory **52**(4), 1289–1306 (2006)

8. Candès, E.: Compressive sampling. In: Proceedings of the International Congress of Mathematicians, pp. 1433–1452. Madrid, Spain, Invited Lectures (2006)

9. Candès, E., Romberg, J., Tao, T.: Robust uncertainty principles: exact signal reconstruction from highly incomplete frequency information. IEEE Trans. Inf. Theory **52**(2), 489–509 (2006)

10. Liao, X., Yuan, Z., Duan, Z., Si, W., Chen, S., Yu, S., Zhao, J.: A robust physics-based 3D soft tissue parameters estimation method for warping dynamics simulation. In: Xiao, T., Zhang, L., Fei, M. (eds.) AsiaSim 2012, Part I. CCIS, vol. 323, pp. 205–212. Springer, Heidelberg (2012)

11. Wu, Z.M.: Radial basis function, scattered data interpolation and the meshless method of numerical solution of partial differential equations. J. Eng. Math. **19**(2), 10–11 (2002)

12. Candès, E., Tao, T.: Near optimal signal recovery from random projections: universal encoding strategies. IEEE Trans. Inf. Theory **52**(12), 5406–5425 (2006)

13. DeVore, R.A.: Deterministic constructions of compressed sensing matrices. J. Complex. **23** (4–6), 918–925 (2007)

14. Baraniuk, R., Davenport, M., DeVore, R., Wakin, M.: A simple proof of the restricted Isometry property for random matrices. Constructive Approximation **28**(3), 253–263 (2008)

15. Elad, M.: Optimized projections for compressed sensing. IEEE Trans. Signal Process. **55** (12), 5695–5702 (2007)

16. Zhao, Y.J., Zheng, B.Y., Chen, S.N.: Adaptive measurement matrix construction in CS. Sig. Process. **28**(12), 1635–1641 (2012)

17. Liu, Z., Zhang, H., Zhang, Y.L.: Image reconstruction algorithm based on weak choosing regularization orthogonal matching pursuit. Acta Photonica Sinica **41**(10), 1217–1221 (2012)

Constructing Learning Maps for Lecture Videos by Exploring Wikipedia Knowledge

Feng Wang[1], Xiaoyan Li[1], Wenqiang Lei[1], Chen Huang[2], Min Yin[1(✉)], and Ting-Chuen Pong[3]

[1] Shanghai Key Laboratory of Multidimensional Information Processing, East China Normal University, 500 Dongchuan Road, Shanghai, China
myin@cs.ecnu.edu.cn
[2] School of Computer Science and Engineering, University of Electronic Science and Technology of China, Chengdu, China
[3] Department of Computer Science and Engineering, Hong Kong University of Science and Technology, Hong Kong, China

Abstract. Videos are commonly used as course materials for e-learning. In most existing systems, the lecture videos are usually presented in a linear manner. Structuring the video corpus has proven an effective way for the learners to conveniently browse the video corpus and design their learning strategies. However, the content analysis of lecture videos is difficult due to the low recognition rate of speech and handwriting texts and the noisy information. In this paper, we explore the use of external domain knowledge from Wikipedia to construct learning maps for online learners. First, with the external knowledge, we filter the noisy texts extracted from videos to form a more precise and elegant representation of the video content. This facilitates us to construct a more accurate video map to represent the domain knowledge of the course. Second, by combining the video information and the external academic articles for the domain concepts, we construct a directed map to show the relationships between different concepts. This can facilitate online learners to design their learning strategies and search for the target concepts and related videos. Our experiments demonstrate that external domain knowledge can help organize the lecture video corpus and construct more comprehensive knowledge representations, which improves the learning experience of online learners.

1 Introduction

With the great advances in Web and multimedia technologies, e-learning has become one of the most important ways of learning new knowledge and skills. In the past few years, MOOC (Massive Open Online Course) has attracted numerous attentions from IT education and multimedia communities. MOOC provides services to learners by sharing educational resources through Internet. It is open accessible, scalable, and free to all social masses. This enables people to learn anywhere at anytime. However, due to the lack of instructions during learning, online learners have to face a mass of course materials and decide the

© Springer International Publishing Switzerland 2015
Y.-S. Ho et al. (Eds.): PCM 2015, Part I, LNCS 9314, pp. 559–569, 2015.
DOI: 10.1007/978-3-319-24075-6_54

suitable learning strategies by themselves. Without any instructions, they may easily get lost in the mass of materials. Knowledge representation has proven an effective way for e-learning [1]. To improve the learning experience, in this paper, we address the construction of learning maps by structuring the course materials to facilitate more efficient browsing for learners.

Among all online course materials, video has been the most commonly used in the current e-learning platforms since it captures the vivid presentation of the domain knowledge from the instructors. In most existing systems, lecture videos are usually presented in a linear manner according to the lecture presentation and recording time. This can be used by learners to linearly browse the lecturer's presentations from the beginning to the end of the course. However, the inner structure of the domain knowledge of a course is usually not linear. For those people who just want to learn a specific domain concept, they prefer to watch only the videos with the related and the prerequisite concepts instead of the whole video corpus. Even for those learners who learn the whole course, a well-organized video corpus can help them to have an overview of the domain knowledge and plan their learning. Thus, it is necessary to re-organize the lecture video corpus to' better represent the knowledge of the course so as to facilitate efficient and effective learning.

Topic threading of video corpus has been actively researched in multimedia field [7,8]. In [8], news video archives are documented by story clustering and threading. To discover the relationship between videos, visual, speech, and text information are frequently employed. However, for lecture videos, visual information is usually less informative where only the lecturer and the whiteboard/slides are present with limited noticeable visual changes between different videos. Instead, the domain concepts that the videos present are the most informative which contain the semantic information of the presentation. In our previous work [3], we propose a framework to construct the knowledge representation by using speech transcript and handwriting texts which contain the domain concepts. However, it is nontrivial to extract the concept words from videos. The speech usually contains many noisy words besides the concepts. The handwriting texts are clearer; however, the recognition rate of handwriting is very low. This reduces the effectiveness of the semantic understanding of the presentations.

Besides video presentations, there are abundant domain knowledge about the courses available from other resources such as Web. In this paper, we explore the external domain knowledge from Wikipedia to aid the analysis of lecture videos. The external knowledge can help the structuring of lecture videos from the following two aspects. First, by identifying the domain concepts, we can filter the noisy information from videos including speech transcripts and handwriting texts. This results in a more elegant and precise representation of the lecture content. We then construct a video map to better represent the relationships between different videos (Sect. 2). Second, by combining videos and the external knowledge, we further construct a concept based knowledge representation to represent the relationships between different concepts and help the learners design their learning strategies (Sect. 3).

The contribution of this paper lies in the following aspects: (i) We propose to employ external domain knowledge to boost the content analysis of lecture videos and the video based knowledge representation. (ii) We propose a revised TF-IDF weighting approach for the similarity measure between lecture videos. (iii) We propose to combine video presentations and academic articles to construct a concept based knowledge representation for online learners.

2 Construction of Video Map by Exploring Domain Knowledge from Wikipedia

The input to our system is a lecture video sequence $V = \{v_1, v_2, \cdots, v_m\}$ for a given course ordered according to the lecture presentation and recording time. Each video captures the lecturer's presentation for one or few domain concepts of the course. In most existing e-learning platforms, the videos are linearly presented to the learners for browsing. In this section, we construct a video map so as to represent the relationships between different videos which can better facilitate the learning and searching of desired videos and concepts by learners. First, we extract the concept words from videos to represent the content of each video by exploring the external domain knowledge from Wikipedia. Second, we revise the TF-IDF (Term Frequency-Inverse Document Frequency) algorithm to measure the similarities between videos based on the extracted concept words. Finally, a video map is constructed as the knowledge representation of the video corpus.

2.1 Extraction of Concept Words

To discover the relationships between different videos, the first step is to understand the content of each video. Since lecture videos are captured to present the domain knowledge of a course, the domain concepts are the most useful information to represent the video content. In [3], speech transcripts and handwriting texts are used to represent the video content. However, both of them suffer from low recognition accuracy. Furthermore, besides the concept words, there are many noisy texts which will affect the similarity measure between videos. To precisely identify the concept words in videos, we employ the domain knowledge of the course from Web to filter the speech transcripts and handwriting texts.

First, for a given online course, we collect the domain concept words from Wikipedia [9] to form a concept word list $C^{(W)} = \{c_1, c_2, \cdots, c_N\}$. Second, for each video in the corpus, the lecturer's speech and handwriting texts are recognized [3], and then filtered by removing those words that are not included in $C^{(W)}$. This results in a more precise and elegant representation of the knowledge content of the video by focusing only on the domain concepts.

2.2 Representing Video Content with Revised TF-IDF

With the filtered concept set, we can measure the video similarity in a more precise way. To this end, the content of each video v_i is represented by a $N - dim$

concept vector $w_i = \{w_{i1}, w_{i2}, \cdots, w_{iN}, \}$ where w_{ij} is the weight of the concept c_j for v_i.

For the weighting of each concept word, TF-IDF (Term Frequency - Inverse Document Frequency) is the most used. However, compared with text documents, the lecture video corpus has its own feature, i.e. the sequence information in the temporal domain. The importance of a concept word in two videos or at two temporal points may be different. For instance, if a concept word c_j appears in a video v_i for the first time (c_j never appears before v_i), it means the lecturer begins to introduce and present this concept. In this case, c_j is the focus of this video and the weight of c_j for video v_i should be increased. Although the concept c_j may be referred when presenting other related concepts in the following videos, the first presence is usually the most important and should be highlighted. Based on this observation, we append the temporal information to TF-IDF by assigning different weights to a given concept according to the time of its presence. More specifically, if a concept c_j has appeared many times before video v_i, its weight for v_i will be reduced, which is defined as

$$w_{ij} = \frac{\alpha}{\alpha + F_{ij}} \cdot \frac{tf_{ij}}{T_i} \cdot \log \frac{m}{df_j} \qquad (1)$$

where tf_{ij} is the term frequency of concept c_j present in video v_i, $T_i = \sum_{p=1}^{N} tf_{ip}$ is the total number of concept words present in video v_i, df_j is the number of videos where c_j is present, $F_{ij} = \sum_{q<i} tf_{qj}$ is the total frequency that c_j is present in all videos before v_i, and α is a smooth factor which is empirically set to be 5 in our experiments. If v_i is the first video where c_j is present, $F_{ij} = 0$. Otherwise, if v_j has already appeared many times before v_i, its importance become minor with the growing of F_{ij}. Finally, by normalizing the weights in Eq. 1 as

$$w'_{ij} = \frac{w_{ij}}{\sum_{j=1}^{m} w_{ij}} \qquad (2)$$

we represent the content of video v_i with a concept vector $a_i = (w'_{i1}, w'_{i2}, \cdots, w'_{iN})$.

2.3   Construction of Video Map by Maximum Spanning Tree

Given a lecture video corpus for a course, we first construct a graph G with the videos as the vertexes. The edge between every two vertexes is weighted with the similarity between the two corresponding videos. Here we employ Cosine similarity to calculate the weight between two videos as

$$sim_{cos}(v_i, v_k) = \frac{a_i \cdot a_k}{||a_i|| \cdot ||a_k||} \qquad (3)$$

where a_i, a_k are the feature vectors for the two videos respectively calculated in Sect. 2.2.

Although the graph G presents the similarities between different videos, as a complete graph, it is less informative and thus learners cannot easily have an overview of the knowledge structure of the video corpus. Next, we simplify

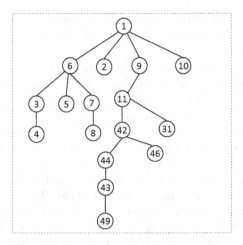

Fig. 1. Part of the video map for the course Chemistry.

the graph to be a tree where the edges with larger weights are preserved and the edges with small weights are removed. This is carried out by employing the Maximum Spanning Tree (MST) algorithm as in our previous work [3]. Figure 1 illustrates an example of the video map for the course *Chemistry*. By representing the video corpus with a tree, users can easily find the learning path by watching the related videos when they are going to explore a specific concept instead of linearly browsing all the videos.

3   Construction of Concept Map by Integrating Wikipedia Knowledge and Lecture Videos

The video based knowledge representation in Sect. 2 shows the relationship between different videos. Most time learners prefer to learn based on the relationships between different concepts and view the video corpus based on concepts. In this section, we construct a learning map between all domain concepts presented in the video corpus. Although videos contain explanations to the concepts, it is difficult to discover the relationship between different concepts based on videos. Each video may contain few different concepts and it is hard to segment a video into clips corresponding to different concepts in order to discover the relationship between them. In our approach, we employ the Wikipedia articles about the domain concepts to discover the relationships between different concepts and construct the learning maps between them. First, we construct an undirected graph to present the relationships between different concepts based on the domain articles. Second, with the lecture videos, we calculate the prerequisite relationship between every two semantically related concepts and generate a directed concept map. If c_a is a prerequisite concept of c_b, the learners need to learn c_a in order to learn c_b. This is represented in our concept by adding a directed edge from c_a to c_b. By identifying the prerequisite relationship between different concepts, learners can easily design their learning strategies according to the concept map.

The construction of concept map has been researched in some previous works. In [5], concept map is generated to find Remedial-Instruction path so as to assist students learn better. The lecture needs to manually pre-set the weights and the threshold during the construction of the concept map. In [4], text information from academic articles is employed to build an domain concept map. The relation strength of two concepts is measured and PCA algorithm is used to construct the concept map. However, the prerequisite relationships between different concepts are not revealed. In [6], a concept map is constructed by comparing the weight between two concepts with a predefined threshold. The prerequisite relationships between different concepts are also not presented. In our approach, lecture videos and the Wikipedia articles are integrated so as to automatically discover the semantic relatedness and the prerequisite relationships between different concepts. More specifically, the academic articles contain the detailed semantic relationship between different concepts, while the video sequence implies the prerequisite relationships between them.

3.1   Constructing Undirected Concept Map with Wikipedia Articles

For each domain concept c_i, we download the corresponding article D_i from Wikipedia, which can be used as an explanatory document for the concept. Next, the semantic content of D_i is represented by an $N-$dimension vector $u_i = (u_{i1}, u_{i2}, \cdots, u_{iN})$ where u_{ij} counts the presence of the concept c_j in D_i. Based on this representation, we calculate the semantic relatedness between every two concepts with Cosine similarity.

A graph $\mathcal{G}$ is then constructed with each concept as a vertex. The edge e_{ab} between two concept c_a and c_b is weighted by the similarity between them. Next, we simplify $\mathcal{G}$ by removing those edges with small weights. An edge e_{ab} is eliminated if its weight $s_{ab} < 3 \cdot \bar{s}$, where $\bar{s}$ is the average weight of all edges in $\mathcal{G}$.

3.2   Constructing Directed Concept Map by Discovering the Prerequisite Relationships Between Concepts

The map constructed in Sect. 3.1 is undirected, which is less useful for learners to decide their learning strategies. In this section, we add directions to the edges to show the prerequisite relationship between two concepts. This is carried out by exploring the temporal information of the concepts presented in lecture videos.

Generally speaking, if c_j is a prerequisite concept of c_k, learners need to learn c_i before learning c_k. In the video corpus, c_j is usually presented before c_k. Here we calculate the prerequisite score for a concept c_j according to its presence time in the video corpus. This is achieved by using the weights calculated in Sect. 2.1 which measures both the temporal information and the importance of a concept in each video. For a concept c_j, we get a ranked video list L in the descending order of the weights for c_j in different videos, i.e. for every p, $1 \leq p < m$, $w'_{L(p),j} > w'_{L(p+1),j}$. The prerequisite score for concept c_j is calculated as

$$Pr(j) = \sum_{p=1}^{M} \frac{L(p) \cdot w'_{L(p),j}}{M} \tag{4}$$

where $L(p)$ is the ordinal number in the video sequence V of the $p - th$ video in L, and M is empirically set to 3 in our experiments. The prerequisite score for a concept c_j actually estimates the ordinal number of the video which presents c_j in the lecture video sequence.

Based on the prerequisite score, we add directions to the edges in the concept map. For each edge connecting two concepts c_j and c_k, we replace it with an directed edge $c_j \rightarrow c_k$ if $Pr(j) \leq (1 + \beta) \cdot Pr(k)$, where β is a smooth factor and set to be 0.05 in our experiment. For those remaining undirected edges, the two related concepts are parallel presented with a high probability. They can be regarded as a concept cluster with very close semantic relationship. By combining lecture videos and academic articles, we can discover both the semantic relatedness and the prerequisite relationship between different concepts which cannot be accomplished by using either single material.

Next, we remove the redundant dependencies in the concept map. An edge from concept c_j to c_k is considered as redundant if there exists another path from c_j to c_k and the length of the path is larger than 1. We traverse the concept map in a depth-first manner to find and remove all the redundant edges.

Finally, in the resulting concept map, we associate each concept c_j with the first three videos from the video list L. With the concept map, learners can view the concept based knowledge structure of the course. If they would like to learn a specific concept, they can find the prerequisite concepts and design the learning path according to the directed map. During learning, they can search a specific concept in the video corpus by clicking the videos associated to it instead of blindly clicking the linearly arranged videos. Figure 2 shows an example of the concept map where each node is a concept from the course *Biology*. For instance, if a learner would like to learn the concept 25, s/he can find its prerequisite concepts and learn them by watching the associated videos along the red paths as shown in Fig. 2.

4   Experiments

Our experiments are carried out on the lecture videos from Khan Academy [10]. Three video corpuses are used for the courses *Chemistry*, *Biology*, and *Physics* respectively. For external domain knowledge, we download the concept list for each course and the explanatory articles for each concept from Wikipedia [9]. After word lemmatization, a concept list is generated for each course. Table 1 shows the statistics of the video corpuses and the concepts for the three courses.

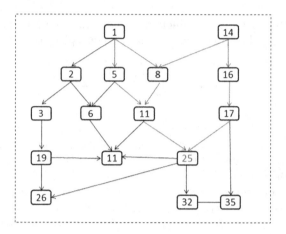

Fig. 2. Part of the concept map for the course Biology.

Table 1. The statistics of the video corpuses for three courses.

Course	Number of videos	Number of concepts
Chemistry	103	240
Biology	72	72
Physics	201	223

4.1 Evaluation of Video Maps

In the first experiment, we evaluate and compare the video based knowledge representation between the approaches in this paper and in [3]. We ask the professors and students majored in the corresponding subject to score the resulting learning maps according to the correctness and comprehensiveness of the knowledge representation. Each judge gives a score ranging from 1 to 5 for the generated maps with 5 meaning the best. Table 2 shows the mean scores for different approaches and courses.

Overall, our video map gets higher scores than the approach in [3]. Most judges comment that the video maps by our approach look more accurate and comprehensive. This is because the external domain knowledge is employed in our approach which enables us to more precisely understand the lecture content and estimate the relationship between different videos. Besides video maps, we provide users with directed concept maps. By referring to both video maps and concept maps, learners can better view the knowledge structure of the course and more efficiently browse the target concepts and videos. Our experiment shows that the concept map is enjoyable for online learners and lecturers.

Table 2. Subjective Evaluation of the Different Approaches.

Course	Video map in [3]	Video map with our approach	Video map + Concept map
Chemistry	3.64	4.25	4.59
Biology	3.87	4.33	4.63
Physics	3.41	4.19	4.42

4.2 Evaluation of Concept Maps

In the second experiment, we evaluate the directed concept maps generated in Sect. 3. The resulting concept maps are evaluated from two aspects: the accuracies of the prerequisite relationship (or the directed edges) and the concept-video association. This is carried out by asking the judges to manually label each directed edge in the concept map and the videos associated to each concept.

Table 3 shows the evaluation results for the prerequisite relationship in the concept map. For each concept, we can find one or few prerequisite concepts. The accuracies of the directed edges are quite encouraging. The judges agree that the concept maps can help a lot as instructions for online learners. Our experiments show that integrating lecture videos and external domain knowledge is a promising way of constructing effective knowledge representation.

Table 3. Accuracies of the prerequisite relationship in the concept map.

Course	# directed edges	# corrected directed edges	Accuracy
Chemistry	544	488	89.7 %
Biology	195	184	94.4 %
Physics	514	472	91.8 %

Table 4 presents the evaluation results of the concept-video association. In the concept map, we associated each node with three videos which present the corresponding concept so that learners can easily find and browse the lecturer's presentations. As shows in Table 4, for most concepts, the first video we provide to learners is the primary presentation by the lecturer. For the remaining concepts, the corresponding presentations can also be found in the second or the third videos. Some concepts are presented in more than one videos and thus we associate each concept with three videos. The concept-video association is mainly based on the weighting of the concept in videos with the revised TF-IDF approach in Sect. 2.1. This experiment validates the effectiveness of our proposed approach.

Table 4. Accuracies of the concept-video association (%).

Course	The first video	The second video	The third video
Chemistry	82.9	16.7	6.3
Biology	90.3	11.1	2.8
Physics	87.4	13.5	8.1

5 Conclusions

We have presented our approach for constructing knowledge representations of online lecture videos. In our approach, two learning maps are provided to online learners as instructions for them to more efficiently browse the course materials and design their learning strategies. Compared with previous works, we explore the external domain knowledge from Wikipedia and integrate it with the recorded presentations from lecturers. This enables us to more precisely understand the semantic content of lecture videos in constructing comprehensive knowledge representations for the courses. Our experiments demonstrate that the resulting learning maps are accurate, comprehensive, and enjoyable by the online learners. For future work, we will explore how to better organize lecture videos and materials from different sources and present them to learners so as to make online learning more efficient, effective, and enjoyable.

Acknowledgments. The work described in this paper was supported by the Science and Technology Commission of Shanghai Municipality under research grant no. 14DZ2260800, the National Natural Science Foundation of China (No. 1103127, No. 61375016), SRF for ROCS, SEM, and the Fundamental Research Funds for the Central Universities.

References

1. Lee, J., Segev, A.: Knowledge maps for e-learning. Comput. Educ. **59**(2), 353–364 (2012)
2. Shaw, B.: A study of learning performance of e-learning materials design with knowledge maps. Comput. Educ. **54**(1), 253–264 (2010)
3. Fan, P.M., Pong, T.C.: Constructing knowledge representation from lecture videos through multimodal analysis. Int. J. Inf. Educ. Technol. **3**(3), 304–309 (2013)
4. Chen, N.S., Kinshuk, Wei, C.W., Chen, H.J.: Mining e-learning domain concept map from academic articles. Comput. Educ. **50**(3), 1009–1021 (2008)
5. Lee, C.H., Lee, G.G., Leu, Y.: Application of automatically constructed concept map of learning to conceptual diagnosis of e-learning. Expert Syst. Appl. **36**(2), 1675–1684 (2009)
6. Tseng, Y.H., Chang, C.Y., Chang, S.N., Rundgren, C.J.: Mining concept maps from news stories for measuring civic scientific literacy in media. Comput. Educ. **55**(1), 165–177 (2010)

7. Pang, L., Zhang, W., Ngo, C.W.: Video hyperlinking: libraries and tools for threading and visualizing large video collection. In: ACM Multimedia Conference (2012)
8. Wu, X., Ngo, C.W., Li, Q.: Threading and autodocumenting in news videos. IEEE Signal Process. Mag. **23**(2), 59–68 (2006)
9. Web Concept Glossary from Wikipedia. http://en.wikipedia.org/wiki/Glossary_of_chemistry_terms, http://en.wikipedia.org/wiki/Glossary_of_biology, http://en.wikipedia.org/wiki/Glossary_of_physics
10. Khan Academy. https://www.khanacademy.org/

Object Tracking via Combining Discriminative Global and Generative Local Models

Liujun Zhao[✉] and Qingjie Zhao

Beijing Key Lab of Intelligent Information Technology,
School of Computer Science, Beijing Institute of Technology, Beijing, China
{zhaolj2011,zhaoqj}@bit.edu.cn

Abstract. In this paper, in order to track objects which undergo rotation and pose changes, we propose a novel algorithm that combines discriminative global and generative local model. Initially, we exploit the wavelet approximation coefficients and completed local binary pattern (CLBP) to represent the object global features. With the obtained global appearance descriptor, we use online discriminative metric learning to differentiate the target object from background. To avoid the drift problem results from global discriminative model, a novel generative spatial geometric local model is introduced. Based on SURF features, the generative local model quantizes the geometric structure information in scale and angle. Then, we combine these global and local models so that they can be benefit each other. Compared with several other tracking algorithms, the experimental results demonstrate that the proposed algorithm is able to track the target object reliably, especially for object pose change and rotation.

Keywords: Object tracking · Global discriminative model · Online metric learning · Generative model · Combined models

1 Introduction

Object tracking is an important task in computer vision and widely applied to many domains such as visual surveillance, human computer interaction. Despite extensive research on this topic, it remains a challenging problem since the appearance of an object suffers different changes such as occlusion, pose change, and rotation.

Numerous tracking algorithms have been proposed to account for appearance variation [1]. These algorithms are divided into discriminative and generative methods. The general tracking method locate the target by seeking the most similarity region in the following frames. Ross et al. [2] propose an incremental visual tracking algorithm which estimated object location by affine transformation of

L. Zhao—This work is partially supported by the Natural Science Foundation of China (No. 61175096, No. 61300082) and Specialized Fund for Joint Building Program of Beijing municipal Education Commission.

Y.-S. Ho et al. (Eds.): PCM 2015, Part I, LNCS 9314, pp. 570–579, 2015.
DOI: 10.1007/978-3-319-24075-6_55

two consecutive frames. Kwon and Lee [3] construct a set of basic observation models to fit the appearance changes. In [4], a algorithm based on local sparse representation histogram is proposed for object tracking where the target is located via model seeking of voting map. Comaniciu et al. [5] propose a kernel-based tracking algorithm using the mean shift-based mode seeking procedure. However, the background information is not used in these algorithms which may improve the tracking accuracy. In contrast, discriminative methods pose the visual tracking as a binary classification problem, and they are more robust and reliable due to the discrimination of foreground and background. Babenko et al. [6] introduce multiple instance learning into online boosting algorithm to adaptive track object. Garbner et al. [7] utilize an online boosting algorithm to select features for object tracking. The CT method [8] formulates the tracking task via a naive Bayesclassifier in compressed domain, which adopts random projection to project a datum in high-dimensional space to a low-dimensional vector. Tsagkatakis et al. [9] utilize a distance metric learning to measure similarity between candidates and the template. Despite much demonstrated success of these discriminative tracking algorithm, tracking drift problem remain to be solved. Recently, several algorithms are presented exploiting the advantages of both generative and discriminative appearance models. The approaches [10, 11] track objects by collaborating this both models with sparse representation. However, the size of sparse dictionary is difficult to determine for object tracking. Beside, the single pattern feature limits their reliability.

Motivated by these previous works, this paper proposes a novel object tracking method, which exploits discriminative global model and generative local model. In order to track objects undergo rotation and pose change, we develop an online discriminative global model which is based on texture feature and wavelet coefficients, and a spatial geometric SURF-based generative local model. The contributions of this paper lie in (1) we extend the object holistic representation with wavelet approximation coefficients and completed local binary pattern(CLBP), and generate a discriminative global object model by online distance metric learning. (2) Our method encodes the geometric structure information of SURF points to form generative local model. The spatial geometric information in local model includes scale and angle, which are able to avoid drift problem and improve the tracking accuracy. (3) A novel object tracking algorithm is proposed to deal with object rotation and pose changes with combined the discriminative global model and generative local model.

The rest of this paper is organized as follows. We introduce the proposed object tracking method that combines the discriminative metric global model and generative local model in Sect. 2. Section 3 shows experiment results and analysis. The conclusion of this paper is in Sect. 4.

2   Proposed Algorithm

In this paper, we track the target object by combining the discriminative global model and generative local model. Figure 1 shows the flow of the proposed

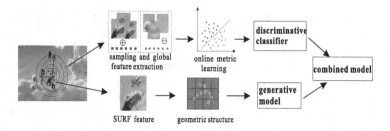

Fig. 1. Overview of the proposed appearance model. ⊕ is the positive sample, ⊖ is the negative sample

appearance model. Given the target location in frame t-1, we collect the positive and negative samples to train discriminative classifier with global appearance, and extract the SURF-based spatial geometric feature to generate local appearance model. In frame t, the combined appearance model can locate the target in the Particle Filter framework. In the following, we will introduce the proposed approach, mainly focusing on the appearance model.

2.1 Discriminative Global Model

Discriminative global model based on online metric learning is used to locate the object in whole image frame. Firstly, the wavelet coefficient and CLBP texture are extracted to present the object appearance. Then, online distance metric learning is introduced to train a classifier with object features.

Global Model: During the object tracking, the major feature energy of tracked object is almost consistent in sequence frames. Wavelet analysis is an useful tool, whose main feature is to fully highlight through frequency transformation. In this paper, we utilize the fast wavelet transformation to extract approximation coefficients as the major energy feature of the target object due to it is less sensitive to appearance change.

In contrast to approximation coefficients, the high coefficients present the texture of object image, but they are not obvious in object tracking due to the interference of image noise. Instead of extracting the high coefficients, we use the CLBP texture [12] as the object detail features. The local difference of CLBP texture is defined by the distance $d_p = s_p * m_p$, which is dived into sign s_p and magnitude m_p. The sign s_p is computed as follows:

$$CLBPS_{P,R} = \sum_{p=0}^{P-1} s(g_p, g_c)2^p, \quad s(x) = \begin{cases} 1 \ x \geq 0, \\ 0 \ x < 0. \end{cases} \tag{1}$$

where g_c is the gray value of center pixel, g_p is value of its neighbors. Similarly, we can obtain the magnitude m_p of the texture feature CLBPM.

$$CLBPM_{P,R} = \sum_{p=0}^{P-1} t(m_p, c)2^p, \quad t(x) = \begin{cases} 1 \ x \geq c, \\ 0 \ x < c. \end{cases} \tag{2}$$

The global appearance descriptor H_g is built by concatenating the histogram CLBPS, CLBPM and wavelet approximation coefficients.

Online Discriminative Metric Learning: Discriminative method views the tracking problem as a binary classification. Based on the appearance feature H_g, this paper learns a distance metric to separate the target object from background for visual tracking. The distance between candidate z_c and template z_t is defined as

$$d(z_c, z_t) = \| z_c - z_t \|^2 = (z_c - z_t)^T Q(z_c - z_t) \tag{3}$$

where z_c is an object sample represented by a feature vector H_g, Q is a distance metric matrix, and it is positive definiteness to guarantee that the distance function will be positive distances.

For object tracking, the location of the target can be achieved by minimizing the object function (3). In order to compute efficiently, we introduce an online distance metric learning (ITML) [13] to solve the optimal problem of (3). Given two vectors u_t and v_t, which consist of two different samples, we first to predict the distance $\hat{y}_t = d_{Q_t}(u_t, v_t)$ with the current distance metric. Then, the distance metric is updated from Q_t to Q_{t+1} by minimizing

$$Q_{t+1} = argmin_{Q \succ 0} D(Q, Q_t) + \eta \ell(d_Q(u_t, v_t), y_t). \tag{4}$$

where $D(Q, Q_t)$ is a regularization function, $\eta > 0$, $\ell(\hat{y}_t, y_t)$ is the loss between the predicted distance $\hat{y}_t$ and target distance y_t, $\ell(\hat{y}_t, y_t) = max(0, \frac{1}{2}b_t(\hat{y}_t - y_t)^2)$ and $\{b_t\} = 1$ or -1. By setting the gradient of (4) to zero, and using Sherman-Morrison inverse formula [14], we can obtain

$$Q_{t+1} = Q_t - \frac{\eta(\overline{y} - y_t)Q_t m_t m_t^T Q_t}{1 + \eta(\overline{y} - y_t)m_t^T Q_t m_t} \tag{5}$$

where $m_t = u_t - v_t$, $\overline{y} = d_{Q_{t+1}}(u_t, v_t)$ is the update distance and is found by

$$\overline{y} = \frac{\eta \hat{y}_t y_t - 1 + \sqrt{(\eta \hat{y}_t y_t - 1)^2 + 4\eta \hat{y}_t^2}}{2\eta \hat{y}_t} \tag{6}$$

During tracking, as shown in Fig. 1, we extract the image regions near the target within radius r_1, extract their global feature vectors as the positive samples. The image regions faraway from the target center within r_2 and r_3 ($0 < r_1 < r_2 < r_3$) are collected and used as negative samples. Then these positive and negative samples are utilized to update discriminative metric matrix according to Eq. 5). By using the learned metric matrix Q the object can be located with the confidences. The confidence of the candidate is defined as:

$$c(H_g) = exp(-\sigma * d(H_g^c, H_g^T)). \tag{7}$$

where H_g^T is the global histogram of template, and σ is the constant.

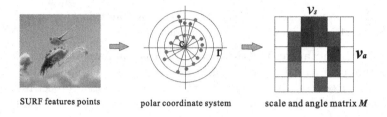

SURF features points   polar coordinate system   scale and angle matrix M

Fig. 2. Spatial geometric structure information

2.2   Generative Local Model

Although the discriminative global model is successful, it is insufficient, because
the different target images may share similarity in holistic view but only differ
in local spatial structure. The relative positions of stable points extracted from
an object in successive image frames almost remain consistent. For this reason,
based on the method in reference [15], we construct the spatial geometric correlation of SURF points, as generative local model, to avoid the estimated error
produced in discriminative global model.

We firstly extract the SURF points $P = \{p\}_{i=1}^n$ from an object image. In
order to obtain the geometric spatial structure of object appearance, as shown
in Fig. 2, we build the polar coordinate system based on the center c of the
normalized image, and quantize the statistics P into $V_s \times V_a$, V_s and V_a present
the different scales and angles respectively. For a region (v_s, v_a), the matrix
$M(v_s, v_a)$ is constructed by counting the number of points in this grid. They are
calculated as follows:

$$M(v_s, v_a) : \begin{cases} v_s = \sqrt{p_x^2 + p_y^2} \\ v_a = arctan(\frac{p_y}{p_x}) \end{cases} \tag{8}$$

where (p_x, p_y) is the location of a SURF point, v_s is scale bin and v_a is angles bin.
So the matrix $M(v_s, v_a)$ is able to preserve the spatial geometric information of
the SURF points P. Then local spatial histogram H_l is presented by the matrix
M, which each bin v indicates the scale and angle of region. In this paper, we
use 6 bins for quantization of scale, and 5 bins for quantization of angle. Next,
based on the local spatial geometric histogram H_l, we compute the conference
i.e. similarity between candidate and template by:

$$c(H_l) = exp(-d_{\chi^2}(H_l^T, H_l^c)) \tag{9}$$

where H_l^c is a local spatial histogram of candidates, d_{χ^2} is a function that computes χ^2−distance.

The final likelihood model p(o/c) which reflects the probability of a candidate
being the target, is utilized to verify the target state. It is constructed as a
combination of global representation H_g and local representation H_l, which is
written as

$$p(o/c) = c(H_g) * c(H_l). \tag{10}$$

2.3   Template Update

The object appearance is changing over time. In order to improve the tracking accuracy, the templates are updated online to account for the appearance change of the target object. Our template update scheme is divided into two respects: global model update and local model update.

For global model updating, we assume that the object state set manually in the first frame is true in the process of tracking. When a new state z_t is coming, the first object state z_1 and previous γ object states are taken into our global model updating scheme as follows:

$$T_g(t+1) = \alpha * z_1 + (1 - \alpha) * \frac{\sum_{i=t-\gamma}^{t} z_i}{\gamma+1}. \tag{11}$$

The update process of the local model update is different from that of the global model. Local template library $T_l(t)$ is composed of three elements, and each one is local spatial geometric histogram. Let S_t be the minimize similarity distance between candidates and templates in the t-th frame. TH is the predefined threshold. If the confidence of the new coming target object state subject to $S_{t-1} < S_t < TH$, the spatial geometric histogram $H_l(t)$ will be pushed into the local template library, and $T_l(t)$ is updated to $T_l(t+1)$ by replacing the oldest element. If $S_t > TH$, this means object is occluded heavily, so the template library T_l is not updated.

3   Experiments

Experiments are carried out on 7 challenging video sequences which are available from [16,17]. Our tracker is implemented in MATLAB 2012b which runs on an Intel Core2 E8400 3.00 GHz with 2 GB memory.

3.1   Implementation Details

In each experiment, the target is manually labeled in the first frame. In the first frame, we randomly select 20 positive samples from $D_1 = \{z| \parallel l(z) - l(\hat{z}_t) \parallel < r_1\}$ and 20 negative samples from $D_2 = \{z| r_2 < \parallel l(z) - l(\hat{z}_t) \parallel < r_3\}$ to train the initial discriminative classifier, where $l(z_t)$ denotes the location of image sample at the t-th frame. In the subsequence of object tracking, the metric matrix is updated online with 10 positive and 10 negative samples respectively. The radius $r_1=4$ for D_1 while $r_2=8$ and $r_3=45$ for D_2. In addition, the parameter σ in Eq. (7) is set to 0.5, α in the global template update strategy Eq. (8) is set to 0.9, γ is set to 5.

In this paper, to demonstrate the performance of the proposed tracker, we compare our approach with other five state-of-the-art methods with the same initialization target. These methods include IVT tracker [2], CT tracker [8], MIL tracker [6], SCM tracker [10], TLD tracker [18]. For fair comparison, these trackers are implemented using their publicly available source code.

3.2   Performance Evaluation

Quantitative Evaluation. We use center error and overlap rate to evaluate the performances of the different algorithms. Center error is a pixel-wise Euclidean distance between the center of tracking target rectangle and center of ground truth. The overlap rate is computed by intersection over union based on the tracking result R_T and ground truth R_G, i.e. $\frac{R_T \cap R_G}{R_T \cup R_G}$. Table 1 summarizes the averaged center location errors over all frames. Figure 3 illustrates the overlap rate of the evaluated trackers. As we can see in Table 1 and Fig. 3, the TLD method does not report the tracking results for a significant number of frames in sequences *bird2*, *bolt*, *tom1* and *panda*. Comparing with IVT, CT, MIL and SCM, our method has less center location error and higher overlap rate, so it is able to track the target more accuracy. In Fig. 3, the SCM tracker performs better than ours in the part frames of sequence *gym*, this due to the tracker is designed to deal with appearance change with trivial templates. However, when the large pose change and rotation occur at the same time, the SCM tracker loses the tracked target. It is obvious that our approach has higher overlap rate in most sequences. Overall, the proposed algorithm performs favorably against the other state-of-the-art methods in seven image sequences.

Table 1. Average center location errors (in pixels). Bold fonts indicates the best results while the *italic* fonts indicate the second best ones

Sequences	IVT	CT	MIL	SCM	TLD	Ours
bird2	55.7	15.5	*14.1*	25.9	-	**10.2**
bolt	387.6	234.4	84.6	*63.7*	-	**11.8**
surfer	*16.7*	47.6	35.1	23.6	28.0	**9.7**
DavidOutdoor	*12.8*	68.5	59.8	231.1	102.2	**8.4**
tom1	51.7	*34.2*	70.3	39.4	-	**9.8**
gym	27.6	46.5	17.9	*17.5*	26.6	**9.1**
panda	*87.6*	115.6	120.6	91.1	-	**6.7**
Average	91.3	80.3	*57.4*	70.3	-	**9.3**

Qualitative Evaluation. Figure 4 gives some tracking results of the sequences. For the sequence *bird2*, it is observed that the trackers IVT, CT, MIL, SCM drift away from bird after rotation and large pose variation occur (#94). In the sequence *bolt*, the target undergoes the non-grid pose variation and camera view changes (#133, #313). In these two sequences, the pose change leads to nonlinear appearance variation of the object where the IVT, CT, MIL and SCM trackers do not deal with robustly. In contrast, our algorithm based on the global discriminative model is able to track more image frames than other methods. The sequence *surfer* presents the case where the object undergoes pose change, rotation and scale changes. We observe that IVT and TLD methods do not

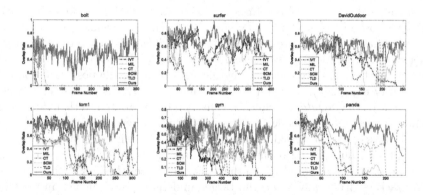

Fig. 3. Overlap rate plots for six test sequences.

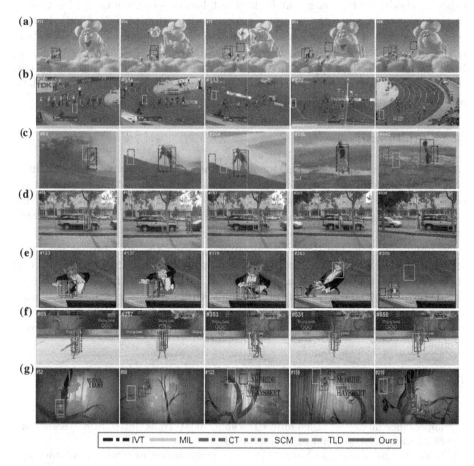

Fig. 4. Screenshots of tracking results of different trackers on video sequences: *bird2*(a), *bolt*(b), *surfer*(c), *DavidOutdoor*(d), *tom1*(e), *gym*(f), *panda*(g).

adapt the scale change of the target (#442), and SCM method suffers drift when the pose of surfer changes (#178, #204). In this scenario, the proposed method achieves the best results. For the *DavidOutdoor* sequence, the target undergoes pose variation and occlusion. The CT and MIL lose track of the target in numerous frames since they have no mechanism to handle pose variation and occlusion (#84, #205), while our method accurately tracks the target through out the entire *DavidOutdoor* sequence. This can be attributed to that the spatial geometric local model is able to distinguish with other similar object. In the *tom1* sequence, as the tracking target, Jerry moves with 360 degrees out-of-plane rotation and partial occlusion. Only our algorithm is able to locate the target accurately compared with others. For instance, the IVT, MIL, CT and SCM trackers drift away when the target undergoes pose change and rotates rapidly (#137, #263, #308). As shown in Fig. 4(f), the *gym* player has large pose change and rotation. The SCM approach tracks the target better than the other methods, except ours, but it does not handle pose change and rotation at the same time, since the limits of sparse presentation (#257, #353). In the sequence *panda*, it is obvious that our tracker performs well whereas the other trackers do not adapt the in-plane rotation and occlusion (#122, #216). This is because that the object appearance can be modeled well by global discriminative model and generative spatial geometric local appearance, where these two models are able to complement well each other.

4   Conclusion

In this paper, a novel object tracking algorithm, which combines the discriminative global model and generative spatial geometric local model, is proposed to cope with the rotation and pose change problems. Based on the wavelet coefficients and CLBP texture features, the discriminative global model locates the target using an online update metric classifier strategy. The spatial geometric local model helps locate the target accurately when the drift problem occurs. In addition, the global and local templates are updated with different scheme to increase the tracking accuracy. Compared with state-of-the-art algorithms, the experiments show that our proposed algorithm performs well on challenging sequences. In the future, the sub-patches feature of the target object could be introduced into tracking for adapting more variation types such as heavy occlusion.

References

1. Li, X., Hu, W., Shen, C., Zhang, Z., Dick, A., van den Hengel, A.: A survey of appearance models in visual object tracking. ACM Trans. Intell. Syst. Technol. **4**, 58 (2013). arXiv preprint arXiv:1303.4803
2. Ross, D., Lim, J., Lin, R., Yang, M.: Incremental learning for robust visual tracking. Int. J. Comput. Vis. **77**(1), 125–141 (2008)
3. Kwon, J., Lee, K.: Visual tracking decomposition. In: IEEE Conference Computer Vision and Pattern Recognition, pp. 1269–1276 (2010)

4. Liu, B., Huang, J., Yang, L., Kulikowsk, C.: Robust tracking usin1g local sparse appearance model and k-selection. In: IEEE Conference Computer Vision and Pattern Recognition, pp. 1313–1320 (2011)

5. Comaniciu, D., Ramesh, V., Meer, P.: Kernel-based object tracking. IEEE Trans. Pattern Anal. Mach. Intell. **25**(5), 564–577 (2003)

6. Babenko, B., Yang, M., Belongie, S.: Robust object tracking with online multiple instance learning. IEEE Trans. Pattern Anal. Mach. Intell. **33**(8), 1619–1632 (2011)

7. Grabner, H., Grabner, M., Bischof, H.: Real-time tracking via on-line boosting. In: British Machine Vision Conference, pp. 47–56 (2006)

8. Zhang, K., Zhang, L., Yang, M.-H.: Real-time compressive tracking. In: Fitzgibbon, A., Lazebnik, S., Perona, P., Sato, Y., Schmid, C. (eds.) ECCV 2012, Part III. LNCS, vol. 7574, pp. 864–877. Springer, Heidelberg (2012)

9. Tsagkatakis, G., Savakis, A.: Online distance metric learning for object tracking. IEEE Trans. Circuits Syst. Video Technol. **21**(12), 1810–1821 (2011)

10. Zhong, W., Li, H., Yang, M.: Robust object tracking via sparsity-based collaborative model. In: IEEE Conference Computer Vision and Pattern Recognition, pp. 1838–1845 (2012)

11. Xie, C., Tan, J., Cheng, P., Zhang, J., He, L.: Collaborative object tracking model with local sparse representation. J. Vis. Commun. Image Represent. **25**(2), 423–434 (2014)

12. Guo, Z., Zhang, D.: A completed modeling of local binary pattern operator for texture classification. IEEE Trans. Image Process. **19**(6), 1657–1663 (2010)

13. Jain, P., Kulis, B., Dhillon, I.S., Grauman, K.: Online metric learning and fast similarity search. In: Advances in Neural Information Processing Systems, pp. 761–768 (2009)

14. Sherman, J., Morrison, W.: Adjustment of an inverse matrix corresponding to a change in one element of a given matrix. Ann. Math. Stat. **21**(1), 124–127 (1950)

15. Yang, F., Lu, H., Yang, M.: Learning structured visual dictionary for object tracking. Image Vis. Comput. **31**(12), 992–999 (2013)

16. Wu, Y., Lim, J., Yang, M.: Online object tracking: a benchmark. In: IEEE Conference Computer Vision and Pattern Recognition, pp. 2411–2418 (2013)

17. Yang, F., Lu, H., Chen, Y.-W.: Human tracking by multiple kernel boosting with locality affinity constraints. In: Kimmel, R., Klette, R., Sugimoto, A. (eds.) ACCV 2010, Part IV. LNCS, vol. 6495, pp. 39–50. Springer, Heidelberg (2011)

18. Kala, Z., Mikolajczyk, K., Matas, J.: Tracking-learning-detection. IEEE Trans. Pattern Anal. Mach. Intell. **34**(7), 1409–1422 (2012)

Tracking Deformable Target via Multi-cues Active Contours

Peng Lv[✉] and Qingjie Zhao

Beijing Key Lab of Intelligent Information Technology, School of Computer Science,
Beijing Institute of Technology, Beijing 100081, People's Republic of China
{plv,zhaoqj}@bit.edu.cn

Abstract. In this study, we present a novel multi-cues active contours based method for tracking target contours using edge, region, and shape information. To locate the target position, a contour based meanshift tracker is designed which combines both color and texture information. In order to reduce the adverse impact of sophisticated background and accelerate the curve motion, we extract rough target region from the coming frame by the proposed target appearance model. What's more, both discriminative pre-learning based global layer and voting based local layer are integrated into our appearance model. For obtaining the detailed target boundaries, we embed edge, region, and shape information into the level sets based multi-cues active contour model (MCAC). Experiments on seven video sequences demonstrate that the proposed method performs better than other competitive contour tracking methods under various tracking environment.

Keywords: Object contour tracking · Active contours · Level sets · Segmentation

1 Introduction

Tracking boundaries of moving objects is an important and challenging task in computer vision. Unlike conventional tracking methods limited to use the rectangular bounding-box or other stationary shapes [1] to represent the target, contour tracking aims to extract the detailed target boundary information. In order to meet this goal, a variety of non-rigid target contour tracking methods have been proposed during last two decades.

In [2], Paragios et al. firstly use the geodesic active contour model [3] to drive the curve to target boundaries during the evolution. Zhang et al. [4] introduce a background mismatching based method for tracking the region of moving target. Niethammer et al. [5] also propose a region based method for target contour tracking. However, due to that only simple edge or region information is used,

---

P. Lv—This work is supported by the National Natural Science Foundation of China (No. 61175096 and No. 61273273) and Specialized Fund for Joint Building Program of Beijing Municipal Education Commission.

© Springer International Publishing Switzerland 2015
Y.-S. Ho et al. (Eds.): PCM 2015, Part I, LNCS 9314, pp. 580–590, 2015.
DOI: 10.1007/978-3-319-24075-6_56

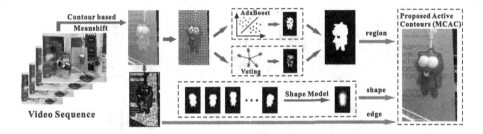

Fig. 1. Framework of the proposed contour tracking method.

these methods could not cope with the complex tracking environment, such as pose variation, occlusion, and sophisticated background. Recently, Bibbly et al. [6] propose a pixel-based contour tracking method which use generative model, however, without the edge information, the tracker may lose precise target boundary information. What's more, in [7], Vaswani et al. use a particle filter based mode tracker to track the target boundaries during the tracking. Combing both region and edge information, Cai et al. [8] propose a contour tracking framework. In [9], Fan et al. introduce an image matting based method to track the target region on a scribble trimap. However, these methods could not capture the various appearance in time, which may result in false segmentation.

In this paper, we propose a novel level sets based framework for tracking non-rigid target boundaries using edge, region and shape information. Our work mainly has the following three-fold contributions: (1) We propose a contour based meanshift target locating algorithm which integrates joint color and texture cues. (2) We propose a novel superpixel based dynamic appearance model using both global and local layers to extract the discriminative rough target region. In our method, a AdaBoost based pre-learning model and a voting algorithm are embedded into the global and local layers, respectively. Besides, for capturing changes of the target under sophisticated background, we update the appearance model dynamically during the tracking. (3) We also propose a multi-cues active contour model (MCAC) which combines edge, discriminative region and shape information for segmenting the target. To avoid the reinitialization procedure and reduce the computing time during the curve evolution, a distance regularization term is added in to the active contour model. With the discriminative region and shape information, the target could be segmented under various environment. The framework of our method is shown in Fig. 1.

The rest of the paper is organized as follows: Sect. 2 describe our target contour tracking framework. We show the qualitative and quantitative results in Sect. 3. In Sect. 4, we summarize the paper.

2   Proposed Method

2.1   Contour Based Meanshift Target Locating

To reduce the impact of complex background and extract the target more effectively, firstly we locate the target region before extracting its boundaries.

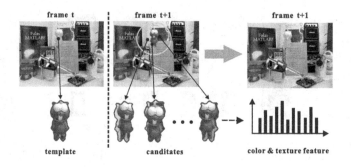

Fig. 2. Illustration of our contour based meanshift tracker.

A natural approach to track and locate the target position would be simply use meanshift, which use the rigid or elliptical region to represent the target. However, this approach has two drawbacks: (1) important target contour information may be lost during the tracking; and (2) the tracker may be interfered by the background in target boundingbox. To cope with these two problems, we use non-rigid region to represent the target in our meanshift tracker.

Considering that the target shape changes continuously during the tracking, which means that the shapes of targets between two continuous frame are highly similar. As shown in Fig. 2, at frame $t+1$, we use the non-rigid target region $I_C(t)$ in frame $I(t)$ as the target template, which provides precise target information. To enable our tracker achieve more robust tracking under various environment, we extract both color and texture information from the target region. For color information, a histogram is extracted in RGB&HSV color space, while for texture information, LBP feature is used:

$$\begin{cases} \mathbf{f}_R = \{\mathbf{f}_{color}, \mathbf{f}_{texture}\} \\ \mathbf{f}_{color} = f_{RGB\&HSV} \\ \mathbf{f}_{texture} = \text{LBP}(I_C(t)) \end{cases} \tag{1}$$

To measure the similarity between template region and candidates, we use the following distance:

$$d(\mathbf{f}_R, \mathbf{f}) = \sqrt{1 - \rho[\mathbf{f}_R, \mathbf{f}]} \tag{2}$$

where $\rho[\cdot]$ is the Bhattacharyya distance between two discrete distributions, which defined as:

$$\rho[\mathbf{f}_R, \mathbf{f}] = \sum_i^N \sqrt{\mathbf{f}_{i,R} \cdot \mathbf{f}_i} \tag{3}$$

Then we use meanshift algorithm to find the target position $\mathbf{y}'_R$ in frame $t+1$ as follows:

$$\mathbf{y}'_R = \frac{\sum_{i=1}^n \mathbf{y}_{i,R} w_i g(\cdot)}{\sum_{i=1}^n w_i g(\cdot)} \tag{4}$$

where w_i is the candidates weights, and $g(\cdot)$ is the kernel, respectively. After several iterations, a new non-rigid target position could be obtained in frame

$I(t+1)$, as shown in Fig. 2. In our method, this non-rigid target region provides important information for our appearance model, which will be described more detailedly in Sect. 2.2.

2.2 Appearance Model Combing Global and Local Layers

Considering that sophisticated background may affect the curve motion during the segmentation procedure, so we propose a appearance model to extract rough target region from the coming frame $I(t+1)$. Some prior works tend to use pixel-based or sparse-based models to represent the target, however, those models usually lost the detailed target boundaries information. In our method, we build a target appearance model based on superpixels, which enables the model to retain both target region and boundary information simultaneously during the tracking. Besides, rather than the single-layer target model in traditional methods, we combine both global and local layers for obtaining the rough region more robustly.

Discriminative Pre-learning Based Global Layer: In global layer, we use discriminative method to extract the global rough target region. For every superpixel sp, a histogram based feature descriptor $\mathbf{s}$ is extracted in RGB and HSV color space. Those feature descriptors are labeled by $+1/-1$ according to the following criteria:

$$\mathbf{s} = \begin{cases} \mathbf{s}^+, & \text{if } (\mathbf{s} \cap \text{Target})/\mathbf{s} \geqslant \eta \\ \mathbf{s}^-, & \text{if } (\mathbf{s} \cap \text{Target})/\mathbf{s} < \eta \end{cases} \tag{5}$$

To extract the rough target region during the tracking, an online AdaBoost classifier is trained and updated dynamically based on the labeled samples.

However, the target appearance may change during the tracking, which may lead to false classification. To avoid this problem, we pre-learning the target appearance from the coming frame $I(t+1)$ before classifying superpixels. Recall that in Sect. 2.1, the meanshift tracker locates the non-rigid target region in frame $I(t+1)$, which enables us to use this information to update the AdaBoost classifier. In pre-learning procedure, we select some unlabeled superpixels randomly from the region $I_C(t+1)$ in next frame as the positive examples to update the classifier. What's more, the internal superpixels have higher probability to be selected than ones closed to the periphery. After pre-learning the target appearance, our model could capture changes of the target. Therefore, in the global layer, a rough target region R_{t+1}^{global} finally obtained, as shown in Fig. 3(e).

Voting Based Local Layer: However, under various tracking conditions, the global layer may miss some local regions, which would result in false segmentation. To reduce the adverse impact of noises caused by global layer, we propose a local voting algorithm to extract the target region. In our model, in order to retain the local features, every unlabeled superpixel in coming frame $I(t+1)$ is voted by the surrounded labeled superpixels in prior frame.

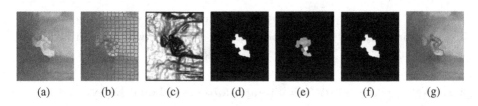

Fig. 3. Illustration of our global and local based appearance model: (a) the located position by our meanshift tracker; (b) superpixel segmentation; (c) the target edge information; (d) discriminative region of the global layer; (e) result of voting in local layer; (f) the final rough target regon; (g) final segmentation result on target region.

In local layer, we use the following distance to measure the similarity between two superpixels:

$$d_{sp}(\mathbf{s}_i, \mathbf{s}_j) = \exp\left(-\frac{\rho^2[\mathbf{s}_i, \mathbf{s}_j]}{\sigma}\right) \tag{6}$$

where $\rho(\cdot)$ is the Bhattacharyya distance given in Eq. 3. For every superpixel in $I(t+1)$, the score which voted by the surrounded superpixels in $I(t)$ is computed by the following formula:

$$\text{score}(sp_{i,t+1}) = \frac{\sum_j^{sp_{j,t}\in\Omega_r} \chi(d_{sp}(\mathbf{s}_i, \mathbf{s}_j))}{\|\chi(d_{sp}(\mathbf{s}_i, \mathbf{s}_j))\|_0} \tag{7}$$

where Ω_r is the region of radius r surrounding the superpixel $sp_{i,t+1}$ in frame $I(t)$. Besides, the kernel function $\chi(\cdot)$ is given by:

$$\chi(d_{sp}(\mathbf{s}_i, \mathbf{s}_j)) = \begin{cases} d_{sp}(\mathbf{s}_i, \mathbf{s}_j), & \text{if } d_{sp}(\mathbf{s}_i, \mathbf{s}_j) \geqslant \zeta \\ 0, & \text{if } d_{sp}(\mathbf{s}_i, \mathbf{s}_j) < \zeta \end{cases} \tag{8}$$

After the voting procedure, the local target region R_{t+1}^{local} is obtained, as shown in Fig. 3 (f).

For obtaining more stable target region, we combine the global and local layers as follows: $R_{t+1} = R_{t+1}^{global} \cup R_{t+1}^{local}$. This rough target region provides important region information for our active contour model, which will be discussed more detailedly in Sect. 2.4. Moreover, in order to reduce the noise cause by those two layers, open operator is used to the expanded rough target region:

$$R_t' = (R_t \ominus B_1) \oplus B_2 \tag{9}$$

where B_1 and B_2 denote the erosion and dilation structuring element, respectively. As shown in Fig. 3 (g) and (h), after integrating both global and local information into the appearance model, target could be extracted accurately.

2.3 Dynamic Shape Model

During the segmentation, various noise such as illumination and target appearance changes may affect the curve evolution, which would results in the false segmentation. What's more, some false negative regions generated by our appearance model may also cause over-segmentation. In order to handle these problems, we build a dynamic shape model to guide curve motion during the evolution.

For representing target shape S_t at time t, a gaussian kernel is applied to target region: $S_t = G(C_t)$, where C_t is the target region mask which is labeled by 1s and 0s. During the tracking, our shape model is updated dynamically as follows:

$$S_t = \sum_{k=1}^{t} p^{t-k} G(C_k) = G(C_t) + \sum_{k=1}^{t-1} p^{t-k} G(C_k)$$

$$= G(C_t) + S_{t-1}. \tag{10}$$

2.4 Multi-cues Active Contours and Curve Evolution

In this section, we will introduce our multi-cues active contour model for segmenting target which combines edge, region, and shape information. Because conventional active contour models [3,10,11] only consider edge or region information, therefore, the curve is vulnerable to be interfered by the complicated background or obvious boundaries and may stop at the false position after evolution. To handle these limitations, in our method, we embed our dynamic appearance model and shape model into active contours.

Edge Information: As many works refer, an edge-detector is defined for extracting the image boundaries: $g(|\nabla I|) = 1/(1 + |\nabla \hat{I}|^2)$. Note that the rough expanded target region R'_t, which is obtained in our appearance model as described in Sect. 2.2, could reduce the negative effect of the background. Therefore, to accelerate the curve evolution, we just let the curve move on the extended rough target region $I'_R(t)$, where $I'_R(t) = R'_t \cdot I(t)$. Then the edge information of the rough target region could be represented as follows:

$$g_{edge} = \frac{1}{1 + |\nabla \widehat{R'_t \cdot I(t)}|^2} = R'_t \cdot g(|\nabla I(t)|) - R'_t + 1. \tag{11}$$

According to the edge information g_{edge}, we define an edge term in our active contour model:

$$\mathcal{F}_1 \triangleq \int_{\Omega} g_{edge} \cdot \delta(\varphi) |\nabla \varphi| dx. \tag{12}$$

Region Information: In many situations, it is hard to extract target boundaries due to the blurred edge or sophisticated background, which would affect the curve motion during the evolution. In order to enable the curve to stop at the target boundaries correctly, target region information is embedded into our active contour model.

Recall that in Sect. 2.2, the rough target region R_t provides important information of target region for the active contour model. To embed the region information into our model, we transform the region R_t into homologous edge information beforehand:

$$g_{region} = g(|\nabla R_t \cdot I(t)|) + g(|\nabla R_t|) - 1$$
$$= R_t \cdot g(|\nabla I(t)|) + g(|\nabla R_t|) - R_t. \tag{13}$$

Therefore, we define the following region term in our active contour model:

$$\mathcal{F}_2 \triangleq \int_\Omega g_{region} \cdot \delta(\varphi)|\nabla \varphi| dx. \tag{14}$$

Shape Information: During the tracking, our appearance model may generate some false negative regions. Caused by the false negative regions information, the curve may move across the target boundaries, and stop at the wrong position. To cope with this problem, we add the target shape information to the active contour model:

$$\begin{cases} g'_{edge} = S_t \cdot g_{edge} \\ g'_{region} = S_t \cdot g_{region} \end{cases} \tag{15}$$

where S_t is the target shape model. Then we use Eq. 15 to update Eqs. 12 and 14, respectively. After integrating with shape information, our active contour model could produce more stable results.

Energy Functional and Curve Evolution: Combining the edge, region, and shape information, we propose a multi-cues active contour model (MCAC):

$$\mathcal{E}(\varphi) = \alpha \mathcal{F}_1(\varphi) + \beta \mathcal{F}_2(\varphi) + \mu \mathcal{R}(\varphi) + \tau \mathcal{A}(\varphi) \tag{16}$$

where $\mathcal{A}(\varphi)$ and $\mathcal{R}(\varphi)$ is area accelerate term and non-reinitialization term to speed up the curve evolution procedure, respectively. These two terms are given by:

$$\mathcal{A}(\varphi) \triangleq \int_\Omega g(|\nabla I(t)|)H(-\varphi)dx \tag{17}$$

$$\mathcal{R}(\varphi) \triangleq \int_\Omega p(|\nabla \varphi|)dx \tag{18}$$

where $H(\cdot)$ is the Heaviside function and $p(\cdot)$ is a potential function define in [11]. By using the finite difference calculation framework, the following gradient flow is obtained to optimize the energy functional $\mathcal{E}(\varphi)$:

$$\frac{\partial \varphi}{\partial t} = \delta_\epsilon(\varphi)\left[\alpha \operatorname{div}\left(S_t \cdot g_{edge} \cdot \mathbf{F}\right) + \beta \operatorname{div}\left(S_t \cdot g_{region} \cdot \mathbf{F}\right)\right]$$
$$+ \mu\operatorname{div}(d_p(|\nabla \varphi|)\nabla \varphi) + \tau g(|\nabla I|)\delta_\epsilon(\varphi), \tag{19}$$

where $\mathbf{F} = \nabla \varphi /|\nabla \varphi|$.

3 Experimental Results

3.1 Experimental Setup

The proposed method is implemented in MATLAB R2010b under Red Hat Enterprise Linux platform on a Intel(R) Core(TM)i7 3.4 GHz processor with 3 GB memory.

Parameters Setting: In Sect. 2.2, the radius r of voting region Ω_r is set to 20, and $\zeta = 0.3$. In Eq. 9, the erosion and dilation structuring element are 5×5 and 12×12, respectively. The updating parameter p in our dynamic shape model is set to 0.6. Besides, we set $\alpha = 1$, $\beta = 3$, $\mu = 1$, and $\tau = 2$ in our active contour model. During the evolution, we set number of the inner and outer iteration steps as 8 and 40, respectively.

Compared Algorithms and Evaluation Criteria: In our experiment, eight target contour tracking algorithms are compared: (a) our method with distance regularized level set evolution (DRLSE) [11]; (b) our method with region-based active contours (G-CV) [10]; (c) our method with edge-based active contours (GAC) [3]; (d) our method without shape information (w/o shape); (e) Scribble tracker which based on matting approach (Scribble tracker) [9]; (f) particle filter based mode tracker (deform PF-MT) [7]; (g) region tracking method based on background mismatch (Mismatch) [4]; and (h) our proposed method (MCAC). Moreover, To evaluate the segmentation performance, mis-tracked pixels rate (MPR) is defined: $MPR_t = |R_t^g \cup R_t - R_t^g \cap R_t| / |R_t^g|$, which indicates the coverage ratio between result and ground truth.

3.2 Qualitative and Quantitative Analysis

Complex Background: We test the methods on video *Lemming*, where the background is sophisticate during target moving, to verify the effectiveness of our method. As shown in Fig. 4, due to the sophisticated background and lacking of target shape information in DRLSE based method, the curve stops at the misplaced boundaries. Because of the accumulated errors during the tracking, Scribble tracker fails to segment the target correctly, yet. In our method, the rough target region extracted by the global and local based appearance model makes the segmentation environment more clear and provides important region information for the active contour model. By combing the region and shape information, our active contour model could cope with the interferences caused by complex background.

Various Appearance: For demonstrating the improvements of our method under various appearance tracking environment, we running the compared methods on video *Seq_sb*. During the tracking, as shown in Fig. 5, due to the changes of target pose and appearance, both deform PF-MT and Mismatch tracker lose the appearance information and fail to extract the target. On the contrary, profiting from the pre-learning procedure, our dynamic appearance model could capture

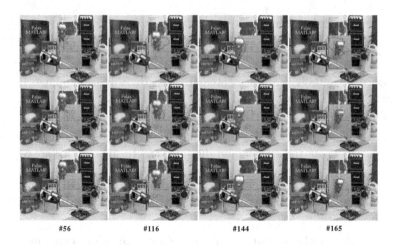

Fig. 4. Tracking results on *Lemming* with three methods (*from top to bottom*): Scribble tracker [9], ours with DRLSE [11], and the proposed method.

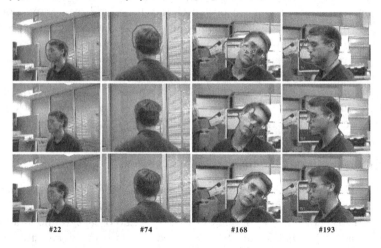

Fig. 5. Tracking results on *Seq_sb* with three methods (*from top to bottom*): Mismatch tracker [4], Deform PF-MT [7], and the proposed method.

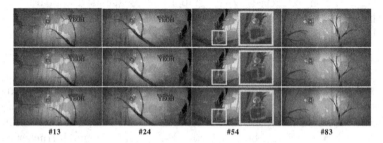

Fig. 6. Tracking results on *Panda* with three methods (*from top to bottom*): Scribble tracker [9], ours w/o shape method, and the proposed method.

Table 1. The mis-tracked pixels rate (MPR) on seven video clips with eight compared methods (the second best results are labeled with red font).

Sequence	w/o shape	G-CV [10]	GAC [3]	DRLSE [11]	Mismatch [4]	Scribble [9]	PF-MT [7]	MCAC
Lemming	0.156	1.425	0.873	0.164	0.443	0.330	0.254	**0.096**
Sufer	0.432	1.995	0.829	0.497	0.664	0.284	0.322	**0.262**
Pedxing1	**0.113**	0.333	0.474	0.216	0.460	0.199	0.328	0.121
Seq_sb	0.269	2.111	0.911	0.504	0.661	0.656	0.459	**0.231**
Seq_dhb	0.439	1.727	0.603	0.326	0.881	0.490	0.405	**0.253**
Seq_dt	0.183	1.343	0.696	0.436	0.494	0.656	0.491	**0.153**
Panda	0.226	2.229	1.698	0.463	1.952	1.327	0.874	**0.181**

the appearance changes promptly, which enables the proposed active contour model to segment the target correctly.

Occlusion: As shown in Fig. 6, in video *Panda*, the target is occluded by a tree and also rotates during its moving. At frame 54, Scribble tracker cannot extract the target region due to the occlusion. Notice that here we also test our method without shape information, as shown in Fig. 6, where we can see that without the shape information, the curve crosses the target boundaries and stops at internal region of the target. Thanks to the shape information in our active contour model, our method could deal with the occlusion during the tracking.

Now we quantify our method. As shown in Table 1, the conventional active contour model based methods (G-CV [10], GAC [3], and DRLSE [11]) usually lose the target during the tracking. That is mainly because the noises from the various tracking environment interfere the curve motion during the evolution, which results in false segmentation. Due to lacking of the dynamic appearance information, Mismatch tracker [4] could not capture the various appearance, and fails to segment the target. Both Scribble [9] and deform PF-MT [7] tracker do better on several tested sequences than conventional tracker, however, these two methods could not cope with the occlusion, as tested on *Panda*. Integrating with edge, region, and shape information, the proposed method performs better than other state-of-the-art methods under various tracking environment.

4 Conclusion

In this paper, we propose a novel level set based target contour tracking method based on multi-cues active contours by combing edge, region, and dynamic shape information for segmenting the target. Qualitative and quantitative results show that our method performs better than other state-of-the-art methods. Further work will aim at developing a more powerful appearance model to represent the target, which may improve the segmentation performance.

References

1. Yilmaz, A.: Kernel-based object tracking using asymmetric kernels with adaptive scale and orientation selection. Mach. Vision Appl. **22**(2), 255–268 (2011)
2. Paragios, N., Deriche, R.: Geodesic active contours and level sets for the detection and tracking of moving objects. IEEE Trans. Pattern Anal. Mach. Intell. **22**(3), 266–280 (2000)
3. Caselles, V., Kimmel, R., Sapiro, G.: Geodesic active contours. In: Proceedings of IEEE International Conference on Computer Vision, pp. 694–699 (1995)
4. Zhang, T., Freedman, D.: Improving performance of distribution tracking through background. IEEE Trans. Pattern Anal. Mach. Intell. **27**(2), 282–287 (2005)
5. Niethammer, M., Tannenbaum, A., Angenent, S.: Dynamic active contours for visual tracking. IEEE Trans. Autom. Control **51**(4), 562–579 (2006)
6. Bibby, C., Reid, I.: Real-time tracking of multiple occluding objects using level sets. In: Proceedings of IEEE Conference on Computer Vision Pattern Recognition, pp. 1307–1314 (2010)
7. Vaswani, N., Rathi, Y., Yezzi, A., Tannenbaum, A.: Deform pf-mt: particle filter with mode tracker for tracking nonaffine contour deformations. IEEE Trans. Image Process. **19**(4), 841–857 (2010)
8. Cai, L., He, L., Yamashita, T., Yiren, X., Zhao, Y., Yang, X.: Robust contour tracking by combining region and boundary information. IEEE Trans. Circuits Syst. Video Technol. **21**(12), 1784–1794 (2011)
9. Fan, J., Shen, X., Ying, W.: Scribble tracker: a matting-based approach for robust tracking. IEEE Trans. Pattern Anal. Mach. Intell. **34**(8), 1633–1644 (2012)
10. Chen, L., Zhou, Y., Wang, Y., Yang, J.: GACV: geodesic-aided c-v method. Pattern Recogn. **39**(7), 1391–1395 (2006)
11. Li, C., Xu, C., Gui, C., Fox, M.D.: Distance regularized level set evolution and its application to image segmentation. IEEE Trans. Image Process. **19**(12), 3243–3254 (2010)

Person Re-identification via Attribute Confidence and Saliency

Jun Liu[1], Chao Liang[1,2]([⊠]), Mang Ye[1], Zheng Wang[1],
Yang Yang[1], Zhen Han[1,2], and Kaimin Sun[3]

[1] School of Computer, National Engineering Research Center for Multimedia
Software, Wuhan University, Wuhan 430072, China
[2] Collaborative Innovation Center of Geospatial Technology,
Wuhan University, Wuhan, China
[3] State Key Laboratory of Information Engineering in Surveying, Mapping,
and Remote Sensing, Wuhan University, Wuhan 430072, China
{jliu_newbee,cliang,yemang,wangzwhu,yangyang0518,
hanzhen,sunkm}@whu.edu.cn

Abstract. Person re-identification is a problem of recognising and associating persons across different cameras. Existing methods usually take visual appearance features to address this issue, while the visual descriptions are sensitive to the environment variation. Relatively, the semantic attributes are more robust in complicated environments. Therefore, several attribute-based methods are introduced, but most of them ignored the diversities of different attributes. We epitomize the diversities of different attributes as two folds: the *attribute confidence* which denotes the descriptive power, and the *attribute saliency* which expresses the discriminative power. Specifically, the attribute confidence is determined by the performance of each attribute classifier, and the attribute saliency is defined by their occurrence frequency, similar to the IDF (Inverse Document Frequency) [1] idea in information retrieval. Then, each attribute is assigned an appropriate weighting according to its saliency and confidence when calculating similarity distances. Based on above considerations, a novel person re-identification method is proposed. Experiments conducted on two benchmark datasets have validated the effectiveness of the proposed method.

Keywords: Person re-identification · Attribute confidence · Attribute saliency

1 Introduction

In recent years, the person re-identification problem, namely matching people across disjoint camera views in a multi-camera systems, has aroused an increasing interest in multimedia analysis communities [2–6]. Person re-identification is still a challenge problem, due to large visual appearance changes caused by variations in viewpoints, lighting, background clutter and occlusion [4,7,8].

© Springer International Publishing Switzerland 2015
Y.-S. Ho et al. (Eds.): PCM 2015, Part I, LNCS 9314, pp. 591–600, 2015.
DOI: 10.1007/978-3-319-24075-6_57

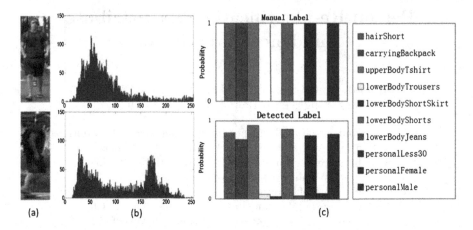

Fig. 1. Illustration of the attribute robustness: (a) Two images of a person under different cameras with 90 degrees viewpoint variation. (b) The histograms of visual features under two cameras. (c) The attribute histograms of the two images. The manual labels were pre-labeled by [9] and the detected labels were extracted by our pre-trained attribute classifies, while the values denote the probabilities to own those attributes. Note that visual features of the two images under different cameras are quite different, while attributes are much invariant, as for those attributes pre-labeled with 1, the corresponding detected label values are far greater than 0.5, for those attributes pre-labeled with 0, the corresponding detected label values are much less than 0.5.

Existing person re-identification methods mainly focus on constructing and selecting various distinctive and stable appearance representations based on various low-level features [10], such as colour feature [10], texture feature [3], and local feature [11], or a combination feature descriptor consisting of color and texture features [8,12]. However, visual features are sensitive to environment variation [7,11]. Relatively, the semantic attributes are more robust in complicated environments *e.g.* viewpoints and lighting [13–15]. An example is shown in Fig. 1. Therefore, we focus on attribute-based method in this paper.

Recently, attribute has been investigated in person re-identification problem. In [13], attributes and the visual features are fused together for person re-identification. Specifically, an attribute-centric representation was learned to describe folks, and a metric for comparing attribute profiles was learned to disambiguate individuals. Ngoc-Bao et al. [16] proposed a re-scoring method for adjusting attribute detection results based on the learned attribute relationship. In [17], Liu et al. presented a novel attribute-restricted latent topic model, which utilized human-specific attributes as restriction priors in a principled generative process. However, these approaches seldom considered that different attributes may have different contributions to person re-identification results (*i.e.* the diversities of attributes). Recently, Layne et al. [18] proposed an Attribute Interpreted Re-identification (AIR) method and a optimized method with weighted attributes (W.AIR). However, W.AIR failed to present quantitative calculation method for the diversities of attributes and the improvement is limited when compared to the original method (AIR).

Two phenomena about the diversities of various attributes were observed. One phenomenon is that, attribute detection accuracies are quite different and those better classified attributes are more reliable (*e.g.* the attribute *"lowerJeans"* detection accuracy is 57.3 %, the attribute *"lowerShortSkirt"* detection accuracy is 96.5 % and the latter is more reliable). Another phenomenon is that, The more frequent an attribute emerge, the less discriminative it is, vice versa. We express the diversities among various attributes by two folds based on the two phenomena, the *attribute confidence* and the *attribute saliency*:

Attribute Confidence. It is introduced to denote the diversity of the descriptive powers of the attributes. The descriptive powers of an attribute has positive correlation with its reliability when describing a image, which is defined by the attribute detection accuracy in this paper, *i.e.* the attribute classifier performance.

Attribute Saliency. It is adopted to express the diversity of the discriminative power of attribute, which is defined by the occurrence frequency of attribute in this paper, as the parallel way referred in [1]. An example is shown in Fig. 2.

Note that, the attribute confidence and saliency are positively related to its descriptive power and discriminative power respectively, they were combined as weight factor of attributes. Thus, in this paper, each attribute was given appropriate weighting according to a positive synthetic of its confidence and saliency.

In this paper, we present a person re-identification method using attributes with appropriate weightings based on the attribute confidence and saliency. The contributions of our work can be summarized in three subjects: (1) the attribute confidence is introduced to denote the descriptive power of attributes, and a quantitative calculation method for the confidence of attribute is presented,

Fig. 2. A example of attribute saliency: the one who wears red shorts is noticeable, and more discriminative as there are eight individuals, seven of them wear long trousers and only one wears red shorts (Color figure online).

(2) the attribute saliency is adopted to express the discriminative power of attributes, and a quantitative calculation method for the saliency of attribute is presented, (3) the combination of the attribute confidence and saliency is applied to illustrate and compute the diversities of attributes, and used as weight factor to promote person re-identification performances. The effectiveness of our proposed method is validated through the experiments developed on VIPeR [19] and PRID [20], two standard datasets for person re-identification.

2   The Proposed Approach

The proposed framework can be divided into three main steps: (1) training the attribute classifiers, (2) calculating the attribute confidence, which is determined by the attribute detection accuracy and the attribute saliency, similar to IDF, which is defined by the occurrence frequency of attribute in this paper, (3) combining the attribute confidence and saliency as weight factor when calculating similarity distances to promote person re-identification performances. The details are discussed in the following.

2.1   Classifiers Training

Attribute Classifiers. We adopted a linear Support Vector Machines (SVM) [21] to train the attributes classifiers. A linear SVM is used because it is both fast to train and fast to test. The training data consists of the visual feature vectors extracted on the training images, together with the pre-labeled attributes contributed by [9]. For each kind of attribute, a linear classifier is trained. We treat the ones contain the attribute as positive samples while the others as negative samples. The classifier is trained using the software package Vlfeat [22], which is optimized for linear classifiers and has complexity linear in the number of training samples.

Gallery Image Attribute Vectors. With the training attributes classifiers, the person image x can be represented by the posteriors $p(a|x)$ from each attribute classifiers as a n dimensional attribute vector: $A(x) = [p(a_1|x), \ldots, p(a_n)|x]$, while $p(a_i|x)$ denotes the output value of the classifiers of the ith attribute, and n is the number of attributes utilized in our study. Based on the training classifiers, attribute vectors can be extracted for the gallery images. Note that the output value of the classifiers for each attribute is from minus infinity to plus infinity. For better distance measure, we transform the original value to [0 1] by sigmoid function [23] as:

$$S(t) = \frac{1}{e^{-t} + 1} \tag{1}$$

where the t denotes the original output of the classifier, $S(t) = [0\ 1]$ is the transformed value that indicates the probability of owing this attribute. Specially, $S(t)$ is more closer to 1 indicates that it will more probably to own this attribute. $S(t) = 0$ expresses that the image do not have this attribute.

2.2   Attribute Confidence and Saliency Calculation

Attribute Confidence Calculation. When we train the classifies, the detection results of attributes are calculated, then we obtain the n dimensional detection accuracy vector $C = [c_1, c_2, \ldots, c_n]$, while n is the number of attributes applied in our method, c_n is the nth accuracy for the corresponding attribute. The attribute confidence, which is adopted to denote the descriptive power of attribute, is determined by the performance of each attribute classifier in our paper. Thus, we can get the attribute confidence based weightings vector $W_C = [w_{c1}, w_{c2}, \ldots, w_{cn}]$, while :

$$w_{ci} = \log_2 (1 + c_i) \tag{2}$$

Attribute Saliency Calculation. Inverse Document Frequency (IDF) is regarded as "term specificity" [24], and obtained by estimating the occurrence frequency of attribute. Thus it is introduced to define the attribute saliency in our paper. The ith attribute's IDF value s_i is estimated as:

$$s_i = \log \frac{M}{N_i} \tag{3}$$

where M is the number of images selected to train the SVM, N_i is the number of images that labeled with the ith attribute. Then, we get the n dimensional IDF vector $S = [s_1, s_2, \ldots, s_n]$, while n is the number of attributes selected in our method. The attribute confidence, which is applied to denote the discriminative power of an attribute, is defined by its occurrence frequency. Thus, we obtain the attribute saliency based weightings vector $W_s = [w_{s1}, w_{s2}, \ldots, w_{sn}]$, while:

$$w_{si} = s_i \tag{4}$$

2.3   Attribute Confidence and Saliency Matching Method

For each attribute, its descriptive power and discriminative power has the positive correlation with its confidence and saliency, respectively. We multiply W_c with W_s to synthesize their ability for expressing the diversities of attributes, which is treated as the weighting factor of attributes when calculating similarity distances. Thus, we obtain the confidence-saliency based weighting factor w_i:

$$w_i = \frac{1}{w_{ci} \times w_{si}} \tag{5}$$

while, $W = [w_1, w_2, \ldots, w_n]$, n is the number of attributes applied in our method.

For our attribute representation, similar to [18], we utilize $L2$-norm distance metric to compute similarity, and define dis_{ATT} as the sum of the original attribute $L2$-distances between images:

$$dis_{ATT}(I_p, I_q) = (A(I_p) - A(I_q))^{\mathrm{T}}(A(I_p) - A(I_q)))$$
$$= \sqrt{\sum_i (p(a_i|I_p) - (p(a_i|I_q)))^2} \tag{6}$$

Considering the diversities of different attributes, a weighted $L2$-norm distance metric is utilized to compute similarity. $dis_{W_{ATT}}$ is defined as the sum of weighted similarity for each attribute i between images, which is calculated by our Attribute Confidence and Saliency Matching method (ACSM):

$$
\begin{aligned}
dis_{W_{ATT}}(I_p, I_q) &= (A(I_p) - A(I_q))^{\mathrm{T}} diag(W)(A(I_p) - A(I_q))) \\
&= \sqrt{\sum_i w_i (p(a_i|I_p) - (p(a_i|I_q)))^2} \\
&= \sqrt{\sum_i \frac{1}{w_{ci} \times w_{si}} (p(a_i|I_p) - (p(a_i|I_q)))^2}
\end{aligned}
\tag{7}
$$

3   Experiments and Results

3.1   Dataset and Evaluation Protocol

Dataset. We choose the publicly representative dataset, VIPeR dataset [19] and PRID dataset [20] to verify our approach. We chose 632 persons in VIPeR and 200 persons in PRID, each of them has two images with drastic appearance difference. For each dataset, its image pairs were randomly split into two equal parts, one for training and acting as the auxiliary dataset, while the other for testing. For those images, each image is pre-labeled with 61 binary attributes by [4]. We chose VIPeR dataset and PRID dataset as both of them provide many challenges occurring in practical surveillance, viewpoint, pose and illumination changes, different backgrounds, low image resolutions, occlusions, etc.

Features. The visual features are extracted for training the SVM classifiers and generate the attribute vector of the gallery images. Similar to [8], a combination feature descriptor consisting of color and texture features is conducted. Specifically, for each image, the RGB and HSV color histograms and LBP descriptor are extracted from overlapping blocks of size 16×16 and stride of 8×8.

Attributes. The attributes we adopted are derived from [9], *i.e.* PETA dataset. This dataset contains 105 attributes. Since some of the attributes are rarely used, we delete them and 53 pre-defined attributes are kept in VIPeR, while 36 pre-defined attributes are kept in PRID.

Evaluation Protocol. For those images in VIPeR dataset and PRID dataset, we select half of them for training and the remaining for testing, while re-identification performance is reported on the held out test portion. We evaluate the re-identification performances based on the comparison of the original attribute detection results, the original attribute detection results with confidence-based weightings, saliency-based weightings and confidence-saliency based weightings separately. To compare performances of various methods, we evaluate the re-identification performance with the averaged Cumulative Matching Characteristic (CMC) curves [25] over 10 trials, and the value of CMC@k indicates the percentage of the real match ranked in the top k.

3.2   Results and Discussions

Attribute Confidence Results. The histogram of some attribute confidence is shown in Fig. 3. It demonstrates the undulation of the descriptive power of attributes. e.g. the greatest ones is 0.681, while the least ones is 0.420, the range is 0.281. Thus, the attribute confidence reflects the diversity of different attributes from the perspective of reliability.

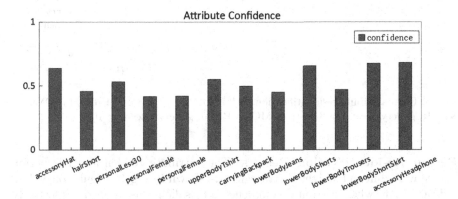

Fig. 3. The histogram of Attribute Confidence on VIPeR.

Attribute Saliency Results: The histogram of some attribute saliency is shown in Fig. 4. It demonstrates the undulation of the discriminative power of attributes. The standard deviation of this set of the attribute confidence is 1.1706, which highlights the diversity of the attribute saliency. Thus, the attribute saliency reflects the diversity of different attributes from the perspective of occurrence frequency.

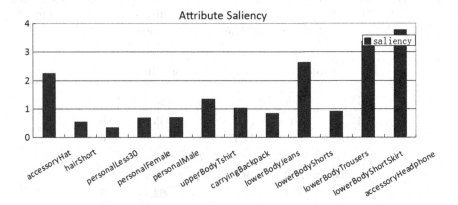

Fig. 4. The histogram of Attribute Saliency on PRID.

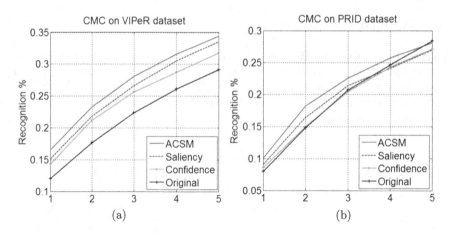

Fig. 5. Final attribute re-identification CMC plots for VIPeR and PRID. (a) CMC on VIPeR, gallery sizes p = 316; (b) CMC on PRID, gallery sizes p = 100.

Person Re-identification Results. The attribute re-identification performance with different weighted strategies are shown in Fig. 5. Results on VIPeR indicate that, when attribute confidence and saliency is applied respectively, the average improvement at rank@1 is 25.5 % and 17.8 % compared to original attributes. Specially, when our ACSM method is applied, the retrieval results are significantly improved. The improvement at rank@1 is 36.6 %, while improved 17.8 % at rank@5. Results on PRID shows that, when we our ACSM method is applied the improvement at rank@1 is 26.3 %. The improvements demonstrate that our proposed ACSM method is effective.

In Table 1, we compared the promote ratios between our ACSM method and W.AIR method in [18]. Our method reaps huger fruits for person re-identification performance. Specially, the improvement at rank@1 is 36.6 % and rank@5 is 17.8 % when our ACSM method is applied. However, corresponding to W.AIR the promote ratio is −12.7 % and 10.8 %. It is strongly suggest that our proposed ACSM method, which epitomizes the diversities of attributes by attribute saliency and attribute confidence is more efficient and reasonable.

Table 1. The promote ratios on VIPeR.

VIPeR	rank@1	5	10	25	50
AIR [18]	5.5	15.7	24.7	45.1	65.6
W.AIR [18]	4.8	17.4	29.0	50.6	68.6
Promote	−12.7 %	10.8 %	17.4 %	12.1 %	4.5 %
Original	12.0	29.1	39.6	56.1	69.8
ACSM	16.4	34.3	45.2	62.1	74.9
Promote	**36.6 %**	**17.8 %**	**14.1 %**	**10.7 %**	**7.3 %**

4 Conclusions

In this paper, a novel attribute confidence and saliency matching method is proposed for person re-identification problem. We applied the attribute confidence and saliency to measure the diversities of different attributes. More specifically, our method contains procedure for calculating confidence-saliency based weighings to evaluate the diversities. It is noteworthy that, to our knowledge, it is the first time to quantitatively describe the diversities of attributes. The improved performances on the VIPeR dataset and PRID dataset illustrated the effectiveness of our proposed method.

Acknowledgement. The research was supported by the National Natural Science Foundation of China (61303114), the Specialized Research Fund for the Doctoral Program of Higher Education (No. 20130141120024), the Nature Science Foundation of Hubei Province (2014CFB712), the China Postdoctoral Science Foundation funded project (2014M562058), the Technology Research Program of Ministry of Public Security (No. 2014JSYJA016), the National Nature Science Foundation of China (No. 61170023), the Internet of Things Development Funding Project of Ministry of industry in 2013 (No. 25), the Fundamental Research Funds for the Central Universities (2042014kf0250, 2042014kf0025), Jiangxi Youth Science Foundation of China (Grant No. 20151BAB217013), The Project Sponsored by the Scientific Research Foundation for the Returned Overseas Chinese Scholars, State Education Ministry ([2014]1685).

References

1. Salton, G., Buckley, C.: Term-weighting approaches in automatic text retrieval. Inf. Process. Manag. **24**, 513–523 (1988)
2. Gheissari, N., Sebastian, T.B., Hartley, R.: Person reidentification using spatiotemporal appearance. In: IEEE Conference on Computer Vision and Pattern Recognition (CVPR) (2006)
3. Farenzena, M., Bazzani, L., Perina, A., Murino, V., Cristani, M.: Person re-identification by symmetry-driven accumulation of local features. In: IEEE Conference on Computer Vision and Pattern Recognition (CVPR) (2010)
4. Zheng, W.S., Gong, S., Xiang, T.: Person re-identification by probabilistic relative distance comparison. In: IEEE Conference on Computer Vision and Pattern Recognition (CVPR) (2011)
5. Leng, Q., Hu, R., Liang, C., Wang, Y., Chen, J.: Person re-identification with content and context re-ranking. In: Multimedia Tools and Applications (MTA) (2014)
6. Liu, X., Song, M., Tao, D., Zhou., X.: Semi-supervised coupled dictionary learning for person re-identification. In: IEEE Conference on Computer Vision and Pattern Recognition (CVPR) (2014)
7. Zhao, R., Ouyang, W., Wang, X.: Learning mid-level filters for person re-identfiation. In: IEEE Conference on Computer Vision and Pattern Recognition (CVPR) (2014)
8. Wang, Y., Hu, R., Liang, C., Zhang, C., Leng, Q.: Camera compensation using feature projection matrix for person re-identification. IEEE Trans. Circ. Syst. Video Technol. (TCSVT) **24**, 1350–1361 (2014)

9. Deng, Y., Luo, P., Loy, C.C., Tang, X.: Pedestrian attribute recognition at far distance. In: ACM International Conference on Multimedia (MM) (2014)

10. Liu, C., Gong, S., Loy, C.C., Lin, X.: Person re-identification: what features are important? In: European Conference on Computer Vision, Workshops and Demonstrations (ECCV) (2012)

11. Gray, D., Tao, H.: Viewpoint invariant pedestrian recognition with an ensemble of localized features. In: Forsyth, D., Torr, P., Zisserman, A. (eds.) ECCV 2008, Part I. LNCS, vol. 5302, pp. 262–275. Springer, Heidelberg (2008)

12. Ye, M., Chao, L., Zheng, W., et al.: Specific person retrieval via incomplete text description. In: International Conference on Multimedia Retrieval (ICMR), Shanghai, China (2015)

13. Layne, R., Hospedales, T.M.,Gong, S.: Towards person identification and re-identification with attributes. In: IEEE Conference on Computer Vision and Pattern Recognition (CVPR) (2014)

14. Zhang, H., Zha, Z.J., Yang, Y., Yan, S., Gao, Y., Chua, T.S.: Attribute-augmented semantic hierarchy: towards bridging semantic gap and intention gap in image retrieval. In: ACM Multimedia (2013)

15. Wang, Z., Hu, R., Liang, C., Leng, Q., Sun, K.: Region-based interactive ranking optimization for person re-identification. In: Ooi, W.T., Snoek, C.G.M., Tan, H.K., Ho, C.-K., Huet, B., Ngo, C.-W. (eds.) PCM 2014. LNCS, vol. 8879, pp. 1–10. Springer, Heidelberg (2014)

16. Nguyen, N.-B., Nguyen, V.-H., Duc, T.N., Le, D.-D., Duong, D.A.: AttRel: an approach to person re-identification by exploiting attribute relationships. In: He, X., Luo, S., Tao, D., Xu, C., Yang, J., Hasan, M.A. (eds.) MMM 2015, Part II. LNCS, vol. 8936, pp. 50–60. Springer, Heidelberg (2015)

17. Liu, X., Song, M., Zhao, Q., Tao, D., et al.: Attribute-restricted latent topic model for person re-identification. Pattern Recogn. (PR) **45**, 4204–4213 (2012)

18. Layne, R., Hospedales, T.M., Gong, S., Mary, Q.: Person re-identification by attributes. In: British Machine Vision Conference (BMVC) (2012)

19. Gray, D., Brennan, S., Tao, H.: Evaluating appearance models for recognition, reacquisition, and tracking. In: IEEE International workshop on performance evaluation of tracking and surveillance (2007)

20. Hirzer, M., Beleznai, C., Roth, P.M., Bischof, H.: Person re-identification by descriptive and discriminative classification. In: Heyden, A., Kahl, F. (eds.) SCIA 2011. LNCS, vol. 6688, pp. 91–102. Springer, Heidelberg (2011)

21. Boser, B.E., et al.: A training algorithm for optimal margin classifiers. In: Proceedings of The 5th Annual ACM Workshop on Copputational Learning Theory, pp. 144–152. ACM Press (1992)

22. Vedaldi, A., Fulkerson, B.: VLFeat: An open and portable library of computer vision algorithms (2008). http://www.vlfeat.org/

23. Mitchell, T.M.: Machine Learning. WCB McGraw-Hill, Boston (1997)

24. Jones, K.S.: A statistical interpretation of term specificity and its application in retrieval. J. Documentation **28**, 11–21 (1972)

25. Wang, X., Doretto, G., Sebastian, T., Rittscher, J., Tu, P.: Shape and appearance context modeling. In: IEEE International Conference on Computer Vision (ICCV) (2007)

Light Field Editing Based on Reparameterization

Hongbo Ao[1], Yongbing Zhang[1]([✉]), Adrian Jarabo[3], Belen Masia[3,4],
Yebin Liu[2], Diego Gutierrez[3], and Qionghai Dai[2]

[1] Graduate School at Shenzhen, Tsinghua University, Shenzhen, China
zhang.yongbing@sz.tsinghua.edu.cn
[2] Department of Automation, Tsinghua University, Beijing, China
[3] Universidad de Zaragoza, Zaragoza, Spain
[4] MPI Informatik, Saarbrücken, Germany

Abstract. Edit propagation algorithms are a powerful tool for performing complex edits with a few coarse strokes. However, current methods fail when dealing with light fields, since these methods do not account for view-consistency and due to the large size of data that needs to be handled. In this work we propose a new scalable algorithm for light field edit propagation, based on reparametrizing the input light field so that the coherence in the angular domain of the edits is preserved. Then, we handle the large size and dimensionality of the light field by using a downsampling-upsampling approach, where the edits are propagated in a reduced version of the light field, and then upsampled to the original resolution. We demonstrate that our method improves angular consistency in several experimental results.

Keywords: Light field · Edit propagation · Reparameterization · Clustering

1 Introduction

In the last years, light fields [10] have gained attention as a plausible alternative to traditional photography, due to its increased post-processing capabilities, including refocus, view shifting or depth reconstruction. Moreover, both plenoptic cameras (e.g. *Lytro* ^{TM} or *Raytrix* ^{TM}) or automultiscopic displays [13] usign light fields have appeared in the consumer market. The wide-spread of this data has created a new need for providing similar manipulation capabilities of traditional images or videos. However, only a few seminal works [6,7] have been proposed to fill this gap.

Editing a light field is challenging for two main reasons: (*i*) the increased dimensionality and size of the light field makes it harder to efficiently edit it, since the edits need to be performed in the full dataset; and (*ii*) angular coherence needs to be preserved to provide an artifact free solution. In this work we propose a new technique to effectively edit light fields based on propagating the edits

© Springer International Publishing Switzerland 2015
Y.-S. Ho et al. (Eds.): PCM 2015, Part I, LNCS 9314, pp. 601–610, 2015.
DOI: 10.1007/978-3-319-24075-6_58

specified in a few sparse coarse strokes. The key idea of our method is to include a novel light field reparametrization that allows us to implicitly impose view-coherence in the edits. Then, inspired in the work by Jarabo et al. [7], we propose a downsampling-upsampling approach, where the edit propagation routine is done in a significantly reduced dataset, and then the result is upsampled to the full-resolution light field.

In comparison to previous work, our results *preserve view-coherence* thanks to the reparametrization of the light field, is *scalable* in both time and memory and is easy to implement on top of any propagation machinery.

2   Related Work

Previous works mainly focus on edit propagation on single images, with some extensions to video. Levin et al. [9] formulate a local optimization to propagate user scribbles to the expected regions in the target image. The method requires a large set of scribbles or very large neighborhoods to propate the edits in the full image. In contrast, An and Pellacini [1] propose a global optimization algorithm by considering similarity between all the possible pixel pairs in a target image; they formulate propagation as a quadratic system and solve it efficiently by taking advantage of its low-rank nature. However, this method scales linearly with the size of the problem, and does not account for view coherence. Xu et al. [15] improve An and Pellacini's method by downsampling the data using a kd-tree in the affinity space, which allows them handling large datasets. However, they scale poorly with the number of dimensions.

Other methods propose to increase the efficiency and generality of the propagation by posing as different energy minimization systems: Li et al. [11] reformulate the propagation problem as an interpolation problem in a high-dimensional space, which could be solved very efficiently using radial basis functions. Chen et al. [5] design a manifold preserving edit propagation algorithm, based on the simple intuition that each pixel in the image is a linear combination of other pixels which are most similar with the target pixel. The same authors later improve this work by propagating first in the basis of a trained dictionary, which is later used to reconstruct the final image [4]. Xu et al. [16] derive a sparse control model to propagate sparse user scribbles successfully to all the expected pixels in the target image. Finally, Ao et al. [2] devise a hybrid domain transform filter to propagate user scribbles in the target image. None of these works are designed to work efficiently with the high-dimensional data of light fields, and might produce inconsistent results between views, that our light field reparametrization avoids.

Finally, Jarabo et al. [7] propose a novel downsampling-upsampling propagation method, which handles the high dimensionality of light fields. We solve our problem efficiently inspired by their approach, although they do not enforce view consistency. This is to the best of our knowledge the only work dealing with edit propagation in light fields, while most previous effort on light field editing have focused on local edits [6,12,14] or light field morphing [3,18].

3   Light Field Editing Framework

The proposed algorithm can be divided into two parts. The first part is light field reparameterization, while the other one is the downsampling-upsampling propagation framework. The latter can be split into three phases: downsampling the light field, propagation on the downsampled light field, and guided upsampling of the propagated data.

We rely on the well-known two-plane parameterization of a light field [10], shown in Fig. 1 (a), in which each ray of light $\mathbf{r}$ in the scene can be defined as a 4D vector which codes its intersection with each of the two planes $\mathbf{r} = [s, t, x, y]$. One of the planes can be seen as the camera plane, where the cameras are located (plane st), and the other as the focal plane (plane xy). Note that the radiance can be reduced to a 4D vector because we assume it travels through free space (and thus does not change along the ray). It is often beneficial to look at the epipolar plane images of the light field. An epipolar volume can be built by stacking the images corresponding to different viewpoints; once this is done, if we fix e.g. the vertical spatial coordinate along the volume, we can obtain an epipolar image or EPI (Fig. 1 (b)).

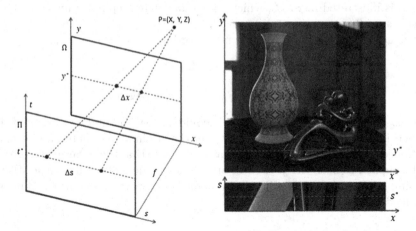

(a) Two-plane parametrization of light field   (b) View (s^*, t^*) and epipolar plane image S_{y^*,t^*}

Fig. 1. (a) Two-plane parametrization of a light field. Plane Π represents the camera plane of the light field, plane Ω represents the focal plane. Each camera location (s^*, t^*) yields a different view of the scene. (b) A view located at (s^*, t^*) and epipolar image S_{y^*,t^*}. We can obtain an epipolar image by fixing a horizontal line of constant y^* in the focal plane Ω and a constant camera coordinate t^* in the camera plane Π.

Once we model the light field with the two-plane parametrization, each pixel in the light field can be characterized by an 8D vector when color and depth information are taken into account. We thus express each pixel $\mathbf{p}$ in the light field as an 8D vector $\mathbf{p} = [r, g, b, x, y, s, t, d]$, where (r, g, b) are the colors of the pixel, (x, y) are the image coordinates on plane Ω, (s, t) are the view coordinates on plane Π and d is the depth information of the pixel. This notation will be used throughout the rest of the paper.

3.1 Light Field Reparameterization

One of the main challenges when doing light field editing is preserving view consistency. Each object point in the light field has a corresponding image point in each of the views of it (excepting occlusions), and these follow a slanted line (with slant related to the depth of the object) in the epipolar images. Here, we exploit this particular structure of epipolar images and propose a well-designed transformation of the light field data that will help preserve this view consistency when performing editing operations.

This transformation amounts to reparameterizing the light field by assigning to each pixel **p** a transformed set of xy coordinates, (x', y'), such that the pixel, in the transformed light field, will be defined by vector $[r, g, b, x', y', s, t, d]$. These new coordinates, will result in a transformed light field in which pixels corresponding to the same object point will be vertically aligned in the epipolar image, that is, will not exhibit spatial variation with the angular dimension; this process is illustrated in Fig. 2, which shows an original epipolar image and the same image after re-parameterization.

The actual computation of these new coordinates is given by Eqs. 1 and 2:

$$x' = \psi(x, y, d) = x - (y - y_c) \cdot (d - 1), \tag{1}$$

$$y' = \phi(x, y, d) = y - (x - x_c) \cdot (d - 1), \tag{2}$$

where x_c and y_c are the coordinates of the middle row and middle column of the epipolar images, respectively, in order to set the origin at the center, and d is, as mentioned, the depth information of that pixel. Note that the reparameterization can be applied to both the $y - t$ slices and the $x - s$ slices of the light field. Using this simple transformation will help in maintaining view consistency within the light field data.

3.2 Downsampling-Upsampling Propagation Framework

To efficiently address the propagation task, we build on the downsampling and upsampling propagation framework proposed by Jarabo et al. [7]. The improved downsampling-upsampling framework implements a three-step strategy to propagate scribbles on the reparameterized light field. To enable efficient calculation, the downsampling-upsampling propagation framework first makes use of k-means clustering algorithm [17] to downsample the light field data in the 8D space. Then a global optimization-based propagation algorithm is applied to the downsampled light field data. Finally, a joint bilateral upsampling method is used to interpolate the propagated data to the resolution of the original light field.

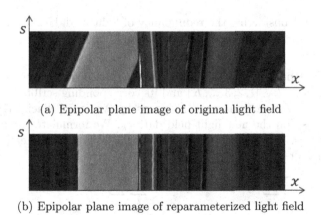

(a) Epipolar plane image of original light field

(b) Epipolar plane image of reparameterized light field

Fig. 2. (a) Epipolar image $S_{y*,t*}$ of the original light field. (b) Reparameterized epipolar image $S'_{y*,t*}$.

Downsampling Phase. To dispose of the unacceptable poor propagation efficiency due to the extremely large size of the light field data, and taking advantage of the large redundancy in it, we use k-means clustering [17] to downsample the original light field data to a smaller size data set. The downsampling phase successfully decreases the data redundancy by representing all the pixels in one cluster with the corresponding cluster center.

Given the original light field data we cluster the $M \times N^1$ 8D data points into K clusters ($K \ll N$), and thus merely need to propagate within the K cluster center points. Each cluster is denoted by C_k, $k \in [1, 2, 3, ..., K]$, and each cluster center is expressed as $\mathbf{c}_k$, $k \in [1, 2, 3, ..., K]$. The set $\mathbf{c}_k$, $k \in [1, 2, 3, ..., K]$, is therefore the downsampled light field.

Original scribbles drawn by the user to indicate the edits to be performed also need to be downsampled according to the cluster results. A weight matrix $\mathbf{D} \in \mathbb{R}^{M \times N}$ is used to record which pixel in the original light field is covered with user scribbles by setting the corresponding element in the matrix to 1 where a user scribble is present, and otherwise set the corresponding element to 0. Assume the original scribbles are expressed as $\mathbf{S} \in \mathbb{R}^{M \times N}$, then the new scribbles of the downsampled light field $\mathbf{s}_k$ can be calculated as follows:

$$\mathbf{s}_k = \frac{1}{M_0} \sum_{(i,j) \in \{(m,n)|\mathbf{p}_{mn} \in C_k\}} D_{ij} * S_{ij}, \tag{3}$$

$$M_0 = \sum_{(i,j) \in \{(m,n)|\mathbf{p}_{mn} \in C_k\}} D_{ij}, \tag{4}$$

where $\mathbf{p}_{mn}$, $m = [1, 2, ..., M]$, $n = [1, 2, ..., N]$ are 8D pixel vectors in the original light field. We get the downsampled scribble set $\{\mathbf{s}_k\}$, $k = [1, 2, ..., K]$, according

---

[1] We can represent the light field as a 2D matrix composed of the different views, as per common practice.

to Eqs. 3 and 4. Considering the redundancy of light field data, a small value K will be good enough to downsample the original light field.

Propagation Phase. After the downsampling phase, we get the downsampled light field data c_k, $k \in [1, 2, 3, ..., K]$ and its corresponding scribble set $\{s_k\}$. We adopt the optimization framework proposed by An and Pellacini [1] to propagate scribbles s_k on the new light field data c_k. We formulate the propagation algorithm in Eqs. 5 and 6, and by optimizing this expression we can acquire the propagated result e_k.

$$\sum_k \sum_j \omega_j z_{kj} (e_k - s_j)^2 + \lambda \sum_k \sum_j z_{kj} (e_k - e_j)^2, \qquad (5)$$

$$z_{kj} = exp(-||(c_k - c_j) \cdot \sigma||_2^2), \qquad (6)$$

where $c_k = (r_k, g_k, b_k, x_k, y_k, s_k, t_k, d_k)$ is pixel vector of the new light field c_k, $k \in [1, 2, 3, ..., K]$; z_{kj} is the similarity between pixel vectors k and j; $\sigma = (\sigma_c, \sigma_c, \sigma_c, \sigma_i, \sigma_i, \sigma_v, \sigma_v, \sigma_d)$ are the weights of each feature in the 8D vector used to compute the affinity and thus to determine the extent of the propagation in those dimensions; and ω_j is a weight coefficient which is set to 1 when s_j is not zero and is otherwise set to 0. For a small number of cluster centers, i.e. a small K, Eq. 5 can be solved efficiently.

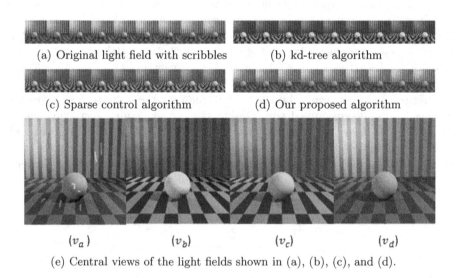

(a) Original light field with scribbles   (b) kd-tree algorithm

(c) Sparse control algorithm   (d) Our proposed algorithm

(v_a)   (v_b)   (v_c)   (v_d)

(e) Central views of the light fields shown in (a), (b), (c), and (d).

Fig. 3. Light field editing result on a 3D light field (horizontal parallax only). We show the initial scribbles drawn by the user and our result compared to that of two other algorithms. Note that (e) shows the central views of the different light fields shown, where the differences can be appreciated. Please refer to the text for details.

Upsampling Phase. Finally, we need to calculate the edited result of all the pixels in the original data set. In the upsampling phase, we utilize the propagated result set $\mathbf{e}_k$ to obtain the resulting appearance of each pixel in the full light field.

For each pixel in the original light field, we find n nearest neighbor cluster centers in the downsampled light field data set $\mathbf{c}_k$ by using a kd-tree for the searching process. Each pixel $\mathbf{p}$ will relate to one nearest neighbor cluster set $\Omega = \{c_j, j = 1, 2, 3, \cdots, m\}$ after the nearest neighbor search procedure. Then joint bilateral upsampling [8] will be used in the upsampling process. More formally, for an arbitrary pixel position p, the filtered result can be formulated as:

$$E(p) = \frac{1}{k_p} \sum_{q\downarrow\in\Omega} e_{q\downarrow} f(||p\downarrow - q\downarrow||) g(||I_p - I_q||), \tag{7}$$

where $f(x)$ and $g(x)$ are exponential functions ($exp(x)$); $q\downarrow$ and $p\downarrow$ are the positional coordinates of the downsampled light field; $e_{q\downarrow}$ is the color of the pixel vector in propagated light field; I_p and I_q are the pixel vectors in the original light field; and k_p is a normalizing factor, which is the sum of the $f \cdot g$ filter weights.

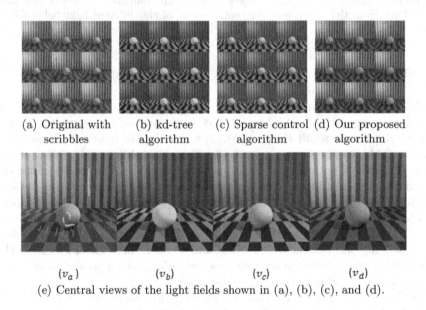

| (a) Original with scribbles | (b) kd-tree algorithm | (c) Sparse control algorithm | (d) Our proposed algorithm |

| (v_a) | (v_b) | (v_c) | (v_d) |

(e) Central views of the light fields shown in (a), (b), (c), and (d).

Fig. 4. Light field editing result on a 4D light field. We show the initial scribbles drawn by the user and our result compared to that of two other algorithms. Note that (e) shows the central views of the different light fields shown, where the differences can be appreciated. Please refer to the text for details.

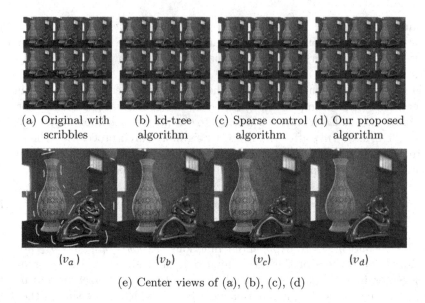

(a) Original with   (b) kd-tree   (c) Sparse control (d) Our proposed
    scribbles   algorithm   algorithm   algorithm

(v_a) (v_b) (v_c) (v_d)

(e) Center views of (a), (b), (c), (d)

Fig. 5. Another light field editing result on a more complex 4D light field. We show the initial scribbles drawn by the user and our result compared to that of two other algorithms. Note that (e) shows the central views of the different light fields shown, where the differences can be appreciated. Please refer to the text for details.

4   Results

In this section we show our results and compare with two state-of-the-art edit propagation algorithms: a kd-tree based method [15], and a sparse control method [16]. In the result shown in Fig. 3, we recolor the light field propagating a few scribbles on the center view of a 1×9 horizontal light field. We show the original light field with user scribbles on the center view, the results of the two previous methods, and our own, as well as larger center views for all for easier visual analysis. Our algorithm (v_d) preserves the intended color of the input scribbles better results, while avoids artifacts such as color bleeding into different areas. In contrast, both the kd-tree (v_b) and sparse control (v_c) methods produce some blending between the colors of the wall and the floor. This blending is also responsible on the change of the user-specified colors, which are darker in v_b and v_c, that our method propagates more faithfully.

In Figs. 4 and 5, we draw some scribbles on the center view of a 3×3 light field. Again, we show the input scribbles, a comparison between the previous methods and ours, plus larger central views. Similar to the results in Fig. 3, our method propagates more faithfully the input colors from the user. In addition, our method results into proper color segmentation based on the affinity of the different areas of the light field, while the results of the kd-tree (v_b) and sparse control methods (v_c) exhibit clear artifacts in form of blended colors, or wrongly propagated areas.

5   Conclusion

We have presented a light field edit propagation algorithm, based on a simple re-parameterization that aims to better preserve consistency between the edited views. We have incorporated it into a downsampling-upsampling framework [7], which allows to handle efficiently the large amounts of data that describe a light field. Our initial results show improvements over other existing edit propagation methods. These are the first steps in a possible direction towards the long-standing goal of multidimensional image editing. Further analysis and developments are needed to exhaustively test the validity of the approach.

Acknowledgements. The project is supported by the National key foundation for exploring scientific instrument No. 2013YQ140517 and partially supported by the National Natural Science Foundation of China under Grants 61170195, U1201255 & U1301257, the Spanish Ministry of Science and Technology (project LIGHTSLICE) and the BBVA Foundation. Diego Gutierrez is additionally supported by a Google Faculty Research Award. Belen Masia is partially supported by the Max Planck Center for Visual Computing and Communication.

References

1. An, X., Pellacini, F.: Appprop: all-pairs appearance-space edit propagation. ACM Trans. Graph. (TOG) **27**(3), 40 (2008)
2. Ao, H., Zhang, Y., Dai, Q.: Image colorization using hybrid domain transform. In: ICASSP, January 2015
3. Chen, B., Ofek, E., Shum, H.Y., Levoy, M.: Interactive deformation of light fields. In: Proceedings of the I3D 2005, pp. 139–146 (2005)
4. Chen, X., Zou, D., Li, J., Cao, X., Zhao, Q., Zhang, H.: Sparse dictionary learning for edit propagation of high-resolution images. In: 2014 IEEE Conference on Computer Vision and Pattern Recognition (CVPR), pp. 2854–2861. IEEE (2014)
5. Chen, X., Zou, D., Zhao, Q., Tan, P.: Manifold preserving edit propagation. ACM Trans. Graph. (TOG) **31**(6), 132 (2012)
6. Jarabo, A., Masia, B., Bousseau, A., Pellacini, F., Gutierrez, D.: How do people edit light fields? ACM Trans. Graph. **33**(4), 146:1–146:10 (2014)
7. Jarabo, A., Masia, B., Gutierrez, D.: Efficient propagation of light field edits. In: Proceedings of the SIACG 2011 (2011)
8. Kopf, J., Cohen, M.F., Lischinski, D., Uyttendaele, M.: Joint bilateral upsampling. ACM Trans. Graph. (TOG) **26**, 96 (2007). ACM
9. Levin, A., Lischinski, D., Weiss, Y.: Colorization using optimization. ACM Trans. Graph. (TOG) **23**, 689–694 (2004). ACM
10. Levoy, M., Hanrahan, P.: Light field rendering. In: Proceedings of the 23rd Annual Conference on Computer Graphics and Interactive Techniques, pp. 31–42. ACM (1996)
11. Li, Y., Ju, T., Hu, S.M.: Instant propagation of sparse edits on images and videos. In: Computer Graphics Forum, vol. 29, pp. 2049–2054. Wiley Online Library (2010)
12. Masia, B., Jarabo, A., Gutierrez, D.: Favored workflows in light field editing. In: CGVCVIP (2014)

13. Masia, B., Wetzstein, G., Didyk, P., Gutierrez, D.: A survey on computational displays: pushing the boundaries of optics, computation and perception. Comput. Graph. **37**, 1012–1038 (2013)
14. Seitz, S.M., Kutulakos, K.N.: Plenoptic image editing. Int. J. Comput. Vision **48**(2), 115–129 (2002)
15. Xu, K., Li, Y., Ju, T., Hu, S.M., Liu, T.Q.: Efficient affinity-based edit propagation using kd tree. ACM Trans. Graph. (TOG) **28**, 118 (2009). ACM
16. Xu, L., Yan, Q., Jia, J.: A sparse control model for image and video editing. ACM Trans. Graph. (TOG) **32**(6), 197 (2013)
17. Žalik, K.R.: An efficient k-means clustering algorithm. Pattern Recogn. Lett. **29**(9), 1385–1391 (2008)
18. Zhang, Z., Wang, L., Guo, B., Shum, H.Y.: Feature-based light field morphing. ACM Trans. Graph. **21**(3) (2002). http://doi.acm.org/10.1145/566654.566602

Interactive Animating Virtual Characters
with the Human Body

Hao Jiang and Lei Zhang[✉]

Shenzhen Key Lab of Broadband Network and Multimedia,
Graduate School at Shenzhen, Tsinghua University, Beijing, China
Jiangh13@mails.tsinghua.edu.cn, zhanglei@sz.tsinghua.edu.cn

Abstract. This paper presents a novel interactive motion mapping system that maps the human motion to virtual characters with different body part size, topology and geometry. Our method is especially effective for characters whose body is disproportional to human structure. To achieve this, we propose an improved Embedded Deformation algorithm to control virtual characters in real-time. In preprocessing stage, we construct the deformation subgraph for each part, and then merge them into a connected deformation graph, these works are entirely automatic and only have to be done once before running. At runtime, we use the Kinect to track human skeletal joints and iteratively solve the rotation matrix and translation vector for each deformation graph node. Then, we update mesh vertices position and normal. We demonstrate the flexibility and versatility of our method on a variety of virtual characters.

Keywords: Embedded deformation · Interactive · Virtual character · Animation

1 Introduction

Suppose that you work at Disney and need to create a 3D animation of the Baymax for the Big Hero 6 sequel. You may think about model rigging, skeleton deformation and key framing. In fact, you do not have to do so. With the development of motion sensing input devices, such as Microsoft's Kinect and Nintendo's Wii, it is possible to animate virtual characters interactively. For example, Kinect tracks human skeletal joints and outputs 3D coordinates of human joints. These coordinates, combined with traditional skeleton based animation techniques, can be used to animate humanoid characters effectively. But it will be failed to animate non-humanoid ones.

Recent years, researchers proposed various methods to animate non-humanoid characters. There are two main solutions: one is based on data-driven method, which maps human motion to the target character in the preprocessing stage, then searching for the proper motion at runtime [1, 2]. Another one is based on mesh deformation, which deforms the mesh directly according to the human skeletal joints [3]. Even though the first approach can map human motion to different topology characters, it inevitably occurs error at runtime. By comparison, mesh deformation is a more flexible and versatile method.

© Springer International Publishing Switzerland 2015
Y.-S. Ho et al. (Eds.): PCM 2015, Part I, LNCS 9314, pp. 611–620, 2015.
DOI: 10.1007/978-3-319-24075-6_59

However, there are three shortcomings of existing mesh deformation method [3]: **Firstly**, it cannot deform the mesh which each body part size is disproportionate to the human's. Take the Baymax as an example, even though we can scale it to the human size, which leg and arm cannot align to the human's well, then the deformation process is fail. **Secondly**, it is sensitive to the initial shape of mesh. If each part of the mesh is close to the other, they will affect each other at runtime. **Thirdly**, the binding process is not very convenient to use. It requires users to keep their skeleton bones intersect with the mesh vertices before running, which is not very natural.

Our paper extends the work of Sumner et al. [4] to animate virtual characters with different body part size, topology and geometry. To achieve this, we present a new method to construct the Embedded Deformation graph model. Specifically, we construct deformation subgraphs for each part of the mesh firstly, all of these subgraphs nodes have K nearest neighbors. Then we present a new method to merge these subgraphs into a connected deformation graph. All of these works only have to be done once at preprocessing stage. At runtime, we track human skeletal joints and iteratively solve affine transformation parameters for deformation graph. Once the process has converged, we update mesh vertices positon and normal, then render it on GPU. Instead of binding user bones to mesh vertices at runtime, we manually select positional constraint vertices which displacement are correspondent to the human limbs and head at preprocessing stage. We demonstrate the performance of our system in Sect. 5.

We summarize our main contributions as follows:

- A new interactive application that maps human motion to virtual characters with different body part size, topology and geometry;
- A new method to construct deformation graph model which is not influenced by the initial shape of input meshes.
- A new method to establish positional constraint relationship, simplifying binding process. Binding five mesh vertices to human skeleton joints is enough to animate humanoid virtual characters.

2   Related Work

Using the inverse kinematics technology to animate humanoid characters has been the research focus in previous decades [8, 10]. Recent years, some researchers combined inverse kinematics with neutral network to control humanoid robotics [13, 14]. Baran et al. proposed an automatic rigging method which combined skeleton embedding with skin attachment to animate humanoid characters [11]. Vlasic et al. presented a method to enhance animation performance by multi-view input images [15].

It is still a challenge work that maps human motion to virtual characters with arbitrary shape and size in real-time. Researchers proposed various approaches to solve it. One is based on data-driven, Seol et al. constructed a motion classifier that maps human motion to characters [1]. Similar to Seol's work, Rhodin et al. devised an interactive motion mapping system which express both motions as dedicated feature spaces and construct mapping functions for these representations [2], which still restricted to the predefine motions. Yamane et al. presented another data-driven method to animate non-humanoid characters, but it cannot do it in real-time [9].

Different from the previously mentioned approaches, mesh deformation is a more versatile and flexible method. Sumner et al. proposed an Embedded Deformation algorithm framework which has a good performance by deforming the subset of mesh vertices directly [4]. Similarly, Sorkine et al. proposed a mesh surface modeling framework to deform mesh vertices [5]. Baran et al. proposed a method to transfer source poses to target poses through semantic transformation [12]. Jacobson et al. proposed a method to animate characters by a subset skeleton bones deformation [16]. Based on Sumner's work, Chen et al. designed a system that scan model firstly and then reconstruct the mesh, once the system run, it can track human motion and map it to virtual characters [3]. Michael et al. presented a new algorithm to reconstruct input models in real-time and can be combined with mesh skinning to animate virtual characters [6]. However, all of these methods are not applicable to our scenario, they cannot interactively control virtual characters with different body part size, topology and geometry. In this paper, we improve deformation graph model and develop an application that maps human motion to these virtual characters.

3   System Pipeline

There are two stages in our system. The first one is sampling process and can be viewed as preprocessing stage, which consists of the following three steps:

1. Select positional constraint vertices manually in each part of the mesh and then construct relationship between constraint vertices and skeleton joints.
2. Traverse mesh vertices and construct deformation subgraph for each part of the mesh. Noticed that each node has K nearest neighbors. Then construct K nearest deformation graph node neighbors for vertices in each part of the mesh.
3. Search for the key graph node in each subgraph and then merge these subgraphs into a connected deformation graph, with the goal of maintaining K neighbors for each graph node.

The above three steps only have to be performed once. After construct the deformation graph, we track human skeletal joints and update mesh vertices position and normal at runtime. Specifically, this stage consists of the following steps:

1. When the user moves, Microsoft's Kinect tracks his skeletal joints and outputs constraint skeleton joints 3D coordinates. Then using these parameters to solve the affine transformation parameters for each deformation graph node. This is a nonlinear least squares problem and can be expressed as an energy formulation, solved by the Gauss-Newton algorithm.
2. Update the graph node position by its affine transformation parameters. Then recomputed mesh vertices position and normal using its K neighbors affine transformation parameters.
3. Upload the new value of mesh vertices position and normal to the GPU and render the mesh in each frame.

4 Methods

4.1 Embedded Deformation [4]

We first review the basic theory of the Embedded Deformation algorithm [4] briefly. The affine transformation parameters for the deformation graph can be expressed by a set of matrices and vectors: $\mathbf{R} = (\mathbf{R}_1, \mathbf{R}_2, \dots, \mathbf{R}_m)$, $\mathbf{t} = (t_1, t_2, \dots, t_m)$, the variable of $\mathbf{R}_j$ is a 3 * 3 matrix and t_j is a 3 * 1 vector, m is the number of graph node. Then a vertex position $\mathbf{v}_i$ can be mapped to a new position $\mathbf{v}_i'$:

$$\mathbf{v}_i' = \sum\nolimits_{j=1}^{K} w_j(\mathbf{v}_i) \left[\mathbf{R}_j \left(\mathbf{v}_i - \mathbf{g}_j \right) + \mathbf{g}_j + t_j \right]. \tag{1}$$

The normal of the vertex $\mathbf{v}_i$ can be mapped to a new value:

$$\mathbf{n}_i' = \sum\nolimits_{j=1}^{K} w_j(\mathbf{v}_i) \mathbf{R}_j^{-1T} \mathbf{n}_i. \tag{2}$$

Where K is the number of nearest graph node around them, we call these nodes as K nearest neighbors, in our system, $K = 4$. The weight $w_j(\mathbf{v}_i)$ is the influence of node $\mathbf{g}_j$ to $\mathbf{v}_i$, its value is decreased with the distance and sum to 1:

$$w_j(\mathbf{v}_i) = \left(1 - \left\| \mathbf{v}_i - \mathbf{g}_j \right\| / d_{max} \right). \tag{3}$$

Here, d_{max} is the distance to the $K + 1$-nearest graph node.

In order to deform the embedded deformation graph precisely, Sumner et al. summarize it as a nonlinear squared optimization problem. The energy equation is consist of three terms. The first one minimizes the rotation error for all of the deformation graph node:

$$E_{rot} = \sum\nolimits_{i=1}^{m} \left\| \mathbf{R}_i^T \mathbf{R}_i - \mathbf{I} \right\|_F^2. \tag{4}$$

Here, $\|\cdot\|_F^2$ is the Frobenius norm and $\mathbf{I}$ is the identity matrix. The second term is a regularization term which minimizes each node neighbor's transformation influences:

$$E_{reg} = \sum\nolimits_{j=1}^{m} \sum\nolimits_{k \in N(j)} \alpha_{jk} \left\| \mathbf{R}_j \left(\mathbf{g}_k - \mathbf{g}_j \right) + \mathbf{g}_j + t_j - \left(\mathbf{g}_k + t_k \right) \right\|_2^2. \tag{5}$$

Here, $N(j)$ is the neighbors of node j. We use $\alpha_{jk} = 1.0$ in the system. The third term is a constraint term minimizes the error of the desired vertex position $\mathbf{q}_l$ and its deformed position $\mathbf{v}_{index(l)}'$:

$$E_{con} = \sum\nolimits_{l=1}^{P} \left\| \mathbf{v}_{index(l)}' - \mathbf{q}_l \right\|_2^2. \tag{6}$$

Here, p is the number of positional constraint vertices. We manually select the desired vertex $\mathbf{q}_l$ as positional constraint vertex in each part of the mesh at preprocessing stage, its desired position in each frame is the sum of original position and correspondent human body displacement. Then the total energy is the weighted sum of three quadratic terms. Similar to [4], we present a simple deformation graph to show these terms effect in Fig. 1(b–f). For the more details, please refer to [4].

4.2 Construct Deformation Graph

In this section, we propose a new method to construct deformation graph. A sparse deformation graph can be viewed as a subset of mesh vertices. Each deformation graph node deform a neighborhood space of a mesh vertex. The layout of the deformation graph is critical to the mesh deformation performance. Since vertices distribution is different in each part of the mesh, we are not construct the deformation graph for the entire mesh at one time. Instead, we first construct a deformation subgraph for each part, then merge these subgraphs into a connected deformation graph which still keep K neighbors for each graph node.

Assuming the deformation graph G is consist of t part, that is $G = \{P_1, P_2, \ldots, P_t\}$. Each variable satisfy conditions:

$$P_i \cap P_j = \emptyset. \ \forall i, j = 1 \ldots t, \ i \neq j \tag{7}$$

$$G = \bigcup_{i=1}^{t} P_i \tag{8}$$

For each subgraph node $g_j \in \mathbb{R}^3$, $j \in 1 \ldots m_j$, m_j is the number of subgraph node belongs to the part of P_i, that is $P_i = \{g_{m_1}, g_{m_2}, \ldots, g_{m_j}\}$. $i = 1 \ldots t$. We construct a sparse deformation subgraph for each part of the mesh by the method of Poisson Disk Sampling. Since vertices distribution is different in each part, we define the sampling distance in the part of P_i as:

$$d_i = \|p - q_2\| / L \tag{9}$$

Here, p, q is the furthest distance of two vertices in P_i, L is proportional to the number of vertices. A reasonable value of L is $\sqrt{S/N}$, here S is the surface area of this part, and N is the sampling number [3].

Since the mesh has to be deformed as a whole, we need to merge these subgraphs into a connected graph. Noticed that each node has K nearest neighbors, we should keep the connected graph with the same topology. We use the following method to generate a connected one: given the part of P_i, P_j, search for the key node $g_i \in P_i$, $g_j \in P_j$ satisfy the condition:

$$\min \left\| g_i - g_j \right\|_2 \tag{10}$$

We add an edge to connect g_i and g_j and break the edge between key node and its Kth neighbor nodes in the part of P_i and P_j. By this means, we generate a connected deformation graph which still keep K neighbors for each node. Noticed that the K neighbors of the key node are not nearest graph node around them now. We show the merge process in Fig. 1(a).

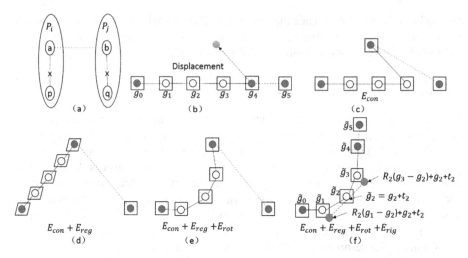

Fig. 1. (a) Connect adjacent subgraphs. Supposed that *a*, **b** is the key node of $\boldsymbol{P}_i, \boldsymbol{P}_j$. *p* is the *Kth* neighbors of *a*, *q* is the *Kth* neighbors of *b*. Then break the edge between *a* and *p*, *b* and *q*, add an edge to connect *a* and *b*. (b–f) A simple example which shows the effectiveness of the four energy terms of the energy equation on a simple deformation graph. The quadrilaterals around each graph node represent the deformation caused by the graph node affine transformation. Noticed that $\boldsymbol{g}_0 \sim \boldsymbol{g}_4$ belong to the same subgraph and $\boldsymbol{g}_5$ belongs to the adjacent subgraph, $\boldsymbol{g}_4$ and $\boldsymbol{g}_5$ is the key node of each subgraph. Manually select the $\boldsymbol{g}_4$ as the positional constraint point and move it to the desired position by constraint term in (b–c). With the regularization term, the deformation of the adjacent graph nodes stay agree with each other in (d–f). We show more details about the regularization term effectiveness on the node $\boldsymbol{g}_2$. The predicted position of neighboring nodes $\boldsymbol{g}_1$ and $\boldsymbol{g}_3$ is shown as gray circle in (f). With the rotation term, the deformation is more natural in (e–f). With the rigidity regularization term, the deformation of the adjacent subgraph nodes stay agree with each other in (f). The actual position of $\tilde{\boldsymbol{g}}_0 \sim \tilde{\boldsymbol{g}}_5$ is influenced by four terms.

In our system, the method of solving *K* nearest neighbor nodes for mesh vertices is different from Sumner's method [4]. Instead of solve it in the entire deformation graph space, we search it in the part which the graph node or the mesh vertex belongs. That is, for the mesh vertex $\boldsymbol{v}_i \in P_j$, its neighbors satisfy the condition $N(j) \in P_j$.

4.3   Optimization

Even though these subgraphs are merged to a connected deformation graph, their connectivity is still weak. In order to reinforce the influence between each subgraph, we add a rigidity regularization term to constraint the influence between adjacent subgraphs [3]:

$$E_{rig} = \sum_{(j,k)} \left\| R_j^T R_k - I \right\|_F^2 + \left\| t_j - t_k \right\|_2^2. \tag{11}$$

Where *j* and *k* is the key node of each subgraph. Then the optimization problem is:

$$\min_{R_1, t_1 \dots R_m, t_m} w_{rot} E_{rot} + w_{reg} E_{reg} + w_{con} E_{con} + w_{rig} E_{rig}. \tag{12}$$

We use the same weights as [3]. Then we use the Gauss-Newton algorithm to solve the unknown affine transformation parameters. For more details about the solving process, please refer to [4].

5 Experiment

5.1 Experimental Setup

We have implemented our method using C++ on a desktop with four-core Inter Core i5-3470 processor running at 3.20 GHz with 4 GB of RAM and an AMD Radeon HD 7400 graphics card. We use Microsoft's Kinect to track human skeletal joints and Eigen library for sparse Cholesky solver.

Since the K nearest neighbors of mesh vertices do not change at runtime, we compute these parameters at preprocessing stage and reuse them in each frame. The main overhead of our system is the Gauss-Newton iteration process, we use the sparse Cholesky decomposition to solve the normal equations in each iteration.

Similar to [3], our system is also a prototype and we cannot formally compare its performance with previous work by specific data. We show the animation result on a variety of virtual characters in the following section and give the statistics and performance data in Table 1. As the number of graph nodes increases, the system performance is getting worse. The fourth column illustrates the minimum frame rate at runtime. In our experiment, we manually select five mesh vertices as positional constraint points which is correspondent to the human head and limbs.

Table 1. Statistics and performance data for the animating characters.

Model	Vertices	Nodes	Minimum FPS	Constraint points
Baymax	1456	93	6.61 fps	5
Kung fu panda	4550	70	8.59 fps	5
Whale	4681	74	8.14 fps	5
Pterosaur	5260	53	12.88 fps	5

5.2 Animating Disproportionate Avatar

Our system can animate virtual characters with different body part size. In Figs. 2 and 3, we animate the Baymax and Kung Fu Panda by human limbs. As mentioned above, Chen's method cannot handle this circumstance. Even though they can scale the Baymax and Kung Fu Panda to the human size, their limbs cannot align to human well.

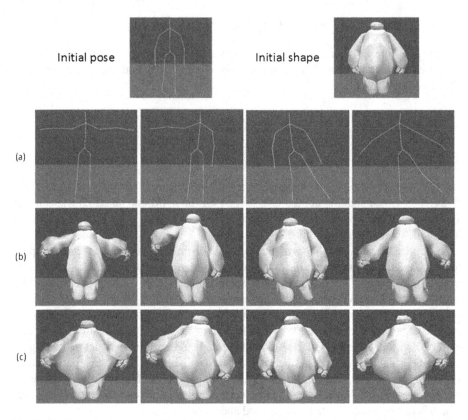

Fig. 2. Animate the Baymax with different methods. (a) Input human pose. (b) Our animation result. (c) Animation result by embedded deformation algorithm which construct the deformation graph by Chen's method.

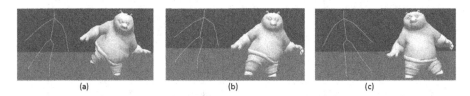

Fig. 3. Animate the kung fu panda with different human poses. (a) Left leg front kick. (b) Left leg back kick. (c) Right leg side kick.

Our system is convenient to use. At preprocessing stage, we bind human limbs to mesh limbs manually. There are only five constraint vertices in Figs. 2 and 3. Compared to Chen's binding process, we reduce the binding skeleton joints count and also animate virtual characters well.

Our system is not sensitive to the initial shape of mesh. We implement Chen's method to construct deformation graph and deform the Baymax. In Fig. 2(b–c), we show four groups of experiment on the same input human pose. By comparison, we can see that our method do not influenced by initial shape of mesh. Once the user moves his hands, the stomach of Baymax would not influenced by its hand in our system.

5.3   Animating Non-humanoid Avatar

Our system works well for virtual characters with different topology and geometry. For different input virtual characters, we only have to specify positional constraint points at preprocessing stage. Our method can animate them without training, it is more versatile and precise than Seol's method [1]. We show the animation result on the whale and pterosaur by human body in Figs. 4 and 5.

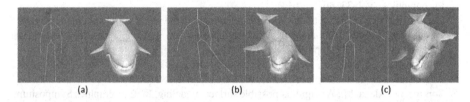

Fig. 4. Animate the whale with different human poses. (a) Initial human pose and initial shape of mesh. (b) Animation result on the tail of whale by right leg. (c) Animation result on the body of whale by right hand.

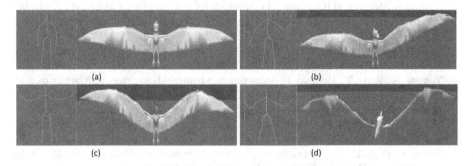

Fig. 5. Animate the pterosaur with different human poses. (a) Initial human pose and initial shape of mesh. (b) Animation result on the right wing of pterosaur by right hand. (c) Animation result on a pair of wings by both hands. (d) Using the human body to change pterosaur direction.

6   Conclusion

In this paper, we improve the Embedded Deformation algorithm to map human motion to virtual characters with different body part size, topology and geometry. We construct subgraph for each part firstly, then merge them into a connected one. Our method is more versatile and precise than previous work. Compared to data-driven based method, our method do not need to training at preprocessing stage and would not occur errors at runtime. Compared to Chen's work, our method can animate virtual characters with disproportionate body size and not sensitive to the initial shape of meshes.

Acknowledgments. This work was partially supported by the National High-tech Research and Development Program of China (2015AA015901) and the National Natural Science Foundation of China (61170195, U1201255, U1301257).

References

1. Seol, Y., O'Sullivan, C., Lee, J.: Creature features: online motion puppetry for non-human characters. In: Proceedings of the 12th ACM SIGGRAPH/Eurographics Symposium on Computer Animation (SCA 2013), pp. 213–221 (2013)
2. Rhodin, H., James,T., Kim, I.K., Varanasi, K., Seidel, H.P., Theobalt, C.: Interactive motion mapping for real-time character control. In: Computer Graphics Forum (Proceedings Eurographics), vol. 33, no. 2 (2014)
3. Chen, J.W., Izadi, S., Fitzgibbon, A.: KinÊtre: animating the world with the human body. In: Proceedings of the 25th Annual ACM Symposium on User Interface Software and Technology (UIST 2012), pp. 435–444 (2012)
4. Sumner, R.W., Schmid, J., Pauly, M.: Embedded deformation for shape manipulation. In: ACM Transactions on Graphics (TOG), vol. 26, article no. 80 (2007)
5. Sorkine, O., Alexa, M.: As-rigid-as-possible surface modeling. In: Eurographics Symposium on Geometry Processing (SGP), pp. 109–116 (2007)
6. Zollhöfer, M., Nießner, M., Izadi, S., Rehmann, C., Zach, C., Fisher, M., Wu, C., Fitzgibbon, A., Loop, C., Teobalt, C., Stamminger, M.: Real-time non-rigid reconstruction using an RGB-D camera. In: ACM SIGGRAPH (2014)
7. Madsen, K., Nielsen, H.B., Tingleff, O.: Methods for non-linear least squares problems. In: Informatics and Mathematical Modeling, 2nd edn. Technical University of Denmark, Lyngby (2004)
8. Shin, H.J., Lee, J., Shin, S.Y., Gleicher, M.: Computer puppetry: an importance-based approach. ACM Trans. Graph. (TOG) **20**, 67–94 (2001)
9. Yamane, K., Ariki, Y., Hodgins, J.: Animating non-humanoid characters with human motion data. In: Proceedings of the 2010 ACM SIGGRAPH/Eurographics Symposium on Computer Animation (SCA 2010), pp. 169–178 (2010)
10. Grochow, K., Martin, S.L., Hertzmann, A., Popović, Z.: Style-based inverse kinematics. ACM Trans. Graph. (TOG) **23**, 522–531 (2004)
11. Baran, I., Popović, J.: Automatic rigging and animation of 3D characters. In: ACM Transactions on Graphics (TOG), vol. 26, article no. 72 (2007)
12. Baran, I., Vlasic, D., Grinspun, E., Popović, J.: Semantic deformation transfer. In: ACM Transactions on Graphics (TOG), vol. 28, article no. 36 (2009)
13. Reinhart, R.F., Steil, J.J: Neural learning and dynamical selection of redundant solutions for inverse kinematic control. In: 11th IEEE-RAS International Conference on Humanoid Robots, pp. 564–569. IEEE Press, Bled (2011)
14. Waegeman, T., Schrauwen, B.: Towards learning inverse kinematics with a neural network based tracking controller. In: Lu, B.-L., Zhang, L., Kwok, J. (eds.) ICONIP 2011, Part III. LNCS, vol. 7064, pp. 441–448. Springer, Heidelberg (2011)
15. Vlasic, D., Baran, I., Matusik, W., Popović, J.: Articulated mesh animation from multi-view silhouettes. In: ACM Transactions on Graphics (TOG), vol. 27, article no. 97 (2008)
16. Jacobson, A., Baran, I., Kavan, L., Popović, J., Sorkine, O.: Fast Automatic Skinning Transformations. In: ACM Transactions on Graphics (TOG), vol. 31, article no. 77 (2012)

Visual Understanding and Recognition on Big Data

Fast Graph Similarity Search via Locality Sensitive Hashing

Boyu Zhang, Xianglong Liu[(✉)], and Bo Lang

State Key Lab of Software Development Environment,
Beihang University, Beijing, China
{byzhang,xlliu,langbo}@nlsde.buaa.edu.cn

Abstract. Similarity search in graph databases has been widely studied in graph query processing in recent years. With the fast accumulation of graph databases, it is worthwhile to develop a fast algorithm to support similarity search in large-scale graph databases. In this paper, we study k-NN similarity search problem via locality sensitive hashing. We propose a fast graph search algorithm, which first transforms complex graphs into vectorial representations based on the prototypes in the database and then accelerates query efficiency in Euclidean space by employing locality sensitive hashing. Additionally, a general retrieval framework is established in our approach. Experiments on three real datasets show that our work achieves high performance both on the accuracy and the efficiency of the presented algorithm.

Keywords: Graph similarity search · Locality sensitive hashing · Graph prototypes · Graph vectorial representation

1 Introduction

Graph search over a large graph dataset plays an important role in many applications, such as multimedia information retrieval [7], pattern recognition [17,18]. For example, to incorporate both region attributes and spatial relationship into estimation, many works represent an image as an attributed graph and transform the image retrieval problem into an attributed graph search problem [7]. Another example is, in chemical analysis graph search can help study properties of a newly synthesized chemical by referring to the existing chemicals with known properties in database. With the increasing amount of graph data, real world databases usually contain thousands of or even millions of graphs [17]. To deal with such large databases efficiently, it is worthwhile to develop fast algorithms to support elementary querying.

In the literature, there are two common types of graph search including subgraph search and similarity search. Along the first direction, mainly concerning the graphs containing the query graph as a subgraph, a number of solutions have been proposed based on subgraph isomorphism. Despite of the successful progress and application, subgraph search suffers from the NP hardness of subgraph isomorphism. The graph similarity search looks for graphs similar to a

© Springer International Publishing Switzerland 2015
Y.-S. Ho et al. (Eds.): PCM 2015, Part I, LNCS 9314, pp. 623–633, 2015.
DOI: 10.1007/978-3-319-24075-6_60

query one, which to some extent avoids the difficulty of pattern match in subgraph search by concentrating on the k or ϵ nearest neighbors (NN). In k-NN query, the k most similar graphs are treated as the retrieved results, while the ϵ one concerns all graphs within a predefined distance ϵ to the query graph.

For graph similarity search, it is important to define a meaningful graph similarity metric that can capture the intrinsic relations between graphs [18], [3]. Graph edit distance (GED) is a most widely adopted one based on the distortion operations (insertions, deletions and substitutions of nodes and edges) on a pair of graphs [4]. However, due to the complex computation of graph edit distance, it is usually infeasible for large database in practice [19]. To tackle this problem, a number of research have attempted to reduce the computation by pruning off unlikely candidates using GED bounds with different assumptions [5,14–16]. The C-Tree method hierarchically treated graph closures and database graphs as interior and leaf nodes in tree based structure, avoiding accessing too dissimilar individual graphs [5]. The k-AT method borrowed the q-gram idea from string similarity problems and approximated the edit distance using k-adjacent tree patterns [15]. In gWT, a recursive search on wavelet trees was performed for efficient many-query search in a very large graph database [14]. SEGOS proposed a two-level indexing and query processing framework to access graphs in increasing dissimilarity with graph punning algorithms [16].

Despite of the aforementioned progress in graph search, there are still several drawbacks suffered in existing research. First, not every subunit in graphs is a meaningful feature as far as the application is concerned. When we break down a graph into tree patterns like k-AT or star structures like SEGOS, we risk losing global information on structure, which may be more important to the real application. Second, the subgraph decomposition and pseudo isomorphism bring expensive computation, which prevents these methods from the successful applications in large-scale databases [5]. Finally, most of existing solutions rely on the contraction of the searching range, which cannot guarantee the desired number of candidate graphs, which will severely limit its applications in practice. For instance, neither k-AT nor SEGOS supports k-NN query, while gWT can only support 1-NN query well.

In this paper, to naturally support k-NN search in large-scale databases, we present a fast graph search framework, which correspondingly addresses aforementioned problems by incorporating the discriminative vectorial representation and fast approximated similarity search. Different from the graph structures that are too complex for the similarity measure, the vectorial representation enjoys sound mathematical properties and efficient algorithmic techniques. Motivated by the representation theory widely used in machine learning, we can efficiently transform complex graphs into vectors in high dimensional space with respect to the selected prototypes in the database using the kernel techniques (e.g., Weisfeiler-Lehman graph kernel [13]). The prototypes serve as the representatives to capture the neighbor structure of database graphs, and approximate the graph similarities based on their distribution for each graph. Therefore, the similarity search can be completed by exhaustive exploiting graphs that share the same prototypes, i.e., the nearest neighbor search in Euclidean space. However, as to large-scale databases, it is still inefficient to conduct exhaustive comparison in Euclidean space. To further speedup the similarity search, locality sensitive hashing (LSH) [6], [8,9] is adopted to index the graph data using hash tables and a multi-probe search strategy [10] is employed for a better performance balancing the efficiency and search accuracy over a large-scale graph database.

To our best of knowledge, this is the first work that achieves fast graph similarity search based on the locality sensitive hashing, which enables low memory and computation cost in practice. The main contributions are two-fold: (1) We efficiently approximate the graph similarity using prototypes based vectorial representation. (2) We further speedup the similarity search in Euclidean space using locality sensitive hashing, which supports fast graph similarity search in large-scale datasets, and meanwhile guarantees the search accuracy.

2 The Proposed Method

It is worth mentioning that all graphs in the paper are assumed as the simple graphs, which have neither self loops nor multiple edges. The simple graphs are the general types of graphs widely used in real applications. In practice, often the simple graph is associated with semantics by labeling its vertex, e.g., the vertex of a chemical compound graph is assigned with atoms as labels. Therefore, in this paper we focus on the undirected *vertex-labeled* simple graph, and for presentation simplicity, it is hereafter abbreviated to a graph.

Given a graph database $D = \{g_1, g_2, ..., g_n\}$ with the semantic class labels, our goal is to retrieve the k nearest graphs $\{g_{i_1}, g_{i_2}, ..., g_{i_k}\}$ to the query graph q, which belong to the same class. To achieve this goal, we will first design a vectorial representation method to transform graphs into high dimensional vectors based on prototypes in the database, and then speedup the large-scale graph similarity search by encoding graphs into hash codes using multiple LSH tables. The overview of our method is described in Fig. 1.

2.1 Vectorial Representation

Graph edit distance (GED) is a powerful metric to measure the similarity between graphs. Its key idea is to define the distances between graphs based on the minimum amount of distortions that are needed to transform one graph into another. A standard set of distortion operations consists of insertions, deletions and substitutions of vertices and edges. We can complete the similarity search directly by explosively ranking all database graphs according to the GED to the query graph. However, the computational complexity of GED is rather expensive, even for the suboptimal algorithms [12]. Consequently, it is infeasible to perform graph similarity search over a large-scale database based on the GED.

Compared to the graph data with complex structures, vector data enjoy series of mathematical properties and computational techniques. To reduce the complexity and difficulty of handling graph data, our straightforward way is to represent graphs in high ($m \ll n$) dimensional vector space using a mapping

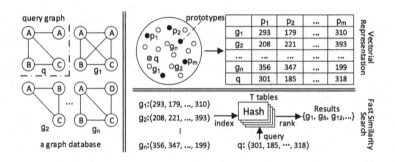

Fig. 1. An overview of our method.

function $\Psi : g \to R^m$, meanwhile preserving the nearest neighbor relations. Since data usually distributed in clusters in the complex space, it is feasible that we can characterize each sample in the space using certain specific prototype corresponding to the cluster structure [2].

Motivated by this fact, we propose a vectorial representation method for graph data based on the prototype graphs in the database. Given a graph database D, a set of m prototypes $P = \{p_1, p_2, ..., p_m\}$ is generated in advance by randomly sampling or clustering. The prototypes serve as the representatives to capture the neighbor structure of database graphs as shown in Fig. 1, and thus similar graphs located around each other will share similar prototypes. By accessing the similarity $s(g, p_i)$ between a graph g and each prototype p_i, a vector of m dimensions with the similarity as an element can be obtained for the graph, and the graph similarity can be approximated based on the Euclidean distances between their vectors. Formally, the mapping function Ψ based on the prototype set can be defined as follows:

$$\Psi(g) = (s(g, p_1), s(g, p_2), ..., s(g, p_m)). \tag{1}$$

Subsequently, the complex similarity search in graph space turns to the nearest neighbor search in Euclidean space.

The mapping Ψ highly depends on s for its capability of neighbor graph preservation. As to the similarity measure between graphs, we can naturally check whether the graphs have identical substructures, e.g. paths, cycles, sub-trees. Possibly one effective measure of similarity of graphs among them is to use subtree patterns, counting all pairs of matching subtree patterns of both graphs, which capture the local neighboring structures of graphs fairly well.

Meanwhile, the mapping Ψ also involves the computation of similarity s between m prototypes, which need to be done fast and with less computational complexity. Given two graphs g and g', a fast way to preserve graph similarity and meanwhile avoid the complexity can be processed by counting all pairs of matching subtree patterns, rooted at the vertices of both graphs, that is what Weisfeiler-Lehman (WL) subtree graph kernel [13] does. For each vertex v in both graphs, a new artificial label is computed by compressing the vertex labels of its neighbors, ensuring that vertices with the same neighboring structure get the same compressed label. The procedure is iterated until we reach the desired subtree height h. By defining $\Sigma^{(h)}$ as the set of all labels that occur as vertex labels at least once in g and g' at the end of the h-th iteration and $f :$ $\{g, g'\} \times \Sigma^{(h)} \rightarrow N$ as a function such that $f(g, \sigma_i)$ is the number of occurrences of the letter $\sigma_i \in \Sigma^{(h)}$ in the graph g, we have:

$$\gamma^{(h)}(g) = (f(g, \sigma_1), f(g, \sigma_2), ..., f(g, \sigma_{|\Sigma^{(h)}|})). \tag{2}$$

The WL subtree kernel uses the above iterative procedures, defining a discriminative s, to efficiently approximate the similarity and represent each graph as a vector with respect to all prototypes. Formally, the WL subtree kernel with h iterations is defined as:

$$s(g, g') = < \gamma^{(h)}(g), \gamma^{(h)}(g') >. \tag{3}$$

2.2 Fast Similarity Search

When the graphs represented as vectors with the mapping function Ψ, one natural way to do the k-NN query is based on the Euclidean distance. The Euclidean distance between vectors can be regarded as an effective measure for similarity in m dimensional vector space. Given a query graph q, the Euclidean distance between the query graph q and each database graph g is computed as

$$||\Psi(q) - \Psi(g)|| = \sqrt{\sum_{i=1}^{m} (s(q, p_i) - s(g, p_i))^2}. \tag{4}$$

We can scan the whole dataset based on the Euclidean distance and return the top k graphs with the minimum distance to the query graph as results.

However, such search method obviously does not scale up when the dataset grows due to the exhaustive comparison in Euclidean space. One possible way to better this situation is to hash the vectorial representations of graphs using several hash functions, ensuring that for each function, the probability of collision is much higher for graphs which are similar to each other than for those which are far apart. Then we can determine near neighbors by hashing the query graph and retrieving similar ones stored in buckets containing that query graph. Consequently, to further speedup the similarity search, the general Euclidean Locality Sensitive Hashing algorithm proposed in [1] is adopted to index the graph data through hash tables. It is based on stable distributions and suitable for L_d, $d \in (0, 2]$. In our case of R^m with L_2 distance, the LSH family for the vectorial representation $\Psi(g)$ of a graph g is defined as follows:

$$H(\Psi(g)) = (h_1(\Psi(g)), h_2(\Psi(g)), ..., h_B(\Psi(g))), \tag{5}$$

$$h_i(\Psi(g)) = \left\lfloor \frac{a_i \cdot \Psi(g) + b_i}{W} \right\rfloor, i = 1, 2, ..., B, \tag{6}$$

where $a_i \in R^m$ is a vector with entries chosen independently from the Gaussian distribution $N(0, 1)$ and b_i is drawn from the uniform distribution $U[0, W]$. Each hash function $h_i : R^m \rightarrow Z$ maps a vectorial representation $\Psi(g)$ onto the set of integers. The parameters B and W control the locality sensitive of the hash function. The index structure is T hash tables with independent hash functions, and the query algorithm is to scan the buckets the query point is hashed to.

Nevertheless, one drawback of the basic LSH is that it requires a large number of hash tables to achieve good search quality. More hash tables to cover most nearest neighbors means more space requirements for the hash tables and more substantial delay to the query process by looking up each hash table. Considering the property of LSH that if a graph is similar to the query graph q but not hashed to the same bucket as q, it is likely to be in a bucket that is close to the bucket where q is hashed, we can try to extend the set of candidate hash buckets to more than one bucket in each hash table by selecting neighboring hash buckets to the query one. The multi-probe LSH [10] is such a search strategy that uses a perturbation-based approach to get a better search performance with fewer hash tables to support the fast similarity search over large-scale graph databases. Given a query graph vector $\Psi(q)$, different from the basic LSH method, which checks the hash bucket $H(\Psi(q))$, the multi-probe LSH applies a hash perturbation vector $\Delta = (\delta_1, \delta_2, ..., \delta_B)$ to probe the hash bucket

$$H(\Psi(q)) + \Delta = (h_1(\Psi(q)) + \delta_1, h_2(\Psi(q)) + \delta_2, ..., h_B(\Psi(q)) + \delta_B). \quad (7)$$

Because the chance of k-th nearest neighbor falling into buckets with $|\delta_i| \geq 2$ is very small, δ_i is restricted to the set $\{-1, 0, 1\}$ in order to simplify the algorithm. To avoid probing a hash bucket more than once, a sequence of perturbation vectors is designed properly so that each vector in the sequence maps to a unique set of hash values. It increases the probability to find a relevant neighbor in a single hash table and consequently reduces the number T of required hash tables.

By processing the multi-probe LSH, the majority of the database graphs that are dissimilar to the query one is filtered out and a candidate set C is returned. In order to precisely locate the top k graphs as the final results, the last step in our approach is to rank the graphs in the candidate set C by the Euclidean distance to the query graph. Since most graphs have been filtered out and only a small part of graphs left in C, the rank procedure does not cost much time.

2.3   Retrieval Framework

We propose a novel retrieval framework for fast graph similarity search, including two main algorithms: building index for graph database and querying procedure. The details are described in Algorithms 1 and 2, respectively.

2.4   Complexity

From [13], we know that the Weisfeiler-Lehman subtree kernel with h iterations on one pair of graphs can be computed in $O(he)$, where e is the number of edges of the graphs. We can regard the time complexity of processing a query in the graph dataset indexed by this multi-probe LSH to $O(1)$.

Assume that the candidate set C returned by the multi-probe LSH contains n' graphs, so the total time of processing one graph similarity search in our approach is bounded by $O(mhe + mn' + n' \log n')$, which consists of (i) $O(mhe)$ time for the vectorial representation of the query graph, (ii) $O(1)$ time for the multi-probe LSH querying procedure, (iii) $O(mn')$ time for computing the Euclidean distance between the query graph and each graph in the candidate set C, and (iv) $O(n' \log n')$ time for ranking the n' Euclidean distances.

Algorithm 1. Building index for graph database.

Input: graph database D.
Output: T hash tables and P.
1: Select a prototype set $P = \{p_1, p_2, ..., p_m\}$ from D.
2: Construct T hash tables, each containing B hash functions, $h_1, h_2, ..., h_B$.
3: **for** each $g_i \in D$ **do**
4:     Compute WL subtree kernel $s(g_i, p_j)$ for all $p_j \in P$.
5:     Generate $\Psi(g_i) \leftarrow (s(g_i, p_1), s(g_i, p_2), ..., s(g_i, p_m))$.
6:     Compute $H(\Psi(g_i))$ in each hash table.
7:     Place $\Psi(g_i)$ into the corresponding hash bucket.
8: **end for**
9: **return** T hash tables and P.

Algorithm 2. Querying procedure.

Input: query graph q, prototype set P, T hash tables and the number of results returned k.
Output: top k graphs.
1: Compute WL subtree kernel $s(q, p_i)$ for all $p_i \in P$.
2: Generate $\Psi(q) \leftarrow (s(q, p_1), s(q, p_2), ..., s(q, p_m))$.
3: Compute $H(\Psi(q))$ in each hash table.
4: Get a candidate set C by the union of all buckets that $H(\Psi(q))$ points to and the multi-probe scheme.
5: Rank C by $\|\Psi(q) - \Psi(g_i)\|$, $g_i \in C$ and get top k graphs.
6: **return** top k graphs.

---

3 Experimental Evaluation

We present an empirical study on real datasets to show that our approach (1) has an effective performance in both the search precision and the efficiency and (2) supports fast graph similarity search in large-scale datasets.

3.1 Experimental Setting

Our method is generic to support the fast similarity search over different types of graph data like part model based visual search. Since in the literature there are rare such image datasets released, we conduct our experiments on the following popular real datasets, described as follows:

Table 1. The search precision (%) and runtime (ms) for different search methods.

k	GED^{rank} precision	time	C-Tree precision	time	gWT precision	time	KVR$^{l_2-rank}$ precision	time	KVR$^{LSH-rank}$ precision	time
NCI1 1	99.60±06.32	8.46×10⁴	98.20±13.31	44.20	99.80±04.47	0.14	99.40±07.73	22.33	99.00±09.96	6.80
5	75.08±22.84	8.46×10⁴	73.00±27.00	70.90	50.00±50.05	7.88	72.92±24.30	23.33	75.24±24.83	6.48
10	69.48±20.48	8.46×10⁴	67.94±27.21	76.10	50.00±50.05	7.88	67.46±22.05	22.83	69.54±22.96	7.55
50	60.71±16.81	8.46×10⁴	59.96±26.61	137.80	50.00±50.05	7.88	59.96±18.26	22.93	60.14±15.63	6.99
NCI109 1	100.00±00.00	6.42×10⁴	99.00±09.96	14.88	100.00±00.00	0.22	100.00±00.00	22.38	100.00±00.00	7.58
5	73.44±23.52	6.42×10⁴	73.12±27.35	45.28	50.00±50.05	7.52	72.16±24.47	23.68	76.12±24.85	7.14
10	66.84±22.13	6.42×10⁴	67.82±27.71	51.82	50.00±50.05	7.52	66.82±22.70	21.98	69.22±22.58	7.52
50	59.36±18.19	6.42×10⁴	58.88±26.21	71.71	50.00±50.05	7.52	60.05±19.28	21.78	61.01±15.77	7.67
MUTAG 1	100.00±00.00	9.72×10⁴	99.00±09.96	23.60	57.20±49.53	0.12	95.00±21.82	26.45	98.00±18.65	11.96
5	75.16±24.28	9.72×10⁴	78.96±24.00	45.15	50.00±10.09	10.58	72.32±24.49	24.75	75.16±24.20	11.25
10	69.66±22.81	9.72×10⁴	73.82±22.11	51.50	49.94±10.16	10.58	68.22±23.01	24.45	69.28±23.17	11.59
50	62.64±19.40	9.72×10⁴	64.52±19.55	100.55	50.02±08.08	10.58	61.61±19.57	24.15	61.34±18.52	9.12

(1) NCI1 and NCI109[1], two balanced subsets of data sets of chemical compounds screened for activity against non-small cell lung cancer and ovarian cancer cell lines, consisting of 4110 and 4127 chemical compounds, respectively.

(2) Mutagenicity (hereinafter referred to as MUTAG) [11], a data set of the molecular compounds of the mutagenicity data, consisting of 4337 elements totally (2401 mutagen elements and 1936 nonmutagen elements).

We make the comparison among the following five different querying methods: (i) the basic ranking solution (GED^{rank} for short) based on approximated GED between all database graphs and the query graph; state-of-the-art graph search methods including (ii) C-Tree [5] and (iii) gWT [14]; (iv) the straightforward way that ranks the whole database using our kernel based vectorial representation in Euclidean space (denoted by KVR$^{l_2-rank}$); (v) the proposed framework incorporating both vectorial representations and LSH based approximated similarity search (KVR$^{LSH-rank}$).

In the experiments, we randomly selected 500 graphs from each dataset as query graphs and varied k to 1, 5, 10, and 50. For each k, we conducted the queries and computed the average of the query results. In order to test the efficiency of our method on large-scale datasets, we expanded the NCI1 dataset up to 10, 100, 1,000 and 10,000 times and processed the same queries on these generated datasets, respectively. Since the high complexity of the optimal GED computation, we used an approximate algorithm known as beam search ($s = 20$) [17] to compute the approximate GED. For C-Tree, we chose the default values, namely, setting the minimum number of child node $m = 20$ and the NBM method [5] was used for graph mapping. For gWT, we set the number of iterations, $h = 10$, for the Weisfeiler-Lehman fingerprints and the similarity threshold was set to 0.8 for 1-NN by default. In order to recall more graphs to do the query for $k = 5$, 10, and 50, we assigned a lower value 0.01 to the threshold. For KVR$^{l_2-rank}$ and KVR$^{LSH-rank}$, we set the same number of iterations $h = 10$ when computing the Weisfeiler-Lehman subtree kernel. In the mapping function, we fixed the number of the prototype graphs $m = 300$ in all the experiments.

---

[1] http://pubchem.ncbi.nlm.nih.gov/.

All the experiments except C-Tree and gWT were run on Windows Server 2008 with Intel Xeon X5650, two 2.66GHz CPUs and 16G memory. C-Tree and gWT were done on the same machine running Linux.

3.2 Experimental Results

We next present our findings on both the effectiveness and the efficiency.

(1) Effectiveness. Table 1 shows the search precision and the query time of the five different search methods performed on the three real datasets. We can see that KVR$^{\text{LSH-rank}}$ outperforms almost all the other methods on the search precision and has a distinct advantage in terms of the query time. Although gWT can also process one query very fast, it only supports 1-NN query well and the precision of gWT gets much worse when k becomes large. We can also notice that KVR$^{\text{LSH-rank}}$ keeps a nearly stable query time, regardless of the value of k, not like C-Tree whose query time gets longer with k increasing. These verify that our approach to tackle graph similarity search is effective.

(2) Efficiency. Figure 2 depicts the query time for processing graph similarity search of the above five search methods on the original and the expanded datasets. As the dataset grows larger, it obviously can be seen that KVR$^{\text{LSH-rank}}$ outperforms all the other methods. Because of the limitation of the experimental environment, we cannot process the query for C-Tree and gWT on the very large datasets. However, according to the growth tendency, we can estimate that KVR$^{\text{LSH-rank}}$ has a significant advantage of the efficiency over all the others within a certain large amount of graph data. Meanwhile, it can also indicate that our method need lower memory cost while C-Tree and gWT needs more when the dataset becomes large. These present that our approach makes a striking speedup for graph similarity search on large-scale datasets.

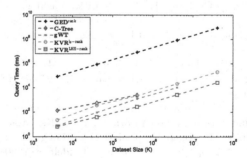

Fig. 2. Efficiency test on large-scale datasets. (The query time for k-NN query with $k = 50$ and NCI1 dataset is duplicated and expanded up to 10, 100, 1000, 10000 times)

4   Conclusions

We provide a fast algorithm based on locality sensitive hashing to support graph similarity search on large-scale databases. First, we transform complex graphs into discriminative vectorial representations based on the prototypes in database. Furthermore, we speedup the similarity search in Euclidean space based on locality sensitive hashing with the search accuracy guaranteed. Finally a general retrieval framework for k-NN query is proposed. Our empirical evaluations verify the effectiveness and the efficiency of our method over real datasets.

Acknowledgements. This work is supported in part by the National Natural Science Foundation of China (61370125 and 61402026), SKLSDE-2014ZX-07 and SKLSDE-2015ZX-04.

References

1. Datar, M., Immorlica, N., Indyk, P., Mirrokni, V.S.: Locality-sensitive hashing scheme based on p-stable distributions. In: SCG, pp. 253–262. ACM (2004)
2. Ding, C., He, X.: K-means clustering via principal component analysis. In: ICML, p. 29. ACM (2004)
3. Fernández, M.L., Valiente, G.: A graph distance metric combining maximum common subgraph and minimum common supergraph. Pattern Recogn. Lett. **22**(6), 753–758 (2001)
4. Gao, X., Xiao, B., Tao, D., Li, X.: A survey of graph edit distance. Pattern Anal. Appl. **13**(1), 113–129 (2010)
5. He, H., Singh, A.K.: Closure-tree: An index structure for graph queries. In: ICDE, pp. 38–38. IEEE (2006)
6. Indyk, P., Motwani, R.: Approximate nearest neighbors: towards removing the curse of dimensionality. In: STOC, pp. 604–613. ACM (1998)
7. Li, C.Y., Hsu, C.T.: Image retrieval with relevance feedback based on graph-theoretic region correspondence estimation. IEEE Trans. Multimedia **10**(3), 447–456 (2008)
8. Liu, X., He, J., Deng, C., Lang, B.: Collaborative hashing. In: IEEE CVPR, pp. 2147–2154. IEEE (2014)
9. Liu, X., He, J., Lang, B.: Reciprocal hash tables for nearest neighbor search. In: AAAI. AAAI Press (2013)
10. Lv, Q., Josephson, W., Wang, Z., Charikar, M., Li, K.: Multi-probelsh: efficient indexing for high-dimensional similarity search. In: VLDB, pp. 950–961. VLDB Endowment (2007)
11. Riesen, K., Bunke, H.: Iam graph database repository for graph basedpattern recognition and machine learning. In: da Vitoria Lobo, N., Kasparis, T., Roli, F., Kwok, J.T., Georgiopoulos, M., Anagnostopoulos, G.C., Loog, M. (eds.) SSPR & SPR 2008. LNCS, vol. 5342, pp. 287–297. Springer, Heidelberg (2008)
12. Riesen, K., Bunke, H.: Approximate graph edit distance computation by means of bipartite graph matching. Image Vis. Comput. **27**(7), 950–959 (2009)
13. Shervashidze, N., Borgwardt, K.M.: Fast subtree kernels on graphs. In: NIPS, pp. 1660–1668 (2009)

14. Tabei, Y., Tsuda, K.: Kernel-based similarity search in massive graph databases with wavelet trees. In: SDM, pp. 154–163. SIAM (2011)
15. Wang, G., Wang, B., Yang, X., Yu, G.: Efficiently indexing large sparse graphs for similarity search. IEEE Trans. Knowl. Data Eng. **24**(3), 440–451 (2012)
16. Wang, X., Ding, X., Tung, A., Ying, S., Jin, H.: An efficient graph indexing method. In: ICDE, pp. 210–221. IEEE (2012)
17. Wang, Y., Xiao, J., Suzek, T.O., Zhang, J., Wang, J., Bryant, S.H.: Pubchem: a public information system for analyzing bioactivities of small molecules. Nucleic Acids Res. **37**(suppl 2), W623–W633 (2009)
18. Yan, X., Yu, P.S., Han, J.: Substructure similarity search in graph databases. In: ACM SIGMOD, pp. 766–777. ACM (2005)
19. Zeng, Z., Tung, A.K., Wang, J., Feng, J., Zhou, L.: Comparing stars: on approximating graph edit distance. VLDB **2**(1), 25–36 (2009)

Text Localization with Hierarchical Multiple Feature Learning

Yanyun Qu[1]([✉]), Li Lin[1], Weiming Liao[1], Junran Liu[1], Yang Wu[2], and Hanzi Wang[1]

[1] Computer Science Department, Xiamen University, Xiamen, China
{quyanyun,linlipj,liaoweimin0909,
ilevanaliu,wang.hz}@gmail.com
[2] Center for Frontier Science and Technology,
Nara Institute of Science Technology, Nara, Japan
wuyang0321@gmail.com

Abstract. In this paper, we focus on English text localization in natural scene images. We propose a hierarchical localization framework which goes from characters to strings to words. Different from existing methods which either bet on sophisticated hand-crafted features or rely on heavy learning models, our approach tends to design simple but effective features and learning models. In this study, we introduce a kind of two level character structure features in collaboration with the Histogram of Gradient (HOG) and the Convolutional Neural Network (CNN) features for character localization. In string localization, a nine-dimension string feature is proposed for discriminative verification after grouping characters. For the final word localization, we learn an optimal splitting strategy based on the interval cues to split strings into words. Experiments on the challenging ICDAR benchmark datasets demonstrate the effectiveness and superiority of our approach.

Keywords: Hierarchical framework · Character structure feature · String feature · Convolutional neural network · Text localization

1 Introduction

With the development of multimedia technology and the popularity of digital imaging devices (such as digital cameras), vast amounts of natural scene images, which carry a wealth of information, are collected and stored. Among all the information contained in an image, text, as a kind of strong high-level semantic resource, provides valuable cues about image content. Actually, text is very important for humans and computers to understand the scenes. Judd [1] proved that people, given an image, tend to fixate on text more than other objects, which suggests the importance of text to humans. Text recognition is also critical in intelligent navigation, movie summation, vision assistance systems, etc. As a result, there is an urgent need to develop the technology of text recognition in natural scene images.

Text recognition is usually divided into two tasks: text localization and word recognition. Text localization is an important prerequisite for word recognition.

© Springer International Publishing Switzerland 2015
Y.-S. Ho et al. (Eds.): PCM 2015, Part I, LNCS 9314, pp. 634–643, 2015.
DOI: 10.1007/978-3-319-24075-6_61

Text localization, as an important task among the renowned competitions held in the International Conference on Document Analysis and Recognition (ICDAR), remains challenging because of various scenes, backgrounds and text appearances. There are two kinds of methods for text localization: heuristics-based and learning-based methods.

Heuristics-Based Methods. The heuristics-based methods are based on heuristics, such as connected component analysis (CCA). Lucas et al. [18, 23] used the Maximally Stable Extremal Regions (MSER) algorithm to detect candidate text and the special text component features for text localization. In [4], they presented a novel approach based on Oriented Stroke Detection to improve the poor performance on noisy images. Another widely used methods are Stroke Width Transform [2] and its extensions such as those in [3]. Epstein et al. [2] proposed a novel stroke filter to improve the detection of text candidate regions. Huang et al. [3] introduced a low-level filter called the Stroke Feature Transform (SFT) by incorporating color cues of text pixels.

Learning-Based Methods. Learning-based methods use the same technology as other object recognition methods. Those methods can be roughly classified into two classes, supervised learning and unsupervised learning. Popular supervised learning algorithms, such as Support Vector Machine (SVM) and AdaBoost, were used for text localization in [4–6]. Unsupervised classification methods were explored for text and non-text classification in [7]. Wang et al. [8, 9] proposed to use Convolutional Neural Network (CNN) to learn the unsupervised features for text recognition.

Among the state-of-the-art methods of text localization, few have focused on character segmentation, because character segmentation is considered as challenging as text localization. In this paper, we present a hierarchical approach for text localization which goes from characters to strings, and to words, in a semantically bottom-up way. Different from existing methods which either bet on a few hand-crafted features [4, 10, 11] or rely on heavy learning models [12], our approach presents a systematic way to integrate various effective features extracted or learned at different semantic levels. We adopt simple learning models, such as kernel SVMs and CNN, and focus more on designing simple yet effective new features. The framework of our approach is shown in Fig. 1. And some visual results are given in Fig. 2.

Character Localization. For character localization, we explore three types of supplementary features: structure, gradient based (HOG), and CNN-based features.

String Localization. For string localization, at first, we group characters by their structure features because the characters in a word often have similar structure feature, such as color, aspect ratio, alignment, etc. Then we design a nine-dimension string feature to learn a SVM model that distinguishes non-text strings efficiently.

Word Localization. In word localization, we use the interval cues of adjacent characters in a word to learn the best strategy to split the candidate strings into words.

Our contributions are as follows: (1) A hierarchical text localization framework which goes from characters to strings and to words is proposed; (2) A group of structure feature combined with HOG and CNN features are implemented for character localization; (3) The structure and string features are designed for string localization.

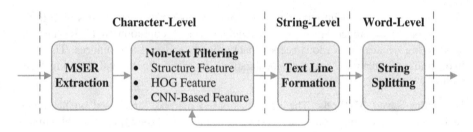

Fig. 1. The framework of the proposed met

(a) MSER Extraction   (b) Non-text Filtering   (c) Text Line Formation   (d) String Splitting

Fig. 2. An example for the visual results obtained at the main steps of the proposed method

The rest of the paper is organized as follows: Character localization is introduced in Sect. 2. String localization and word localization are proposed in Sects. 3 and 4, respectively. Experimental results are described in Sect. 5. Conclusions are given in Sect. 6.

2 Character Localization

2.1 Structure Features

Similar to many methods [13–15], we use MSER to extract the connected components as the candidate character regions. [11, 14] show that MSER-based methods have a high capability to detect most text components in images. Then, to filter out the discernible negative regions, two-level structure features are proposed, as shown in Tables 1 and 2.

In the first level, there are five features, which are special to characters. These character region features usually lie within reasonable ranges, but non-character region features do not. By means of the simple but discriminative features, we remove the non-character regions whose structures are different from the character regions.

Some regions, such as character-like noise regions and distorted character regions which are very different from normal character regions, cannot be easily classified as characters or non-characters only according to their own features. For those regions, we extract the second-level structure features of two adjacent regions, which will be used to form the text line (see Sect. 3). The two character region features (see Table 2)

Table 1. The first-level structure features

Feature	Denotation	Description
Bounding box	$[x, y, w, h]$	(x, y) is the coordinates of the top-left pixel. w, h denote the width and height of a region, respectively.
Area	a	The number of pixels in a region
Centroids	$[cx, cy]$	The coordinates of the centroid in a region
Color	c	Mean pixel value of a region in R,G, B channel images
Aspect Ratio	ar	The ratio between the width and height of a region

Table 2. The second-level structure features

Feature	Definition	Feature	Definition
Difference of color	$\left|c_i - c_j\right|$	Ratio of width	$\frac{\min(w_i, w_j)}{\max(w_i, w_j)}$
Horizontal distance of centroids	$\left|cx_i - cx_j\right|$	Ratio of height	$\frac{\min(h_i, h_j)}{\max(h_i, h_j)}$
Vertical distance of centroids	$\left|cy_i - cy_j\right|$	Ration of aspect ratio	$\frac{\min(ar_i, ar_j)}{\max(ar_i, ar_j)}$

usually satisfy certain constraints; however, the features of most non-character regions do not show strong regularity. Based on this, we successfully add those ambiguous regions into character groups or non-character groups.

2.2 HOG Feature

The structure filter achieves high recall but low precision (see Fig. 4). To filter more negative regions, we use HOG to represent the candidate regions and learn a SVM classifier with the RBF (Radial Basis Function) kernel. To train the classifier, we collect 7,396 character samples and 8,073 non-character samples cropped from the training dataset of ICDAR2003. We apply the SVM classifier on the candidate regions and divide them into two groups according to their confidence scores. The regions whose confidence scores are higher than a threshold are placed into a high-confidence group. The others are placed into a low-confidence group. The HOG feature works well on differentiating those regions that cannot be classified by the structure feature.

2.3 CNN-Based Features

Besides the HOG feature, we also use CNN to learn unsupervised features. The flowchart is shown in Fig. 3. The structure of our CNN network is similar to those in [8, 9] including two convolutional layers, each of which has a convolution step and a pooling step. The number of filters for the two layers is 96 and 256, respectively. To learn the filters, we first normalize the training image patches to 32×32, from which

8×8 patches are randomly extracted. Next, each 8×8 patch is contrast normalized and ZCA whitened [16] to form a 64-dimension vector. We then use K-means described in [8] to learn a set of filters. As is common in CNNs, we adopt the average pooling over the convolution layer response map. The output feature of the second layer is then used to train a SVM classifier with the Polynomial kernel. The CNN-based feature, as the supplement for the structure feature and HOG feature, leads to more sophisticated character and non-character binary classifications.

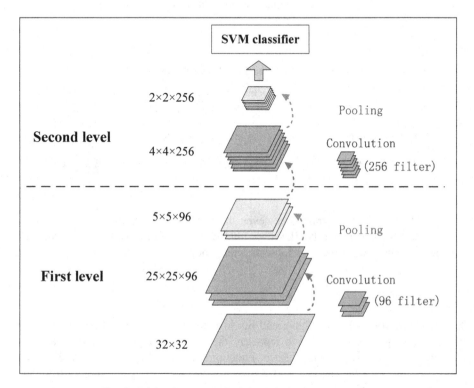

Fig. 3. Bi-level unsupervised CNN feature representation

3 Text Line Formation

As shown in Table 2, the characters in a word often have similar structure features. We use those structure features to group the candidate regions. In detail, if two neighbor regions satisfy all structure feature constraints, they are treated as a pairwise region. Then we search all pairwise regions. Furthermore, if two pairwise regions share one common region and have similar direction, they are merged. This operation is repeated until no character pairs can be merged. Each merging set of characters forms a candidate string. We also filter out a string which contains only a single character, because it is assumed that a text string has at least three characters.

For the candidate strings, we learn a model to distinguish non-text strings. Here, as shown in Table 3, we design a nine-dimension string feature. For SF3, two pairwise adjacent regions, having similar direction, form a text line. A rotation angle is defined as the angle between every two text lines. We use the nine-dimension features to train a SVM classifier with the RBF kernel. In our experiments, we collect 649 word patches as positive samples and 287 non-word patches as negative samples which are cropped from the training dataset of ICDAR 2003.

Table 3. The nine-dimension string feature

Feature	Description
SF1	The number of characters
SF2	The average probability of characters
SF3	The average rotation angle
SF4	The size variation of characters
SF5	Variation of the distances between two centroids of adjacent characters
SF6	The average bias between the direction of a character and the text line
SF7	The average axial ratio of the major axis to the minor axis of characters
SF8	The average density ratio of the number of pixels to the area of the bounding box
SF9	The average color similarity defined by the cosine similarity of two characters' color histograms

4 String Splitting

Many of the candidate strings include multiple words. To solve this problem, we use the interval cues to split the candidate strings into words. Generally speaking, the intervals of two adjacent characters in a word are almost equal, and the inter-word intervals are usually greater than the intra-word intervals.

First, within each string, we denote the intervals between successive characters as $D_k, k \in [1, n-1]$, where n is the number of characters in the string. Then, we calculate the average of the intervals between successive characters (denoted as $\bar{D}$) and the average width of the characters (denoted as $\bar{W}$). We split a string region if D_k satisfies the conditions defined by the following inequality (1), where α_i and β_i are the weights for $\bar{D}$ and $\bar{W}$, respectively. A is the set of all possible combinations of the weights.

$$D_k > \max_{(\alpha_i, \beta_i) \in A} (\alpha_i * \bar{D} + \beta_i * \bar{W}) \tag{1}$$

To learn the optimal splitting strategy, we apply our text line formation to the ground truth words. By analyzing the interval cues of the ground truth strings, we learn the best combinations of weights. After splitting, each sub-region is regarded as a word region.

5 Experimental Results

In this section, we apply our approach on the ICDAR2003 and ICDAR2011 datasets. As shown in Fig. 5 below, the visual results of text localization achieved by our method demonstrate its robustness to challenges such as illumination changes, cluttered background, text font variations, and so on.

Feature Performance. Different from the state-of-the-art methods, the proposed method is based on character localization. By the definition, a detected character is correct if the ratio of the intersection to the union of the detected region and the ground truth region is greater than 0.5. We use the recall, precision and F-measure as the criteria following [17]. Precision is defined as the number of correct estimates divided by the total number of estimates. Recall is defined as the number of correct estimates divided by the total number of targets. As shown in Fig. 4, recall for the three stages remains constant, but the precision shows improvement, indicating that the proposed method filtered out more noise regions, yet reserved character regions, which demonstrates the supplementary of the structure, HOG, and CNN features.

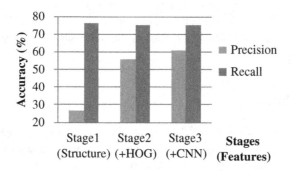

Fig. 4. The performance of the proposed method for character localization on the ICDAR2003

Table 4. The performance comparison of different methods on the ICDAR2003

Methods	Year	Precision	Recall	F-measure
Our method	2015	0.83	0.67	0.74
Huang *et al.* [14]	2014	**0.84**	0.67	**0.75**
SFT-TCD [3]	2013	0.81	**0.74**	0.72
Kim's [13]	2013	0.78	0.65	0.71
Neumann and Matas [18]	2011	0.72	0.62	0.67
Epshtein *et al.* [2]	2010	0.73	0.60	0.66
Results only on the words with more than two characters				
Our method*	2015	**0.86**	**0.75**	**0.80**
Yi's Method* [19]	2013	0.71	0.62	0.63

*The asterisk are evaluated only on those words with more than three characters.

The data with bold fonts are the best results.

SVM Classifier Performance. There are three SVM classifiers in our experiment: HOG-SVM, CNN-SVM, and STRING-SVM. Evaluated by cross-validation, they have achieved classification accuracies of 92.78 %, 96.4 %, and 91.6 %, respectively.

Text Localization Performance. We evaluate the text localization performance of our method on the ICDAR2003 (Table 4) and ICDAR2011 (Table 5) datasets by the standard criteria used in the competitions of ICDAR2003 and ICDAR2005 [10, 17].

Table 5. The performance comparison of different methods on ICDAR2011

Methods	Year	Precision	Recall	F-measure
Our method	2015	79.4	65.8	**72.0**
Shi *et al.* [20]	2013	63.1	83.3	71.8
Gao *et al.* [21]	2013	61.3	**83.9**	70.8
Kim's method [22]	2011	**83.0**	62.5	71.3
Neumann's method [11]	2012	64.7	73.1	68.7
Results only on the words with more than two characters				
Our method*	2015	**83.4**	**71.6**	**77.0**
Yi's method* [19]	2013	76.0	68.0	67.0

* The asterisk are evaluated only on those words with more than three characters.
The data with bold fonts are the best results.

Fig. 5. Some examples of the text localization results obtained by the proposed method on ICDAR2003. The boxes in blue are detection results and the boxes in red are the ground truth (Color figure online).

As shown in Tables 4 and 5, the performances of the proposed method on IC-DAR2003 and ICDAR2011 are very competitive to the other state-of-the-art methods. Note that in [19], the authors selected images in which a text string contains at least three characters. Similarly, we also ignore all words that contain two or fewer characters, because the text line formation step of the proposed method is based on the assumption that a string has at least three characters. For comparisons in Tables 4 and 5, the methods marked with an asterisk are evaluated only on those words with more than three characters. The data with bold fonts are the best results.

It is noticing that all of our learning models are trained on ICDAR2003, but the good performance on ICDAR2011 in Table 5 demonstrates the generalization ability of the proposed method. References [15] and [23] follow the standard protocol of the ICDAR2011 [22], which not only gives a score of "1" for one-to-one matching and a score of "0" for no matching, but also gives a score of "0.8" for one-to-many matching, which definitely increases precision without decreasing recall.

6 Conclusions

In this paper, we proposed a hierarchical approach for text localization in a natural scene image, which goes from characters to strings and to words. We designed three types of features for character detection: the character structure feature, HOG feature, and unsupervised feature based on CNN. Furthermore, we used the structure feature to group characters into words. To distinguish the non-text strings, we proposed the nine dimension word string feature. We compared our method with several state-of-the-art methods in terms of precision, recall, and F-measure. The experimental results demonstrate that the proposed method is competitive with the state-of-the-art methods in terms of accuracy. However, the proposed method is not only simple but also free of heavy learning steps, which is considered to be important for real applications.

Acknowledgments. This work was supported by the National Natural Science Foundations of China under Grants 61373077, 61472334 and 61170179, the Natural Science Foundation of Fujian Province of China Under Grant 2013J01257, the Fundamental Research Funds for the Central Universities under Grant 20720130720, the 2014 national college students' innovative and entrepreneurial training project, and the Scientific Research Foundation for the Introduction of Talent at Xiamen University of Technology YKJ12023R.

References

1. Judd, T., Ehinger, K., Durand, F., Torralba, A.: Learning to predict where humans look. In: 2009 IEEE 12th International Conference on Computer Vision, pp. 2106–2113. IEEE (2009)
2. Epshtein, B., Ofek, E., Wexler, Y.: Detecting text in natural scenes with stroke width transform. In: 2010 IEEE Conference on Computer Vision and Pattern Recognition (CVPR), pp. 2963–2970. IEEE (2010)
3. Huang, W., Lin, Z., Yang, J., Wang, J.: Text localization in natural images using stroke feature transform and text covariance descriptors. In: 2013 IEEE International Conference on Computer Vision (ICCV), pp. 1241–1248. IEEE (2013)
4. Jain, A.K., Yu, B.: Automatic text location in images and video frames. Pattern Recogn. 31, 2055–2076 (1998)

5. Jung, C., Liu, Q., Kim, J.: Accurate text localization in images based on SVM output scores. Image Vis. Comput. **27**, 1295–1301 (2009)
6. Chen, X., Yuille, A.L.: A time-efficient cascade for real-time object detection: with applications for the visually impaired. In: 2005 IEEE Computer Society Conference on Computer Vision and Pattern Recognition-Workshops, CVPR Workshops, pp. 28–28. IEEE (2005)
7. Liu, C.M., Wang, C.H., Dai, R.W.: Text detection in images based on unsupervised classification of edge-based features. Proceedings of the Eighth International Conference on Document Analysis and Recognition, vols. 1 and 2, pp. 610–614 (2005)
8. Coates, A., Carpenter, B., Case, C., Satheesh, S., Suresh, B., Wang, T., Wu, D.J., Ng, A.Y.: Text detection and character recognition in scene images with unsupervised feature learning. In: 2011 International Conference on Document Analysis and Recognition (ICDAR), pp. 440–445. IEEE (2011)
9. Wang, T., Wu, D.J., Coates, A., Ng, A.Y.: End-to-end text recognition with convolutional neural networks. In: 2012 21st International Conference on Pattern Recognition (ICPR), pp. 3304–3308. IEEE (2012)
10. Lucas, S.M.: ICDAR 2005 text locating competition results. In: 2005 Proceedings of the Eighth International Conference on Document Analysis and Recognition, pp. 80–84. IEEE (2005)
11. Neumann, L., Matas, J.: Real-time scene text localization and recognition. In: 2012 IEEE Conference on Computer Vision and Pattern Recognition (CVPR), pp. 3538–3545. IEEE (2012)
12. Pan, Y.-F., Hou, X., Liu, C.-L.: Text localization in natural scene images based on conditional random field. In: 2009 10th International Conference on Document Analysis and Recognition. ICDAR 2009, pp. 6–10. IEEE (2009)
13. Koo, H.I., Kim, D.H.: Scene text detection via connected component clustering and nontext filtering. IEEE Trans. Image Process. **22**, 2296–2305 (2013)
14. Huang, W., Qiao, Y., Tang, X.: Robust scene text detection with convolution neural network induced MSER trees. In: Fleet, D., Pajdla, T., Schiele, B., Tuytelaars, T. (eds.) ECCV 2014, Part IV. LNCS, vol. 8692, pp. 497–511. Springer, Heidelberg (2014)
15. Yin, X.-C., Yin, X., Huang, K., Hao, H.-W.: Robust text detection in natural scene images. IEEE Trans. Pattern Anal. Mach. Intell. **36**, 970–983 (2014)
16. Hyvärinen, A., Oja, E.: Independent component analysis: algorithms and applications. Neural Netw. **13**, 411–430 (2000)
17. Lucas, S.M., Panaretos, A., Sosa, L., Tang, A., Wong, S., Young, R.: ICDAR 2003 robust reading competitions. In: 2003 International Conference on Document Analysis and Recognition (ICDAR), pp. 682–682. IEEE Computer Society (2003)
18. Neumann, L., Matas, J.: Text localization in real-world images using efficiently pruned exhaustive search. In: 2011 International Conference on Document Analysis and Recognition (ICDAR), pp. 687–691. IEEE (2011)
19. Yi, C., Tian, Y.: Text extraction from scene images by character appearance and structure modeling. Comput. Vis. Image Underst. **117**, 182–194 (2013)
20. Shi, C., Wang, C., Xiao, B., Zhang, Y., Gao, S.: Scene text detection using graph model built upon maximally stable extremal regions. Pattern Recogn. Lett. **34**, 107–116 (2013)
21. Gao, S., Wang, C., Xiao, B., Shi, C., Zhang, Y., Lv, Z., Shi, Y.: Adaptive Scene Text Detection Based on Transferring Adaboost. In: 2013 12th International Conference on Document Analysis and Recognition (ICDAR), pp. 388–392. IEEE (2013)
22. Shahab, A., Shafait, F., Dengel, A.: ICDAR 2011 robust reading competition challenge 2: Reading text in scene images. In: 2011 International Conference on Document Analysis and Recognition (ICDAR), pp. 1491–1496. IEEE (2011)
23. Neumann, L., Matas, J.: Scene text localization and recognition with oriented stroke detection. In: 2013 IEEE International Conference on Computer Vision (ICCV), pp. 97–104. IEEE (2013)

Recognizing Human Actions by Sharing Knowledge in Implicit Action Groups

RuiShan Liu, YanHua Yang, and Cheng Deng[✉]

Video and Image Processing System Lab, School of Electronic Engineering,
Xidian University, Xi'an 710000, China
rsliu.xd@gmail.com, chdeng@mail.xidian.edu.cn

Abstract. Most of the current action recognition approaches learn each action category separately. An important observation is that many action categories are correlated and could be clustered into groups, which are always ignored to decreasing the recognition accuracy. In this paper, we employ a multi-task learning framework with group-structured regularization to share knowledge in category groups. First, we employ Fisher Vector, concatenated by gradients with respect to mean vector and covariance matrix of GMM, to represent action data. Intuitively, the action categories in the same group are prone to have a closer relationship with the same Gaussian components. The proposed method uses one-vs-one SVM margin to measure the degree of similarity between each pair of categories and obtain the implicit group structure by Affinity Propagation Clustering. In order to encourage the categories in the same group to share dimensions feature from the same Gaussian component and vice versa, the implicit group structure is used as the prior regularization in multi-task learning. Our experiments on large and realistic dataset HMDB51 show that the proposed method has achieved the comparative even higher accuracy with less dimensions of feature over several state-of-the-art approaches.

Keywords: Action recognition · Implicit group structure · Multi-task learning · Fisher vector

1 Introduction

Human action recognition [1,8] in videos has been an active research area in computer vision due to its widely applications in surveillance, human-computer interaction and so forth. Earlier work mainly focused on simple and small datasets such as Weizmann dataset [5] recorded in a controlled environment. However, due to background clutter, viewpoint changes, occlusion and motion style variation, the great challenge still remains in this area especially when focusing on large and more realistic datasets (e.g., UCF101 dataset [15], HMDB51 dataset [10]).

Recently, action recognition methods using local hand-crafted features, including Dense Trajectories [16] and Improved Dense Trajectories [17] have obtained the state-of-the-art performance on many datasets cooperating with

© Springer International Publishing Switzerland 2015
Y.-S. Ho et al. (Eds.): PCM 2015, Part I, LNCS 9314, pp. 644–652, 2015.
DOI: 10.1007/978-3-319-24075-6_62

Fig. 1. Action categories from HMDB51 could be clustered into different groups. Action categories in the same group are prone to share more dimension features from the same Gaussian.

Bag-of-Visual-Words [11] or its variants. Among its variants, Fisher Vector [12], concatenated by gradients with respect to mean vector, and covariance matrix of each Gaussian component, is especially outstanding. Intuitively, each Gaussian component captures specific motion patterns and feature characteristic and the correlated action categories are inclined to have a closer relationship with the same Gaussian component. For example, action chew and eat in HMDB51 are highly correlated and prone to share similar information captured by the same Gaussian component. In addition, different from the traditional multi-task leaning based on the assumption that all of the action categories are correlated or irrelevant, we observe the fact that in dataset action categories which are highly correlated could be clustered into groups and categories in the same group are prone to share the dimensions feature from the same Gaussian component and vice versa (Fig. 1).

Motivated by the above analysis, we employ a multi-task learning framework [2] with group-structured constraint to exploit the relationship between action categories and Gaussian components. Specifically, we employ the SVM margin between two categories to measure the category-level similarity. Then we use Affinity Propagation Clustering [4] to yield the implicit group structure of categories based on the similarity matrix. Finally, we employ a multi-task learning framework with the group structure used as the prior regularization. Multi-task learning is known as its capability to improve the generalization of each single task by learning a common feature representation. With the group structure served as the prior regularization, our model could constrain knowledge sharing only exists in each category group and among the groups this sharing is prohibited. The framework of the proposed work is shown in Fig. 2.

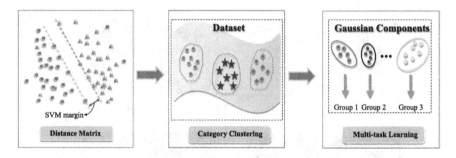

Fig. 2. The framework of the proposed work. Firstly we train one-vs-one classifier for every pair of action categories in dataset to compute the category-level similarity. Then based on the similarity matrix, implicit group structure could be obtained by Affinity Propagation Clustering. Finally, a multi-task learning is employed with group-structure constraint to encourage the action categories in the same group to share more dimensions feature from the same Gaussian component.

The contribution of this paper are three-fold: (1) we explore the implicit group structure of action categories embedding in datasets which is neglected by all of the former work; (2) we propose an effective multi-task learning framework to encourage knowledge transferring in the category groups; (3) we conduct experiments on HMDB51 dataset and the results confirms that the proposed work could achieve a comparative even better recognition accuracy compared to the state-of-the-art methods with less dimensions feature.

2   The Proposed Method

In this section, we first review Fisher Vector briefly. Then we explore the implicit group structure by using the SVM margin to measure the degree of similarity among categories. Finally, we introduce the multi-task learning objective with the group structure employed as a prior regularization.

2.1   Fisher Vector

Fisher Vector coding method, derived from Fisher Kernel, is based on the assumption that generation process of local descriptor $\mathbf{X}$ can be modelled by a probability density function $p(,:\theta)$. Since the gradient of the log-likelihood could describe the way that parameters contributes to the generation process of $\mathbf{X}$. Then the sample could be describe as:

$$G_\theta^{\mathbf{X}} = \frac{1}{N}\nabla_\theta log p(\mathbf{X};\theta). \tag{1}$$

Note that the dimensionality of this vector depends on the number of parameters in θ. The probability density function is usually modelled by Gaussian Mixture Model, and $\theta = \{\pi_1, \mu_1, \sigma_1, ..., \pi_K, \mu_K, \sigma_K\}$ are the model parameters denoting the mixture weights, means and diagonal covariance respectively. In [12] an improved Fisher Vector is proposed as follows:

$$\rho_k = \frac{1}{\sqrt{\pi_k}} \gamma_k \left(\frac{x - \mu_k}{\sigma_k} \right), \tag{2}$$

$$\tau_k = \frac{1}{\sqrt{\pi_k}} \gamma_k \left(\frac{(x - \mu_k)^2}{\sigma_k^2} - 1 \right), \tag{3}$$

Where γ_k is the weight of local descriptor x to k^{th} Gaussian component,

$$\gamma_k = \frac{\pi_k \mathcal{N}(x; \mu_k, \sigma_k)}{\sum_{i=1}^{K} \pi_i \mathcal{N}(x; mu_i, \sigma_i)}. \tag{4}$$

The final Fisher Vector is concatenated by this two gradients as:

$$S = [\rho_1, \tau_1, ..., \rho_K, \tau_K] \tag{5}$$

2.2   Exploring the Implicit Group Structure

Supposing that there are C action categories existing in the dataset. For the c-th action category, the corresponding training data are $\{x_{c,i}, y_{c,i}\}_{i=1}^{N_c} \subset \mathbb{R}^{D \times 1} (c = 1, ..., C)$, where i and N_c denote the index and the number of video belonging to the c-th action category, respectively. $x_{c,i}$ is concatenated as Eq. (5) mentioned above.

To measure the degree of similarity of two action categories, we train one-vs-one SVM classifiers using the subset of the training data as in [6]. Intuitively, more similar the two action categories are, shorter the SVM margin between them is. Then the similar action categories are prone to share more dimensions feature from the same Gaussian. The similarity degree could be computed as follows:

$$s_{y_i, y_j} = \sum_{\forall (i,j) \in y_p \cup y_q} \alpha_i \alpha_j (-1)^{I[c(i), c(j)]} K(x_{p,i}, x_{q,i}), \tag{6}$$

Where i and j are all pairs of samples from the union of p-th and q-th categories. α is the coefficient for support vectors which could be yielded by the one-vs-one SVM classifiers. $c(.)$ returns the category of the given sample and $I[.]$ is the indicator function as follows:

$$I_{i,j} = \begin{cases} 0 & c(i) == c(j) \\ 1 & otherwise, \end{cases} \tag{7}$$

$K(., .)$ is the kernel function. In our experiment, we use the linear kernel which shows a good performance cooperating with Fisher Vector. Since each classifiers use only a tiny subset of the training data ignoring test/train splits in this phase, they are quick to train $\binom{2}{C}$ one-vs-one classifiers in total.

In [6], the category-level similarity is also computed using SVM margins. However, our work is different from their work in application and formulation. Their work focus on identifying pairs of categories that are close to learn discriminative category-level features while our goal is to explore the group structure embedding in the dataset. Furthermore, we employ a multi-task learning framework with group-structure regularization to encourage knowledge sharing in category groups.

Finally, we get a symmetric similarity matrix $S \in \mathbb{R}^{C \times C}$, then the implicit group structure of category could be obtained by using Affinity Propagation Clustering [4]. We don't have to determine the number of clusters in advance and it could be obtained based on the similarity matrix S owing to the use of Affinity Propagation Clustering. It is notable that the there is not an optimal group structure for an action dataset. The number of groups and the specific classes in each group changed with the type of descriptors. Specifically, for different types of descriptors, (e.g., Histograms of Oriented Gradients (HOG) and Histograms of Optical Flow (HOF)) we obtained different SVMs and similarity matrix S, then the resulted group structures are also different, respectively. An example of the group structure is showed in Sect. 3.2.

2.3 Model Learning

Obtaining the action categories groups, we employ a multi-task learning with group-structure regularization [7] to learn a binary linear classifier for each category simultaneously. This approach could encourage features competition inter-groups and feature sharing intra-groups with structured sparisity as follows:

$$\min_{W} \sum_{c=1}^{C} \sum_{i=1}^{N_c} \frac{1}{2} \left[\max(0, 1 - y_{c,i} w_c x_{c,i}) \right]^2 + \lambda_1 \sum_{d=1}^{D} \sum_{l=1}^{L} \|w^{d,g_l}\|_2 + \lambda_2 \|W\|_F^2, \quad (8)$$

The first term is the squared hinge loss function. $W \in \mathbb{R}^{N \times D}$ denotes the parameters of all the categories and w^{d,g_l} denotes the parameters of c-th category. In the second term, w^{d,g_l} denotes subset of action categories belonging to the group g_l. The group-structure regularization takes advantage of both ℓ_1 lasso and ℓ_{21} lasso to encourage feature sharing intra-groups and feature competition inter-groups. The last term constrained by Frobenius norm helps W to avoid overfitting. λ_1 and λ_2 are scalar regularization parameters balancing sparsity against classification loss.

Obviously, the group-structure regularization imposed on the objective function Eq. (8) is a mixed norm regularization. Even though it is convex, but non-smooth and non-trivial to optimize at the same time. As in [7,9], firstly we reformulate the objective by representing the 2-norm in the regularizer in its dual form, then we employ the smoothing proximal gradient descent [3] to optimize a smooth approximation of the resulting objective.

A binary linear classifier for each action category could be obtained by Eq. (8) after learning the after learning the model parameters matrix W. When coming a new test video represented as the Fisher Vector x_n, the corresponding label y_n could be obtained as follows:

$$y_n = \arg\max_{c} w_c x_n. \quad (9)$$

In [7], the group-structure multi-task learning framework is also employed to learn attribute classifiers, but compared with their methods, the proposed work obtain the group structure based on the data while in [7] it is prespecified which is too strong for many practical problems. Furthermore our work is equipped with the ℓ_2 regularization to prevent overfitting caused by the high dimension of Fisher Vector.

3   Experiment

In this section, we firstly introduce the implementations details. Then we present the recognition results on the HMDB51 datasets with the proposed method. Besides, a comparison to the state-of-the-art methods is given in this section.

3.1   Experimental Setup

Dataset. We conduct the experiments on HMDB51 dataset [10] which is collected from a wide range of sources such as: movies, Youtube. It is consisted of 6766 videos from 51 action categories and each category has more than 100 videos. Our experiments follow the original evaluation protocol using three training and testing splits and the average accuracy over the three splits is given to measure the performance.

Features. We evaluate our approach with improved dense trajectories [17] sampling local descriptors. Each video is represented by four types of descriptors: HOG, HOF, MBHx and MBHy. As the steps in [18], we first reduce the dimensionality by a factor of two by performing PCA on each type of descriptor. Then we randomly sample a subset of 256,000 features and learn the GMM with $K = 512$ components via the EM algorithm. We also whiten each types of descriptors and apply L_2 and intra normalization to them. Finally, four types of descriptors are combined by concatenating to form final representation.

Parameters. The scalar parameters $\lambda_1 4$ and λ_2 are chosen by a cross validation procedure. The optimal number of action category groups could be computed by Affinity Propagation Clustering.

Baseline. We compare the proposed method with three baselines. They are multi-task learning without regularization, multi-task learning with ℓ_1 rgularization, multi-task learning with ℓ_{21} regularization. In all multi-task learning methods, we use hinge-loss as the loss function and choose scalar parameters by cross validation.

3.2   Experiments on HMDB51 Dataset

Firstly, we show the group structure obtained by the proposed method on HMDB51. 51 action categories is divided into 7 groups by Affinity Propagation Group1: *turn, walk, shake hands, hug and kiss.* Group2: *brush hair, clap, wave, shoot gun, draw sword, sword, climb, climb stairs and drink, eat, pick, sit, stand, pour, shoot bow and pullup.* Group3: *chew, smile, laugh, smoke and talk.* Group4: *push up, sit up.* Group5: *cartwheel, flic flac, hand stand, somersault, push, ride bike and ride horse.* Group6: *catch, hit, swing baseball, throw, dive, fall floor, jump, run, kick ball, fencing, kick, sword exercise and punch.* Group7: *dribble, shoot ball and golf.* The result shows that action categories which are similar in appearance and motion trajectories are clustered into the same group. For example, Actions chew, smile, laugh, smoke and talk which are all describe the subtle facial expressions are in the same group.

Table 1. Comparison of the baselines with our method on the HMDB51 dataset.

Feature/Method	Without regularizer	ℓ_1 lasso	ℓ_{21} lasso	Proposed work
HOG	40.22	44.05	43.84	**44.54**
HOF	49.00	50.04	50.09	**50.62**
MBHx	40.24	42.81	42.61	**44.09**
MBHy	47.13	48.60	48.62	**49.01**
Combined	60.15	60.38	60.31	**60.54**

Then we compare the proposed method with all the baselines in Table 1. The results shows that the proposed work outperforms multi-task learning method with each type of features and their combination. As mentioned in [7], when the number of groups is set to only one, the group-structure regularization degenerates to l_{21} lasso. In other word, all of the action categories are highly correlated. When each action category belongs to its own singleton group, the group-structure regularization degenerates to l_1 lasso. Above all, the group-structure regurlarization take advantage of both l_{21} and l_1 lasso.

Finally, we compare the proposed method with some state-of-the-art methods on the HMDB51 dataset. As demonstrated in Table 2, the proposed method even outperforms the approach [14] based on Deep Learning.

Table 2. Comparison of the baselines with our method on the HMDB51 dataset.

Methods	Accuracy
Sadanand and J. Corso [13]	26.9 %
Yang et al. [19]	53.9 %
Hou et al. [6]	57.88 %
K. Simonyan and A. Zisserman [14]	59.5 %
Proposed method	**60.54%**

4   Conclusion

In this work, we proposed a new method to recognize human action in videos by exploring the implicit group structure embedding in the dataset based on the assumption that action categories in the same group are prone to share more dimensions feature from the same Gaussian component in Fisher Vector and vice versa. The group structure were used as a prior regularization cooperating with a multi-task learning framework, encouraging related knowledge transferring as well as reducing the dimension of Fisher Vector. We evaluate the performance the proposed method on a complex and large-scale action datasets: HMDB51 dataset. Experimental results demonstrate that our work outperforms several baselines and some state-of-the-art methods.

Acknowledgements. This work is supported by the National High Technology Research and Development Program of China (2013AA01A602), the Program for New Century Excellent Talents in University (NCET-12-0917), the Fundamental Research Funds for the Central Universities (No. K5051302019), the Key Science and Technology Program of Shaanxi Province, China (2014K05-16).

References

1. Aggarwal, J.K., Ryoo, M.S.: Human activity analysis: a review. ACM Comput. Surv. **43**(3), 16 (2011)
2. Argyriou, A., Evgeniou, T., Pontil, M.: Convex multi-task feature learning. Mach. Learn. **73**(3), 243–272 (2008)
3. Chen, X., Lin, Q., Kim, S., Carbonell, J.G., Xing, E.P.: Smoothing proximal gradient method for general structured sparse learning. CoRR abs/1202.3708 (2012)
4. Frey, B.J., Dueck, D.: Clustering by passing messages between data points. Science **315**, 972–976 (2007)
5. Gorelick, L., Blank, M., Shechtman, E., Irani, M., Basri, R.: Actions as space-time shapes. Trans. Pattern Anal. Mach. Intell. **29**(12), 2247–2253 (2007)
6. Hou, R., Roshan Zamir, A., Sukthankar, R., Shah, M.: DaMN – discriminative and mutually nearest: exploiting pairwise category proximity for video action recognition. In: Fleet, D., Pajdla, T., Schiele, B., Tuytelaars, T. (eds.) ECCV 2014, Part III. LNCS, vol. 8691, pp. 721–736. Springer, Heidelberg (2014)
7. Jayaraman, D., Sha, F., Grauman, K.: Decorrelating semantic visual attributes by resisting the urge to share. In: CVPR (2014)
8. Ke, S.R., Hoang, L.U.T., Lee, Y.J., Hwang, J.N., Yoo, J.H., Choi, K.H.: A review on video-based human activity recognition. Computers **2**(2), 88–131 (2013)
9. Kim, S., Xing, E.P.: Tree-guided group lasso for multi-task regression with structured sparsity. In: ICML (2010)
10. Kuehne, H., Jhuang, H., Garrote, E., Poggio, T., Serre, T.: HMDB: a large video database for human motion recognition. In: ICCV (2011)
11. Peng, X., Wang, L., Wang, X., Qiao, Y.: Bag of visual words and fusion methods for action recognition: comprehensive study and good practice. CoRR abs/1405.4506 (2014)

12. Perronnin, F., Sánchez, J., Mensink, T.: Improving the fisher kernel for large-scale image classification. In: Daniilidis, K., Maragos, P., Paragios, N. (eds.) ECCV 2010, Part IV. LNCS, vol. 6314, pp. 143–156. Springer, Heidelberg (2010)
13. Sadanand, S., Corso, J.J.: Action bank: a high-level representation of activity in video. In: ICCV (2012)
14. Simonyan, K., Zisserman, A.: Two-stream convolutional networks for action recognition in videos. In: NIPS (2014)
15. Soomro, K., Zamir, A.R., Shah, M.: UCF101: a dataset of 101 human actions classes from videos in the wild. CoRR abs/1212.0402 (2012)
16. Wang, H., Kläser, A., Schmid, C., Liu, C.L.: Action recognition by dense trajectories. In: CVPR (2011)
17. Wang, H., Schmid, C.: Action recognition with improved trajectories. In: ICCV (2013)
18. Wang, H., Ullah, M.M., Klser, A., Laptev, I., Schmid, C.: Evaluation of local spatio-temporal features for action recognition. In: BMVC (2009)
19. Yang, X., Tian, Y.L.: Action recognition using super sparse coding vector with spatio-temporal awareness. In: Fleet, D., Pajdla, T., Schiele, B., Tuytelaars, T. (eds.) ECCV 2014, Part II. LNCS, vol. 8690, pp. 727–741. Springer, Heidelberg (2014)

Human Parsing via Shape Boltzmann Machine Networks

Qiurui Wang[1,2], Chun Yuan[2(✉)], Feiyue Huang[3], and Chengjie Wang[3]

[1] Department of Computer Science, Tsinghua University, Beijing, China
wangqr12@mails.tsinghua.edu.cn
[2] Graduate School at Shenzhen, Tsinghua University, Shenzhen, China
yuanc@sz.tsinghua.edu.cn
[3] BestImage Team, Tencent, Shanghai, China
{garyhuang,jasoncjwang}@tencent.com

Abstract. Human parsing is a challenging task because it is difficult to obtain accurate results of each part of human body. Precious Boltzmann Machine based methods reach good results on segmentation but are poor expression on human parts. In this paper, an approach is presented that exploits Shape Boltzmann Machine networks to improve the accuracy of human body parsing. The proposed Curve Correction method refines the final segmentation results. Experimental results show that the proposed method achieves good performance in body parsing, measured by Average Pixel Accuracy (aPA) against state-of-the-art methods on Penn-Fudan Pedestrians dataset and Pedestrian Parsing Surveillance Scenes dataset.

Keywords: Segmentation · Human parsing · Shape Boltzmann machine

1 Introduction

A basic issue in objects segmentation is how to discriminate the different semantic parts of human, which can provide significant assistance for several tasks, such as gesture and attire recognition. This is a challenge task, because the situation of appearances and poses of a figure are totally different, occasionally blurry or obstructed.

Much research has been done on this challenging problem. Existing studies of human parsing [1–4] and other segmentation methods can be generally divided into three categories: template matching based methods, Bayesian inference methods and deep learning network methods. Rauschert and Collins [5] used a Bayesian framework to estimate articulated body pose and parts in pixel-level segmentation, treating the human body as a collection of geometrically linked regions. This approach used color and posed skeletons to express the structure of the object. Eslami and Williams [6] used a Shape Boltzmann Machine (ShapeBM) [7] employing a MCMC process to simulate the multivariate distributions of the human body and infer all the parts. Luo et al. [8] proposed a

© Springer International Publishing Switzerland 2015
Y.-S. Ho et al. (Eds.): PCM 2015, Part I, LNCS 9314, pp. 653–663, 2015.
DOI: 10.1007/978-3-319-24075-6_63

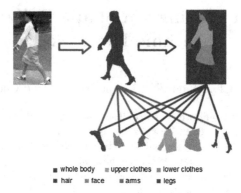

Fig. 1. The architecture of our model. There are three steps in our proposed method: (a) Feature extraction and whole human body segmentation (blue arrow), (b) Human body parsing (blue lines), (c) Output optimization (green arrow and green lines). The blue lines represent the interaction optimization process between the whole body and all its parts, the green lines stand for curve correction optimization process (Color figure online).

three level Deep Decompositional Network (DDN) to express a labeled map by distinguishing occlusion areas and non-occlusion areas at the body part scale and whole body scale. These works further motivate the researchers to focus on body parsing problem, a very difficult task in this area, using a deep learning network.

In the reported experiments, it was found that although the results are accurate, the edges are ragged and the masks of objects or body parts are irregular. Combining edge and shape based clustering methods [1], it is proposed Curve Correction approach can further improve the performance.

This study generally provides a Shape Boltzmann Machine framework for human body parsing, which can be simultaneously accomplished. The reported Curve Correction method is a good example of how to refine segmentation results. Experimental results show that the proposed methods achieve good performance in body parsing against state-of-the-art methods.

2   Model Design

2.1   Model Structure for Human Parsing

It is clear that different models of objects are obtained by training different objects, even the parts of some object. But if we train and parse the parts of human body separately, it is at great computation cost and we may lose the relationship between parts and its whole body. In this paper, each human is regarded as being taken as a portion of the map of the whole body, while the whole body can be treated as the joint mapping of all its parts. This is concept behind the model shown in Fig. 1.

The process shown in Fig. 1 assumes the following steps. First, each part is treated as an independent channel task, including the whole body, hair, face, arms, upper clothes, lower clothes and legs. Each channel provides its output by one layer of ShapeBM, which is Restricted Boltzmann Machine (RBM) in fact, making the task simpler, excluding interruptions by other body parts. This process is similar to a Switchable Deep Network [8] but only in this first step. Next, the individual parts separately connect the whole body part. Thus in the second layer, there are some connections between the components, and the whole body channel can be treated as the top layer, which now makes it a three layer Boltzmann Machine network. After several interactions, some accurate segmentation output is obtained for both the whole body and its parts, which are optimized in the final results by the proposed Curve Correction method as is described in Sect. 2.3.

2.2 Multi Channel Segmentation by Shape Boltzmann Machine Network

In the RBM, the hidden variables $h \in \{0,1\}^n$ are used to obtain global dependencies between visible variables. The energy function takes the form $E(y,h) = -y^T W h - b^T y - c^T h$. We only need to train and learn the parameters b, c, W in RBMs. The defect of RBM in segmentation is that it requires a large set of hidden variables and numerous training samples to model a complex distribution of object, which is difficult for some tasks like human body parsing since the labeled images are scarce, such as the datasets reported by Penn-Fudan Pedestrians (PFP) [12] and Pedestrian Parsing Surveillance Scenes (PPSS) [13] datasets.

However, Eslami and Williams [6] proposed a part-connected Boltzmann Machine within a layer in a two-layer structured network. The first layer of the hidden variables h^1 is divided into several disconnected subsets. Each of these has a range, which only connects to a local patch of the object silhouette. Hence, the first layer can be exploited as parts of the body although it is difficult to map them directly. The second layer of the hidden variables h^2 connects to all the variables h^1 of the first layer with no connection to each other. Yang et al. [14] ameliorated ShapeBM by connecting images to a hidden layer as a full generative model Max Margin Boltzmann Machines (MMBM). MMBM provides a better expression, though is hard to train, which can be solved by optimizing the variational upper bound of the log-likelihood $log\,p(y|x)$ using the EM algorithm.

Generative RBM and ShapeBM do well in segmentation of human body shape modeling, but the computation cost is high for inferring a part silhouette of the human body y from an image x. MMBM has a better capability to express of $\{0,1\}^n$ as previously mentioned, because it constructs a fully generative model between human body images and their silhouette mask $p(y,x)$. Human body contours can be inferred from the image by the conditional distribution $p(y|x)$.

This study presents the silhouette as a set of visible variables, but it is difficult to construct the joint distribution of a human body, since the conditional distributions of body parts are various and ambiguous. Similar to MMBM, we attempt to train the conditional probability distribution of each part of the human body directly, instead of a joint distribution for estimating the silhouette. But compared to MMBM, our variances are: (a) our task is multi-target recognition and segmentation for each part of human body while MMBM is used only for the whole body, (b) the variables of one patch in hidden layer h1 in MMBM cant connect the variables in other patches, but the proposed model differs. With regard to (a), we use only a one layer Boltzmann Machines where visible variables are connected with the features of the images for which we have other optimization methods, instead of two hidden layers in MMBM, otherwise the computation cost is high. For (b), we need to refine the whole body or one part of body.

The reason we can use each component of the human body to refine the model is each part of the body is not overlapped with the other referring back to ground image, which means all the pixels in the image should belong to one type of body component because the whole body contains more features and is unambiguous while some parts, such as hair, arms, legs, are too small to be recognized and may be very similar to the background, we use the whole body to refine all the semantic parts. We achieve this process by an interaction optimization process as explained in Fig. 1. First, our proposed network is constructed similar to MMBM. Next, a curve correction described in Sect. 2.3 is used during interaction optimization process.

We try to train the conditional probability distribution of each part of human body directly instead of the joint distribution for estimating the silhouette. The energy function of $p(y, h|x)$ is represented as:

$$E(y, h, x) = -y^T W h - h^T (V^1 x + c) - y^T (V^0 x^0 + b) \tag{1}$$

Since the whole body and its protential semantic parts are relative to each other, we can model one part with respect to the other parts and the whole body silhouette. It is concluded that these can have the form

$$E(y_w, y_p(i), h, x) = - y_w W^{1*} y_p(i) - y_p(i) W h - y_p(i)^T (V^1 x^1 + c^1)$$
$$- h^T (V x + c) - y_w (V^0 x^0 + b) \tag{2}$$

where y_w is the whole body silhouette and $y_p(i)$ is the i-th channel of the model, which is a part of the whole body. x^0 is the features extracted directly from image. x^1 and x^2 are part of x^0, representing the relative features to some part of $y_p(i)$ and the whole body yw respectively. V^0, V^1 and V show the structure the part. W is the mapping from the features with respect to a body part while W^{1*} defines the relationship between the whole body and all the parts, which can be considered as the gesture of human body.

Considering that the network has two layers, the energy function is a high-order form of y, which can hardly get maximum a posteriori (MAP) estimation of it. Instead, the log-likelihood $log\, p(y|x)$ is optimized to obtain the value for the maximum by using the EM algorithm. However, reaching the maximum is still problematic when the other parameters are fixed, so we approximate it by optimizing

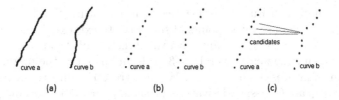

Fig. 2. Curve Similarity Measure. (a) Two curves a and b, which are similar but not the same. (b) The curves are expressed by fragments. (c) Compare the similarity of fragments and find out the best matching fragment pair.

$$\{y_w, y_p(i), h_w, h_p(i)\} = argmax \, p(y_w, y_p(i), h_w, h_p(i)|x) \tag{3}$$

where h_w, $h_p(i)$ can be optimized independently by $exp(-F(y, x)/Z)$. And according to [14], we have:

$$-F(y, x) \propto c_j + y^T W_{.j} + V_{j.}^1 x^1 \tag{4}$$

where $.j$ and $j.$ present fixing the j-th column and row of W respectively, and partition function Z is a constant.

In this next step, we prove that h_w and $h_p(i)$ can be learned in the model to yield their initial values. Then y_w, $y_p(i)$, h_w and $h_p(i)$ can be optimized iteratively as Mixed Iterative MAP Inference, as is seen in Algorithm 1.

Algorithm 1. Mixed Iterative MAP Inference

1: Initialize $h_w, h_p(i)$
2: **while** do not converge **do**
3:   $y_w \leftarrow max \, p(y_w | \sum_{i \in \theta} y_p(i), h_w, x)$
4:   $y_p(i) \leftarrow max \, p(y_p(i)|y_w, \sum_{j \in \theta, j! = i} y_p(j), h_p, x)$
5:   $h_w \leftarrow max \, p(h_w|y_w, x)$
6:   $h_p(i) \leftarrow max \, p(h_p(i)|y_p(i), x)$
7: **end while**

2.3 Similarity Measurement of the Curve and Curve Correction

Similarity Measurement of the Curve. In the section, a method to correct inaccurate edges of object is discussed. Existing methods, such as UCM [15], Canny, Sobel. focus on curve extraction; Some methods attempt to compare the entire difference between contours of objects, such as Shape Contexts [16,17]. All these efforts are done at object level, which is not adequate for correcting the existing curves of segmentation results with convincing curves.

The proposed approach is inspired by Shape Contexts, but is totally different. Assume that there are two curves a and b, shown in Fig. 2(a). The purpose is to use curve a to correct curve b, but it will be found that the differences is hard to describe whether in Euclidean distance or in curvature. Since any curve is a set of points, we can use a few points of the curve to express all the curve point shown in Fig. 2(b), and accompany with the direction of the curve.

We divide an image into several 3×3 grids and quantize the fragment of the curve within a grid as a single point. We also trace the direction changes in a curve. I_x and I_y are Gaussian derivatives in x and y directions at some point of a curve. $I_\theta = I_y/I_x$ is the direction vector of Gaussian derivative and is quantized to 8 binned directions. For each adjacent point pair, if I_θ is the same as the former or changes one-binned degrees in clockwise or anti-clockwise, they can be grouped together as one curve. Otherwise they are separated into different curves. We record all the directions and allow there are at most two directions existing in one curve.

Then any curve can be parameterized by a vector $C = \{I_\theta, f(p)\}$, where now I_θ represents the overall direction and $f(p)$ stands for a fragment of the curve, p is the index. The curves can be compared and matched using the direction and fragments. For all the same direction curves, we use minimum Euclidean distance of the fragments to find the best matching one from different curves: $f_* = argmin \, d(f_a(i), f_b(j))$, where $f_a(i)$ and $f_b(j)$ are fragments from curve a and b respectively and d is Euclidean distance of the two curve fragments. We dont allow the minimum distance beyond the threshold value d_t, otherwise they will not match.

We define all the qualified fragments between two curves and give a score:

$$S(a,b) = \sum I(I_\theta, f_*) \tag{5}$$

where $I(I_\theta, f_*)$ is an indicator function, measuring if the fragment pair are matching under the same direction. So $S(a,b)$ presents the number of matching fragment pairs. The higher the score is, the more similar the curves are.

Curve Correction. By using the proposed curve similarity measurement method, we can correct rough curves and out of position curves. To do this, first, we divide an image into several regular rectangle regions. Then, the edges of whole body (WE), parts of body (SE) and potential edges (PE) are matching in each region, where WE and EE are the curves extracted from the output contours of whole human body and each part of the body. Although HO-CC [18] performs well in regional clustering, it requires marginal training based on structured support vector machines, which extracts a heavy computational cost from our system, because we have used a Boltzmann Machine network. So we simply take OWT-UCM as the contours and extract the curves as PEs. If there are no WE, we use SE and PE.

Finally, we compute the similarity of the three types of edges using a similarity measure algorithm of curves as previously mentioned to check if they have a sufficient score (Eq. 5). The selective priority order in the candidate set is PE, WE, SE, meaning that we will trust PE more and use it as the contour of object if they are similar. The reason for the priority order is (a) PE reflects the nature

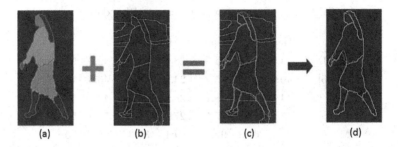

Fig. 3. Curve correction example. (a) An example of our multi channel segmentation result and the contours of each part of human body. (b) Edges extracted from OWT-UCM. (c) Comparison for curves in (a) and (b). Blue curves are our segmentation results, yellow curves are from OWT-UCM. (d) The results after curve correction. Blue curves are original ones, and yellow ones are corrected curves. From the results we can see that curve correction makes the result more fit to the real contours of human parts (Color figure online).

edge of all objects more, but is more likely to contain background edges, (b) WE is more accurate than SE in our experience, since, after learning, the whole body model is more reliable than the parts models. For example, small parts of the body sometimes are too small to distinguish from the background, such as arms, legs, or are sometimes too various, like different colors and shapes of hair. If there is no matching pair in a candidate set, we simply use SE (Fig. 3).

2.4   Overlap Regions and Missing Regions

To exit from wrong mapping between the whole body and the parts learned by our BM model, the silhouettes from different parts may overlap with each other (overlap regions) and some parts of whole body silhouette may be unlabeled (missing regions).

For overlap areas, as we know the number of each part, like one face and two arms, we can group all un-overlapped regions and overlap regions, compute each region center and simply use K-means algorithm to label each overlap region again. After grouping the overlap regions, we can also take the similar steps to label the missing regions by K-means.

3   Experiments

3.1   Dataset and Implements

We take PFP [12] and PPSS [13] as the benchmark datasets to verify our proposed method. Similar to [1,5,6,13], we regard the human body segmentation as six regions: hair, face, arms, upper (upper clothes), lower (lower clothes) and legs.

Experiments are implemented by Intel Xeon CPU E5-2620 with 128 G memory. We use HumanEva [19] as pre-training dataset and take 2,000 random images from PPSS as our training set, the residual 1,922 images and the whole

Table 1. Segmentation accuracies on PFP dataset

	Hair	Face	Upper	Arms	Lower	Legs	Aver
SBP [1]	44.9	60.8	74.8	26.2	71.2	42.0	53.3
P&S [5]	40.0	42.8	75.2	24.7	73.0	46.6	50.4
DDN [13]	44.7	54.2	78.1	25.3	75.0	49.8	54.7
DL [13]	43.2	57.1	77.5	27.4	75.3	52.3	56.2
BM	67.1	67.2	92.3	54.1	89.3	54.6	70.8
BM_curve	**69.5**	**68.9**	**93.7**	**64.5**	**89.7**	**66.0**	**75.4**
BM*	86.9	85.9	93.1	54.9	93.2	65.5	79.9
BM_curve*	**89.3**	**87.1**	**94.5**	**70.0**	**95.3**	**76.1**	**85.6**

Table 2. Segmentation accuracies on PPSS dataset

	Hair	Face	Upper	Arms	Lower	Legs	Aver
DDN [13]	22.0	29.1	57.3	10.6	46.1	12.9	30.0
DL [13]	35.5	44.1	68.4	17.0	61.7	23.8	41.8
BM	77.3	**74.1**	**90.6**	**54.6**	**87.1**	**56.8**	**73.4**
BM_curve	**77.4**	73.8	**90.6**	54.5	86.9	56.6	73.3

(334) of PFP consist of the testing set for the first situation. Then, in order to get rid of the influence on PFP by PPSS, we use random 200 images of PFP as the training set and the residual 134 images as testing set, as is regarded as the second situation. We normalize the size of all the images to 32×64 for feature extraction and 64×128 for input and output. All the images are horizontally flipped in the pre-training and training datasets. The total time of training and testing is 28 h for the first situation in average and 13 h for the second. For both training and testing, we spend around 30 s to extract features (especially UCM) from a single image, which is our main bottleneck.

3.2   Results and Performances

Average Pixel Accuracy (aPA) is used to measure the results on the two datasets. We use BM represents for the proposed Multi Channel Shape Boltzmann Machine network while BM_curve stands for the results with Curve Correction. For PFP dataset, if they are with a asterisk symbol, it means the results of the second situation as is described in Sect. 3.1, otherwise it is the first's. We compare our results with SBP [1], P&S [5], DDN and DL [13]. For PPSS dataset, results are listed with DDN and DL [13].

Figure 4 gives some samples of our final output under the condition that training list and testing list are from the same dataset. In Fig. 4, the left 4 columns are from PFP dataset and the others show the PPSS dataset's results.

Fig. 4. Results of our method on PFP and PPSS datasets. The rows from top to below are the original images, ground truths of human parts and our predicted label silhouettes.

Tables 1 and 2 also give the details on results, showing our proposed method provides the best performance on accuracies. For PFP dataset, Curve Correction improves the accuracy rates of all the parts under both of the two situations, though performances are better if we use its own images as training set. For PPSS, the reason that Curve Correction doesn't improve all the results of human parts can be that the images in PPSS dataset are blurry and it is difficult to extract proper OWT-UCM curves. In our experiments, the total training time is about 20 h, the testing time for each image is about 48 s, both of the time mainly cost in feature extraction.

4   Conclusion

We present Shape Boltzmann Machine Networks to model the whole, the parts of human body and their relationship, optimize silhouettes of them interactively. After this process, a more realistic edge can be found by Curve Correction. Finally, all the overlap and missing regions are mapped again. Our proposed method outperforms the state-of-the-art methods on PFP and PPSS datasets.

Acknowledgements. This work is supported by the National High Technology Development Plan (863 Plan) under Grant No. 2011AA01A205, the National Significant Science and Technology Projects of China under Grant No. 2013ZX01039001-002-003, the NSFC project under Grant No. U1433112 and No. 61170253. We also thank the support from the academic program of Tencent Inc.

References

1. Fowlkes, C.C., Bo, Y.: Shape-based pedestrian parsing. In: IEEE Conference on Computer Vision and Pattern Recognition (2011)
2. Bourdev, L., Malik, J.: Poselets: body part detectors trained using 3D human pose annotations. In: IEEE International Conference on Computer Vision (2009)
3. Lin, L., Yang, W., Luo, P.: Clothing co-parsing by joint image segmentation and labeling. In: IEEE Conference on Computer Vision and Pattern Recognition (2014)
4. Luis, K., Ortiz, E., Berg, T.L., Yamaguchi, K., Hadi, M.: Parsing clothing in fashion photographs. In: IEEE Conference on Computer Vision and Pattern Recognition (2012)
5. Rauschert, I., Collins, R.T.: A generative model for simultaneous estimation of human body shape and pixel-level segmentation. In: Fitzgibbon, A., Lazebnik, S., Perona, P., Sato, Y., Schmid, C. (eds.) ECCV 2012, Part V. LNCS, vol. 7576, pp. 704–717. Springer, Heidelberg (2012)
6. Williams, C., Ali Eslami, S.M.: A generative model for parts-based object segmentation. In: Advances in Neural Information Processing Systems, pp. 272–281 (2012)
7. Williams, C.K.I., Winn, J., Eslami, S.M.A., Heess, N.: The shape boltzmann machine: a strong model of object shape. In: IEEE Conference on Computer Vision and Pattern Recognition (2012)
8. Wang, X., Tang, X., Luo, P., Tian, Y.: Switchable deep network for pedestrian detection. In: IEEE Conference on Computer Vision and Pattern Recognition (2014)
9. Salakhutdinov, R., Fidler, S., Zhu, Y., Urtasun, R.: Segdeepm: exploiting segmentation and context in deep neural networks for object detection. In: IEEE Conference on Computer Vision and Pattern Recognition (2015)
10. Girshick, R., Malik, J., Hariharan, B., Arbelez, P.: Hypercolumns for object segmentation and fine-grained localization, eprint (2014). arXiv:1411.5752
11. Darrell, T., Malik, J., Girshick, R., Donahue, J.: Rich feature hierarchies for accurate object detection and semantic segmentation. In: IEEE Conference on Computer Vision and Pattern Recognition (2014)
12. Wang, L.-M., Shi, J., Song, G., Shen, I.-F.: Object detection combining recognition and segmentation. In: Yagi, Y., Kang, S.B., Kweon, I.S., Zha, H. (eds.) ACCV 2007, Part I. LNCS, vol. 4843, pp. 189–199. Springer, Heidelberg (2007)
13. Tang, X., Wang, X.: Pedestrian parsing via deep decompositional network. In: IEEE International Conference on Computer Vision (2013)
14. Simon, M., Yang, J., Safar, Y.: Max-margin boltzmann machines for object segmentation. In: IEEE Conference on Computer Vision and Pattern Recognition (2014)
15. Fowlkes, C., Malik, J., Arbelez, P., Maire, M.: Contour detection and hierarchical image segmentation. In: IEEE Transaction on Software Engineering (2011)
16. Puzicha, J., Belongie, S., Malik, J.: Shape matching and object recognition using shape contexts. In: IEEE International Conference on Computer Science and Information Technology (2002)
17. Liu, S., Guo, X., Lin, L., Cao, X., Zhang, H.: Sym-fish: a symmetry-aware flip invariant sketch histogram shape descriptor. In: IEEE International Conference on Computer Vision (2013)

18. Nowozin, S., Kim, S., Yoo, C.D., Kohli, P.: Image segmentation using higher-order correlation clustering. IEEE Trans. Pattern Anal. Mach. Intell. **36**(9), 1761–1774 (2014)
19. Balan, A.O., Sigal, L., Black, M.J.: Humaneva: synchronized video and motion capture dataset for evaluation of articulated human motion. In: IEEE International Conference on Computer Vision (2006)

Depth-Based Stereoscopic Projection Approach for 3D Saliency Detection

Hongyun Lin[1], Chunyu Lin[1(✉)], Yao Zhao[1], Jimin Xiao[2], and Tammam Tillo[2]

[1] Institute of Information Science, Beijing Key Laboratory of Advanced Information Science and Network, Beijing Jiaotong University, Beijing, China
{14120323,cylin,yzhao}@bjtu.edu.cn
[2] Department of Electrical and Electronic Engineering, Xian Jiaotong-Liverpool University, Suzhou, China
{jimin.xiao,tammam.tillo}@xjtlu.edu.cn

Abstract. With the popularity of 3D display and the widespread using of depth camera, 3D saliency detection is feasible and significant. Different with 2D saliency detection, 3D saliency detection increases an additional depth channel so that we need to take the influence of depth and binocular parallax into account. In this paper, a new depth-based stereoscopic projection approach is proposed for 3D visual salient region detection. 3D images reconstructed with color and depth images are respectively projected onto XOZ plane and YOZ plane with the specific direction. We find some obvious characteristics that help us to remove the background and progressive surface where the depth is from the near to the distant so that the salient regions are detected more accurately. Then depth saliency map (DSM) is created, which is combined with 2D saliency map to obtain a final 3D saliency map. Our approach performs well in removing progressive surface and background which are difficult to be detected in 2D saliency detection.

Keywords: 3D saliency detection · Depth-based stereoscopic projection · Salient region

1 Introduction

Human Visual System (HVS) is able to identify region of interest (ROI) from the scene accurately and rapidly. Saliency detection is to simulate the human eyes' visual processing and find region of interest in images accurately and efficiently. Two mechanisms of visual attention are usually distinguished: bottom-up and top-down [1]. The bottom-up attention is fast, subconscious, data-driven and independent of a particular viewing task; whereas the top-down attention is slow, voluntary and dependent both on the viewing task and the semantic information [9]. These two kinds of mechanisms interact with each other and affect the human visual behavior [2–4]. Most of previous work focuses on bottom-up visual process since the complexity of the top-down process depends on the individual differences.

© Springer International Publishing Switzerland 2015
Y.-S. Ho et al. (Eds.): PCM 2015, Part I, LNCS 9314, pp. 664–673, 2015.
DOI: 10.1007/978-3-319-24075-6_64

The existing 2D saliency detection algorithms give priority to local contrast algorithm, in which Itti's model [5] is representative. Achanta et al. [6] calculate the pixel saliency value through the superposition of multiple-scale regional contrast to create a full-resolution saliency map. Goferman et al. [7] calculate saliency value with the view that novel features and high colors are more salient, and concentrated rather than dispersed in the image. Spectral residual approach [8] is based on the global contrast and builds saliency map by analyzing the characteristics of a salient area in the frequency domain.

Several computational models of 3D visual attention are investigated in previous literatures and they are divided into three categories: Depth-weighting models (DW), depth-saliency models and stereo-vision models [9]. DW models treat the depth as weight to multiply with 2D saliency map [10]. Depth-saliency models take depth saliency as additional information [11, 12]. Depth features such as relative depth, pop-out effect and depth gradient are firstly extracted from the depth map to create additional feature maps, which are then used to generate depth saliency maps (DSM). Finally these depth saliency maps are combined with 2D saliency maps to obtain a final 3D saliency map. Different from the previous two models which use depth map directly, stereo-vision models take into account the mechanisms of the stereoscopic perception in the HVS [13].

Although the existing 3D saliency detection approaches utilize the various characteristics of the depth, most of them are just simple extension of 2D schemes. On the contrast, our approach solves the problem in three-dimensional view. By projecting the 3D images with specific direction, the background and progressive surface that are clearly not the salient regions can be detected and removed easily, whereas the previous work cannot differentiate these regions. Progressive surface is a region in which the depth is progressive. Experimental results demonstrate the effectiveness of the proposed scheme. In summary, our main contributions are shown in the following:

- We introduce a new approach for 3D saliency detection, which is able to remove the background and progressive surface.
- Our approach efficiently output full resolution saliency maps which uniformly highlight whole salient regions and establish well-defined boundaries of salient objects.
- A comparison of our approach against several typical methods is presented and verifies the effectiveness of depth information in improving 3D saliency detection.

The paper is organized as follows. Section 2 introduces our novel saliency method. The qualitative and quantitative experimental results are given in Sect. 3. Finally, the conclusions are presented in Sect. 4.

2 Depth-Based Stereoscopic Projection Approach

The diagram of the proposed model is presented in Fig. 1 in which the key step is highlighted with the thick box. Firstly, the 3D image is reconstructed with the color and depth images. With the 3D image, an object can be observed from multiple visual angles, which makes it easy to find salient objects in 3D. We project 3D image onto XOZ plane and YOZ plane respectively (O denotes the origin point, X and Y represent respectively horizontal and vertical axis, Z is the direction perpendicular to XOY). The projection of background and progressive surface is a curve or a straight line on XOZ and YOZ plane by viewing the projected images, so the background and progressive surface can be removed easily in order to get the salient regions. Then we create a depth saliency map (DSM) based on the depth-bias and center-bias rules. This depth saliency map is finally combined with 2D saliency map to obtain a final 3D saliency map.

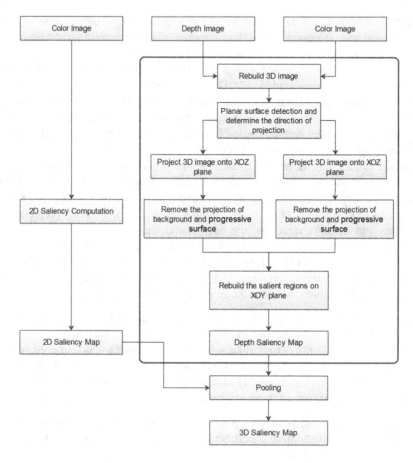

Fig. 1. The diagram of the proposed model.

2.1 Three-Dimensional Reconstruction and Stereographic Projection

We can reconstruct 3D image by using the color and depth images as shown in Fig. 2, where Fig. 2(a) and (d) are two typical color images, Fig. 2(b) and (e) are their corresponding depth maps, Fig. 2(c) and (f) are their 3D images reconstructed with color and depth images. The color plus depth images are acquired from the Middlebury image dataset [15].

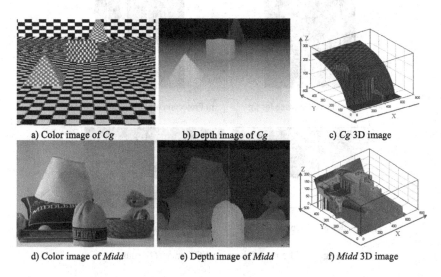

| a) Color image of *Cg* | b) Depth image of *Cg* | c) *Cg* 3D image |

| d) Color image of *Midd* | e) Depth image of *Midd* | f) *Midd* 3D image |

Fig. 2. Reconstructed 3D image.

We can discover different characteristics by different perspectives, thus the key step is the appropriate selection of visual angles that makes the characteristics obvious and useful. The selection of projection angle depends on the direction of background plane and progressive surface. Thus we detect the background plane and progressive surface with depth-map driven planar surfaces detection [16]. The projection of background and progressive surface along with planar direction is a straight line or curve as shown in Fig. 3.

In general case, the background and progressive surface are perpendicular to XOZ and YOZ plane. Therefore an example of the three most typical directions is given in the following, which are from the Z axis direction, X axis direction and Y axis direction. The rebuilt 3D images from Fig. 2 are projected onto XOZ plane and YOZ plane, as shown in Fig. 3. Then we calculate the number of the points on the projection planes respectively and get the results in XOZ and YOZ presented in Fig. 3. The objects on color image respectively correspond to its projection on XOZ and YOZ plane. The projection of progressive surface and background is a curve or a straight line on YOZ plane as shown in Fig. 3(a) and (b). The projection of background is a horizontal line on XOZ plane as shown in Fig. 3(c) and (d).

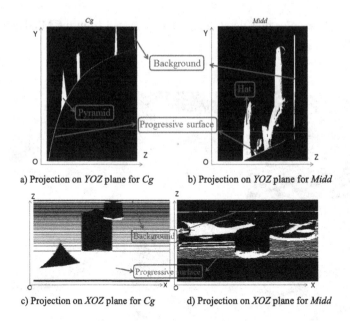

a) Projection on *YOZ* plane for *Cg*   b) Projection on *YOZ* plane for *Midd*

c) Projection on *XOZ* plane for *Cg*   d) Projection on *XOZ* plane for *Midd*

Fig. 3. The projections on *YOZ* and *XOZ* plane.

2.2   Processing Based on Characteristics of Projected Images

By observing images on XOZ and YOZ projection planes in Fig. 3, we can find the
projection of progressive surface and background easily. The details are given as
follows:

1. **On YOZ projection plane:** Firstly, the projection of progressive surface is a curve
   or a straight line. The number of projection points can be up to the width of image,
   but in fact, because the progressive surface may not be perpendicular to the YOZ
   plane, it will lead to scatter points on a few adjacent lines, namely getting a relatively
   thick line. Secondly, the projection of background is nearly a vertical line. The
   vertical thickness depends on whether the background is in the same depth.
2. **On XOZ projection plane:** Firstly, progressive surface is tiled on the XOZ plane
   commonly. The number of projection points is 1 generally. Secondly, the projection
   of background is nearly a horizontal line. The above characteristics can be noticed
   from Fig. 3.

   The curve and straight line that are represented as the projection of progressive
surface and background can be removed easily on YOZ projection plane as shown in
Fig. 4. If the number of projection points is one on XOZ plane, we directly eliminate
the projection of progressive surface, and then eliminate the horizontal line that is the
projection of background, which can be observed in Fig. 5. Finally, we get the projected
drawings where there is no projection of progressive surface and background. The rest
areas are the projection of more salient regions on XOZ and YOZ plane as shown in
Figs. 4 and 5(b).

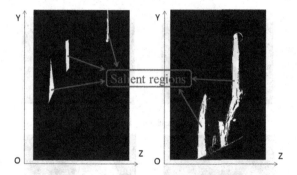

Fig. 4. Removing the projection of progressive surface and background on YOZ plane.

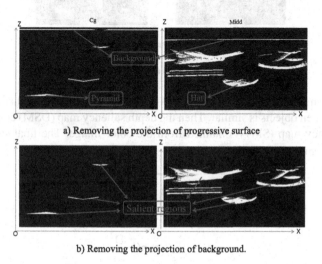

a) Removing the projection of progressive surface

b) Removing the projection of background.

Fig. 5. Processing on XOZ plane.

2.3   Generating Depth Saliency Map and 3D Saliency Map

The projection of more salient regions has been found as shown in Figs. 4 and 5(b). Thus we can rebuild depth saliency map with the help of the processed images as shown in Fig. 6(a). The creation of depth saliency map (DSM) base on depth-bias, that closer pixel is more salient.

Depth saliency map is combined with the existing 2D saliency computation models to create 3D saliency map. But our pooling way is different from the typical ways [9]. The input image of 2D saliency computation model is the segmented color image by using DSM as shown in Fig. 6(b). Our approach is based on the ideas that what attracts our attention is mainly the pop-out effect, and saliency mainly comes from dramatic change on depth and color. Hence, depth maps already carry a large amount of information, such as the contour and size of the object and so on. Our DSM performs well

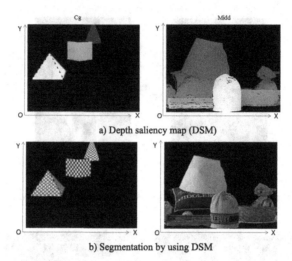

Fig. 6. Depth saliency map and the segmented images.

in predicting more salient areas of 3D images in most cases except that the relative distance between objects is similar. Then the depth saliency map (DSM) combines with the 2D saliency map (SM_{2D}) obtained from the color image. The final saliency map (SM_{3D}) is the sum of two maps.

$$D(x,y) = \sqrt{(\frac{x - X_C}{0.5\text{width}})^2 + (\frac{y - Y_C}{0.5\text{height}})^2} \tag{1}$$

$$W_C(x,y) = N(\frac{1}{\sqrt{2\pi\sigma}} \exp(-\frac{D^2}{2\sigma^2})) \tag{2}$$

$$SM_{3D} = W_C \left(\omega_1 \times DSM + \omega_2 \times SM_{2D}\right) \tag{3}$$

where X_C and Y_C denote the coordinates of center, D represents the relative distance between each pixel and center. $N(.)$ indicates a function in which the value is normalized to $[0.5, 1]$. σ is set as 1. W_C denotes the weight based on center-bias. $\omega_1 = 0.5$, $\omega_2 = 0.5$.

3   The Experimental Results and Analysis

In order to assess the proposed computational model, a set of experimental results is presented in Fig. 7. For comparison, several classical saliency maps obtained by IT [5], FT [6], SR [8] and GB [14] are provided against their editions added with DSM, which are denoted as DSM + IT, DSM + FT, DSM + SR and SDM + GB. We can see that our 3D saliency maps can not only accurately find the salient regions, but also provide the clear boundary that classical 2D approaches are unable to generate. Specifically, our

3D saliency maps have more uniform saliency values in object and maintain the integrity of object because the pop-out effect and progressive direction of depth are used to generate the clear boundary in our approach. The color plus depth images are acquired from the Middlebury image dataset [15]. Here, just the scenes with progressive surface and background are provided to demonstrate the effectiveness of the proposed scheme. As for the other scenes, the proposed scheme can produce the same performance as that of 2D saliency scheme at least.

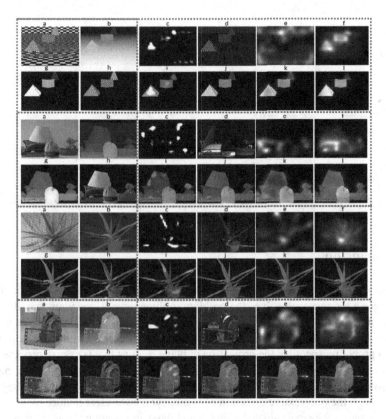

Fig. 7. 3D saliency map comparison with different 2D models. (a) Color image; (b) depth image; (c) Itti's model [5] alone; (d) FT [6] alone; (e) SR [8] alone; (f) GB [14] alone; (g) depth saliency map (DSM); (h) the processed color image by DSM; (i) DSM + IT; (j) DSM + FT; (k) DSM + SR; (l) DSM + GB.

To provide a quantitative evaluation, the PR (Precision-Recall) curve is used as the metric. The PR curve can indicate how well the saliency maps coincide with the ground truth maps for different binary saliency thresholds. The definition of precision and recall are as following:

$$precision = \frac{\sum S \cap G}{\sum S}, recall = \frac{\sum S \cap G}{\sum G} \tag{4}$$

where S denotes the binary saliency pixels for different thresholds, G denotes the ground truth pixels and Σ refers to sums of all pixels. The experimental results are presented in Fig. 8, it shows that our proposed approach significantly improves the precision and recall. The PR (Precision-Recall) curve demonstrates a great improvement of DSM on 3D saliency map.

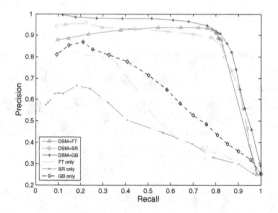

Fig. 8. Precision-recall curves comparison for different saliency detection scheme.

4   Conclusion

In this paper, we propose a depth-based computational model of visual attention for 3D still images. Our approach solves the problem in three dimensions view that allows us to utilize much additional information. There are a large number of flat surfaces that are not salient in artificial scene, such as the ground, wall, ceiling and the desktop. If we choose a proper perspective, the projection of plane is a line that is easy to process. This idea can be used not only for 3D saliency detection but also for segmentation, motion estimation, target tracking and so on. For example, our method can achieve the segmentation through the progressive direction of plane, even if the clear texture such as black alternating with white exists in color image and the depth is not pop-out. Our approach performs well in removing progressive surface and the background in the distance and finding salient regions. The key of our approach is to find a proper visual angle in which the projection of plane is a line. In addition, the proposed model introduces relative little computation. Last but not least, it is compatible with any 2D saliency detection schemes.

Acknowledgement. This work was supported by National Natural Science Foundation of China (no.61402034, no.61210006 and no.61202240, no. 61272051), supported by Beijing Natural Science Foundation (4154082) and SRFDP (20130009120038), supported by the Fundamental Research Funds for the Central Universities (2015JBM032).

References

1. Yarbus, A.L.: Eye Movements and Vision. Plenum Press, New York (1967)
2. Wang, J., Chandler, D.M., Callet, P.L.: Quantifying the relationship between visual salience and visual importance. In: SPIE Human and Electronic imaging (HVEI) XV, p. 75270K (2010)
3. Einhäuser, W., Rutishauser, U., Koch, C.: Task-demands can immediately reverse the effects of sensory-driven saliency in complex visual stimuli. J. Vis. **8**(2), 1–19 (2008)
4. van Zoest, W., Donk, M.: Bottom-up and top-down control in visual search. Percept. London **33**, 927–937 (2004)
5. Itti, L., Koch, C., Niebur, E.: A model of saliency-based visual attention for rapid scene analysis. IEEE Trans. Pattern Anal. Mach. Intell. **20**(11), 1254–1259 (1998)
6. Achanta, R., Hemami, S., Estrada, F., Susstrunk, S.: Frequency-tuned salient region detection. In: IEEE Conference on Computer Vision and Pattern Recognition, pp. 1597–1604 (2009)
7. Goferman, S., Zelnik-Manor, L., Tal, A.: Context-aware saliency detection. IEEE Trans. Pattern Anal. Mach. Intell. **34**(10), 1915–1926 (2011)
8. Hou, X., Zhang, L.: Saliency detection: a spectral residual approach. In: IEEE Conference on Computer Vision and Pattern Recognition, pp. 1–8 (2007)
9. Wang, J., et al.: Computational model of stereoscopic 3D visual saliency. IEEE Trans. Image Process. **22**(6), 2151–2165 (2013)
10. Chamaret, C., Godeffroy, S., Lopez, P., Meur, O.L.: Adaptive 3D rendering based on region-of-interest. In: Proceedings of SPIE 7524, Stereoscopic Displays and Applications XXI, p. 75240V (2010)
11. Ouerhani, N., Hugli, H.: Computing visual attention from scene depth. In: 15th IEEE International Conference on Pattern Recognition, vol. 1, pp. 375–378 (2000)
12. Potapova, E., Zillich, M., Vincze, M.: Learning what matters: combining probabilistic models of 2D and 3D saliency cues. In: Crowley, J.L., Draper, B.A., Thonnat, M. (eds.) ICVS 2011. LNCS, vol. 6962, pp. 132–142. Springer, Heidelberg (2011)
13. Bruce, N., Tsotsos, J.: An attentional framework for stereo vision. In: The 2nd Canadian Conference on Computer and Robot Vision, 2005. Proceedings, pp. 88–95 (2005)
14. Harel, J., Koch, C., Perona, P.: Graph-based visual saliency. In: Advances in Neural Information Processing Systems, pp. 545–552 (2006)
15. Scharstein, D., Hirschmüller, H., Kitajima, Y., Krathwohl, G., Nešić, N., Wang, X., Westling, P.: High-Resolution Stereo Datasets with Subpixel-Accurate Ground Truth. In: Jiang, X., Hornegger, J., Koch, R. (eds.) GCPR 2014. LNCS, vol. 8753, pp. 31–42. Springer, Heidelberg (2014)
16. Jin, Z., Tillo, T., Cheng, F.: Depth-map driven planar surfaces detection. In: Visual Communications and Image Processing Conference, pp. 514–517 (2014)

Coding and Reconstruction of Multimedia Data with Spatial-Temporal Information

Revisiting Single Image Super-Resolution Under Internet Environment: Blur Kernels and Reconstruction Algorithms

Kai Zhang, Xiaoyu Zhou, Hongzhi Zhang, and Wangmeng Zuo[✉]

Center of Computational Perception and Cognition,
School of Computer Science and Technology,
Harbin Institute of Technology, Harbin 150001, China
cswmzuo@gmail.com

Abstract. Due to limited network bandwidth, the blurred and down-sampled high-resolution images in the spatial domain are inevitably used for transmission over the internet, and so single image super-resolution (SISR) algorithms would play a vital role in reconstructing the lost spatial information of the low-resolution images. Recently, it has been recognized that the blur kernel is crucial to the SISR performances. As most of the existing SISR methods typically assume the blur kernel is known, and in fact the blur kernel is either fixed with the scaling factor or unknown, it thus would be of high value to investigate the relationship between blur kernels and reconstruction algorithms. In this paper, we first propose a fast and effective SISR method based on mixture of experts and then give an empirical study on the sensitivity of different SISR algorithms to the blur kernels. Specially, we find that different algorithms have different sensitivity to the blur kernels and the most suitable blur kernels for different algorithms are different. Our findings highlight the importance of the blur models for SISR algorithms and may benefit current spatial information coding methods in multimedia processing.

Keywords: Image super-resolution · Blur kernel · Spatial information · Image downsamling · Mixture of experts

1 Introduction

Single image super-resolution (SISR), with the goal of generating a visually pleasant high-resolution (HR) image from a given low-resolution (LR) input, has been successfully applied in various real applications for different kinds of reasons such as cost of camera, storage limitation, insufficient computational power and limited bandwidth [1]. Particularly, the transmission of blurred and downsampled HR images over the internet is inevitable due to limited network bandwidth. Fortunately, with the powerful computational efficiency of modern computers and advanced algorithms, the SISR algorithms are practicable to effectively and efficiently recover the lost spatial information from the LR images.

© Springer International Publishing Switzerland 2015
Y.-S. Ho et al. (Eds.): PCM 2015, Part I, LNCS 9314, pp. 677–687, 2015.
DOI: 10.1007/978-3-319-24075-6_65

In general, the relationship between LR image y and HR image x can be mathematically modeled by

$$y = (x * k) \downarrow_s,$$

where $*$ is a convolution operator, k is a blur kernel, $\downarrow_s$ is a downsampling operator with scaling factor s. Reconstructing the HR image from LR image is a typically ill-posed problem since the lost spatial information is usually several times larger than that of the existing one. One classical solution to SISR is the simple and efficient bicubic interpolation method. However, it tends to generate over-smoothed results when the scaling factor is relative large, thus it is far enough to fulfill real applications.

With the purpose of seeking more effective SISR methods, in the past decade, researchers mainly focus on exploiting image priors (e.g., gradient profile prior [2], sparsity prior [3,4] and non-local similarity prior [5,6]) or/and exemplar images [7,8] to alleviate the inherent ill-posedness, thereby enriching the details of the super-resolved image. Nevertheless, little attention has been paid to the blur kernel estimation in existing SISR methods. Instead, two types of blur kernels, i.e., bicubic kernel [4,9,10] (default setting of Matlab function *imresize*) and squared Gaussian kernel with a suitable standard deviation σ (e.g., 7×7 Gaussian kernel with $\sigma = 1.6$ for upscaling factor of 3 [3]) are commonly assumed to be the default kernel because bicubic kernel can produce visually pleasant LR outputs and is widely used to synthesize LR images in many state-of-the-art SISR algorithms while the Gaussian kernel fits well to real camera blur [11].

In practice, the blur kernel used for generating LR images is either fixed with the scaling factor (such as bicubic kernel) or unknown. Correspondingly, the assumed kernel of SISR methods are either matched or mismatched to the true one. A related work concerning the blur kernel match is the one by Yang et al. [12] who gave an empirical study on the sensitivity of different SISR algorithms to various assumed Gaussian blur kernels. However, they did not consider the case with widely-used bicubic kernel, as well as the case of blur kernel mismatch. For the blur kernel mismatch, Efrat et al. [13] pointed out that the super-resolved images may deteriorate seriously when the assumed blur kernel deviates from the true one and an accurate blur kernel is more important than a well-designed image prior. An alternative to tackle this is to incorporate the blur kernel estimation and super-resolution into a unified framework [14]. However, this will result in unnecessary computational load for the estimation of blur kernel if the coding of LR images under the internet environment has uniform standards. As a result, it is of high value to give a thoroughly analysis on the sensitivity of SISR methods to various blur kernels, thus providing an insight into the coding methods for internet images and state-of-the-art SISR methods for real applications.

In this paper, we first propose a fast and effective SISR method based on mixture of experts [15] and then give an empirical evaluation on the sensitivity of different SISR algorithms to the blur kernels. In particular, we focus on the bicubic kernel and Gaussian kernel and the evaluation is split into two parts. In the first part, we aim to evaluate the best blur kernel so that the SISR methods can recover the lost spatial information as accurate as possible. To this end, various blur kernels with a certain scaling factor are separately used as assumed kernel to train the SISR model and then the synthesized LR images with the same kernel are super-resolved by the learned model. In the second part, we focus on the case of blur kernel mismatch, thus giving a sensitivity analysis on SISR methods to mismatched blur kernels.

The rest of the paper is organized as follows. Section 2 presents the evaluated SISR methods along with a new SISR method based on mixture of experts. Section 3 details the evaluation settings, including blur kernels, scaling factors, evaluated datasets and features. In Sect. 4, the sensitivity analysis on SISR methods to various blur kernels with qualitative and quantitative evaluation is presented. Section 5 finally concludes the paper.

2   Evaluated SISR Methods

In general, there are three main types of SISR methods, namely, interpolation-based, reconstruction-based, and learning-based ones. Interpolation-based methods, such as bicubic interpolation and bilinear interpolation, are fast and simple but not effective. Reconstruction-based methods which usually incorporate a certain prior information to tackle the inherently ill-posed problems can be effective but not computationally efficient [3]. Learning-based methods which attempt to learn correspondence between LR and HR information could be the most popular type for their excellent performance. Among them, neighbor embedding [8] and sparse coding [4] approaches are two representative methods. Recently, anchor-based local learning approaches which learn a set of anchor points from training data to divide the LR space into a number of regions and then learn a local model for each region have attracted intensive attention for its high computational efficiency without compromise of visual quality. Timofte et al. [9,10] proposed a fast anchored neighborhood regression approach based on the neighbor embedding and sparse coding approaches. Yang and Yang [16] proposed a simple functions (SF) based SISR method where the cluster centers of K-means are adopted as the anchors, and the least square technique is employed to learn the linear transformation for each anchor point.

2.1   A Fast and Effective SISR Method Based on Mixture of Experts

Despite the high computational efficiency and competitive quality results of the anchors based local learning methods, there still exists some drawbacks, e.g., the partition is separately learned from local models which not only makes the learned partition not tailored for subsequent local model learning but also results in an abundant local regressors, thus increasing the memory and computational complexity for real applications. By pursuing a joint learning of the partition and local models, we found that the plain mixture of experts (MoE) model [15] which consists of a gating network and several expert networks meets our goal.

To be specific, the gating network plays the same role of anchor points, i.e., partitioning the data into local regions in the training phase and determining which expert (regressor) is best suited for an input in the testing phase. With such a joint learning strategy, better partition and local regressors can be learned and thus comparable or even better results can be achieved with much fewer regressors. As a result, the computational cost could be highly reduced. In this paper, the number of regressors is set to 384 as it achieves good trade-off between image quality and testing speed and outperforms Yang's SF method with 4096 regressors.

Taking both visual quality and computational efficiency into consideration, we evaluate six SISR methods, including the classic bicubic interpolation method, one sparse coding method proposed by Zeyde et al. [17], one neighbor embedding approach proposed by Chang et al. [8], Timofte's global regression approach and adjusted anchored neighborhood regression approach which are referred to as GR [9] and A+ [10], respectively, and our proposed method based on mixture of experts.

3   Experimental Settings

There are several settings for the evaluation of the SISR methods against various blur kernels, such as the blur kernels and scaling factors. In this section, we give the experimental settings for the evaluation. All the experiments are carried out in the Matlab (R2014a) environment running on a modern computer with Intel(R) Xeon(R) CPU 3.30 GHz and 32 GB memory.

3.1   Blur Kernels and Scaling Factors

As mentioned in Sect. 1, we adopt two types of blur kernels, i.e., bicubic kernel and Gaussian kernel. For Gaussian kernel, we use kernel width (i.e., σ) ranging from 0.2 to 2.4 with a step size of 0.1. Such a wide range of widths for Gaussian kernel can guarantee that the best kernel widths for different scaling factors are included. In this paper, we set the scaling factors to 2, 3 and 4. By a combination of different blur kernels and scaling factors, various LR images which contain different spatial information from an HR image can be synthesized. In fact, the richness of information preserved in the LR images from the HR images is determined by the blur kernel and the best kernels for different scaling factors are different. In addition, smoother kernel would result in smoother LR images. Taking the Gaussian kernel as an example, the local information will lose during downsampling if the kernel width is too small while the high-frequency information will be smoothed if the kernel width is too large (see Fig. 1 for an illustration). Thus, only using the proper kernel can preserve the spatial information to the uttermost. However, in the context of SISR, another central factor that can affect the SISR performance is the applied reconstruction algorithm.

(a) BIC   (b) 0.2   (c) 0.6   (d) 1.0   (e) 1.4   (f) 1.6   (g) 2.0   (h) 2.4

Fig. 1. Synthesized LR images by bicubic kernel (referred to as "BIC") and Gaussian kernel with different σ. The scaling factor is 3.

(a)   (b)   (c)   (d)   (e)   (f)   (g)   (h)   (i)   (j)

Fig. 2. First row: Some training samples from the 91 images. **Second row:** The 10 testing images: (a) baby (b) bird (c) butterfly (d) lena (e) pepper (f) ppt3 (g) comic (h) monarch (i) woman (j) zebra.

3.2   Datasets and Features

We use a training set containing 91 natural images as in [4, 10] for a fair comparison. Due to the fact that the quality inconsistency between training and testing images can influence the performance of SISR methods, we use 10 widely-used testing images which are quality-consistent with the training images for a fair evaluation study. Some training samples and the testing images are shown in Fig. 2.

Since human vision system is much more sensitive to the luminance components than the chrominance components for color images, we convert the original color images into the YCbCr space and extract features on the luminance components Y. Same to many other SISR works, the bicubic interpolation is applied to the color components and only the luminance component of an image is upscaled by the main algorithms. However, different from most of the SISR methods which first upscale the LR images into the image of desired size by bicubic interpolation and then extract features based on the interpolated image (see [10] for more details), in our proposed method we extract the features of each LR/HR patch pair by taking the luminance values and subtracting the mean value of LR patch. We set the size of LR patch to be 5×5 which is different from the 7×7 LR patch size of SF method and thus the memory and computational time can be further reduced. Considering that the boundary pixels of each HR patch do not convey much mutual information with respect to the counterpart pixels, we simply use the central region for HR patches which means the sizes of HR patch are 6×6, 9×9 and 12×12 for scaling factor 2, 3 and 4, respectively.

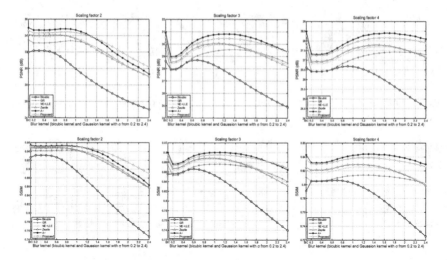

Fig. 3. The average PSNR and SSIM results of different methods with respect to various matched blur kernels (bicubic kernel and Gaussian kernel with σ from 0.2 to 2.4) and scaling factors (2, 3 and 4). Results are better viewed in zoomed mode.

4   Evaluation Results

In order to give a thorough evaluation on the sensitivity of SISR algorithms to various blur kernels, we spit our evaluation into two parts. The first part is to evaluate the performance of SISR methods with respect to matched blur kernels (the assumed kernel is same to the true one) while the second part focuses on the performance of SISR methods with respect to mismatched blur kernels (the assumed kernel is different to the true one). Since SISR algorithms aim to not only reconstruct the HR images but also generate visually pleasant output, we adopt two image quality metrics, i.e., PSNR and SSIM [18]. Correspondingly, the PSNR is used to measure the reconstruction accuracy and the SSIM is used for visual quality evaluation.

4.1   SISR Methods w.r.t. Mismatched Blur Kernels

In this subsection, for each evaluated SISR method (except for the bicubic interpolation method), the synthesized training LR images by various blur kernels and scaling factors are used to train different SISR models and then each learned model with a certain blur kernel and scaling factor is applied to super-resolving the testing LR images which are synthesized by the same blur kernel and scaling factor. Figure 3 shows the average PSNR and SSIM results of different methods with different blur kernels (bicubic kernel and Gaussian kernel with σ from 0.2 to 2.4) and scaling factors (2, 3 and 4). Figure 4 shows some visual results of super-resolved "monarch" images by different SISR methods with matched blur kernels. Intuitively, we can obtain the observations and results as follows:

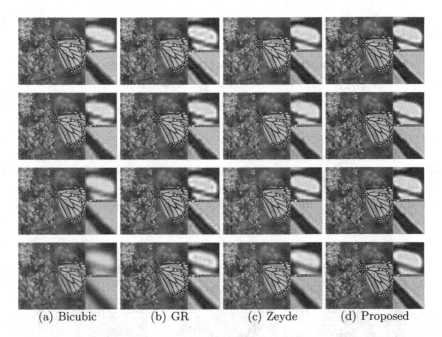

(a) Bicubic   (b) GR   (c) Zeyde   (d) Proposed

Fig. 4. Visual results of super-resolved "monarch" images by different SISR methods with various matched blur kernels. Rows correspond to the blur kernels (from first to last: bicubic kernel and Gaussian kernel with σ=0.6, 1.2 and 2.4). The scaling factor is 3. Results are better viewed in zoomed mode.

1. The bicubic interpolation method is more sensitive to the blur kernels than other methods. Even with a single regressor, the GR method can outperform the bicubic method significantly. The NE+LLE method and Zeyde's method have similar results while our proposed method is slightly worse than A+. The bicubic interpolation method generates over-smoothed results when the Gaussian kernel width σ is large (say 2.4) whereas the other methods do not have such a big visual difference when the matched kernel varies.

2. The best kernel for bicubic interpolation method could be a Gaussian kernel rather than the bicubic kernel, e.g., the Gaussian kernel with σ=0.6 is better than bicubic kernel for the scaling factor of 3. However, this will result in jaggy artifacts as shown in Fig. 4, thus it does not mean the best Gaussian kernel for bicubic interpolation method is also the best for other SISR methods. In fact, the best Gaussian kernel widths for other methods are larger than that of bicubic interpolation method.

3. The best Gaussian kernel widths for different SISR methods are different, e.g., with a scaling factor of 3, the best Gaussian kernel width for Zeyde's method is 0.9 whereas the best Gaussian kernel width for A+ is 1.1.

4. Overall, the best blur kernel for the SISR methods are bicubic kernel and Gaussian kernel with σ of ranges [0.2, 0.8], [0.8, 1.6] and [1.2, 1.8] for scaling factors 2, 3, and 4, respectively.

Table 1. The computational time (s) of different SISR methods on the "butterfly" and "lena" images with a scaling factor of 3.

Images	Size (LR)	GR	NE+LLE	Zeyde	A+	Proposed
butterfly	85×85	0.14	0.83	0.59	0.20	0.05
lena	170×170	0.54	3.14	2.39	0.74	0.18

Table 1 shows the computational time of different SISR methods on the "butterfly" and "lena" images with a scaling factor of 3. As one can see, our proposed method is at least two times faster than the other methods. From the aspects of visual quality and computational efficiency, our proposed method is more practical for real applications.

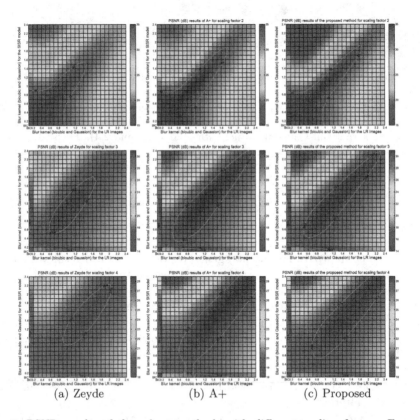

(a) Zeyde   (b) A+   (c) Proposed

Fig. 5. PSNR results of three best methods with different scaling factors. For each sub-figure, columns correspond to the true blur kernel for the LR images while rows indicate the assumed blur kernel for the SISR methods. Results are better viewed in zoomed mode.

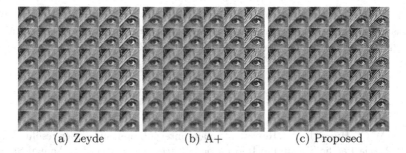

(a) Zeyde   (b) A+   (c) Proposed

Fig. 6. Some visual results of three best methods with scaling factor of 3. For each sub-figure, rows and columns correspond to the blur kernels (from first to last: bicubic kernel and Gaussian kernel with $\sigma=0.2$, 0.6, 1.0, 1.6 and 2.4) for the LR images and SISR models, respectively. Results are better viewed in zoomed mode.

4.2   SISR Methods w.r.t. Mismatched Blur Kernels

For the case that the assumed blur kernel of learned model is mismatched with the true blur kernel of LR images, we use all the learned models to super-resolve the LR images synthesized by various blur kernels. Figure 5 shows the average PSNR results of three best methods with different scaling factors. Note that here we do not show the results in terms of SSIM due to its consistence with PSNR metric. From this figure, we can have several observations. First, the PSNR results deteriorate seriously when the assumed blur kernel greatly deviates from the true one. Second, the sensitivities of different SISR methods to the mismatched kernels are different. Specially, although Zeyde's method does not provide the best PSNR results when the blur kernels are matched, it has less sensitivity to the mismatched blur kernels than A+ and our proposed method. This merit of Zeyde's method may be attributed to the flexibility and adaptivity of sparse coding. Third, the PSNR results are not symmetric about the secondary diagonal where the blur kernels are matched. Compared with the lower-right of each sub-figure, the upper-left presents lower PSNR results. Some visual results are illustrated in Fig. 6. As one can see, high frequency ringing artifacts along edges appear when the assumed blur kernel is sharper than the true one. On the other hand, the reconstructed image is blurred when the assumed blur kernel is smoother than the true one.

5   Conclusions

In this paper, we first propose a mixture of experts based SISR method which not only delivers an effective results but also shows promise in computational efficiency, and then study the sensitivity of different SISR methods to various blur kernels. By the empirical study, we show that different algorithms have different sensitivity to the blur kernels and the most suitable blur kernels for different algorithms are different. We also give a suggestion about the best blur kernels for image coding. Moreover, we present the possible properties of visual results with various combination of assumed blur kernels and true blur kernels. Our findings highlight the importance of the blur models for SISR algorithms and may benefit current spatial information coding methods in multimedia processing.

References

1. Siu, W.C., Hung, K.W.: Review of image interpolation and super-resolution. In: 2012 Asia-Pacific Signal and Information Processing Association Annual Summit and Conference (APSIPA ASC), pp. 1–10. IEEE (2012)
2. Sun, J., Xu, Z., Shum, H.Y.: Image super-resolution using gradient profile prior. In: IEEE Conference on Computer Vision and Pattern Recognition, pp. 1–8 (2008)
3. Dong, W., Zhang, D., Shi, G., Wu, X.: Image deblurring and super-resolution by adaptive sparse domain selection and adaptive regularization. IEEE Trans. Image Process. **20**(7), 1838–1857 (2011)
4. Yang, J., Wright, J., Huang, T.S., Ma, Y.: Image super-resolution via sparse representation. IEEE Trans. Image Process. **19**(11), 2861–2873 (2010)
5. Dong, W., Zhang, L., Shi, G.: Centralized sparse representation for image restoration. In: IEEE International Conference on Computer Vision, pp. 1259–1266 (2011)
6. Romano, Y., Protter, M., Elad, M.: Single image interpolation via adaptive nonlocal sparsity-based modeling. IEEE Trans. Image Process. **23**(7), 3085–3098 (2014)
7. Freeman, W.T., Jones, T.R., Pasztor, E.C.: Example-based super-resolution. IEEE Comput. Graph. Appl. **22**(2), 56–65 (2002)
8. Chang, H., Yeung, D.Y., Xiong, Y.: Super-resolution through neighbor embedding. IEEE Conf. Comput. Vis. Pattern Recogn. **1**, 275–282 (2004)
9. Timofte, R., De Smet, V., Van Gool, L.: Anchored neighborhood regression for fast example-based super-resolution. In: IEEE International Conference on Computer Vision, pp. 1920–1927 (2013)
10. Timofte, R., De Smet, V., Van Gool, L.: A+: adjusted anchored neighborhood regression for fast super-resolution. In: IEEE Asian Conference on Computer Vision (2014)
11. Deng, H., Zuo, W., Zhang, H., Zhang, D.: An additive convolution model for fast restoration of nonuniform blurred images. Int. J. Comput. Math. **91**(11), 2446–2466 (2014)
12. Yang, C.-Y., Ma, C., Yang, M.-H.: Single-image super-resolution: a benchmark. In: Fleet, D., Pajdla, T., Schiele, B., Tuytelaars, T. (eds.) ECCV 2014, Part IV. LNCS, vol. 8692, pp. 372–386. Springer, Heidelberg (2014)
13. Efrat, N., Glasner, D., Apartsin, A., Nadler, B., Levin, A.: Accurate blur models vs. image priors in single image super-resolution. In: IEEE International Conference on Computer Vision, pp. 2832–2839. IEEE (2013)
14. Michaeli, T., Irani, M.: Nonparametric blind super-resolution. In: IEEE International Conference on Computer Vision (ICCV), pp. 945–952. IEEE (2013)
15. Jordan, M.I., Jacobs, R.A.: Hierarchical mixtures of experts and the EM algorithm. Neural Comput. **6**(2), 181–214 (1994)
16. Yang, C.Y., Yang, M.H.: Fast direct super-resolution by simple functions. In: IEEE International Conference on Computer Vision, pp. 561–568. IEEE (2013)

17. Zeyde, R., Elad, M., Protter, M.: On single image scale-up using sparse-representations. In: Boissonnat, J.-D., Chenin, P., Cohen, A., Gout, C., Lyche, T., Mazure, M.-L., Schumaker, L. (eds.) Curves and Surfaces 2011. LNCS, vol. 6920, pp. 711–730. Springer, Heidelberg (2012)
18. Wang, Z., Bovik, A.C., Sheikh, H.R., Simoncelli, E.P.: Image quality assessment: from error visibility to structural similarity. IEEE Trans. Image Process. **13**(4), 600–612 (2004)

Prediction Model of Multi-channel Audio Quality Based on Multiple Linear Regression

Jing Wang[✉], Yi Zhao, Wenzhi Li, Fei Wang, Zesong Fei, and Xiang Xie

School of Information and Electronics, Beijing Institute of Technology, Beijing, China
{wangjing,feizesong,xiexiang}@bit.edu.cn

Abstract. Perceived audio quality is an important metric to measure the perception degradation of multi-channel audio signals especially for coding and rendering systems. Conventional objective quality measurement such as PEAQ (Perceptual Evaluation of Audio Quality) is limited to describe both the basic audio quality and the spatial impression. A novel prediction model is proposed to predict the subjective quality of 5.1-channels audio systems. Two attributes are included in the evaluation including basic quality and surround effects. Multiple Linear Regression (MLR) combined with Principal Component Analysis (PCA) is used to establish the prediction model from the objective parameters to subjective audio quality. Data set for model training and testing is obtained from formal listening tests under different coding conditions. Preliminary experiment results with 5.1-channels audio show that the proposed model can predict multi-channel audio quality more accurately than the conventional PEAQ method considering both the basic audio quality and the surround effects.

Keywords: Multi-channel audio · Prediction model · Objective audio quality · Subjective audio quality · Multiple linear regression

1 Introduction

Multi-channel audio can provide enhanced hearing experience with high audio quality and good spatial effect for applications such as realistic video conference, 3D movie appreciation, and 3D game playing. With the demand of convenience in the storage, transmission and distribution, research on multi-channel audio coding and rendering techniques becomes more significant especially in multimedia interactive applications like live broadcasting and Internet streaming. Successful techniques, such as MP3 [1], AAC [2] and MPEG Surround [3], has been widely used to compress multi-channel audio signals in order to reduce the storage and transmission requirements particularly for 2D multi-channel audio over the entire horizontal (2D) plane such as ITU-5.1 audio formats [4], which includes 5 channels of standard loudspeakers plus a low frequency effects channel. Lately, more efforts are made to compress and reproduce 3D multi-channel audio [5, 6] over the whole 3D space with large multiple channel like NHK 22.2 channels and even more.

In order to compare and optimize the performance of multi-channel audio coding and rendering systems, it is very important to research on the perceived audio quality

© Springer International Publishing Switzerland 2015
Y.-S. Ho et al. (Eds.): PCM 2015, Part I, LNCS 9314, pp. 688–698, 2015.
DOI: 10.1007/978-3-319-24075-6_66

evaluation methods. There are two kinds of audio quality test methods including subjective and objective ones. Some subjective listening test standards such as ITU-R BS.1116 [7], ITU-R BS.1285 [8] and ITU-R BS.1534 [9] have been widely used to assess the stereo and multi-channel sound systems under specific experimental conditions. The sound attributes specific to monophonic, stereophonic and multi-channel evaluations are different and for the 5.1-channels audio evaluation the attributes include basic audio quality, front image quality and impression of surround quality. On the other hand, there are very few studies on objective quality assessment methods of multi-channel audio. In fact, the existing techniques, mainly containing ITU-R BS.1387-1 PEAQ [10], cannot handle multi-channel audio and even the stereo signals. In the case of multi-channel, most researchers perform PEAQ algorithm in the same manner and independently for each channel and average the final objective difference grade (ODG) values. That is to say, no spatial cues are taken into consideration in PEAQ. Lately, more efforts have been done to evaluate the objective quality of multi-channel audio from the view of spatial hearing information. Reference [11] proposed an objective measurement method of perceived auditory quality in multi-channel audio compression coding systems which firstly converts multi-channel signals into binaural signals using Head related transfer functions (HRTFs). Reference [12] extend and modified the standard PEAQ algorithm to measure two-channels stereo audio signals more accurately by introducing binaural hearing model. Reference [13] proposed a mapping model to assess multi-channel audio quality by employing a binaural auditory model as front-end to provide perceptually relevant binaural features for the reference and test audio signal.

In order to provide an automatic and proper assessment of spatial audio quality with multiple channels, this paper proposes a novel prediction model to measure both the basic audio quality and the spatial effects based on a series of objective parameters including PEAQ scores and spatial cues through the mapping method of Multiple Linear Regression (MLR) combined with Principal Component Analysis (PCA). Basic audio quality is indicated by PEAQ scores and surround effects can be described by the distortion of binaural space cues. These two kind of attributes will also be included in the subjective evaluation to obtain the subjective spatial audio quality. Considering the limitation of formal listening environment, this paper only designs the mode based on the test of 2D multi-channel audio, e.g. 5.1-channels audio, which has been widely used in current digital surround systems. The signal with 5.1 channels has 6 channels including FL (front left), FR (front right), FC (front center), LFE (low frequency effects), BL (back left) and BR (back right). The prediction model is trained and verified with formal listening tests of 5.1-channels audio under different test conditions.

2 Objective Measurement of Spatial Audio Quality

2.1 Motivation of the Prediction Model

The purpose of the prediction model is to automatically measure multi-channel audio quality using the extracted objective parameters from the input signal and to better evaluate the spatial audio quality than the conventional PEAQ algorithm considering both the basic quality and the space perception. Therefore, three conditions should be satisfied: objective parameters should be extracted to indicate the two attributes; mapping from objective parameters to subjective quality should be established; subjective evaluation data should be collected to train the prediction model.

Considering that PEAQ is limited in describing space perception, the distortion of spatial cues between the neighboring channels are computed to indicate the surround effects. These two kinds of attributes form the necessary objective parameters of the prediction model. Then the objective parameters are pre-processed to form a new set of variables and then the principle components are extracted by PCA. Prediction model is established using MLR to get the MAQ (Multi-channel Audio Quality) scores. The next step is to obtain subjective test data including the original clean audio, the distorted audio and the subjective scores. The subjective test data is used to generate the training and testing data in which audio samples are used to extract the model's input parameters and the scores are used to be the model's output for training and testing.

The whole diagram of the predicting procedure is shown in Fig. 1.

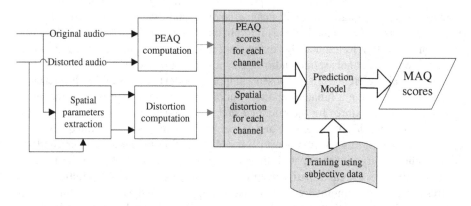

Fig. 1. Diagram of the predicting procedure

2.2   Extracting the Objective Parameters

Basic Audio Quality. PEAQ is a known objective measurement tool recommended by ITU-R BS.1387-1 [10] which has been used widely in audio coding quality assessment. It is designed only for monaural and stereo sounds and cannot be used to assess multi-channel audio systems because no inter-channel cues are taken into consideration. Most of the research on multi-channel spatial audio uses PEAQ to be the objective measurement that leads to a poor matching between the estimated quality and the perceived subjective spatial quality. PEAQ is limited in describing the fidelity of spatial information but can indicate the basis audio quality of each channel. In this paper, the basic quality of each channel is computed by PEAQ tool to get the objective difference grad (ODG) values.

The ODG values range from 0 to –4, where 0 corresponds to an imperceptible impairment and –4 to an impairment judged as very annoying. The negative values of ODG are firstly converted to the positive values range from 5 to 1, where 5 stands for very good and 1 stands for very bad. Thus we get the positive PEAQ value by the operation $PEAQ = 5 + ODG$. The basic audio quality of multi-channel audio are defined by **B** vector for one sample. As to the 5.1-channels audio, there are six values in **B**.

$$\mathbf{B} = [PEAQ_1, PEAQ_2, \dots, PEAQ_6] \tag{1}$$

Extracting Spatial Parameters. Human's brain perceives audio space through hearing by analyzing mainly the three ear entrance signal properties including Inter-aural Phase Difference (IPD), Inter-aural Level Difference (ILD) and Inter-aural Coherence (IC) that are important to the positioning of spatial sound image according to the analysis of the effectiveness of the various spatial factors [14]. The directions of sound sources in the horizontal (2D) plane can be expressed by ILD and IPD, and the distance of sound sources can be expressed by ILD and intensity of sound sources [5]. These factors belong to the inter-aural difference. Because the inter-channel difference corresponds to the inter-aural difference for the multi-channel audio signal, the former parameters are denoted as Inter-Channel Phase Difference (ICPD), Inter-Channel Level Difference (ICLD), and Inter-Channel Coherence (ICC), respectively. Therefore, it can restore the original spatial information based on these factors for multi-channel audio signal. These three space cues have been successfully used in BCC coding [15] and MPEG Surround [3]. In this paper, we use ICPD, ICLD and ICC to describe the surround effects of 5.1-channels audio.

In this paper, a set of spatial parameters (ICID, ICPD, ICC) are extracted between adjacent channels of each frame from the original and the distorted multi-channel audio signal respectively, in order to describe the surround effects. For 5.1-channels audio, the combining forms of X_1 and X_2 belong to the set of FL-FR, BL-BR and FC-LFE which finally generate nine kinds of spatial parameters P^{ij} for each audio sample.

$$P^i \in \{ICID, ICPD, ICC\}, \quad i = 1, 2, 3$$
$$P^j \in \{FL - FR, BL - BR, FC - LFE\}, \quad j = 1, 2, 3 \tag{2}$$

Finally, the average spatial distortion between the original and the distorted signal is calculated using the Euclidean distance as follows.

$$D^{ij} = \frac{1}{T} \sum_{t=1}^{T} \left\| P_1^{ij}(t) - P_2^{ij}(t) \right\|_F \tag{3}$$

P_1 stands for the spatial parameters of the original clean audio and P_2 stands for the distorted one. D is the average distortion of T frames. The indicators of spatial distortions can be expressed by the distortions of these three kinds of binaural space cues which are defined by **S** vector for one sample.

$$\mathbf{S} = \left[D^{11}, D^{12}, D^{13}, D^{21}, D^{22}, D^{23}, D^{31}, D^{32}, D^{33} \right] \tag{4}$$

2.3 Design of Subjective Listening Test

The subjective evaluation is used to collect the data set for the model training and testing. In this paper, formal listening tests based on ITU-R BS.1116-1 [7] are designed to come as close as possible to a reliable estimate of the judgment of the audio quality. The test procedure refers to the "double-blind triple-stimulus with hidden reference" in BS.1116. Experienced listeners are selected in order to evaluate the basic audio quality and the surround effects more accurately. Pre-screening and Post-screening of listening subjects are carried out respectively before and after the formal test. Every listener gets to hear three stimuli ("A", "B" and "C"). The known Reference Signal is always available as source "A". The hidden Reference Signal and the Signal Under Test are simultaneously available but are "randomly" assigned to "B" and "C". The listener is asked to assess the sound attributes' impairments on "B" compared to "A", and "C" compared to "A", according to the continuous five grade impairment scale given in Table 1.

Table 1. Five grade impairment scale

Grade	Impairment
5.0	Imperceptible
4.0	Perceptible, but not annoying
3.0	Slightly annoying
2.0	Annoying
1.0	Very annoying

Normally, only one attribute, "Basic Audio Quality", is used in the multi-channel audio test. In our experiments, both the basic audio quality and the surround effects (or spatial impression) are assessed simultaneously. The listening test is conducted in an acoustic shielding room meeting requirements of BS.1116. The 5.1-channels audio signals are reproduced from five full-band and one low frequency band Genelec studio monitors which are connected with one high quality sound card. The test conditions includes different audio codec at different coding rates. The listening scores are collected and averaged among different listeners under different conditions.

3 The Specific Structure of the Model

3.1 Data Pre-processing

There are 15 parameters computed from every audio sample, denoted as

$$\{\mathbf{Parameters_m}\}_1^n = [\mathbf{B_m S_m}] \tag{5}$$

Where m is the audio sample index, n is the total number of the collected audio data, also named as sample capacity. As to 5.1-channels audio, **B** vector have six values indicating the basic audio quality shown in Eq. (1) and **S** vector have nine values indicating the spatial distortions shown in Eq. (4). Further data analysis, including data preprocessing and PCA, should be performed after the establishment of 15 parameters in terms of data mining. MLR is adopted to establish the mapping model.

Table 2 shows that high correlation of the six PEAQ values for the six channels statistically. In order to reduce the complexity of mapping model, a new parameter called *Mean_PEAQ* is computed from *PEAQ* values, where

$$\left\{Mean\_PEAQ_m\right\}_1^n = \frac{1}{6}\left(PEAQ_{m1} + PEAQ_{m2} + \cdots + PEAQ_{m6}\right) \tag{6}$$

Table 2. The correlation coefficient among PEAQ values

	PEAQ1	PEAQ2	PEAQ3	PEAQ4	PEAQ5	PEAQ6
PEAQ1	1.000					
PEAQ2	0.909	1.000				
PEAQ3	0.868	0.954	1.000			
PEAQ4	0.539	0.689	0.723	1.000		
PEAQ5	0.978	0.905	0.865	0.524	1.000	
PEAQ6	0.941	0.978	0.947	0.632	0.946	1.000

The value '*Mean_PEAQ*' instead of **B** vector will be used as the predictor related with basic audio quality.

3.2 Principle Components Extraction

The data have a larger dimension after pre-processing, which makes the analysis of relationship among the pre-processed data difficult in multidimensional space. Besides, the parameters have very strong correlations with each other, which lead to cross impacts on the audio quality, and it will be difficult to analyze and present this cross effect. Principal component analysis (PCA) is introduced to solve the problem. Specifically, we use PCA to calculate the principle components of spatial parameters. The principle components are shown in the following equation.

$$\begin{cases} \mathbf{PC}_1 = [PC_{11}PC_{12} \cdots PC_{1n}] \\ \mathbf{PC}_2 = [PC_{21}PC_{22} \cdots PC_{2n}] \\ \cdots \\ \mathbf{PC}_9 = [PC_{91}PC_{92} \cdots PC_{9n}] \end{cases} \tag{7}$$

In Eq. (7), the element $\{PC_k\}_1^9$ is defined as

$$\mathbf{PC}_k = e_{k1} \times S_1 + \ldots + e_{km} \times S_m + \ldots + e_{km} \times S_n = \mathbf{S} \times \mathbf{e}_k \qquad (8)$$

Where $\mathbf{e}_k$ are defined as

$$\mathbf{e}_k = [e_{k1} \; e_{k2} \cdots e_{kn}]^{\mathrm{T}} \qquad (9)$$

The symbol $\mathbf{e}_k$ denotes the eigenvector corresponding to eigenvalue related to covariance coefficient matrix of $\mathbf{S}$ vector, which is arranged in descending order according to the value of corresponding eigenvalue.

3.3 Quality Measurement with MLR

The input parameters of the mapping model is denoted as $\mathbf{x}$ vector after data preprocessing and PCA.

$$\mathbf{x} = \begin{bmatrix} PC_1 \; PC_2 \ldots PC_9 \; Mean\_PEAQ \end{bmatrix} \qquad (10)$$

MLR is adopted to further investigate the relationship between the input parameters and the audio quality. The general mapping model is given in the following equation, by which the predicted Multi-channel Audio Quality (MAQ) is calculated.

$$MAQ = f\left(\mathbf{x}|\{a_k\}_0^{10}\right)$$
$$= a_0 + a_1 \times PC_1 + \ldots + a_9 \times PC_9 + a_{10} \times Mean\_PEAQ \qquad (11)$$

In the Eq. (11), $\{a_k\}_0^{10}$ are the fitting coefficients in MLR mapping model. Firstly, the preliminary least-squares fitting of the input data and the audio quality values would be taken, and then the test of significance (e.g. the F-test and T-test) will apply. An F-test is used to determine whether the liner relationship of the equation is significant, while a T-test is used to determine whether the impact of each variable is significant, leading to some variables excluded according to the result.

4 Performance Analysis

4.1 Training and Testing Data Set

Eight kinds of 5.1-channels audio samples are chosen to generate the original and the distorted signals. The test conditions in the subjective tests include different coding rates and different parametric coding methods. Totally, forty-five samples with a sampling rate 48 kHz are selected to get the training and testing data set. The coding bit rates range from 128 kbps to 256 kbps. PEAQ scores and the spatial distortion parameters are calculated between the original and the distorted signals. Subjective scores of different samples are obtained by averaging the scores from different listeners through BS1116 listening test. The predictors are computed from the audio samples itself. Outliers will firstly be excluded from the data sets which are mainly caused by the shortcomings of PEAQ computation and the abnormal subjective scores. In our experiments, there are only three outliers that were excluded. The original data will be randomly separated into two parts, i.e., training and test sets. The training data set is used to fit the model, while the test set is used for assessing the accuracy of the finally chosen model. Typically, 75 % of original data would be selected for training and the others for testing. Finally, the size of training set is 32 and the size of testing set is 10.

4.2 Algorithm Evaluation

Traditionally, prediction performance is measured by Pearson's correlation coefficient ρ and root mean squared error (RMSE). They are usually used to interpret the relationship between the predicted quality data **x** and the subjective quality data **y**. Pearson's correlation coefficient and RMSE are computed based on the following equations.

$$\rho = \frac{\sum_{k=1}^{N}\left[\left(x_k - \bar{x}\right)\left(y_k - \bar{y}\right)\right]}{\sqrt{\sum_{k=1}^{N}\left(x_k - \bar{x}\right)^2 \sum_{k=1}^{N}\left(y_k - \bar{y}\right)^2}} \in [-1 1] \tag{12}$$

$$RMSE = \sqrt{\frac{1}{N}\sum_{k=1}^{N}\left(x_k - y_k\right)^2} \tag{13}$$

Where $\bar{x}$ is the average of x_k and $\bar{y}$ is the average of y_k. N is the data size.

The proposed prediction model is compared with the standard PEAQ method through the two measures listed in Table 3. It can be shown that the prediction model achieves higher correlation compared to the PEAQ method which has a very low correlation with the subjective quality under the condition of the selected data. The negative correlation coefficient's in the table states that PEAQ method cannot express the subjective quality correctly under the testing data and will lead to unnormal results.

Table 3. Fitting accuracy for training and test sets

Measure / Data	ρ		RMSE	
	MAQ	Mean_PEAQ	MAQ	Mean_PEAQ
Training data	0.778	0.545	0.47	1.96
Testing data	0.668	-0.129	0.45	2.26

Figure 2 shows the scattering points of the objective MAQ scores vs. subjective listening scores for different audio samples under test data. Figure 3 shows the performance of PEAQ. From the figures, it can be seen that the scattering points in Fig. 2 are aligned with the diagonal line more closely than those in Fig. 3. The experimental results show that the proposed mapping model can predict the objective scores better.

Table 4 shows an example of the scores by different test methods in our experiment. We can see that PEAQ results does not consist with the subjective scores and cannot

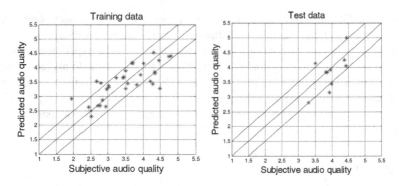

Fig. 2. Scatter-plots of the prediction model vs. subjective tests

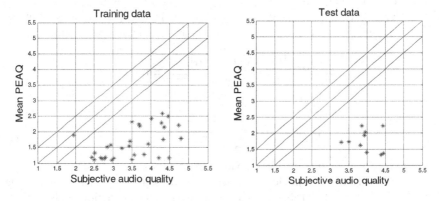

Fig. 3. Scatter-plots of the PEAQ method vs. subjective tests

indicate the subjective quality under the parametric multi-channel coding. By introducing three kinds of spatial parameters, the prediction model is able to correctly identify the subjective quality. However, the correlation is not very high within the limited train and test data. PEAQ may be not a reliable method under the parametric coding schemes which has also been shown in Reference [12]. In another way, we can further improve the PEAQ tool by designing the listening tests only measuring the basic audio quality and meanwhile more efforts could be tried to generate more proper spatial parameters considering the sound image and spatial effects.

Table 4. An example of the scores

Test sample / Test method	Sample1	Sample2
Subjective test score	4.8063	2.5100
Mean_PEAQ	1.7898	1.3275
MAQ score	4.4039	2.3068

5   Conclusions

This paper introduces a prediction model which can be used to predict the perceived quality in 5.1-channels audio systems. The method takes into account distortions in both basic audio quality and spatial quality. PEAQ is used to measure the basic audio quality and binaural space cues are used to indicate the spatial quality. The input parameters of the prediction model are PEAQ scores of the six channels and the nine spatial distortion values. Multiple linear regression is used to map the input parameters into the subjective score and the model parameters are trained through subjective listening databases with the test method of BS.1116. This paper gives a preliminary try by using a limited 5.1-channels data set considering the difficulty of formal listening tests. Experimental results show that the inclusion of spatial information into the audio quality prediction, compared with the standard PEAQ, can generate a higher correlation between the objective and the subjective scores. To further improve the correlation, more formal listening tests should be done to increase the number of samples used for training and testing and complicated prediction model can be designed with other spatial cues and data mining models.

References

1. ISO/IEC 13818-3: Information technology-Generic coding of moving pictures and associated audio information – Part 3: Audio (1998)
2. ISO/IEC 13818-7: Information technology - Generic coding of moving pictures and associated audio information-Part 7: Advanced Audio Coding (AAC) (2006)
3. ISO/IEC JTC1/SC29/WG11: Information technology - report on the verification tests of MPEG-D MPEG surround (2007)
4. ITU-R BS.775-2: Multichannel stereophonic sound system with and without accompanying picture (2006)
5. Cheng, Y., Ruimin, H., Liuyue, S., et al.: A 3D audio coding technique based on extracting the distance parameter. In: IEEE International Conference on Multimedia and Expo, pp. 1–6. IEEE Press, California (2014)
6. Bin, C., Christian, R., Ian, S.B., Xiguang, Z.: A general compression approach to multi-channel three-dimensional audio. IEEE Trans. Audio Speech Lang. Process. **21**(8), 1676–1688 (2013)
7. ITU-R BS.1116-1: Methods for the subjective assessment of small impairments in audio systems including multichannel sound systems, Geneva, Switzerland (1997)
8. ITU-R BS.1285: Pre-selection methods for the subjective assessment of small impairments in audio systems, Geneva, Switzerland (1997)
9. ITU-R BS.1534: Method for the subjective assessment of intermediate quality level of coding systems, Geneva, Switzerland (2001)
10. ITU-R BS.1387-1: Method for objective measurements of perceived audio quality, Geneva, Switzerland (2001)
11. Inyong, C., Shinn-Cunningham, B.G., Sang, B.C., Sung, K.-M.: Objective measurement of perceived auditory quality in multichannel audio compression coding systems. J. Audio Eng. Soc. **56**, 3–17 (2008)

12. Schafer, M., Bahram, M., Vary, P.: An extension of the PEAQ measure by a binaural hearing model. In: International Conference on Acoustics, Speech and Signal Processing, pp. 8164–8168. IEEE Press, Vancouver (2013)
13. Smimite, A., Beghdadi, A., Chen, K., Jafjaf, O.: A new approach for spatial audio quality assessment. In: International Conference on Telecommunications and Multimedia, pp. 46–51. IEEE Press, Greece (2014)
14. Jeroen, B., Par, S.V.D., Armin, K., Erik, S., Jeroen, B., Erik, S.: Parametric coding of stereo audio. EURASIP J. Adv. Signal Process. **9**, 1305–1322 (2005)
15. Faller, C., Baumgarte, F.: Binaural cue coding: a novel and efficient representation of spatial audio. In: International Conference on Acoustics, Speech and Signal Processing, pp. 1841–1844. IEEE Press, Florida (2002)

Physical Properties of Sound Field Based Estimation of Phantom Source in 3D

Shanfa Ke[1,2], Xiaochen Wang[1,2](✉), Li Gao[1,2], Tingzhao Wu[1,2], and Yuhong Yang[1,2]

[1] National Engineering Research Center for Multimedia Software, School of Computer, Wuhan University, Wuhan 430072, China
[2] Research Institute of Wuhan University in Shenzhen, Shenzhen, China
{kimmyfa,clowang}@163.com, {gllynnie,wutz01}@126.com, ahka_yang@yeah.net

Abstract. 3D spatial sound effects can be achieved by amplitude panning with several loudspeakers, which can produce the auditory event of phantom source at arbitrary location with loudspeakers at arbitrary locations in 3D space. The estimation of the phantom source is to estimate the signal and location of a sound source which produce the same perception of auditory event with that of phantom source by loudspeakers. Several methods have been proposed to estimate the phantom sources, but these methods couldn't ensure the conservation of sound energy at listening point in sound field, which including kinetic energy (particle velocity) and potential energy (sound pressure), so estimated errors were caused. A new method to estimate phantom source signal and the position is proposed, which is based on the physical properties (particle velocity, sound pressure) of the listening point in the sound field by loudspeakers. Moreover, the proposed method could be also appropriate for arbitrary asymmetric arranged loudspeakers. Experimental results showed that compared with current methods, estimated distortions of the location of phantom source and the superposed signal by loudspeakers with proposed method have been reduced obviously.

Keywords: 3D · Amplitude panning · Phantom sources estimating · Particle velocity · Sound pressure

1 Introduction

With the rapid development of 3D video, 3D audio also develop rapidly. Especially in 3D listening scene, people want to reconstruct phantom source which can be perceived at any desired position. So several kinds of methods have been

S. Ke – The research was supported by National Nature Science Foundation of China (61201169); National High Technology Research and Development Program of China (863 Program)(2015AA016306); National Nature Science Foundation of China (61231015); Science and Technology Plan Projects of Shenzhen (ZDSYS201405091-6575763); National Nature Science Foundation of China(61201340); the Fundamental Research Funds for the Central Universities (2042015kf0206).

© Springer International Publishing Switzerland 2015
Y.-S. Ho et al. (Eds.): PCM 2015, Part I, LNCS 9314, pp. 699–710, 2015.
DOI: 10.1007/978-3-319-24075-6_67

proposed by researchers, such as binaural methods [1], synthesis methods [2] and auditory perception of virtual image based reconstruction methods [3]. Methods based on auditory perception of virtual image, such as amplitude panning, generate phantom source by controlling the loudspeaker signals, it is based on virtual auditory properties and psychoacoustic spatial perception properties to create same auditory event, which is generated by a real source at the position of phantom source. As this kind of methods is simple and low demand for hardware, it became the common methods [3,4].

Amplitude panning methods have a hypothesis, that the sound field generated by loudspeaker signals can replace the sound field of a real sound source, which at the position of phantom source generated by loudspeaker signals. The position of phantom source can be controlled by the relation of gains of loudspeaker signals. The first model for frontal pair-wise amplitude panning was the law of sine [5], it is based on a simple geometrical model of the head. Different path to ear cause interaural time difference (ITD), then the azimuth of phantom source is estimated using ITD. In 1959, Lea et al. proposed the tangent law [3], it is also based on a simple geometrical model of the head, Compared to the sine law, this method works when head turns to phantom source. Because these two models are only suitable for two loudspeakers configuration, in 1997, Pulkki et al. proposed Vector Based Amplitude Panning (VBAP) [6]. This method can estimate the azimuth of phantom source which created by three loudspeaker. For the configuration of more than three loudspeakers, Multiple-Direction Amplitude Panning (MDAP) [7] has been proposed, for a desired panning angle, MDAP superimposes the results of VBAP for N panning directions uniformly distributed around the desired panning direction within a spread of $\pm\phi_{MDAP}$. Velocity vector method [8] also works for more than three loudspeakers, this method estimates the velocity vector of phantom source by a linear summation of the weighted loudspeaker direction, the direction of this vector is assumed to correspond to the localization of low frequencies. The result of velocity vector method is identical to the result of VBAP for the configuration of three loudspeakers, but this method only works for low frequencies. In 1992, Ger et al. proposed energy vector [8] based on the idea of velocity vector, and it is expected to model the localization direction for higher frequencies or broadband signals. VBAP has became the most popular method of amplitude panning as it's simple.

The goal of estimating phantom source is to determine a real sound source (in the following contents of this paper, this real sound source is called as real substituted source) located at the position of phantom source and with the gain of it, so that the energy, which generated by the real substituted source, can conserve with the energy of sound field generated by loudspeaker signals at the central listening point. But the existing methods, such as VBAP, just maintain the conservation of sound pressure generated between a real substituted source and loudspeaker signals, the energy of sound field is determined by the kinetic energy (particle velocity) and potential energy (sound pressure) [9], so they can't maintain conservation of energy at central listening point. Moreover, these methods don't work for the asymmetric arrangement of loudspeakers. Therefore, this work carried out the relative research based on the physical properties of sound field aiming at this goal.

Firstly, this work derived the condition of keeping the hypothesis of amplitude panning established, it's based on the conservation of sound pressure and particle velocity between loudspeaker signals and the real substituted source at listening point. Then we proposed methods to estimate the phantom source for loudspeakers symmetric arrangement case, which refer as loudspeakers contributed on the surface of the same sphere, the angles and gain were estimated respectively based on the conservation of particle velocity and the conservation of energy (PCEC) at listening point, the above conservations are all between the loudspeaker signals and the real substituted source. At last, this work proposed method for loudspeakers asymmetric arrangement case to estimate the phantom source for arbitrary listening point. Results of the experiment have shown the validity of the proposed PCEC, comparing with the existing methods, the real substituted source of phantom source generated more similar experience of auditory perception as generated by loudspeaker signals at the listening point.

2   Proposed Physical Properties Based Method

The hypothesis of amplitude panning is that the sound field of loudspeaker signals can replace the sound field of a real substituted source which generated at the central listening point. The physical properties of sound field including sound pressure and particle velocity, the energy of sound field is determined by the kinetic energy (particle velocity) and potential energy (sound pressure) [9]. The existing methods can't maintain the conservation of energy of sound field respectively generated by loudspeaker signals and a real substituted source. In this section, we firstly derived the condition of keeping the hypothesis of amplitude panning established. Then we proposed methods for estimating the phantom source based on physical properties of sound field for symmetric arrangement case. At last, method for estimating the phantom source for asymmetric arrangement case was proposed.

This paper denoted the azimuth, elevation, radius of the position of sound signal and the gain of the signal respectively as θ_l, φ_l r_l, g_l. So the particle velocity of sound signal at the central listening point are denoted as [10,11]: $(g_l \frac{e^{-ikr_l}}{r_l} \cos\theta_l \cos\varphi_l, g_l \frac{e^{-ikr_l}}{r_l} \sin\theta_l \cos\varphi_l, g_l \frac{e^{-ikr_l}}{r_l} \sin\varphi_l)$. The sound pressure of sound signal at the central listening point are denoted as [10,11]: $g_l \frac{e^{-ikr_l}}{r_l}$. l denotes the index of sound signals, $k = \frac{2\pi f}{c}$ denotes the number of wave, f denotes the frequency of signal, and c denotes sound velocity. Figure 1 has shown the sketch of environment arrangement for the horizontal plane with two loudspeakers.

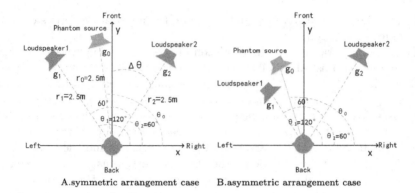

A.symmetric arrangement case   B.asymmetric arrangement case

Fig. 1. Sketch of environment arrangement.

2.1 Verification of Panning Law Based on Physical Properties of Sound Field

At the central listening point, assuming the superposed sound pressure and particle velocity generated by loudspeaker signals is equal to these generated by the real substituted source, then these equations can be achieved:

$$
\begin{aligned}
\sum_{l=1}^{L} g_l \frac{e^{-ikr_l}}{r_l} \cos\theta_l \cos\varphi_l &= g_0 \frac{e^{-ikr_0}}{r_0} \cos\theta_0 \cos\varphi_0 \\
\sum_{l=1}^{L} g_l \frac{e^{-ikr_l}}{r_l} \sin\theta_l \cos\varphi_l &= g_0 \frac{e^{-ikr_0}}{r_0} \sin\theta_0 \cos\varphi_0 \\
\sum_{l=1}^{L} g_l \frac{e^{-ikr_l}}{r_l} \sin\varphi_l &= g_0 \frac{e^{-ikr_0}}{r_0} \sin\varphi_0 \\
\sum_{l=1}^{L} g_l \frac{e^{-ikr_l}}{r_l} &= g_0 \frac{e^{-ikr_0}}{r_0}
\end{aligned}
\tag{1}
$$

L denotes the number of loudspeaker, $l = 1, \cdots, L$ denotes lst loudspeaker. $l = 0$ denotes phantom source. The existing methods assume all loudspeakers distributed on the surface of the same sphere, namely $r_0 = r_1 = \cdots = r_L$, so Eq. (1) can be transformed as:

$$
\sum_{l=1}^{L} g_l \cos\theta_l \cos\varphi_l = g_0 \cos\theta_0 \cos\varphi_0
\tag{2}
$$

$$
\sum_{l=1}^{L} g_l \sin\theta_l \cos\varphi_l = g_0 \sin\theta_0 \cos\varphi_0
\tag{3}
$$

$$
\sum_{l=1}^{L} g_l \sin\varphi_l = g_0 \sin\varphi_0
\tag{4}
$$

$$
\sum_{l=1}^{L} g_l = g_0
\tag{5}
$$

When $L = 2, \varphi_0 = \varphi_l = 0$, as Fig. 1A, we derived the equation by Eqs. (2)–(5) as:

$$
\begin{aligned}
(g_1 \cos\theta_1 + g_2 \cos\theta_2)^2 + (g_1 \sin\theta_1 + g_2 \sin\theta_2)^2 &= (g_1 + g_2)^2 \\
2g_1 g_2 (\cos(\theta_1 - \theta_2) - 1) &= 0
\end{aligned}
\tag{6}
$$

Due to the different position of loudspeakers as Fig. 1A, that to say $\theta_1 \neq \theta_2$, so the Eq. (6) isn't established. So under the configuration of two loudspeaker in symmetric arrangement in horizontal plane, the sound pressure and particle velocity of the real substituted source can't conserve with these of loudspeaker signals, therefore, the law of sine and the law of tangent are partial solutions. Such as the law of sine estimates the azimuth just using Eqs. (3) and (5), the law of tangent using Eqs. (2) and (3) to estimate.

For the configuration of three and four loudspeakers in symmetric arrangement case, we derived the conditions for the conservation of both sound pressure and particle velocity, which generated by loudspeaker signals and the real substituted source at central listening point:

$$
\begin{aligned}
&g_1 g_2 (\cos \varphi_1 \cos \varphi_2 \cos(\theta_1 - \theta_2) + \sin \varphi_1 \sin \varphi_2 - 1) + \\
&g_1 g_3 (\cos \varphi_1 \cos \varphi_3 \cos(\theta_1 - \theta_3) + \sin \varphi_1 \sin \varphi_3 - 1) + \\
&g_2 g_3 (\cos \varphi_2 \cos \varphi_3 \cos(\theta_2 - \theta_3) + \sin \varphi_2 \sin \varphi_3 - 1) = 0
\end{aligned}
\tag{7}
$$

$$
\begin{aligned}
&g_1 g_2 (\cos \varphi_1 \cos \varphi_2 \cos(\theta_1 - \theta_2) + \sin \varphi_1 \sin \varphi_2 - 1) + \\
&g_1 g_3 (\cos \varphi_1 \cos \varphi_3 \cos(\theta_1 - \theta_3) + \sin \varphi_1 \sin \varphi_3 - 1) + \\
&g_1 g_4 (\cos \varphi_1 \cos \varphi_4 \cos(\theta_1 - \theta_4) + \sin \varphi_1 \sin \varphi_4 - 1) + \\
&g_2 g_3 (\cos \varphi_2 \cos \varphi_3 \cos(\theta_2 - \theta_3) + \sin \varphi_2 \sin \varphi_3 - 1) + \\
&g_2 g_4 (\cos \varphi_2 \cos \varphi_4 \cos(\theta_2 - \theta_4) + \sin \varphi_2 \sin \varphi_4 - 1) + \\
&g_3 g_4 (\cos \varphi_3 \cos \varphi_4 \cos(\theta_3 - \theta_4) + \sin \varphi_3 \sin \varphi_4 - 1) = 0
\end{aligned}
\tag{8}
$$

For more loudspeakers case, the condition can be easily achieved according to the principle shown by the above two equation.

It has shown that the conservation of both sound pressure and particle velocity can be achieved only when the number of loudspeakers is not less than three, this is consistent with the conclusion of the existing literature. Moreover, this conservation can be achieved only when the condition shown in Eqs. (7) or (8) is satisfied, otherwise, the solution, which is solved by methods discussed above, is also just a approximate solution, such as the configuration of two loudspeakers.

2.2 Estimation of Phantom Source for Symmetric Arrangement Case

Due to the above deduction and analysis, to maintain the least distortion of energy generated between loudspeaker signals and the real substituted source at listening point, the paper proposed methods of estimating phantom source that based on physical properties of sound field. Firstly, in 3D scene, the azimuth and elevation angle of the phantom source can be estimated based on the conservation of particle velocity by using Eqs. (2)–(4):

$$
\theta_0 = \arctan(\sum_{l=1}^{M} g_l \sin \theta_l \cos \varphi_l / \sum_{l=1}^{M} g_l \cos \theta_l \cos \varphi_l)
\tag{9}
$$

$$
\varphi_0 = \arctan(\sum_{l=1}^{L} g_l \sin \varphi_l / \sqrt{(\sum_{l=1}^{l} g_l \cos \theta_l \cos \varphi_l)^2 + (\sum_{l=1}^{L} g_l \sin \theta_l \cos \varphi_l)^2})
\tag{10}
$$

Equations (9) and (10) have shown that the way of estimating angles of the proposed PCEC is identical to the way of VBAP when the number of loudspeakers is three, but the way of estimating the gain of the real substituted source is based on the conservation between energy of the real substituted source and energy of loudspeaker signals at central listening point using Eqs. (2)–(5), while VBAP just maintains the conservation between sound pressure:

$$g_0 = \sqrt{((\sum_{l=1}^{L} g_l \cos\theta_l \cos\varphi_l)^2 + (\sum_{l=1}^{L} g_l \sin\theta_l \cos\varphi_l)^2 + (\sum_{l=1}^{L} g_l \sin\varphi_l)^2 + (\sum_{l=1}^{L} g_l)^2)/2} \qquad (11)$$

Equations (9)–(11) have shown that, the angles and gain of phantom source depend on the angles and gains of loudspeaker signals, the frequency of loudspeaker signal and the radius of loudspeaker have no effect on them. The energy of sound field is decided by the kinetic energy and potential energy, because the proposed method computed angles and gain of phantom source based on physical properties of sound field, so comparing with the existing methods, the energy distortion generated by the real substituted source and loudspeaker signals at central listening point is smaller. Moreover, the PCEC method can maintain the conservation of energy between loudspeaker signals and real substituted source at the central listening point.

2.3   Estimation of Phantom Source for Asymmetric Arrangement Case

For the case that listening point located at the origin of coordinate with a symmetric arrangement of given loudspeakers, the distance between listening point and each loudspeaker is the same. But when the listening point isn't origin of coordinate, or loudspeakers aren't symmetrical configured related to the origin of coordinate, these distance are no longer the same, which can be called as asymmetric arrangement, the existing amplitude panning methods can't work for this case. But the PCEC method still can estimate the position and gain of phantom source for this case. Firstly, we estimated the azimuth and elevation of phantom source based on the conservation of particle velocity, namely using the first three sub-equation of Eq. (1):

$$\theta_0 = \arctan(\sum_{l=1}^{M} g_l \frac{e^{-ikr_l}}{r_l} \sin\theta_l \cos\varphi_l / \sum_{l=1}^{M} g_l \frac{e^{-ikr_l}}{r_l} \cos\theta_l \cos\varphi_l) \qquad (12)$$

$$\varphi_0 = \arctan(\sum_{l=1}^{L} g_l \frac{e^{-ikr_l}}{r_l} \sin\varphi_l / \sqrt{(\sum_{l=1}^{l} g_l \frac{e^{-ikr_l}}{r_l} \cos\theta_l \cos\varphi_l)^2 + (\sum_{l=1}^{L} g_l \frac{e^{-ikr_l}}{r_l} \sin\theta_l \cos\varphi_l)^2}) \qquad (13)$$

We transformed the left of the fourth sub-equation of Eq. (1) as:

$$\sum_{l=1}^{L} \frac{g_l e^{-ikr_l}}{r_l} = \sqrt{(\sum_{l=1}^{L} \frac{g_l}{r_l} \cos(kr_l))^2 + (\sum_{l=1}^{L} \frac{g_l}{r_l} \sin(kr_l))^2} e^{i \arctan(\frac{\sum_{l=1}^{L} \frac{g_l}{r_l} \sin(kr_l)}{\sum_{l=1}^{L} \frac{g_l}{r_l} \cos(kr_l)})} \qquad (14)$$

So the radius and gain of phantom source can be computed by equations:

$$r_0 = -\frac{1}{k} \arctan(\sum_{l=1}^{L} \frac{g_l}{r_l} \sin(kr_l) / \sum_{l=1}^{L} \frac{g_l}{r_l} \cos(kr_l)) \qquad (15)$$

$$g_0 = r_0 \sqrt{(\sum_{l=1}^{L} \frac{g_l}{r_l} \cos(kr_l))^2 + (\sum_{l=1}^{L} \frac{g_l}{r_l} \sin(kr_l))^2} \qquad (16)$$

3   Experiment and Analysis

In order to evaluate the difference of different methods, and evaluate whether the physical properties of sound field of a real substituted source are consistent with the one of loudspeaker signals at the central listening point, especially the energy of central listening point, and for verifying whether the phantom source generated by loudspeaker signals has the same quality as an auditory event caused by the real substituted source. We performed the objective simulation experiments and subjective listening test.

3.1   Objective Experiments

We simulated the free sound field to preform the objective experiment. The azimuth estimated by the subjective method proposed by Lee. H et al. [12] is chosen as the reference, the summation of energy of loudspeaker signals is set to 1, the ratio of gains of loudspeaker signals is determined by making the reference azimuth increasing by $1°$. Firstly, azimuth of phantom source and distortion of azimuth compared to the reference were computed. Then the distortion of energy for different methods at the central listening point and at ear were computed. At last, sound field generated respectively by loudspeaker signals and the real substituted sources estimated by different methods were drawn. The following equation was used to compute the distortion of energy, E_P and E_l is referred to the energy of sound field at central listening point, which respectively generated by the real substituted source and loudspeaker signals.

$$E_P = (g_0 \frac{e^{-ikr_0}}{r_0} \cos\theta_0 \cos\varphi_0)^2 + (g_0 \frac{e^{-ikr_0}}{r_0} \sin\theta_0 \cos\varphi_0)^2 + (g_0 \frac{e^{-ikr_0}}{r_0} \sin\varphi_0)^2 + (g_0 \frac{e^{-ikr_0}}{r_0})^2$$

$$E_l = (\sum_{l=1}^{L} g_l \frac{e^{-ikr_l}}{r_l} \cos\theta_l \cos\varphi_l)^2 + (\sum_{l=1}^{L} g_l \frac{e^{-ikr_l}}{r_l} \sin\theta_l \cos\varphi_l)^2 + (\sum_{l=1}^{L} g_l \frac{e^{-ikr_l}}{r_l} \sin\varphi_l)^2 + (\sum_{l=1}^{L} g_l \frac{e^{-ikr_l}}{r_l})^2$$

$$Distortion = \frac{E_l - E_P}{E_l} \times 100\%$$

Results of Symmetric Arrangement Case

Firstly, the experiments set the radius of sphere on which loudspeakers distributed to 1 m, 1.5 m, 2.5 m, 3.5 m, the frequency of signal is set to 500 Hz, 1 kHz, 3 kHz and 5 kHz, the distance between the center of head and binaural is set to 0.05 m, 0.06 m, 0.07 m. The experiments chosen four aperture angle of loudspeakers for the case of horizontal plane ($30°$, $60°$, $90°$,$140°$). Here we have shown the result for aperture angle is $60°$,radius of binaural is 0.07, frequency is 1 kHz, radius of sphere is 2.5 m.

Figure 2A has shown the azimuth of phantom source estimated by different methods with different gains between two loudspeakers. The azimuth of phantom source estimated by proposed methods is equal to the one estimated by the law of tangent. In symmetric arrangement case, the azimuth of phantom source is frequency-independent, radius-independent, it depend on the aperture angle of loudspeakers. Figure 2B has shown the distortion between azimuth of phantom source by above methods and the reference.Both the proposed method and the law of tangent have the same distortion, its maximum is about $6°$. Compared

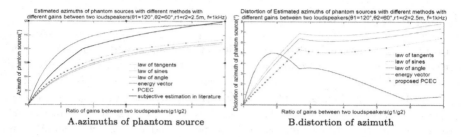

A.azimuths of phantom source   B.distortion of azimuth

Fig. 2. Azimuths and distortion of azimuth of phantom source estimated by different methods for horizontal plane

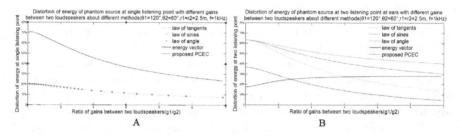

A   B

Fig. 3. Distortion of energy between of loudspeaker signals and of the real substituted source. A. the distortion at central listening point; B. the distortion at two listening point at ears

to the law of sine, the distortion of proposed methods is reduced by about 2°. While the ratio of gains between two loudspeakers is less than 2.3, the distortion of proposed methods is smaller than of methods of energy vector. But when the ratio is more than 2.3, the azimuth of phantom source estimated by energy vector is more close to the reference value than others. Moreover, the distortion of the law of sine has the most distortion of azimuth, the maximum is 8°.

The distortion of energy between of a real source located at the position of phantom source and of loudspeaker signals at central listening point and at ear have been shown in Fig. 3. Figure 3A has shown that the proposed method achieved the conservation of energy at central listening point. The energy distortion of energy vector at central listening point has arrived at maximum 50 %. Distortion of other methods has arrived at about 10 %. So the proposed method is more suitable for the characteristic of auditory perception. Figure 3B shown the average distortion of energy of phantom source compared to the energy of loudspeaker signals at ear. The average distortion of energy of the proposed PCEC has reduced at 40 % compared to the one of the law of sine at ear, and also has reduced at 20 %–30 % compared to other methods. The energy of sound field generated by real source of phantom source estimated by the proposed method is more similar to the one generated by the two loudspeaker signals at ear.

Figure 4 has shown the sound field of both loudspeaker signals and phantom source which estimated by the above discussed methods, it shown that the sound

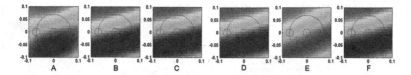

Fig. 4. Sound field of sound signals. A. generated by loudspeaker signals with $\theta_1 = 120°$, $\theta_2 = 60°$; B. sound field with the law of tangent; C. sound field with the law of sine; D. sound field with the angle law; E. sound field with energy vector; F. sound field with proposed PCEC

field of phantom source which estimated by the proposed methods is more similar to the sound field of loudspeaker signals both at central listening point and at ears.

We also performed the experiment in 3D scene, as shown in Fig. 5. Figures 5 A and B have shown the angles of phantom source estimated by different methods for three loudspeakers configuration in 3D scene. it has shown the position of phantom source estimated by the proposed method is identical to the one estimated by VBAP. Figure 5 C and D have shown the distortion of energy between of a real substituted source of estimated phantom source and of loudspeaker signals at single listening point and at two listening point at ears in 3D scene, the proposed PCEC method achieved the conservation of energy at central listening point. It can be shown that the proposed PCEC method generated less distortion of energy compared to VBAP and energy vector method in 3D scene.

Results of Symmetric Arrangement Case

The existing methods don't work for asymmetric arrangement case, the proposed method solved this problem. In this section, we estimated the azimuth of phantom source for asymmetric arrangement case for the horizontal plane as Fig. 1B. the aperture angle is chosen as $60°$, the radius of position of loudspeaker1 is set as 2.16 m, 2.5 m, 2.84 m, 3.18 m; The result is shown in Figs. 6 and 7.

Figure 6 has shown that the larger the distance between loudspeaker1 and the central listening point, the closer the phantom source to loudspeaker2 for a

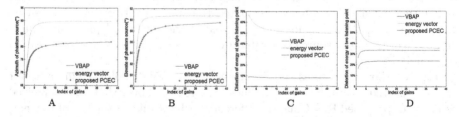

Fig. 5. Angles of phantom source estimated by different methods and distortion of energy of sound field in 3D scene. A. azimuth of phantom source; B. elevattion of phantom source; C. distortion of energy at central listening point; D. distortion of energy at two listening point at ears

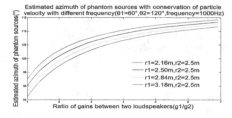
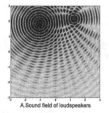
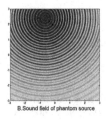
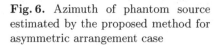

Fig. 6. Azimuth of phantom source estimated by the proposed method for asymmetric arrangement case

Fig. 7. Sound field of sound source with for asymmetric arrangement case

given gains, as the gain increasing, the azimuth of phantom source increasing too. Moreover, the frequency has no effect on the estimation of the azimuth of phantom source under the configuration of this experiment. Figure 7 has shown the sound field generated by two loudspeaker signals with different radius and the sound field generated by the phantom source estimated under the configuration of different radius of loudspeakers. We can see the sound field of phantom source is consistent with the sound field of loudspeaker signals.

3.2 Subjective Experiment

As a supplement to the objective metrics, subjective experiment was conducted, a multi-stimulus test with hidden reference and anchor (MUSHRA) [13] was chosen as the subjective experiment platform. Four 0.4 s test sequences have been used in the subjective experiment, there are three single frequency signals (frequency is 0.5 k,1 k,5 k) is synthesised by Adobe Audition 6.0, and a speech sequence es01.wav which is chosen from the MEPG standard test data. There were 14 subjects participating in the experiment, they are graduated students engaged in audio research with age in 20–25. The symmetric environment arrangement was chosen as Fig. 1A, azimuths of loudspeakers were $60°$ and $120°$, gains of loudspeaker signals were $g_1 = 1.154$ and $g_2 = 0.818$, the azimuth of the real substituted source estimated by different methods under this configuration of loudspeaker signals. Subjects evaluated the similarity between the position and intensity of the real substituted source of the estimated phantom source and these of phantom source generated by loudspeaker signals, and then scored this similarity.

Figure 8 shows the scores and the corresponding 95 % confidence intervals for real substituted source for different methods. It has shown that, the perceived azimuth for proposed PCEC has no difference with the one for the law of tangent, but it was more close to the perceived azimuth generated by loudspeaker signals compared with the energy vector method. The perceived energy for the law of tangent is much larger than the one generated by loudspeaker signals, and the perceived energy for energy vector is very small. Compared to them, The

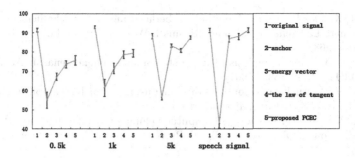

Fig. 8. Average scores and corresponding 95 % confidence intervals of the scores for different methods with different signals

perceived energy for the proposed PCEC is more close to the perceived energy of loudspeaker signals. Figure 8 has indicated the perceived score of similarity for the proposed PCEC is higher than for other methods, and it is more close to the perceived score of original loudspeaker signals. Here we just conducted a simple subjective experiment in 2D scene, for 3D scene, it need more sophisticated subjective experiments to verify the proposed PCEC method.

4   Conclusion

This paper proposed a new method to estimate phantom sources in 3D, it's based on physical properties of the sound field. Firstly, this paper derived the condition of maintaining the conservation of energy between of loudspeaker signals and of the real substituted source. Then we proposed method for estimating the azimuth and gain of phantom source, which based on physical properties of sound field at the central listening point, so that maintaining the conservation of energy between of loudspeaker signals and of the real substituted source at the central listening point. At last, this work proposed the method to estimate the phantom source for asymmetric arrangement case. The experiments has shown that the proposed PCEC has maintained the conservation of energy at central listening point, the distortion of energy of between of real substituted source and loudspeaker signals at ear has reduced by maximum 40 % compared to the existing methods.

References

1. Laitinen, M., et al.: Influence of Resolution of Head Tracking in Synthesis of Binaural Audio. in Audio Engineering Society Convention. Audio Engineering Society, 132 (2012)
2. Spors, S., Rabenstein, R., Ahrens, J.: The theory of wave field synthesis revisited. In: 124th AES Convention (2008)
3. Leakey, D.M.: Some measurements on the effects of interchannel intensity and time differences in two channel sound systems. J. Acoust. Soc. Am. **31**(7), 977–986 (1959)

4. Wendt, K.: Das Richtungsh? ren bei der berlagerung zweier Schallfelder bei Inten- sit? ts-und Laufzeitstereophonie. Rheinisch-Westf lische Technische Hochschule Aachen (1963)

5. Bauer, B.B.: Phasor analysis of some stereophonic phenomena. J. Acoust. Soc. Am. **33**(11), 1536–1539 (1961)

6. Pulkki, V.: Virtual sound source positioning using vector base amplitude panning. J. Audio Eng. Soc. **45**(6), 456–466 (1997)

7. Pulkki, V.: Uniform spreading of amplitude panned virtual sources. In: 1999 IEEE Workshop on Applications of Signal Processing to Audio and Acoustics, IEEE (1999)

8. Gerzon, M.A.: General metatheory of auditory localisation. In: Audio Engineering Society Convention. Audio Engineering Society, 92 (1992)

9. Fahy, F.J.: Measurement of acoustic intensity using the cross-spectral density of two microphone signals. J. Acoust. Soc. Am. **62**(4), 1057–1059 (1977)

10. Pierce, A.D.: Acoustics: An Introduction to its Physical Principles and Applica- tions. McGraw-Hill, New York (1981)

11. Ando, A.: Conversion of multichannel sound signal maintaining physical properties of sound in reproduced sound field. IEEE Trans. Audio Speech Lang. Process. **19**(6), 1467–1475 (2011)

12. Lee, H., Rumsey, F.: Level and Time Panning of Phantom Images for Musical Sources. J. Audio Eng. Soc. **61**(12), 978–988 (2013)

13. ITU-R: Recommendation BS.1534-1. Method for the subjective assessment of inter- mediate quality levels of coding systems. International Telecommunication Union (2003)

Non-overlapped Multi-source Surveillance Video Coding Using Two-Layer Knowledge Dictionary

Yu Chen[1], Jing Xiao[1,2(✉)], Liang Liao[1], and Ruimin Hu[1,2]

[1] National Engineering Research Center for Multimedia Software, School of Computer, Wuhan University, Wuhan, China
{cynercms,jing,liaoliangwhu,hrm}@whu.edu.cn
[2] Collaborative Innovation Center of Geospatial Technology, Wuhan, China

Abstract. In multi-source surveillance videos, a large number of moving objects are captured by different surveillance cameras. Although the regions that each camera covers are seldom overlapped, similarities of these objects among different videos still result in tremendous global object redundancy. Coding each source in an independent way for multi-source surveillance videos is inefficient due to the ignoring of correlation among different videos. Therefore, a novel coding framework for multi-source surveillance videos using two-layer knowledge dictionary is proposed. By analyzing the characteristics of multi-source surveillance videos in large scale of spatio and time space, a two-layer dictionary is built to explore the global object redundancy. Then, a dictionary-based coding method is developed for moving objects. For any object in multi-source surveillance videos, only some pose parameters and sparse coefficients are required for object representation and reconstruction. The experiment with two simulated surveillance videos has demonstrated that the proposed coding scheme can achieve better coding performance than the main profile of HEVC and can preserve better visual quality.

Keywords: Multi-source surveillance video · Global object redundancy · Knowledge dictionary

1 Introduction

Intelligent video systems have been used in wide series of applications, such as intelligent transportation, public security and so on. Due to the trend of high definition in intelligent video systems, a growing amount of surveillance video data is produced every day, bringing an extremely high demand for data storage and compression which is difficult to meet.

Surveillance cameras in urban intelligent video systems are well arranged. Videos taken by these cameras, namely multi-source surveillance videos, capture a large number of moving objects. Although the regions that each camera covers are seldom overlapped, similarities of these objects among different videos give rise to another kind of redundancy named global object redundancy. It becomes the most influential kind of redundancy with a dramatically increasing number of moving objects

© Springer International Publishing Switzerland 2015
Y.-S. Ho et al. (Eds.): PCM 2015, Part I, LNCS 9314, pp. 711–720, 2015.
DOI: 10.1007/978-3-319-24075-6_68

when taking all the videos in a large scale of space and time as a whole. Therefore, utilizing the spatial and temporal correlation of objects among different videos to eliminate global object redundancy becomes a key issue for multi-source surveillance video coding.

Previous efforts have been made on single source video coding. The H. 26x series [1, 2] adopt intra/inter frame prediction to explore similarities in spatial/temporal domain. Although most of local redundancy within one video can be effectively eliminated by prediction, the same treatment for static background and dynamic foreground results in suboptimal performance for coding surveillance video data. In order to make full use of the characteristics of surveillance video, several coding schemes have been proposed, which can be classified into two categories: background coding and foreground coding. The former aims at generating a long-term reference frame of high quality for background that can significantly improve the prediction accuracy [3–5]. The latter focuses on utilizing the 2D shapes or 3D models of foreground objects to help with the motion estimation and compensation process [6, 7]. However, the above mentioned methods measure similarities on pixel level, which are prone to be affected by illumination and object pose changes. Multi-source surveillance videos are often discontinuous in both spatial and temporal domains. The lack of consistency on pixel level makes it difficult to explore the global redundancy in multi-source surveillance videos.

In recent years, multi-source image coding schemes have been proposed and have drawn lots of attention. Yue et al. [8] utilized similar images retrieved from cloud to reconstruct the original one. Only a few features extracted from the original image and its down sampled version need to be transmitted. Shi et al. [9] generated a feature-based minimum spanning tree to establish the prediction relationships for each subset that contains images of the same scene or object, then feature-based prediction was adopted. Thanks to its invariant property to rotation and robustness to illumination changes, SIFT feature can build the correlations among different images thus achieving inter-image prediction to explore redundancies among multi-source images. Nevertheless, in a larger scale of space and time, it always happens that the pose of object changes dramatically, for example, the same object shows the front side in one frame but the back side in another. Under this circumstance, the feature based methods will fail for the reason that there are no matched features. Prior knowledge about objects is robust to the notable pose change and taking advantage of it can provide a considerable coding gain.

Therefore, we build a two-layer knowledge dictionary to extract the common information in large scale of spatio and time space from multi-source surveillance videos for object representation and reconstruction. In this way, both the correlations among different videos and the prior knowledge about objects are efficiently utilized to explore the global object redundancy. Then, a dictionary-based coding method is developed to make full use of the two-layer knowledge dictionary for multi-source surveillance videos.

The rest of this paper is organized as follows. In Sect. 2, detailed analysis of global object redundancy and the purpose of building the two-layer knowledge dictionary are given. Our proposed coding framework is in Sect. 3. Section 4 presents the experimental results. Conclusion and future work of this paper are described in Sect. 5.

2   Global Object Redundancy

2.1   Analysis of Global Object Redundancy

In surveillance videos, there is a large quantity of moving objects. According to [10, 11], on average around 99 % bitrate is used to record moving objects, and 77 % for moving vehicles when coding the surveillance videos. Further removing the global object redundancy caused by moving objects will lead to a considerable improvement of the compression ratio. Vehicles not only play an important role among moving objects in surveillance videos, but also have rigid geometric structure that is convenient to be modeled. As a consequence, our efforts will focus on moving vehicles.

For most of the surveillance videos, there are two phenomena that can be easily observed as followings: First, the appearances of one vehicle in different videos may be various, but they correspond to only one model. Second, the appearance changes of one vehicle in different videos over large scale of space and time have correlations. Based on the above observations, we divide the global object redundancy into two components: model consistency and residual similarity (shown in Fig. 1).

- Model consistency: model consistency refers to the uniform model that all the appearances of one vehicle share.
- Residual similarity: residual similarity is the common information of the difference between model and real appearances in videos.

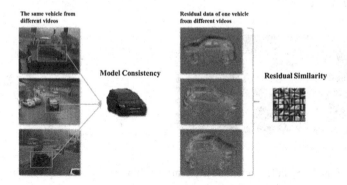

Fig. 1. Global object redundancy in multi-source surveillance videos

By extracting the model consistency and the residual similarity of objects as the common information for object representation and reconstruction, global object redundancy can be effectively eliminated, and then achieve better coding performance.

2.2   Two-Layer Knowledge Dictionary for Eliminating Global Object Redundancy

According to the analysis, global object redundancy consists of model consistency and residual similarity, i.e. common information of model and residual data. It is widely acknowledged that dictionary learning is an effective method to obtain common structures of images or signals, thus, a two-layer dictionary is proposed. The first layer of knowledge dictionary consists of 3D models with textures, representing the primitive knowledge about objects. The second layer of knowledge dictionary is learned from the difference between the appearances in real videos and the reconstruction ones from the first layer, representing the knowledge about factors that affect the appearances of objects. This two-layer knowledge dictionary simulates the real process that the appearances of original objects change over space and time with different factors.

A preliminary experiment has been conducted on several real surveillance videos of different crossroads in CIF format (shown in Fig. 2(a)) to prove the second observation mentioned in the above section and testify the effectiveness of the two-layer dictionary. Image blocks from the same object and their residual data are extracted as the training samples. Then original image blocks and residual blocks are used to train the normal dictionary and the two-layer dictionary respectively. Figure 2(b) shows the average PSNR of reconstruction blocks using various number of dictionary atoms. The reconstruction performance of two-layer dictionary is 4 dB better on average than that of normal dictionary when using the same number of atoms, indicating that residual data of the same object has more common information and stronger correlations than the original one. Moreover, it is noticeable that only a few atoms are needed to reach a relatively high PSNR value (using 5 atoms can achieve nearly PSNR of 36 dB on average). Therefore, the two-layer knowledge dictionary is proven to have the ability to represent objects and eliminate global object redundancy in large scale of spatio and time space efficiently.

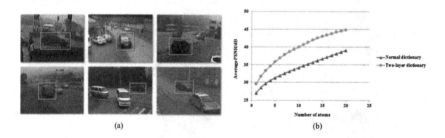

(a)   (b)

Fig. 2. Sample frames and performance comparison using top k atoms

3   Proposed Method

In previous section, we have analyzed the characteristics of a special kind of redundancy, i.e. global object redundancy, in multi-source surveillance videos and presented the two-layer knowledge dictionary as an effective way to deal with it. In the following sections, the learning procedure of the two-layer dictionary and how to utilize it to achieve high-efficiency coding for multi-source surveillance videos are described in details.

3.1   Two-Layer Dictionary Learning

Two-layer knowledge dictionary is used for object representation and converts the image blocks of object into sparse coefficients at the encoding side. A suitable dictionary adaptive to video content can preserve better visual quality whereas using less number of atoms. Therefore, the compression ratio that proposed coding framework can reach is closely related to the dictionary. Learning procedure of the dictionary is described as follows (shown in Fig. 3):

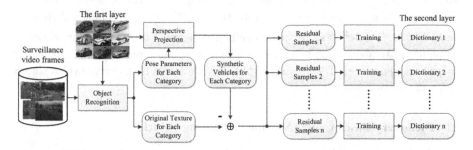

Fig. 3.   Two-layer dictionary learning procedure

- The first layer

    The first layer reflects the primitive knowledge about objects and simulates the original condition without any factors acting on them. It is composed of real 3D models of vehicles that appear in the multi-source surveillance videos. We can collect these models with textures on them through working with vehicle manufacturers or downloading from the Internet.

- The second layer

    The second layer represents the environmental factors, such as illumination, weather, that affect the appearance of real model in different videos. It is composed of the difference between 3D model and real appearance in videos.

Initialization: A portion of multi-source videos are chosen as samples to learn the dictionary. Vehicles are extracted from samples through vehicle recognition process. Then, residual data r_i for the ith vehicle can be derived by subtracting the textures from the corresponding 3D model:

$$r_i = V_i - P_i \cdot M_i \tag{1}$$

where V_i and M_i are original video data and textures from 3D model for the ith vehicle respectively. P_i is the projection matrix related to the pose parameters which are estimated during the vehicle recognition process [12]. The residual data is then divided into 16×16 blocks as training data.

Training: Considering the adaptability to video content, a specific dictionary is built for each category of vehicle. Residual blocks are classified according to the vehicle recognition result. For each category j, the dictionary updates through optimizing the following problem:

$$\arg \min_{D_j, C_j} \left\| R_j - D_j C_j \right\|_2 + \lambda \left\| C_j \right\|_0 \tag{2}$$

where $R_j \in \mathbb{R}^{N \times K_j}$ is the residual matrix whose columns are vectorized residual blocks $r_k, k = 1, 2, 3 \cdots K_j$ having a size of N pixels. $D_j \in \mathbb{R}^{N \times M}$ is the overcomplete dictionary where $N < M$. $C_j \in \mathbb{R}^{M \times K_j}$ is the sparse coefficient matrix. λ is a constant that balances the representation error and the sparsity. This optimizing problem can be efficiently solved by adopting the K-SVD algorithm [13–15]. Moreover, the l_0 norm can be replaced by l_1 norm for its convexity.

3.2 Dictionary-Based Coding Scheme for Moving Vehicles

To make full use of two-layer knowledge dictionary, a dictionary-based coding scheme for moving vehicle is proposed.

The encoding procedure is presented in Fig. 4. Vehicles in each frame are extracted beforehand and numbered by the index i. The same dictionaries are stored in both encoding and decoding side for representation and reconstruction respectively.

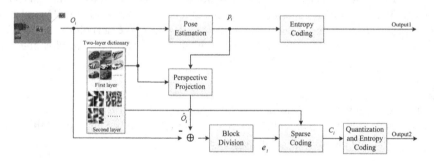

Fig. 4. Proposed coding scheme for moving vehicles

When coding the ith vehicle O_i, 3D model from the first layer of dictionary is used to estimate the pose parameters p_i meanwhile textures of the 3D model is projected to the image plane by perspective projection with extracted pose parameters to form a synthetic vehicle $\hat{O}_i$. Then, the residual data of the vehicle e_i is derived by $O_i - \hat{O}_i$ and further divided into blocks. Only those blocks related to the vehicle need to be sent to sparse coding module to compute the sparse coefficients C_i using OMP algorithm [16]. At this stage, the corresponding dictionary in the second layer is chosen according to the category of the vehicle in the first layer. Finally, pose parameters and sparse coefficients are coded to form the bitstream.

At the decoding side, a basic vehicle is synthesized by the decoded pose parameters and the 3D model from the first layer of dictionary. Then, residual blocks are reconstructed by the corresponding dictionary for specific category from the second layer. By combining these two components, the final reconstructed vehicle is obtained.

3.3 Overview of the Coding Framework

In this section, a coding framework for multi-source surveillance videos using two-layer knowledge dictionary is proposed to eliminate the global object redundancy.

As shown in Fig. 5, moving objects are first extracted from multisource surveillance videos. Then, vehicle recognition is adopted to classify moving vehicles into several categories. For each category, vehicles are coded by our proposed dictionary-based coding scheme. The background frames with remaining foreground objects together are coded by the surveillance profile of IEEE 1857-2013 standard. In the decoding procedure, vehicles are synthesized by two-layer knowledge dictionary and project back to their original positions. The decoded background frames and other objects are added to form the reconstruction videos.

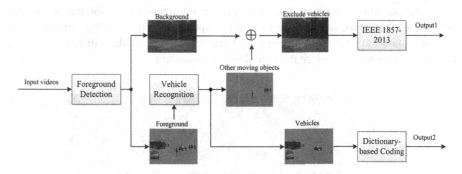

Fig. 5. Overall coding framework

In our method, spatial and temporal information is used for building the correlations of the same object in different videos with the help of prior knowledge during the recognition process. Only with the correlations we can further efficiently utilize the common information and explore the global object redundancy in large scale of spatio and time space, thus achieving better coding performance.

4 Experimental Results

In order to evaluate the performance of our proposed coding scheme, the main profile of HEVC (implemented in HM 16.2) is employed as the comparison method under our simulated environment. The configuration is shown in Table 1. In our proposed coding framework, the surveillance profile of IEEE 1857-2013 standard (implemented in SM2_1.6) adopts the same configuration as the main profile of HEVC except that the QP of G-picture is set to 10 to ensure the quality of the background frames used for referencing.

The two-layer dictionary is built for representing the red vehicle in the video clips. The 3D model and the corresponding texture of the vehicle obtained beforehand from the Internet are stored in the first layer. A total number of 10000 image blocks with the size of 16×16 are extracted as the training samples from 300 images of the red vehicle taken under various environmental conditions with the changing poses. The second layer is trained off-line using the training strategy described in Sect. 3.1. The size of second layer is set to 600 to ensure its overcompleteness. The maximal number of atoms used to represent the vehicle L is set to 20.

Table 1. The experimental configuration of HEVC and IEEE 1857-2013

Items	Value	Items	Value
Frame rate	30	Rate control	Disable
QP	22,27,32,37	Search range	64
Intra period	−1	Ref. Num.	5

Two video clips (shown in Fig. 6) with the resolution of 640×480 and the length of 90 s are taken in the pre-designed scenario, capturing the non-overlapped regions to simulate multi-source surveillance videos. Vehicles in the two video clips are the same, for the convenience to explore the global object redundancy.

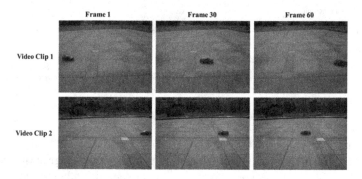

Fig. 6. Sample frames extracted from two video clips

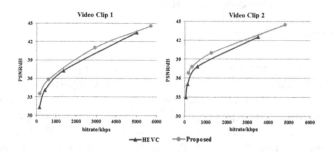

Fig. 7. The rate distortion curves for test video clips

The encoding performance of the HEVC and our proposed method are compared. As shown in Fig. 7, our method outperforms the HEVC within the range of low bitrate but has less PSNR increment with the increasing bitrate. This is mainly because that we fix the maximal number of atoms used to represent the vehicle in this experiment. Under low bitrate condition (less than 5000 kbps in our experiment), the PSNR of vehicle reconstructed by our method is better than the one of HEVC. However, due to the constraint of maximal number of atoms, the PSNR of vehicle reconstructed by our method will reach the maximum value at one point whilst the one of HEVC continuously increases with the increment of bitrate.

To further compare the ability to eliminate the redundancy of moving objects, only bits for coding the vehicle and PSNR of the area that vehicle covers are taken into consideration. For HEVC bitstream, we utilize the bitstream analyzer named CodecVisa to obtain the bits used for coding the vehicle. The bitrate for vehicles used in our proposed method is obtained from entropy coding for object parameters such as pose parameters and sparse coefficients. For both methods, image blocks that cover the vehicle are extracted to calculate the object PSNR. In this experiment, the maximal number of atoms used to represent the vehicle L in our method is set to 5, 15, 20, 25, 30 while the QP for HEVC is set to 22, 27, 32, 37 to draw the rate distortion curves.

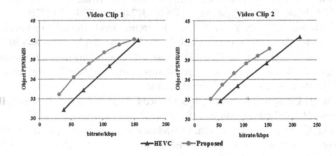

Fig. 8. The rate distortion curves for the vehicle

As presented by Fig. 8, our method achieves PSNR gain of 2.38 dB and 2.24 dB for video clip 1 and video clip 2 respectively in low bitrate condition (under 150 kbps with the average resolution of 80 × 35 for the vehicle in video clip 1 and 200 kbps with the average resolution of 100 × 45 for the vehicle in video clip 2). Nevertheless, with the increase of bitrate, the quality of reconstructed object of HEVC is very high. To reach the same quality, much more atoms in dictionary are needed, thus generating a large number of sparse coefficients, resulting in a performance drop.

5 Conclusion

In this work, a novel coding scheme for multi-source surveillance videos is proposed. We first analyze the global object redundancy among different surveillance videos in large scale of spatio and temporal space. Then, based on its characteristics, a two-layer knowledge dictionary is built to represent the moving objects and eliminate global object redundancy. Finally, we developed a dictionary-based coding method for moving vehicles. The method is proven to be more effective than HEVC in low bitrate. To improve the performance of our method in high bitrate, further efforts will be spared on the prediction method for compressing the sparse coefficients.

Acknowledgments. This work was partly supported by the China Postdoctoral Science Foundation (2014M562058), Scientific Research Foundation for the Returned Overseas Chinese Scholars, State Education Ministry ([2014]1685), Fundamental Research Funds for the Central Universities (2042014kf0025, 2042014kf0286), EU FP7 QUICK project (PIRSES-GA-2013-612652).

References

1. Wiegand, T., Sullivan, G.J., Bjøntegaard, G., Luthra, A.: Overview of the H.264/AVC video coding standard. IEEE Trans. Circuits Syst. Video Technol. **13**(7), 560–576 (2003)
2. Sullivan, G.J., Ohm, J., Han, W., Wiegand, T.: Overview of the high efficiency video coding(HEVC) standard. IEEE Trans. Circuits Syst. Video Technol. **22**(12), 1649–1668 (2012)
3. Paul, M., Lin, W. et al.: A Long-term reference frame for hierarchical B-picture-based video coding. IEEE Trans. Circuits Syst. Video Technol. **24**(10), 1729–1742 (2014)
4. Zhang, X., Huang, T., Tian, Y., Gao, W.: Background-modeling-based adaptive prediction for surveillance video coding. IEEE Trans. Image Process. **23**(2), 769–784 (2014)
5. Zhang, X., Tian, Y., Huang, T., Dong, W., Gao, W.: Optimizing the hierarchical prediction and coding in HEVC for surveillance and conference videos with background modeling. IEEE Trans. Image Process. **23**(10), 4511–4526 (2014)
6. Ng, K., Wu, Q., Chan, S., Shum, H.: Object-Based Coding for plenoptic videos. IEEE Trans. Circuits Syst. Video Technol. **20**(4), 548–562 (2010)
7. Tsai, T., Lin, C.: Exploring contextual redundancy in improvingobject-based video coding for video sensornetworks surveillance. IEEE Trans. Multimedia **14**(3), 669–682 (2012)
8. Yue, H., Sun, X., Yang, J., Wu, F.: Cloud-based image coding for mobile devices—toward thousands to one compression. IEEE Trans. Multimedia **15**(4), 845–857 (2013)
9. Shi, Z., Sun, X., Wu, F.: Feature-based image set compression. In: IEEE International Conference on Multimedia and Expo, pp. 1–6 (2013)
10. Xiao, J., Chen, Y., Hu, J., Hu, R.: Global coding of multi-source surveillance video data. In: Data Compression Conference, pp. 33–42 (2015)
11. Xiao, J., Liang, L., Hu, J., Chen, Y., Hu, R.: Exploiting global redundancy in big surveillance video data for high efficient coding. J. Cluster Comput. (2014)
12. Tan, T.N., Sullivan, G.D., Baker, K.D.: Model-based localization and recognition of road vehicles. Int. J. Comput. Vis. **27**(1), 5–25 (1998)
13. Aharon, M., Elad, M., Bruckstein, A.: K-SVD: an algorithm for designing over-complete dictionaries for sparse representation. IEEE Trans. Sig. Process. **54**(11), 4311–4322 (2006)
14. Liu, Q., Liang, D., Song, Y., et al.: Augmented Lagrangian based sparse representation method with dictionary updating for image deblurring. SIAM J. Imaging Sci. **6**(3), 1689–1718 (2013)
15. Liu, Q., Wang, S., Ying, L., et al.: Adaptive dictionary learning in sparse gradient domain for image recovery. IEEE Trans. Image Process. **22**(12), 4652–4663 (2013)
16. Tropp, J.A., Gilbert, A.C.: Signal recovery from random measurements via orthogonal matching pursuit. IEEE Trans. Inf. Theory **53**(12), 4655–4666 (2007)

Global Motion Information Based Depth Map Sequence Coding

Fei Cheng[1]([⊠]), Jimin Xiao[1], Tammam Tillo[1], and Yao Zhao[2]

[1] Department of Electrical and Electronic Engineering,
Xian Jiaotong-Liverpool University (XJTLU), 111 Ren Ai Road, SIP, Suzhou,
Jiangsu Province 215123, People's Republic of China
{fei.cheng,jimin.xiao,tammam.tillo}@xjtlu.edu.cn
http://www.mmtlab.com
[2] Beijing Jiaotong University Institute of Information Science, Beijing, China
yzhao@bjtu.edu.cn

Abstract. Depth map is currently exploited in 3D video coding and computer vision systems. In this paper, a novel global motion information assisted depth map sequence coding method is proposed. The global motion information of depth camera is synchronously sampled to assist the encoder to improve depth map coding performance. This approach works by down-sampling the frame rate at the encoder side. Then, at the decoder side, each skipped frame is projected from its neighboring depth frames using the camera global motion. Using this technique, the frame rate of depth sequence is down-sampled. Therefore, the coding rate-distortion performance is improved. Finally, the experiment result demonstrates that the proposed method enhances the coding performance in various camera motion conditions and the coding performance gain could be up to 2.04 dB.

Keywords: Depth map · Global motion information · Down-sampling

1 Introduction

Due to the rapid development of the range and depth sensing technology, depth cameras such as Microsoft Kinect and SwissRange SR4000 [1] have been developed. Depth maps are widely employed in the texture-plus-depth representation for 3D video coding.

A depth map, which represents the distance from the objects in the scene to the capturing camera, together with its aligned texture, have been exploited to describe 3D scenes. Multi-view Video plus Depth (MVD) format is a promising way to represent 3D video content, and recently extensions supporting for the MVD format have been introduced [2,3]. With the MVD format, only a small number of texture views associated with their depth views are required. At the

F. Cheng –This work was supported by National Natural Science Foundation of China (No.61210006, No.60972085).

© Springer International Publishing Switzerland 2015
Y.-S. Ho et al. (Eds.): PCM 2015, Part I, LNCS 9314, pp. 721–729, 2015.
DOI: 10.1007/978-3-319-24075-6_69

decoder or display side, Depth-Image-Based Rendering (DIBR) [4,5] is used to synthesize additional viewpoint video.

In the DIBR based 3D video coding scheme, depth map is represented as a gray-scale image, which is encoded independently. A texture and its corresponding depth map describe the features of the same scene in terms of content and distance respectively. The correlation between them should be exploited by an encoder to reduce the redundancy. In [6], the coding performance of depth maps is improved by taking into account the Motion Vector (MV) of texture. This can reduce the time of the Motion Estimation (ME) for depth map encoding due to the reduced coding complexity. Furthermore, it is proposed to add the 3D search to expand the ME of depth map.

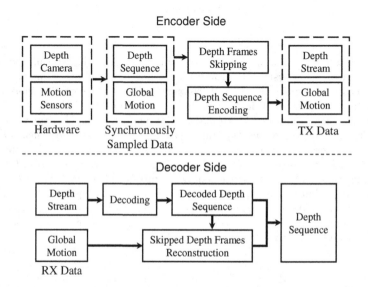

Fig. 1. The diagram of proposed depth map sequence encoding and decoding method

A depth map contains position information of each object, which can be exploited to project the neighboring frames by using the camera global motion information. In this paper, we propose a novel global motion information based depth map sequence coding method. As shown in Fig. 1, the synchronously sampled global motion information of a depth camera has been exploited to improve the depth map sequence coding performance. We intentionally skip some depth map frames or blocks during encoding according to the amount of camera global motion. Then, the skipped parts are projected from their neighboring frames using the global motion information between frames. As the bitrate is reduced, the proposed method achieves the global rate-distortion performance gain.

We develop the hardware prototype to simulate the camera global motion and produce different sequences to test the proposed method under different conditions. In order to simplify the experiment, we test the proposed method under the H.264/AVC standard. The experimental results demonstrate that the proposed method can improve the coding performance compared to H.264/AVC. The average gain could be up to 2.04 dB. It is worth mentioning the proposed method is independent of the coding tools, which means that it can be used with other video coding standards.

The rest of this paper is organized as follows. In Sect. 2, the details of the proposed method are described. Then, the experimental methods of the proposed scheme and the results are presented in Sect. 3. Lastly, Sect. 4 concludes this work.

2 Proposed Method

In many video capturing and depth map sampling scenarios, the camera is moving. The camera global motion leads to the change of image content. As the depth information is represented as gray-scale map after quantization, the gray level changes with the change of depth value. According to the imaging principle, the impact of global camera motion on depth map can be pictorially presented in Fig. 2. A cube and a cylinder as examples are captured by a depth camera. The cylinder is farther away from the camera than the cube. The depth map is quantized linearly, while the gray levels (black to white) represent the distance (near to far). The depth camera samples a new depth image after dolly and tracking. With the dolly motion, the depth camera moves forwards. Therefore, the two objects are enlarged with different scales. The scale of the cube is larger than that of the cylinder as it is closer to the camera. At the same time, the gray levels of two objects become darker. For camera tracking, the position of each object is shifted, meanwhile the relative distance between them is also changed. But the gray level of each object dose not change.

The motion information of the depth camera could be obtained using the motion sensor. The depth information together with the relative position can be exploited to project the neighboring frames to the current position. The projected depth map should be similar with the real depth frame.

With the similar principle, some depth map frames or some blocks of one depth frame can be skipped for encoding. Instead, only the global motion information is transmitted to the decoder side. The skipped frames or blocks are reconstructed by projecting from neighboring depth map frames. By reducing the encoding frames or blocks, the total bitrate decreases. Finally, the overall rate-distortion performance is improved.

In summary, two key procedures have to be implemented in order to achieve the proposed depth map sequence coding method. Firstly, some depth map frames or blocks should be skipped for the proposed method. Secondly, at the decoder, skipped frames or blocks are projected from the neighboring frames. It is worth mentioning that the transmitted global motion information can also benefit many other applications, such as deblurring and background extraction for moving cameras. The details of each procedure are introduced as follows.

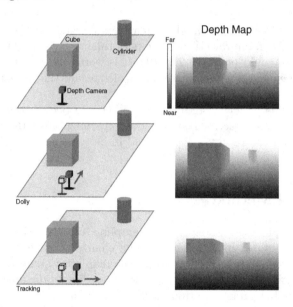

Fig. 2. The impact of camera global motion on depth map

2.1   Depth Map Skipping

Whether the skipped depth frame could be properly projected from the neighboring depth frames is related to two main factors. One factor is the amount of camera global motion. The smaller the motion, the more similar the current frame to the neighboring frames, which means it is not difficult for the decoder to project the skipped depth frames. Therefore, the depth frame skipping can be decided dynamically by the amount of global motion. The threshold of skipping is evaluated based on the content of the sequence.

Another factor is the change of contents in the scene. The moving objects are difficult to be projected from their neighboring frames. Therefore, blocks containing moving objects are segmented and encoded separately, whereas, the rest blocks of the current frames are skipped. It is worth mentioning that both sides of the neighboring frames should be projected to the current position. The decision on whether the frames or blocks are skipped or not is based on the differences between the current frame and the projected frames (blocks). Large difference means that the frame (block) is skipped, and vice versa.

An example of depth map frame skipping is described in Fig. 3. The depth frames D_{n+1}, D_{n+3} and D_{n+4} are skipped at the encoder side. Then, D_{n+1} is projected from D_n and D_{n+2}, while D_{n+3} and D_{n+4} are projected from D_{n+2} and D_{n+5}. Finally, each frame of the depth sequence is reconstructed at the decoder side.

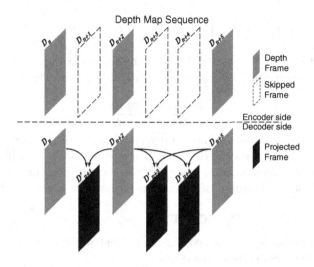

Fig. 3. An example of depth map frame skipping and projection

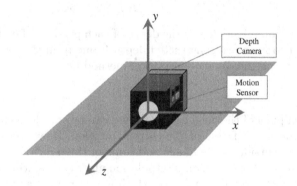

Fig. 4. The 3D coordinate system definition for the camera

2.2 Depth Map Projection

Each pixel in the depth map needs to be converted from a 2D point into 3D coordinate space for 3D projection. First, the 3D coordinate system needs to be defined as shown in Fig. 4. The x-axis and the y-axis are parallel to the image, while the z-axis represents the depth.

Let $\mathbf{P} = [w, h]$, where w and h represent the horizontal and vertical coordinate of a pixel in the image. As the depth z of each pixel has been quantized to an integer n in the depth map frame, then it needs to be dequantized by:

$$z = Q^{-1}(n), \tag{1}$$

where Q is the quantization method of depth map sequence. The 3D homogeneous coordinate of a pixel can be converted by:

$$\mathbf{C} = [x, y, z, 1] = \left[K(\frac{W}{2} - w) \cdot z, K(\frac{H}{2} - h) \cdot z, z, 1 \right], \tag{2}$$

where W and H are horizontal and vertical resolution of the depth map respectively. K is the intrinsic parameters of the depth camera, which is represented as:

$$K = \frac{\tan(\phi_w)}{W} = \frac{\tan(\phi_h)}{H}, \tag{3}$$

where ϕ_w and ϕ_h are the horizontal and vertical angles of the view respectively. The 4×4 projective transformation matrix is represented by $\mathbf{T}$, which is related to the translation and rotation from the neighboring position to the current position. The new coordinate of a pixel on the project frame can be obtained by:

$$\mathbf{C}_p = \mathbf{C} \times \mathbf{T} = [x_p, y_p, z_p, 1]. \tag{4}$$

The 3D coordinate of each pixel in the current position has to be inversely converted to 2D coordinate in the depth map:

$$\mathbf{P}_p = [w_p, h_p] = \left[\frac{W}{2} - \frac{x_p}{Kz_p}, \frac{H}{2} - \frac{y_p}{Kz_p} \right]. \tag{5}$$

As the global motion might change the depth of each pixel, n' of each pixel needs to be quantized to form a reconstructed depth frame from the projected depth information by using the same quantization method Q:

$$n' = Q(z_p). \tag{6}$$

Finally, each pixel of the neighboring frames can be transformed to a new 2D coordinate in the depth frame of the current position. However, some of them might be located outside of the image and some of them might not be integers, which leads to holes. Therefore, an interpolation algorithm is utilized to fill holes and smooth the projected depth frame, which is represented as $\mathbf{D}'$.

3   Experimental Method and Results

To the best of our knowledge, there is no standard sequence where the proper (not estimated) global motion information is available. Consequently, we produced some sequences with synchronously sampled global motion information using our platforms. We tested the proposed scheme using the sequences we produced. These sequences are available for download at http://mmtlab.com/dmcmb.

3.1   Data Acquisition and Prototypes

In this paper, we developed a programmable track slider as shown in Fig. 5 (the texture camera is not used in this paper). A shaft encoder is employed as the motion sensor to get the accurate translational distance. The depth camera is a Mesa Imaging SwissRanger SR4000.

3.2   Experiments and Results

In the experiment, we used forward movement (dolly) and the right movement (tracking) as examples to test the proposed method. The resolution of depth map generated by SR4000 is QCIF (176×144) @ 25 fps. The depth map quantization method is uniform, where white presents farthest distance and black presents nearest distance. To evaluate the proposed method, H.264/AVC JM reference software 14.1 [7] is used. The projection program is developed based on the principle mentioned in Sect. 2. To span a reasonable range of bitrate, the QP was differently set for the proposed method and the standard H.264/AVC. As the frame skipping scheme in the proposed method decreases the bitrate, we have to adjust QP values in order to obtain similar range of bitrate. In the forward dolly experiment, the QP value for the proposed method test is set from 24 to 50, while for standard H.264/AVC it is set from 26 to 50. In the right tracking experiment, the QP value for the proposed method test is set from 26 to 50,

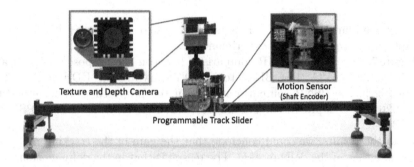

Fig. 5. The customized prototype for the translational motion

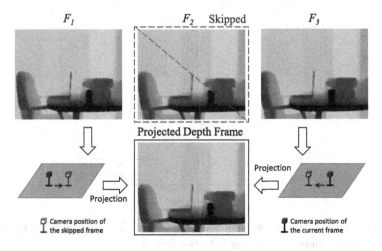

Fig. 6. An example of the processing of the tracking right motion experiment

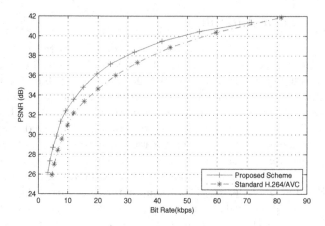

Fig. 7. PSNR versus bitrate for the proposed scheme and standard H.264/AVC in dolly motion; skipping one frame from two frames

while for standard H.264/AVC it is set from 28 to 50. Figure 6 illustrates an example of the right tracking experiment.

Figure 7 presents the PSNR comparison between the proposed method and standard H.264/AVC. The BD-Rate is -29.7 %, while the BD-PSNR is 1.29 dB. In Fig. 8, the PSNR comparison between the proposed method and standard H.264/AVC is presented. The BD-Rate is -41.12 %, while the BD-PSNR is 2.04 dB.

From the results, we could conclude that the gain of the tracking motion is larger than that of the dolly motion. The reason is that in the dolly motion, each

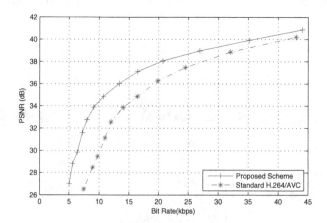

Fig. 8. PSNR versus bitrate for the proposed scheme and standard H.264/AVC in tracking motion; skipping one frame from two frames

object of the depth map is scaled, and the interpolation needs to be employed. This would reduce the accuracy of the reconstructed frames.

4   Conclusions

This paper has introduced a novel depth map sequence coding method using the camera motion information. Compared with the existing H.264/AVC standard, the proposed scheme is able to improve the coding performance up to 2.04 dB. It is noticed that the accuracy of the depth and motion information affects the performance of the proposed method. In the future, we will improve the data precision and test more types of motion, such as rotation and combined motion.

References

1. MESA IMAGING, SR4000. http://www.mesa-imaging.ch/products/sr4000/
2. Hannuksela, M., Chen, T.Y., Suzuki, J.-R.O., Sullivan, G. (ed.).: Avc draft text 8. JCT-3V document JCT3V-F1002, vol. 16 (2013)
3. Chen, Y., Hannuksela, M.M., Suzuki, T., Hattori, S.: Overview of the mvc+ d 3D video coding standard. J. Vis. Commun. Image Representation **25**(4), 679–688 (2014)
4. Merkle, P., Smolic, A., Muller, K., Wiegand, T.: Multi-view video plus depth representation and coding. In: 2007 IEEE International Conference on Image Processing (ICIP 2007), pp. I–201, IEEE (2007)
5. Fehn, C.: Depth-image-based rendering (dibr), compression, and transmission for a new approach on 3D-tv, in electronic imaging. Int. Soc. Optics Photonics **2004**, 93–104 (2004)
6. Lee, P-J., Huang, X-X.: 3D motion estimation algorithm in 3D video coding. In: 2011 International Conference on System Science and Engineering (ICSSE), pp. 338–341, June 2011
7. HHI Fraunhofer Institute, H.264/AVC Reference Software. http://iphome.hhi.de/suehring/tml/download/

Author Index

Printed in the United States
By Bookmasters